Undergraduate Lecture Notes in Physics

Undergraduate Lecture Notes in Physics (ULNP) publishes authoritative texts covering topics throughout pure and applied physics. Each title in the series is suitable as a basis for undergraduate instruction, typically containing practice problems, worked examples, chapter summaries, and suggestions for further reading.

ULNP titles must provide at least one of the following:

- An exceptionally clear and concise treatment of a standard undergraduate subject.
- A solid undergraduate-level introduction to a graduate, advanced, or non-standard subject.
- A novel perspective or an unusual approach to teaching a subject.

ULNP especially encourages new, original, and idiosyncratic approaches to physics teaching at the undergraduate level.

The purpose of ULNP is to provide intriguing, absorbing books that will continue to be the reader's preferred reference throughout their academic career.

More information about this series at http://www.springer.com/series/8917

R. Prasad

Analog and Digital Electronic Circuits

Fundamentals, Analysis, and Applications

 Springer

R. Prasad
Emeritus Professor of Physics
Aligarh Muslim University
Aligarh, India

ISSN 2192-4791   ISSN 2192-4805   (electronic)
Undergraduate Lecture Notes in Physics
ISBN 978-3-030-65131-2   ISBN 978-3-030-65129-9   (eBook)
https://doi.org/10.1007/978-3-030-65129-9

This Springer imprint is published by the registered company Springer Nature Switzerland AG
The registered company address is: Gewerbestrasse 11, 6330 Cham, Switzerland

Preface

A course on electronics may broadly be divided into three components: circuit analysis, analog circuits and digital electronics. Books on the above-mentioned individual components are available; however, the present book covers all three components in sufficient details. Since different topics of the three components are covered in different semesters of the graduate course, a single book covering all topics will be of much utility for students so far as the continuity of the subject is concerned. My own experience and the valuable experience of my other colleagues who teach these courses at different Engineering/Physics institutes indicated the need for such a comprehensive book.

The book is divided into three parts: Part I, consisting of three chapters, covers the fundamentals of circuit analyses. Chapter 1 introduces the basic circuit elements, their properties, classification and network theorems that simplify circuit analyses. The application of network theorems has been explained by a large number of solved examples.

A special feature of the book is the self-assessment questions that are evenly distributed in all chapters. These short and crisp questions are meant for the reader to test his/her understanding till that point of the text. It is expected that a sincere reader will satisfy himself/herself by answering these questions before proceeding further.

Chapter 2, Part I, is devoted to the Laplace technique of circuit analysis. Since most problems of Physics in general and of electrical circuits, in particular, can be represented by differential equations, the Laplace transform technique, which basically is a method of solving differential equations, may be used to solve electrical circuits. The transformation of a complicated electrical network from the time domain to s-domain using the method of Laplace transform converts the complicated circuit into a circuit containing only impedances in s-domain. Use only of Kirchhoff's laws then solves the circuit. In view of the emerging importance of the Laplace transform method, a complete chapter having a sufficient number of solved examples is devoted to the topic. Remaining topics, like first-order and second-order circuits, transients, phasor representation, Fourier analysis, etc. are covered in Chap. 3 of Part I.

Part II covers analog electronics/semiconductor physics. It has five chapters, from Chaps. 4 to 8. The basic electrical properties of solids, their classification on the basis of these properties, intrinsic and extrinsic semiconductors, Fermi energy, its importance, etc. are discussed in Chap. 4, while Chap. 5 covers the physics of p-n junction. Junction diodes and their applications are discussed in this chapter. Both Junction and Field Effect Transistors are discussed in Chap. 6. Transistor amplifiers and their different small-signal equivalents are also described in this chapter.

Feedback plays a very important role in amplifiers. Chapter 7 (Part II) is devoted to the topic of feedback in amplifiers. Properties of the two-port networks are also reviewed in this chapter.

An operational amplifier is one of the most versatile devices of semiconductor electronics. Chapter 8 (Part II) covers the operation and applications of operational amplifiers.

Topics of digital electronics are covered in Part III, which is divided into three chapters. Chapter 9 starts with the introduction of discrete-time digital signals, basics of A/D and D/A conversion, number systems, logic gates and covers the combinational logic circuits. Sequential logic circuits, Latches, Flip-flop, Multiplexer, Counters, etc. are discussed in sufficient detail in Chap. 10. The last chapter of the book, Chap. 11, deals with special circuits and devices like semiconductor memories, sample and hold circuits, architecture of Analog to Digital and Digital to Analog converters, Arithmetic Logic Unit, etc.

Distinguishing features of the book are as follows:

(i) It targets the bottom 50% students of the class, and, therefore, very simple language is used in the book and difficult concepts are explained by giving a large number of examples from everyday life.

(ii) To provide the reader an opportunity to test his/her understanding of the subject while reading the book at successive stages, crisp and short self-assessment questions based on the matter covered up to that stage have been uniformly distributed over the text of each chapter. It is hoped that a sincere reader will satisfy himself/herself by answering these questions before proceeding further.

(iii) Each topic in the book is developed from the very basics; a short review of already known parameters is prefixed to every chapter. Efforts have been made to make the discussion as complete as possible so that the reader may not have to refer to any other book/material.

(iv) A large number of solved examples illustrating the method of approach and supported by relevant circuit diagrams are provided in each chapter.

(v) A sufficient number of problems on topics covered in the chapter is given at the end of each chapter. Answers to problems are also provided for readers to check their solutions.

(vi) A bank of multiple-choice questions, some of which have more than one correct alternatives, are provided at the end of each chapter. An answer is treated complete only if all correct alternatives are marked. The correct answers to these questions are also given.

(vii) A special feature is the list of short answer questions given at the end of each chapter. In each chapter, sample answers to a few short answer questions have also been included to provide a template or specimen for the reader.

The author will be happy to receive any suggestions for further improvement in the book.

R. Prasad
Retired Professor of Physics
Ex-Dean Faculty of Science
Aligarh Muslim University
Aligarh, India

Acknowledgements

I start my acknowledgements first thanking the youngest, my Grandson, Ansh. Ansh suggested some better substitutes for words that I used in the text, for example, same batch instead of same lot etc. My Granddaughter Antra also helped me in many ways, including re-setting my small laptop that often goes out of order. They both are unending sources of joy and happiness. I sincerely thank both of them.

Members of my research group, who are also my colleagues, deserve acknowledgements. Professor B. P. Singh, (Chairman, Department of Physics, AMU, Aligarh) and Dr. Manoj Sharma (Assistant Professor, S.V. Degree College, Aligarh) needs special mention as both of them have helped me at each step in the development of this text. Dr. Sharma, specially emphasised the need of such a book, and collected some literature for me. I thank all members of the group including Prof. Singh, Prof. M. M. Mustapha, Dr. Sharma, Dr. Sunita, Dr. Unnati, Dr. P. P. Singh, Dr. D. P. Singh, Dr. Abhisheik Yadav, Dr. Shuaib and others for their help in this project.

I take this opportunity to thank Aligarh Muslim University, Aligarh (India) where I served as Lecturer, Reader, Professor and the Dean Faculty of Science spanning over more than 40 years.

Last but not the least, I wish to thank all my family members, in particular my wife Sushma Mathur, without their support this project was not possible. I dedicate this book to my family.

Kellyville, Australia   R. Prasad

Contents

Part I
Circuit Analyses

Electrical Network Theorems and Their Applications

Abstract

Network theorems are tools to convert complicated electric networks into simpler equivalents. Important network theorems, their proof and applications in different situations are discussed in this chapter.

1.1 Objective

An arrangement of interconnected basic units (or elements) of a system is called a network. There may be several different types of networks that may differ from each other on the basis of the type of interconnections and the type of basic units or both. Initially, the term network was used to describe the interconnected electrical elements like resistances, capacitors, inductors, sources of electric potential difference, etc. and was then extended also to include basic elements of electronics, like diodes, transistors, etc. The concept of the network has now been extended to include other systems, like thermal networks, and biological systems like neuron networks, etc.

A typical network may be made up of several branches, meshes, loops, etc. A network may have one or more than one 'Input' port(s) where some physically measurable and defined quantity, called 'signal', is applied. A signal is often produced by some energy source. The applied signal while travelling through different branches, loops and meshes of the network divides and recombines at several points in the network and thus get modified. It is often desired to know the strength and other characteristic parameters of the signal at a particular location in the network, called the 'output' port. The process of obtaining the relationship between the input and output signals is termed network analysis. Network analysis is easy if the network contains only a few elements, however, it becomes tedious for complicated multi-branch, multi-element networks. Network theorems are used to reduce complicated networks to simpler less-complicated networks. In this section, we will

© The Author(s), under exclusive license to Springer Nature Switzerland AG 2021
R. Prasad, *Analog and Digital Electronic Circuits*, Undergraduate Lecture Notes in Physics, https://doi.org/10.1007/978-3-030-65129-9_1

study electrical network theorems with the objective of applying these theorems to reduce complicated networks and obtain relationships between the input and output signals.

1.2   Some Definitions

Before going into the study of electrical network theorems, let us refresh the definitions and explanations of some terms frequently used in electrical/electronic circuit theory.

(a) Electrical quantities
(i) Electric charge: It is the most fundamental quantity for circuit theory. Atom, the building block of matter, is overall electrically neutral because it contains both negative and positive charges in equal amounts. The charge is generally represented by the letter q or Q and is measured in the unit of Coulomb, denoted by C. The magnitude of the negative charge on an electron is: $(-)\ 1.602 \times 10^{-19}$ C.

(ii) Potential difference or Voltage: The amount of work done (or the energy required) in moving a unit positive charge from one point (say A) to another point (say B) across a circuit element is called the potential difference between the two points (A and B) and is often represented by letters v, V or E. Potential difference is measured in the unit of Volt, denoted by symbol V. The change in the work done dw (or in energy) when a charge of magnitude q is moved across a potential difference dv is given as $dw = qdv$.

(iii) Current: The rate of flow of charge through a circuit element is called current and is denoted by letters i or I. Current is measured in the unit of Ampere, represented by symbol A; a flow of 1 Coulomb of charge per second through a circuit element constitutes 1A of current. The relationship between charge q and current i in mathematical form may be written as $i = \frac{dq}{dt}$ and $q = \int idt$.

(iv) Energy: Energy is the capacity to perform work and generally it is denoted by letters w or W. Its unit is Joule which is represented by the symbol J. For electric networks, it may be written as $W = \int dw = \int vdq = \int vidt$.

(v) Power: Rate of doing work or rate of change in energy is defined as power which is represented by letter p or P. In case of electrical/electronic networks, $P = \frac{dW}{dt} = vi$. The power is measured in the unit of Watt which is denoted by W.

(b) Electrical circuit elements
Basic building blocks/basic units or electrical circuit elements may be classified into two types: active and passive. This classification is based on the property of the element whether it can convert other form(s) of energy into electrical energy or not.

Elements that cannot convert other forms of energy like heat energy or chemical energy, etc. into electrical energy (that is into electric current or electric potential difference) are termed as Passive elements. Resistor, Capacitor, and Inductor are examples of passive elements. It is worth mentioning that passive elements are not capable of producing electric current/voltage of their own but when electrical energy is imposed on them by some external source, the passive element may convert this electrical energy into some other form, may convert it into heat or magnetic field, etc. Circuit elements that may produce electrical energy (electric current or potential difference) by converting energy from some other form are called Active elements. Cells, batteries, DC and AC current generators, photo diodes, etc. are examples of active elements.

(I) Passive circuit elements

(a) Resistor: As already mentioned, passive elements do not generate electrical energy, which essentially means that passive elements cannot convert energy from any other form to electrical form. A Resistor is denoted by R and is represented in a circuit diagram by the symbol —$\wedge\wedge$—. General-purpose resistors are made using wires of metallic alloys of high resistivity. Since large lengths of wires are required to produce resistors of high values, they are coiled to save space. A coil made of some conducting material like metallic alloy also possesses a finite non-zero value of inductance and capacitance. As such any general-purpose resistor is essentially equivalent to a combination of a resistor, a capacitor and an inductor. However, in simple treatments the inductance and the capacitance associated with resistance are neglected, assuming their magnitudes to be negligibly small as compared to the magnitude of the resistance. Ohm's law describes the behaviour of pure resistance. According to Ohm's law, when some electric current of instantaneous value $i(t)$ at instant 't' is made to pass through an ideal or pure resistor of magnitude R, an instantaneous potential difference $v(t)$ is produced across the two ends of the resistor which is directly proportional to the magnitude of the current $i(t)$, as shown in Fig. 1.0. The law further states that the end A through which the current enters the resistor is at a higher potential than the end B where the current leaves it. Arrows in this figure show the direction of the flow of current and the direction of the increase of the electrical potential.

Hence from Ohm's law,

$$v(t) \propto i(t) \text{ or } v(t) = Ri(t) \tag{1.1}$$

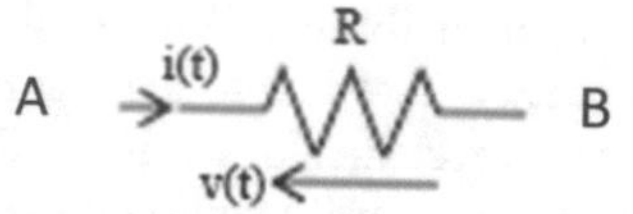

$$\text{R}$$

A $\xrightarrow{i(t)}$ ⟋⟍⟋⟍⟋⟍ B

$v(t)$ ⟵———

Fig. 1.0 Direction of the current and the potential difference across an ideal or pure resistance

Here, R is the magnitude of the resistance in ohm (Ω) when $v(t)$ is in Volt (V) and $i(t)$ is in Ampere (A). One may also write

$$i(t) = Gv(t) \tag{1.2}$$

where $G = 1/R$ is called the conductance and is measured in the unit of Siemen. The electric energy in the circuit $w(t)$ is given by

$$w(t) = \int v(t)i(t)dt = \int R(i(t))^2 dt = \int G(v(t))^2 dt \tag{1.3}$$

This shows that the energy in the circuit is always positive and increases with time. Energy consumption in the circuit, which is the rate of depletion of energy and is given by $\frac{dw}{dt}$, becomes

$$\frac{dw}{dt} = p(t) = R(i(t))^2 = G(v(t))^2 \tag{1.4}$$

The above equation shows that energy is continuously consumed by pure resistance if some current is passed through it and that energy consummation does not depend on the direction of flow of the current. A resistor continuously converts electric energy into heat and dissipates it to the surroundings.

(b) Inductor: The inductance of an inductor (a cylindrical coil having several turns) is denoted by the letter L and is measured in the unit of Henry (H). In electrical circuits, it is represented by the symbol —⟋⟍⟋⟍⟋⟍—.

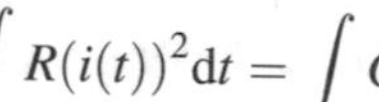

Like a resistor, actual inductance also has associated resistance and capacitance that are neglected in a simple analysis. The behaviour of an inductor is described by Faraday's law of electromagnetism. If current $i(t)$ is passed through an inductor, a magnetic flux $\varphi(t)$ gets linked to the inductor. The flux $\varphi(t)$, for a given inductor, is proportional to the magnitude of the current $i(t)$. If the current changes with time, increases or decreases, the magnetic flux linked to the inductor also changes. According to Faraday's law, the change in the flux generates a potential difference, say $'e(t)'$ across the terminals of the inductor, which is proportional to the rate of

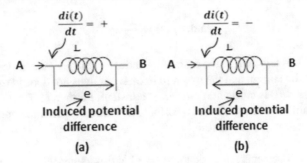

Fig. 1.1 Direction of the induced potential difference across an inductor when **a** current increases and **b** current decreases

change of the magnetic flux $\varphi(t)$. The nature of the induced potential difference is such that it opposes the source of the change in current in the inductor. For example, in Fig. 1.1a, for the case when current $i(t)$ increases, the magnitude of the induced potential difference 'e' will be proportional to the rate of the change in current but it will be lower at point A and higher at B. On the other hand, if the current in the inductor decreases, figure (b), the induced potential will fall from A to B.

Therefore,

$$e(t) \propto \frac{di(t)}{dt}$$
$$\text{or} \quad e(t) = L\frac{di(t)}{dt} \tag{1.5}$$

Here L is the inductance of the inductor. It is evident from (1.5) that for any inductor if $\frac{di(t)}{dt}$ is infinite, i.e. rate of change of current is infinite, then the induced potential difference $e(t)$ will also be infinite. This is not possible. An infinite rate of change of current may occur when a step change in current takes place, that is, the current is suddenly made zero from some finite value or vice versa. Hence, step changes in currents are not possible through an inductor. Current through an inductor must change slowly. When current passing through an inductor is switched off, the large emf induced in the circuit produces a spark which allows a small current to flow for a short time after the switch is off and thus reduces the rate of decrease of current.

Also, energy w(t) in the system is given by

$$w(t) = \int e(t).i(t)dt = \int L\frac{di}{dt}.i(t)dt = \int Li(t)dt = \frac{1}{2}Li^2 \tag{1.6}$$

$$\text{And}\quad p(t) = \frac{dw}{dt} = 0 \tag{1.7}$$

Equation (1.6) tells that in the case of an inductor, the net energy depends on the magnitude of the current and the magnitude of L but does not depend on time. For a given inductor, w(t) is constant if current is constant. Further, (1.7) tells that there is no consumption of energy by the inductor. An inductor stores the applied electric energy in electromagnetic form.

(c) Capacitor: The capacitor is denoted by C and is represented in circuits by the symbol—|⊢—. When some electric potential difference $v(t)$ is applied across a capacitor, charge $q(t)$ gets accumulated on the plates of the capacitors. The charge $q(t)$ is proportional to the applied potential difference $v(t)$. Hence,

$$q(t) \propto v(t) \text{ or } q(t) = Cv(t) \tag{1.8}$$

In the above equation, C is the capacity of the capacitor and is measured in the unit of farad (F). On differentiation with respect to time 't', (1.8) yields

$$i(t) = \frac{dq}{dt} = C\frac{dv(t)}{dt} \tag{1.9}$$

It may be observed from (1.9) that if there is a sudden or step change in the potential $v(t)$ across the capacitor, i.e. $\frac{dv(t)}{dt}$ is infinite; the current $i(t)$ will also become infinite, which is not possible. This essentially means that a step change or sudden change of potential difference across a capacitor is not possible.

And the energy in the system is given by

$$w = \int i(t).v(t)dt = \int \left\{ C\frac{dv(t)}{dt} \right\} v(t)d(t) = \int Cv(t)dv(t) - \frac{1}{2}C(v(t))^2 \tag{1.10}$$

Equation (1.10) indicates that the energy contained in the capacitor at a given instant is proportional to the magnitude of the capacitance C and the square of the instantaneous potential difference $v(t)$ across the capacitor. For a given capacitor, if $v(t)$ is increased, the energy stored in the capacitor increases and the energy decreases on lowering the potential difference. If $v(t)$ remains fixed, the energy associated with the capacitor also remains constant. This shows that a capacitor stores energy (in electrostatic form) and there is no loss or dissipation of energy by a (an ideal or pure) capacitor.

It may thus be concluded that an inductor does not allow a step change of current through it while a capacitor does not allow a step change of potential difference across it. Further, both the pure inductor and the pure capacitor store electrical energy, former as electromagnetic and latter as electrostatic energy, and they do not dissipate or lose energy to the surroundings. Contrary to that, a pure resistor dissipates energy to the surroundings in the form of heat, even when the current or the voltage drop across it is kept constant.

(II) Alternating current

In DC current, the magnitude of the current may vary but the direction of flow of the current does not change with time, as shown in Fig. 1.2a. In alternating current (AC), both the magnitude and the direction of flow of the current change with time with some fixed linear frequency f, Fig. 1.2b.

The linear frequency f is related to the cyclic frequency ω by the relation $\omega = 2\pi f$. In simple electrical/electronic networks, one deals either with DC or sinusoidal AC. The sinusoidal AC may be represented by a sine function as $i_{ac} = I_{max}\sin\omega t$ (or $i_{ac} = I_{max}\cos(\omega t + \varphi)$). Here, I_{max} is the amplitude or the maximum value of the current and φ the phase angle. As may be observed in Fig. 1.2b, the

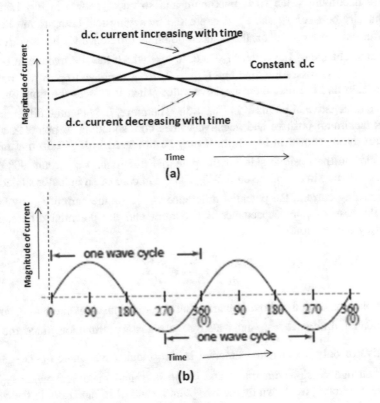

Fig. 1.2 Direct and sinusoidal alternating currents

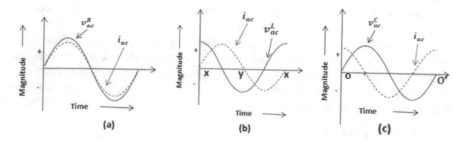

Fig. 1.3 Phase relationship between the applied sinusoidal alternating current and the potential difference induced across **a** pure resistor, **b** pure inductor and **c** pure capacitor

time axis may also be marked in terms of the phase angle. In the present presentation AC always refers to sinusoidal AC, unless specified otherwise.

When AC passes through a pure resistance, it develops an alternating potential difference across the two terminals of the resistance. This alternating potential difference $v_{ac}^R (= R i_{ac})$ follows the applied alternating current i_{ac} , which means that v_{ac}^R is a maximum when i_{ac} is maximum and so on as shown in Fig 1.3a. This simultaneity between i_{ac} and v_{ac}^R is expressed by saying that both are in phase.

In the case of an inductor, the induced potential difference v_{ac}^L is proportional to the rate of change of the current i_{ac} , i.e. to $\frac{di_{ac}}{dt}$ which has the maximum positive value when i_{ac} crosses the time line $(i_{ac} = 0)$ from negative to positive (points X in Fig. 1.3b and the maximum negative value when it crosses the time line from positive to negative at points Y in Fig. 1.3b. Therefore, the potential difference v_{ac}^L attains maximum positive and negative values corresponding to points X and Y, respectively, as shown in Fig. 1.3b. It may be observed in Fig. 1.3b that the first maximum on the positive side for the potential difference v_{ac}^L occurs 90° earlier than the first maximum of the current i_{ac}. Thus, in case of an inductor subjected to an alternating current, the potential difference v_{ac}^L leads the current i_{ac} by 90°.

In the case of pure capacitance, the current and the potential difference are related by the relation:

$$\frac{dq}{dt} = i_{ac} = C \frac{dv_{ac}^C}{dt} \tag{1.11}$$

It is obvious from the above equation that if i_{ac} is sinusoidal alternating current; the potential difference v_{ac}^C must also be a sinusoidal function since the time derivative of only a sinusoidal function is sinusoidal. Also, since $i_{ac} \propto \frac{dv_{ac}^C}{dt}$, the rate of change of v_{ac}^C with time must be positive and maximum when i_{ac} is a positive maximum, as shown by the two points o and o' in Fig. 1.3c. In the case of the capacitance, the first maximum of v_{ac}^C on the positive side occurs after a time

Fig. 1.4 Phase relationship
of potential drops across pure
resistance, pure inductance
and pure capacitance to the
sinusoidal alternating current

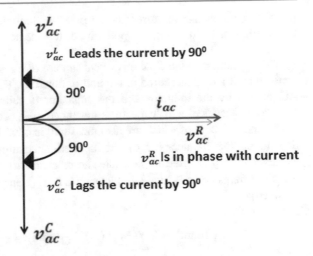

corresponding to 90° with respect to the positive maximum of the current i_{ac}. Hence in this case, the potential difference v_{ac}^C lags behind the current i_{ac} by 90°.

Figure 1.4 shows the phase relationship between v_{ac}^R, v_{ac}^L and v_{ac}^C with respect to i_{ac} on a single graph.

Power p(t) in the circuit at each instant may be calculated by multiplying the instantaneous values of current i_{ac} with the corresponding potential difference, with their appropriate negative/or positive signs. In case of pure resistance p(t), is always positive indicating that power is continuously dissipated by a resistor to the surroundings in the form of heat, while in cases of a pure inductor and pure capacitor power fluctuates every quarter of a period from positive to negative, indicating that power is either stored in the device or is given back to the circuit. There is no dissipation of power in pure L and C circuits.

Self-assessment question: Draw graphs showing the variation of power p(t) with time for a pure capacitor and a pure inductor circuits subjected to sinusoidal alternating currents.

(a) **Resistance, Reactance, Impedance, Admittance, Conductance and Susceptance.**

The current in conductors is generated by the instantaneous motion of electrons in a particular direction. Conventional current, which is assumed to be due to the motion of a fictitious positive charge, flows in a direction opposite to the direction of flow of real electrons. All circuit elements, like resistor, inductor and capacitor, oppose the flow of alternating current through them. In the case of a resistor, the opposition to the flow of current (both DC and AC) arises essentially from the collision of moving electrons with the lattice of the resistor material. Though a pure inductor does not oppose the flow of steady (constant magnitude) DC but both the inductor and the capacitor oppose the flow of sinusoidal AC, the inductor on account of the fact that rapid change in current is not possible through it and the capacitance for the reason that rapid change of potential difference across it are not possible.

The opposition to the flow of current in case of a resistor is called 'resistance', denoted by R and is measured in the unit of ohm (Ω). The opposition to the flow of AC offered by the inductor and the capacitor is called 'reactance', specifically 'inductive reactance' and 'capacitive reactance', respectively, and are represented by X_L and X_C, both of which are also measured in ohm. However, both X_L and X_C depend on the frequency f of the applied alternating current. The following expressions may be used to calculate the reactance X_L and X_C in ohms provided the linear frequency f is in Hz, inductance L in H (Henry) and the capacitance C in F (farad):

$$X_L(ohm) = 2\pi f(Hz)L(H); X_C(ohm) = \frac{1}{2\pi f(Hz)C(F)} \tag{1.12}$$

In the case of a pure resistance, both for the DC and AC, the current i_{ac} and the potential difference v_{ac}^R are always in phase. In the case of inductance and the capacitor, however, there is always a phase difference between i_{ac} and v_{ac}^L or v_{ac}^C. Thus, while dealing with AC quantities apart from the magnitude of the quantity another parameter, the phase difference with respect to the phase of some other standard quantity (maybe i_{ac}) comes into play. To include this additional parameter of phase difference, which is required to completely define an AC quantity, AC quantities are often represented by complex numbers. If an electrical quantity is represented by a complex number A, such that

$$A = B + jC \tag{1.13}$$

then the magnitude of the quantity

$$|A| = \sqrt{(B^2 + C^2)} \tag{1.14}$$

And the phase angle ϕ with respect to the real part B is given by

$$\phi = \tan^{-1}\left(\frac{C}{B}\right) \tag{1.15}$$

Equation (1.13) is the rectangular representation of the complex quantity and is convenient if the addition or subtraction of complex quantities is done; for example, the sum/difference of A and $A' = B' + jC'$ are given by

$$A \mp A' = (B \mp B') + j(C \mp C') \text{ and } \phi^\mp = \tan^{-1}\frac{(C \mp C')}{(B \mp B')} \tag{1.16}$$

The quantity A may also be represented in polar form as

$$A = |A| \angle \phi_A \qquad (1.17)$$

The polar form is convenient if the multiplication or division of complex quantities is required,

$$A x A' = (|A| x |A'|) \angle (\phi_A + \phi_{A'}); \frac{A}{A'} = \frac{|A|}{|A'|} \angle (\phi_A - \phi_{A'}) \qquad (1.18)$$

Often in electrical circuits, all the three passive elements, R, L and C, are present simultaneously. As a result, the total opposition offered to AC current by such circuits contains both resistive and reactive components. For example, in case of the following series combination of pure resistance R (=100 Ω), pure inductance L (=10 mH) and pure capacitance C (=5 μF) if an AC current of frequency $f = (159.15 Hz)$ is passed, the total opposition to the current offered by the circuit will be

(a) R = (100 + j 0) Ω by the resistance,
(b) $X_L = (0 + j(2\pi \times 159.15 \times 10 \times 10^{-3}))) = (0 + j10)\Omega$ by the inductance and
(c) $X_L = (0 - j(\frac{1}{2\pi \times 159.15 \times 5 \times 10^{-6}})) = (0 - j200)\Omega$ by the capacitor.

A general term for the total opposition to the flow of current offered by a circuit or an electrical element is **impedance**. Impedance is denoted by the letter Z and in general it is a complex quantity, particularly in the case where AC flows through the circuit. The real part of impedance is the resistance R offered by resistive elements, and the imaginary part X is the reactance present in the circuit on account of the inductance and or capacitance.

For the series combination of Fig. 1.5, the impedance Z of the circuit may be written as

$$Z = R + jX_L - jX_C = (100 + j0) + (0 + j10) + (0 - j200) = (100 - j190)\Omega$$

Impedance Z is like a vector of magnitude (Fig. 1.6)

$$|Z| = \sqrt{(100)^2 + (190)^2} = 214.71\Omega \text{ and } \phi = -\tan^{-1}\frac{190}{100} = -62.24^0$$

An advantage of impedance representation is that Ohm's law can be applied between i_{ac}, v_{ac} and Z, such that $v_{ac} = Z i_{ac}$, here v_{ac} is the total potential difference between the two ends of the series combination and i_{ac} is the AC current of frequency f (=159.15 Hz) through the circuit. This form of Ohm's law is applicable to any type of circuit operated through a direct or alternating current/voltage source. It may, however, be kept in mind that here all the three quantities v_{ac}, Z and i_{ac} are in general complex numbers. It may thus be observed that with the concept of an impedance function Z, that is a complex number, all laws like Kirchhoff's law, etc.

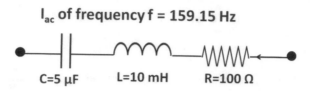

I_{ac} of frequency f = 159.15 Hz

C=5 µF   L=10 mH   R=100 Ω

Fig. 1.5 A series combination of resistor R, inductor L and capacitor C subjected to the AC current

Fig. 1.6 Representation of impedance Z

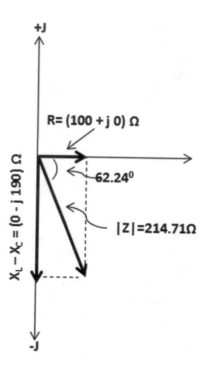

that are valid for DC circuits may well be applied to AC circuits. The only limitation arises when calculating power in the circuit. It is because reactive components, unlike resistive components, do not consume power, but rather store it and release it for parts of the input cycles. The following solved examples will demonstrate how Ohm's law may be applied to AC circuits using complex numbers for current, potential difference (also called voltage) and impedance.

Solved example (SE1.1) The following circuit is energized by a 120 V, 50 Hz source. Calculate the current passing through the circuit and the potential drops against each component (Fig. SE1.1)

Solution: *Impedance offered by resistance R,* $Z_R = (250 + j0)\Omega = 250\angle0\Omega$.
  Impedance offered by inductance L,

Fig. SE1.1 Circuit for
(SE1.1)

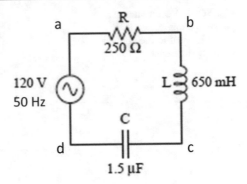

$$Z_L = \left(0 + j\, 2\pi \times 50 \times 650 \times 10^{-3}\right)\Omega = (0 + j204.2)\Omega = 204.2\angle 90^0\ \Omega$$

Impedance offered by capacitance C,

$$Z_C = \left(0 - j\frac{1}{2\pi \times 50 \times 1.5 \times 10^{-6}}\right)\Omega = (0 - j2122.2)\Omega = 2122.2\angle - 90^0\Omega$$

$$\text{Total impedance } Z_{\text{total}} = \left(Z_R + Z_L + Z_C\right) = (250 - j1918.0)\Omega$$
$$= 1934.22\angle - 82.57^0\Omega$$

The current $i_{ac} = \frac{v_{ac}}{Z_{\text{total}}} = \frac{120\angle 0}{1934.22\angle -82.57^0\Omega} = 0.062\angle 82.57^0 A$.

Thus, current of 62 mA flows through the circuit and the current leads the applied potential difference by 82.57°.

The potential drop (pd) across each component may be obtained by using the expression $v_{ac} = Z_{ac} i_{ac}$ *where* v_{ac} *is the pd against an element of impedance* Z_{ac}.

$$v_R = Z_R i_{ac} = (250\angle 0)(0.062\angle 82.57^0) = 15.5\angle 82.57^0 V$$
$$= Z_L i_{ac} = (204.2\angle 90^0)(0.062\angle 82.57^0) = 12.66\angle 172.57^0 V$$

$$v_C = Z_C i_{ac} = (2122.2\angle - 90^0)(0.062\angle 82.57^0) = 131.58\angle - 7.43^0 V$$

It may sound strange that the potential drop against the capacitance $(131.58\angle -7.43^0 V)$ *is larger than the applied potential difference of 120 V. However, this is quite possible in circuits with AC currents and is due to the phase difference between currents in different elements. Further,* v_R, v_L *and* v_C *are all vectors and their vector sum must be equal to the applied pd* v_{ac} *as shown in Fig. SE1.2.*

Admittance: The reciprocal of the impedance, $\frac{1}{Z}$, is called admittance or admittance operator. It is denoted by the letter Y and is measured in the unit of Siemen. It

Fig. SE1.2 Vector sum of potential drops across R, L and C is equal to the applied potential difference

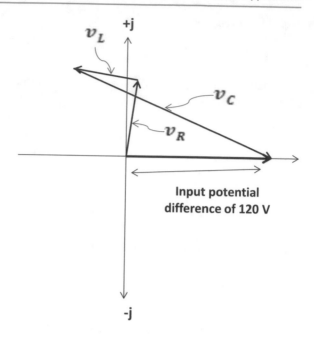

Input potential difference of 120 V

is also a vector in a two-dimensional plane made of a real axis and an imaginary axis. As impedance is a measure of how much a circuit or any circuit element impedes the flow of alternating current, admittance is a measure of how much the circuit admits the alternating current. In rectangular vector notation the real part of admittance, denoted by G, is called conductance. In impedance there are two parts, the resistive part R and reactive part X; conductance G is the counterpart of R and susceptance, denoted by B, is the counterpart of reactance X in the formulation of admittance Y, as given below

$$Y = \frac{1}{Z} = G + JB \qquad (1.19)$$

The pairs of 'resistance' and 'conductance', and 'reactance 'and 'susceptance' have some linguistic logic associated with them. While 'resistance' is a measure of the opposition to the flow of current offered by the circuit, conductance is a measure of the ease with which current can pass through the circuit. Similarly, while reactance is a measure of how much a circuit reacts to the change in current over a period of time, susceptance is a measure of how much the circuit is susceptible to the changes in current over a period of time. Ohm's law in admittance form may be written as

$$i_{ac} = Y v_{ac} \qquad (1.20)$$

Solved example (SE1.2) Compute the conductance G and susceptance B for the network given in (SE1.1).

Solution: *The total impedance Z_{total} for the circuit of Fig. SE1.1 as calculated in the earlier solved example is $Z_{total} = (250 - j1918.0)\Omega$, hence the admittance 'A' of the network is*

$$A = \frac{1}{Z_{total}} = \frac{1}{(250 - j1918.0)} \text{Siemens} = \frac{(250 + j1918.0)}{(250 - j1918.0(250 + j1918.0))}$$

$$= \frac{(250 + j1918.0)}{\left[(250)^2 + (1918)^2 \right]} = \frac{250}{3741224} + j\frac{1918}{3741224}$$

$$= \left[(6.68 \times 10^{-5}) + j(5.13 \times 10^{-4}) \right] \text{Siemens}$$

Hence, Conductance $G = (6.68 \times 10^{-5})$ Siemens.
And Susceptance $B = (5.13 \times 10^{-4})$ Simens.

Self-assessment question (SAQ): Calculate the admittance and hence the conductance and susceptance of a pure capacitor of 100 μF capacity when AC current of 10 mA and 100 Hz is applied across it.

(III) Active circuit elements
Those components of an electric circuit or network that may generate electrical energy by converting some other form of energy into electrical energy are called active elements or components or simply (electrical energy) sources. Energy sources may be of two types: (i) voltage source and (ii) current source.

(a) **Voltage source**: An ideal voltage source delivers a constant potential difference or voltage across its output terminals and the voltage does not change with the load in the circuit. The internal impedance of an ideal voltage source is, therefore, zero and the output voltage is the same as the internal potential difference. However, an ideal voltage source is nonexistent and an actual voltage source has some small internal impedance. An actual voltage source in electrical circuits is represented by a circle with a small series impedance as shown in Fig. 1.7. The terminal potential difference $v(t)$ of an actual voltage

Fig. 1.7 A practical voltage source

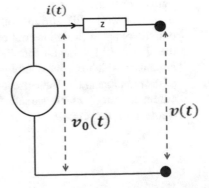

Fig. 1.8 A practical current
source

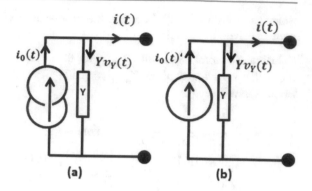

source is given by $v(t) = v_0(t) - Zi(t)$, where $v_0(t)$ is the internal potential
difference, Z internal impedance and $i(t)$ is the instantaneous current through
the circuit.

(b) **Current source**: An ideal current source is one which delivers current of
constant magnitude to the load, and the current does not vary with the magnitude of the load. The admittance of an ideal current source is zero and,
therefore, the external current is the same as the internal current of the source.
The actual current source, however, has a small admittance which may be
represented as a shunt in parallel to the source terminals. An actual or practical
current source may be represented as shown in Fig. 1.8a, b. In an actual current
source, the terminal current $i(t)$ is slightly less than the internal current $i_0(t)$ as a
small current $i_Y(t)$ passes through the internal admittance Y. The current $i_Y(t)$
produces a potential drop $v_Y(t)$ across the admittance Y. Therefore,
$i(t) = i_0(t) - Yv_Y(t)$.

(c) **Dependent and Independent energy sources**: Energy sources may be classified into two categories: independent and dependent sources. Independent
sources are those for which the output voltage or current does not depend on
any other parameter of the network. A Battery or a Cell is an example of an
independent DC voltage source. The dependent sources may further be classified as (i) voltage-controlled voltage source, a voltage source the output
voltage of which is controlled by the potential drop across some other element
of the network, (ii) current-controlled voltage source, a voltage source the
output voltage of which is controlled by current through some other element of
the circuit, (iii) voltage-controlled current source, a current source the output
current of which is controlled by the potential drop across some other element
of the circuit, and (iv) the current-controlled current source, a current source the
output current of which is controlled by the current passing through some other
element of the circuit. An operational amplifier in a certain configuration

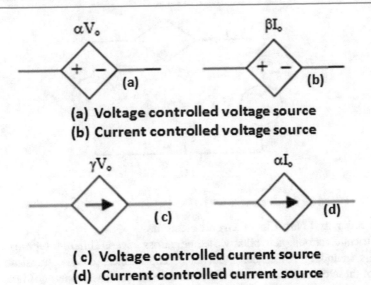

(a) Voltage controlled voltage source
(b) Current controlled voltage source

(c) Voltage controlled current source
(d) Current controlled current source

Fig. 1.9 Voltage- and current-controlled sources. V_0 and I_0 are either potential drop or current in some element of the circuit. α, β, γ *and* δ are constants of proportionality

behaves like a voltage-controlled voltage source. The dependent sources are represented by diamond figures in circuit diagrams as indicated in Fig. 1.9.

Solved example (SE1.3) Calculate the current I and the pd V_0 for the circuit shown in Fig. SE1.3.

Solution: *The circuit shown in* Fig. SE1.3 *contains a voltage-controlled voltage source. Now applying Ohm's law to the resistance of 1Ω, one gets*

$$1 x I = -V_0 \tag{SE1.3.1}$$

Also by applying Kirchhoff's law to the circuit,

$$10 = 4I + 4V_0 + 5 - V_0 \tag{SE1.3.2}$$

Solving (SE1.3.1) and (SE1.3.2), one gets

$$I = 5A \text{ and } V_0 = 5V$$

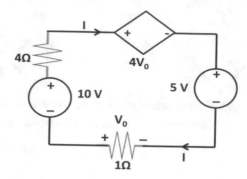

Fig. SE1.3 Circuit for solved example SE(1.3)

(IV) **Linear and Non-Linear circuit elements**

In electrical/electronic networks, we come across several different types of circuit elements including resistance, inductance, capacitance, diodes, transistors, etc. Some of these elements are linear and some non-linear. The property of linearity of any element or system is defined through the relationship that exists between the input and the output signals. Let the box in Fig. 1.10 represent an element or a system. When some input X_1 is applied to the system (or element), a signal y_1 is obtained in the output, as shown for the topmost box in the figure. Similarly, when some signal X_2 is applied to the input, a signal y_2 is obtained in the output. Since X_1 and X_2 are independent of each other, output signals y_1 and y_2 are also not related to each other. Now if a signal of strength $(aX_1 + bX_2)$, where 'a' and 'b' are real coefficients, is applied at the input and a signal of strength $(ay_1 + by_2)$ is obtained at

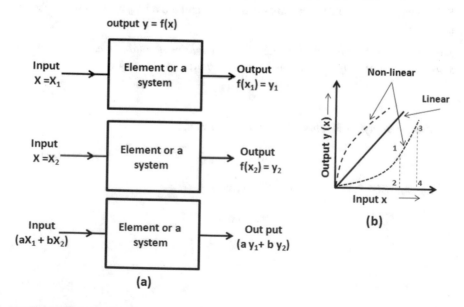

Fig. 1.10 Defining linearity of an element or a system

the output, then the system is said to be a linear system. The output (y) versus input (x) characteristic (also called transfer characteristic) of a linear system (or element) is a straight line as shown in part (b) of the figure. The transfer characteristics of non-linear systems may have any shape other than a straight line, as indicated by dashed lines in figure (b).

Most of the systems and elements in nature are non-linear. However, non-linear systems may be treated as linear over some portions/parts of the transfer characteristics (see the transfer characteristic between points marked 1 and 3 in the figure). When some potential difference V is applied across a pure resistance, it obeys Ohm's law and the current I, which is proportional to V flows through it. Thus, a pure resistance initially behaves like a linear element. However, with time the temperature of resistance increases and the magnitude of the resistance changes, turning a heated resistance non-linear. In spite of this non-linearity, pure (or ideal) resistance, pure inductance and pure capacitance are treated as linear elements. Diodes, transistors and electromagnetic relays, on the other hand, are examples of some non-linear elements.

Self-assessment question (SAQ): Draw a rough sketch for the V-I characteristics of a junction diode and show that it is a non-linear element. Also, show the part of the V-I curve for which the diode may be treated as a linear element.

(V) Bidirectional and Unidirectional elements and circuits

A bidirectional circuit or circuit element is one the (electrical) properties of which do not change if the direction of current or voltage across it is reversed keeping the magnitude same. For example, the resistance of an ideal resistance and the heat dissipated by it per unit time do not change when the direction of flow of current through the resistance is reversed keeping the magnitude the same. Resistance, inductance and capacitance are bidirectional (or bilateral) elements and circuits that contain only these elements are bidirectional. Circuits or circuit elements the electrical properties of which change with the change in the direction of current or potential difference applied to them are termed as unidirectional or unilateral components. A diode or a transistor, for example, offers different amounts of resistance to the flow of current when they are forward biased and when reversed biased. Diode, transistors, logic gates, etc. are examples of unidirectional elements. Any circuit that contains any one or more than one unidirectional element becomes a unidirectional circuit.

(VI) Lumped and distributed elements and circuits

In conventional circuit theory, one deals with the performance of ideal circuit elements and circuits formed by interconnecting them. In such cases, the 'cause' and the 'effect' are regarded as time coincident. For example, when ideal or pure resistance is connected to a battery, the flow of current through the resistance (the cause) and the potential drop across the resistance (the effect) are assumed to occur simultaneously; it is assumed that no time is taken by the current to reach and pass through the resistance and that the potential drop gets generated as soon as the current passes through the resistance. This fundamental concept of time simultaneity of 'cause' and effect is a characteristic of lumped behaviour. This fundamental concept is extended to include the world of the real elements, i.e. the physical hardware where the electrical performance over its useful range approximates closely to that of the ideal elements. The physical dimensions of the element, for example, of the resistance R do not affect the simultaneity of current i , the 'cause' and the potential difference v, the 'effect'. The physical and mechanical dimensions of the resistance may affect the magnitude R of the resistance. So far as the relation $v = Ri$ is concerned, the mechanical dimensions of the resistance, of connecting wires and of the battery, are irrelevant; it means that the components are considered as 'lumped'. The concept of lumped networks is not limited to electrical or electronic networks but can be applied in many other fields like that of the study of heat transfer, behaviour of biological systems, etc.

In case of electrical circuits, the assumption of a lumped network may be justified as the speed of current flow that is almost equal to the velocity of light $\left(\approx 3 \times 10^8 \mathrm{m/s}\right)$ and, therefore, any electrical disturbance at one end of 1 cm long resistance will not take more than 0.03 ns to be felt at the other end. This delay may be regarded as negligible for pulse signals having rise and fall times of a few nanoseconds. The analysis of lumped networks is comparatively simple as the building blocks of the lumped circuit are assumed to be concentrated at a singular point in space and, therefore, the behaviour of the lumped networks may be described using either **linear algebra** or **differential equations** involving only time. For example, the current through an ideal capacitor is related to the time derivative of the potential drop across it, $i = C\frac{dv}{dt}$. No space dependence is involved in the lumped analysis.

A circuit may be considered as 'distributed' rather than 'lumped' if the delay between the cause and effect is significant. In such cases, the behaviour of the circuit requires both the space and time information of the variables. Therefore, the behaviour of the distributed network may be described using **calculus** of **partial differential equations** involving partial derivatives both with respect to time and space. The distributed approach assumes that all circuit attributes like resistance, capacitance, inductance, etc. are infinitesimally small and are distributed throughout the spatial dimensions of the circuit. Connecting wires are not assumed to be perfect conductors but they have resistance, capacitance and inductance per unit length distributed over their lengths. The distributed approach further assumes that currents through branches and voltages at nodes are non-uniform. The distributed network model is often used for the study of transmission lines and transistors/IC, etc. at high frequencies. A thumb rule for the use of the distributed model is that it is applicable if the physical dimensions of the network are equal or larger than the wavelength corresponding to the frequency of the applied signal.

(VII) Ohm's law in vector notation

German scientist Prof. Georg Simon Ohm, in 1827, published an article entitled 'Die galvanische Kette mathematisch bearbeitet' meaning 'The Galvanic circuit

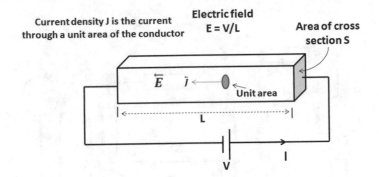

Fig. 1.11 Current density and the electric field through a conductor

investigated mathematically' in which he for the first time put Ohm's law in vector form. In order to understand this law, let us refer to Fig. 1.11. As shown in the figure, a potential difference of V volt is applied across a rod of a uniform area of cross section S of some conducting material. The application of the potential difference produces two effects: (i) an electric field E of strength $|E| = |V/L|$ directed from a positive end to a negative end of the rod gets established and (ii) and current $I = V/R$ flows through the circuit. Here, R is the resistance of the rod.

Ohm argued that the current density $\overrightarrow{J}$ (current per unit area) and the electric field $\overrightarrow{E}$, both of which have the same direction, are proportional to each other, i.e.

$$\overrightarrow{J} = \sigma \overrightarrow{E} \tag{1.21}$$

Here σ is the constant of proportionality. Equation (1.21) is the vector form of Ohm's law. A word of caution about σ is necessary; for non-isotropic conductors σ may be a tensor, having different values in different directions. Using the above expression, it is easy to obtain the following relations:

$$I = \overrightarrow{j}.\overrightarrow{S} = \sigma \overrightarrow{E}.\overrightarrow{S} = \sigma SE = \sigma S \frac{V}{L} = \left(\frac{\sigma S}{L}\right)V \tag{1.22}$$

(The surface areas 'S' may be taken as a vector directed normally to the surface.)

or

$$V = \left(\frac{L}{\sigma S}\right)I = RI \tag{1.23}$$

Here R $\left(= \frac{L}{\sigma S}\right)$ is the resistance, $\left(\frac{\sigma S}{L}\right)$ the conductance G and $\frac{1}{\sigma} = \rho$ is the conductivity of the material. Equation (1.23) represents the well-known scalar form of Ohm's law.

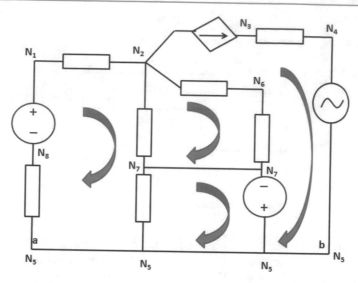

Fig. 1.12 A typical network

(VIII) **Electrical/electronic network, branch, node, loop and mesh**
A group of active and/or passive circuit elements connected through conducting wires in different ways constitute a circuit or a network. A typical network or circuit is shown in Fig. 1.13, where rectangular blocks represent passive circuit elements like resistance, capacitance, inductance, diode, transistor, etc. An arm that contains only one circuit element is called a '*branch*'. In the circuit shown in the figure, N_1 N_2; $N_2 N_3$; $N_3 N_4$; $N_4 N_5$, etc. are branches of the circuit. The point where two or more than two branches join together is called a '*Node*'. N_1, N_2, ... N_7 are all nodes. Any close path starting from a node and passing through different active and passive elements reaching back to the starting node constitutes a '*loop*'. Starting from N_1 passing to N_2, N_3, N_4, N_5, N_8 and back to N_1 constitutes a loop. N_1, N_2, N_7, N_5, N_8 is another closed path and hence a loop. Similarly,

N_2, N_6, N_7, N_2 is another loop in the circuit. It may be observed that it is possible that another loop is nested within a loop. For example, within the loop N_1, N_2, N_3, N_4, N_5, N_8, N_1, several loops are nested (or lying). Thus in general, there may be one or more than one smaller loops or closed paths within a loop. A loop that does not contain any other loop or branch of the circuit is called a '*mesh*'. Four meshes are indicated in the figure with curly arrows.

Self-assessment question: Fig. 1.12 ab is a branch or not? Give your reason.

Self-assessment question: Count the total number of loops and meshes in the circuit of Fig. 1.12.

Self-assessment question: A transmission line works at a frequency of 10 MHz. What must be the length of the transmission line in order to treat it as a distributed network?

Fig. 1.13 Application of Kirchhoff's voltage law

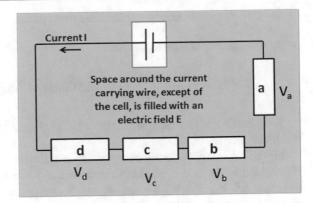

(IX) Kirchhoff's Laws

Gustav Robert Kirchhoff, a German Physicist and Mathematician, in 1845, embodied two laws on electric circuits. The first law, called Kirchhoff's current law (KCL), states that at any node of a circuit the amount of current reaching the node is exactly equal to the amount of the current leaving the node. This essentially means that in a circuit, there cannot be an accumulation of charge at a node. This is understandable as the accumulation of charge at a node will stop the further flow of current through the conductor. The second law, called Kirchhoff's voltage law (KVL), states that the total potential drop around a close path or loop in a circuit is zero. This means that the algebraic sum of all potential drops, taken with proper signs, in a closed loop of a network is zero.

James Clerk Maxwell, a Scottish scientist, synthesized the electric and magnetic fields and developed a sound theoretical background of electromagnetism. He showed that the behaviour of electric and magnetic fields is contained in four differential equations called Maxwell's equations given below:

$$\nabla.E = \frac{\rho}{\varepsilon} \quad \text{(Gauss's law)} \tag{1.24}$$

$$\nabla.B = 0 \quad \text{(Gauss's law for magnetism)} \tag{1.25}$$

$$\nabla \times E = -\frac{\partial B}{\partial t} \text{(Faraday's law of induction)} \tag{1.26}$$

$$\nabla \times B = \mu\left(\varepsilon\frac{\partial E}{\partial t} + J\right) \quad \text{(Ampere's law)} \tag{1.27}$$

Here, E and B are respectively the electric and the magnetic fields, j the total current density, ρ the charge density, ε the permittivity and μ the permeability. The magnetizing field H is related to the magnetic field B by $B = \mu H$ and the displacement field D is related to the electric field E by $D = \varepsilon E$. With these substitutions, Gauss's law becomes

$$\nabla . D = \rho \quad \text{(Modified form of Gauss's law)} \tag{1.24a}$$

And Ampere's law takes the form:

$$\nabla \times H = \left(\frac{\partial D}{\partial t} + J \right) \quad \text{(Modified form of Ampere's law)} \tag{1.27a}$$

Kirchhoff's laws for electric circuits may be obtained using Maxwell relations as discussed below.

Figure 1.13 shows a circuit energized by a cell with current I passing through the circuit elements a, b, c, d, etc. Let us assume that potential drops across these circuit elements are V_a, V_b, V_c, V_d and so on. Some of these potential drops may be negative and some positive. The net external potential drop which is the algebraic sum of Va, ... Vd may be represented by $V = V_a + V_b + V_c \ldots \ldots$ The current I in the circuit produces an electric field E in the space around the conducting wires, leaving the space inside the cell. The line integral $\left(\int Edl \right)$ of the electric field from the negative terminal of the cell to the positive terminal, i.e. in the direction of the potential rise, is equal to the emf V_{emf} or the terminal voltage of the cell. Therefore,

$$V_{emf} = V \text{ or } V - V_{emf} = 0$$

i.e. algebric sum of all voltage drops in a closed circuit or loop is zero.

This establishes Kirchhoff's voltage law.

To obtain Kirchhoff's current law, let us consider a node in the circuit, an enlarged view of which is shown in Fig. 1.14. As may be seen in the figure, n number of conducting wires with areas of cross section s_1, s_2, s_3, ... s_k ... s_n and current densities j_1, j_2, j_3, ... j_k,... j_n, respectively, join at the node. It may be observed that some current densities are positive, leading towards the node, and some negative, leaving the node. The total current through the kth conductor wire, for example, is $I_k = \left(\vec{j}_k . \vec{n}_k \right)$. Here, n_1, n_2,...etc. indicate the unit vector normal to the respective surface. The current will have a negative sign for those conductors in which the direction of current density is opposite to the unit normal vector.

Let us enclose the node in some imaginary volume V and let the total surface of the volume be S.

Using the divergence theorem one may obtain

$$\iiint \nabla . j dV = \iint j . n dS \tag{1.28}$$

Fig. 1.14 Enlarged view of the node in a circuit. Several conductor wires of different areas of cross section, carrying currents of different current densities, join at the node

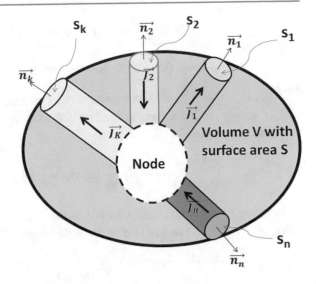

In the above equation, the integral on the left-hand side is over the imaginary volume V while the integral on the right is over the surface S of the volume V. The integral on the right side may be broken into two parts, one covering the surfaces of the conductors and the other covering that part of the surface where there is no conductor, i.e. surface $S' = \left[S - (s_1 + s_2 + s_3 \ldots + .s_k \ldots + s_n) \right]$.

$$\iiint \nabla.j dV = \sum_{1}^{n} \iint j_k.n_k ds_k + \left(\iint J.n dS' \right)_{S'} \qquad (1.29)$$

or

$$\iiint \nabla.j dV = \sum_{1}^{n} I_k + 0 \qquad (1.30)$$

The first surface integral on the right side of (1.29) gives the algebraic sum of currents through conductor wires. The second integral on the right-hand side over the surface S' has the value zero as there are no currents through the remaining surface s'. Thus, it follows from (1.24a) that in order to obtain the algebraic sum of current it is required to evaluate the volume integral $\int \int \int \nabla.j dV$.

From (1.27a) of modified Ampere's law, it follows that

$$\frac{\partial D}{\partial t} + J = \nabla \times H \qquad (1.31)$$

Applying gradient operator on both sides of the above equation, one obtains

$$\frac{\partial(\nabla.D)}{\partial t} + \nabla.J = \nabla.(\nabla \times H) = 0 \tag{1.32}$$

since $\nabla.(\nabla \times H)$ for any field vector is always zero.

Also, $(\nabla.D) = \rho$ *from* (1.24a) , hence (1.32) reduces to
$\nabla.J = -\frac{\partial \rho}{\partial t}$, taking volume integral on both sides and assuming that ρ and $\frac{\partial \rho}{\partial t}$ are continuous, one gets

$$\iiint \nabla.j dV = -\iiint \frac{\partial \rho}{\partial t} dV = -\frac{d}{dt}\iiint \rho dV = \frac{dQ}{dt} = 0 \tag{1.33}$$

In (1.33), Q is the total charge, and for steady current $\frac{dQ}{dt}$ is zero. Substituting the value of $\int \int \int \nabla.J dV = 0$ in (1.30), one gets Kirchhoff's current law,

$$\sum_1^n I_k = 0$$

(X) Planner and Non-planner circuits

A planner circuit is a circuit that may be drawn on a piece of paper, i.e. circuit elements are all arranged in only two dimensions. However, there may be networks where elements are distributed in three directions. The circuits where elements are arranged in three dimensions are called non-planner and are more involved so far as their analysis is concerned.

1.3 Circuit Analysis

Circuit analysis essentially means to find out the magnitudes and directions of currents and potential drops at various locations in the circuit. In principle, network analysis may be done using Kirchhoff's voltage and current laws. However, in case of complicated circuits having several sources and passive elements, the use of some shortcut methods simplify the application of Kirchhoff laws. Two such methods, namely Node voltage method and Mesh current method are frequently used for circuit analysis. However, the node voltage and mesh current methods are applicable only to planner circuits. Brief reviews of these methods are presented here.

(a) Node voltage method

This method allows the calculation of voltages at nodes of the network. In order to reduce the number of calculations, it is customary to select one particular node and assign it to zero potential so that the voltages of other nodes are calculated with reference to this zero potential node. Once the potentials of the nodes are known, currents in different branches may be calculated. The method leads to a set of simultaneous equations that may be solved to get currents and voltages in the circuit. The following steps are generally taken in nodal voltage analysis.

1. To identify all the nodes in the circuit and to select a particular node with respect to which the voltages of all other nodes are to be measured. Also, do away with the non-essential nodes, so far as possible, to reduce the number of nodes.
2. Identify the node for which the potential is known because of some voltage source.
3. Assign voltages to identified notes and currents to different branches.
4. Apply Kirchhoff's voltage and current laws to obtain a set of simultaneous equations.
5. Solve these equations to obtain currents and voltages at different branches and at different nodes.

The following solved examples will further clarify the method of application of the nodal voltage method.

Solved example (SE1.4) Determine the magnitude and direction of currents I_1, I_2 and the voltage at node B of the given circuit using the node voltage method (Fig. SE1.4).

Solution: *As may be observed in the figure, the circuit has six nodes: A, B, B', C, D, E and F. Nodes B and B' are the same as there is no circuit element between them. Let us choose node D to be our reference node and let us assume that the potential of this node is zero, i.e. $V_D = 0$. The potentials of all other nodes with respect to node D are denoted as V_A, V_B, V_C... The circuit contains three voltage sources of 20 V, 10 V and 5 V, four resistances each of 10Ω and one independent current source of 15 A. Voltages at nodes A and E are known as they are connected directly to voltage sources the other ends of which are connected to node D at $V_D = 0$.*

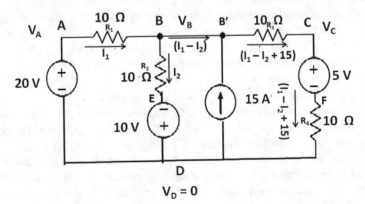

Fig. SE1.4 Circuit for example (SE1.4)

We assume current I_1 flows from 20 V source towards the resistance R_1 as shown in the circuit diagram. At node B, current I_1 divides into two parts I_2 going down through R_2 and $(I_1 - I_2)$ flowing through towards node B' where 15 A current from the current source joins it. Therefore, the current of magnitude $(I_1 - I_2 + 15)A$ passes through resistors R_3, R_4 and the voltage source of 5V. From the circuit, it is already known that voltages at nodes A and E with respect to node D are

$$V_A = 20\,V \text{ and } V_E = -10\,V$$

Using Ohm's law, we can calculate the voltage V_B of node B through three independent paths: (i) DAB, (ii) DEB and (iii) DFCB. This will result in three simultaneous equations involving three unknown quantities I_1, I_2 and V_B .

$$V_B = (20 - 10I_1) \quad \text{Path } (i)\ DAB \qquad\qquad \text{(SE1.4.1)}$$

$$V_B = (20 - 10I_2) \quad \text{Path } (ii)\ DEB \qquad\qquad \text{(SE1.4.2)}$$

$$V_B = [20(I_1 - I_2 + 15) + 5] \quad \text{Path } (iii)\ DFCB \qquad\qquad \text{(SE1.4.3)}$$

The above three simultaneous equations can be easily solved to get

$$I_1 = -4.5\,A; I_2 = +7.5; (I_1 - I_2 + 15) = +3\,A \text{ and } V_B = 65V$$

The negative sign in front of the value of current I_1 indicates that the current is flowing in the direction opposite to the direction assumed.

(b) Mesh analysis

Another powerful technique of analysing complicated circuits is the mesh analysis, also called Maxwell's cyclic current method. The method is based on identifying meshes in the network and assigning consistently clockwise or anti-clockwise cyclic currents to each mesh. Kirchhoff's voltage law (KVL) is then applied to these meshes to obtain simultaneous equations involving assigned currents as unknown variables. Since the number of unknown quantities is equal to the number of meshes and each mesh gives one equation corresponding to the application of KVL, the solution of these simultaneous equations yields the values of the currents assigned to different meshes. Care is taken to assign positive or negative signs to voltage drops against different circuit elements and voltage sources.

As a thumb rule, voltage drops across resistances are taken positive if traversing the resistance along the direction of the flow of current and negative if opposite to the direction of current flow. In case of voltage sources, if the assumed cyclic current enters the source at the positive terminal and goes out at the negative terminal, a potential drop of the source is taken as positive. On the other hand, if the cyclic current enters the source at its negative terminal and goes out at the positive terminal, the source voltage is taken as negative. The sign convention for voltage

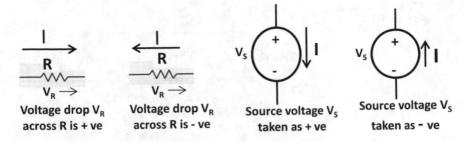

Fig. 1.15 Sign convention for voltage drops

drops is shown in Fig. 1.15. The following worked out example will further clarify the use of the mesh technique.

Solved example (SE1.5) Compute the currents in different branches of the following circuit (Fig. SE1.5a).

Solution: *The circuit shown in the figure has three meshes marked 1, 2, 3. Cyclic currents I_1, I_2 and I_3 in the clockwise direction have been assigned, respectively, to meshes 1, 2 and 3. Applying KVL to mesh 1 yields the following equation:*

$$R_3 I_1 + R_4(I_1 - I_3) + R_2(I_1 - I_2) = 0$$
$$\text{or} \quad 50I_1 + 40(I_1 - I_3) + 10(I_1 - I_2) = 0 \tag{SE1.5.1}$$
$$\text{or} \quad 100I_1 - 40I_3 - 10I_2 = 0$$

Similarly, the application of KVL to mesh 2 yields

$$R_2(I_2 - I_1) + V_S + R_1 I_2 = 0$$
$$\text{or} \quad 10(I_2 - I_1) + 50 + 30I_2 = 0 \tag{SE1.5.2}$$
$$\text{or} \quad 50I_2 - 10I_1 + 50 = 0$$

Fig. SE1.5a Circuit for example (SE1.5)

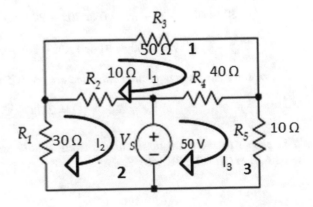

Fig. SE1.5b Network of
example (SE1.5)

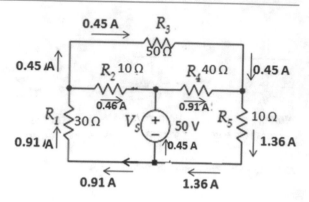

Similarly, the application of KVL to mesh-3 gives

$$50I_3 - 40I_1 - 50 = 0 \qquad\qquad \text{(SE1.5.3)}$$

On solving the three simultaneous equations (SE1.5.1), (SE1.5.2) and (SE1.5.3), one gets

$$I_1 = 0.45A; I_2 = -0.91A; \text{ and } I_3 = 1.36A$$

Currents in different branches of the circuit may be obtained from the values of mesh currents and are shown in Fig. (SE1.5b).

Solved example (SE1.6) Use the mesh analysis method to obtain the voltage of the voltage-controlled voltage source and currents in different branches of the given network (Fig. SE1.6a).

Solution: *The circuit contains two meshes 1 and 2. A voltage-controlled voltage source is connected in mesh-2. The controlling voltage V_x is the voltage drop against the resistance R_2 of 20 Ω. The gain factor of the source is 5. Let us assign cyclic currents x and y, respectively, to mesh-1 and mesh-2. It is simple to write KVL equations for the two meshes as*

$$50x + 20(x - y) - 50 = 0 \text{ or } 70x - 20y - 50 = 0 \qquad \text{(SE1.6.1)}$$

$$20(y - x) - 5V_x + 100y = 0 \text{ or } 120y - 20x - 5V_x = 0 \qquad \text{(SE1.6.2)}$$

Also, since the voltage of the controlled source is 5 times V_X where

$$V_X = R_2(x - y) = 20(x - y) \qquad\qquad \text{(SE1.6.3)}$$

it is simple to solve the three simultaneous equations (SE1.6.1), (SE1.6.2) and (SE1.6.3) to get (Fig. SE1.6b)

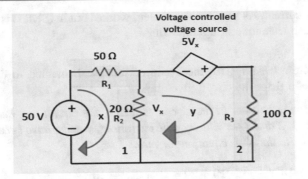

Fig. SE1.6a Circuit for example (SE1.6)

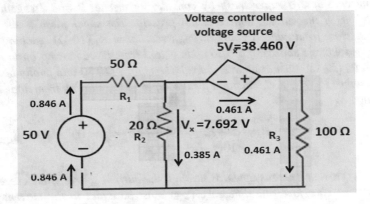

Fig. SE1.6b Current distribution and the voltage of the controlled source are shown in the figure

$$x = 0.846A; y = 0.461A \text{ and } V_X = 7.692V$$

(c) Super mesh and current source

The mesh analysis becomes a little involved if there is a current source in the circuit. It is because in the mesh analysis, one applies KVL for each mesh and the voltage drop across the current source is not known, therefore, is not possible to write the KVL equation for the mesh containing the current source. In such cases involving current sources, one assigns cyclic currents in all meshes as is always done in the mesh analysis but the KVL equation is not written for the mesh containing the current source. Instead, one creates a super mesh by removing the current source as if it is not in the circuit and writes the KVL equation for this super mesh. An additional equation is obtained by writing the current passing through the current source in terms of the cyclic mesh currents and equating it to the given value of the source current. As an example problem worked out in (SE1.7) is solved using the super mesh technique in the following.

Solved example (SE1.7) Use the cyclic current mesh method to determine the voltage of node B for the circuit given below.

Solution: *As may be seen in the circuit diagram, the circuit can be divided into three meshes, 1, 2 and 3 for which cyclic currents x, y and z have been assigned. It is simple to write the KVL equation for mesh-1 as*

$$10x + 10(x - y) - 10 - 20 = 0 \text{ or } 20x - 10y - 30 = 0 \qquad (SE1.7.1)$$

However, it is not possible to write KVL equations both for mesh-2 and mesh-3 because the voltage drop against the current source is not known. To overcome this, one considers a super mesh by considering meshes 2 and 3 together and assuming as if the current source is not present. The super mesh is the closed circuit starting from node B going through resistance R_3 (10 Ω) reaching C and moving via voltage source of 5 V to node F and from there moving on to node D through R_4 (10 Ω), then to node E via voltage source (10 V) and finally back to B via resistance R_2. The KVL equation for the super mesh, starting from node B, may be written as

$$R_3z + 5 + R_4z + 10 + R_2(y - x) = 0$$
$$\text{or} \quad 20z + 10(y - x) + 15 = 0 \qquad (SE1.7.2)$$

Finally, current z passes upwards through the current source while current y is going down. Therefore, the difference (z-y) must be equal to the current of the source.
  So

$$z - y = 15 \qquad (SE1.7.3)$$

The solution of simultaneous (SE1.7.1), (SE1.7.2) and (SE1.7.3) gives

$$x = -4.5\,A; \ y = -12.0A \text{ and } z = 3.0\,A$$

The magnitude of current x is 4.5 A and its direction is opposite to the assumed direction (clockwise). Current x, therefore, flows through R_1 from B to A. It means that the potential of B is more than the potential at A by the amount $(R_1.x) = 45$ V. Node A is at +20 V with respect to node D. Therefore, the potential of node B with respect to node D is (45 + 20 =) 65 V.

Self-assessment question (SAQ): Compute the currents in all branches of the circuit given in Fig. SE1.7.

Fig. SE1.7 Circuit diagram
for problem (SE1.7)

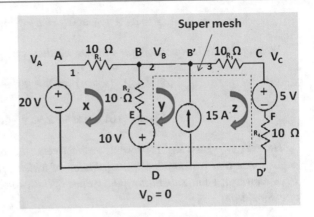

The mesh method of analysing is also applicable to AC networks. However, in AC networks an additional parameter of phase difference has to be taken into account at each step. As such, mesh currents for AC networks are taken as complex numbers like (x + j y), and values of the real component 'x' and the imaginary component 'y are both determined by solving the KVL equations for different meshes. The following solved example will outline the steps involved in the application of the mesh analysis to AC networks.

Solved example (SE1.8) Using the mesh analysis method, calculate currents through the three elements of the circuit of Fig. SE1.8.

Solution: *Since the circuit contains AC sources, the assumed cyclic currents x and y in the two meshes are complex numbers having a real part and an imaginary part. Let the currents x and y be represented as*

Fig. SE1.8 Network for
example (SE1.8)

$$X = (a_x + j\ b_x)$$
$$y = (a_y + j\ b_y)$$

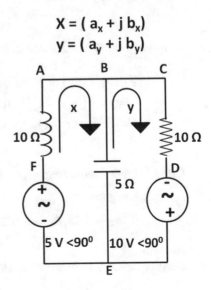

$$x = a_x + jb_x \text{ and } y = a_y + jb_y \tag{SE1.8.1}$$

Writing the KVL equations for the two meshes, one gets

$$Z_L.x + Z_C(x - y) - 5V < 90^0 = 0 \tag{SE1.8.2}$$

$$\text{or } Z_R.y - 10V < 90^0 + Z_C.(y - x) = 0 \tag{SE1.8.3}$$

Here Z_L, Z_C, and Z_R represent, respectively, the impedances of inductance, capacitance and resistance the magnitudes of which are given as 10Ω, 10Ω and 10Ω. However, while substituting the values of these impedances in KVL equations, one should put

$$Z_L = (0 + j10\Omega), Z_C = (0 - j5) \text{ and } Z_R = (10 + j0) \tag{SE1.8.4}$$

Also,

$$5V < 90^0 = (0 + j5) \text{ and } 10V < 90^0 = (0 + j10) \tag{SE1.8.5}$$

Substituting the values of x and y in terms of a and b from (SE1.8.1) and of impedances from (SE1.8.4) along with source potentials from (SE1.8.4), in (SE1.8.2) one may obtain

$$(0 + j10)(a_x + jb_x) + (0 - j5)\left[(a_x - a_y) + j(b_x - b_y)\right] - (0 + j5) = 0$$
$$\text{or } j10a_x - 10b_x - j5a_x + j5a_y + 5b_x - 5b_y = 0 + j5$$
$$\text{or } j5(a_x + a_y) - 5(b_x + b_y) = 0 + j5$$
$$\text{or } (a_x + a_y) = 1 \text{ and } (b_x + b_y) = 0$$

$$\tag{SE1.8.6}$$

Similarly from (SE1.8.3), one gets

$$(10 + j0)(a_y + jb_y) + (0 - j5)\left[(a_y - a_x) + j(b_y - b_x)\right] = (0 + j10)$$
$$\text{or } \left[10a_y + 5(b_y - b_x)\right] + j\left[10b_y - 5(b_y - b_x)\right] = (0 + j10)$$
$$\text{or } 10a_y + 5(b_y - b_x) = 0 \text{ and } 10b_y - 5(b_y - b_x) = 10$$

$$\tag{SE1.8.7}$$

Using the four relationships contained in (SE1.8.6) and (SE1.8.7), it is easy to get

$$a_x = 1.25, b_x = -0.25. a_y = -0.25 \text{ and } b_y = 0.25$$

Therefore,
Current through Inductance $L = X = (a_x + jb_x) = (1.25 - j0.25)A$

Current through the resistance R=y=$(a_y + jb_y) = (-0.25 + j0.25)A$
Current through the capacitance $C = (X-Y) = (1.50 - j0.5)A$

Self-assessment question (SAQ): In networks that contain current source, one considers super mesh to bypass the current source. What is the reason for that, what is it that is not known about the current source?

(d)   Conversion of a voltage source into a current source and vice versa
Two types of independent sources are generally used to energize electrical and electronic circuits: The constant voltage source and the constant current source. An ideal voltage source is one which delivers a voltage or potential difference V_S at its terminals that remain constant for an infinitely long time and that (V_S) does not change with the load connected to the source. The load is the combination of various elements, external to the source that is connected to the two terminals of the source and is powered by it. Similarly, an ideal current source is one that delivers current I_S of constant magnitude for an infinitely long time, and the current does not change with the load. Under ideal conditions, the internal resistance of a voltage source is zero and that of a current source is infinite. An ideal voltage or current source is only conceptual. A real or practical voltage/current source delivers a nearly constant voltage/current for a reasonable period of time and the output voltage/current of the source changes, by a small amount, with load. There are two reasons for the deviation of a practical source from the ideal one. Firstly, the process by which some other form of energy is converted into electrical energy within the source depletes with time. For example, in most of the cells some chemical reaction produces electrical energy and with time, concentrations of the energy-producing chemicals deplete resulting in the corresponding change in the voltage/current of the source. The second reason is the non-zero internal resistance of the voltage source and finite resistance of the current source. These resistances arise from the materials used in the fabrication of these sources. A practical voltage source is, therefore, represented with a small internal resistance R_{int} in series with the source and a practical current source with resistance in parallel with the source.

Figure 1.16 shows a practical voltage source. As shown in this figure, the electromotive force (emf) or the open-circuit potential difference and the internal resistance of the source are, respectively, V_s and R_{int}. The internal structure of the source is assumed to be contained within the dotted lines. The maximum current that may be drawn from this voltage source, $(I_{max})_V = V_S/R_{int}$. This maximum current will be drawn when the load is zero, i.e. the terminals of the source are short-circuited. Now if it is required to convert this voltage source into a current source, obviously the maximum current that the corresponding current source will be able to deliver will be equal to $(I_{max})_V = V_S/R_{int}$. This maximum current will be drawn when the terminals of the current source are open.

The equivalent current source is shown in (b) part of the figure. For example, if there is a voltage source of 10 V and internal resistance 0.5Ω, then it is equivalent to a current source of 20 A with resistance of 0.5Ω in parallel to it. It needs to be remembered that a voltage source is used in parallel to the load while a current

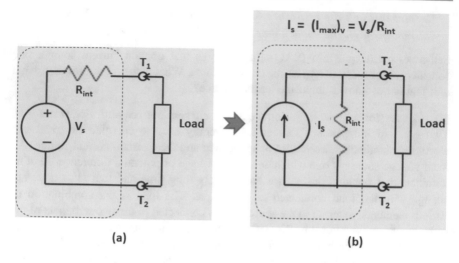

Fig. 1.16 Converting a practical voltage source into a current source

source is always in series to the load. Therefore, if a load of infinite impedance is connected to the current source which does not draw any current from the source (which means that terminals T_1 and T_2 of the current source, figure (b), are open), the terminal voltage across the current source will be ($I_S \cdot R_{int} =$) V_s, which is the emf of the voltage source.

Similarly, a current source of current I_s with parallel resistance R_S is equivalent to a voltage source of voltage $V_S = I_S R_S$ with resistance in series of value R_s. A 10 A current source with parallel resistance of 2Ω is equivalent to a voltage source of voltage $V_S = 10 \times 2 = 20V$ with a series resistance of 2Ω.

Self-assessment question (SAQ): In what respect a practical current or voltage source differs from an ideal source and why?

1.4 Network Theorems

The node voltage method and the cyclic current mesh method are two shortcuts for solving electrical networks. Network theorems are additional tools to simplify complicated electrical networks. Theorems, by definition, are the statements that are true and can be proved. An essential difference between a postulate and a theorem is that a postulate cannot be proved while a theorem can be proved. Fundamental network theorems and their applications will be discussed in the following.

1.4.1 Superposition Theorem

The superposition theorem is the most fundamental theorem that is applicable to both DC and AC networks. The theorem applies to those networks which have many sources distributed in the network at different nodes. The theorem states that:

The current through or the voltage across any passive circuit element in a network having many sources distributed in the network is equal to the algebraic (or vector) sum of the current or voltage produced in the given element independently by one source at a time.

The theorem essentially resolves a complicated network having many, say N, sources (current and/or voltage) distributed in the network into N simple networks each with only one source. The current through or voltage across a particular circuit element can be computed for each of these N single-source networks, and the desired final value of the current or voltage may be obtained by taking the algebraic sum of the N values obtained earlier.

In this prescription of the theorem, the obvious question that arises is: **what should be done to (N-1) other sources when analysing the circuit with only one particular source?** The answer is simple:

When a voltage source is removed it is replaced by a short circuit, if it has no internal resistance otherwise by resistance equal in value to the internal resistance of the removed source.

Similarly,

While removing a current source it is replaced by an open circuit if it has no internal resistance, otherwise, it is replaced by resistance equal in value to the internal resistance of the source.

It may be realized that a current source has an internal resistance in parallel to the source while a voltage source has an internal resistance that is in series with the source. The above-stated technique of removing voltage and current sources may be explained by the circuit diagrams shown in Fig. 1.17.

As shown in the left side of Fig. 1.17, when a voltage source is removed it is replaced by its internal resistance R_{int}, and if the internal resistance is zero then by a

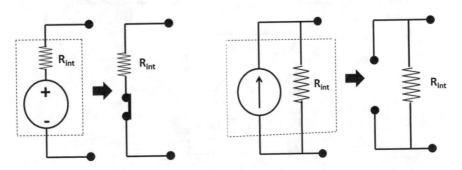

Fig. 1.17 Replacing voltage and current sources for the application of Superposition theorem

short circuit. A current source is replaced by an open circuit, if it has infinite parallel impedance or by the parallel impedance R_{int}, as shown in the right side of Fig. 1.17.

The superposition theorem says that each energy source has an independent effect on each element of the network and that these effects do not interfere with each other, meaning thereby that the effect of a particular source S_1 on a specific element R_X does not depend on the effect of source S_2 (or of any other source) on R_X. The net cumulative effect on a particular element is the algebraic (or vector) sum of the effects independently produced by different sources.

Limitation: It is important to note that the superposition theorem applies only to those effects that are linear with regards to the element, like current through and voltage across resistance or impedance. The theorem cannot be applied for non-linear quantities like power dissipated by a resistor, etc. or energy stored in an inductor or a capacitor.

The following solved examples will demonstrate the application of the superposition theorem.

Self-assessment question (SAQ): Is it possible to calculate the amount of total power dissipated by resistance by summing up the power dissipated using one source at a time invoking the superposition theorem? Give reasons for your answer.

Solved example (SE1.9) Fig. SE1.9a shows the original circuit of the problem and figs (b) and (c) the single source equivalents. Calculate using the superposition theorem the current through 10Ω resistance. Also, calculate the power supplied to the resistance individually by the two sources

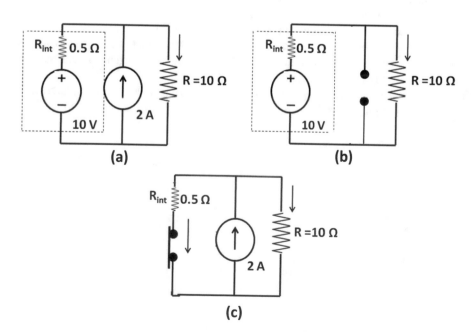

Fig. SE1.9 Circuits for example (SE1.9)

Solution: *The two single-source reductions of the given circuit as per the superposition theorem are shown in figures (b) and (c). In figure (b), the current source has been removed and since it has no internal resistance in parallel to the current source, an open pair of terminals is left. The total resistance in the circuit of figure (b) is* $R_{total} = R_{int} + R = (0.5 + 10)\Omega = 10.5\Omega$.

Hence, the current through the circuit or the resistance R is $I_{R1} = \dfrac{V_S}{R_{total}} = \dfrac{10V}{10.5\Omega} = 0.9524A$.

And the power dissipated per second by resistance R is $P_1 = R(I_{R1})^2 = 10 \times (0.9524)^2 = 9.07W$.

Figure (c) shows the circuit after removing the voltage source. The internal resistance of the source which is taken to be in series with the voltage source is retained after short-circuiting the source. Let us assume that current I_{R2} *passes through resistance R (=10 Ω) and the remaining current* $(2 - I_{R2})A$ *passes through the internal resistance* R_{int} *(=0.5Ω). Since both R and* R_{int} *are connected to the terminals of the current source, the potential difference against the two resistances must be the same, hence*

$$R.I_{R2} = R_{int}.(2 - I_{R2}) \text{ or } 10xI_{R2} = 0.5x2 - 0.5I_{R2}$$

or $10.5I_{R2} = 1.0$ *and* $I_{R2} = 0.095$ A, *and the energy dissipated by* $R = RI_{R2}^2 = 0.090$ W

Since for both cases (b) and (c) the current through resistance R is in the same direction, the algebraic sum of currents $I = I_{R1} + I_{R2} = 1.047A$ *and algebraic sum of energy dissipated* = (9.0 + 0.09) = 9.09 W.

Now to test the validity of the superposition theorem, let us compute the current through R when both sources are present simultaneously as shown in figure (a). Let us use the mesh current method for this computation. Let x and y be the assumed cyclic currents in two meshes as shown in figure (d). Since there is a current source, we take the super mesh and write the following KVL equations (Fig. SE1.9).

$$10.y - 10 + 0.5.x = 0 \text{ and } y - x = 2$$

which gives x = −0.952 A and y = 1.047 A. Since current y flows through R, the result of the superposition theorem method matches from the current obtained

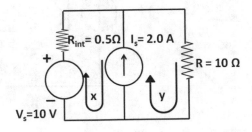

Fig. SE1.9 Circuits for example (SE1.9)

*using the mesh method. Now let us further calculate energy dissipation in resis-
tance R when current of 1.047 A passes through it. W = 10 × (1.047)² = 10.962 W.
It is very clear that the total energy dissipation calculated assuming the superposition theorem does not match with the value obtained using the mesh method. This
was expected because power dissipated is not a linear function.*

Solved example (SE1.10) Calculate the magnitude and the direction of the current
through resistor R_2 using the superposition theorem for the network and data
provided in Fig. SE1.10a.

Solution: *The given network has two energy sources, a voltage source of 20 V with
no internal resistance and a current source of 5A also with no parallel internal
resistance. Figure (b) shows the circuit left after removing the current source. As
may be seen in* Fig. SE1.10b, *the resistances R_1 and R_2 are connected in series and
this series combination is connected to the two terminals of the voltage source of
emf 20 V. The current through this series combination is given by
$I_1 = \frac{V_S}{R_1+R_2} = \frac{20}{30} = 0.66A$. This current will flow through R_2 from the positive
terminal of the voltage source towards the negative terminal that is from A to B.*

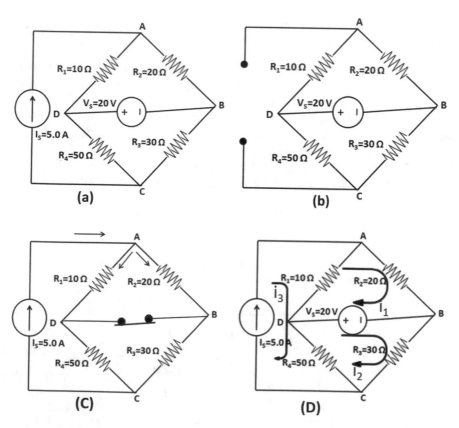

Fig. SE1.10 Circuit diagrams for example (SE1.10)

Next, in Fig. SE1.10c *the voltage source is removed and is replaced by a short circuit. Now the resistances R_1 and R_2 are in parallel and 5.0 a current from the current source will enter at A. This 5.0 A current will divide into two parts, one part say X goes through R_2 and the remaining (5-x) will go through R_1. The currents X and (5-x) will be in the inverse ratio of the resistances, more current through the smaller resistance. Therefore,*

$$\frac{X}{(5-x)} = \frac{R_1}{R_2} = \frac{10}{20} \text{ or } 2x = 5 - x \text{ or } 3x = 5 \text{ and } x = 1.66 \ A$$

The current X will also flow from A to B through R_2.

Thus, current $I_1(= 0.66A)$ flows through R_2 from A to B when the current source is removed, and another current X flows from A to B through R_2 when the voltage source is removed. According to the superposition theorem, the total current through R_2 will be equal to the algebraic sum of I_1 and x. Since both the currents flow in the same direction, the current I through R_2 when both the sources are present is

$$I = I_1 + x = (0.66 + 1.66) = 2.32A$$

Let us verify if the current I obtained using the superposition theorem is the correct value of current when both the sources are present. For that, we use the mesh method to find the current through R_2. As shown in Fig. (d), currents I_1, I_2 and I_3 are assigned to the three meshes.

The KVL equations for the two meshes are

$$20I_1 - 20 + 10(I_1 - I_3) = 0$$
$$\text{And} \quad 30I_2 + 50(I_2 - I_3) + 20 = 0$$

Now for the third mesh that contains the current source, we do not write the KVL equation but use the fact that $I_3 = 5.0A$

The three simultaneous equations may be solved to get the values:

$$I_1 = 2.33A; I_2 = 2.88A; I_3 = 5.0A$$

It may be observed that Current I through R_2 obtained by the superposition theorem (I = 2.32 A) matches well with the value ($I_1 = 2.33A$) obtained using the mesh analysis, within errors of rounding off.

Solved example (SE1.11) Compute the current passing through each element of the network of Fig. (SE1.11a) using the superposition theorem. The frequency of both voltage sources is 100 Hz.

Solution: Figures (SE1.11b, c) *show the circuits obtained by removing the 10 V and 5V sources, respectively. Let us first consider the circuit of figure (b). As may be seen in this figure, the 10Ω resistance and 5Ω capacitors are in parallel. Let us denote the impedance of this parallel combination by $Z_{R\|C}$ that is given by*

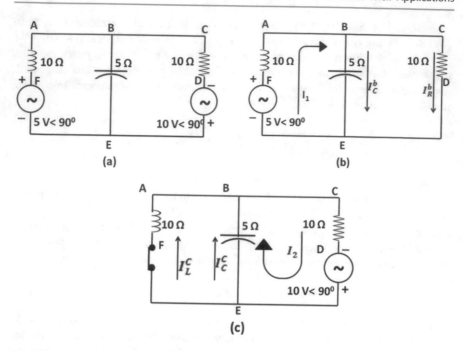

Fig. SE1.11   Circuits for example (SE1.11)

$$\frac{1}{Z_{R||C}} = \frac{1}{Z_R} + \frac{1}{Z_C} = \frac{1}{10} + \frac{1}{(-j5)} = 0.1 + j\frac{1}{5} = 0.1 + j(0.2)S$$

$$\text{And} \quad Z_{R||C} = \frac{0.1 - j(0.2)}{(0.1)^2 + (0.2)^2} = \frac{0.1 - j(0.2)}{0.05} = 2.0 - j(4.0)\Omega$$

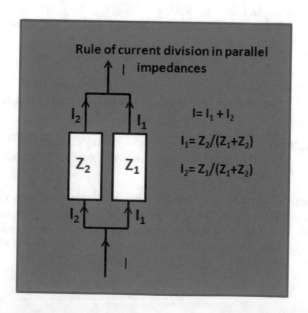

This $Z_{R||C}$ *is in series with* $Z_L = (0+j10)\Omega$. *Therefore, the total impedance across the 5V voltage source is*

$$Z_{Total}^b = (2.0 - j4.0) + (0+j10.0) = (2.0+j6.0)\Omega = \left(2.0^2 + 6.0^2\right)^{\frac{1}{2}} < tan^{-1}\frac{6.0}{2.0}$$

$$= 6.32\Omega < 71.56^0$$

Therefore, the current through the series combination of resistance and $Z_{R||C}$ *is*

$$I_1 = \frac{5\ V < 90^0}{6.32\Omega < 71.56} = 0.79\ A < 18.43^0 = 0.75 + j(0.25)A \qquad (SE1.11.1)$$

Thus, the current through the 10Ω *inductance for the circuit of figure (b) is*

$$I_L^b = I_1 = 0.79A < 18.43^0 = 0.75 + j(0.25)A$$

Now from the law of current division in parallel impedances, the current I_R^b *through the resistance is given by*

$$I_R^b = \frac{Z_C}{Z_C + Z_R}I_1 = \frac{-j5}{10-j5}\left(0.79A < 18.43^0\right) = (0.2 - j0.4)\left(0.79A < 18.43^0\right)$$

or $\quad I_R^b = (0.45A < -63.43)(0.79A < 18.43^0) = 0.36A < -45^0 = (0.25 - j0.25)A$

And the current I_C^b *through the capacitor is given by*

$$I_C^b = \frac{Z_R}{Z_C + Z_R}I_1 = \frac{10}{10-j5}\left(0.79A < 18.43^0\right) = (0.8 + j0.4)\left(0.79A < 18.43^0\right)$$

$$= \left(0.89A < 26.56^0\right)\left(0.79A < 18.43^0\right) = 0.70A < 45^0 = (0.49 + j0.49)A$$

Therefore, for the circuit represented by figure (b):

$$I_L^b = I_1 = 0.79\ A < 18.43^0 = 0.75 + j(0.25)A.\ \text{From } F \text{ to } A$$

$$I_C^b = 0.70A < 45^0 = (0.50 + j0.50)A.\ \text{From } B \text{ to } E$$

$$I_R^b = 0.36A < -45^0 = (0.25 - j0.25)A.\ \text{From } C \text{ to } D$$

Next, we analyse the circuit shown in figure (c) obtained by removing the 5 V source. In this case, the inductance (of impedance $+j10\Omega$) *and capacitance (of impedance* $-j\ 5\ \Omega$) *are in parallel. The impedance* $Z_{L||C}$ *of the parallel combination is given by*

$$\frac{1}{Z_{L\|C}} = \frac{1}{+j10} + \frac{1}{-j5} = (-j0.1 + j0.2) = (0 + j0.1)S = 0.1S < 90^0$$

Hence, $Z_{L\|C} = \frac{1}{0.1 < 90} = -j10\Omega = 10 < -90^0$.

The $Z_{L\|C}$ and the reactance of the resistance are in series in figure (c), hence the total impedance in the circuit of figure (c) is

$$Z^c_{Total} = Z_{L\|C} + Z_R = -j10 + 10 = 14.14\Omega < -45^0$$

The current through the series combination of resistance and $Z_{L\|C}$ is given by

$$I_2 = \frac{10V < 90^0}{Z^c_{Total}} = \frac{10V < 90^0}{14.14\Omega < -45^0} = 0.71A < 135^0 = (-0.50 + j0.50)A$$

Currents through the inductance L and the capacitance C may be calculated using the law of current distribution in parallel combinations as

$$I^c_L = \frac{Z_c}{Z_L + Z_c} = \frac{-j5}{j10 - j5}I_2 = -1(0.71A < 135^0) = -1\{-0.5 + j0.5\} = (0.5 - j0.5)A$$

From F to A(same direction as I^b_L)

$$I^c_C = \frac{Z_L}{Z_L + Z_c} = \frac{j10}{j10 - j5}I_2 = 2I_2 = 1.42A < 135^0 = (-1.0 + j1.0)A.$$

From E to B (opposite to I^b_C)

And $I^c_R = I_2 = 0.71A < 135^0 = (-0.5 + j0.5)$ A From C to D (same direction as I^b_R)

Therefore, using the superposition theorem, currents through the circuit elements of the network given in figure (a) may be obtained by taking the algebraic sum of currents through the corresponding element for circuits of figure (b) and figure (c).

$$I^a_R = I^b_R + I^c_R = (0.25 - j0.25)A + (-0.5 + j0.5)A = (-0.25 + j0.25)A$$

$$I^a_C = I^b_C + I^c_C = (0.5 + j0.5)A - (-1.0 + j1.0)A = (1.5 - j0.5)A$$

$$I^a_L = I^b_L + I^c_L = (0.75 + j0.25 + (0.5 - j0.5)A = (1.25 - j0.25)A$$

The network of Fig. SE1.11 has been solved using the mesh cyclic current method in the worked out example (SE1.8). Results obtained by the application of *the superposition theorem method and the cyclic current method agree with each other proving the superposition theorem.*

(ii) Network with a dependent energy source

The method of tackling networks which contain dependent energy sources is illustrated in the following by taking an example.

Solved example (SE1.12) Figure SE1.12 shows a network that has a voltage-controlled voltage source. Compute the currents through the two resistances and discuss the possibility of using the superposition theorem to calculate currents in the circuit.

Solution: *Circuits having dependent sources, in general, could not be resolved using the superposition theorem. It is because controlled sources are not themselves sources of energy, that is, they do not generate energy; their energy output depends on the value of the voltage or current through some other element of the network. As such, they could not be replaced by an open circuit (in case of current source) or by a short circuit (in case of a voltage source). For example, in the circuit of Fig. SE1.12 the voltage-controlled voltage source cannot be replaced by a short circuit, unless voltage V across R_1 is zero. The voltage V will be zero only when current I_s from the current source is zero. If I_s is zero, it means that the circuit is not energized and then there will be no currents anywhere in the circuit. Therefore, the superposition theorem is not applicable in circuits that have dependent sources.*

To analyse such circuits, the conventional method of using KVL and KCL equations is employed. In the given circuit, the current I_1 is given by

$$I_1 = \frac{V}{R_1}$$

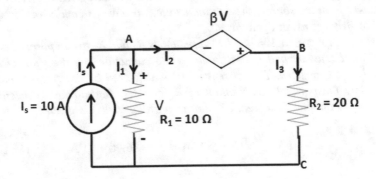

Fig. SE1.12 Circuit for example (SE1.12)

And the current I_3 through the resistance R_2 is not equal to I_2, since the dependent voltage source adds some additional potential at point B. The potential of point B with respect to point C is

$$V_B = V + \mu V = V(1 + \beta)$$

Therefore, $\quad I_3 = \frac{V_B}{R_2} = \frac{V(1+\beta)}{R_2}$

Now let us calculate V in terms of the current I_s. The total current delivered by the source I_s must be equal to the sum of $I_1 + I_3$. i.e.:

$$I_S = I_1 + I_3 = \frac{V}{R_1} + \frac{V(1+\beta)}{R_2} = V\left[\frac{R_2 + R_1(1+\beta)}{R_1 R_2}\right]$$

Hence, $\quad V = \frac{R_1 R_2}{[R_2 + R_1(1+\beta)]} I_S$

Substituting the given values of R_1, R_2, and I_s, one gets

$$V = \frac{200}{(20 + 10(1+\beta)} x10 = \frac{200}{3+\beta} V$$

The Current $\quad I_3 \frac{200}{3+\beta} x \frac{(1+\beta)}{20} = 10x\frac{(1+\beta)}{(3+\beta)} A$

And Current $\quad I_1 = \frac{20}{3+\beta} A$

Solved example (SE1.13) Figure SE1.13 shows a network that is energized by two DC and one AC sources. Assuming that all circuit elements are ideal having reactances as given in the figure, use the superposition theorem to compute the current through resistance R_1 of the network.

Solution: *Since all circuit elements are ideal, the inductance acts as a short circuit or a conducting wire and a capacitor acts as an open circuit when DC current is made to pass through them.*

Figure SE1.13a *shows the modified network when the DC current source is retained and the other voltage sources are replaced by short circuits. In case of the DC current from the source, the two inductances behave as conductors, while the capacitor behaves as an open circuit.*

As shown in Fig. SE1.13a, *the two resistances R_1 and R_2 are in parallel and the part of the circuit beyond C that includes the capacitor and the short-circuited battery is not connected. The current of 10 A delivered by the current source is divided into the parallel combination of R_1 and R_2. The current I_1 through R_1 (according to the law of current division in parallel loads) is given by*

$$I_{R1}^1 = \frac{R_2}{R_1 + R_2} I_s = \frac{20}{30} x10 = 6.66A < 0^0 = (6.66 + j0)A \qquad (SE1.13.1)$$

Fig. SE1.13 Network for
example SE(1.13)

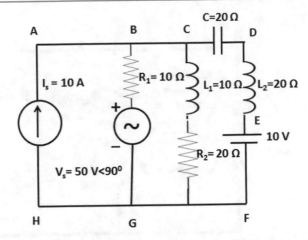

Fig. SE1.13a Modified
circuit for (SE1.13)

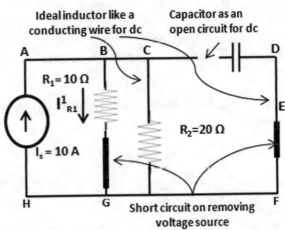

And the current I_{R1}^1 flows in the direction from B to G (Fig. SE1.13b).

*The circuit obtained by removing the DC current and DC voltage sources is
shown in figure (b). Now an AC voltage source is energizing the circuit and hence
inductors and capacitors offer their specified reactance. As may be seen in the
figure, the capacitor C and L_2 are in series, and the net impedance of this series
combination Z_{s1} is given as*

$$Z_{s1} = Z_C + Z_{L2} = (0 - j10) + (0 + j20) = (0 + j10)\Omega \qquad (SE1.13.2)$$

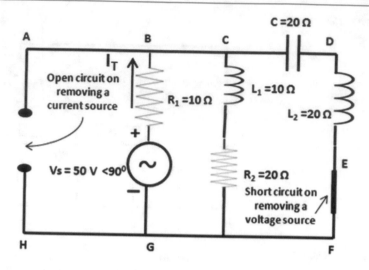

Fig. SE1.13b Modified circuit of example (SE1.13)

Similarly, the inductor L_1 and resistance R_2 are in series and their joint impedance is given by

$$Z_{s2} = Z_{R2} + Z_{L1} = (20 + j0) + (0 + j10) = (20 + j10)\Omega \qquad \text{(SE1.13.3)}$$

Now, Z_{s1} and Z_{s2} are in parallel and if $Z_{||}$ denotes the impedance of this parallel combination then,

$$\frac{1}{Z_{||}} = \frac{1}{Z_{s1}} + \frac{1}{Z_{s2}} = \frac{1}{(0 + j10)} + \frac{1}{(20 + j10)} = \frac{(0 + j10) + (20 + j10)}{(0 + j10)(20 + j10)} = \frac{1 + j1}{-5 + j10}$$

Therefore $Z_{||} = \frac{-5 + j10}{1 + j1} = \frac{5 + j15}{2} = (2.5 + j7.5)\Omega$

The series combination of R_1 and $Z_{||}$ is now connected across the voltage source V_s. As such, the total load across V_s is

$$Z_T = Z_{R1} < +Z_{||} = (10 + j0) + (2.5 + j7.5) = (12.5 + j7.5)\Omega$$
$$= 14.58\Omega < 30.96^0 \qquad \text{(SE1.13.4)}$$

The current drawn from source V_s by the load Z_T is

$$I_T = \frac{V_S}{Z_T} = \frac{50\ V < 90^0}{14.58\ \Omega < 30.96^0} = 3.43A < 59.04^0$$

Since R_1 is in series with the source, the current through R_1 in figure (b) is

$$I_{R1}^2 = I_T = 3.43A < 59.04^0 = (1.76 + j2.94)A \qquad (SE1.13.5)$$

And current $I_{R1}^2 (= I_T)$ flows in the direction from G to B.

Figure SE1.13c *shows the circuit obtained after removing the DC current source and the AC voltage source. The only source present is the 10 V batteries. It is to be noted that the capacitor blocks any DC to flow to the resistor R_1. Hence in this case, no current flows through resistance R_1.*

Thus, it may be seen that currents respectively I_{R1}^1 and I_{R1}^2 flow through the resistance R_1 when all the three sources of energy are present. The current through R_1, according to the superposition theorem, is the algebraic sum of I_{R1}^1 and I_{R1}^2. Since the direction of the flow of the two currents is opposite of each other, the current through R_1 is

$$I_{R1} = I_{R1}^1 - I_{R1}^2 = (6.66 + j0) - (1.76 + j2.94)A$$
$$= (4.90 - j2.94)A = 5.71A < -30.96^0$$

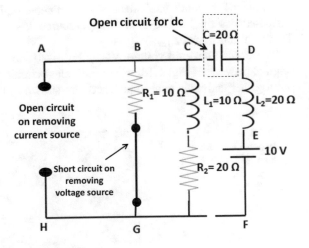

Fig. SE1.13c Modified circuit for example (SE1.13)

1.4.2   Thevenin's Theorem

Thevenin's theorem is another powerful tool to simplify complicated networks. The theorem is applicable to linear bidirectional DC and AC networks. Thevenin's theorem may be stated as follows:

"A linear bidirectional (DC or AC) network with respect to two specified terminals A and B of the network may be replaced by a voltage source V_{Thev} in series with impedance Z_{Thev} where V_{Thev} is the open-circuit potential difference between terminals A and B; and Z_{Thev} is the impedance of the network looked back into the network from terminals A and B".

For DC networks, where resistances and energy sources are the circuit elements, the impedance Z_{Thev} means the resistance of the network looked back into the network from the reference terminals A and B.

Figure 1.18 shows a typical linear bidirectional network that contains an energy source and some impedance distributed in the network. Suppose terminals A and B are chosen as reference terminals and it is required to apply Thevenin's theorem with respect to A and B. Let there be a load Z_L which when connected between terminals A and B, current I_L flows through it as shown in the figure. According to Thevenin's theorem, the network lying in the shaded area may be replaced by single energy source V_{Thev} in series with impedance Z_{Thev}, such that the same current I_L in the same direction flows through the load Z_L if connected between terminals A' and B' of the new Thevenin equivalent circuit as shown in Fig. 1.19. Since the simple circuit shown in Fig. 1.19 produces the same affect on the load as is produced by the original network of Fig. 1.18, the circuit shown in Fig. 1.19 is called Thevenin's equivalent between points A and B of the original circuit.

To draw Thevenin's equivalent of a given network, it is required to compute the characteristics of the Thevenin voltage source V_{Thev} and the magnitude of the Thevenin impedance Z_{Thev}.

To obtain Thevenin voltage V_{Thev}: Fig. 1.20 shows a simplified version of the original network of Fig. 1.18, where individual circuit elements in each branch are put together and their total impedances are shown in the boxes. According to the

Fig. 1.18 A typical linear bidirectional network to be replaced by the Thevenin equivalent between terminals A and B

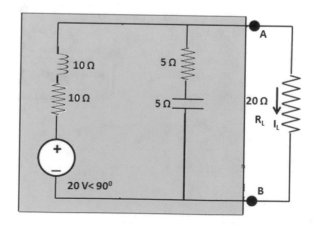

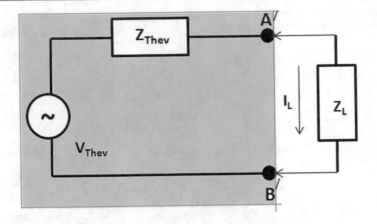

Fig. 1.19 Thevenin equivalent circuit between terminals A and B of network

Fig. 1.20 Simplified representation of the original network

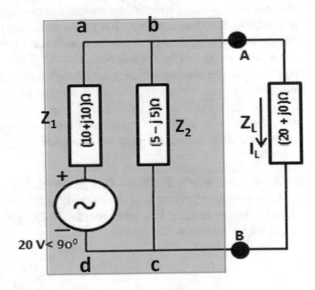

statement of Thevenin's theorem, V_{Thev} is the open-circuit potential difference between terminals A and B in the original network. In Figure 1.20, V_{Thev} is equal to the potential difference between points b and c or between terminals A and B. Therefore,

$$V_{Thev} = (\text{Potential drop } V_{Z2} \text{ across } Z_2) = \frac{(Z_2)(20V < 90^0)}{(Z_1 + Z_2)}$$

$$= \left\{ \frac{(5 - j5)(0 + j20)}{(10 + j10) + (5 - j5)} \right\} V = \frac{(100 + j100)}{(15 + j5)} = \frac{141.42 < 45^0}{15.81 < 18.43^0} = 8.94V < 26.56^0$$

$$(1.34)$$

Fig. 1.21 Voltage division
across two series impedances

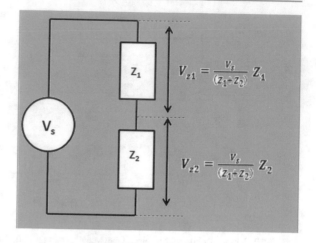

In driving equation (1.34), the principle of voltage division as shown in Fig. 1.21 has been used. It may, however, be pointed out that any other convenient method depending on the specific problem like the mesh current method or the node voltage method, etc. may be used to obtain the open-circuit voltage V_{Thev}.

To obtain Thevenin impedance: The Thevenin impedance Z_{Thev} is the impedance of the network looked back through terminals A and B. Further, while computing Z_{Thev} all current sources are replaced by an open circuit, and all voltage sources are replaced by a short circuit. Care is to be taken to retain the internal resistances of the sources, if present in the original network.

The Thevenin impedance may be calculated using one of the three possible methods.

(i) As the first step, replace the voltage sources by short circuits and current sources by an open circuit as shown in Fig. 1.22a. Now it may be observed that the impedances Z_1 and Z_2 are in parallel. If $Z_{\parallel}$ represents the impedance of the parallel combination, then

$$\frac{1}{Z_{\parallel}} = \frac{1}{(10+j10)} + \frac{1}{(5-j5)} = 0.15 + j0.05 \text{Siemen}$$

Hence, $Z_{\parallel} = \frac{(0.15 - j0.05)}{((0.15)^2 + (0.05)^2)} = (6.0 - j2.0)\Omega = (6.32\Omega < -18.43^0)$.

Since $Z_{\parallel}$ is the total impedance of the network looked back into the network from terminals A and B, it is equal to Z_{Thev}.

Therefore,

$$Z_{Thev} = (6.0 - j2.0)\Omega = (6.32\Omega < -18.43^0) \tag{1.35}$$

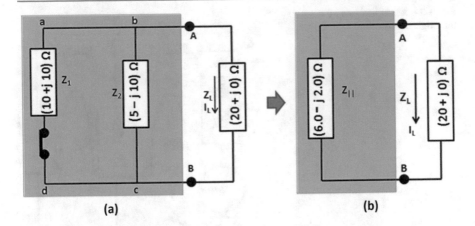

Fig. 1.22 a Voltage sources are replaced by short circuits and current source by open circuit. **b** $Z_{\parallel}$ represents the impedance of the parallel combination of Z_1 and Z_2

Figure 1.23 shows the final Thevenin's equivalent circuit with respect to terminals A and B of the original network of Fig. 1.18. It is simple to calculate the current I_L^{Thev} through the load obtained using the Thevenin equivalent as given below

$$I_L^{Thev} = \frac{V_{Thev}}{Z_{Thev} + Z_L} = \frac{8.94V < 26.56^0}{(6 - j2) + (20 + j0)} = \frac{8.94 < 26.56^0}{(26.08\Omega < -4.4^0)}$$
$$= 0.34A < 30.96^0 \tag{1.36}$$

To prove Thevenin's theorem, let us calculate the load current in the original network of Fig. 1.18, a simplified circuit of which is provided in Fig. 1.24. As may be seen in this figure, there are two distinct meshes in the network to which cyclic

Fig. 1.23 Thevenin equivalent of the original network

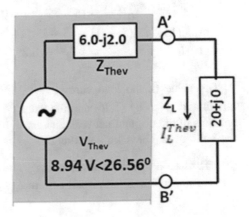

Fig. 1.24 Mesh analysis of
the original network

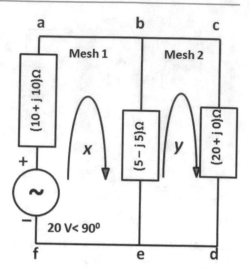

currents x and y have been assigned. It may further be noted that the load current I_L is the same as the mesh current y. Therefore, we write KVL equations for the two meshes and solve them to obtain the value of current y in mesh-2.

The KVL equation for mesh-1 is

$$(10+j10).x + (5-j5)(x-y) - (0+20j) = 0$$
$$\text{or}\quad (15+j5).x - (5-j5)y = (0+j20) \tag{1.37}$$

The KVL equation for mesh-2 may be written as

$$(5-j5)(y-x) + (20+j0)y = 0$$
$$\text{or}\quad (25-j5)y - (5-j5).x = 0 \; Or \; x = \frac{(5-j1)}{(1-j1)}y$$

Substituting the above value of x in (1.37), one gets

$$\left[\left\{\frac{(15+j5)(5-j1)}{(1-j1)}\right\} - (5-j5)\right]y = (0+j20)$$
$$\text{or}\quad [(15+j5)(5-j1) - (5-j5)(1-j1)]y = (0+j20)(1-j1) \tag{1.38}$$
$$\text{or}\quad I_L = y = \frac{(1+j)}{(4+j1)} = \frac{(5+3j)}{17} = 0.294 + j0.176 = 0.342A < 30.96^0$$

The value of the load current I_L^{Thev} calculated using Thevenin's equivalent circuit and given by (1.36) and the value of the load current obtained by the mesh analysis of the original network and given by (1.38) agree with each other proving the validity of Thevenin's theorem.

(ii) The second method of calculating Thevenin's impedance Z_{Thev} is by using the concept of short-circuit current I_{short}. I_{short} is the current that will flow

Fig. 1.25 Original network
with terminals A and B
short-circuited

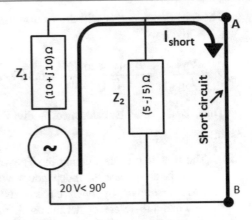

in the original network if terminals A and B of the network are connected together or short-circuited. In the present case, the original network with its terminals A and B short-circuited is shown in Fig. 1.25.

It is obvious from Fig. 1.25 that the impedance Z_2 is now redundant as it is in parallel to a short circuit of zero resistance. Hence, now current will flow through Z_2. The short-circuit current I_{short} is therefore given by

$$I_{short} = \frac{\text{Source voltage}}{\text{Impedance in the circuit}} = \frac{20V < 90^0}{Z_1} = \frac{20V < 90^0}{(10+j10)\Omega} = \frac{20V < 90^0}{14.14\Omega < 45^0}$$
$$= 1.41A < 45^0$$

$$(1.39)$$

Now we go back to the Thevenin equivalent circuit (Fig. 1.23) and demand that current I_{short} flows from the equivalent network when its terminals A' and B' are short-circuited. With this condition, we may calculate the impedance required in the circuit as follows (Fig. 1.26):

Fig. 1.26 Impedance Z that
will draw current I_{short} from
Thevenin's source

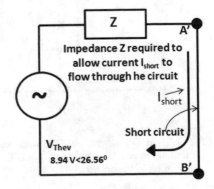

Impedance Z required to
allow current I_{short} to
flow through he circuit

I_{short}

Short circuit

V_{Thev}
8.94 V<26.56^0

The impedance Z that will draw current equal to I_{short} from Thevenin's source V_{Thev} is given as

$$Z = \frac{V_{Thev}}{I_{short}} = \frac{8.94V < 26.56^0}{1.41A < 45^0} = 6.34\Omega < -18.44^0 = (6.0 - j2.0) = Z_{Thev}$$

Thus, Z_{Thev} may be calculated by dividing V_{Thev} by the short-circuit current I_{short}.

(iii)   The first step of calculating the Thevenin impedance Z_{Thev} for a given network is to replace all independent voltage sources by a short circuit and
current sources by open circuits. Next, connect a voltage source V_{trial}
between the reference terminals A and B and calculate the current I_{trial}
through the modified network. The Thevenin impedance Z_{Thev} is then given
by V_{trial}/I_{trial}.

As shown in Figure 1.27a, b, when an external source V_{trial} is connected between
terminals A and B, current I_{trial} flows through the circuit which is given by

$$I_{trial} = \frac{V_{trial}}{Z_{\parallel}}, \text{ and therefore,}$$

$$\frac{V_{trial}}{I_{trial}} = Z_{Thev} = Z_{\parallel} = (6 - j2)\Omega$$

Thus, it may be observed that the first step for the application of Thevenin's
theorem is to find the open-circuit potential difference V_{Thev} between the reference
terminals A and B and then to find the Thevenin impedance Z_{Thev} using one of the

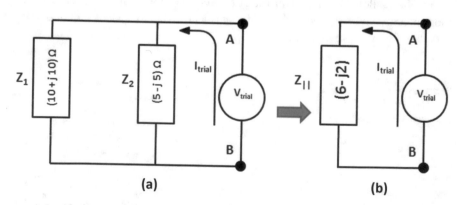

Fig. 1.27 a The original network after replacement of voltage source by a short circuit and
connecting source V_{trial} between A and B. **b** The network after the reduction of the parallel
combination of impedances

above-mentioned three methods. It may further be noted that Thevenin's equivalent circuit for a given network changes with the change of the reference terminals. The following worked out examples will further enhance the understanding of Thevenin's theorem and its applications.

Note: In case of AC networks, the impedance of circuit elements depends on the frequency of the applied AC source. Thevenin's theorem in AC networks holds good at each individual frequency. That means that the magnitudes of V_{Thev} and Z_{Thev} may have different values at different source frequencies.

Solved example (SE1.14) Obtain Thevenin's equivalent circuit for the DC network of Figure SE1.14 and hence calculate the current through the load resistance R_L.

Solution: *As may be seen in* Fig. SE1.14, *the four resistances R_1, R_2, R_3 and R_4 are arranged along the arms of a rectangle abcd. A battery of 10 V is connected at two opposite corners a and c of the rectangle. The reference terminals A and B are the other two corners d and b of the rectangle. In step one, let us calculate the potential difference V_{bd} between the two corners b and d when load R_L is not connected. This V_{bd} will be equal to the V_{Thev}.*

Simplified versions of the given network after removing the load resistance R_L are shown in Fig. SE1.14.1a, b. *As may be seen in these figures, the battery supplies current I to the parallel combinations of (R_1+R_2) and (R_3+R_4), i.e. to the parallel combination of $(25+20)\Omega \parallel (15 + 30)\Omega$. It may be noted that the*

Fig. SE1.14 Network for example (SE1.14)

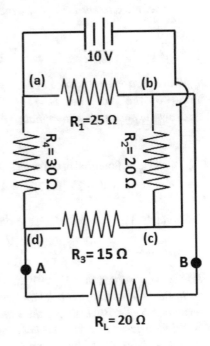

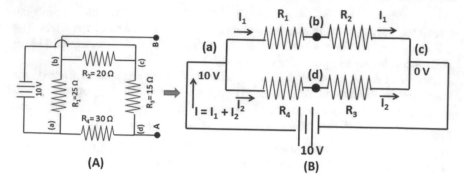

Fig. SE1.14.1 **a** Original network without load resistance. **b** Simplified version of the network of figure a

magnitudes of each resistance of the parallel combination are 45Ω. Therefore, the net resistance of the parallel combination $Z_{||}$ is given by

$$Z_{||} = \frac{45}{2}\Omega = 22.5\Omega$$

Hence, the current I drawn by the parallel combination is

$$I = \frac{10}{22.5} = 0.44A$$

Since the total resistance of each parallel arm is 45Ω, the current I will divide into two equal parts, i.e. $I_1 = I_2 = \frac{0.44}{2} = 0.22A$.

Now the potentials of terminals b and d if represented respectively by V_b and V_d, then

$$V_b = \text{Potential at terminal } a - \text{potential drop around resistance } R_1$$
$$= 10V - 25.(0.22)V = 4.5V$$

Similarly, V_d = potential of vterminal a − potential drop around resistance R_4
$$= 10V - 30.(0.22)V = (10 - 6.6) = 3.4V$$

It may be observed that $V_b > V_d$ and $V_{bd} = V_b - V_d = 1.2V = V_{Thev}$.

In the second step, we calculate the resistance of the network looked back from terminals A and B which are the same as terminals d and b of the network.

While determining the Thevenin resistance R_{Thev}, it is required that all sources of voltage be replaced by a short circuit and all current sources be replaced by an open circuit. Since in the present problem there is only one voltage source, the

Fig. SE1.14.2 Reduction of
networks for calculating
Thevenin resistance

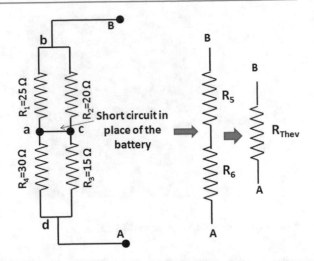

following network is obtained when that voltage source is replaced by a short
circuit (Fig. SE1.14.2).

As may be seen in the figure, the resistance between terminals A and B is a series
combination of two resistances R_5 and R_6, where R_5 and R_6 are themselves parallel
combinations respectively of $R_1\|R_2$ and $R_3\|R_4$. Now,

$$R_5 = \frac{R_1 R_2}{R_1 + R_2} = \frac{500}{45} = 11.11\Omega \text{ and } R_6 = \frac{R_\# R_4}{R_3 + R_5} = \frac{450}{45} = 10.0\Omega$$

Therefore, $R_{\text{Thev}} = 21.11\Omega$.
The Thevenin equivalent circuit of the problem is given in Fig. SE1.14.3.
The current I_L through the load R_L is

$$I_L = \frac{1.2}{41.11} = 0.029A = 29\,\text{mA}$$

Fig. SE1.14.3 Thevenin
equivalent circuit

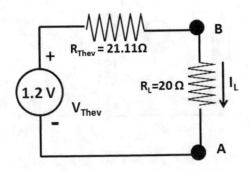

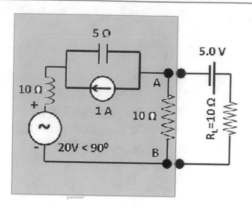

Fig. SE1.15 Network for example (SE1.15)

Solved example (SE1.15) Draw the Thevenin equivalent circuit with reference to nodes A and B for the shaded part of the network shown in Fig. SE1.15 and hence find the current in the load resistance R_L.

Solution: *As the first step, we calculate the Thevenin voltage V_{Thev} for the given network. V_{Thev} may be calculated if the magnitude and direction of flow of current through resistance between terminals A and B (of 10 Ω) are known. One may find out the current through the ten-ohm resistance in two different ways: (i) making use of the Superposition theorem and (ii) using the mesh analysis. We will use both these methods to obtain the direction and magnitude of the current.*

(i) *Using the superposition theorem: According to the superposition theorem, the current through the ten-ohm resistance will be equal to the algebraic sum of currents through it with one source at a time and replacing the other source either by an open circuit or a short circuit, depending on whether the other source is a current source or a voltage source. In Fig. SE1.15.1b, the current source is replaced by an open circuit while the voltage source is retained. If the currentthrough resistance of ten ohms, in this case, is designated by I_1, then*

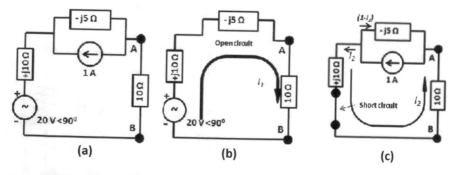

Fig. SE1.15.1 **a** original network, **b** network after replacing current source by open circuit and **c** network with voltage source replaced by a short circuit

$$I_1 = \frac{\text{Voltage}}{\text{Total impedance}} = \frac{20V < 90^0}{(+j10-j5+10)} = \frac{20V < 90^0}{(10+j5)}$$

$$\text{or} \quad I_1 = \frac{20V < 90^0}{11.18 < 26.56^0} = 1.79A < 63.44^0 = (0.8+j1.6)A \text{ from } A \text{ to } B$$

The network on replacement of the voltage source by a short circuit and retaining the current source is shown in Fig. SE1.15.1c. As may be observed in this figure, the current of 1 A delivered by the current source gets divided into two parts, part I_2 passes through the series combination of the impedance of $(+j10)$ Ω and the ten-ohm resistance, while $(1-I_2)$ current passes through the impedance of $(-j5)$ Ω. One may calculate the current I_2 using the law of the division of currents in parallel impedances.

$$I_2 = \frac{(-j5)}{(10+j10)+(-j5)} = \frac{-j5}{(10+j5)} = (-0.2 - j0.4)A \text{ from } B \text{ to } A$$

The net current I through the Ωresistance is the algebraic sum of I_1 and I_2, and since I_2 is flowing in the opposite direction. Hence,

$$I = I_1 - I_2 = [(0.8+j1.6) - (-0.2 - j0.4)] = (1.0+j2.0)A \text{ from } A \text{ to } B.$$

(ii) **Using mesh analysis** The original network with impedance values of circuit elements is shown in Fig. SE1.15.1d, where two meshes are clearly visible in the network. Cyclic currents x and y are assigned to the meshes as shown in the figure. Since there is a current source in the network, one uses the concept of super mesh and treats the current source as if it is not present. The KVL equation for the super mesh may be written as

$$(+ j10)x + (- j5)y + (10)x - (+ j20) = 0 \qquad \text{(SE1.15.1)}$$
$$\text{or} \quad (10 + j10)x - (j5)y = + j20$$

$$\text{Also} \quad y - x = 1 \text{ or } y = 1 + x$$

Fig. SE1.15.1d Mesh analysis of the network

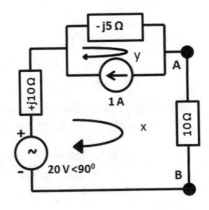

Fig. SE1.15.1e Network for
the calculation of Thevenin
resistance

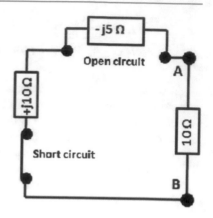

On substituting this value of y in (1.69), one gets

$$(10+j5)x = j25 \text{ or } x = \frac{25j}{(10+j5)}A = (1+j2)A \text{ from } A \text{ to } B$$

As expected, current values obtained by two different methods agree with each other.

The Thevenin voltage V is equal to the potential drop across the resistance of 10Ω, hence,

$$V_{Thev} = (10\Omega)(\text{current } I \text{ or } x) = (10)(1+j2)V = (10+j20)V$$

The second step is to find out the Thevenin impedance Z_{Thev} which is the impedance of the network looked back from terminals A and B, after replacing the current source by an open circuit and the voltage source by a short circuit. The circuit for the calculation of Z_{Thev} is shown in Fig. SE1.15.1e, where it is obvious that the impedance between terminals A and B (i.e. Z_{Thev}) is equal to the value of the parallel combination of Ω resistance and the series combination of $(-5j)$ and $(+10j)$ impedances.

$$\text{Therefore, } Z_{Thev} = (j10 - j5)||(10\Omega) = \frac{(j5)(10)}{(10+j5)} = (2+j4)\Omega$$

$$= 4.47 < 63.43^0 \Omega$$

The Thevenin equivalent of the given network is shown in Fig. SE1.15.1f. The required current I_L through the load resistance is given by

$$I_L = \frac{\text{Total voltage in the circuit}}{\text{Total impedance}} = \frac{(10+j20)+(5)}{(2+j4)+10} = \frac{15+j20}{12+j4}$$

$$= (1.62+j1.12)A$$

Fig. SE1.15.1f Thevenin
equivalent of the given
network

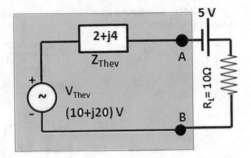

Solved example (SE1.16) Draw the Thevenin equivalent of the network given in
Fig. SE1.16a with respect to terminals A and B.

Solution: *In Step-1, we determine the Thevenin voltage V_{Thev}. It is obvious that
V_{Thev} is equal to the voltage drop across the resistance of 10 Ω connected between
terminals A and B. If I ampere is the current through this ten-ohm resistance then,*
   $V_{Thev} = 10IV$
   In order to find the current I, mesh analysis is used. As shown in Fig. SE1.16b,
*currents x, y and I are assigned to the three meshes of the circuit. The KVL
equation of the X mesh may be written as*

$$(j10 - j5)x + (-j5 + j10)(x - y) - (10V < -90^0) - (20V < 90^0) = 0$$
or $(j10)x - (j5)y - (-j10) - (j20) = 0$ or $(j10)x - (j5)y - j10 = 0$
or $(j2)x - (j)y = 0 + j2$

$$(SE1.16.1)$$

   *Since a current source is present between the x and I meshes, the KVL equation
for the super mesh may be given as*

Fig. SE1.16a Network of
example (SE1.16)

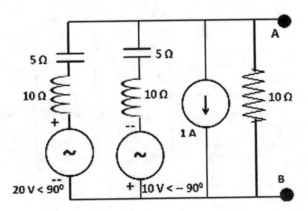

Fig. SE1.16b Assignment of
currents to different meshes

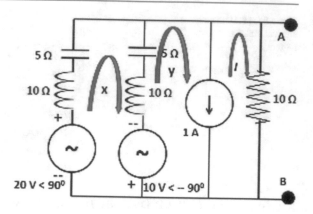

$$(j5)(y - x) + (10)I + (10 < -90^0) = 0 \text{ or} (-j5)x + (j5)y + 10I = 0 + j10$$
$$\text{or} \quad (-j)x + (j)y + (2)I = 0 + j2$$

$$(SE1.16.2)$$

Since the current source is delivering current of 1 A, therefore,

$$(y - I) = 1A \qquad (SE1.16.3)$$

Equations (SE1.16.1), (SE1.16.2) *and* (SE1.16.3) *may be solved to get the value
of current I through the resistance of 10 Ω, which gives*

$$I = 0.29 + j1.17$$
$$\text{And} \quad V_{Thev} = (10)(0.29 + j1.17)V = (2.9 + j11.7)V = 12.05V < 76^0$$

In Step-II, the Thevenin impedance Z_{Therv} *needs to be calculated. For these
calculations, all voltage sources are replaced by a short circuit and the current
source by an open circuit as is shown in the leftmost circuit in* Fig. SE1.16c. *As may
be seen in this figure, two equal impedances each of* $j5$ *(+$j10$ − $j5$ =+$j5$) are in*

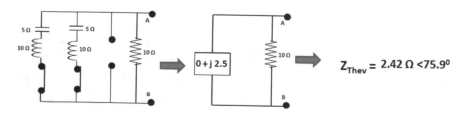

Fig. SE1.16c Circuit after simplification

Fig. SE1.16d Thevenin's
equivalent circuit

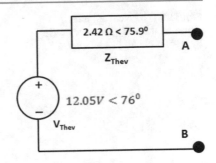

parallel, and therefore the net impedance is $(j2.5)$ Ω. This $(j2.5$ $\Omega)$ is now in parallel with 10 Ω impedance of the resistance. The $Z_{||}$ is then given by

$$Z_{||} = (j2.5)||(10) = \frac{(j2.5)(10)}{(j2.5)+(10)} = \frac{(j25)(10-j2.5)}{(10+j2.5)(10-j2.5)} = 0.59 + j2.35$$
$$= 2.42\Omega < 75.9^0$$

The desired Thevenin equivalent circuit is given in Fig. SE1.16d.

Solved example (SE1.17) Draw Thevenin's equivalent circuit between terminals A and B of the network shown in Fig. SE1.17a.

Solution: *The network shown in* Fig. SE1.17a *has a dependent current source of current* $50\,I_1$, *where* I_1 *is the current through resistance* R_2. *In the first step, we shall find out the magnitude and direction of the Thevenin voltage* V_{Thev}. *The current* $50\,I_1$ *while flowing through the resistance* R_3 *produces the output voltage which is the required* V_{Thev}. *Since the current flows from terminal 'b' to terminal 'a', the voltage* V_{Thev} *has B as a positive terminal and terminal A as negative. The magnitude of current* I_1 *is given by*

Fig. SE1.17a Network for
example (SE1.17)

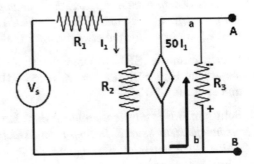

Fig. SE1.17b Network after
replacing the voltage source
by a short circuit

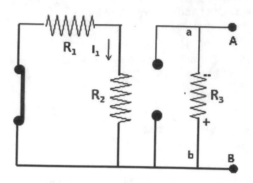

Fig. SE1.17c Thevenin's
equivalent of the given
network

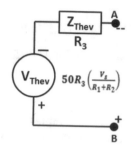

$$I_1 = \frac{V_S}{R_1 + R_2} \text{ and therefore, } V_{Thev} = R_3(50I_1) = 50R_3\left(\frac{V_s}{R_1 + R_2}\right) \quad \text{(SEA.17.1)}$$

In the second step, let us obtain the value of the Thevenin impedance Z_{Thev}.
To determine Z_{Thev}, voltage sources in the original network are to be replaced
by a short circuit. As such when Vs is removed and substituted by a short circuit,
current I_1 becomes zero. Therefore, no current flows through the dependent current
source also. The dependent current source behaves as an open circuit. The network
now looks like the one given in Fig. SE1.17b.

As is evident from this figure, the impedance (the resistance in the present case)
between terminals A and B is R_3, which is then equal to Z_{Thev}. The Thevenin
equivalent circuit for the given network is shown in Fig. SE1.17c.

Solved example (SE1.18) Draw the Thevenin equivalent circuit for the network
shown in Fig. SE1.18a for $Z_1 = (5+j4)\Omega, Z_2 = (0+j6)\Omega, Z_3 = (10+j0)\Omega$
and $\lambda = 2.0$, V=10 V.

Solution: *This is a typical problem which has only a dependent voltage source and
no independent source. In Step-I, let us calculate the Thevenin voltage V_{Theve}
which is the potential drop against the impedance Z_2. Therefore, the current I_1
through Z_2 is given by*

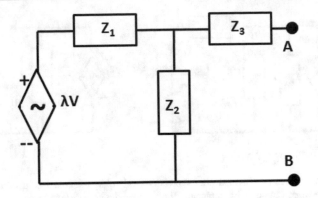

Fig. SE1.18a Network of example (SE1.18)

$$I_1 = \frac{\lambda V}{Z_1 + Z_2} \text{ and hence } V_{Thev} = Z_2 \left(\frac{\lambda V}{Z_1 + Z_2} \right) \qquad (SE1.18.1)$$

In Step-II, we calculate the Thevenin impedance Z_{Thev}. It may be calculated in the following three different ways:

(a) If the dependent voltage source is replaced by a short circuit, the circuit shown in Fig. SE1.18b results. The resultant impedance between terminals A and B is then given by

$$Z_{Thev} - \left(Z_{\parallel}, \text{ the parallel combination of } Z_1 \text{ and } Z_2 \right) \mid Z_3$$

where $Z_{\parallel} = \frac{Z_1 Z_2}{Z_1 + Z_2}$.

$$\text{Therefore,} \quad Z_{Thev} = \frac{Z_1 Z_2}{Z_1 + Z_2} + z_3 \qquad (SE1.18.2)$$

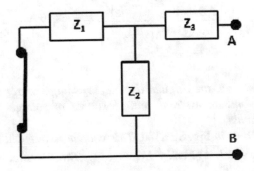

Fig. SE1.18b Network after replacing the voltage source by a short circuit

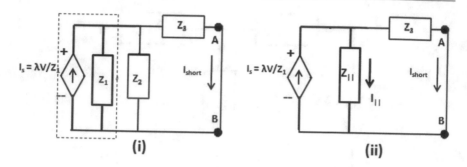

Fig. SE1.18c i Network obtained after conversion of voltage source into a current source. **ii** Simplified network obtained by replacing the parallel combination of impedances by single impedance

(b) *The dependent voltage source may be converted into a dependent current source with source current $I_\mathrm{S}(=\frac{\lambda V}{Z_1})$ and impedance Z_1 in parallel with the current source as shown by the dotted part in* Fig. SE1.18c(i). *With this conversion, the original network assumes the circuit shown in* Fig. SE1.18c(i). *The parallel combination of impedances Z_1 and Z_2 may be lumped together to get $Z_\parallel$ which is given by $Z_\parallel = \frac{Z_1 Z_2}{Z_1 + Z_2}$. Also, the open-circuit voltage between terminals A and B, denoted by V_ABopen, is given by $V_\mathrm{ABopen} = Z_\parallel I_\mathrm{S} = \frac{Z_1 Z_2}{Z_1 + Z_2}\left(\frac{\lambda V}{Z_1}\right)$.*

If terminals A and B of the modified circuit are shorted, current I_short will flow through the short circuit AB and current $I_\parallel$ through the impedance $Z_\parallel$ such that $I_\mathrm{S} = I_\parallel + I_\mathrm{short}$. Further, one may use the law of current division in parallel impedances to get

$$I_\parallel = \frac{Z_3}{Z_\parallel + Z_3} I_\mathrm{S} \text{ and } I_\mathrm{short} = \frac{Z_\parallel}{Z_\parallel + Z_3} I_\mathrm{S}$$

Now, $Z_\mathrm{Thev} = \dfrac{V_\mathrm{ABopen}}{I_\mathrm{short}} = \dfrac{[Z_\parallel I_\mathrm{S}]}{\left(\frac{Z_\parallel}{Z_\parallel + Z_3} I_\mathrm{S}\right)} = (Z_\parallel + Z_3) = \dfrac{Z_1 Z_2}{Z_1 + Z_2} + Z_3.$

$$(SE1.18.3)$$

(c) *In the third method, the original dependent voltage source is replaced by a short circuit, and an external voltage source of voltage V_ex is connected between terminals*
A and B as shown in Fig. SE1.18d. *This voltage source will establish current in the network which is given by $I_\mathrm{ex} = \frac{V_\mathrm{ex}}{Z_\parallel + Z_3}$. The Thevenin impedance is now defined as*

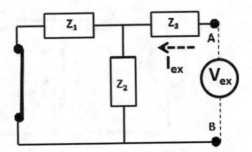

Fig. SE1.18d Original voltage source is replaced by a short circuit and an external voltage source is connected at the output AB

$$Z_{Thev} = \frac{V_{ex}}{I_{ex}} = \frac{V_{ex}}{\left(V_{ex}/(Z_{\|} + Z_3)\right)} = Z_{\|} + Z_3 = \frac{Z_1 Z_2}{Z_1 + Z_2} + Z_3 \qquad (SE1.18.4)$$

It may be observed from equations (SE1.18.2), (SE1.18.3) and (SE1.18.4) that the same value for Z_{Thev} is obtained by three different methods of calculation.

The following values for Z_{Thev} and V_{Thev} are obtained when the given numerical values of impedances, etc. are substituted.

$$Z_{Thev} = \frac{Z_1 Z_2}{Z_1 + Z_2} + Z_3 = \frac{(5+j4)(0+j6)}{(5+j4)+(0+j6)}\,\Omega + (10+j0)\Omega$$
$$= (1.44 + j3.12) + (10 + j0) = (11.44 \mid J3.12)\Omega$$
$$= 11.86\Omega < 15.26^0$$

And

$$V_{Thev} = Z_2 \frac{(\lambda V)}{(Z_1 + Z_2)} = (0+j6)\left(\frac{2x10}{(5+j4)+(0+j6)}\right)V = (9.6 + j4.8)V$$
$$= 10.73V < 26.57^0$$

Thevenin's equivalent of the original network is shown in Fig. SE1.18e.

Fig. SE1.18e Thevenin's equivalent

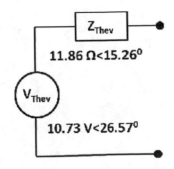

Fig. SE1.19a Network for
example (SE1.19)

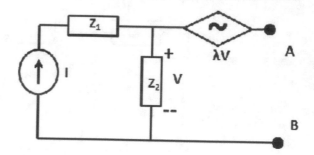

Solved example (SE1.19) Draw the Thevenin equivalent circuit with respect to
terminals A and B for the network given in Fig. SE1.19a.

Solution: *Step-I. Calculation of Thevenin's voltage* V_{Thev}.

V_{Thev} *is the potential difference between terminals A and B when no load is
connected between them. It is obvious from Fig. SE1.19a that the current I supplied
by the current source will flow only through impedances Z_1 and Z_2 and that no
current will pass through the dependent voltage source. As such the potential drop
V across Z_2 will be $Z_2 I$, and the potential drop across the dependent voltage source
will be equal to* $\lambda V = \lambda(Z_2 I)$. *Therefore, the total potential difference between
terminals A and* $B = V + \lambda V = Z_2 I + \lambda(Z_2 I)$. *Hence,*

$$V_{\text{Thev}} = Z_2 I + \lambda(Z_2 I) = Z_2 I (1 + \lambda) \qquad (\text{SE1.19.1})$$

Step-II. Calculation of Thevenin's impedance Z_{Thev}.

*As shown in Fig. SE1.19b, current I_{ex} flows in the indicated loop when an
external voltage source is applied between terminals A and B. In this case, the
potential drop V across impedance Z2 is given by* $V = Z_2 I_{ex}$ *and the voltage
across the dependent source* $\lambda V = \lambda Z_2 I_{ex}$. *Therefore, the total voltage drop in the*

Fig. SE1.19b Network after
replacing current source by an
open circuit and connecting
an external voltage source
Vex between terminals
A and B

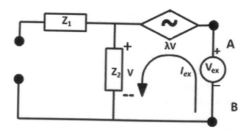

Fig. SE1.19c Thevenin's
equivalent circuit of the given
network

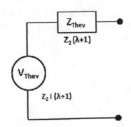

closed loop is $(V + \lambda V) = Z_2 I_{ex}(1 + \lambda)$. *This total voltage drop in the loop must be equal to the externally applied potential difference* V_{ex} *(Fig. SE1.19c).*

$$\text{Hence,} \quad V_{ex} = Z_2 I_{ex}(1 + \lambda) \text{ and } \frac{V_{ex}}{I_{ex}} = Z_2(1 + \lambda).$$

$$\text{But} \quad Z_{Thev} = \frac{V_{ex}}{I_{ex}} = Z_2(1 + \lambda).$$

Solved example (SE1.20) The network of Fig. SE1.20a has only a dependent current source; draw the Thevenin equivalent of the circuit between terminals A and B. Use Thevenin's equivalent circuit to find the current through the load resistance R_L.

Solution: *Since the network for which Thevenin's equivalent is desired does not have any independent source, either of voltage or of current, Thevenin's voltage* V_{Thev} *must be zero. Now the only parameter required to calculate is the Thevenin impedance* Z_{thev}.
    To calculate Z_{Thev}, *we replace the parallel combination of the dependent source of current (of* $2I_x$*) and* 5Ω *resistance into a voltage source of voltage* $(5x\ 2I_x = 10I_x)$ *V, in series with resistance of* 5Ω. *The modified network is shown in* Fig. SE1.20b. *We now connect between terminals A and B an external voltage source of voltage* V_{ext} *that supplies energy to the network. Let current x be drawn from this external voltage source as shown in the figure.*

Fig. SE1.20a Network for
example (SE1.20)

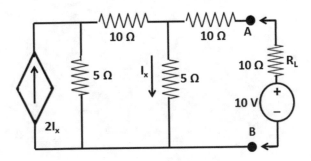

Fig. SE1.20b Assignment of
currents to meshes

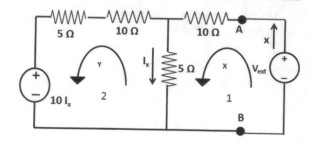

The modified network has two meshes marked 1 and 2. We assume currents x and y flow through these meshes as shown and write KVL equations for these meshes.

$$\text{Mesh 1} \quad 10x + 5(x - y) = V_{ext} \quad \text{or} \quad 15x - 5y = V_{ext} \qquad \text{(SE1.20.1)}$$

$$\text{Mesh 2} \quad (10 + 5)y + (10I_x) + 5(y - x) = 0 \text{ or } 20y - 5x = -10I_x \quad \text{(SE1.20.2)}$$

$$\text{And} \quad x - y = I_X \qquad \text{(SE1.20.3)}$$

The above three equations may be solved to get

$$17.5\,x = V_{ext} \text{ and } Z_{Thev} = \frac{V_{ext}}{x} = 17.5\Omega$$

Thevenin's equivalent circuit with load is shown in Fig. SE1.20c. Since the original network has no independent source, the equivalent circuit has only resistance of the value 17.5 Ω. When the load section is connected to Thevenin's equivalent circuit, a current I will flow through the load resistance R_L which is given by

Fig. SE1.20c Thevenin's
equivalent circuit with load

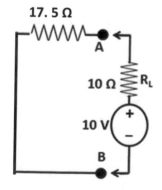

$$I_{RL} = \frac{10\ V}{(10+17.5)\Omega} = 0.36\ A$$

Self-assessment question: While calculating Thevenin's impedance of an active network, why are all independent voltage sources replaced by short circuits and all independent current sources by open circuits?

1.4.3 Norton's Theorem

Norton's theorem states that "A linear bidirectional DC or AC network with respect to any two terminals A and B of the network may be replaced by a current source of current I_{Nort} in parallel with impedance Z_{Nort}, where I_{Nort} is the current through terminals A and B when they are shorted and Z_{Nort} is the impedance of the network looked back through terminals A and B, while all independent voltage and current sources are replaced respectively by a short circuit and an open circuit". It is obvious that in case of a DC network, Z_{Nort} is the resistance of the network looked back through A and B, independent voltage sources replaced by short circuits and current sources by open circuits.

Figure 1.28 shows a schematic representation of Norton's theorem. It is to be noted that like Thevenin's theorem, Norton's theorem for AC networks holds separately at each frequency of the energy source. Moreover, for a given network Z_{Nort} and Z_{Thev} are same. Further, Norton's theorem is complimentary to Thevenin's theorem since the series combination of the voltage source V_{Thev} and Thevenin's impedance Z_{Thev} of Thevenin's equivalent circuit can always be transformed into a current source in parallel with the impedance as shown in Fig. 1.29.

Like the case of Thevenin's theorem, in the case of obtaining Norton's equivalent circuit for a complicated network, there are three possibilities: the given network contains (i) only independent sources of energy, (ii) both independent and dependent sources and (iii) only dependent sources.

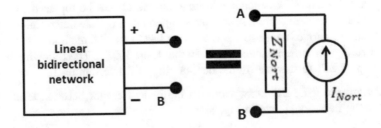

Fig. 1.28 Schematic representation of Norton's theorem

Fig. 1.29 Equivalence of Thevenin's and Norton's equivalent circuits

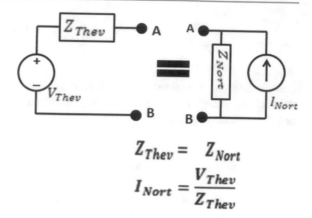

$$Z_{Thev} = Z_{Nort}$$

$$I_{Nort} = \frac{V_{Thev}}{Z_{Thev}}$$

In the last case, when the network does not have any independent energy source, the current I_{Nort} is zero and the parallel impedance Z_{Nort} may be determined either by connecting an external voltage (V_{ext}) or a current source of known strength (I_0) at the two reference terminals. The network may then be analysed by using some convenient tool like the mesh cyclic current method, nodal analysis method or any other method to obtain current (I_{ext}) drawn from the external voltage source (in case an external voltage source is connected) or the potential difference (V_0) between the reference terminals (if the current source is connected). The ratio, (V_{ext}/I_{ext}) or (V_0/I_0), as the case may be, will give the value of impedance Z_{Nort}.

In (ii) case, when both independent and dependent sources are distributed in the network, the short-circuit current I_{Nort} may be determined by analysing the circuit obtained after shorting the reference terminals and replacing the independent voltage sources by short circuits and independent current sources by open circuits. However, the Norton impedance Z_{Nort} may be determined by the method of connecting external voltage or current sources at the reference terminals, as detailed earlier.

In the case when the network contains only independent sources, any of these methods may be applied to determine I_{Nort} and Z_{Nort}. The important point to remember is that while it is possible to set independent voltage sources as short circuits and independent current sources to open circuits, the same cannot be done for dependent sources. A dependent voltage source cannot be replaced by a short circuit and a dependent current source by an open circuit while determining I_{Nort} or Z_{Nort}. Hence in case of dependent sources, one has to use the method of connecting external voltage or current source to obtain the desired Z_{Nort}. The following worked out examples will further clarify the methodology of determining I_{Nort} and Z_{Nort}.

Solved example (SE1.21) Draw Norton's equivalent circuit between terminals A and B of the network shown in Fig. SE1.21a.

Solution: *The given network has two independent sources and therefore any method may be used.*

Fig. SE1.21a Network for example (SE1.21)

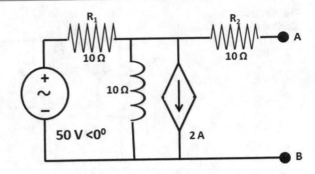

Fig. SE1.21b Mesh analysis of the modified network

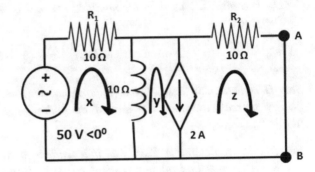

Step-I *To calculate* I_{Nort}: *The reference terminals A and B are short-circuited to find Norton's current as shown in* Fig. SE1.21b. *The modified network has three distinct meshes to which cyclic currents, respectively, x, y and z are assigned. The required* I_{Nort} *is equal to the cyclic current z. Since meshes two and three have the current source between them, one has to consider the super mesh made by joining the two meshes. The KVL equations for the meshes are given as*

Mesh $- 1 : 10x + (0+j10)(x-y) = 50 + j0 \text{ or} (10+j10)x - j10y = 50 + j0$

$$\text{or} \quad (1+j1)x - j1\,y = 5 + j0 \qquad\qquad \text{(SE1.21.1)}$$

$$\text{Mesh} - 2 : \quad (y-z) = 2\,A \qquad\qquad \text{(SE1.21.2)}$$

Super mesh: $10z + (j10)(z-y) = 0$

$$\text{or} \quad (1+j1)z - (j1)y = 0 \qquad\qquad \text{(SE1.21.3)}$$

The above three simultaneous equations may be solved to get

$$x = (2.5 - j0.5)A; \; y = (2+j2)A; z = (0+j2)A$$
$$\text{and } (x-y) = (0.5 - j2.5)A \qquad\qquad \text{(SE1.21.4)}$$

Fig. SE1.21c Current
distribution in the modified
network

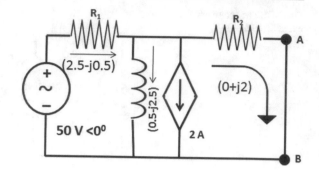

The current distribution in the modified network is shown in Fig. SE1.21c.
Therefore,

$$\text{Norton current } I_{Nort} = z = (0+j2)A \qquad (\text{SE1.21.5})$$

Step-II *To determine* Z_{Nort}: *Since the original network contains only independent energy sources, one may use any of the following three possible methods of obtaining* Z_{Nort}.

(a) *In order to determine* Z_{Nort}, *all independent voltage sources are replaced by short circuits and all independent current sources by open circuits, which leaves the original network as is shown in Fig. SE1.21d.*
    An inspection of the reduced network shows that looking back through terminals A and B, the total impedance of the reduced network Z_{Nort} *is given by*

$$Z_{Nort} = \text{series combination of } R_2 \text{ and } Z_{\|}$$

where

$$Z_{\|} = \text{parallel combination of } R_1 \text{ and } L = \frac{R_1 L}{(R_1+L)} = \frac{10(j10)}{(10+j10)} = (5+j5)\Omega.$$

Therefore, $\boldsymbol{Z_{Nort} = 10\Omega + (5+j5)\Omega = (15+j5)\Omega.}$

Fig. SE1.21d Reduced
network for the calculation of
Z_{Nort}

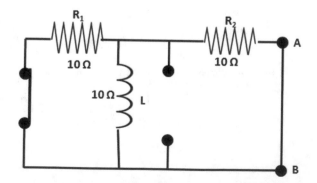

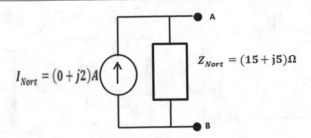

Fig. SE1.21e Norton's Equivalent of the original network

    Norton's equivalent of the original network is given in Figs. SE1.21d *and* SE1.21e.

(b) *In the second method to determine* Z_{Nort}, *one connects a voltage source of voltage* V_x *at terminals A and B and analyses the circuit to get the current* I_x *drawn from the voltage source. As shown in* Fig. SE1.21f, *there are two meshes in the reduced network. It is assumed that cyclic current x flows in the right mesh and y in the left mesh. Writing the KVL equations for the two meshes, one gets*

$$\text{Right mesh}: \quad 10x - V_X + (j10)(x - y) = 0$$

$$\text{or} \quad (10 + j10)x - j10y = V_X \qquad \text{(SE1.21.6)}$$

$$\text{Left mesh}: \quad (j10)(y - x) + 10y = 0 \qquad \text{(SE1.21.7)}$$

    The above two equations may be solved to get

$$x = \frac{V_X}{(15 + j5)}$$

Fig. SE1.21f Mesh analysis of the reduced circuit

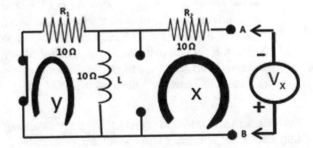

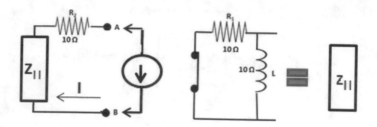

Fig. SE1.21g Reduced network being fed by an external current source

Thus, the current I_x drawn from V_x is same as x, and the ratio $\frac{V_X}{I_X} = \frac{V_X}{x} = (15 + j5)\Omega.$

Thus, $Z_{\text{Nort}} = (15 + j5)\Omega$, the value obtained by method (a).

(c) In the third method of finding the Norton impedance, a current source of known strength, say, I is connected between the terminals A and B. The current fed by the current source gets distributed in different branches of the reduced network. Then from the analysis of the reduced network, one finds the potential difference V across terminals A and B. The ratio V/I gives Norton's impedance. Figure SE1.21g shows the reduced network with a current source.

The current I fed by the external source passes through the parallel combination of resistance R_1 and L and then through the resistance R2 as shown in the figure.

Resistance R_1 and inductance L form a parallel combination; the resultant impedance of this parallel combination is $Z_{||}$ given by

$$Z_{||} = \frac{(10)(j10)}{(10 + j10)} = (5 + j5)\Omega$$

The potential difference V from B to A is given by

$$V = Z_{||}I + R_2I$$

And the ration $\frac{V}{I} = Z_{\text{Nort}} = Z_{||} + R_2 = (5 + j5)\Omega + 10\Omega = (15 + j5)\Omega .$

It may thus be noted that in case of only independent sources in the network, any one of the three methods may be used to calculate the Norton impedance. However, if the network contains dependent sources, then either the (b) or the (c) method should be used to obtain the value of Norton's impedance. This was the case with Thevenin's theorem also.

Self-assessment question (SAQ): While calculating Norton's or Thevenin's impedance for a linear bidirectional active network, is it possible to replace a dependent voltage source by a short circuit and a dependent current source by an open circuit? Give reasons for your answer.

Fig. SE1.22a Circuit for
example (SE1.22)

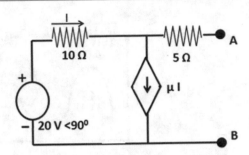

Solved example (SE1.22) Obtain the Norton current I_{Nort} and the Norton impedance Z_{Nort} with respect to terminals A and B for the circuit shown in Fig. SE1.22a.

Solution: *Step-I. To calculate I_{Nort}: The modified circuit with terminals A and B shorted is shown in Fig. SE1.22b. The modified circuit contains two meshes which are separated by a dependent current source. It is assumed that cyclic current I flows through the left-side mesh and current I_N through the right mesh. Obviously, $I_N = I_{Nort}$. Since there is a current source separating the two meshes, we have to consider the super mesh to apply KVL. The KVL equation for the super mesh is*

$$10I + 5I_N = j20 \qquad \text{(SE1.22.1)}$$

$$\text{And} \quad I = \mu I + I_N \text{ or } I = \frac{I_N}{(1-\mu)} \qquad \text{(SE1.22.2)}$$

From the last two equations,

$$I_{Nort} = I_N = j\frac{4(1-\mu)}{(3-\mu)}A \qquad \text{(SE1.22.3)}$$

Step-II: To obtain Z_{Nort} The reduced circuit with an independent voltage source replaced by a short circuit and an external voltage source connected at terminals A and B is shown in Fig. SE1.22c. It is assumed that currents I_{ex} and I flow

Fig. SE1.22b Modified
circuit with terminals A and B
shorted

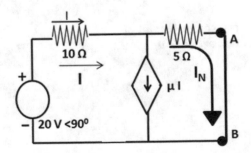

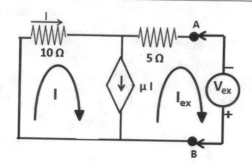

Fig. SE1.22c Reduced circuit with independent voltage source replaced by a short circuit and an external voltage source connected at terminals A and B

through the two meshes of the reduced circuit as shown in the figure. The KVL equation for the super mesh is written as

$$10I + 5I_{ex} = V_{ex} \tag{SE1.22.4}$$

$$\text{And} \quad I - I_{ex} = \mu I \text{ or } I = \left(\frac{1}{1-\mu}\right)I_{ex} \tag{SE1.22.5}$$

It follows from equations (SE1.22.4) and (SE1.22.5) that

$$Z_{Nort} = \frac{V_{ex}}{I_{ex}} = \frac{(15 - 5\mu)}{(1 - \mu)}\,\Omega$$

Self-assessment question (SAQ): For a given network and for the same terminals, how is V_{Thev} related to I_{Nort} ?

Solved example (SE1.23) Obtain the Thevenin and the Norton equivalent circuit parameters $V_{Thev}, Z_{Thev}, I_{Nort}$ *and* Z_{Nort} between terminals A and B of the network given in Fig. SE1.23.

Fig. SE1.23 Network of example (SE1.23)

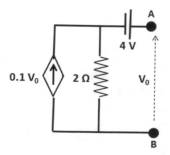

Fig. SE1.23a Circuit
obtained after transforming
the current source into a
voltage source

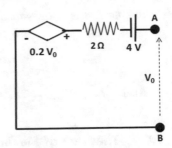

Solution: *The circuit contains an independent voltage source and a dependant current source with resistance in parallel. This parallel combination of dependent current source and 2Ω resistance may be converted into a dependent voltage source of (0.1 x 2=) 0.2 V in series with resistance of 2Ω, as shown in Fig. SE1.23a.*

Since terminals A and B are open, no current is flowing through the network and V_0 is, therefore, the open-circuit potential difference. Hence,

$$V_{Thev} = V_0 \qquad\qquad (SE1.23.1)$$

And $\quad (0.2V_0 + 4) = V_0$ or $0.8V_0 = 4$ or $V_0 = 5V = V_{Thev}$ $\qquad$ (SE1.23.2)

If now terminals A and B are short-circuited, the potential V_0 becomes zero and then the potential of the dependent voltage source which is 0.2 V_0 will also become zero. Under these conditions, the dependent voltage source may be replaced by a short circuit and the circuit of Fig. SE1.23a reduces to the circuit shown in Fig. SE1.23b.

The short-circuit current

$$I_{shor} = \frac{4}{2} = 2A. \qquad\qquad (SE1.23.3)$$

Fig. SE1.23b Reduced
circuit after short-circuiting
terminals A and B

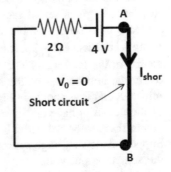

But the Thevenin Impedance

$$Z_{\text{Thev}} = \frac{V_{\text{Thev}}}{I_{\text{shor}}} = \frac{5\ V}{2\ A} = 2.5\Omega. \qquad \text{(SE1.23.4)}$$

Also, $I_{\text{Nort}} = I_{\text{shor}} = 2A$, and $Z_{\text{Nort}} = Z_{\text{Thev}} = 2.5\ \Omega;\ V_{\text{Thev}} = 5\ V.$

1.4.4 Theorem of Maximum Power Transfer

It is well known that all cell- or battery-operated devices work well till their batteries are good but stop working when their batteries get old. It is interesting to investigate what happens to the battery or the cell when they get old and get exhausted. If one measures the output voltage of an exhausted cell or battery, surprisingly in most cases it is found that the output voltage is nearly the same as is specified for the new battery. However, when current is drawn from an exhausted battery or cell, its output voltage decreases sharply with the increase in the current. This happens because with sustain use the internal resistance of the cell (or battery) increases. So when current is drawn from the exhausted cell, considerable voltage drops against its internal resistance, and the net output voltage decreases to a low value. As such the exhausted battery could not deliver the required power to the device which stops working. The cell or the battery or any other active device that delivers power/current to some other device is often called the source and the other device the load. Every load needs a specified power to work properly. Power P of a source is defined as the multiplication of constant voltage V at which the current I is delivered to the load, i.e. P = V I. The maximum values of voltage V and current I, i.e. of the maximum power of a source is fixed. For example, a typical AAA cell may deliver up to 750 mA of current at 1.5 V for an hour or 75 mA current at a constant voltage of 1.5 V for 10 hours. In batteries, energy is produced by converting chemical energy into electrical energy. However, electronic power supplies that run on electricity may deliver a fixed amount of power continuously for considerably longer times or, till they remain connected to the electricity line.

Every active or passive electrical/electronic system has got some impedance with respect to a given set of terminals. Since any given system or network may be replaced by a series combination of Thevenin's voltage source V_{Thev} and Thevenin's impedance Z_{Thev}, it is always possible to determine the impedance of any source or the load as its Thevenin's impedance. Let Z_S and Z_L represent respectively the impedances of the source and the load with respect to connecting terminals A and B as shown in Fig. 1.30. Further, if V_0 and I_0 are the maximum values of the output voltage and currents of the source, then theoretically the maximum power that can be delivered by the source is $V_0 I_0$. However, this is not possible. Suppose Z_L is very large approaching infinity (as if open circuit towards load), then the source output voltage will be almost V_0 but the current delivered to the load $(I = V_0/Z_L)$ will be very small tending to zero and therefore, negligible power will

Fig. 1.30 Transfer of power
from the source to the load

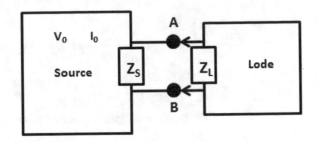

be delivered to the load. Similarly, for the other extreme, if the impedance Z_L is very small (that will mean as if terminals A and B are shorted), the current through the load may be almost as large as I_0, but the voltage across AB will be negligibly small and so again very little power will be delivered to the load. It may thus be observed that both for very large and for very small load impedances, minimal power is delivered to the load. It is obvious that between these two extremes, there must be one particular value of Z_L for which maximum power will be delivered to the load.

Summing up, it may be said that a power source may be capable of supplying a certain amount of power but how much power will actually be drawn by the load depends on the output impedance of the source and the input impedance of the load. The theorem of maximum power transfer specifies the relationship between these two impedances for maximum power transfer from the source to the load.

The theorem of maximum power transfer states:

"An active network (source) of output impedance $Z_S = (R_S + jX_S)$ *will deliver maximum power to the load if the input impedance* Z_L *of the load is conjugate of* Z_S, *i.e.* $Z_L = (R_S - jX_S)$*".*

Let us discuss the implications of the theorem in three different situations.

(a) In the case when both source and load have pure resistive impedances, the theorem tells that maximum power transfer will take place when $R_S = R_L$. For purely resistive systems, the current through the load I_L is given by

$I_L = \dfrac{V_0}{\left(R_S + R_L\right)}$. And the voltage across the load V_L is given by $V_L = I_L . R_L$.

The power delivered to the load

$$P_L = V_L I_L = R_L I_L^2 = R_L \cdot \left\{ \left(\frac{V_0}{R_S + R_L} \right)^2 \right\}. \tag{1.40}$$

In order to find the value of load resistance R_L for maximum power transfer, one may differentiate P_L with respect to R_L and equate it to zero, treating R_L as a variable.

$$\frac{dP}{dR_L} = \frac{d[\left(\frac{V_0}{R_S + R_L}\right)^2 . R_L]}{dR_L} = \frac{V_0^2}{(R_S + R_L)^4}\left[(R_S + R_L)^2 . 1 - 2(R_S + R_L) . R_L\right] = 0$$

or $R_S + R_L - 2R_L = 0$

And $R_S = R_L$ (for maximum power transfer).   (1.41)

One may calculate the magnitude of the maximum power P_L^{max} transferred to the load by putting $R_L = R_S$ in (1.40):

$$P_L^{max} = R_L\left(\frac{V_0}{2R_L}\right)^2 = \frac{1}{4}\left(\frac{V_0^2}{R_L}\right) = \frac{1}{4}\left[V_0\left(\frac{V_0}{R_S}\right)\right] = \frac{1}{4}\{V_0 I_0^{max}\}$$

$$= \frac{1}{4}(\text{maximum power of the source})$$ (1.42)

Thus, it may be observed that under the most favourable condition of maximum power transfer, only one-fourth of the maximum power possessed by the source is transferred to the load.

Further, the potential drop against the load $\left(V_L^{max}\right)$ when maximum power is delivered is given as

$$V_L^{max} = R_L I_L^{max} = R_L\left(\frac{V_0}{2R_L}\right) = \frac{V_0}{2}$$ (1.43)

Equation (1.43) tells that at maximum power transfer, the potential drop against the load is only half of the maximum voltage; the remaining half of the input voltage V_0 drops against the output resistance $R_S (=R_L)$ of the source. It means that at the maximum power transfer, the same amount of power $\left[\frac{1}{4}\left(\frac{V_0^2}{R_S}\right)\right]$ is consumed by the source. In the case of pure resistive networks, at the instant of maximum power transfer, as much power is dissipated by the source as is transferred to the load. In other situations when $R_L \neq R_S$, much more power is dissipated by the source than what is delivered to the load. Moreover, in the most favourable condition of maximum power transfer, only one half of the total maximum power associated with the source $\left(V_0 I_0^{max}\right)$ is available for consumption, one-fourth of which is consumed by the source itself and the remaining one-fourth is passed on to the load for utilization by it. Figure 1.31 shows the variation of power transferred to load P_L as a function of load resistance R_L for given values of V_0 and R_S.

(b) Suppose both the source and the load have complex impedances $Z_S = (R_S + jX_S)$ and $Z_L = (R_L + jX_L)$, respectively. Now the current through the lode I_L is given as

Fig. 1.31 Power delivered to
the load as a function of load
resistance

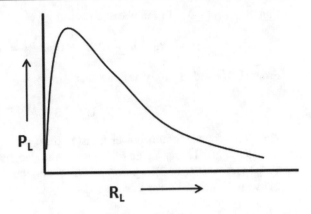

$$I_L = \frac{V_0}{(R_S + R_L) + j(X_S + X_L)}$$

And the magnitude of the load current is given by

$$|I_L| = \frac{V_0}{\sqrt{(R_S + R_L)^2 + (X_S + X_L)^2}}$$

The power delivered to the load P_L is

$$P_L = R_L \cdot (|I_L|)^2 - R_L \left[\frac{V_0{}^2}{(R_S + R_L)^2 + (X_S + X_L)^2} \right] \qquad (1.44)$$

Now there are two independent variables R_L and X_L and to maximize P_L, partial derivatives with respect to these variables have to be equated to zero, i.e. $\frac{\partial P_L}{\partial X_L} = 0$ and $\frac{\partial P_L}{\partial R_L} = 0$. These conditions give

$$\frac{\partial P_L}{\partial X_L} = \frac{R_L V_0{}^2}{\left[\left(R_S + R_L \right)^2 + \left(X_S + X_L \right)^2 \right]^2} \left[-2(X_S + X_L).1 \right] = 0$$

That is

$$X_L = -X_S \qquad (1.45)$$

Also,

$$\frac{\partial P_L}{\partial R_L} = \frac{V_0{}^2}{\left[(R_S + R_L)^2 + (X_S + X_L)^2 \right]^2} \left[\left\{ \left[(R_S + R_L)^2 + (X_S + X_L)^2 \right].1 \right\} - 2R_L (R_S + R_L) \right] = 0$$

Putting $X_L = -X_S$ in the above equation gives

$$R_S + R_L - 2R_L = 0 \ or \ R_L = R_S \tag{1.46}$$

From (1.45) and (1.46), it follows that

$$Z_L = R_L + jX_L = R_S - jX_S = Z_S^* \tag{1.47}$$

Therefore, for maximum power transfer from the source to the load, the load impedance should be conjugate of the source impedance. Further, with $Z_L = Z_S^*$, $Z_L + Z_S$ becomes $2R_S$ or $2R_L$, purely resistive, and the discussion under subheading (a) above becomes valid for this case also.

(c) It often happens that a source of output impedance $Z_S(= R_S + jX_S)$ is required to supply power to a load of impedance $Z_L(= R_L + jX_L)$ but of fixed power factor $f\left(= \dfrac{R_L}{\sqrt{R_L^2 + X_L^2}}\right)$. This, for example, happens when the source is used to run an induction motor of fixed power factor $f \approx 0.8 lag$ and in such a situation, the power factors of load and source may have different values. However, as shown here, the power factor f establishes a relationship between R_L and X_L:

$$f^2 = \frac{R_L^2}{R_L^2 + X_L^2} \ or \ X_L = R_L \left(\frac{1}{f^2} - 1\right)^{\frac{1}{2}} = \mu R_L, \ \ where \ \mu = \left(\frac{1}{f^2} - 1\right)^{\frac{1}{2}}. \tag{1.48}$$

Hence, the current I_L through the load is given by

$$I_L = \frac{V_0}{Z_S + Z_L} = \frac{V_0}{(R_S + jX_S) + (R_L + j\mu R_L)}$$

And the magnitude of current $|I_L| = \dfrac{V_0}{\sqrt{\left(R_S + R_L\right)^2 + \left(X_S + \mu R_L\right)^2}}$.

Power delivered to load $P_L = R_L \cdot (|I_L|)^2 = R_L \cdot \dfrac{V_0^2}{\left(R_S + R_L\right)^2 + \left(X_S + \mu R_L\right)^2}$.

To find the condition for the maximum value of P_L, one must differentiate P_L with respect to R_L and equate it to Zero. That is,

$$\frac{dP_L}{dR_L} = \frac{d\left[R_L \cdot \frac{V_0^2}{(R_S + R_L)^2 + (X_S + \mu R_L)^2}\right]}{dR_L} = 0$$

or $\dfrac{v_0^2}{\left[(R_S+R_L)^2+(X_S+\mu R_L)^2\right]^2}\left[\left\{(R_S+R_L)^2+(X_S+\mu R_L)^2\right\}.1-R_L\{2(R_S+R_L)+2\mu(X_S+\mu R_L)\}\right]=0$

or $(R_S+R_L)^2+(X_S+\mu R_L)^2-2\left[R_L R_S+R_L^2+\mu X_S R_L+\mu^2 R_L^2\right]=0$

$R_S^2+R_L^2+2R_S R_L+X_S^2+\mu^2 R_L^2+2\mu X_S R_L-2[R_L R_S+R_L^2+\mu X_S R_L+\mu^2 R_L^2]=0$

or $R_S^2+X_S^2-R_L^2-\mu^2 R_L^2=0$ or $R_S^2+X_S^2=R_L^2+\mu^2 R_L^2$

But $\mu^2 R_L^2=X_L^2$, and so $R_S^2+X_S^2=R_L^2+X_L^2$ but $R_S^2+X_S^2=Z_S^2$ and $R_L^2+X_L^2=Z_L^2$

Therefore, for maximum power transfer $Z_S^2=Z_L^2$ or $|Z_S|=|Z_L|$.

Thus, in cases where the power factor of the load has a fixed value different from that of the source, the magnitude of the load impedance should be equal to the magnitude of the source impedance for maximum power transfer to the load.

Figure 1.32 demonstrates the method of solving problems associated with power transfer. The first step is to carry out the Thevenin analysis of both the source and the load networks to obtain Thevenin impedances and Thevenin voltages. The source must be an active network having at least one independent energy source and, therefore, $Z_S{}^{\text{Thev}}$ must have a finite non-zero value. However, the Thevenin voltage for the load may be zero if the load is a passive network. Once Thevenin impedances are obtained, the appropriate condition of maximum power transfer may be used to solve the problem.

(d) Both the power source and the load have their own impedances, respectively, $Z_S=R_S+X_S$, and $Z_L=R_L+X_L$. For maximum power transfer, either $Z_L=Z_S{}^*$ or $|Z_L|=|Z_S|$. However, sometimes it is not possible to adjust the

Fig. 1.32 Schematic representation of the methodology of solving power transfer problems

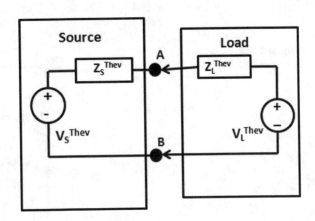

reactance X_L so as to make it a conjugate of X_S. In such situations, it is still possible to transfer a sufficiently high amount of power to the load by adjusting the load resistance R_L. Without going into mathematical derivation it may be said that for fixed values of R_S, X_S and X_L, the best adjusted value of the load resistance R_L^{Adj} is given by

$$R_L^{Adj} = \sqrt{R_S^2 + (X_S + X_C)^2} \qquad (1.49)$$

And if one defines

$$R_{Mean} = \frac{R_S + R_L^{Adj}}{2} \qquad (1.50)$$

then the maximum power transferred to load is given by

$$P_L^{max} = \frac{V_S^2}{4R_{Mean}} \qquad (1.51)$$

It is obvious that in the case where the Thevenin equivalent circuit is used as a power delivering source, V_S is replaced by V_{Thev}, R_S by R_{Thev} and so on.

Solved example (SE1.24) Determine the load that may draw maximum power from the network shown in Fig. SE1.24a when connected at port AB. Also, obtain the magnitude of the maximum power.

Solution: *As the first step, one has to find Thevenin parameters Z_{Thev} and V_{Thev} for the given network. To get Z_{Thev} we use the method of external voltage source*

Fig. SE1.24a Network for example (SE1.24)

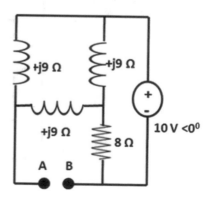

Fig. SE1.24b Modified
network on short-circuiting
the voltage source

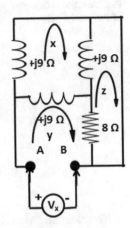

being connected at terminals A and B and the original voltage source being
replaced by a short circuit as shown in the modified network in Fig. SE1.24b.

The modified network has three meshes for which cyclic currents x, y and z have
been assigned. Using the mesh analysis, the value of current y will be determined to
calculate Z_{Thev} using the expression $Z_{\text{Thev}} = \frac{V_X}{y}$. The three mesh equations may
be written as

For mesh x : $(j9)x + (j9)(x - z) + (j9)(x - y) = 0$
or $3x - y - z = 0$ (SE1.24.1)

For mesh y : $(j9)(y - x) + 8(y - z) - V_X = 0$
or $(8 + j9)y - (j9)x - 8z = V_X$ (SE1.24.2)

For mesh z : $8(z - y) + (j9)(z - x) = 0$
or $(8 + j9)z - 8y - (j9)x = 0$ (SE1.24.3)

Equations (SE1.24.1), (SE1.24.2) and (SE1.24.3) may be solved to get

$$\frac{(-27 + 48j)}{(8 + 6j)} y = V_X$$

And $Z_{Thev} = \frac{V_x}{y} = (0.72 - j5.46)\Omega = 5.51\Omega < -82.48^0$

For maximum power transfer to the load, the load impedance Z_{Load} must be
the complex conjugate of Z_{Thev}. Therefore, $Z_{Load} = (0.72 + j5.46)\Omega =$
$5.51\Omega < 82.48^0$.

Fig. SE1.24c Mesh analysis
of the source network

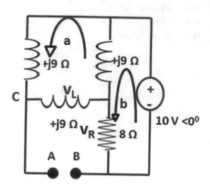

To calculate the maximum power delivered to the load, $P_{\text{Load}}^{\max}$, it is required to know V_{Thev} of the given source network. Z_{Thev} the open-circuit potential difference between A and B is (Fig. SE1.24c) given by $Z_{\text{Thev}} = v_L + v_R$, where v_L and v_R are respectively the potential drops across 9 Ω inductance and 8 Ω resistance. Also, if cyclic mesh currents a and b are assigned to the two closed meshes in the circuit, then $v_L = (9j.a)V$ and $v_R = (8.b)$ V . The two mesh equations using KVL may be written as

$$\text{For mesh}(a) \quad (j9).a + (j9).a + (j9)(a - b) = 0$$
$$\text{or} \qquad\qquad 3a - b = 0 \ or \ b = 3a \qquad\qquad (SE1.24.4)$$

$$\text{For mesh}(b) \quad (j9)(b - a) + 8b - 10 = 0$$
$$\text{or} \qquad\qquad (8 + j9)b - (j9)a = 10 \qquad\qquad (SE1.24.5)$$

Substituting b=3a in (SE1.24.5), one gets

$$(24 + j18)a = 10 \ or \ a = (0.27 - j0.2)A \ \text{and} \ b = (0.81 - j0.6)A$$

Therefore,
$$v_L = (j9)(a) = (1.8 + j2.43)V = \text{and} \ v_R = (8)(b) = (6.48 - j4.8)V.$$

$$\text{And} \quad V_{Thev} = (8.28 - j2.3)V = 8.59V < -15.6^0$$

$$\text{Also,} \quad P_L^{max} = \frac{V_{Thev}^2}{4R} = \frac{(8.59)^2}{4x(0.72)} = 25.62 \ W$$

Solved example (SE1.25) Which series combination A: (R + L) or B: (R + C) of the load be connected between terminals 1 and 2 of the source network of Fig. SE1.25a so that maximum power is transferred to the load. Calculate the values of the load elements (R, L or C) and the maximum power delivered to the load. How much power is consumed by the load and where?

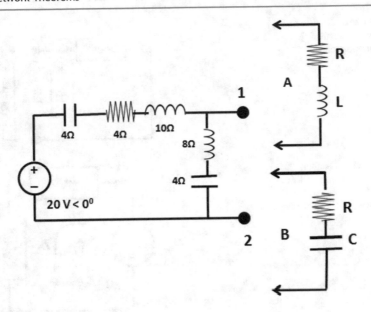

Fig. SE1.25a Network for example (SE1.25)

Solution: *For maximum power transfer, the load impedance must be a complex conjugate of the source impedance. Therefore, the first step is to find the Z_{Thev} of the source network between terminals 1 and 2. For that, the voltage source is replaced by a short circuit and with that the source network reduces to the one shown in Fig. SE1.25b.*

As may be observed in the figure, Thevenin's impedance of the network between terminals 1 and 2 is equal to the parallel combination of impedances Z_1 and Z_2. Therefore,

$$Z_{Thev} = \frac{1}{Z_1} + \frac{1}{Z_2} = \frac{Z_1 Z_2}{Z_1 + Z_2} = \frac{(4+j6)(j4)}{(4+j6)+(j4)} = \frac{(-24+j16)}{(4+j10)} = \frac{(-24+j16)(4-j10)}{16+100}$$

$$Z_{Thev} = (0.55 + j2.62)\Omega$$

$$(SE1.25.1)$$

Now for maximum power transfer to the load, the load impedance should be conjugate of Z_{Thev}, i.e. $Z_{Load} = Z^*_{Thev} = (0.55 - j2.62)\Omega$. A series combination of resistance $R=0.55\Omega$ with capacitance of $X_C = 2.62\Omega$ will make Z^*_{Thev}. Such a combination in Fig. SE1.25a is represented by B.

To calculate the maximum power delivered to a load of impedance $Z_{Load}(= 0.55 - j2.62)$, it is required to obtain the V_{Thev} of the given source network, as shown in Fig. SE1.25d. The current i_0 is given by (Fig. SE1.25c)

Fig. SE1.25b Reduced
network for calculating
Thevenin impedance

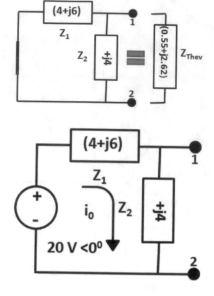

Fig. SE1.25c Original
network in a compact form

$$i_0 = \frac{20V < 0^0}{(4+j10)} = \frac{20V < 0^0}{10.77 < 68.2^0} = 1.86A < -68.2^0$$

Now, V_{THEV} is the potential drop across impedance $Z_2 = Z_2.i_0$

$$= (4\Omega < 90^0)(1.86A < -68.2^0) = 7.44 \ V < 21.8^0$$

The current, I, through the circuit shown in Fig. SE1.25d is given by

$$I = \frac{7.44V < 21.8^0}{(0.55+j2.62)+(0.55-j2.62)} = \frac{7.44 \ V < 21.8^0}{1.1\Omega} = 6.76A < 21.8^0$$

(SEA.25.2)

Fig. SE1.25d Thevenin's
equivalent of the source
network with a load that
draws maximum power
connected to it

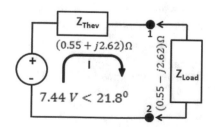

It is important to note that now the total load is $Z_{\text{Thev}} + Z_{\text{Load}} = 1.1\Omega$, *which is purely resistive and, therefore, in the circuit of* Fig. SE1.25d, *there is no phase difference between the voltage and current. The phase angle of 21.80 is relative to the voltage source (20V<0°) of the original network.*

In general, if R denotes the resistive component of both the Thevenin impedance of the source and that of the load impedance for maximum power transfer, (in the present problem R=0.55Ω) then maximum power $P_{\text{Load}}^{\max} = RI^2 = R\dfrac{\left(|V_{\text{Thev}}|\right)^2}{(R+R)^2} = \dfrac{\left(|V_{\text{Thev}}|\right)^2}{4R}.$

Hence, $P_{\text{Load}}^{\max} = \dfrac{(7.44)^2}{4.(0.55)} = 25.16W.$

The load dissipates 25.16 W of power as heat via its resistive component of 0.55Ω.

Solved example (SE1.26) A load consisting of a series combination of resistance R and capacitive reactance of 4Ω is connected between terminals 1 and 2 of the network shown in Fig. SE1.25a. Obtain the most suitable value of R so that maximum power is transferred to the load. Also, calculate the maximum power transferred to the load and compare it with the power transferred to the load for which $Z_{\text{Load}} = Z^*_{\text{Thev}}$.

Solution: *As has been calculated in (SE1.25), the load for maximum power transfer is (0.55-j2.62) Ω and that maximum power transferred to the load is 25.16 W. However, in the present problem it is given that the load is (R-j4) Ω. It is required to calculate the most suitable value of the load resistance R so that maximum power is delivered to this mismatched load.*

The most suitable value of load resistance is given by

$$R = \sqrt{R_{\text{Thev}}^2 + \left(X_{\text{Thev}} + X_{\text{L}}\right)^2}$$

or $R = \sqrt{(0.55)^2 + (+j2.62 - j4)^2} = \sqrt{0.30 + 1.90} = 1.48\Omega$

For calculating the power transferred to the load (1.48-j4), it is required to get the value of R_{Mean}, *which is given by* $R_{\text{Mean}} = \dfrac{R_{\text{Thev}} + R}{2} = \dfrac{0.55 + 1.48}{2} = 1.01~\Omega.$

And the power transferred to the load $P_{\text{L}} = \dfrac{V_{\text{Thev}}^2}{4R_{\text{Mean}}} = \dfrac{7.44^2}{4 \times 1.01} = 13.70W.$

It may be pointed out that with a fixed (-j4)Ω capacitive reactance in the load, if a 1.01 Ω resistance is added in series to the capacitance then the load draws 13.70 W of power which is almost half of 26.16 W drawn by a perfectly matched load.

1.4.5 Reciprocity or Reciprocality Theorem

The theorem of reciprocity or reciprocality is applicable to linear, bidirectional networks that have only one source of energy, either an independent voltage or

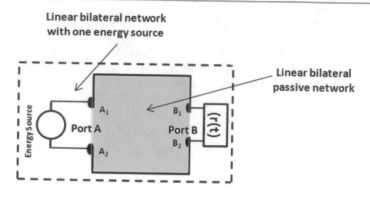

Fig. 1.33 Schematic representation of reciprocity theorem

current source. The theorem is not applicable to linear and bilateral networks that contain dependent sources. The statement of the theorem may be framed in two different ways, as given below:

"In any linear bilateral network having only one energy source, if the energy source at port A produces a response 'r(t)' at another port B of the network, then the same response 'r(t)' will appear at port A when the energy source is instead connected at port B".

A port is a pair of terminals where some circuit element may be connected and the response 'r(t)' in electrical circuits may mean a current or a voltage drop.

Figure 1.33 shows a linear bilateral network with a single energy source which is connected through terminals A_1, A_2 at port A. At some other location in the network, some measuring instrument is connected through terminals B_1, B_2 of port B which records the response 'r(t)'. The theorem says that the same response 'r(t)' will be recorded if the source is connected at port B and the measuring instrument at port A.

An active network with only one energy source may be considered as a combination of an energy source and a passive network. Thus, the linear bilateral active network of Fig. 1.33 may be treated as a combination of an energy source plus a linear bilateral passive network (shaded part). The reciprocity theorem tells that:

"In a linear bidirectional passive network, an excitation 'e(t)' by a source at some port A of the network produces a response 'r(t)' at another port B, then the same response 'r(t)' will be observed at port A if the same excitation 'e(t)' is applied at port B".

The following worked out examples will further help in understanding the theorem.

Solved example (SE1.27) Prove the reciprocity theorem with respect to ports A and B of the network shown in Fig. SE1.27a.

Solution: *To prove the theorem of reciprocity, let a voltage source of voltage V_s (V $<0°$) be connected at port A and a short circuit at port B, as shown in* Fig. SE1.27b

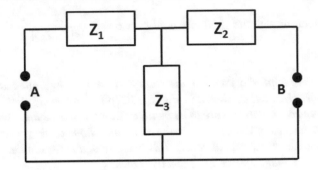

Fig. SE1.27a Network of example (SE1.27)

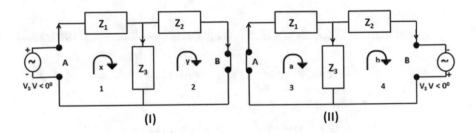

Fig. SE1.27b **I** Source at port A. **II** source at port B

(I). The direction of current flow in different branches of the network is indicated in the figure. To determine the magnitude of the current at the short circuit at port B, one may carry out the mesh analysis. There are two meshes 1 and 2, to which cyclic currents x and y have been assigned. It may be noted that the directions of the cyclic currents are chosen to match with the direction of flow of currents in different branches.

$$\text{KVL for mesh 1} \quad Z_1 x + Z_3(x - y) = V_S \; or \; (Z_1 + Z_3)x - Z_3 y = V_S \quad \text{(SE1.27.1)}$$

$$\text{KVL for mesh 2} \quad (Z_2 + Z_3)y - Z_3 x = 0 \quad\quad\quad\quad \text{(SE1.27.2)}$$

From (SE1.27.2,) one gets $x = \frac{(Z_2 + Z_3)}{Z_3} y$.
Substituting this value of x in (SE1.27.1), one gets

$$(Z_1 + Z_3)\frac{(Z_2 + Z_3)}{Z_3}y - Z_3 y = V_S \; or \; [Z_1(Z_2 + Z_3) + Z_2 Z_3]y = Z_3 V_S$$

or   current through the short circuited port $B = y = \dfrac{Z_3 V_S}{[Z_1(Z_2 + Z_3) + Z_2 Z_3]}$.

$$(SE1.27.3)$$

To prove the reciprocity theorem, the voltage source is now shifted to port B and port A is short-circuited as shown in Fig. SE1.27b(ii). However, while connecting the voltage source at port B, care must be taken that the source must send currents to different branches of the network in the same directions as they were flowing in the case when the source was at port A. It is because of this reason that the positive terminal of the voltage source is connected down.

In Fig. SE1.27b(ii) also, there are two meshes 3 and 4 to which currents 'a' and 'b', respectively, are assigned. The KVL equations for the two meshes may be written as

For mesh 3 $(Z_1 + Z_3)a - Z_3 b = 0$ *or* $b = \dfrac{(Z_1 + Z_3)}{Z_3} a$ $(SE1.27.4)$

For mesh 4 $(Z_2 + Z_3)b - Z_3 a = V_S$; *substituting the value of b from (SE1.27.4), one gets* $(Z_2 + Z_3)\frac{(Z_1 + Z_3)}{Z_3} a - Z_3 a = V_S$

or $[Z_1(Z_2 + Z_3) + Z_2 Z_3]a = Z_3 V_S$

And $a = \dfrac{Z_3 V_S}{[Z_1(Z_2 + Z_3) + Z_2 Z_3]}$ $(SE1.27.5)$

It is, therefore, observed that current y is equal both in magnitude and direction to current a. This proves the theorem of reciprocity.

1.4.6 Compensation Theorem

In electrical/electronic circuits, it often happens that we analyse the network and at the end find that some minor change in some network parameter is required to get the desired results. Normally, in such cases one changes the required parameter in small steps and repeats the circuit analysis over and over again after every change so as to find the appropriate value of the parameter. In such cases, the compensation theorem helps in estimating the required change in the desired parameter without sacrificing accuracy and carrying out a detailed analysis each time. It results in the considerable saving of time and effort.

Before coming to the statement of the theorem let us consider a linear, bilateral active network a particular branch, say AB, of which carries current I and has impedance Z with voltage drop against Z as V. Now suppose that the impedance Z is changed by a small amount ΔZ; this change of impedance from Z to $(Z + \Delta Z)$ will change current and voltage drop in branch AB as well as in other branches of the network. Let the new value of current in branch AB be $(I + \Delta I)$ and the new voltage drop across $(Z + \Delta Z)$ be $(V + \Delta V)$.

$$\text{Now,} \quad V = ZI.$$
$$\text{And} \quad (V + \Delta V) = (Z + \Delta Z)(I + \Delta I) \tag{1.52}$$
$$= ZI + I\Delta Z + (Z + \Delta Z)(\Delta I)$$

In (1.52), ZI on the right-hand side gets cancelled with V on the left and, therefore,

$$[\Delta V - (Z + \Delta Z)(\Delta I)] = I(\Delta Z) \tag{1.53}$$

In (1.53) on the left-hand side, there are the changes that have taken place in arm AB of the network due to the change in the impedance Z to $(Z + \Delta Z)$ and on the right-hand side is the term $I(\Delta Z)$ that has dimensions and units of voltage. Therefore, (1.53) may be interpreted to mean that all changes in branch AB are due to a voltage source of magnitude $I(\Delta Z)$ in this branch. This argument may be extended to include the whole network; all changes in each branch of the network are due to a voltage source of strength $(-I(\Delta Z))$ in branch AB. The negative sign is included to take into account the fact that in our discussion both ΔZ and ΔI are taken as positive, which is physically not possible; if Z increases, I will decrease and vice versa. With this background, it is now possible to state the compensation theorem as follows:

"In any linear bilateral active network, if any branch (say AB) carrying current 'I' has its impedance changed from Z to (Z+ΔZ), the corresponding changes that occur in all branches of the network are the same as those that would have occurred by the addition of a voltage source (-IΔZ) in the modified branch AB".

Figure 1.34 shows (i) the original active linear bilateral network with branch AB having current I, impedance Z and voltage drop against the impedance V, (ii) now the impedance of branch AB is changed to $(Z+\Delta Z)$ so that current in this branch becomes $(I+\Delta I)$ and the voltage drop against the impedance becomes $(V+\Delta V)$. The compensation theorem says that changes occurring in all branches of the network including branch AB (due to the change of Z to $(Z+\Delta Z)$) may be considered to have been produced by the voltage source $(-I\Delta Z)$ that is included in branch AB (the branch for which Z has been changed), as shown in Fig. 1.34(iii). The following worked out example will further explain the method of applying the compensation theorem and the advantage of using it.

Solved example (SE1.28) Calculate currents in branches be, cd and af of the network shown in Fig. SE1.28a and use the compensation theorem to calculate the change in the value of resistances R_1, R_2 and R_3, respectively, to increase currents through them by 20%.

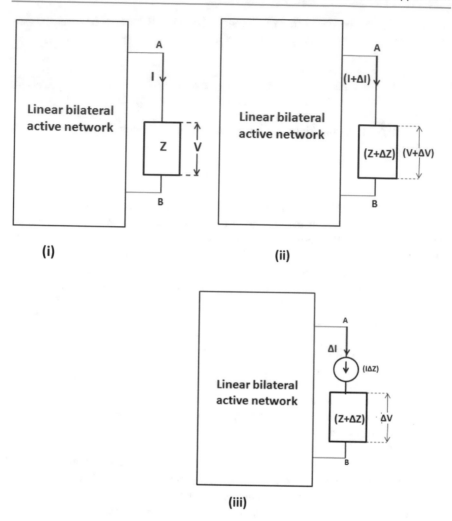

Fig. 1.34 i the original network with branch AB having current I and impedance Z, **ii** network with impedance of branch AB changed to $(Z + \Delta Z)$ and **iii** additional voltage source $(-I\Delta Z)$ in branch AB

Solution: *Refer to* Fig. SE1.28b; *the current* I_1 *drawn from the battery may be obtained by dividing the voltage 50 V by the total circuit resistance* $R_T = [30 + (50||100)]\Omega = \frac{190}{3}\,\Omega$.

$$\text{Hence,} \quad I_1 = \frac{50}{(190/3)} = 0.79\ A \text{ from } a \text{ to } b.$$

Fig. SE1.28a Network for example (SE1.28)

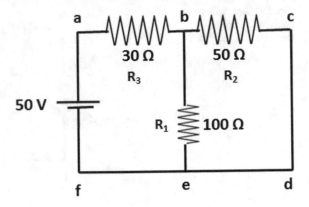

Fig. SE1.28b Current distribution in the network

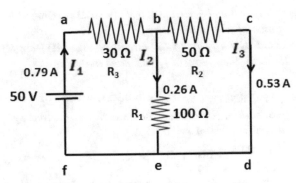

Currents I_2 and I_3 may be obtained using the law of current division in parallel impedances as $I_2 = \frac{50}{100+50} \times 0.79 = 0.26$ A and $I_3 = \frac{100}{100+50} \times 0.79 = 0.53$ A

Next, it is required to calculate the change in the resistance of the particular branch so that current in that branch increases by 20% of its initial value. We consider branch af that contains resistance R_3 of 30Ω. The initial value of current in this branch is 0.79 A. 20% of the initial value is 0.16 A. Therefore, it is required to calculate the change in the magnitude of resistance $R_3=30$Ω so that the current increases by 0.16 A. Let us assume that a change of (ΔR_3) in the value of R_3 produces the desired change. According to the compensation theorem, the desired change may be brought about by putting a voltage source of magnitude $(-I\Delta R_3)$ in the modified branch af as shown in Fig. SE1.28c.

The required change in current I_1 is 20%, i.e. $(\Delta I)= 0.16$ A. New voltage source $\Delta V = (-I\Delta R_3)$ V. New resistance in the branch $R'_1 = [(30+\Delta R_3)+(100\|50)]$. According to the compensation theorem,

$$+ (\Delta I) = \frac{\Delta V}{R'_3} \text{ Or } 0.16 = \frac{-(I\Delta R_3)}{(30+\Delta R_3)+(50\|100)} = \frac{-0.79\Delta R_3}{(30+\Delta R_3)+\frac{50 \times 100}{150}}$$

or $(0.16)[(30+\Delta R_3)+33.33] = -0.79(\Delta R_3)$

or $-0.95(\Delta R_3) = 10.12$

or $\Delta R_3 = -10.65$Ω

Fig. SE1.28c Circuit on
applying compensation
theorem to branch af

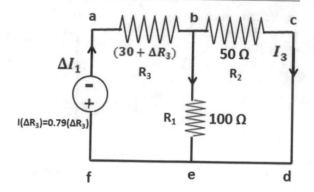

It may thus be seen that if R_3 (=30Ω) is reduced by 10.65 Ω, then the current in branch af will increase by 20%.

Let us verify it. In the circuit of Fig. SE1.28a, R_3 has the value (30–10.65=) 19.35 Ω, then current I_1 will have the value

$I_1' = \frac{50}{19.35 + 100\|50} = \frac{50}{52.68} = 0.95A$, which is 20% more than 0.79 A

Similarly, for branch 'be', which has current $I_2 = 0.26$ A required change in current = 20% of I_2= 0.05 A.

So, $0.05 = \frac{-0.26(\Delta R_1)}{(100 + (\Delta R_1)) + 50\|30}$

or $\Delta R_1 = -19.15\Omega$

Therefore, resistance of (100–19.15=) 80.85 Ω instead of 100 Ω will increase current by 20% in branch be.

Finally, for branch bc or cd, where the initial current is 0.53 A, a 20% rise will mean a change of 0.1 A in the current. Using the compensation theorem and assigning the change in R_2 as (ΔR_2), one gets

$$(\Delta I_3) = \frac{-I_3(\Delta R_2)}{(R_2 + \Delta R_2) + 100\|30}$$

Substituting the values $(\Delta I_3) = 0.1$, $R_2 = 50$, and $I_3 = 0.53$ in the above equation, the following is obtained:

$$0.1 = \frac{-0.53(\Delta R_2)}{(50 + \Delta R_2) + \frac{100\times30}{130}} = \frac{-0.53(\Delta R_2)}{73.07 + \Delta R_2}$$

or $\Delta R_2 = -11.60\Omega$

Therefore, if resistance R_2 is reduced by 11.6Ω, i.e. resistance of only 38.4Ω instead of 50Ω is kept, the current in branch bc will increase by 20%.

Self-assessment question (SAQ): In example (SE1.28), verify that reducing R_2 by 11.6Ω and R_1 by 19.15Ω increases current in respective branch by 20%.

Self-assessment question (SAQ): How will the mathematical steps in example (SE1.28) change if it is required to reduce the current by 20%?

Self-assessment question (SAQ): Draw the circuit diagram after applying the compensation theorem at branch bc in example (SE1.28).

Solved example (SE1.29) Calculate currents in different arms of the network shown in Fig. SE1.29a and apply the compensation theorem to calculate the change in R_5 required to reduce the current in arm DE by 0.1 A. How will this change affect currents in the other arms of the network?

Solution: *To find currents in various branches of the given network, we carry out the mesh analysis, assigning cyclic currents x, y and z to the three meshes as shown in Fig. SE1.29b. The three mesh equations, obtained on applying KVL to the meshes, are*

$$3x - y = 0;\ 3y - 2x + 5 = 0 \text{ and } 3z - y - 5 = 0$$

The solution of the above simultaneous equations gives

$$x = -0.71\ A;\ y = -2.14\ A \text{ and } z = 0.95\ A$$

Figure SE1.29c *shows the current distribution in various arms of the network. The current in arm DE is 0.95 A; to reduce it by 0.1 A, we have to increase the resistance R_5 by, say, ΔR. Then according to the compensation theorem, all changes in the network may be assumed to have been produced by a voltage source of voltage $(-I\Delta R)$ put in arm DE in series with $(R+\Delta R)$, and all other sources of potential replaced by a short circuit, as shown in* Fig. SE1.29d.
Now the required change in current in arm DE is -0.1 A. The negative sign is included to indicate the reduction in current. Using the compensation theorem, one gets

$$Current = \frac{Voltage}{Total\ resistance} = -0.1 = \frac{-I\Delta R}{(20+\Delta R)+5.71} = \frac{-0.95\Delta R}{25.71+\Delta R}$$

or $2.571 + 0.1\Delta R = 0.95\Delta R\ Or\ 0.85\Delta R = 2.571\ Or\ \Delta R = 3.02\Omega$

Fig. SE1.29a Network for example (SE1.29)

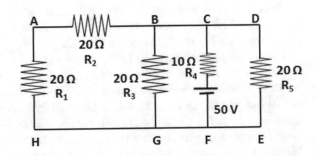

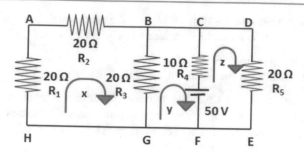

Fig. SE1.29b Mesh analysis of the network

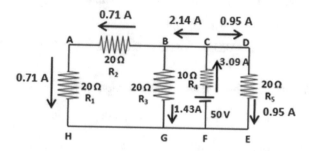

Fig. SE1.29c Current distribution in different branches

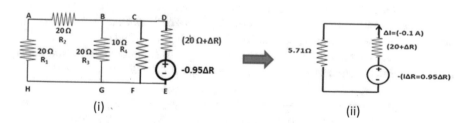

Fig. SE1.29d (i) Network on the application of Compensation theorem on arm DE. (ii) Reduced network

Thus, the compensation theorem tells that to reduce current by 0.1A, the resistance of 20Ω in branch DE should be increased by 3.0220Ω. Therefore, the total resistance in arm DE becomes 23.02Ω.

Let us now find out the effect of this change of resistance from 20Ω to 23.02Ω in arm DE on the currents in other branches. As is indicated in Fig. SE1.29d, the net result of resistance change is that an additional 0.1A current is pushed into the network at point D or C, as shown in Fig. SE1.29e. Now, this current will be distributed in branches CF, BG and BAH. Let currents I_1, I_2 and I_3 be assumed to pass through these branches, respectively. Since current distribution in a parallel combination of resistances is inversely proportional to the resistance, I_1 will be maximum

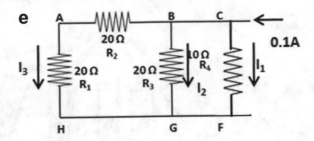

Fig. SE1.29e Effect of the increase of resistance in branch DE on currents in other branches

(as it passes through 10Ω resistance), I_2 will be half of I_1 (as it passes through 20Ω resistance) and I_3 will be ¼ of I_1 (as it passes through 40Ω resistance). Therefore,

$$\tilde{I}_1 + I_2 + I_3 = 0.1A = I_1 + \frac{1}{2}I_1 + \frac{1}{4}I_1 \, Or \, 1.75I_1 = 0.1$$
$$and \, I_1 = 0.057 \, A; I_2 = 0.028A, I_3 = 0.014A$$

Going back to Fig. SE1.29c, *the initial current distribution in different branches was as follows:*

Branch DE: *0.95 A from D to A, now the effect of the increase in resistance may be simulated as if current of 0.1A flows in the opposite direction so as to reduce current in the branch to 0.85A.*
Branch CF: *3.09A from F to C, now additional current of 0.057A passes from C to F; the net current in this branch is (3.09–0.057=) 3.033A flows from F to C.*
Branch BG: *1.43A from B to G, now additional current of 0.028 from B to G, the net current in this branch is (1.43 + 0.028=) 1.458A.*
Branch BAH: *0.71A from A to H, now additional current of 0.014A from A to H, net current in this branch is (0.71 + 0.014=) 0.724A.*

It may thus be observed that changes in current/voltage distributions in the network may be easily calculated without carrying out a detailed analysis using the compensation theorem when some parameter of the network is changed by a small amount.

1.4.7 Millman's Theorem

The theorem deals with networks where several admittances are connected at one terminal. The theorem may be stated as follows:
If in a network N admittances Y_1, Y_2, Y_3, … Y_P, … Y_N are connected to a common point M and the potentials of the other ends of these admittances V_{1G},

Fig. 1.35 Schematic representation of Millman's theorem

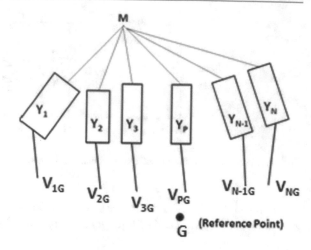

V_{2G}, V_{3G} ... V_{PG}... V_{NG} *are known with respect to some reference G, then according to Millman's theorem the potential of the common point M with respect to reference G,* V_{MG}*, is given by* $V_{MG} = \dfrac{\sum_{k=1}^{N} Y_k V_{kG}}{\sum_{k=1}^{N} Y_k}$ (Fig. 1.35).

As an example, Fig. 1.35a shows a network where four admittances, respectively, of values 4 S, 3 S, 1 S and 7 S (S stands for Siemen, the unit of admittance) have one of their two terminals connected to a common point M and the other terminals have 3V, 9V, 5V and 2V potential differences with respect to ground. The potential of the common point M with respect to ground is not known. Millman's theorem provides the formula to obtain the potential V_{MG} of the common point with respect to ground as

$$V_{MG} = \frac{\sum_{P=1}^{4} Y_P V_{PG}}{\sum_{P=1}^{4} Y_P} = \frac{4x3 + 3x9 + 1x5 + 7x2}{4+3+1+7} = \frac{58}{15} = 3.87V$$

Fig. 1.35a An example of the application of Millman's theorem

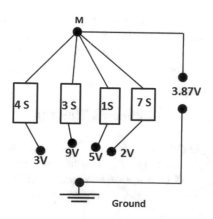

It is easy to prove Millman's theorem using Kirchhoff's current law (KCL). Knowing the magnitude of the admittance and the potential difference at its two terminals, it is possible to calculate the current through the admittance, for example, in case of the Pth admittance the current I_P through it is given by $I_P = \frac{V_{PK} - V_{MK}}{Z_P}$; here V_{PK} is the potential of one end terminal of admittance P with respect to some reference K, V_{MK} is the potential of the other terminal of the admittance with respect to the same reference point K (V_{MK} is not known) and Z_P is the impedance which is reciprocal of admittance Y_P. Now, the algebraic sum of all currents reaching the node M, from KCL, must be zero, hence,

$$\sum_{P=1}^{N} I_P = 0 \; Or \; \sum_{P=1}^{N} \frac{V_{PK} - V_{MK}}{Z_P} = \sum_{P=1}^{N} Y_P(V_{PK} - V_{MK}) = 0$$

$$\text{or} \quad \sum_{P=1}^{N} Y_P V_{PK} = \sum_{P=1}^{N} Y_P V_{MK} = V_{MK} \sum_{P=1}^{N} Y_P$$

$$\text{or} \quad V_{MK} = \frac{\sum_{P=1}^{N} Y_P V_{MK}}{\sum_{P=1}^{N} Y_P}$$

Solved example (SE1.30) One terminal of five admittances, respectively, of values 2 S, 5 S, 6 S, 1 S and 3 S is connected to a point M which has a potential of 6.7 V with respect to ground. The other ends of the admittances are, respectively, at 2 V, 5 V, 6 V, 1V and V_0 V with respect to the ground. Determine the current through the admittance of 3 S.

Solution: Using Millman's expression for the voltage at node M, we have

$$\text{Also,} \quad 6.7 = \frac{\sum_{P=1}^{5} Y_P V_P}{\sum_{P=1}^{5} Y_P} = \frac{2x2 + 5x5 + 6x6 + 1x1 + 3V_0}{2+5+6+1+3} = \frac{66 + 3V_0}{17}$$

$$\text{or} \quad 3V_0 = 47.9 \; \text{or} \; V_0 = 15.96V$$

The current through 3S admittance $= Y(V_0 - V_M) = 3(15.96 - 6.7) = 27.8A$.

1.4.8 Equivalent Generator Theorem

This theorem which is an extension of Millman's theorem may be stated as follows:
A system of n voltage sources operating in parallel may be replaced by a single equivalent voltage source of voltage V_{eq} in series with impedance Z_{eq} where

$$V_{eq} = \frac{\sum_{k=1}^{n} \frac{V_k}{Z_k}}{\sum_{k=1}^{n} \frac{1}{Z_k}} = \frac{\sum_{k=1}^{n} Y_k V_k}{\sum_{k=1}^{n} Y_k} \quad \text{and} \quad \frac{1}{Z_{eq}} = \sum_{k=1}^{n} \frac{1}{Z_k} \quad \text{or} \; Y_{eq} = \sum_{k=1}^{n} Y_k.$$

The theorem may also be looked at as the application of Thevenin's theorem to a combination of parallel voltage sources feeding a load (Fig. 1.36).

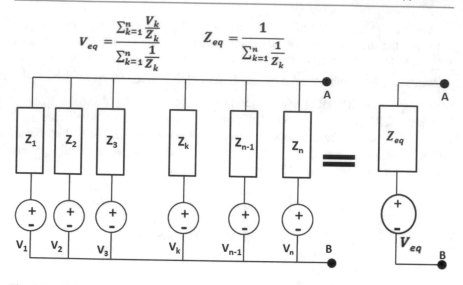

$$V_{eq} = \frac{\sum_{k=1}^{n} \frac{V_k}{Z_k}}{\sum_{k=1}^{n} \frac{1}{Z_k}} \qquad Z_{eq} = \frac{1}{\sum_{k=1}^{n} \frac{1}{Z_k}}$$

Fig. 1.36 Schematic representation of equivalent generator theorem

Solved example (SE1.31) Use generator equivalent theorem to calculate current through the capacitance C of Fig. SE1.31a.

Solution: *In the given network,* $Z_1 = 20\Omega; Z_2 = +j40\Omega; V_1 = 100\ V; V_2 = 50\ V; Z_L = -j40\Omega.$

Using the generator equivalent theorem, $V_{eq} = \frac{\frac{V_1}{Z_1} + \frac{V_2}{Z_2}}{\frac{1}{Z_1} + \frac{1}{Z_2}} = \frac{\frac{100}{20} + \frac{50}{j40}}{\frac{1}{20} + \frac{1}{j40}} = \frac{5 - j1.25}{0.05 - j0.025} = \frac{5.15 < -14.03^0}{0.06 < -26.56^0}$

or $V_{eq} = 85.83 < 12.53^0 V$

Also $\frac{1}{Z_{eq}} = \frac{1}{Z_1} + \frac{1}{Z_2} = 0.06 < -26.56^0$ (SE1.31.1)

Fig. SE1.31a Network for example (SE1.31)

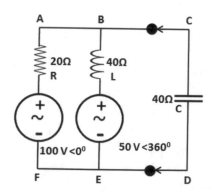

Hence, $Z_{eq} = 16.67\Omega < 26.56 = (14.91 + j7.45)\Omega$ (SE1.31.2)

Total   impedance $Z_{Total} = Z_{eq} + X_c = (14.91 + j7.45)\Omega + (-j40)\Omega = (14.91 - j32.54)\Omega$

or $Z_{Total} = (14.91 - j32.54)\Omega = 35.79\Omega < -65.38^0$ (SE1.31.3)

The current through capacitance $C = \dfrac{V_{eq}}{Z_{Total}} = \dfrac{85.83 < 12.53^0 V}{35.79\Omega < -65.38^0} = 2.40A < 77.92^0.$

Solved example (SE1.32) Determine the equivalent generator voltage V_{eg} and equivalent impedance Zeq for the circuit of Fig. SE1.32a.

Solution: *There are four independent voltage sources connected in parallel, three of which have the same polarity, while the fourth has inverse polarity. This inverse polarity may be taken care of by assigning a negative sign to the voltage of the fourth voltage source. Let us first calculate the impedances and admittances of each branch.*

$$Z_1 = 10\ \Omega;\ Z_2 = +j10\Omega;\ Z_3 = -j10\Omega;\ Z_4 = (+j5 - j5 + 5)\Omega = 5\Omega$$

Hence, $Y_1 = 0.1\ S;\ Y_2 = -j0.1\ S;\ Y_3 = +j0.1\ S;\ Y_4 = 0.2\ S$

Now, $V_{eq} = \dfrac{\sum_{k=1}^{4} Y_k V_k}{\sum_{k=1}^{4} Y_k} = \dfrac{20x0.1 + 100x(-j0.1) + 50x(+j0.1) - (50j)(0.25)}{0.1 - j0.1 + j0.1 + 0.2} = \dfrac{2 - j17.5}{0.3}$

or $V_{eq} = (6.67 - j58.33)V = 58.71\ V < -83.47^0$

Also $Y_{eq} = \sum_{k=1}^{4} Y_k = (0.1 - j0.1 + j0.1 + 0.2)S = 0.3S$

Therefore, $Z_{eq} = \dfrac{1}{Y_{eq}} = \dfrac{1}{0.3} = 3.33\Omega$

Fig. SE1.32a Network for example (SE1.32)

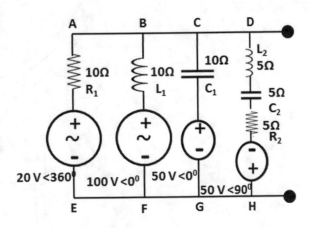

1.4.9   Nodal–Mesh Transformation or Rosen's Theorem

A nodal network is a network of impedances where one terminal of each impedance is connected to a common node and the other terminals are connected at various other nodes of the network, as shown in Fig. 1.37. On the other hand, in a mesh network impedances are connected with each other making a mesh structure, Fig. 1.37a. Rosen's theorem provides a pathway to convert a nodal network into a mesh network without disturbing the outer structure of current flow that means if current I_1 enters at node 1 of the nodal network, the same current I_1 will flow through the corresponding port 1 of the mesh network.

For the nodal network of Fig. 1.37, according to Millman's theorem,

$$V_{MG} = \frac{\sum_{k=1}^{N} Y_k V_{kG}}{\sum_{k=1}^{N} Y_k}$$

So for any node q

$$I_q = Y_q \left(V_{qG} - V_{MG} \right) = Y_q V_{qG} - Y_q \frac{\sum_{k=1}^{N} Y_k V_{kG}}{\sum_{k=1}^{N} Y_k} = \frac{Y_q \left[V_{qG} \sum_{k=1}^{N} Y_k - \sum_{k=1}^{N} Y_k V_{kG} \right]}{\sum_{k=1}^{N} Y_k}$$

or $\quad I_q = \frac{Y_q \left[\sum_{k=1}^{N} Y_k V_{qG} - \sum_{k=1}^{N} Y_k V_{kG} \right]}{\sum_{k=1}^{N} Y_k} = \frac{Y_q \left[\sum_{k=1}^{N} Y_k (V_{qG} - V_{kG}) \right]}{\sum_{k=1}^{N} Y_k} = \sum_{k=1}^{N} \frac{Y_q Y_k}{\sum_{k=1}^{N} Y_k} (V_{qG} - V_{kG})$

Fig. 1.37 A nodal network

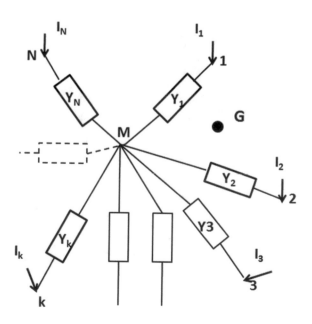

Fig. 1.37a A mesh network

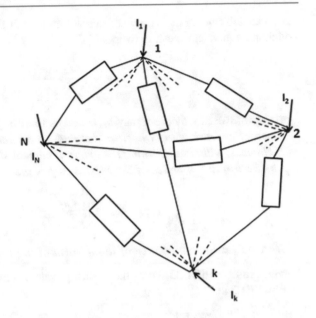

Since V_{qG} is the potential of node q with respect to reference G and V_{kG} is the potential of node k with respect to the same reference G, therefore, $\left(V_{qG} - V_{kG}\right)$ is the potential difference $(V_q - V_G)$, between node q and node k. Hence,

$$I_q = \sum_{k=1}^{N} \left[\frac{Y_q Y_k}{\sum_{k=1}^{N} Y_k}\right](V_q - V_k) \tag{1.54}$$

Now in a mesh network from Kirchhoff's current law, the current between two nodes q and k is given by

$$I_{qk} = Y_{qk}(V_q - V_k) \tag{1.55}$$

Here Y_{qk} is the admittance between the qth and k^{th} nodes. Since in a mesh network each node is connected to several other nodes, the total current I_q through node q may be obtained by summing (1.55) over all other nodes k. Therefore,

$$I_q = \sum_{k=1}^{N} I_{qk} = \sum_{k=1}^{N} Y_{qk}(V_q - V_k) \tag{1.56}$$

If (1.54), which is true for the Nodal network, is compared to (1.56) which is true for a mesh network, then one finds that the outer current structures of the two networks will remain the same, i.e. same currents will flow through nodes both in nodal and equivalent mesh networks if

$$Y_{qk} = [\frac{Y_q Y_k}{\sum_{k=1}^{N} Y_k}] \qquad (1.57)$$

As such, Rosen's theorem may be stated as follows:

A nodal network with N open nodes may be converted into a mesh network with N nodes, without disturbing the external current structure, if the admittance Y_{qk} between q and k nodes of the mesh satisfies the following condition:

$$Y_{qk} = [\frac{Y_q Y_k}{\sum_{k=1}^{N} Y_k}]$$

Here, Y_q is the admittance of the qth arm of the nodal network.

Solved example (SE1.33) Draw the mesh equivalent network for the nodal network of Fig. SE1.33a.

Solution: *The given nodal network has five free nodes, numbered 1 to 5 in the network. Since the outer current structure of the network must not change in the transformation from a nodal to a mesh structure, the equivalent mesh network must*

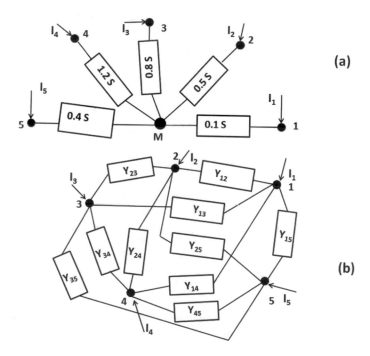

Fig. SE1.33 **a** Nodal network of problem (SE1.33). **b** Equivalent mesh network

also have five nodes. Further, in a mesh network different nodes are connected with each other through appropriate admittances; the values of these interconnecting admittances are obtained using Rosen's theorem, (1.57).

In the given nodal network, there are five admittances, $Y_1 = 0.1\ s, Y_2 = 0.5\ s, Y_3 = 0.8\ s, Y_4 = 1.2\ s$ and $Y_5 = 0.4\ s$. According to Rosen's theorem, the magnitudes of admittances of the equivalent mesh network are given by the relation

$Y_{qk} = [\frac{Y_q Y_k}{\sum_{k=1}^{N} Y_k}]$, it may, however, be kept in mind that $Y_{qk} = Y_{kq}$

Hence, $Y_{12} = \frac{Y_1 Y_2}{Y_1+Y_2+Y_3+Y_4+Y_5} = \frac{0.1x0.5}{0.1+0.5+0.8+1.2+0.4} = \frac{0.05}{3.0} = 0.017S$

$$Y_{13} = \frac{Y_1 Y_3}{Y_1+Y_2+Y_3+Y_4+Y_5} = \frac{0.1x0.8}{0.1+0.5+0.8+1.2+0.4} = \frac{0.08}{3.0} = 0.027S$$

$$Y_{14} = \frac{Y_1 Y_4}{Y_1+Y_2+Y_3+Y_4+Y_5} = \frac{0.1x1.2}{0.1+0.5+0.8+1.2+0.4} = \frac{0.12}{3.0} = 0.04S$$

$$Y_{15} = \frac{Y_1 Y_4}{Y_1+Y_2+Y_3+Y_4+Y_5} = \frac{0.1x0.4}{0.1+0.5+0.8+1.2+0.4} = \frac{0.04}{3.0} = 0.013S$$

$$Y_{23} = \frac{Y_2 Y_3}{Y_1+Y_2+Y_3+Y_4+Y_5} = \frac{0.5x0.8}{0.1+0.5+0.8+1.2+0.4} = \frac{0.40}{3.0} = 0.133S$$

$$Y_{24} = \frac{Y_2 Y_4}{Y_1+Y_2+Y_3+Y_4+Y_5} = \frac{0.5x1.2}{0.1+0.5+0.8+1.2+0.4} = \frac{0.60}{3.0} = 0.20S$$

$$Y_{25} = \frac{Y_2 Y_5}{Y_1+Y_2+Y_3+Y_4+Y_5} = \frac{0.5x0.4}{0.1+0.5+0.8+1.2+0.4} = \frac{0.20}{3.0} = 0.067S$$

$$Y_{34} = \frac{Y_3 Y_4}{Y_1+Y_2+Y_3+Y_4+Y_5} = \frac{0.8x1.2}{0.1+0.5+0.8+1.2+0.4} = \frac{0.96}{3.0} = 0.32S$$

$$Y_{35} = \frac{Y_3 Y_5}{Y_1+Y_2+Y_3+Y_4+Y_5} = \frac{0.8x0.4}{0.1+0.5+0.8+1.2+0.4} = \frac{0.32}{3.0} = 0.107S$$

$$Y_{45} = \frac{Y_4 Y_5}{Y_1+Y_2+Y_3+Y_4+Y_5} = \frac{1.2x0.4}{0.1+0.5+0.8+1.2+0.4} = \frac{0.48}{3.0} = 0.16S$$

Figure SE1.33b *shows the equivalent mesh network with numerical values of impedances obtained as above.*

Self-assessment question (SAQ): Why is it necessary to have the number of nodes in a nodal and its mesh equivalent network to be the same?

Self-assessment question: A nodal network has 4 free nodes, what will be the number of branches in the mesh equivalent network of this nodal network?

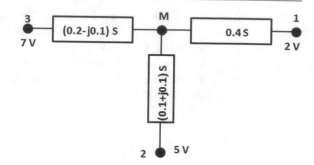

Fig. SE1.34a Nodal network of example (SE1.34)

Solved example (SE1.34) Draw an equivalent mesh network for the nodal network of Fig. SE1.34a. If potentials with respect to ground of nodes 1, 2 and 3 are, respectively, 2 V, 5 V and 7 V, calculate the potential of node M with respect to ground and hence show that the nodal to mesh transformation does not change the external current configuration of the network.

Solution: *The potential of the common node M with respect to ground is given by*

$$V_{MG} = \frac{\sum_{k=1}^{3} Y_k V_k}{\sum_{k=1}^{3} Y_k} = \frac{0.4x2 + (0.1+j0.1)x5 + (0.2-j0.1)x7}{0.4 + (0.1+j0.1) + (0.2-j0.1)} = \frac{2.7 - j0.2}{0.7}$$
$$= (3.85 - j0.28)A$$

(SE1.34.1)

External current configuration of the nodal network
The potential of node M is larger than the potential of node 1, therefore, current I_1 through node 1 will flow from M to 1.

$$I_1 = (V_{MG} - V_{1G})xY_1 = [(3.85 - j0.28) - 2](0.4) = (0.74 - j0.11)A$$

(SE1.34.2)

Similarly, $I_2 = (V_{MG} - V_{2G})xY_2 = [(3.85 - j0.28) - 5](0.1+j0.1)A$
or $I_2 = -(0.087 + j0.143)A$

(SE1.34.3)

The negative sign in the above equation means that I_2 flows from node 2 (which is at a higher potential) towards M (at lower potential).

Also $I_3 = (V_{MG} - V_{3G})xY_2 = [(3.85 - j0.28) - 7](0.2 - j0.1)A$
or $I_3 = -(0.66 - j0.26)A$

(SE1.34.4)

Again the negative sign tells that the current is flowing towards node M.

Fig. SE1.34b Voltage and current distributions in nodal network

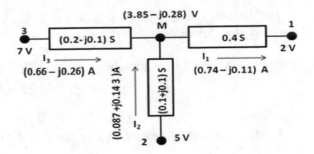

The current and voltage distribution in the nodal network is shown in Fig. SE1.34b.

The mesh equivalent network is shown in Fig. SE1.34c. The magnitude of the impedances in the mesh network is given by

$$Y_{qk} = \frac{Y_q Y_k}{\sum_{k=1}^{3} Y_k}, \text{ therefore,}$$

$$Y_{12} = \frac{Y_1 Y_2}{Y_1 + Y_2 + Y_3} = \frac{0.4 x (0.1 + j0.1)}{0.4 + (0.1 + j0.1) + (0.2 - j0.1)} = (0.057 + j0.057)S$$

$$(SE1.34.5)$$

$$Y_{13} = \frac{Y_1 y_3}{Y_1 + Y_2 + Y_3} = \frac{0.4 x (0.2 - j0.1)}{0.4 + (0.1 + j0.1) + (0.2 - j0.1)} = (0.114 - j0.057)S$$

$$(SE1.34.6)$$

And

$$Y_{23} = \frac{Y_2 Y_3}{Y_1 + Y_2 + Y_3} = \frac{(0.1 + j0.1) x (0.2 - j0.1)}{0.4 + (0.1 + j0.1) + (0.2 - j0.1)} = (0.042 + j0.014)S$$

$$(SE1.34.7)$$

Fig. SE1.34c Current distribution in a mesh network

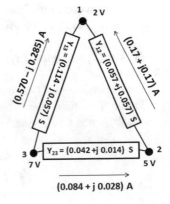

Now currents between different branches of the mesh network may be calculated easily as given hereunder:

Current I_{12} in branch 12 is given by $I_{12} = (V_2 - V_1)Y_{12} = (5 - 2)(0.057 + j0.057)$
$$I_{12} = (0.17 + j0.17)A \; from \; 2 \; to \; 1$$

$$(SE1.34.8)$$

Similarly, current I_{13} branch 13 is given by $I_{13} = (V_3 - V_1)Y_{13} = (7 - 2)$ $(0.114 - j0.057)$

or $I_{13} = (0.57 - j0.28)A$ from 3 to 1   (SE1.34.9)

And $I_{23} = 2x(0.042 + j0.014) = (0.084 + j0.028)A$ from 3 to 2   (SE1.34.10)

Current distribution in a mesh network is shown in Fig. SE1.34c.
It may be observed from Fig. SE1.34c that total current going out from node 1

$$I_{12} + I_{13} = (0.17 + j0.17)A + (0.57 - j0.28)A = (0.74 - j0.11)$$

It may be seen that $(I_{12} + I_{13})$*is equal to I_1given by* (SE1.34.2).
Similarly, current through node 2 is $I_2 = I_{21} - I_{32} = (0.17 + j0.17) -$ $(0.084 + j0.028)$
or $I_2 = (0.086 + j0.142)A$ *which agrees well with the value obtained fusing the method of the node network and is given by* (SE1.34.3).
The calculation of current at node 3 is left as an exercise. It may, however, be pointed out that current values calculated for the nodal network may slightly differ from the values obtained from the analysis of the mesh network because of the rounding off errors.

Thus, it is proved that the external configuration of currents at nodes 1, 2 and 3 does not change when the nodal network is transformed into an equivalent mesh network.

(i) T- and π-Transformation

Often in networks, one finds impedances/admittances arranged in typical formations making some well-known figures. Some of the common formations are

(a) Tee (T) formation, (b) Pi-($\prod$) formation, (c) Star formation and (d) Delta (Δ) formation.

Let us consider the T-section and $\prod$-section circuits that are shown in Fig. 1.38.
In the T-section, the impedance between terminals A and B, Z_{AB}, between terminals A and C, Z_{AC}, and between terminals B and C, Z_{BC} are given by

$$Z_{AB} = (Z_2 + Z_3), Z_{AC} = (Z_1 + Z_2), and \; Z_{BC} = (Z_3 + Z_1) \qquad (1.55)$$

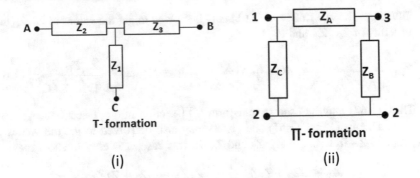

Fig. 1.38 **i** T-formation (or section) and **ii** $\prod$-formation (or section)

In the $\prod$-section, on the other hand, the effective impedance between n terminals 1 and 2, Z_{12}, is

$$Z_{12} = [Z_C || (Z_A + Z_B)] = \frac{Z_C(Z_A + Z_B)}{(Z_A + Z_B) + Z_C} = \frac{Z_A Z_C + Z_B Z_C}{Z_A + Z_B + Z_C} \quad (1.56)$$

Similarly,

$$Z_{13} = [Z_A || (Z_B + Z_C)] = \frac{Z_A(Z_B + Z_C)}{(Z_A | Z_B) + Z_C} = \frac{Z_A Z_B + Z_A Z_C}{Z_A + Z_B + Z_C} \quad (1.57)$$

$$Z_{23} = [Z_B || (Z_A + Z_C)] = \frac{Z_B(Z_A + Z_C)}{(Z_A + Z_B) + Z_C} = \frac{Z_A Z_B + Z_B Z_C}{Z_A + Z_B + Z_C} \quad (1.58)$$

If the T-section and the $\prod$-sections are equivalent, then $Z_{AB} = Z_{13}; Z_{AC} = Z_{12} and Z_{BC} = Z_{23}$.
Therefore,

$$Z_{AB} = Z_{13} or\ (Z_2 + Z_3) = \frac{Z_A Z_B + Z_B Z_C}{Z_A + Z_B + Z_C} \quad (1.59)$$

$$Z_{AC} = Z_{12} Or\ (Z_1 + Z_2) = \frac{Z_A Z_C + Z_B Z_C}{Z_A + Z_B + Z_C} \quad (1.60)$$

And

$$Z_{BC} = Z_{23} Or\ (Z_3 + Z_1) = \frac{Z_A Z_B + Z_A Z_C}{Z_A + Z_B + Z_C} \quad (1.61)$$

Equations (1.59), (1.60) and (1.61) may be solved to get the values of Z_1, Z_2 and Z_3 in terms of Z_A, Z_B and Z_C :

$$Z_1 = \frac{Z_B Z_C}{Z_A + Z_B + Z_C}; Z_2 = \frac{Z_A Z_C}{Z_A + Z_B + Z_C}; and\ Z_3 = \frac{Z_A Z_B}{Z_A + Z_B + Z_C} \quad (1.62)$$

Thus, (1.62) may be used to transform a $\prod$-section into a T-section.

Simultaneous (1.56), (1.57) and (1.58) may also be solved to get the values of Z_A, Z_B and Z_C in terms of Z_1, Z_2 and Z_3. In this case, it is easy to show that

$$Z_A = \frac{Z_1 Z_2 + Z_2 Z_3 + Z_1 Z_3}{Z_1}, Z_B = \frac{Z_1 Z_2 + Z_2 Z_3 + Z_1 Z_3}{Z_2}, Z_C$$
$$= \frac{Z_1 Z_2 + Z_2 Z_3 + Z_1 Z_3}{Z_3} \quad (1.63)$$

The set of equations contained in (1.63) may be used to convert a T-section into a $\prod$-section.

Solved example (SE1.35) Draw the T-section that is equivalent to the $\prod$-section of Fig. SE1.35a and show that the two sections are equivalent to each other.

Solution: *The sum of the three impedances is $(10 + j10) + (5 - j5) + (10 - j5) = (25 + j0)\ \Omega$. The impedances Z_1, Z_2 and Z_3 of the equivalent T-section are given as*

$$Z_1 = \frac{(5 - j5)(10 - j5)}{(25 + j0)} = (1 - j3)\Omega; Z_2 = \frac{(5 - j5)(10 + j10)}{(25 + j0)} = (4 + j0)\Omega$$

And $Z_3 = \frac{(10 + j10)(10 - j5)}{(25 + j0)} = (6 + j2)\Omega$

The two sections will be equivalent to each other if the same voltage source is connected between the corresponding terminals of the two sections, then the same amount of current is drawn by the two sections. Let us connect a voltage source of voltage V volts between terminals 1 and 2 of the given $\prod$-section, and let current $I_{\prod}$ be drawn from the voltage source by the section.

Fig. SE1.35a $\prod$-section of example (SEA.35)

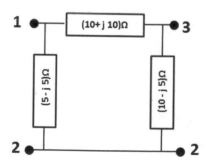

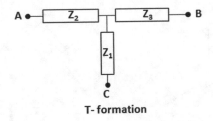

T- formation

Fig. SE1.35b Equivalent T-section

$$I_\pi = \frac{V}{\text{effective impedance between terminals 1 and 2}}$$
$$= \frac{V}{(5-j5)\|[(10+j10)+(10-j5)]} = \frac{V}{(5-j3)}A \qquad (\text{SE1.35.1})$$
$$\text{or} \quad I_\pi = \frac{V(5+j3)}{34}A$$

Next, let us connect the same voltage source of voltage V between the corresponding terminals A and C of the T-section. The current I_T in this case is given by (Fig. SE1.35b)

$$I_T = \frac{V}{\text{effective impedance between terminals } A \text{ and } C}$$
$$= \frac{V}{Z_1+Z_2} = \frac{V}{(1-j3)+(4+j0)} = \frac{V}{(5-j3)} \qquad (\text{SE1.35.2})$$
$$\text{or} \quad I_T = \frac{V(\% +j3)}{34}A$$

From (SE1.35.1) *and* (SE1.35.2), *it is clear that* $I_T = I_\pi$ *which proves the equivalence of the two sections.*

(ii) Star-Delta (Δ)Transformation

A typical pair of star and delta formations is shown in Fig. 1.39. It may be noted that while the star section is a nodal network, the delta section is a mesh network. Therefore, Rosen's theorem for the transformation from nodal to mesh may be used for the star to delta transformation. According to Rosen's theorem,

$$Y_{AB} = \frac{1}{Z_{AB}} = \frac{Y_{AS}Y_{BS}}{Y_{AS}+Y_{BS}+Y_{CS}} = \frac{\frac{1}{Z_{AS}}\frac{1}{Z_{BS}}}{\frac{1}{Z_{AS}}+\frac{1}{Z_{BS}}+\frac{1}{Z_{CS}}}$$
$$= \frac{Z_{CS}}{Z_{AS}Z_{BS}+Z_{BS}Z_{CS}+Z_{CS}Z_{AS}}$$

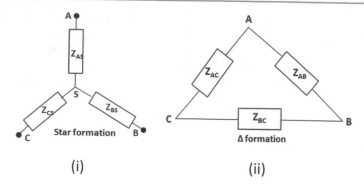

Fig. 1.39 **i** A three-node star network **ii** a three-node delta network

$$\text{or} \quad Z_{AB} = \frac{Z_{AS}Z_{BS} + Z_{BS}Z_{CS} + Z_{CS}Z_{AS}}{Z_{CS}} \tag{1.64}$$

$$\text{Similarly,} \quad Z_{AC} = \frac{Z_{AS}Z_{BS} + Z_{BS}Z_{CS} + Z_{CS}Z_{AS}}{Z_{BS}} \tag{1.65}$$

$$\text{And} \quad Z_{BC} = \frac{Z_{AS}Z_{BS} + Z_{BS}Z_{CS} + Z_{CS}Z_{AS}}{Z_{AS}} \tag{1.66}$$

Further, a Delta section may also be transformed into a star section; for that, let us consider the two networks of Fig. 1.39 and equate the effective impedances between different pairs of nodes in the two sections.

The effective impedance between terminals A and B is

$$(Z_{AS} + Z_{BS})_{\text{Star}} = [Z_{AB}||(Z_{AC} + Z_{BC})]_{\text{Delta}} = [\frac{Z_{AB}(Z_{AC} + Z_{BC})}{Z_{AB} + Z_{AC} + Z_{BC}}]_{\text{Delt}} \tag{1.67}$$

Similarly, for effective impedance between terminals A and C, we have

$$(Z_{AS} + Z_{CS})_{\text{Star}} = [Z_{AC}||(Z_{AB} + Z_{BC})]_{\text{Delta}} = [\frac{Z_{AC}(Z_{AB} + Z_{BC})}{Z_{AB} + Z_{AC} + Z_{BC}}]_{\text{Delt}} \tag{1.68}$$

And for effective impedance between terminals B and C, we have

$$(Z_{CS} + Z_{BS})_{\text{Star}} = [Z_{BC}||(Z_{AC} + Z_{AB})]_{\text{Delta}} = \left[\frac{Z_{BC}(Z_{AC} + Z_{AB})}{Z_{AB} + Z_{AC} + Z_{BC}}\right]_{\text{Delt}} \tag{1.69}$$

The above three simultaneous equations may be solved to get

$$Z_{AS} = \frac{Z_{AB}Z_{AC}}{Z_{AC}+Z_{AB}+Z_{BC}}; Z_{BS} = \frac{Z_{AB}Z_{BC}}{Z_{AC}+Z_{AB}+Z_{BC}}; Z_{CS}$$

$$= \frac{Z_{BC}Z_{AC}}{Z_{AC}+Z_{AB}+Z_{BC}} \qquad (1.70)$$

Note: It may be marked that in each of the above equalities defined by (1.70), if one wants to find the particular star element value, say Z_{AS}, one has to multiply together the two delta elements joined at node A, i.e. Z_{AB} and Z_{AC} and divide by the sum of the delta impedances.

Solved example (SE1.36) In the network of Fig. SE1.36a, calculate the effective resistance between pairs of terminals AB, BC and AC.

Solution: *In order to find the desired resistances, one has to simplify the given network. Let us start with the innermost delta formation with resistances 5 Ω, 3 Ω, and 2 Ω. As shown in* Fig. SE1.36b, *the delta section may be transformed into a star formation using the transformation* (1.70).

The 1.5 Ω resistance in series with 6.5 Ω makes 8 Ω, 1 Ω in series with 9 Ω resistance makes 10 Ω, and 9.4Ω resistance with 0.6 Ω resistance results in 10 Ω resistance. The star formation of 8 Ω, 10 Ω and 10 Ω resistances may again be converted into an equivalent Δ-section (see Fig. SE1.36c),

$$\text{with} \quad R_4 = \frac{R_1R_2 + R_1R_3 + R_2R_3}{R_3} = \frac{8x10 + 8x10 + 10x10}{10} = \frac{260}{10} = 26\Omega$$

$$R_5 = \frac{R_1R_2 + R_1R_3 + R_2R_3}{R_2} = \frac{8x10 + 8x10 + 10x10}{10} = \frac{260}{10} = 26\Omega$$

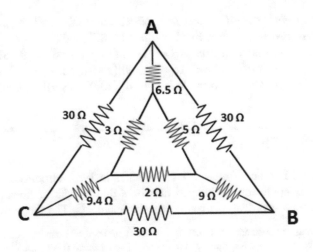

Fig. SE1.36a Network for problem (SE1.36)

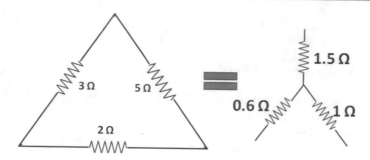

Fig. SE1.36b Delta to star conversion

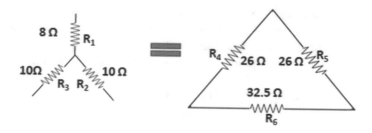

Fig. SE1.36c Star to delta transformation

$$And \quad R_6 = \frac{R_1R_2 + R_1R_3 + R_2R_3}{R_1} = \frac{8x10 + 8x10 + 10x10}{8} = \frac{260}{8} = 32.5\Omega$$

Thus, the original circuit reduces to a delta section with each having a parallel combination of two resistances, as shown in Fig. SE1.36d.
The effective resistance in arm $AB = [13.9||(13.9 + 15.6)] = 9.45\Omega.$
The effective resistance in arm $AC = [13.9||(13.9 + 15.6)] = 9.45\Omega.$
The effective resistance in arm $BC = [15.6||(13.9 + 13.9)] = 9.99\Omega.$

1.5   Tellegen Theorem

Dutch Electrical Engineer Bernard D. H. Tellegen in 1952 introduced this theorem. The theorem is applicable to lumped networks that consist of linear, non-linear, time-variant, time-invariant, active and passive elements. The theorem states,

> The summation of instantaneous powers for the n number of branches in an electrical network is zero provided Kirchhoff's current (KCL) and voltage (KVL) laws hold for each branch.

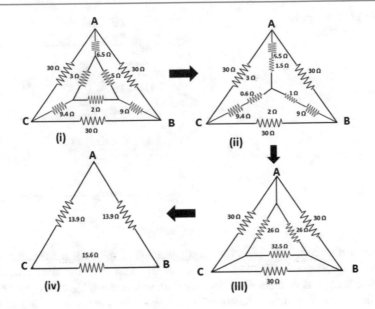

Fig. SE1.36d Stages of circuit simplification

In mathematical language, the theorem for a network of k branches may be written as

$$\sum_{k}^{n} v_k i_k = 0$$

Here, v_k, and i_k are respectively the instantaneous current and voltage drops in the kth branch of the network and Kirchhoff's current (KCL) and voltage (KVL) laws hold for each branch of the network.

Though the theorem at first glance appears obvious and, therefore, not exploited too much, it is of great importance and many other theorems may be derived using this theorem.

The theorem may be proved for any network where currents and voltages in each branch obey Kirchhoff's law. The following solved example will prove the theorem.

Solved example (SE1.37) Figure SE1.37 shows a network where currents and voltages at different branches are marked. Obtain the values of unknown currents and voltages in the circuit and show that the Tellegen theorem holds for the network.

Fig. SE1.37 Network for
example (SE1.37)

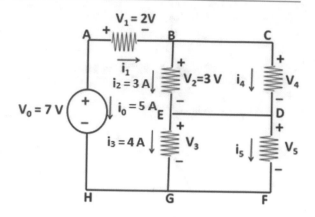

Solution: An essential condition for the application of the Tellegen theorem is that currents and voltages in different branches of the network must obey Kirchhoff's laws of currents and voltages. We first calculate voltages V_3, V_4 and V_5 that are not given in the problem using Kirchhoff's voltage law. Let us apply KVL to the closed-loop ABEGHA to get

$$V_1 + V_2 + V_3 = V_0 \text{ or } 2 + 3 + V_3 = 7; \text{Hence } V_3 = 2V$$

Similarly from closed-loop BCDEB, $V_4 = V_2 = 3V$.
And from closed-loop EDFGE, $V_5 = V_3 = 2V$.
Having obtained the values of unknown voltages, we now apply KCL at the following nodes to get the values of unknown currents. For that, let us assume that current i_0 is 5 A so that from Kirchhoff's law $i_1 = -5$ A; and at node B if $i_2 = 3$ A; then $i_4 = -8$ A. Further if at node E it is assumed that $i_3 = 4$ A, then $i_5 = -9$ A.

So now we have voltages at all branches that obey KVL and currents in all branches that obey KCL. Substituting these values in the expression for the Tellegen theorem, one gets

$$\sum_{k}^{n} v_k i_k = 7x5 + 2(-5) + 3x3 + 2x4 + 3x(-8) + 2x(-9) = 0$$

Hence, the Tellegen theorem gets verified.

Problems

P1.1   Calculate the magnitude of an inductor and of a capacitor that offers
impedance of 10 Ω each at a linear frequency of 100 Hz. Express, both in
complex and vector forms, the impedance of a series combination of a
capacitor, an inductor and an ideal resistance each of impedance 10 Ω.

Answer: [15.9mH; 159 μF; (10 + j0); (10Ω < 0°)].

P1.2   An AC voltage source gives a voltage of (100 V<60°). Write this voltage
in rectangular form.

Answer: [50+j86.60] V.

P1.3   Three AC voltage sources are cascaded as shown in Fig. P1.3. Compute
the output voltage between A and B.

Answer: [121.98 V<62.10°].

P1.4   State the superposition theorem and discuss the conditions for its
application. For the network shown in Fig. P1.4, find the current through
the load Z_L (series combination of inductance and the resistance) between
terminal A and B using (a) Superposition theorem and (b) Thevenin's
equivalent circuit with reference to terminals A and B.

Answer: [12.19 A<62.10°].

Fig. P1.3 Circuit for
problem (P1.3)

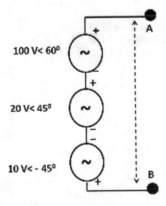

100 V< 60°

20 V< 45°

10 V< - 45°

A

B

Fig. P1.4 Circuit for
problem (1P.4)

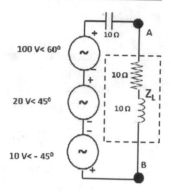

P1.5   Using the superposition theorem, determine the current through the 10
Ωresistance between terminals A and B of the network of Fig. P1.5.

Answer: [current 1.79A<63.44° by the voltage source and 0.45A
<63.44° by the current source in the direction from A to B, total
current through 10Ω resistance 2.23 A<63.44° from A to B].

P1.6   Three energy sources supply power to the network shown in Fig. P1.6;
use the superposition theorem to obtain currents through the ten-ohm
resistance established by each source individually. Also, confirm your
result using the mesh analysis.

Answer: [currents respectively by S_1, S_2 and S3: $(0.24 + j0.94)$
A *from A to B*; $(0.11 + j0.46)A$ *from A to B*; $(0.06 + j0.24)A$ *from B to A*].

Fig. P1.5 Circuit for
problem (P1.5)

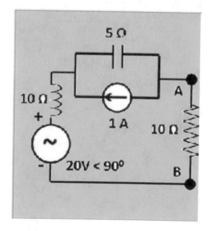

Fig. P1.6 Circuit for
problem (P1.6)

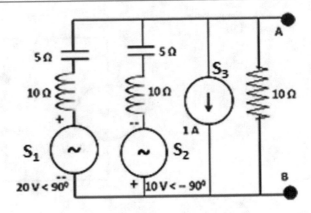

Fig. P1.7 Network for
problem (1P.7)

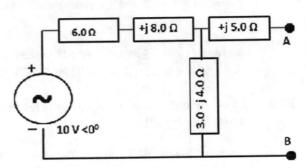

P1.7 For the circuit given in Fig. P1.7, obtain the values of the Thevenin
voltage and the Thevenin impedance between terminals A and B.

Answer: $[V_{Thev} = (5.08 \ V < -77.09^0); Z_{Thev} = (4.64 + j2.94)\Omega].$

P1.8 For the network given in Fig. P1.8, determine the current in each branch
of the network, the total current drawn from the voltage source,

Fig. P1.8 Network for
problem (P1.8)

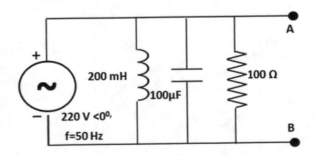

Fig. P1.9 Network for
problem (P1.9)

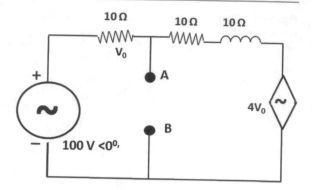

Thevenin's voltage and Thevenin's impedance between terminals A and
B of the network.

Answer: $[\,[I^L = (3,50\,A < -90^0); I^C = (6.91\,A < 90^0), I^R = (2,2\,A < 0^0), I^{total} = (4.06\,A < 57.17^0), V_{Thev} = (220V < 0^0); Z_{Thev} = (54.19 < -57.17^0)\,\Omega]$.

P1.9   Draw Thevenin's equivalent circuit between terminals A and B of the
network of Fig. P1.9.

Answer: Figure P1.9a shows Thevenin's equivalent circuit for problem
(P1.9).

P1.10   Determine the Thevenin voltage V_{Thev} and the Thevenin impedance Z_{Thev}
with respect to terminals A and B for the network shown in Fig. P1.10.

Answer: $[V_{Thev} = 7.5\,V, Z_{Thev} = 4.0\Omega$. Note: For determining
Z_{Thev}, replace the independent voltage source by a short circuit and
connect an external voltage source of voltage V_{ex} between terminals
A and B and calculate current I_{ex} drawn from V_{ex}. The ratio
V_{ex}/I_{ex} will give $Z_{Thev}]$.

Fig. P1.9a Thevenin
equivalent circuit

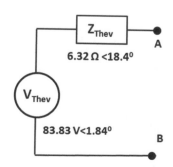

Fig. P1.10 Network for
problem (P1.10)

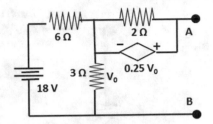

Fig. P1.11 Network for
problem (P1.11)

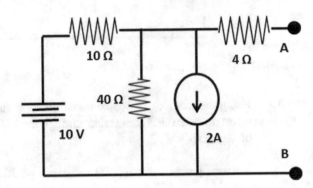

P1.11   Draw Thevenin's equivalent circuit, between terminals A and B of the
network given in Fig. P1.11.

Answer: Thevenin's equivalent circuit is given in Fig. P1.11a.

P1.12   Obtain the Thevenin impedance and voltage between terminals a and b of
the network given in Fig. P1.12. Also, use the superposition theorem to
obtain currents in each branch of the circuit due to individual sources.

Answer: [V_{Thev}= 1 V ,Z_{Thev}= 1.56 Ω, current due to 3V battery in
all branches = 0.33A, from a to c through R_1,c to b through R_2, b to d
through R_3 .Currents due to current source: Through R_1 1.33 A from c
to a and 0.67A through R_2 and R_3 from c to b to d].

Fig. P1.11a Thevenin
Equivalent

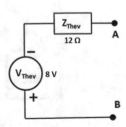

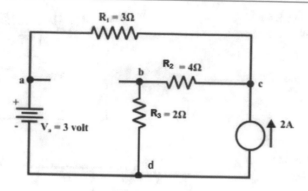

Fig. P1.12 Network for problem (P1.12)

P1.13   Using the methods of source transformation and the superposition
theorem, determine the current through the 7Ω resistance in the network
of Fig. P1.13.

Answer: [current through 7Ω resistance =1.78 A].

P1.14   Use the Thevenin equivalent circuit between terminals A and B of the
network.Shown in Fig. (P1.14a) to determine the current through the 6Ω
load resistance R_L.

Answer: [I_L = 2.65 A].

P1.15   Find the values of the Norton current and the Norton impedance with
reference to terminals A and B for the circuit of Fig. P1.15.

Answer: $\left[I_{Nort} = (5.36\ A < -\ 10.3^0)\ and\ Z_{Nort} = (7.5 + j2.5)\Omega \right]$.

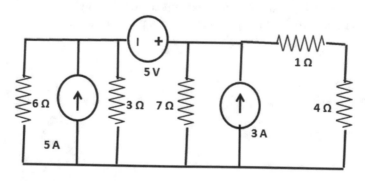

Fig. P1.13 Network for problem (P1.13)

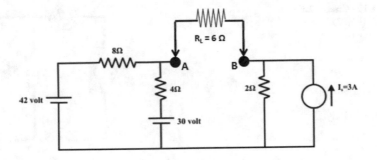

Fig. P1.14a Circuit for problem (P1.14)

Fig. P1.15 Circuit for
problem (P1.15)

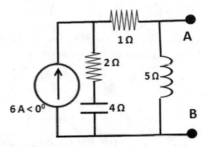

P1.16   Obtain the values of the parameters for the Norton and the Thevenin
equivalent circuits for the network of Fig. P1.16 with reference to
terminals A and B.

    Answer: $[V_{Thev} = 155\ V; Z_{Thev} = Z_{Nort} = 55\Omega; I_{Nort} = 2.82\ A]$.

P1.17   Show that Norton's and Thevenin's equivalent circuits with respect to
terminals A and B for the network of Fig. P1.17 are identical. Draw one of
them.

    Answer: [*Since the circuit has no independent energy source, both*
V_{Thev} *and* I_{Nort} *are zero,* $Z_{Thev} = Z_{Nort} = 1.2\Omega$]. Thevenin's or
Norton's equivalent circuit is shown in Fig. P1.17a.

Fig. P1.16 Network for
problem (P1.16)

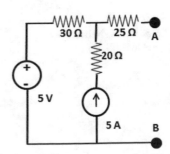

Fig. P1.17 Circuit for
problem (P1.17)

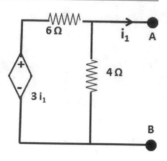

Fig. P1.17a Norton or
Thevenin equivalent for
problem (P1.17)

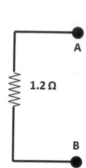

P1.18   Calculate the Load that must be connected at terminals 1 and 2 so that
maximum power is transferred to the load from the source circuit. Also,
calculate the maximum power that may be delivered by the source
network to the load (Fig. P1.18).

Answer: (P1.18) $\left[Z_{Load} = 2.52 - j2.81;\ P_{Load}^{max} = 0.60\ kW\right]$.

P1.19   Norton's equivalent circuit parameters for a network are $I_{Nort} = 5.3A < -10^0$ and $Z_{Nort} = (7.5 + j2.5)\Omega$. . Calculate the load that may
be connected to Norton's equivalent circuit to absorb maximum power
and the amount of maximum power absorbed by such a load.

Answer: $\left[Z_{Load} = (7.5 - j2.5)\Omega;\ P_{Load}^{max} = 58.43\ W\right]$.

Fig. P1.18 Network for
problem (P1.18)

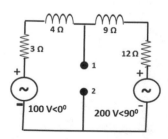

P1.20 (a) For a source network, Norton's Current I_{Nort} is $(4+j3)A$ and Thevenin's potential difference V_{Thev} is $(8+j2)V$. Calculate the value of the load that may draw maximum power from the network and the magnitude of the maximum power delivered by the network.(b) In case the load has a fixed inductance of 2Ω reactance, calculate the value of the resistance R that must be put in series with the inductance for optimal power transfer to the load, and also the power transferred in this situation.

Answer: (a) [$Z_{Load} = (1.6 + j0.64)\Omega$; $P_{Load}^{max} = 10.61\ W$];
(b)[R = 2.1 Ω, $P_{Load}^{optimum} = 9.17\ W$].

P1.21 Prove the reciprocity theorem for ports A and B of the network shown in Fig. P1.21 and calculate the current through the 100 Ω resistor.

Answer: [0.157 A].

P1.22 Calculate the change in the impedance of branch CD of the network of Fig. P1.22 so that the current in this branch reduces by 0.1A.

Answer: [Increase in the impedance **of branch** CD by $(2.53-j1.26)\Omega$].

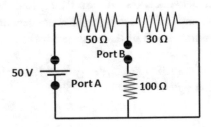

Fig. P1.21 Network for problem (P1.21)

Fig. P1.22 Circuit for problem (P1.22)

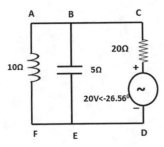

Fig. P1.23 Circuit for
problem (P1.23)

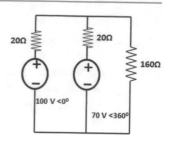

P1.23   Use the generator equivalent theorem to find the equivalent voltage source
V_{eq} and the equivalent impedance Zeq for the circuit of Fig. P1.23,
assuming 160 Ω resistance as load.

   Answer: [V_{eq}=85 V; Z_{eq}=10Ω].

P1.24   One terminal each of three impedances of $(1 + j1)\Omega$, $(2 + j2)\Omega$ and $(3 - j3)\Omega$ is connected to a terminal P, and the potential difference of the other
terminals of these impedances with respect to ground are respectively, 4
V, 3 V and 2 V. What is the potential of terminal P with respect to
ground? (Fig. P1.24)

   Answer: [3.61V<−5.46°].

P1.25   For the nodal network shown in Fig. P1.25, find the current I_{Y1} through
impedance Y_1 and draw the equivalent mesh network. What is the current
I_{AB} through branch AB of the mesh network?

   Answer: $[V_{MG} = (7.18 - j0.036)V;\ I_{Y1} = (4.18 - j0.036)A\ ;$
   $I_{AB} = (0.36 + j0.036)A]$ (Fig. P1.25b).

Fig. P1.24 Circuit for
problem (P1. 24)

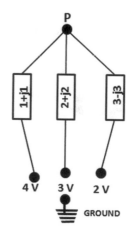

Fig. P1.25 Nodal network of problem (P1.25)

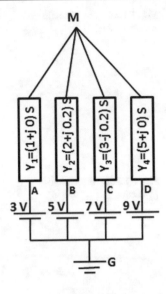

Fig. P1.25b Equivalent mesh network for problem (P1.25)

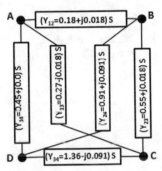

P1.26   Find the equivalent ∏-section for the following T-section (Fig. P1.26).

  Answer: The equivalent ∏-section is given below in Fig. P1.26a.

P1.27   Compute the resistance of the given network (Fig. P1.27a) between terminals X and Y using delta–star conversion for the section enclosed in the dotted rectangle.

  Answer: [5.65 Ω] .

P1.28   Use ∏–T and star–Delta conversions to simplify the network shown in Fig. P1.28 and get the resultant capacitance between terminals A and B.

  Answer: [8 μF].

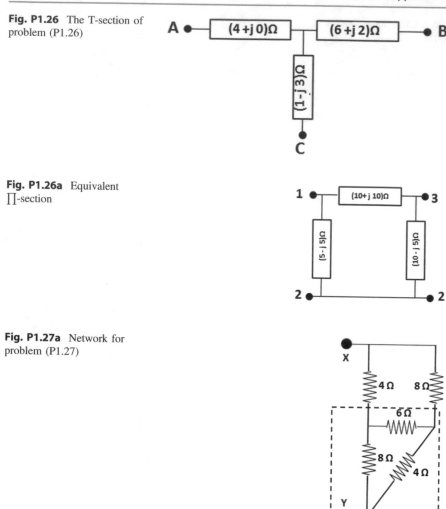

Fig. P1.26 The T-section of problem (P1.26)

Fig. P1.26a Equivalent ∏-section

Fig. P1.27a Network for problem (P1.27)

P1.29   Use Star–Delta transformations to simplify the network of Fig. P1.29 and thus find the effective resistance between terminal A and B of the network.

Answer: [2.49 Ω].

P1.30.   Find the impedance between terminals A and C of the network of Fig. P1.30.

Answer: [3 Ω].

Fig. P1.28 Circuit for problem (P1.28)

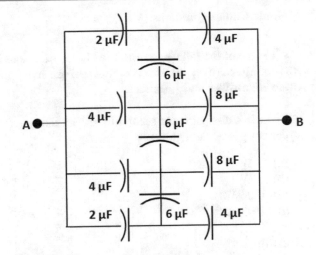

Fig. P1.29 Network for problem (P1.29)

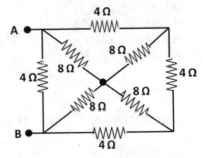

Fig. P1.30 Network for problem (P1.30)

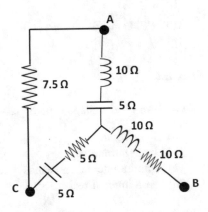

Multiple Choice Questions (MCQ)

Note: **Some of the following multiple-choice questions may have more than one correct alternative; all correct alternatives must be ticked/selected for the complete answer and full marks.**

MC1.1 Heat dissipated per second by the impedance $(5\Omega < 36.86^0)$ when 10 mA current passes through it is

(a) 4 W
(b) 3 W
(c) 0.04 W
(d) 0.03 W.

ANS: (c).

MC1.2 Current of $(4 + j3)A$ in polar notation may be written as

(a) **(5 A< 36.86°)**
(b) **(5 A<−36.86°)**
(c) **(5 A<53.13°)**
(d) **(5 A<−53.13°)**

ANS: (a).

MC1.3 Pick the vector form of Ohm's law:

(a) $(\vec{\nabla} \times \overrightarrow{E}) = -\frac{\partial \overrightarrow{B}}{\partial t}$
(b) $(\vec{\nabla}.\overrightarrow{E}) = \frac{\rho}{\varepsilon}$
(c) $(\vec{\nabla}.\overrightarrow{D}) = \rho$
(d) $\vec{J} = \sigma\vec{E}.$

ANS: (d).

MC1.4 An ideal voltage source has

(a) Large emf
(b) large current capacity
(c) infinite internal resistance
(d) zero internal resistance.

ANS: (d).

MC1.5 What is required to be done to dependent current and voltage sources present in a network while applying the superposition theorem:

(a) Replace all dependent sources by short circuits.
(b) Replace all dependent sources by open circuits.
(c) Keep the dependent sources in their original form without replacing them.
(d) Replace dependent voltage sources by short circuits and dependent current sources by open circuits.

ANS: (c).

MC1.6 In which of the following types of networks is it possible to separately treat its elements like resistance, capacitance and inductance for analysis?

(a) Distributed network
(b) Linear network
(c) Lumped network
(d) Bidirectional network.

ANS: (c).

MC1.7 In determining the magnitude and polarity of the potential drop across impedance, it is necessary to know:

(a) The magnitude of the impedance
(b) The magnitude of the current
(c) The direction of flow of current
(d) the total emf in the network.

ANS: (a), (b), (c).

MC1.8 Nodal and Mesh cyclic current methods of analysis are applicable only to

(a) Unidirectional networks
(b) Planner networks
(c) Bidirectional networks
(d) DC networks.

ANS: (b).

MC1.9 The superposition theorem can be applied only to networks that are

(a) Resistive
(b) unidirectional

(c) bilateral
(d) linear and bilateral.

ANS: (d).

MC1.10 Which of the following theorems help in simplifying a parallel combination of current sources into a single current source?

(a) Norton's
(b) Thevenin's
(c) Millman's
(d) Rosen's.

ANS: (c).

MC1.11 Reciprocity theorem is applicable to

(a) Any network
(b) only to a planner network
(c) linear network
(d) linear bilateral network.

ANS: (d).

MC1.12 Kirchhoff's laws are not applicable to network with

(a) active elements
(b) passive elements
(c) lumped parameters
(d) distributed parameters.

ANS: (d).

MC1.13 Efficiency of power transfer in the case when load and source impedances are equal is

(a) 50%
(b) 70%
(c) 90%
(d) 100%.

ANS: (a).

MC1.14 Which of the following is a non-linear element?

(a) metallic resistance
(b) diode
(c) transistor
(d) ideal capacitor.

ANS: (a), (b), (c).

MC1.15 In the mesh analysis of a network, the concept of super mesh is used when the mesh has a

(a) Current source
(b) Voltage source with internal resistance
(c) Voltage source with no internal resistance
(d) Dependent voltage source.

ANS: (a).

MC1.16 Pick all correct statements from the following:

(a) In a nodal network, one end of each impedance is connected to a common node.
(b) In a mesh network, several impedances may be connected at each node.
(c) While calculating Thevenin's impedance Z_{Thev} both independent and dependent voltage sources are replaced by open circuits.
(d) Rosen's theorem helps in the conversion of a nodal network into an equivalent mesh network.
(e) Millman's theorem is applicable only to mesh networks.
(f) Millman's theorem is applicable to nodal networks.
(g) It is possible to transform a mesh network into an equivalent nodal network using Rosen's theorem.
(h) The equivalent generator theorem is based on Rosen's theorem.

ANS: [(a), (b), (d), (f), (g)].

MC1.17 If Z_{Thev} and V_{Thev} are, respectively, Thevenin's impedance and Thevenin's voltage for a given network corresponding to two terminals A and B and Z_{Nort} & I_{Nort} are Norton's impedance and Norton's current for the same network with respect to terminals A and B, then

(a) $Z_{Thev} = Z_{Nort}; I_{Nort} = \dfrac{V_{Thev}}{Z_{Thev}}$

(b) $Z_{Thev} = Z_{Nort}; I_{Nort} = \dfrac{Z_{Thev}}{V_{Thev}}$

(c) $Z_{Thev} = \frac{Z_{Nort}}{V_{Thev}}$; $I_{Nort} = \frac{Z_{Thev}}{V_{Thev}}$

(d) $Z_{Thev} = \frac{I_{Nort}}{V_{Thev}}$; $Z_{Nort} = \frac{Z_{Thev}}{V_{Thev}}$

ANS: [(a)].

MC1.18 *"In any linear bilateral active network, if any branch (say AB) carrying current 'I' has its impedance changed from Z to (Z+ΔZ), the corresponding changes that occur in all branches of the network are the same as those that would have occurred by the addition of a voltage source (-IΔZ) in the modified branch AB"* is the statement of

(a) Norton's theorem
(b) Theorem of maximum power transfer
(c) Compensation theorem
(d) Rosen's theorem.

ANS: (c).

MC1.19 A mesh is

(a) A part of a network which has at least four nodes.
(b) A closed circuit made by several branches of the network.
(c) A closed circuit in a network which does not contain any other branch inside it.
(d) A closed circuit in a network with at least one nested closed circuit inside.

ANS: (c).

MC1.20 The impedance of a series combination of a 10 Ω ideal resistance, an ideal capacitance and an ideal inductance each of reactance 10 Ω is

(a) 10 Ω
(b) (10 + j 20) Ω
(c) (+ j 10) Ω
(d) (-j10) Ω

ANS: (a).

MC1.21 A nodal and a mesh network are equivalent when

(a) Power consumed by the nodal and mesh networks are equal.
(b) Currents flowing through the corresponding branches of the two networks are equal.

(c) The effective resistance between two corresponding branches of the network are equal.
(d) The external configuration of current flow through the corresponding nodes of two networks is the same.

ANS: (d).

MC1.22 A current source of strength 10 mA and parallel admittance of 100×10^{-6} S is equivalent to a voltage source of

(a) 100 Volt; series impedance 10 kΩ
(b) 100 Volt; series impedance 1 kΩ
(c) 10 Volt; series impedance 10 kΩ
(d) 10 Volt; series impedance 1 kΩ

ANS: (a).

Short Answer Questions (SAQ)

Note: As the name suggests, short answer questions require brief and to the point replies. Often, short answer questions are asked as notes or parts of long answer questions. Though they may be framed in different ways, however, some short answer questions are given here as notes, and an answer to a randomly selected question is also provided as an example.

SA1.1   Write a brief note on

(a) Impedance and its components
(b) Rectangular and polar representation of impedance
(c) Super mesh and its application
(d) Dependent sources
(e) Conversion of a voltage source into a current source
(f) Time relationship between current and voltage across inductance subjected to sinusoidal voltage
(g) Time relationship between current and voltage across a capacitor subjected to sinusoidal voltage
(h) Statement and limitations of the superposition theorem
(i) Compensation theorem and its application
(j) Transformation of Thevenin's equivalent into Norton's equivalent
(k) Need for impedance matching for power transfer
(l) Conversion from nodal to mesh network and Rosen's theorem
(m) Tellegen's theorem
(n) Millman's theorem

Sample Answer to SA1.1 (c)

Brief Note on Super Mesh and its Application

The mesh analysis using the cyclic current method is a powerful tool to analyse planner networks. However, one faces some difficulty in using this method if a current source is imbedded in the network, for example, the network shown in Fig. SA1.1C has a current source connected in arm CF of the network.

As shown in the figure, the network can be broken into three distinct meshes, 1, 2 and 3 to which cyclic currents a, b, and c may be assigned. It is simple to write the KVL equation for mesh-1 as

$$Z_1 a + Z_2(a - b) - V_S = 0 \text{ or } (Z_1 + Z_2)a - Z_2 b = V_S \qquad \text{(SA1.1(c).1)}$$

However, it is not possible to write KVL equations for mesh-2 and mesh-3 since the magnitude of the voltage difference across the current source is not known. In such situations one uses the concept of super mesh; for a moment, it is assumed as if the current source does not exist in the circuit and, therefore, the KVL equation for the super mesh BCDEFGB may be written as

$$Z_2(b - a) + Z_3 b + Z_4 c = 0 \qquad \text{(SA1.1(c).2)}$$

Further, in view of the fact that current through arm CF must be equal to the current I provided by the current source, an additional equation may be written as

$$(c - b) = I \qquad \text{(SA1.1(c).3)}$$

The above three simultaneous equations have three unknown variables a, b, and c, and their values may be obtained by solving these equations.

Thus, it is possible to use the concept of a super mesh in the mesh analysis where there are current sources.

Fig. SA1.1C Mesh and super mesh

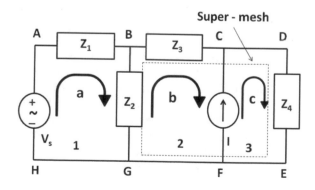

Sample Answer to SA1.1(i)

Brief note on the Compensation theorem and its application

The compensation theorem states "In any linear bilateral active network, if any branch (say AB) carrying current 'I' has its impedance changed from Z to (Z +ΔZ), the corresponding changes that occur in all branches of the network are the same as those that would have occurred by the addition of a voltage source (- IΔZ) in the modified branch AB".

As is evident from the statement of the theorem, the theorem may be used to calculate changes in the currents/voltages in different branches of the network, without a detailed analysis and loss of accuracy, when the magnitude of some element in a given branch of the network is changed by a small amount. The utility of the theorem lies in the considerable saving of effort and time that would have otherwise gone in the re-analysis of the whole network if it is required to change the magnitude, by a small amount, of some element in one of the several branches of the network. Further, the theorem may be used to estimate the change required to be made in the magnitude of the impedance of a particular branch so as to change the current in that branch by a certain factor. To illustrate this point, let us take a very simple example of the circuit shown in Fig. SA1.1(i).1, where it is simple to calculate that currents of 1A and 0.5 A flow, respectively, through branches AB, BE and CD of the circuit.

Suppose it is now required that current in branch CD should be 0.45A instead of the present value 0.5 A. The easiest way to reduce the current in arm CD (from 0.5 A to 0.45 A) is to increase the resistance in this branch from 5Ω to, say, (5 + ΔR) Ω. The magnitude of ΔR may be easily determined using the compensation theorem, without further analysis of the circuit.

The compensation theorem says that all changes in the currents/voltages in the network may be assumed to originate from a voltage source of magnitude −(IΔR) present in branch CD. This is shown in Fig. SA1.1(i).2. Now if it is required to reduce the current by 0.05A (from 0.5 to 0.45 A) in branch CD, i.e. ΔI = -0.05 A, then according to the compensation theorem, this change in current may be assumed to have originated due to a voltage source of magnitude (−IΔR) in branch CD, i.e.:

Fig. SA1.1(i).1 Circuit for
example SAQ.A1

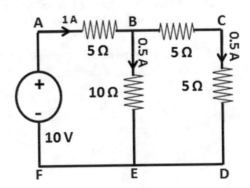

Fig. SA1.1(i).2 Modified
circuit

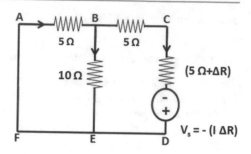

$$\Delta I = \frac{voltage\ source}{total\ resistance\ in\ the\ circuit} = \frac{-I\Delta R}{(5||10)+5+(5+\Delta R)} = \frac{-(0.5)(\Delta R)}{3.33+10+\Delta R}$$

$$or \quad \Delta I = -0.05 = \frac{-0.5\Delta R}{13.33+\Delta R}$$

$$or \quad 0.45\Delta R = 0.667\ or\ \Delta R = 1.48\ \Omega$$

*Thus, using the compensation theorem it is found that if the resistance in arm
CD is increased by 1.48Ω, which means if resistance of 6.48Ω instead of 5Ω is put
in branch CD the current in this branch will reduce by 0.05 A.*

Long Answer Questions (LAQ)

Long answer questions on topics of network analysis mostly contain two parts, a
part based on a short answer question (given earlier) and another part containing
some numerical problem, a large number of which are provided in the text. Some
long answer questions may also be framed like the following:

LA1.1   State the theorem of maximum power transfer and prove it taking a
suitable example.
LA1.2   Discuss the superposition theorem and conditions of its applicability.
Prove the theorem using an example.
LA1.3   State and prove Millman's theorem.
LA1.4   Distinguish between a nodal and a mesh network. State Rosen's theorem
and show how it can be used to transform a nodal network into a mesh
network.
LA1.5   State the equivalent generator theorem and show that it may be derived
from Millman's theorem.

Circuit Analyses Using the Laplace Transform

2

Abstract

The Laplace transformation is essentially a method of solving differential equations; however, it may also be used for analysing electrical networks. Applications of the Laplace transform method for solving electrical networks and its advantage over other conventional methods of circuit analysis will be discussed in this chapter. A number of solved examples given in the chapter will help in a complete understanding of the method.

2.1 Introduction

In principle, the Laplace transform method is a very general method for solving differential equations. It is known that a large number of processes, happenings, systems, etc. in different branches of science may be described by appropriate differential equations and, therefore, the Laplace transform method may be used for such cases. The Laplace transform method, therefore, finds applications in almost all branches of science, be it electrical circuits, dynamical problems, mechanical systems, nuclear physics, chemical reactions and so on. One may ask the question: why is this method of the Laplace transforms superior or better than the other known methods of solving differential equations? The answer is simple; while most of the other conventional methods of solving differential equations are applicable, each to a specific class of differential equations, one has to know beforehand which of the conventional methods should be used to solve a given differential equation. On the other hand, the Laplace transform method may be applied to solve any differential equation of any class, in general. As a result, the Laplace transform method has wide applicability. Another practical advantage of using the Laplace transform method is that it does not require the knowledge of complicated and

© The Author(s), under exclusive license to Springer Nature Switzerland AG 2021   147
R. Prasad, *Analog and Digital Electronic Circuits*, Undergraduate Lecture
Notes in Physics, https://doi.org/10.1007/978-3-030-65129-9_2

advanced techniques of mathematical manipulations; basic knowledge of calculus and algebra is enough to workout the Laplace transform procedure.

The method of solving differential equations using the Laplace transform technique may be divided into three steps: (i) Write the Laplace transform for the given differential equation. (ii) Carry out algebraic simplifications in the Laplace transform. (iii) Write the inverse Laplace transform of the expression obtained by simplifying the Laplace transform in step-II, and interpreting the results.

In order to fully appreciate the Laplace transform technique, in the first step, the Laplace transform will be defined, and the expressions for the Laplace transforms for a few core or essential functions will be derived. These functions frequently pop up in problems associated with different branches of science. In the second step, the Inverse Laplace transform will be defined and some properties of both the Laplace transform and the Inverse Laplace transform will be discussed. The method of writing the Laplace transform for differential equations will be covered in the next step. The application of the Laplace transform technique to an electrical circuit will be discussed in the final step.

Self-assessment question: Why do conventional methods of solving differential equations pose problems?

2.2   Laplace Transform

In many different branches of science, we often come across functions that depend on time. For example, in case of electrical/electronic circuits, we deal with time-dependent voltages, currents, electronic signals, etc. A time-dependent function is often represented as $f(t)$, where 't' in bracket tells that the function f may have different magnitudes at different times, i.e. the 'domain' over which the function may have different values is time, which is denoted by 't'.

The Laplace transform of a function $f(t)$ that operates in domain 't' is another function $F(s)$ that operates in another domain 's'. The process of obtaining the Laplace transform of a function is often denoted by the curly letter $\mathcal{L}$, which is also called the Laplace operator. Thus,

$$\mathcal{L}[f(t)] \rightarrow F(s) \equiv \int_{t=0}^{t=\infty} e^{-st} f(t) dt \qquad (2.1)$$

Equation (2.1) tells that the Laplace operator $\mathcal{L}$ operating on function $f(t)$ produces the Laplace transform which is function $F(s)$ defined by the integral shown at the extreme right of the equation. It may be noted the final expression for $F(s)$ that will be obtained by solving the integral and putting the upper and lower limits of integration will be a function of variable 's' and will not contain any 't' dependence, i.e. $F(s)$ will operate exclusively in domain 's'.

In case of electrical circuits, domain 's' is called the frequency domain and may be complex; it has a real component and an imaginary component. We shall be more specific regarding s when we discuss electrical circuits. With the definition of the Laplace transform as given by (2.1), we now proceed to calculate expressions for the Laplace transforms for some core or essential functions.

2.2.1 Laplace Transform for an Exponential Function

As an example, let $f(t) = e^{\alpha t}$, where α is a constant. By definition, the Laplace transform for this function may be written as

$$\mathcal{L}[e^{\alpha t}] = F(s) = \int_{t=0}^{t=\infty} e^{-st} f(t) dt = \int_{t=0}^{t=\infty} e^{-st} e^{\alpha t} dt$$

$$= \int_{t=0}^{t=\infty} e^{-(s-\alpha)t} dt \tag{2.2}$$

Equation (2.2) may be easily integrated to get

$$\mathcal{L}[e^{\alpha t}] = F(s) = \left[-\frac{1}{s-\alpha} e^{-(s-\alpha)t} \right]_{t=0}^{t=\infty} = -\frac{1}{(s-\alpha)} \left[e^{-\infty} - e^0 \right] = \frac{1}{(s-\alpha)} \tag{2.3}$$

Thus, $\mathcal{L}[e^{\alpha t}] = \dfrac{1}{(s-\alpha)}$ **for** $s > \alpha$

The condition $s > \alpha$ ensures that the denominator of expression (2.3) does not become zero.

Further, in the special case when constant $\alpha = 0$, (2.3) reduces to

$$\mathcal{L}[e^{0t}] = \mathcal{L}[1] = \frac{1}{s} \textbf{ for } s > 0 \tag{2.4}$$

2.2.2 Laplace Transform for Function f(t) = t^n

The Laplace transform for function $f(t) = t^n$, where n is a number, may be written as

$$\mathcal{L}[t^n] = \int_{t=0}^{t=\infty} e^{-st} f(t) dt = \int_{t=0}^{t=\infty} e^{-st} t^n dt. \tag{2.5}$$

Equation (2.5) may be evaluated using the method of integration by parts, with the substitutions:

$$u = t^n; du = nt^{(n-1)}dt; dv = e^{-st}dt \text{ that gives } v = -\frac{1}{s}e^{-st}$$

$$\text{so that } L[t^n] = \int_{t=0}^{t=\infty} e^{-st}t^n dt = \int_{t=0}^{t=\infty} udv = [uv]_{t=0}^{t=\infty} - \int_{t=0}^{t=\infty} vdu$$

$$= \left[-\frac{t^n}{s}e^{-st}\right]_{t=0}^{t=\infty} - \int_{t=0}^{t=\infty} \left\{-\frac{1}{s}e^{-st} \cdot nt^{(n-1)}dt\right\} \tag{2.6}$$

$$= \left[-\left\{\frac{\infty^n e^{-\infty}}{s} - \frac{0^n e^{-0}}{s}\right\}\right] + \frac{n}{s}\int_{t=0}^{t=\infty} e^{-st}t^{(n-1)}dt$$

The first term on the right in (2.6) goes to zero under the given limits and the term under the integration sign is nothing but the Laplace transform of function $t^{(n-1)}$. Therefore, (2.6) represents a recursive relation between the Laplace transforms of functions t^n and $t^{(n-1)}$. Hence,

$$\mathcal{L}[t^n] = \frac{n}{s}\int_{t=0}^{t=\infty} e^{-st}t^{(n-1)}dt = \frac{n}{s}\mathcal{L}\left[t^{(n-1)}\right]$$

$$= \frac{n}{s}\cdot\frac{(n-1)}{s}\mathcal{L}\left[t^{(n-2)}\right] = \frac{n}{s}\cdot\frac{(n-1)}{s}\cdot\frac{(n-2)}{s}\mathcal{L}\left[t^{(n-3)}\right]$$

$$\cdots\cdots\cdots\cdots\cdots\cdots\cdots \tag{2.7}$$

$$= \frac{n}{s}\cdot\frac{(n-1)}{s}\cdot\frac{(n-2)}{s}\cdots\cdots\frac{1}{s}\mathcal{L}[t^0] = \frac{n!}{s^n}\mathcal{L}(1) = \frac{n!}{s^n}\left(\frac{1}{s}\right)$$

$$\text{or } \mathcal{L}[t^n] = \frac{n!}{s^{(n+1)}}$$

In obtaining expression (2.7), we have made use of the relation $\mathcal{L}(1) = \frac{1}{s}$ as given by (2.4).

2.2.3 Laplace Transforms for Cosine and Sine Functions

Cosine and sine functions pop up frequently while dealing with sinusoidal variables. As will be seen, the Laplace transform for the sine function will automatically appear while deriving expressions for the Laplace transform for the cosine function.

Let $f(t) = \cos\beta t$, where β is a constant.

$$\mathcal{L}[\cos\beta t] = \int_{t=0}^{t=\infty} e^{-st}f(t)dt = \int_{t=0}^{t=\infty} e^{-st}\cos\beta t dt \tag{2.8}$$

Integral at the extreme right in (2.8) may be evaluated by using the method of integration by parts. Let us make the following substitutions for that

$u = e^{-st}$ and $dv = \cos\beta t$; so that $du = -se^{-st}dt$ and $v = \int \cos\beta t = \frac{1}{\beta}\sin\beta t$

With the above substitutions, (2.8) reduces to

$$\mathcal{L}[\cos\beta t] = \int_{t=0}^{t=\infty} udv = [uv]_{t=0}^{t=\infty} - \int_{t=0}^{t=\infty} vdu \tag{2.9}$$

It is easy to show that $[uv]_{t=0}^{t=\infty} = 0$, when limits are put on the values of u and v. Equation (2.9) reduces to

$$\mathcal{L}[\cos\beta t] = -\int_{t=0}^{t=\infty} vdu = -\int_{t=0}^{t=\infty} \left(\frac{1}{\beta}\sin\beta t\right)(-se^{-st}dt)$$

$$\text{or } \mathcal{L}[\cos\beta t] = \frac{s}{\beta}\int_{t=0}^{t=\infty} e^{-st}\sin\beta dt = \frac{s}{\beta}\mathcal{L}[\sin\beta t] \tag{2.10}$$

Equation (2.10) gives a relation between the Laplace transforms of $\cos\beta t$ and $\sin\beta t$.

Let us proceed further to derive the expression for the Laplace transform of $\cos\beta t$ using the expression

$$\mathcal{L}[\cos\beta t] = \frac{s}{\beta}\int_{t=0}^{t=\infty} e^{-st}\sin\beta dt \tag{2.11}$$

The integral in the above expression may be evaluated using the method of integration by parts. Let us make the following substitutions in the above expression.

$u = e^{-st}$ so that $du = -se^{-st}dt$ and $dv = \sin\beta t$ so that $v = -\frac{1}{\beta}\cos\beta t$

With these substitutions, (2.11) becomes

$$\mathcal{L}[\cos\beta t] = \frac{s}{\beta}\int_{t=0}^{t=\infty} udv = \frac{s}{\beta}\left\{[uv]_{t=0}^{t=\infty} - \int_{t=0}^{t=\infty}\left(-\frac{1}{\beta}\cos\beta t\right)(-se^{-st}dt)\right\} \tag{2.12}$$

or

$$\mathcal{L}[\cos \beta t] = \frac{s}{\beta}\left[(e^{-st})\left(-\frac{1}{\beta}\cos \beta t\right)\right]_{t=0}^{t=\infty} - \frac{s^2}{\beta^2}\int_{t=0}^{t=\infty} e^{-st}\cos \beta t dt$$

$$\text{or} = \frac{s}{\beta}\left[\left\{e^{-\infty}\left(-\frac{1}{\beta}\cos \infty\right)\right\} - \left\{e^0\left(-\frac{1}{\beta}\cos 0\right)\right\}\right] - \frac{s^2}{\beta^2}\mathcal{L}[\cos \beta t]$$

$$= \frac{s}{\beta}\left[\{0\} - \left\{-\frac{1}{\beta}\right\}\right] - \frac{s^2}{\beta^2}\mathcal{L}[\cos \beta t] \tag{2.13}$$

$$\text{or}\left(1 + \frac{s^2}{\beta^2}\right)\mathcal{L}[\cos \beta t] = \frac{s}{\beta^2}$$

$$\text{or } \mathcal{L}[\cos \beta t] = \frac{s}{(s^2 + \beta^2)}$$

Equation (2.13) gives the expression for the Laplace transform of function $\cos\beta t$. For the Laplace transform of function $\sin\beta t$, we revert to (2.10), according to which

$$\mathcal{L}[\cos \beta t] = \frac{s}{\beta}\mathcal{L}[\sin \beta t]$$

$$\text{And } \mathcal{L}[\sin \beta t] = \frac{\beta}{s}\mathcal{L}[\cos \beta t]$$

On substituting the value of $\mathcal{L}[\cos\beta t]$ from (2.13) in the above expression, one gets

$$\mathcal{L}[\sin \beta t] = \frac{\beta}{(s^2 + \beta^2)} \tag{2.14}$$

With these examples, it is obvious that it is possible to calculate the Laplace transform of any well-behaved function using the laws of calculus.

2.2.4   Inverse Laplace Transform and Properties of Transform and Inverse Transforms

The Laplace transform operator (denoted by $\mathcal{L}$) converts a function $f(t)$ of domain 't' into a corresponding function $F(s)$ of domain 's'. The inverse Laplace transform operator, denoted by $\mathcal{L}^{-1}$, turns back the function $F(s)$ of domain 's' into the original function $f(t)$ of domain 't'. Thus, the two operators are reciprocal or inverse of each other. Thus,

$$\mathcal{L}[f(t)] \rightarrow F(s) \equiv \int_{t=0}^{t=\infty} e^{-st}f(t)dt \tag{2.15}$$

And

$$\mathcal{L}^{-1}[F(s)] \to f(t) \equiv \int_{s=0}^{s=\infty} e^{st}F(s)ds \qquad (2.16)$$

The method of solving problems represented by differential equations in t-domain is to transfer them in corresponding equations in s-domain using the Laplace transform. Because of the special way of transformation from t to s-domain, it has been found that the simplification of expressions of s-domain is much easier; one has just to carry out some simple algebraic manipulations and the expressions of s-domain, in most cases, reduce to standard forms. The s-domain expressions in standard form are then converted back to corresponding expressions in t-domain using the inverse Laplace operation. The main advantage of the Laplace technique lies in the ease of simplifying s-domain functions.

Self-assessment question: What is the main advantage of working in s domain?

(a) **Properties of Laplace and Inverse Laplace transforms**
  (i) **Linearity**
      Operators $\mathcal{L}$ and $\mathcal{L}^{-1}$ are both linear. Mathematically, the property of linearity of these operators may be expressed by the following equations.

$$\mathcal{L}[K_1f_1(t) + K_2f_2(t) + K_3f_3(t) + \cdots \ldots] = K_1\mathcal{L}[f_1(t)] + K_2\mathcal{L}[f_2(t)] + K_3\mathcal{L}[f_3(t)]\ldots$$
$$(2.17)$$

And

$$\mathcal{L}^{-1}[K_1f_1(t) + K_2f_2(t) + K_3f_3(t) + \cdots \ldots]$$
$$= K_1\mathcal{L}^{-1}[f_1(t)] + K_2\mathcal{L}^{-1}[f_2(t)] + K_3\mathcal{L}^{-1}[f_3(t)]\ldots \qquad (2.18)$$

K_1, K_2, K_3, etc. appearing in these equations are simply constants and $f_1(t)$, $f_2(t)$, $f_3(t)$, etc. are functions defined in domain 't'.

The linearity of these operators makes it easy to break their operation over an expression containing the sum of several terms into operations on individual terms and then taking the sum of these individual operations. This linearity property goes a long way in writing down the Laplace transforms and the Inverse Laplace transforms for complicated differential equations.

  (ii) **First shift theorem or frequency shift**
      $$\mathcal{L}[e^{at}f(t)] = F(s-a) \text{ and } \mathcal{L}^{-1}[F(s-a)] = e^{at}f(t)$$

(iii) **Time shift property**

$$\mathcal{L}[f(t-a)] = F(s)e^{-as} \text{ and } \mathcal{L}^{-1}[F(s)e^{-sa}] = f(t-a)$$

It is important to observe the symmetry or duality between time shift and frequency shift.

(iv) **Transform of integral**

$$\mathcal{L}\left[\int_0^t f(t)dt\right] = \frac{1}{s}F(s)$$

(v) **Initial and final value theorems**

$$f(0) = \lim_{t \to 0} f(t) = \lim_{s \to \infty} \frac{F(s)}{1} \text{ and } f(\infty) = \lim_{t \to \infty} f(t) = \lim_{s \to 0} F(s)$$

Initial and final value theorems are of great significance; they are very often used to check the results obtained by the Laplace technique, both in t and in s domains.

(vi) **Scaling property**

$$\mathcal{L}[f(at)] = \frac{1}{a}F\left(\frac{s}{a}\right) a > 0$$

The above equation tells that a time expansion of the signal by a factor a in t domain causes compression of the signal by the same factor in s domain.

Some important properties of the Laplace operator are listed in Table (2.1).

Table 2.1 Some properties of Laplace transform

Property	Function $f(t)$	Laplace transform $F(s)$
Linearity	$K_1 f_1(t) + K_2 f_2(t) + \dots$	$K_1 F_1(s) + K_2 F_2(s) + \dots$
Scaling	$f(at)$	$\frac{1}{a}F\left(\frac{s}{a}\right)$
Frequency shift	$e^{-at}f(t)$	$F(s+a)$
Time shift	$f(t-a)u(t-a)$	$e^{-as}F(s)$

2.2.5 Tables of Laplace and Inverse Laplace Transforms

In the last few sections, expressions for the Laplace transforms for some frequently used functions have been derived. This was done to develop the skill of obtaining the Laplace transforms. However, there are many more functions, different from the standard ones, which pop up, though less frequently, in some problems. Therefore, tables of the Laplace transforms and the inverse Laplace transforms have been developed for ready reference and use. A table of some useful Laplace transform pairs is given below. The double arrow ($\leftrightarrow$) between terms of the table indicates their equivalence and relation with the inverse operator (Table 2.2).

2.2.6 Solution of Ordinary Differential Equations Using Laplace Transforms

A general differential equation may be defined as a relationship between a finite set of functions and its derivatives. Depending on the domain of the functions involved, we have ordinary differential equations or ODE, when only one variable appears in the equation (or there are full derivatives) and the partial differential equations, in

Table 2.2 Useful Laplace transform pairs

$$\mathcal{L}[\delta(t)] = 1 \leftrightarrow \mathcal{L}^{-1}[1] = \delta(t)$$

$$\mathcal{L}[e^{\lambda t}] = \frac{1}{(s-\lambda)} \leftrightarrow \mathcal{L}^{-1}\left[\frac{1}{(s-\lambda)}\right] = e^{\lambda t}$$

$$\mathcal{L}[1] = \mathcal{L}[u(t)] = \frac{1}{s} \leftrightarrow \mathcal{L}^{-1}\left[\frac{1}{s}\right] = 1$$

$$\mathcal{L}[K.1] = K\mathcal{L}[u(t)] = \frac{K}{s} \leftrightarrow \mathcal{L}^{-1}\left[\frac{K}{s}\right] = K \quad \text{Here K is any constant.}$$

$$\mathcal{L}[\sqrt{t}] = \frac{\sqrt{\pi}}{2s^{3/2}} \leftrightarrow \mathcal{L}^{-1}\left[\frac{\sqrt{\pi}}{2s^{3/2}}\right] = \sqrt{t}$$

$$\mathcal{L}[t^n] = \frac{n!}{s^{n+1}} \leftrightarrow \mathcal{L}^{-1}\left[\frac{n!}{s^{n+1}}\right] = t^n \text{ or } \mathcal{L}^{-1}\left[\frac{1}{s^n}\right] = \frac{t^{(n-1)}}{(n-1)!}$$

$$\mathcal{L}[t^n e^{-at}] = \frac{n!}{(s+a)^{n+1}} \leftrightarrow \mathcal{L}^{-1}\left[\frac{n!}{(s+a)^{n+1}}\right] = t^n e^{-at}$$

$$\mathcal{L}[\cos\beta t] = \frac{s}{s^2+\beta^2} \leftrightarrow \mathcal{L}^{-1}\left[\frac{s}{s^2+\beta^2}\right] = \cos\beta t \text{ or } \mathcal{L}^{-1}\left[\frac{1}{s^2+\beta^2}\right] = \frac{1}{s}\cos\beta t$$

$$\mathcal{L}[\sin\beta t] = \frac{\beta}{s^2+\beta^2} \mathcal{L}^{-1}\left[\frac{\beta}{s^2+\beta^2}\right] = \sin\beta t \text{ or } \mathcal{L}^{-1}\left[\frac{1}{s^2+\beta^2}\right] = \frac{1}{\beta}\sin\beta t$$

$$\mathcal{L}[\sin(\beta t+\theta)] = \frac{s\sin\theta + \beta\cos\theta}{(s^2+\beta^2)} \leftrightarrow \mathcal{L}^{-1}\left[\frac{s\sin\theta + \beta\cos\theta}{s^2+\beta^2}\right] = \sin(\beta t+\theta)$$

$$\mathcal{L}[\cos(\beta t+\theta)] = \frac{s\cos\theta - \beta\sin\theta}{s^2+\beta^2} \leftrightarrow \mathcal{L}^{-1}\left[\frac{s\cos\theta - \beta\sin\theta}{s^2+\beta^2}\right] = \cos(\beta t+\theta)$$

$$\mathcal{L}[e^{-at}\sin\beta t] = \frac{\beta}{(s+a)^2+\beta^2} \leftrightarrow \mathcal{L}^{-1}\left[\frac{\beta}{(s+a)^2+\beta^2}\right] = e^{-at}\sin\beta t$$

$$\mathcal{L}[e^{-at}\cos\beta t] = \frac{s+a}{(s+a)^2+\beta^2} \leftrightarrow \mathcal{L}^{-1}\left[\frac{s+a}{(s+a)^2+\beta^2}\right] = e^{-at}\cos\beta t$$

$$\mathcal{L}[t\sin(\alpha t)] = \frac{2as}{(s^2+\alpha^2)^2} \mathcal{L}^{-1}\left[\frac{2as}{(s^2+\alpha^2)^2}\right] = t\sin(\alpha t)$$

$$\mathcal{L}[t\cos(\alpha t)] = \frac{s^2-a^2}{(s^2+\alpha^2)^2} \leftrightarrow \mathcal{L}^{-1}\left[\frac{s^2-a^2}{(s^2+\alpha^2)^2}\right] = t\cos(\alpha t)$$

short PDE, where more than one variable is involved. The method of writing the Laplace transform for ODE will be described in the following.

There are different ways of denoting the derivative of a function, for example, the first derivative of a function $x(t)$ of time may be denoted as $\frac{dx(t)}{dt}$ or as $x'(t)$ or as D_x. Similarly, higher derivatives may be denoted as $\frac{d^2x(t)}{d^2}, \frac{d^3x(t)}{d^3}, \frac{d^4x(t)}{d^4} \ldots$ or $x''(t), x'''(t), \ldots$ and $D_x^2, D_x^3, \ldots$ For the sake of uniformity, we shall be using the last symbol D_x^n to denote the nth derivative with time 't' of function $x(t)$.

The Laplace transform for the derivative of a function is defined as

$$\mathcal{L}\left[x'(t)\right] = \mathcal{L}[D_x^1] \equiv \int\limits_{t=0}^{t=\infty} e^{-st} D_x^1 \text{ and } \mathcal{L}[D_x^n] \equiv \int\limits_{t=0}^{t=\infty} e^{-st} D_x^n \qquad (2.19)$$

In most cases, the integral on the right, defining the Laplace transform, may be solved using the method of integration by parts, however, tables of the Laplace transforms for the derivatives of different order are available for ready use. Expressions for the Laplace transforms of the first few derivatives of function $x(t)$ as well as the general expression for its nth derivative are given below:

$$\mathcal{L}\left[x'(t)\right] = \mathcal{L}[D_x^1] = s\mathcal{L}[x(t)] - x(0) \qquad (2.20)$$

$$\mathcal{L}[x''(t)] = \mathcal{L}[D_x^2] = s^2\mathcal{L}[x(t)] - sx(0) - x'(0) \qquad (2.21)$$

$$\mathcal{L}[x'''(t)] = \mathcal{L}[D_x^3] = s^3\mathcal{L}[x(t)] - s^2x(0) - sx'(0) - x''(0) \qquad (2.22)$$

$$\cdots\cdots\cdots\cdots\cdots\cdots\cdots\cdots$$

$$\mathcal{L}[D_x^n] = s^n\mathcal{L}[x(t)] - s^{n-1}x(0) - s^{n-2}D_x^1(0) - s^{n-3}D_x^2(0) + \cdots \qquad (2.22a)$$

A definite pattern in the expressions for the Laplace transforms of successively higher orders of derivatives is visible in the above expressions. Terms like $x(0)$, x'(0) and $x''(0)$ represent, respectively, the value of function $x(t)$ at time $t = 0$, value of the first derivative of $x(t)$ at time 0 (initial time) and value of the second derivative of the function $x(t)$ at initial time $t = 0$. It may be recalled that the above-mentioned terms define the boundary conditions of the problem. A first-order ODE needs only one boundary condition to uniquely define the function, second-order ODE two boundary conditions, third-order ODE three boundary conditions and so on. It may be observed that the Laplace transform of the first derivative given by (2.20) contains only one boundary condition: the value of the function at $t = 0$. Similarly, the Laplace transform of the third derivative, (2.22), contains three boundary conditions: the initial value of the function at $t = 0$, value of the first derivative at time $t = 0$ and the value of the second derivative at $t = 0$.

Let us consider a second-order differential equation with constant coefficients a and b that represents a certain process:

$$x''(t) + ax'(t) + bx = u(t) \tag{2.23}$$

Here, u(t) is the input deriving force, which may be electric current or voltage in the case of electrical circuits, and $x(t)$ is the output or response of the circuit. Let $x(0) = K_1$ and $x'(0) = K_2$ be the two boundary conditions. We now take the Laplace transform of the two sides of the equation to get

$$\left[s^2 X(s) - sx(0) - x'(0) \right] + a[sX(s) - x(0)] + bX(s) = U(s) \tag{2.23a}$$

Equation (2.23a) is called the subsidiary equation. Collecting $X(s)$ terms, one gets

$$(s^2 + as + b)X(s) = (s + a)x(0) + x'(0) + U(s)$$

or $\tag{2.23b}$

$$X(s) = [(s + a)x(0) + x'(0)]Q(s) + U(s)Q(s)$$

where $Q(s) = \frac{1}{(s^2 + as + b)}$ is called the transfer function.

In the special case when $x(0) = x'(0) = 0$, (2.23b) becomes

$$X(s) = U(s)Q(s) \tag{2.23c}$$

And the transfer function $Q(s)$ may be written as the quotient,

$$Q(s) = \frac{X(s)}{U(s)} = \frac{Laplace\ transform\ for\ input\ x(t)}{Laplace\ transform\ of\ output\ u(t)} \tag{2.23d}$$

Self-assessment question: Is it possible to encounter derivatives like $\frac{dF(s)}{ds}$ or $\frac{d^2 F(s)}{ds^2}$, etc. in expressions of s domain obtained by the Laplace transformation?

2.2.7 Partial Fractions

The solution $X(s)$ of the subsidiary equation of a differential equation usually comes as a quotient of two polynomials,

$$X(s) = \frac{P(s)}{Q(s)}$$

In order to determine x(t), one takes the inverse Laplace transform of $X(s)$. It is easy to get the inverse Laplace transform if $X(s)$ is in the form of the sum of partial fractions. The form of the partial fractions depends on the types of factors in the product form of $Q(s)$. Often, the following types of factors are found:

(i) Unrepeated factors like $(s-a)(s-b)$; in that case, $X(s)$ may be expressed as

$$X(s) = \frac{A}{(s-a)} + \frac{B}{(s-b)}$$

In such cases, it is easy to take the value of the inverse Laplace transform for each partial fraction from the table and add them to get $x(t)$,

$$x(t) = \mathcal{L}^{-1}X(s) = \mathcal{L}^{-1}\left[\frac{A}{(s-a)}\right] + \mathcal{L}^{-1}\left[\frac{B}{(s-b)}\right]$$

(ii) Repeated factors like $(s-a)^2$

$$X(s) = \frac{A}{s} + \frac{Bs+C}{(s+a)^2}$$

And $x(t) = \mathcal{L}^{-1}X(s) = \mathcal{L}^{-1}\left[\frac{A}{(s)}\right] + \mathcal{L}^{-1}\left[\frac{Bs+C}{(s-b)^2}\right]$

(iiii) Complex factors like $[s-(\alpha+j\beta)][s-(\alpha-j\beta)]$

$$X(s) = \frac{As+B}{[s-(\alpha+j\beta)][s-(\alpha-j\beta)]} = \frac{As+B}{(s-\alpha)^2+\beta^2}$$

And $x(t) = \mathcal{L}^{-1}X(s) = \mathcal{L}^{-1}\left[\frac{As+B}{(s-\alpha)^2+\beta^2}\right]$

The method of finding the values of constants A, B, C, etc. will be discussed while solving such cases.

Having developed the main body of Laplace's methodology, we now proceed to solve some problems using the Laplace transforms.

Solved example (SE2.1) A function $X(t)$ is given as $X(t) = cos2t + e^{-5t}$ *for* $t \geq 0$.

Find the Laplace transform of the function.

Solution: *Function $X(t)$ has two terms; we use the property of linearity and write the Laplace transform of the function as the sum of the Laplace transforms of individual terms. Next from the table of Laplace transforms, one may find the desired values and sum them as shown here,*

$$\mathcal{L}[X(t)] = \mathcal{L}[\cos 2t] + \mathcal{L}[e^{-5t}] = \frac{s}{s^2+4} + \frac{1}{s+5} = \frac{2s^2+5s+4}{(s^2+4)(s+5)}$$

The desired Laplace transform is $\mathcal{L}[X(t)] = \frac{2s^2+5s+4}{(s^2+4)(s+5)}$.

Solved example (SE2.2) Find the Laplace transform for the function $Y(t)$ given below

$$Y(t) = t^3 e^{-5t} + \cos(\omega t + \theta) + \sin(\omega t + \theta)\, t \geq 0$$

Solution: *The given function contains three terms, and invoking the linearity property, the Laplace transform of the function may be found by taking the sum of the Laplace transforms of each term separately and adding them up. It is easy to find the Laplace transform for each term from the table as*

$$\mathcal{L}[t^3 e^{-5t}] = \frac{3!}{(s+5)^4} = \frac{6}{(s+5)^4}; \mathcal{L}[\cos(\omega t + \theta)] = \frac{s\cos\theta - \omega\sin\theta}{(s^2+\omega^2)};$$

$$\textit{and} \quad \mathcal{L}[\sin(\omega t + \theta)] = \frac{s\sin\theta + \omega\cos\theta}{(s^2+\omega^2)}$$

Adding the three terms, one gets

$$\mathcal{L}[y(t)] = \frac{6}{(s+5)^4} + \frac{s\cos\theta - \omega\sin\theta}{(s^2+\omega^2)} + \frac{s\sin\theta + \omega\cos\theta}{(s^2+\omega^2)} = \frac{6}{(s+5)^4} + \frac{(s-\omega)\sin\theta + (s+\omega)\cos\theta}{(s^2+\omega^2)}$$

$$\textit{or } \mathcal{L}[y(t)] = \frac{6}{(s+5)^4} + \frac{(s-\omega)\sin\theta + (s+\omega)\cos\theta}{(s^2+\omega^2)}$$

Solved example (SE2.3) Obtain the inverse Laplace transform for

$$F(s) = 1 + \frac{7}{s+5} - \frac{10\,s}{s^2+25} + \frac{1}{s^5}$$

Solution: *Using the linearity property of the inverse Laplace transform, one may write*

$$\mathcal{L}^{-1}[F(s)] = f(t) = \mathcal{L}^{-1}[1] + 7\mathcal{L}^{-1}\left[\frac{1}{s+5}\right] - 10\mathcal{L}^{-1}\left[\frac{s}{s^2+5^2}\right] + \mathcal{L}^{-1}\left[\frac{1}{s^5}\right] t \geq 0$$

$$\text{(SE2.2)}$$

From the table of the inverse Laplace transforms, one gets

$$\mathcal{L}^{-1}[1] = \delta(t); 7\mathcal{L}^{-1}\left[\frac{1}{s+5}\right] = 7e^{-5t}; 10\mathcal{L}^{-1}\left[\frac{s}{s^2+5^2}\right] = 10\,cos5t \text{ and } \mathcal{L}^{-1}\left[\frac{1}{s^5}\right] = \frac{t^4}{4!}$$

Substituting these values in (SE2.2), one gets the desired inverse Laplace transform as

$$f(t) = \delta(t) + 7e^{-5t} - 10\,cos5t + \frac{t^4}{4!}\,for\,t \geq 0$$

Solved example (SE2.4) The Laplace transform $F(s)$ of a function $f(t)$ is given as

$$F(s) = \frac{s^2 + 14}{s(s+1)(s+2)}, \text{ find the function f(t)}$$

Solution: *The given F(s) has a denominator that contains the multiplication of three terms. If any one of these three terms becomes zero, i.e. if s=0, or (s+1) = 0, or (s+2) = 0, the function F(s) becomes infinite. Those values of s for which F(s) becomes infinite are called the poles of function F(s). The given function in this case has three poles at s = 0, s = −1 and s = −2.*

In order to find the inverse Laplace transform of this function, it is required that the given function must be converted into a combination of additive and/or sub-tractive terms so that the inverse Laplace transform of each term is obtained and added. It means that the function is converted in the following form:

$$F(s) = \frac{A}{s} + \frac{B}{(s+1)} + \frac{C}{(s+2)},$$

where A, B and C are the coefficients of different terms. In the present case, these coefficients may be determined in two different ways: (i) using the algebraic method and (ii) using the residue method as worked out here.

(i) *Algebraic method*:

We have

$$F(s) = \frac{s^2 + 14}{s(s+1)(s+2)} = \frac{A}{s} + \frac{B}{(s+1)} + \frac{C}{(s+2)}$$

$$Or\,s^2 + 14 = A(s+1)(s+2) + Bs(s+2) + Cs(s+1)$$

$$= A(s^2 + 3s + 2) + B(s^2 + 2s) + C(s^2 + s)$$

$$Or\,s^2 + 14 = (A + B + C)s^2 + (3A + 2B + C)s + (2A)$$

Comparing the coefficients of s^2, s and constant terms, one gets

$$(A + B + C) = 1; (3A + 2B + C) = 0 \text{ and } (2A) = 14$$

That gives $A = 7; B = -15; C = 9$.

With the substitution of these values of A, B and C, F(s) becomes

$$F(s) = \tfrac{7}{s} + \tfrac{(-15)}{(s+1)} + \tfrac{9}{(s+2)}$$
$$And f(t) = \mathcal{L}^{-1}[F(s)] = 7\mathcal{L}^{-1}\left[\tfrac{1}{s}\right] - 15\mathcal{L}^{-1}\left[\tfrac{1}{s+1}\right] + 9\mathcal{L}^{-1}\left[\tfrac{1}{s+2}\right] \qquad (SE2.4)$$
$$or\ f(t) = 7u(t) - 15e^{-t} + 9e^{-2t}\ t \geq 0$$

The desired function is given by equation (SE2.4).

(ii) **Residue method of finding coefficients**

Coefficients A, B and C may also be obtained by the method of residues. There are three terms, s, (s+1) and (s+2) in the denominator of function F(s). Each of these three terms has a corresponding residue. The residue corresponding to s is $[s\ F(s)] = \frac{s^2+14}{(s+1)(s+2)}$, similarly, the residue corresponding to term (s+1) is $[(s+1)F(s)] = \frac{s^2+14}{s(s+2)}$ and the residue corresponding to term (s+2) is $[(s+2)F(s)] = \frac{s^2+14}{s(s+1)}$.

Without going into details, it can be shown that the coefficients may be obtained by evaluating the residues of function F(s) at values of s corresponding to poles of the function.

$$A = [sF(s)]_{s=0} = \left[\frac{s^2+14}{(s+1)(s+2)}\right]_{s=0} = \left[\frac{0+14}{(0+1)(0+2)}\right] = \frac{14}{2} = 7$$

$$B = [(s+1)F(s)]_{s=-1} = \left[\frac{s^2+14}{s(s+2)}\right]_{s=-1} = \frac{(-1)^2+14}{(-1)(-1+2)} = \frac{15}{-1} = -1$$

$$C = [(s+2)F(s)]_{s=-2} = \left[\frac{s^2+14}{s(s+1)}\right]_{s=-2} = \frac{(-2)^2+14}{(-2)(-2+1)} = \frac{18}{2} = 9$$

Note: The above method can be applied only if there are no terms with higher powers in the denominator of F(s); it is not applicable if the denominator has terms like $(s + a)^2$ or $(s + b)^3$, etc. It is also not applicable if the domain s is complex. Methods used in such cases will be studied separately when we come across such problems.

Solved example (SE2.5) The Laplace transform of a function $f(t)$ is given as $F(s) = \frac{4-2s}{(s^2+1)(s-1)^2}$. Find f(t)

Solution: *Function F(s) has two terms in its denominator. One of the terms contains the s^2 term and the other term has the power of 1 of s but it is squared. Intuition tells that if it is converted into additive components, then the enumerator of this term should be of type (As + B), i.e. the corresponding term will be like $\frac{As+B}{(s^2+1)}$; here A and B are constants. The term corresponding to $(s-1)^2$ may be like $\frac{C}{(s-1)^2}$, but there may be an extra term like $\frac{D}{(s-1)}$ where D is constant. Adding all these terms, the total expression for F(s) becomes;*

$$F(s) = \frac{4-2s}{(s^2+1)(s-1)^2} = \frac{As+B}{(s^2+1)} + \frac{C}{(s-1)^2} + \frac{D}{(s-1)}$$

$$= \frac{(A+D)s^3 + (B-2A-D+C)s^2 + (A-2B+D)s + (B-D+C)}{(s^2+1)(s-1)^2}$$

$$or\ 4-2s = (A+D)s^3 + (B-2A-D+C)s^2 + (A-2B+D)s + (B-D+C)$$

$$(SE5.1)$$

On comparing coefficients of s^3, s^2, s terms and constants on the two sides of the equation, one gets
$A = 2, B = 1, C = 1$ *and* $D = -2$; *substituting these values in (SE5.1), one gets*

$$F(s) = \frac{2s+1}{(s^2+1)} + \frac{1}{(s-1)^2} - \frac{2}{(s-1)}$$

$$= 2\frac{s}{(s^2+1)} + \frac{1}{(s^2+1)} + \frac{1}{(s-1)^2} - \frac{2}{(s-1)}$$

Taking the inverse Laplace transform of each term on the right side, one gets

$$f(t) = 2\cos(t) + \sin(t) + e^{-t}t - 2e^{-t}\ for\ t \geq 0$$

Self-assessment question: What is the reason for including an extra term $\frac{D}{(s-1)}$?

2.2.8 Convolution Theorem

The fraction expansion method of finding the inverse Laplace transform at times could be quite complex. In such cases, one may use the convolution theorem to find the inverse Laplace transform.

The convolution theorem is based on the convolution of two functions f(t) and g (t); according to the definition, the convolution of f(t) and g(t) is written as (f(t)*g(t)) and is given by

$$f(t) * g(t) = \int_0^t f(\tau)g(t-\tau)d\tau \qquad (2.23e)$$

If one substitutes $\tau_1 = (t-\tau)$ in the above equation, it becomes

$$f(t) * g(t) = \int_0^t f(\tau)g(t-\tau)d\tau = -\int_0^t f(t-\tau_1)g(\tau_1)d\tau_1 = g(t) * f(t) \quad (2.23f)$$

The convolution theorem, therefore, states that

$$\mathcal{L}[f(t) * g(t)] = F(s)G(s) \quad\quad (2.23g)$$

And,

$$\mathcal{L}^{-1}[F(s)G(s)] = f(t) * g(t) = \int_0^t f(\tau)g(\tau)d\tau \quad\quad (2.23h)$$

The following example will further clarify the application of the convolution theorem for obtaining the inverse Laplace transform. Let us determine the inverse Laplace transform of the following s domain function $Y(s)$:

$$Y(s) = \frac{1}{(s^2 + a^2)(s^2 + b^2)}$$

One may write function $Y(s)$ as a product of two terms $F(s)$ and $G(s)$ where

$$F(s) = \frac{1}{(s^2 + a^2)} ; G(s) = \frac{1}{(s^2 + b^2)}$$

And,

$$\mathcal{L}^{-1}[Y(s)] = \mathcal{L}^{-1}[F(s)G(s)]$$
$$= \mathcal{L}^{-1}[F(s)]\mathcal{L}^{-1}[G(s)] = \mathcal{L}^{-1}\left[\frac{1}{(s^2 + a^2)}\right]\mathcal{L}^{-1}\left[\frac{1}{(s^2 + b^2)}\right] \quad (2.23i)$$
$$= \frac{1}{a}\sin(at)\frac{1}{b}\sin(bt)$$

But from (2.23h),

$$\mathcal{L}^{-1}[F(s)G(s)] = f(t) * g(t)$$

Comparing (2.23h) and (2.23i), one gets

$$f(t) = \frac{1}{a}\sin(at) \text{ and } g(t) = \frac{1}{b}\sin(bt)$$

And, convolution $f(t) * g(t) = \frac{1}{ab} \int_0^t \sin(a\tau)\sin(b\{t - \tau\})d\tau$

or convolution $f(t) * g(t) = \frac{1}{ab(a^2 + b^2)} [a\sin(bt) - b\sin(at)]$

And $y(t) = \text{convolution} f(t) * g(t) = \frac{1}{ab(a^2 + b^2)} [a\sin(bt) - b\sin(at)]$

2.3   Application of Laplace Transformation Technique for Circuit Analysis

Having understood the application of the Laplace transformation to the solution of differential equations, we shall now proceed to apply the technique for the analysis of electrical/electronic circuits. The procedure may be divided into three steps as follows.

1. **Transform the given circuit from t (time) domain to s domain using the technique of the Laplace transform. The special property of s domain is that all passive circuit elements behave like impedances; there are no derivatives of electric current or voltage in s domain. Therefore, Ohm's law and all other laws based on Ohm's law like KVL, KCL, mesh analysis, etc. can be applied on the s domain equivalent circuit.**
2. **Solve the circuit in s domain using any tool of circuit analysis like nodal analysis, mesh analysis, superposition theorem, source transformation, or any other technique, etc.**
3. **Take the inverse Laplace transform of the simplified circuit in s domain to get back the required solution in the time domain.**

2.3.1   Transformation of the Circuit from Time Domain to s Domain

Electrical circuits are made of basic circuit elements including current and voltage sources and passive circuit elements like resistance R, capacitor C and inductor L. For the transformation of a given circuit from t domain to s domain, it is required that each of these circuit elements is replaced by its s domain equivalent. S-domain equivalent of these passive circuit elements are derived in the following. Quantities represented by lower case letters refer to t-domain values and those with upper case letters to s-domain values.

(i) **Resistance R in t and s domains**
Potential difference $v(t)$ across resistance R and the current $i(t)$ through the resistance in t-domain are related by the equation

$$v(t) = Ri(t) \tag{2.24}$$

To write the equivalent equation in s-domain, we operate both sides of the above equation by the Laplace transform operator $\mathcal{L}$ to get

$$\mathcal{L}[v(t)] = R\mathcal{L}[i(t)] \tag{2.25}$$

We now define $\mathcal{L}[v(t)] \equiv V(s)$ and call $V(s)$ the **voltage in s-domain**. Similarly, $\mathcal{L}[i(t)] \equiv I(s)$ and is called **current in s-domain**. It is obvious that units of $V(s)\,and\,I(s)$ are not volts and amperes. Quantities in s-domain do not have the same units as the corresponding quantities of t-domain. The normal procedure is to retain the numerical values of circuit parameters given in their **basic units** in t-domain as such in the equivalent circuit of s-domain. It is because finally the result is obtained back in t-domain by taking the inverse Laplace transform. Therefore, one generally does not much care about the units of s-domain quantities. However, before undertaking the transformation from t to s domain, the units of quantities in t domain must be converted into their basic unit, for example, if a capacitor of 1 μF is given then the numerical value of this capacitor that will be carried to s domain will be 10^{-6} F; similarly, 1 MΩ resistance will be converted to 10^6 Ω resistance in s domain, 1 mH inductance will be carried as inductance of 0.001H to the s domain and so on. The relation in s-domain corresponding to relation (2.24) of t-domain may now be written as

$$V(s) = RI(s) \tag{2.34}$$

It may be observed that in the case of resistance R, there is one-to-one correspondence between the expressions of the t and s domain quantities. The pictorial representation of this equivalence is shown in Fig. 2.1.

One may also define the impedance of an element in s-domain as the ratio of voltage V(s) in s-domain to the current I(s) in s-domain. In case of resistance, the impedance z_R^s, from (2.26), may be given as

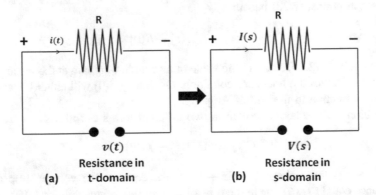

Fig. 2.1 Resistance in t- and s-domains

$$z_R^s = R$$

From the way the impedance is defined in s-domain, it follows that the voltage drop V(s) against any impedance may be obtained by multiplying the impedance with current I(s) through it. This is in analogy to the definition of impedance in t-domain. However, there is one special character of s-domain, which will be apparent later, that every circuit element of t-domain, be it R or L or C, turns into impedance when transformed in s-domain.

Self-assessment question A2SA: Why does every circuit element behave as impedance in s-domain?

(ii) Inductance L in t and s domains

The potential difference $v(t)$ across an ideal inductor at the instant of time t is equal to the value of the inductance L of the inductor multiplied by the instantaneous value of the rate of change of current $i(t)$ through the inductor. In mathematical terms, it may be written as

$$v(t) = L\frac{di(t)}{dt} \tag{2.27}$$

Equation (2.27) is a mathematical representation or definition of inductance L in t domain. In order to describe the inductor in s-domain, we take the Laplace transform of both sides of (2.27) to get

$$\mathcal{L}[v(t)] = \mathcal{L}\left[L\frac{di(t)}{dt}\right] = L\mathcal{L}\left[\frac{di(t)}{dt}\right] \tag{2.28}$$

However, by definition $\mathcal{L}[v(t)] = V(s)$ and using the expression from the table for the first-order derivative, one may substitute $L\mathcal{L}\left[\frac{di(t)}{dt}\right] = L[sI(s) - i(0)]$. With these substitutions, (2.28) becomes

$$V(s) = L[sI(s) - i(0)] \tag{2.29}$$

Factor $i(0)$ in (2.29) refers to the value of current in t-domain at the instant $t = 0$. That is, it specifies the boundary condition. Obviously, i(0) will either be zero or a number or fraction in the unit of Ampere.

Equation (2.29) may be written in two different forms as follows.

$$V(s) = (sL)I(s) - Li(0)$$

This shows that voltage V(s) in s-domain is equal to the voltage drop across impedance equal to (sL) due to the flow of current I(s), minus voltage $Li(0)$. It may be noted that the equivalent of L (of t-domain) becomes impedance of magnitude

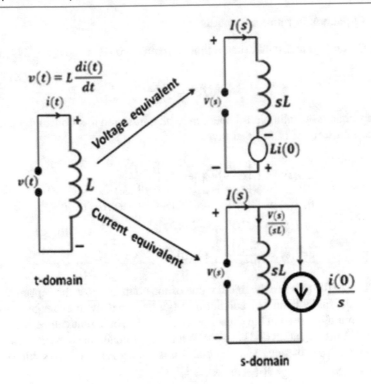

Fig. 2.2 Inductor in t- and s-domains

(sL) in s-domain through which current I(s) flows minus a voltage source of magnitude Li(0). Here, i(0) is the value of current through the inductance in t domain at time t = 0. Obviously, the magnitude of i(0) depends on the initial boundary conditions of the original circuit in t domain. Therefore, inductance L of t domain is replaced by impedance of value (sL) through which current I(s) flows and a voltage source of value L i(o) which opposes the voltage $(sL)I(s)$ when transformed in s domain.

Therefore, inductance in s-domain is equivalent to a series combination of impedance (sL) and a voltage source of voltage (-L i(0)). This series combination is referred to as the voltage equivalent of inductance.

(b) $I(s) = \frac{V(s)}{(sL)} + \frac{i(0)}{s}$

Equation (b) represents the current source equivalent and tells that two currents $\frac{V(s)}{(sL)}$ through (sL) and $\frac{i(0)}{s}$ in parallel to it constitute the total current I(s) in s-domain. The voltage and current source equivalents in s-domain for inductance are shown in Fig. 2.2. It may be observed that an inductor behaves as if it has impedance equal to (sL) in s-domain.

(iii) **Capacitor in t and s domains**

Current i(t) at instant t (in the time domain) through a capacitor C is given by

$$i(t) = C\frac{dv(t)}{dt} \qquad (2.30)$$

The capacitance operating in t-domain may be transferred to s-domain by taking the Laplace transform of (2.30) that gives

$$\mathcal{L}[i(t)] = \mathcal{L}\left[C\frac{dv(t)}{dt}\right] = C\mathcal{L}\left[\frac{dv(t)}{dt}\right]$$

$$\text{or } I(s) = C\{sV(s) - v(0)\} = (sC)V(s) - Cv(0)$$

$$\text{or } I(s) = (sC)V(s) - Cv(0) \qquad (2.31)$$

$$\text{or } I(s) = \frac{V(s)}{1/(sC)} - Cv(0)$$

In (2.31), $v(0)$ refers to the initial value of the voltage across the capacitor at time t = 0, i.e. the initial boundary condition in t-domain. $v(0)$ will either be zero, if no energy is initially stored in the capacitor, or it will be some finite numerical value in the unit of Volt. Equation (2.31) also tells that in s-domain, the capacitor behaves as if it has impedance of value 1/sC, so that if it is put across a voltage source of value $V(s)$, the current through it becomes $\frac{V(s)}{1/(sC)}$.

On rearranging equation (2.31), one gets

$$V(s) = \left(\frac{1}{sC}\right)I(s) + \frac{v(0)}{s} \qquad (2.32)$$

Equation (2.32) may be interpreted saying that voltage V(s) is the sum of the voltage drop due to current I(s) through impedance of value 1/sC and a voltage source of value $\frac{v(0)}{s}$.

The two equivalent circuits, current equivalent and voltage equivalent, for a Capacitor in s-domain are shown in Fig. 2.3.

As mentioned earlier, the three elements R, L and C when transformed in s-domain behave as if they are impedances of values R, sL and 1/sC, respectively. The reason why there are no elements like inductors or capacitors in s-domain is that in s-domain, there are no derivatives; when a derivative of t-domain is transformed in s-domain, it gets changed to a function with boundary conditions. Since elements like capacitor and inductor require derivatives of current or voltage, which do not exist in s-domain, therefore, there are no such elements.

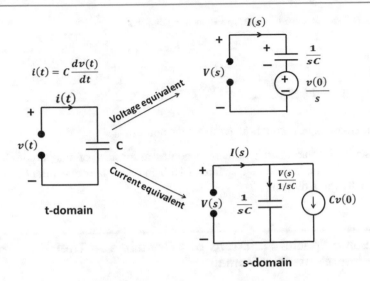

Fig. 2.3 Capacitor in t and s-domains

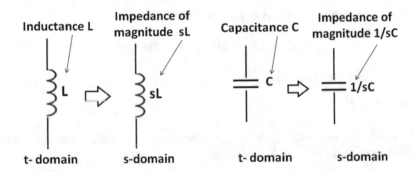

Fig. 2.4 Inductance and capacitance equivalents in s-domain with initial zero condition

In case when the initial values of current $i(t = 0)$ and or voltage $v(t = 0)$ in a circuit are zero, the circuit is said to have zero initial conditions. In that situation, the s-domain equivalents of inductance and capacitance reduce as shown in Fig. 2.4.

Initial boundary conditions play a crucial role in the solution of a problem. The total time domain may be divided into three components: (i) before $t = 0$, i.e. $t < 0$, (ii) at $t = 0$ and (iii) at any time after $t = 0$, i.e. for $t > 0$. The value of a given parameter in t domain is often specified as $t(0^-)$, which means just earlier than $t = 0$, or at $t < 0$. Similarly, $t(0^+)$ refers to the time just after $t = 0$. The following example will clarify the importance of boundary conditions.

With the knowledge of the Laplace technique acquired so far, we are now ready to solve some problems. Table (2.3) summarizes some t to s domain equivalents for ready reference.

Table 2.3 Table of some t domain to s domain equivalents

Parameter	t-domain	s-domain
Inductance	L	sL (as impedance)
Capacitance	C	1/sC (as impedance)
Constant like voltage, etc.	V(t)	V/s

Transformation of a constant from t to s domain

A constant of value X in t domain becomes a constant of magnitude X/s in s domain. For example, a voltage source of 10 V in t domain is equivalent to a source of value 10/s in s domain.

2.4   Some Special Functions of t Domain and Their Equivalents in s Domain

i   **Unit Step Function:** $f(t) = u(t)$

Step function u(t) is defined in t domain as

$$u(t) = 0 \ for \ t < 0; u(t) = 1 \ for \ t \geq 0$$

The Laplace transform $\mathcal{L}[u(t)] = 1/s (in \ s \ domain)$; and $\mathcal{L}^{-1}[1/s] = u(t)$.

ii   **Function** $f(t) = \{u(t) - u(t-1)\}$

Function f(t) =[u(t) – u(t-1)] is shown in the figure; the Laplace transform of this function is given as (Figs. 2.5a and 2.5b)

$$\mathcal{L}[\{u(t) - u(t-1)\}] = \mathcal{L}[u(t)] - \mathcal{L}[u(t-1)] = \left[\frac{1}{s} - \frac{1}{s}e^{-s}\right] = \frac{1}{s}\{1 - e^{-s}\}$$

iii   **Function:** $[f(t) = 4u(t) - u(t-1) - u(t-2) - u(t-3) - u(t-4]$

$$\mathcal{L}[4u(t) - u(t-1) - u(t-2) - u(t-3) - u(t-4] = \frac{1}{s}[4 - e^{-s} - e^{-2s} - e^{-3s} - e^{-4s}]$$

iv   **Function unit impulse** $\delta(t)$: Unit impulse function has zero value for all times $t > 0$ or $t < 0$; at t = 0, it has undefined value such that $\int\limits_{-\infty}^{+\infty} \delta(t)dt = 1$

$$\mathcal{L}[\delta(t)] = 1 \ and \ \mathcal{L}^{-1}[1] = \delta(t)$$

Fig. 2.5a Step function u(t)

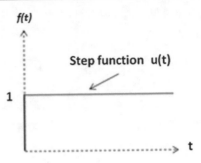

Fig. 2.5b Step function [u(t)
– u(t−1)]

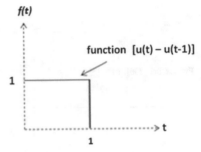

It is easy to realize that the impulse function is the derivative of the unit step function (Fig. 2.5c).

v   **Ramp**: A function that increases linearly with time as shown in Fig. 2.5d is called a ramp function. Unit ramp function is the integral of the unit step function.

A ramp function may have different values of its slope, for example, the function shown in Fig (2.5e) has a slope of 1.

$$\mathcal{L}[t] = \frac{1}{s^2}$$

In the general case when the ramp function has a slope α, (either positive or negative),

$$\mathcal{L}[\alpha t] = \alpha \frac{1}{s^2}$$

vi   **Some other functions**: Figure (2.5f) shows three functions; $Y_1(t)$, which is a ramp function with slope +0.5, may be written as $Y_1(t) = 0.5\,t$ and, therefore, its Laplace transform $F_1(s)$ may be written as

Fig. 2.5c Step function

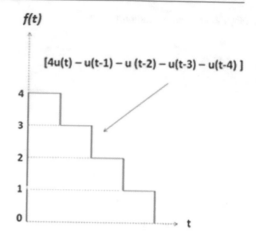

$f(t)$

$[4u(t) - u(t-1) - u(t-2) - u(t-3) - u(t-4)]$

Fig. 2.5d Delta function

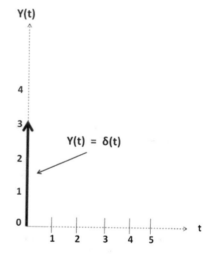

$Y(t)$

$Y(t) = \delta(t)$

$$\mathcal{L}[Y_1(t)] = F_1(s) = 0.5 \cdot \frac{1}{s^2}$$

Function $Y_2(t)$ has a magnitude zero till t = 2 and then it becomes a ramp of slope −0.5; the Laplace transform for this function $F2(s)$ is given as

$$\mathcal{L}[Y_2(t)] = F_2(s) = -0.5e^{-2s}\frac{1}{s^2}$$

Function $Y_3(t)$ is a triangular pulse that contains a ramp of slope +1 from $t = 0$ to $t = 3$, and it assumes value zero for $t > 3$. Function $Y_3(t)$ may be thought of as the sum of three functions: (a) a ramp of slope 1 up to $t = 3$. (b) It may be observed that

Fig. 2.5e Ramp function

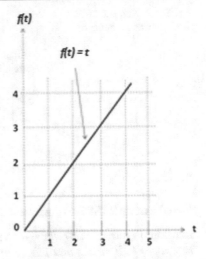

Fig. 2.5f Figures of three different functions

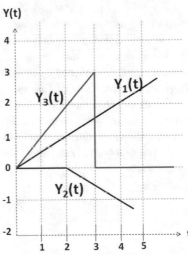

at $t = 3$, the slope of the curve has become zero; a zero slope may be obtained by adding at $t = 3$ another ramp of slope -1.0. (c) With the addition of two ramps given by (a) and (b), a negative discontinuity will crop up at $t = 3$; to overcome that, a step function of height -1 needs to be added to the sum of (a) and (b). Finally,

$Y_3(t) = 1t - 1(t - 3) - u(t - 3)$; and the corresponding Laplace transform $F_3(s)$ is given by

$$\mathcal{L}[Y_3(t)] = F_3(s) = \frac{1}{s^2} - e^{-3s}\frac{1}{s^2} - e^{-3s}\frac{1}{s}$$

Fig. SE2.6 Series LCR
circuit in t and s domains

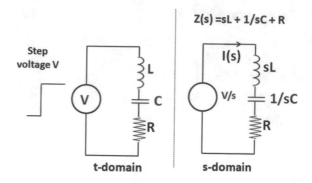

t-domain   s-domain

Solved example (SE2.6) Analyse a series LCR circuit using the Laplace transform.

Solution: *Figure (SE2.6) shows a series LCR circuit subjected to a step voltage V at instant t = 0. It means that just before t = 0, v(t) was zero or v(t = 0⁻) = 0. And v (t=0⁺) = v(t= ∞) = V which may be a number like 10 V or 20 V, etc.. Therefore, f (t=0+) = f(any positive value of t).*

The equivalent circuit in s-domain is shown in the right half of the figure. It may be noted that in s-domain equivalent, the voltage V is transformed into V/s; inductance L is replaced by impedance of magnitude (sL), and capacitance by the equivalent impedance of magnitude (1/sC) while the resistance remains as such.

It is to be noted that the initial conditions in t-domain tell that voltage v(t = 0-) is zero, therefore, each circuit element is replaced by its equivalent impedance in s domain   without any current or voltage source. Since the three impedances in s-equivalent are in series, they may be added to get the total impedance Z(s) of the s-equivalent.

$$Z(s) = sL + \frac{1}{sC} + R \qquad\qquad (SE2.6.1)$$

And

$$I(s) = \frac{V(s)}{Z(s)} = \frac{\frac{V}{s}}{\left(sL + \frac{1}{sC} + R\right)} = \frac{V}{L\left(s^2 + \frac{1}{LC} + \frac{R}{L}s\right)} = \left(\frac{V}{L}\right)\frac{1}{\left(s^2 + \frac{R}{L}s + \frac{1}{LC}\right)}$$

$$(SE2.6.2)$$

We now make the following substitution:

$$\left(s^2 + \frac{R}{L}s + \frac{1}{LC}\right) = (s+a)^2 + b^2$$

so that on comparing the coefficients of s and the constant terms, one gets

$$2a = \frac{R}{L} \text{ and } \left(a^2 + b^2 = \frac{1}{LC}\right) \text{ or } a = \frac{R}{2L} \text{ and } b = \sqrt{\frac{1}{LC} - \left(\frac{R}{2L}\right)^2} \quad \text{(SE2.6.3)}$$

With this substitution, (SE2.6.2) becomes

$$I(s) = \left(\frac{V}{L}\right) \frac{1}{(s^2 + \frac{R}{L}s + \frac{1}{LC})} = \left(\frac{V}{L}\right) \frac{1}{\left\{(s+a)^2 + b^2\right\}} \quad \text{(SE2.6.4)}$$

We take the inverse Laplace transform of (A2SE.6.4) to revert back in t-domain,

$$i(t) = \left(\frac{V}{L}\right) \mathcal{L}^{-1}\left[\frac{1}{\left\{(s+a)^2 + b^2\right\}}\right] = \left(\frac{V}{L}\right) \frac{1}{b} e^{-at} \sin(bt) \quad \text{(SE2.6.5)}$$

The above equation gives the required expression for current $i(t)$ at any time $t > 0$ in the series LCR circuit. At $t = 0$, the exponential term becomes 1, and the expression reduces to

$$i(t) = i(t = 0^+) = \left(\frac{V}{L}\right) \frac{1}{b} \sin(bt)$$

$$= \left(\frac{V}{L}\right) \frac{1}{\sqrt{\frac{1}{LC} \left(\frac{R}{2L}\right)^2}} \sin\left(\{\sqrt{\frac{1}{LC} - \left(\frac{R}{2L}\right)^2}\}t\right) \quad \text{(SE2.6.6)}$$

It is not difficult to identify $\frac{1}{\sqrt{LC}}$ with the resonance frequency ω of the circuit. Equation (SE2.5.6) then reduces to the familiar form

$$i(t) = \frac{V}{L} \left(\frac{1}{\sqrt{\omega^2 - a^2}}\right) \sin\left(\sqrt{\omega^2 - a^2}\right)t, \text{ where } a = \frac{R}{2L} \quad \text{(SE2.6.7)}$$

As expected the current in the circuit in t domain is a sine wave. It may be remembered that in the present case, the applied voltage is step voltage which remains constant for all positive values of time. However, if it is assumed that after some time say t_1 the input voltage is switched off, then expression (SE2.5.7) will hold for $t > 0+ < t_1$ and after time t_1, current $i(t > t_1)$ will decay exponentially due to the factor e^{-at} with a time constant $\tau = \frac{1}{a} = \frac{2L}{R}$.

Solved example SE2.7 Draw the s domain equivalent circuit of the network given in Fig. (SE2.7a), and calculate the current through the 42 Ω resistance. Assume that there is no initial energy stored in the network.

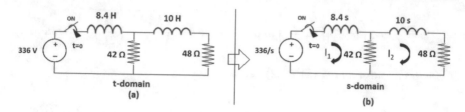

Fig. SE2.7 Given network **(a)** and its s equivalent

Solution: *It is easy to draw the s-equivalent circuit by replacing 336 V by 336/s, 10 H inductance by 10 s, 8.4 H inductance by 8.4 s and retaining the resistances as such. The s-equivalent circuit is shown in Fig .SE2.7a. It may be noted that the two inductances are replaced simply by multiplying s to their numerical values without including any current source like Li(0) in the circuit. It is done as it is given that no energy was stored in the circuit initially, which means i(0) = 0.*

Next, it is required to calculate the current through the 42Ω resistance. To obtain it, we carry out a mesh analysis assuming currents I_1 and I_2 in the two meshes as shown in the figure. Two mesh equations may be written by applying KVL to each mesh as follows:

$$-\frac{336}{s} = (8.4s + 42)I_1 - 42I_2 \qquad\qquad (SE2.7.1)$$

And

$$(10s + 90)I_2 - 42I_1 = 0 \qquad\qquad (SE2.7.2)$$

Therefore, $I_1 = \frac{(10s + 90)}{42}I_2$; putting this value of I_1 in (SE2.7.1), one gets

$$(8.4s + 42)\frac{(10s + 90)}{42}I_2 - 42I_2 = -\frac{336}{s}$$

or

$$I_2 = \frac{168}{s^3 + 14s^2 + 24s} = \frac{168}{s(s+2)(s+12)} = \frac{A}{s} + \frac{B}{(s+2)} + \frac{C}{(s+12)} \qquad (SE2.7.3)$$

The coefficients A, B and C may be obtained by the method of residues as

$$A = [sI_2]_{s=0} = \left[\frac{168}{(s+2)(s+12)}\right]_{s=0} = \frac{168}{2x12} = 7$$

$$B = [(s+2)I_2]_{s=-2} = \left[\frac{168}{s(s+12)}\right]_{s=-2} = \frac{168}{(-2)(10)} = -8.4$$

$$And\ C = [(s+12)I_2]_{s=-12} = \left[\frac{168}{s(s+2)}\right]_{s=-12} = \frac{168}{(-12)(-10)} = 1.4$$

Substituting these values of A, B and C back in (SE2.7.3)

$$I_2 = \frac{7}{S} - \frac{8.4}{(s+2)} + \frac{1.4}{(s+12)} \tag{SE2.7.4}$$

Taking the inverse Laplace transform of each term gives

$$i_2(t) = 7u(t) - 8.4e^{-2t}u(t) + 1.4e^{-12t}u(t)\,Ampere\,(A)\,for\,t \geq 0 \tag{SE2.7.5}$$

u(t) in each term of the above expression appears because one can always associate a multiplicative factor 1 with each term of (SE2.7.4) and when the inverse Laplace transform is taken as $\mathcal{L}^{-1}[1] = u(t)$, it gives u(t).

The value of current I_1 may be obtained by substituting the value of I_2 in (SE2.7.2) and is given as

$$I_1 = \frac{40s+360}{s(s+2)(s+12)} = \frac{E}{s} + \frac{F}{(s+2)} + \frac{G}{(s+12)} \tag{SE2.7.6}$$

E, F and G in the above expression are constants, which may either be found by the method of residues or by comparing the coefficients of identical terms. It may be shown that E = 5, F = −14 and G = −1. Therefore, I_1 is given by

$$I_1 = \frac{40s+360}{s(s+2)(s+12)} = \frac{15}{s} - \frac{14}{(s+2)} - \frac{1}{(s+12)}$$

On taking the inverse Laplace of each term, one gets the t-domain value of $i_1(t)$ as

$$i_1(t) = \left[15 - 14e^{-2t} - e^{-12t}\right]u(t) \tag{SE2.7.7}$$

Checking the result

The correctness of the solution, i.e. of the result obtained above may be verified by two methods.

1. *By finding out the values of the parameters obtained by the analysis, in the present case, the two currents $i_1(t)$ and $i_2(t)$, at times t = 0 and at t = ∞, both*

Fig. SE2.7.1 Circuit at t = ∞

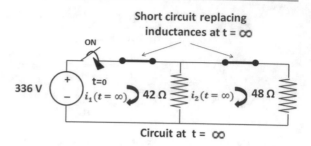

Short circuit replacing
inductances at t = ∞

Circuit at t = ∞

by circuit analysis and from the result obtained by the s domain analysis. If the
two values agree then the derivation and analysis are correct. Let us use this
method to verify the correctness of $i_1(t)$. From (SE2.7.7), that contains the
result of theoretical analysis, theoretical values $i_1^{The0}(t = 0)$ and $i_1^{Theo}(t = \infty)$
may be obtained as

$$i_1^{The0}(t = 0) = 0,$$

Since the boundary condition says that initially there was no energy in the
system.

And $i_1^{Theo}(t = \infty) = 15A$, obtained by putting $t = \infty$ in $(A2SE - 7.7)$.   it   is
interesting to see that at $t = \infty$ the value of i_1 is constant, i.e. the system has
reached a steady state and there is no change in the value of currents, etc. in the
circuit. If current is not changing then the inductance behaves like a short circuit,
and the equivalent circuit of the network looks as shown in Fig. SE2.7.1 .

Now the equivalent resistance of the parallel combination of resistances 42 Ω
and 48 Ω is 22.4 Ω, and current i_1^∞ is given as

$$i_1^\infty = \frac{336}{22.4} = 15A$$

The theoretical value of i_1, obtained earlier $(i_1^{Theo}(t = \infty) = 15A)$, agrees with
the value of i_1^∞ obtained by circuit analysis. The same may be tested for i_2^∞.

2. The other method of verifying the correctness of the derived result is based on
   the application of the initial and final value theorems. According to the initial
   and final value theorems,

$$\lim_{t=0} f(t) = \lim_{s \to \infty} sF(s) \text{ and } \lim_{t \to \infty} f(t) = \lim_{s \to 0} sF(s)$$

Let us first find the current $i_1^{t=0}$ using the initial value theorem,

Fig. SE2.8 Network for
example (SE2.8)

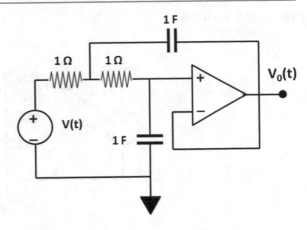

$$\lim_{t\to 0} i_1(t) = \lim_{s\to\infty} sI_1(s) = \lim_{s\to\infty}\left[s\left(\frac{40s + 360}{s^3 + 14s^2 + 24s} \right) \right]$$

$$= \lim_{s\to\infty}\left[\left(\frac{40/s + 360/s^2}{1 + 14/s + 24/s^2} \right) \right] = 0$$

Thus with the use of the initial value theorem, the initial value of current i_1 at t = 0 comes out to be zero, as expected from the boundary condition. Next, let us apply the final value theorem to get the value of current i_1 at infinite time.

$$\lim_{t\to\infty} i_1(t) = \lim_{s\to 0} sI_1(s) = \lim_{s\to 0}\left[s\left(\frac{40s + 360}{s^3 + 14s^2 + 24s} \right) \right]$$

$$\lim_{t\to\infty} i_1(t) = \lim_{s\to 0} sI_1(s) = \lim_{s\to 0}\left[s\left(\frac{40s + 360}{s^2 + 14s^s + 24} \right) \right] = \frac{360}{24} = 15A$$

It is the same value obtained earlier. Verifying the value of current i_2 using both these methods is left as an exercise. (Fig. SE2.8)

Solved example (SE 2.8) Use the Laplace transform method to obtain the output voltage $V_0(t)$ for the following op-amp amplifier.

Solution: *Let us first draw the equivalent circuit in s domain and arbitrarily assign currents x(s), y(s), z(s) and a(s) in different branches of the given network as shown in* Fig. SE2.8.1.
   Let V_P be the potential at node P, and the potential of node Q will be $V_o(s)$ because of the property of the op amp.
   Now currents x(s), y(s), a(s) and z(s) are given as
$$x(s) = \frac{V(s) - V_P}{1}; y(s) = \frac{V_P - V_0(s)}{1}; z(s) = \frac{V_0(s) - 0}{\left(\frac{1}{s}\right)} \text{ and } a(s) = \frac{V_0(s) - V_P}{(1/s)}$$

We now apply KCL at node P.
   At node P : $y(s) = x(s) + a(s)$ or $y(s) = \frac{V_P - V_0(s)}{1} = \frac{V(s) - V_P}{1} + \frac{V_0(s) - V_P}{(1/s)}$, which
gives

Fig. SE2.8.1 Equivalent circuit in s domain

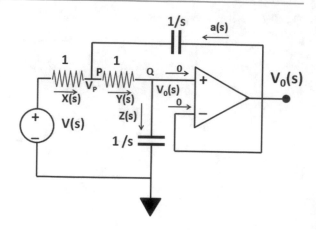

$$V(s) = (2+s)V_P - (1+s)V_0(s) \tag{SE8.1}$$

Similarly, applying KCL at node Q gives

$$y(s) = z(s) \; Or \; \frac{V_P - V_0(s)}{1} = \frac{V_0(s) - 0}{\left(\frac{1}{s}\right)} \; Or \; V_P = (s+1)V_0(s) \tag{SE8.2}$$

Substituting this value of V_P in (SE8.1), one gets

$$V(s) = (2+s)(s+1)V_0(s) - (1+s)V_0(s) = \left[(s^2+3s+2) - (1+s)\right]V_0(s)$$

or $V(s) = (s+1)^2 V_0(s)$

Hence the voltage gain $A_v(s)$ of the op amp is $A_v(s) = \frac{V_0(s)}{V(s)} = \frac{1}{(s+1)^2}.$

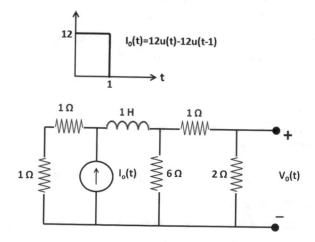

Fig. SE9 Network for example (SE2.9)

Taking the inverse Laplace transform of $A_v(s)$ *gives the op-amp gain* $A_v(t)$ *in t domain.*

Hence, $A_v(t) = \mathcal{L}^{-1}[A_v(s)] = \mathcal{L}^{-1}\left[\frac{1}{(s+1)^2}\right] = te^{-t}.$

The output voltage $V_0(t)$ *in t domain is given as*

· *Ans:* $V_0(t) = A_v(t)V(t) = te^{-t}V(t).$

Solved example (SE2.9) Assuming that there was no energy in the system initially, obtain the value of the output voltage $V_0(t)$ in the network of Fig. SE9.1. The input current source $I_0(t)$ has the form as shown in the figure.

Solution: *The important point to pay attention to is the input source is a current source that is represented by the function* $i_0(t) = 12u(t) - 12u(t-1).$

Figure SE9.1 shows the function 12 u(t) in which a step function has magnitude zero for t < 0, assumes value 12 at t = 0 and the magnitude remains 12 for all times t > 0. Function 12 u(t−1) has a magnitude 0 for all times t < 1, assumes a value 12 at t = 1 and the magnitude remains 12 for t > 1. The difference {12 u(t) −12 u(t−1)} is a function that has magnitude 0 for t <0 and t > 1; =12 for t = 0 to t = 1, as shown at the extreme right of the figure. Further, the Laplace transform of {12 u(t)−12 u(t−1)} is given as

$$I_0(s) = \mathcal{L}[\{12u(t) - 12u(t-1)\}] = \mathcal{L}[12]\mathcal{L}[u(t)]\mathcal{L}[u(t-1)] = \frac{12}{s}\{(1 - e^{-s})\}$$

(SE9.1)

We draw the equivalent circuit of the given network in s domain and simplify it as shown in Fig. SE9.2. As may be seen in Fig. SE9.2, inductance has been replaced by its equivalent impedance sL in the s domain and resistances by their impedances equal to their resistance values.

We next change the current source $I_0(s)$ *into equivalent voltage source* $V_E(s)$ *by absorbing the parallel impedance* $(Z_R = 2)$, *such that* $V_E(s) = 2I_0(s)$ *with a series impedance of 2 as shown in the bottom circuit diagram of the figure. Total current* $I_1(s)$ *in the circuit is given by*

$$I_1(s) = \frac{V_E(s)}{(s+4)}$$

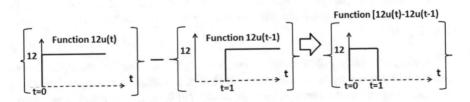

Fig. SE9.1 Pictorial representation of function 12{u(t)-u(t-1)}

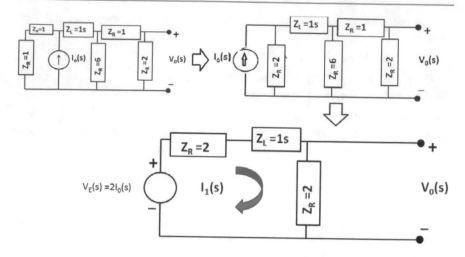

Fig. SE9.2 Network in s domain

And the output voltage $V_0(s) = 2I_1(s) = \frac{2V_E(s)}{(s+4)} = \frac{4I_0(s)}{(s+4)} = \frac{4}{(s+4)}I_0(s)$

Substituting the value of $I_0(s)$ from (SE9.1), one gets

$$V_0(s) = \frac{4}{(s+4)}I_0(s) = \frac{4}{(s+4)}\frac{12}{s}\{(1-e^{-s})\} = \frac{48}{(s+4)s}[1-e^{-s}] = \frac{A}{s} + \frac{B}{(s+4)}$$

$$(SE9.2)$$

where coefficient A and B are given by

$A = |\frac{48}{(s+4)}[1-e^{-s}]|_{s=0} = 0$ and $B = |\frac{48}{s}[1-e^{-s}]|_{s=-4} = -12(1-e^4)$

So, $V_0(s) = 0 - \frac{12(1-e^4)}{s+4} = -12(1-e^4)$. Taking the inverse Laplace of the

equation, one gets

$$V_0(t) = \mathcal{L}^{-1}[V_0(s)] = -12[\mathcal{L}^{-1}[1] - \mathcal{L}^{-1}[e^4]] = -12[u(t)] + 12e^{-4(t-1)}u(t)$$

ANS: $V_0(t) = -12[u(t)] + 12e^{-4(t-1)}u(t)$.

Solved example (SE2.10) The switch of the circuit was made off at t = 0, Find the initial conditions and calculate the current through the circuit i(t) for t > 0 (Fig. SE2.10).

   Solution: *To find the initial conditions at t = 0, let us consider the circuit at a time t < 0 or at t = 0 when the switch is still closed. In that case, the capacitor is no more in the circuit and a constant current, say, $i_L(0)$ is flowing in the circuit. For a*

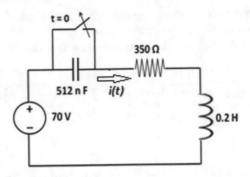

Fig. SE2.10 Network for example (SE2.10)

constant current, the inductance behaves as a short circuit and, therefore, the current $i_L(0)$ through the inductor is given by $i_L(0) = \frac{70V}{350\Omega} = 0.2\,A$.

Having found the initial current through the inductance, the circuit with the switch turned off at $t = 0$ is considered when the capacitor is in the circuit. The circuit in t domain at $t > 0$, its equivalent circuit in s domain and the reduced circuit in s domain are shown in Fig. SE2.10a. As shown in this figure, the s domain circuit contains impedances Z_c, Z_R and Z_L corresponding, respectively, to capacitance, resistance and inductance. Apart from that, the inductance may be replaced by its impedance (sL) in parallel with a current source of magnitude $i_L(o)$

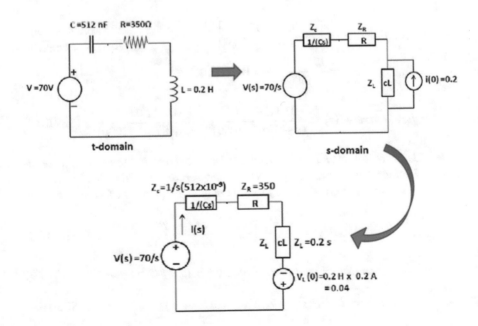

Fig. SE2.10a Circuit in t and s domains

*or by a voltage source (with opposite polarity) of magnitude L $i_L(0)$. The voltage
equivalent is shown at the bottom of the figure.*

The current I(s) through the circuit in s domain is given by

$$I(s) = \frac{V(s) + V_L(0)}{Z_C + Z_R + Z_L} = \frac{\frac{70}{s} + 0.04}{\frac{1}{s(512 \times 10^{-9})} + 350 + 0.2s} = \frac{70 + 0.04s}{0.2s^2 + 350s + 1953125} = \frac{70 + 0.04s}{0.2[s^2 + 1750s + 9765625]}$$

$$or \ I(s) = \frac{0.2s + 350}{[s^2 + 1750s + 9765625]}$$

$$(SE2.10.1)$$

*The denominator of (SE2.10.1) cannot be factorized into real factors, hence, it
has been broken into two complex factors:*

$$[s^2 + 1750s + 9765625] = (s + 875 - J3000)(s + 875 + j3000)$$

$$I(s) = \frac{0.2s + 350}{[s^2 + 1750s + 9765625]} = \frac{0.2s + 350}{(s + 875 - j3000)(s + 875 + j3000)}$$

$$= \frac{A}{(s + 875 - j3000)} + \frac{B}{(s + 875 + j3000)}$$

The coefficient A $= \left[\frac{0.2s + 350}{(s + 875 + j3000)}\right]_{(s = -875 + j3000)} = \frac{-175 + j600 + 350}{(-875 + j3000 + 875 + j3000)} = \frac{175 + j600}{(j6000)}$

$= -j0.03 + 0.1 = 0.01\angle - 0.02°$

And the coefficient B $= \left[\frac{0.2s + 350}{(s + 875 - j3000)}\right]_{(s = -875 - j3000)} = \frac{-175 - j600 + 350}{(s + 875 - j3000)}$

$= \frac{175 - j600}{j6000} = 0.01\angle 0.02°.$

Therefore, current I(s) $= \frac{0.01\angle - 0.02°}{(s + 875 + j3000)} + \frac{0.01\angle 0.02°}{(s + 875 - j3000)}$ $(SE2.10.2)$

*In order to obtain current i(t) corresponding to current I(s), one has to take the
inverse Laplace transform of I(s). Now whenever any variable like current or
voltage in s domain is given in the form of complex conjugates like the following:*

$$F(s) = \frac{M\angle Y^0}{(S + a - jb)} + \frac{M\angle - Y^0}{(S + a + jb)}$$ $(SE2.10.3)$

then it can be shown that the corresponding function f(t) in t domain is given by

$$f(t) = \mathcal{L}^{-1}[F(s)] = 2(M)e^{-at}\cos(bt + Y)$$ $(SE2.10.4)$

Comparing Expression for I(s) given by (SE2.10.2) and (SE2.10.3), one finds
  M = 0.01; Y = 0.02; a = 875; and b = 3000.
  Therefore, using (SE2.10.4) one can write the current in the time domain as

$$i(t) = 2(0.01)e^{-875t}\cos(3000t + 0.02)$$

Solved example (SE2.11) Use the Laplace transform method to find the voltage $V_A(t)$ at node A in the given circuit. It is given that the initial voltage across the capacitor $V_c(0)$ was 5 V (Fig. SE2.11).

Solution: *The s domain equivalent circuit of the given network is shown in Fig. SE2.11a. It may be observed that the given voltage source $10\,e^{-t}u(t)$ has been replaced by its Laplace transform $10/(s+1)$, while the current source $2\delta(t)$ by 2. In s domain, all passive circuit elements behave like corresponding impedances which are shown in the figure. Since the initial voltage across capacitance at $t = 0$ is given as 5 V, the capacitor in s domain may be replaced by its impedance $1/sF = 10/s$ and a current source of magnitude $cV = 0.1 \times 5 = 0.5$ A.*

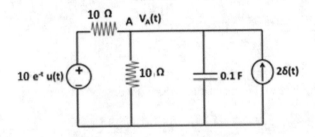

Fig. SE2.11 Circuit for example (SE2.11)

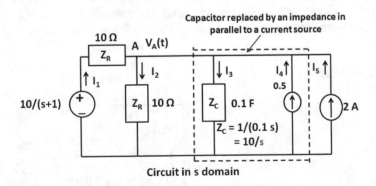

Fig. SE2.11a Equivalent circuit in s domain

The s domain network consists of five currents, I_1, I_2, I_3, I_4 and I_5. Applying KCL at node A, one gets

$$I_1 + I_4 + I_5 = I_2 + I_3, \text{ where} \qquad\qquad (SE2.11.1)$$

$I_1 = \frac{\frac{10}{s+1} - V_A(s)}{20}$; $I_2 = \frac{V_A(s)}{20}$; $I_3 = \frac{V_A(s)}{10/s}$; $I_4 = 0.5$; *and* $I_5 = 2$. *Substituting these values of currents in (SE2.11.1), one gets*

$$\frac{\frac{10}{s+1} - V_A(s)}{10} + 2 + 0.5 = \frac{V_A(s)}{10} + \frac{V_A(s)}{10/s}$$

$$or \quad \frac{1}{s+1} + 2.5 = \frac{2V_A(s)}{10} + \frac{sV_A(s)}{10} = \frac{1}{10}[(2s+1)V_A(s)]$$

$$or \quad V_A(s) = \frac{25s + 35}{(s+1)(s+2)} = \frac{K_1}{(s+1)} + \frac{K_2}{(s+2)}$$

Constants K_1 and K_2 may be found either by comparing the coefficients of powers of s or by the method of residues as

$$K_1 = [(s+1)V_A(s)]_{s=-1} = \frac{25(-1)+35}{(-1+2)} = 10; K_2 = [(s+2)V_A(s)]_{s=-2}$$

$$= \frac{25(-2)+35}{(-2+1)} = 15$$

Hence, $V_A(s) = \frac{10}{(s+1)} + \frac{15}{(s+2)}$.
Taking the inverse Laplace of the above gives

$$V_A(t) = \mathcal{L}^{-1}[V_A(s)] = \mathcal{L}^{-1}\left[\frac{10}{(s+1)}\right] + \mathcal{L}^{-1}\left[\frac{15}{(s+2)}\right] = \left(10e^{-t} + 15e^{-2t}\right)u(t)$$

Solved example (SE2.12) Calculate current i(t) drawn from the source $v(t) = e^{-0.6t}\sin(0.8t)V$ at a time $t > 0$ by the elements of the circuit shown in Fig. SE2.12. It may be assumed that there are no currents and voltages in the circuit at $t < 0$.

Fig. SE2.12 Circuit for example (SE2.12)

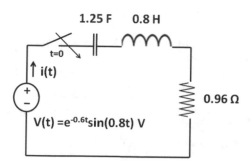

Fig. SE2.12a Equivalent
circuit in s domain at t > 0

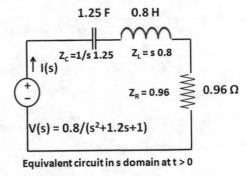

Equivalent circuit in s domain at t > 0

Solution: *In order to draw the equivalent circuit in s domain at time t > 0, it is required to write the Laplace transforms of all parameters given in the original circuit.*

$$V(s) - \mathcal{L}[v_i(t)] = \mathcal{L}\left[e^{-0.6}\sin(0.8t) = \frac{0.8}{(s+0.6)^2 + (0.8)^2}\right] = \frac{0.8}{(s^2 + 1.2s + 1)}$$

$$Z_c = \frac{1}{Cs} = \frac{1}{1.25s} = \frac{0.8}{s}; Z_L = sL = 0.8s; Z_R = R = 0.96$$

The equivalent s domain circuit at t > 0 is shown in Fig. SE2.12a.
The current I(S) is given b,

$$I(s) = \frac{V(s)}{Z_c + Z_L + Z_R} = \frac{\left(\frac{0.8}{s^2 + 1.2S + 1}\right)}{\frac{0.8}{s} + 0.8s + 0.96}$$

$$or\ I(s) = \frac{0.8s}{(0.8s^2 + 0.96s + 0.8)((s^2 + 1.2s + 1))} = \frac{0.8s}{0.8(s^2 + 1.2s + 1)^2} = \frac{s}{(s^2 + 1.2s + 1)^2}$$

$$or\ I(s) = \frac{s}{(s^2 + 1.2s + 1)^2}$$

$$(SE2.12.1)$$

It may be observed that the denominator $(s^2 + 1.2\ s + 1)$ cannot be factorized into real factors, therefore, it is broken into complex factors that are complex conjugates of each other, as follows:

$$(s^2 + 1.2s + 1) = \{(s + 0.6) + j0.8\}\{(s + 0.6) - j0.8\}$$

Hence,

$$I(s) = \frac{s}{(s^2 + 1.2s + 1)^2} = \frac{s}{(s+0.6-j0.8)^2(s+0.6+j0.8)^2}$$

$$= \frac{A}{(s+0.6-j0.8)^2} + \frac{B}{s+0.6-j0.8} + \frac{C}{(s+0.6+j0.8)^2} + \frac{D}{s+0.6+j0.8}$$

where

$$A = \left[\frac{s}{(s+0.6+j0.8)^2}\right]_{s=-0.6+j0.8} = \frac{-0.6+j0.8}{(-06+j0.8+0.6+j0.8)^2} = \frac{-0.6+j0.8}{(j1.6)^2} = 0.234 - j0.313$$

$$or\ A = \left[(0.234)^2 + (0.313)^2\right]^{1/2} tan^{-1}\left(-\frac{0.313}{0.234}\right) = 0.390 tan^-(-1.337) = 0.39\angle - 53.2$$

In order to calculate B, we differentiate the function and then evaluate it at the value of s = −0.6 + j0.8

$$B = \frac{d}{ds}\left[\left[\frac{s}{(s+0.6+j0.8)^2}\right]\right] =$$

$$= \left[\left(\frac{1}{(s+0.6+j0.8)^2}\right) - \left\{\frac{2s}{(s+0.6+j0.8)^3}\right\}\right]_{s=-0.6+j0.8} = 0.29\angle 90$$

Other constants C and D will be the complex conjugates of these values.

Hence, $I(s) = \frac{0.39\angle-53.2}{(s+0.6+j0.8)^2} + \frac{0.29\angle90}{(s+0.6+j0.8)} + \frac{-0.39\angle53.2}{(s+0.6-j0.8)^2} + \frac{-0.29\angle-90}{(s+0.6-j0.8)}.$

Taking the inverse Laplace of I(s),

$$i(t) = \mathcal{L}^{-1}[I(s)]$$
$$= [2(0.39)te^{-0.6t}\cos(0.8t - 53.2) + 2(0.29)e^{-0.6t}\cos(0.8t + 90)]u(t)$$

Solved example (SE2.13) The switch of the circuit is turned around at t = 0; calculate the current through the circuit at a time t > 0 (Fig. SE2.13).

Solution: *Figure (SE2.13.1) shows (a) the circuit at time t < 0, when a 60 V battery is connected through resistance of 3 kΩ to a capacitor of 1μF (=10⁻⁶ F). Since in steady state a capacitor behaves like an open circuit, the total voltage of 60 V appears across the two ends of the capacitor. This gives the initial value of voltage against the capacitor at time t = 0, $V_C(0)$ = 60 V. Now at t = 0, the switch is turned*

Fig. SE2.13 Network for example (SE2.13)

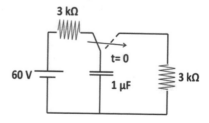

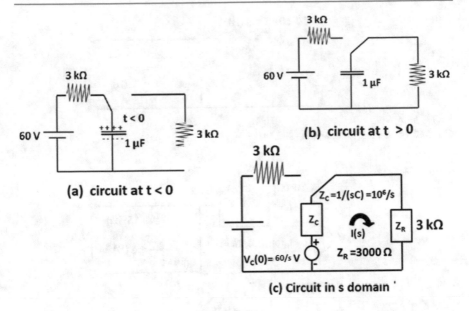

Fig. SE2.13.1 a Circuit at $t < 0$; **b** circuit at $t > 0$; **c** Circuit in s domain

the other way which leaves the circuit for $t > 0$ as shown in Fig. SE2.13.1b; the s domain equivalent of which is shown in part (c) of the figure. It is required to find out the current $I(s)$, which is given as

$$I(s) = \frac{60/s}{Z_c + Z_R} = \frac{60/s}{\frac{1000000}{s} + 3000} = \frac{60}{3000s + 1000000} = \frac{60}{3000}\left(\frac{1}{s + 1000/3}\right)$$

or $I(s) = 0.02\left(\frac{1}{s + 333.33}\right)$

The current in t domain may be obtained by taking the inverse Laplace of $I(s)$,

$$i(t) = \mathcal{L}^{-1}[I(s)] = \mathcal{L}^{-1}0.02\left[\frac{1}{s + 333.33}\right] = 0.02e^{-333.33t}u(t)$$

As may be observed from the above expression for current in t domain , the current decays exponentially with a time constant $\tau = 1/333.33 = 0.003s^{-1}$.

Solved example (SE2.14) For the circuit given in the figure, obtain current $i_{L1}(0)$ and $i_{L2}(0)$ and then draw the equivalent circuit in s domain to determine the output voltage $V_o(t)$ (Fig. SE2.14).

Solution: *Figure (SE2.14.1a) shows the circuit at a time t < 0, when the switch is short-circuiting the part of the circuit on the right which may be left out of con- sideration. Further, the circuit at t < 0 is shown in figure (b) where steady current*

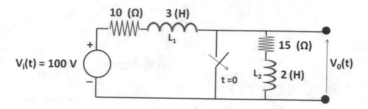

Fig. SE2.14 Circuit for example (SE2.14)

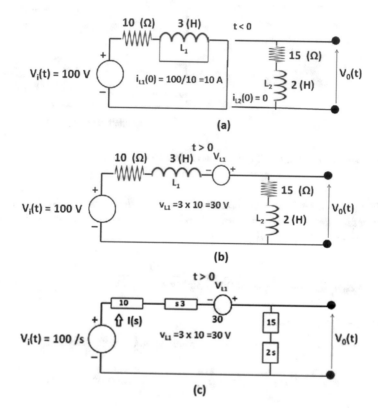

Fig. SE2.14.1 Equivalent circuits

flows through the left-side part of the circuit and inductance L_1 behaves as a short circuit. The current through this short circuit $i_{L1}(0) = v_i(t) /R = 100/10 = 10$ A. Since the right-side part of the circuit is not connected to any energy source, the current through inductance L_2 is $i_{L2}(0) = 0$.

The equivalent circuit in s domain is shown in figure (c). Here, all passive elements have been replaced by their equivalent impedances. Inductance L_1 is replaced by its impedance $sL_1 = s3$ in series with a voltage source (of opposite polarity) of value $(L_1 \times i_{L1}(0) = 3 \times 10) = 30$. Since there was no initial current $i_{L2}(0) = 0$, the second inductance has no series voltage source.

Current $I(s) = \dfrac{Total\ voltage}{Total\ impedance} = \dfrac{\frac{100}{s}+30}{10+3s+15+2s} = \dfrac{1}{s}\left(\dfrac{100+30s}{5s+25}\right)$

And the output voltage $V_0(s) = I(s)\times(15+2s) = \dfrac{1}{s}\left(\dfrac{100+30s}{5s+25}\right)(15+2s) = $

$\dfrac{(6s+20)(2s+15)}{s(s+5)}$

or $V_0(s) = \dfrac{A}{s} + \dfrac{B}{(s+5)}$. Constants A and B may be calculated by the method of residues

where

$A = \left[\dfrac{(6s+20)(2s+15)}{(s+5)}\right]_{s=0} = \dfrac{300}{5} = 60\ and\ B = \left[\left[\dfrac{(6s+20)(2s+15)}{s}\right]\right]_{s=-5} = \dfrac{-50}{-5} = 10.$

Therefore, $V_0(s) = \dfrac{60}{s} + \dfrac{10}{(s+5)}.$

Hence, $v_0(t) = \mathcal{L}^{-1}[V_0(s)] = \mathcal{L}^{-1}\left[\dfrac{60}{s}\right]+\mathcal{L}^{-1}\left[\dfrac{10}{(s+5)}\right]$

or $v_0(t) = 60u(t)+106e^{-5t}u(t)$

Solved example (SE2.15) As shown in the figure, a parallel combination of resistance and capacitance is excited by an impulse current source $10\delta(t)$. Calculate the current i(t) and the voltage $v_0(t)$ at $t > 0$.

Solution: Since the delta function $\delta(t)$ has zero value for $t < 0$ and $t > 0$, therefore, $i_g(t = 0^-)$ is zero and hence $i_C(0^-) = 0$. No energy is stored in the circuit just before t $= 0$. At $t = 0$, ig(t=0^+) becomes finite and some energy $(=1/2\ C\ v(0^+))$ gets stored in the circuit. The equivalent circuit in s domain is shown in Fig. SE2.15a where Z_C and Z_R are, respectively, the impedances offered by resistance and capacitance.

From s domain circuit, it follows that

$I(s) = \dfrac{10}{s}$. Also, the equivalent impedance of the parallel combination of Z_R and Z_C is

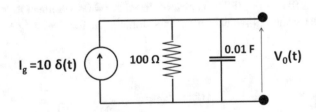

Fig. SE2.15 Circuit for example (SE2.15)

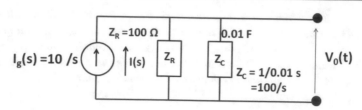

Fig. SE2.15a Circuit in s domain

$$Z_{Eq} = Z_C \| Z_R = \frac{Z_C Z_R}{Z_C + Z_R} = \frac{\frac{100}{s} 100}{\frac{100}{s} + 100} = \frac{1}{100(s+1)}$$

Now $V_0(s) = Z_{Eq}I(s) = \frac{1}{100(s+1)} \cdot \frac{10}{s} = 0.1 \frac{1}{s(s+1)} = 0.1 \left[\frac{A}{s} + \frac{B}{s+1}\right].$
We use the residue method to obtain the values of A and B:

$$A = \left[\frac{1}{s+1}\right]_{s=0} = 1 \ and \ B = \left[\frac{1}{s}\right]_{s=-1} = -1$$

Therefore, $V_0(s) = 0.1 \left[\frac{A}{s} + \frac{B}{s+1}\right] = 0.1 \left[\frac{1}{s} - \frac{1}{s+1}\right].$
Now $v_0(t) = \mathcal{L}^{-1}[V_0(s)] = \mathcal{L}^{-1}\left[0.1\frac{1}{s}\right] - \mathcal{L}^{-1}\left[0.1\frac{1}{s+1}\right] = 0.1[u(t) - e^{-t}u(t)]$
or $v_0(t) = 0.1[u(t) - e^{-t}u(t)]$

Problems on Chapter 2

P2.1 For the circuit given in the figure. show that the current I(s) is given by
(Fig. P2.1)

$$I(s) = 15 \left[\frac{1 + sRC}{s^3 RLC + s^2(R^2 C + L) + Rs}\right]$$

P2.2 Switch SW in Fig (P2.2) is pushed down at $t = 0$ to connect terminals A and
C. Draw the equivalent circuit in s domain at for $t > 0$.

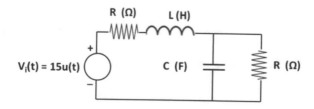

Fig. P2.1 Circuit for problem (P2.1)

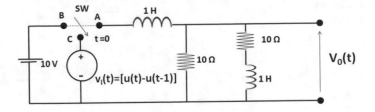

Fig. P2.2 Circuit for problem (P2.2)

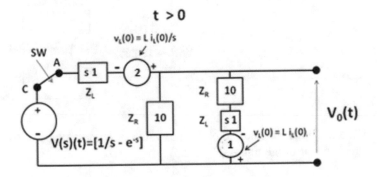

Fig. P2.2a Equivalent s domain circuit

Answer: Figure (P2.2a) shows the s domain equivalent circuit. The magnitudes of the two voltage sources shown in the figure have been obtained
from the initial conditions at $t < 0$.

P2.3   Derive expression for the output voltage $V_0(t)$ in terms of the parameters
provided in the circuit (Fig. P2.3).

Answer: $[v_0(t) = (C_1 V_1)/(C_1 + C_2)]$.

P2.4   Solve the following differential equation using the method of Laplace
transformation. The initial conditions are given as $y(0) = 1$ and $dy/dt\,(0) = -2$

$$\frac{d^2y}{dt^2} + 6\frac{dy}{dt} + 8y = 2u(t)$$

Answer: $y(t) = \frac{1}{4}(1 + 2e^{-2t} + e^{-4t})u(t)$.

Fig. P2.3 Circuit for
problem (P2.3)

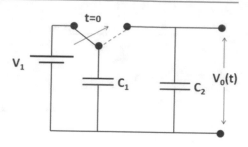

Fig. P2.5 Circuit for
problem (P2.5)

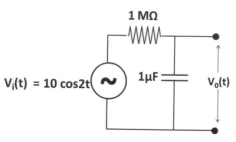

P2.5   Use the Laplace method to obtain the value of output voltage $v_0(t)$ for the
circuit given in the figure (Fig. P2.5).

Answer: $[v_0(t) = 5 \sin 2t]$.

P2.6   A parallel combination of 1 kΩ resistance and 1 mH inductance is excited
by a current source $i_i(t) = 10\ u(t)$. Calculate the current $i_L(t)$ through the
inductance at time t > 0.

Answer: $\left[i_L(t) = \left(1 - e^{-10^6 t} \right) u(t) \right]$.

Fig. P2.7 Circuit for
problem (P2.7)

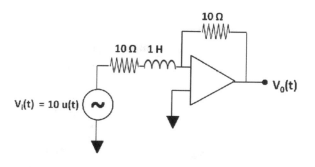

Fig. P2.8 Circuit for
problem (P2.8)

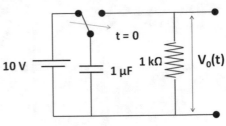

Fig. P2.9 Circuit for
problem (P2.9)

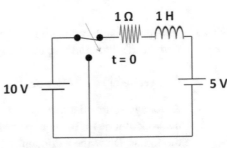

P2.7 Draw equivalent circuit in s domain for the circuit of Fig (P2.7) and
calculate the voltage $v_0(t)$ at the output of the op amp.

Answer: $[v_0(t) = -100(1 - e^{-t})u(t)]$.

P2.8 For the circuit of Fig (P2.8) draw the s domain circuit at $t > 0$ and obtain the
value of output voltage $v_0(t)$.

Answer: $v_t(t) = 10e^{10^{-3}t}u(t)$.

P2.9 Calculate current i(t) through the circuit at $t > 0$ (Figs. P2.9 and P2.10)

Answer: [5 u(t)].

Fig. P2.10 Circuit for
problem (P2.10)

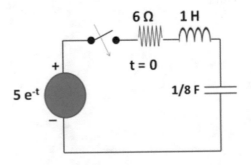

Fig. P2.11 Circuit for
problem (P2.11)

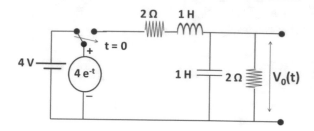

P2.10   Assuming that there is no initial energy stored in the circuit, calculate
current i(t) through the circuit (Figs. P2.11 and MC2.5).

Answer: $[i(t) = -\frac{5}{3}e^{-t} + 5e^{-2t} - \frac{10}{3}e^{-4t}]$.

P2.11   A voltage source defined by $v_i(t) = 4e^{-t}$ *for t > 0* gets connected to the
network at time $t = 0$ when the switch is thrown the other side as shown in
the figure. Draw the s domain equivalent circuit and calculate the output
voltage $v_0(t)$ using the Laplace transform method.

Answer: $[v_0(t) = 8e^{-t} - 6e^{-\frac{5}{4}t}cos\left(\frac{\sqrt{7}}{4}t\right) + \frac{2}{\sqrt{7}}e^{-\frac{5}{4}t}sin(\frac{\sqrt{7}}{4}t)]$.

P2.12   The value of a particular voltage V(s) in equivalent s domain circuit is
given as $V(s) = \frac{2s^2 + 15s + 24}{s(s+2)(s+5)}$; calculate the value of the corresponding
voltage $v(t)$ in t domain.

Answer: $v(t) = \left(\frac{12}{5} - \frac{1}{3}e^{-2t} - \frac{1}{15}e^{-5t}\right)u(t)$.

Multiple Choice Questions

Note: *Some of the multiple-choice questions have more than one correct alternatives; all correct alternatives must be picked up for complete answers in such
cases.*

MC2.1   The function shown in the figure is described by

(a) 8u(t+4)
(b) 8u(t+2)
(c) 4u(t-8)
(d) 8u(t-4)

   ANS: {a}.

MC2.2   The value of integral $\int\limits_{-10}^{+10} (8t^2 + 5)\delta(t - 3)dt$ is

(a) 27
(b) 30
(c) 50
(d) 77

   ANS: {d}.

MC2.3   The Laplace transform of $\sqrt{t}$ is

(a) $\frac{2s^{3/2}}{\sqrt{\pi}}$
(b) $\frac{\sqrt{\pi}}{2s^{3/2}}$
(c) $\frac{\sqrt{\pi}}{2s^{5/2}}$
(d) $\frac{2s^{5/2}}{\sqrt{\pi}}$

   ANS: (b)

MC2.4   The Laplace inverse transform of $\frac{s^2-a^2}{(s^2+a^2)^2}$ is

(a) $t\cos(at)$
(b) $t\sin(at)$
(c) $\sin(at) - at\cos(at)$
(d) $\cos(2at)$

   ANS: (a).

MC2.5   Part of the triangular curve from $t = 0$ to $t = 1$ may be represented by the expression

(a) $4t[u(t)-u(t+1)]$
(b) $4t[u(t)+u(t-1)]$
(c) $4t[u(t)-u(t-1)]$
(d) $4t[u(t)+u(t+1)]$

   ANS: (c).

MC2.6   The Laplace transform of a function $g(t)$ is given as

(a) $\int\limits_{-\infty}^{\infty} g(t)e^{st}$

(b) $\int\limits_{-\infty}^{\infty} g(t)e^{-st}$

(c) $\int\limits_{-\infty}^{0} g(t)e^{-st}$

(d) $\int\limits_{0}^{\infty} g(t)e^{-st}$

   ANS: (d).

MC2.7   The inverse Laplace transform of $\frac{1}{s(s+2)}$ is given by

(a) $[1+e^{2t}]u(t)$
(b) $[1-e^{2t}]u(t)$
(c) $\frac{1}{2}[1+e^{-2t}]u(t)$
(d) $\frac{1}{2}[1-e^{-2t}]u(t)$

   ANS: (d).

MC2.8   Step signal $u(t)$ is made to undergo 50 convolution operations to get the function $f(t) = u(t)*u(t)*u(t)...........50$ terms. The Laplace transform of (ft) is

(a) 1
(b) s^{50}
(c) s^{-50}
(d) $50s$

   ANS: (c).

MC2.9   The Laplace transform for function f(t) = sin(2t)cos(2t) is

(a) $\frac{1}{s^3}$

(b) $\left(\frac{1}{2} - \frac{s}{s+1}\right)$

(c) $\frac{1}{2s} + \frac{s}{s^2+2}$

(d) $\frac{2}{s^2+16}$

   ANS: (d).

MC2.10   A circuit contains a capacitor C, an inductor L and resistance R. It is energeized by a source $v_i(t) = 10\,[u(t)-u(t-1)]$. The s domain equivalent circuit will contain elements like

(a) 1/sC, L, R, $v_L(0)$, $v_C(0)$

(b) 1/sc, sL, R, $v_c(0)$, $v_L(0)$, 10/s , (10/s) e^{-1s}

(c) 1/sc, sL, R, $i_c(0)$, $i_L(0)$, 10/s , (10/s) e^{-1s}

(d) 1/sC, 1/sL, R, 10/s

   ANS: [(b) and (c)].

MC2.11   The Laplace transform of the delta function $\delta(t)$ is

(a) 1/s

(b) 1

(c) s/s+1

(d) s+1/s

   ANS: (b).

MC2.12   The inverse Laplace transform of 1/s is

(a) The unit step function

(b) sin(at)

(c) u(t)

(d) $\delta(t)$

   ANS: (a) and (c).

Short Answer Questions

SA2.1   What is the advantage of working in s domain?

SA2.2   An electric circuit is given in t domain; eloborate the steps you will take to draw its s domain equivalent circuit.

SA2.3   Why do networks in s domain not contain derivatives of voltage and current?

SA2.4   How can one check the correctness of a result in the Laplace analysis?

SA2.5   Discuss the role of initial conditions in drawing an equivalent circuit in s domain.

SA2.6   Describe the voltage and current equivalents of an inductor in s domain.

SA2.7   Explain why a capacitor behaves as an open circuit and an inductor as a short circuit in steady state.

SA2.8   Is it possible to use the superposition theorem for simplifying circuits of s domain ? Give reasons for your answer.

SA2.9   Show that inductance of value L in t domain is equivalent to impedance of value (sL) in s domain.

SA2.10 What is the equivalent of t domain function [u(t0 –u(t-1)]?

SA2.11 Derive the Laplace transform of ramp function f(t)= t dt.

SA2.12 Using the definition of the Laplace transformation, derive the Laplace transform of the delta function.

Fig. MC2.5 Figure for MCQ (2.5)

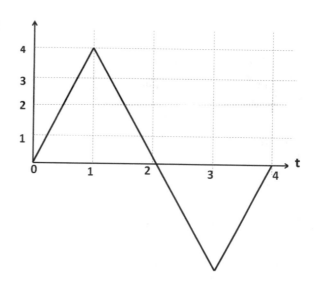

Sample answer for SA2.6

Describe the voltage and current equivalents of an inductor in s domain.

Answer: *True or ideal inductance does not allow sudden or step changes of current through it. It happens because a changing current through the inductance generates a reverse or back emf that opposes the source that changes the current through it. As such the instantaneous voltage* $v_i(t)$ *across the inductance of magnitude L and the rate of change of current* $\frac{di(t)}{dt}$ *are related by the expression*

$$v(t) = L\frac{di(t)}{dt} \tag{SA2.6}$$

The above equation may be considered as the equation that defines the electrical property of inductance. One important conclusion that may be drawn from (SA2.6) is that if di(t)/dt is zero, vi(t) is also zero. This means that in case of a steady current, voltage across the inductance is zero or for constant DC current inductor behaves as a short circuit.

Coming back to equation (SA2.6), it is obvious that when it is required to transport inductance from t domain to s domain, one should take the Laplace transform of (SA2.6), which defines inductance in t domain. Therefore, taking the Laplace transform of both sides of (SA2.6), one gets

$$\mathcal{L}[v(t)] = \mathcal{L}\left[L\frac{di(t)}{dt}\right] = L\mathcal{L}[\frac{di(t)}{dt}] \tag{SA2.6.1}$$

Now $\mathcal{L}[v(t)] = V(s)$ *defines in s domain the quantity equivalent of the voltage of the time domain. The Laplace transformation of the first derivative of a quantity, say i(t), is given as*

$$L\mathcal{L}\left[\frac{di(t)}{dt}\right] = sLI(s) - Li(0) \tag{SA2.6.2}$$

Thus, (SA2.6) that defines the electrical property of an inductor in t domain may look like the following in s domain :

$$V(s) = (sL)I(s) - Li(0) \tag{SA2.6.3}$$

Assuming that V(s) and I(s) behave like voltage and current in s domasin, (SA2.6.3) may be interpreted as the voltage across an inductor in s domain that is the sum of two voltages, (sL) I(s) + {- Li(0)}; the first term indicates a voltage drop across impedance of magnitude (sL) due to the flow of current I(s) and a voltage of opposite polarity of fixed magnitude Li(o), where i(o) is the current through the inductor v at t = 0 in t domain.Obviously, i(o) is the current that flowed through the inductance just at the instant t = 0 or at $(t = 0^-)$ *and depends on the previous history of the circuit in t domain. I(o) is defined by the boundary condition of the problem.*

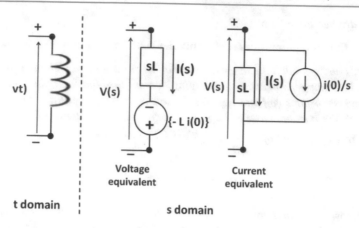

Fig. SA2.6 t and s domain equivalents of an inductor

Equation (SA2.6.3) defines the electric property of an inductor in s domain ; it is equivalent to impedance (sL) in series with a voltage source (−Li(0)). Thus, L in s domain may be represented as a series circuit as shown in Fig. SA2.6. The current equivalent of the inductor in s domain is shown at the extreme right of the figure. If i (0) is zero , i.e.there was no energy stored in the inductor at time t = 0, the current and voltage sources vanish and the inductor v is equal to impedance sL in s domain.

Long Answer Questions

LA2.1 Define the Laplace transform of a function f(t) and using it, obtain the value of the Laplace transform of f(t) =t.

LA2.2 Starting from the definition of the Laplace transform obtain expressions of Laplace transforms of sin(t) and cos(t).

LA2.3 State the Convolution theorem and explain its physical significance.

LA2.4 List the important properties of the Laplace transform and discuss two of them in detail.

LA2.5 List the steps that are taken to solve a circuit using the Laplace method and explain the procedure taking an example.

LA2.6 Discuss the method of transporting resistance, capacitance and inductance from t domain to s domain. Derive equivalent expressions for passive circuit elements in s domain.

LA2.7 Write a detailed note on the Laplace transform method of carrying out electrical network analysis. What are the advantages of this method? Explain with a suitable example.

First- and Second-Order Circuits, Phasor and Fourier Analysis

3

Transient and Steady States

Abstract

Electrical circuits may be represented mathematically by time-dependent differential equations. Classification of electrical circuits as first- and second-order circuits and specific methods of simplifying them to obtain their transient and steady-state solutions are discussed in this chapter. Phasor representation of electrical parameters which is useful for further simplifying networks is also presented. Details of Fourier analysis that transforms the given electrical system from time domain to frequency domain are also provided in this chapter.

3.1 Introduction

A general electrical circuit consists of some sources of energy, either voltage or current sources or both of them coupled to a combination of passive elements like resistances, inductances and/or capacitances. Out of the three passive elements, an ideal resistance consumes electrical energy and converts it into heat energy continuously so long as it is energized. When current $i(t)$ flows through an ideal resistance of magnitude R, it converts the applied electrical energy into heat at the rate of $H = R[i(t)]^2$. The dissipation of electrical energy by a resistance stops as soon as the flow of current is stopped. Thus, the response of an ideal resistance to the energy source is instantaneous. Let us consider the case of a resistance that is connected to a current source through a switch as shown in Fig. 3.1a.

We consider the response of these elements at three different times; (i) just before the switch is put on, i.e. at $t = (0^-)$, (ii) at $t = 0$, when the switch was operated and (iii) at $t = (0^+)$, just after the switch was operated. As shown in the figure, it does not matter whether we consider a moment 1 h before the time $t = 0$ when the switch was put on or only one microsecond before $t = 0$, the condition of the circuit does not depend on that. One may, therefore, say that the system was in a steady state for all times before $t = 0$ and that the system remains in steady state from

© The Author(s), under exclusive license to Springer Nature Switzerland AG 2021   203
R. Prasad, *Analog and Digital Electronic Circuits*, Undergraduate Lecture
Notes in Physics, https://doi.org/10.1007/978-3-030-65129-9_3

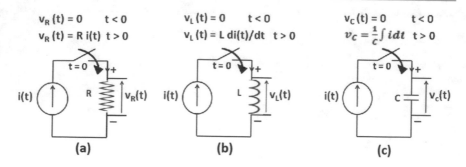

Fig. 3.1 Energizing of **a** resistance; **b** inductor; **c** capacitor by a current source

$t = -\infty$ to a time $t = (0^-)$, which is infinitely small time before $t = 0$. Now in case of a resistance, instantly at time $t = 0$ when switch is put on, current starts to flow through the resistance and a voltage $v_R(t) = R.i(t)$ develops across the resistance. It may be noted that at $t = (0^-)$, just before the switch was operated, the voltage across the resistance $v(0^-) = 0$; but at $t = 0$, at $t = (0^+)$, and up to $t = \infty$, $v_R(t) = R\,i(t)$. Thus, changes of current and voltages across an ideal resistance may take place suddenly or abruptly. Further, so long as the current $i(t)$ remains constant the voltage $v_R(t)$ also remains constant. When currents and voltages in a circuit do not change with time, the circuit is said to be in steady state. For an isolated resistance, it was in a steady state from $t = -\infty$, up to $t = (0^-)$, with zero voltage across it and it remains in a steady state again from $t = 0$ to $t = (0^+)$ and for all times up to $t = +\infty$, with constant current $i(t)$ and a constant voltage $v_R(t)$. It is worth noting that in case of purely resistive circuit, the circuit remains in a steady state of zero voltage and zero current before the operation of the switch and it goes to another steady state with constant current and constant value of voltage drop across it after the switch is operated. It is, however, obvious that the two steady states of the resistive circuit are different from each other.

When switches are used in circuits, it is often required to consider the status of the circuit just before the switch was closed, i.e. at $t = (0^-)$, when switch was just closed, i.e. at $t = (0)$ just after the switch closure, i.e. at $t = (0^+)$ and after a considerably long time $t = \infty$. If suppose the switch is closed (or opened) at instant $t = 0$, then one denotes the instant just before the switch closure by $t = (0^-)$ and just after the switch closure by $t = (0^+)$.

On the other hand, an ideal inductance and an ideal capacitance respond differently to a source of energy. An ideal inductance does not allow a sudden change in the magnitude of current through it. As shown in part (b) of the figure, $i\,(0^-) = 0$ and, therefore, initially there was no current through the inductor. That is to say that the inductor was in a steady state before $t = 0$. However, when at $t = 0$, switch was made on, suddenly current tries to rush through the inductor. Inductor does not allow this to happen, how does the inductor oppose this flow of current? It generates a back emf which opposes the flow of current, so the current gets reduced from $i(t)$ to some value $i_1(t')$ at some time $t' > t$. The average rate of change of current through the inductor may be given as $\frac{\mathrm{d}i(t)}{\mathrm{d}t} = \frac{i(t) - i(t')}{t' - t}$, and the back emf generated in the inductor is proportional to this rate of current change. Initially, just at the instant $t = 0$, the rate of change of current is maximum and, therefore, the back emf is largest which considerably slows down the growth of current through the inductor. The rate of change of current decreases with time and slowly current increases through the inductance and after some time steady current almost equals to $i(t)$ flows through the inductor. As soon as a steady current starts flowing through the inductor, it loses the property of opposing the change of current and the inductor behaves as a short circuit or simply a conducting wire. It may be observed that while a resistor switches from a steady state at $t = (0^-)$ to another steady state at $t = 0$, an inductor starting from a steady state at $t = (0^-)$ (with zero current) attains a second steady state after some time laps when a constant current flows through it again. The inductive circuit passes through a **transient state** between the two steady states. In a transient state, currents and voltages in the circuit are not constant they keep changing. Now suppose, after say one hour after turning the switch on when the system has attained second steady state, the switch is again made off. In case of a resistance the current through it will instantly stop. However, inductor will again pass through a transient state before attaining the final steady state of current zero. In contrast to a resistance an ideal inductor does not dissipate electrical energy, rather it stores it. During the transient period an inductor stores energy in the form of magnetic field which is always associated with an inductor.

Story of an ideal capacitor is also similar. At time $t = (0^-)$, there was no current and no voltage across the two ends of the capacitor, and it was in a steady state. However, at $t = 0$, when switch is put on, a sudden change of voltage occurs across the capacitor which it does not allow. Voltage through capacitor cannot change suddenly. A charging current denoted by $i_C(t)$ passes through the capacitor which charges the capacitor and voltage across the two ends of the capacitor rise with the accumulation of more and more charge on it. Since the charging current $i_C(t)$ decreases and the voltage across capacitor $v_C(t)$ increases with time, both the current and the voltage across the capacitor keep changing for some time immediately after $t = 0$ when switch was operated. As such the capacitive circuit also passes through a transient. Like an inductor, a pure or ideal capacitor also does not dissipate energy, it stores energy in the form of electric field which is always associated with the accumulation of charges. Ultimately, after a given time determined by the circuit parameters, capacitive circuit attains steady state when the voltage across the capacitor becomes equal to the voltage of the applied source and charging current becomes zero. **In a steady state, capacitor behaves as an open circuit** (since no charging current passes through it), while **inductance behaves as a short circuit**. Further, both the inductor and the capacitor carry some energy that gets stored in them during the transient period.

In summary, it may be said that a pure resistive circuit may pass from one steady state to another steady state without passing through any transient state. This happens because an ideal resistance does not store energy. On the other hand, both the inductor and the capacitor circuits pass through transient states when switched from one steady state to another steady state. Transient states are the outcome of the energy storage property of these elements. Because of their property of storing energy, capacitor and inductor are also called dynamic (passive) elements.

3.2 First- and Second-Order Circuits

An electric circuit that contains energy source(s) along with resistances and only one type of dynamic passive element, i.e. apart from resistances only either capacitance or inductance are called first-order circuits. It is because the behaviour of these circuits may be represented by first-order differential equation. In principle, RC and RL circuits are the only first-order circuits.

When an electric circuit apart from resistances also contains two dynamic elements, i.e. both L and C, the circuit is called a second-order circuit as the response of such circuits may be represented by a second-order differential equation. A second-order circuit may have L, C and R in parallel or two of them in parallel and third in series and so on.

3.2.1 Analysis of First-Order Circuits

(a) Solution of first-order linear differential equation

Since the response of first-order circuits may be represented by a first-order differential equation, we first find out the general solution of a first-order linear differential equation. A general first-order linear differential equation may be written as

$$a_1 \frac{dX}{dt} + a_0 X = f(t). \tag{3.1}$$

Here, a_1 and a_0 are the constant coefficients. The solution X of this equation may be written as

$$X = X_n + X_f, \tag{3.2}$$

where X_n and X_f represent, respectively, the natural and force responses of the system.

(i) Natural response

The natural response X_n is the solution of the homogeneous equation,

$$a_1 \frac{dX}{dt} + a_0 X = 0. \tag{3.3}$$

The functional form of X_n is

$$X_n = Ke^{st}. \tag{3.4}$$

Here, K and s are constants (symbol s used here has nothing to do with the s domain of Laplace transformations. s used here is just a symbol for a constant). The value of 's' may be found by substituting the functional form given by (3.4) back in the homogeneous equation (3.3) to get

$$a_1 \frac{d(Ke^{st})}{dt} + a_0 Ke^{st} = 0,$$

$$a_1 Kse^{st} + a_0 Ke^{st} = 0 \quad or \quad a_1 s + a_0 = 0 \text{ or } s = -\frac{a_0}{a_1}. \tag{3.5}$$

The value of the constant K is determined from the boundary conditions and since boundary conditions do not apply only to X_n but they (boundary conditions) apply to the total function X; K may be determined only after determining the forced solution X_f.

(ii) Forced response

In order to determine the forced response solution X_f, some trial function is chosen and substituted in the original equation to determine the parameters of trial function. The choice of the trial function depends on the type of the function $f(t)$ that occurs on the RHS of the original (3.1). For example, if the function $f(t) = a$, some constant than the trial function should also be a constant like A; if $f(t) = at + b$, the trial function may be $A + Bt$; if $f(t)$ is $at^n + bt^{n-1} + \ldots$, then the trial function may be $At^n + Bt^{n-1} + \ldots$; if $f(t) = a \cos(wt) + b \sin(wt)$, the trial function may be $A \cos(wt) + B \sin(wt)$. The coefficients A, B, etc. may be found from actual substitution of trial functions in original (3.1). Following solved examples may further illustrate the method of obtaining the forced response solution.

Solved example (SE3.1) Solve the first-order differential equation given below with boundary condition $y(t = 0) = 0$.

$$\frac{dy(t)}{dt} + 8y(t) = 2t + 1$$

Solution: *First, we find the natural solution $y_n(t)$ of the given equation, for which the homogeneous equation becomes*

$$\frac{dy(t)}{dt} + 8y(t) = 0. \tag{SE3.1.1}$$

We try a solution $y_n(t) = Ke^{st}$ and substitute it in (SE3.1)
$$\frac{d(Ke^{st})}{dt} + 8(Ke^{st}) = 0 \Rightarrow Kse^{st} + 8Ke^{st} = 0 \Rightarrow s = -8.$$
The natural response

$$y_n(t) = Ke^{-8t}. \tag{SE3.1.2}$$

Now, we proceed to find the forced response $y_f(t)$.

Since in this case it is given that f(t) =2t + 1, we chose the trial function $y_f(t)$. as (A t + B) and substitute it in the original equation to get

$$\frac{d(At + B)}{dt} + 8(At + B) = 2t + 1 \Rightarrow A + 0 + 8(At + B) = 2t + 1.$$

Comparing the coefficients of t and constant terms on the two sides of the equation one gets

$$8A = 2 \text{ and } A + 8B = 1; \text{ therefore,}$$
$$A = \frac{2}{8} = \frac{1}{4}; \text{ and } 8B = 1 - \frac{1}{4} = \frac{3}{4}; B = 3/32.$$

The forced response $y_f(t) = \frac{1}{4}t + 3/32$.
The total response

$$y(t) = y_n(t) + y_f(t) = Ke^{-8t} + \frac{1}{4}t + 3/32. \tag{SE3.1.3}$$

We now apply boundary condition that $y(0) = 0$ and put this condition on (SE3.1.3)

$$0 = Ke^{-8\times0} + \frac{1}{4} \times 0 + \frac{3}{32} \Rightarrow K + \frac{3}{32} = 0 \Rightarrow K = -\frac{3}{32}.$$

Final value of the total response may be given as

$$y(t) = y_n(t) + y_f(t) = -\frac{3}{32}e^{-8t} + \frac{1}{4}t + \frac{3}{32}.$$

(b) **Natural response of a RL circuit:** In general, a series RL circuit may have several resistances in the network but only one inductance. It is always possible to reduce all the resistances in a single equivalent resistance R_T in series (or in parallel) with the inductance as shown in Fig. 3.2a. The equivalent simplified circuit of network of figure (a) is shown in figure (b). Further, since we are **considering natural response**, there is no source of energy in the circuit.

Assuming that current $i_L(t)$ is flowing in the circuit both through R_T and through the inductance L, and writing the voltage drop $v_L(t)$ against the inductance as $\left(L\frac{di_L(t)}{dt}\right)$, one may apply KVL to the closed circuit to obtain

$$v_{R_T}(t) + v_L(t) = 0, \quad or \quad R_T i_L(t) + L\frac{di_L(t)}{dt} = 0. \qquad (3.6)$$

Equation (3.6) is a first-order differential equation, the natural response for which may be obtained by substituting a trial function $i_L(t) = Ke^{st}$ in (3.6)

$$L\frac{d(Ke^{st})}{dt} + R_T(Ke^{st}) = 0 \Rightarrow sL + R_T = 0 \Rightarrow s = -\frac{R_T}{L}.$$

Therefore, the natural response

$$i_L^n(t) = Ke^{-\left(\frac{R_T}{L}\right)t}. \qquad (3.7)$$

Let us now consider the boundary condition that $i_L^n(t = 0) = I_0$ which means that at time $t = 0$, the initial time, there was some current I_0 through the inductor. Putting this boundary condition on (3.7) one gets

$$i_L^n(t = 0) = I_0 = Ke^{\left(-\frac{R}{L}x0\right)} \Rightarrow K = I_0.$$

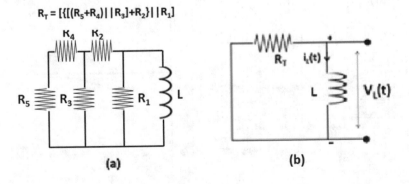

$$R_T = [\{[(R_5+R_4)||R_3]+R_2\}||R_1]$$

(a)   (b)

Fig. 3.2 a Series LR circuit; **b** a parallel LR circuit

Therefore, the complete natural response function

$$i_L^n(t) = I_0 e^{-\frac{R_T}{L}t}. \tag{3.8}$$

Let us now discuss the nature of the natural response $i_L^n(t)$ of a RL network given by expression (3.8). The response may be written as

$$i_L^n(t) = I_0 e^{-\frac{R_T}{L}t} = I_0 e^{-\frac{t}{\tau}},$$

where $\tau = \frac{L}{R_T}$ is called the time constant and decides the rate at which the exponential curve representing the natural response of a RL circuit decays with time.

Exponential decay curve of the inductor current with time is shown in Fig. 3.3a. It may be observed that after each time constant τ the value of the inductor current falls by a factor of $1/e$. After 1τ the value of current becomes 36.8% of its maximum value, after 2τ, 13.5% I_0, after 3τ, 5% of I_0 and so on. Though theoretically the value of current will become zero only after an infinite time, however, for all practical purposes the value of current will become negligible (less than 1% of the original value I_0) after a time lapse of 5τ. The time gap of $\approx 5\tau$ is, therefore, called the settling time or the time after which the system will attain a steady state.

Self-assessment question: Write the natural response of a RL circuit for the boundary condition $i_L(t = t_0) = I_0$.

It is easy to find the voltage drop v_L^n across the inductor using the fact that the voltage drops across the inductor L and the resistor R are equal in magnitudes but are of opposite polarity.

$$i.e., v_L^n(t) = -v_L^n(t) = -Ri_L^n = -RI_0 e^{-\frac{t}{\tau}}.$$

Figure 3.3b shows the variation of the inductor voltage with time.

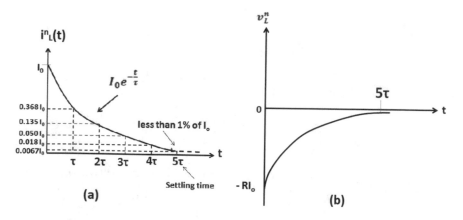

Fig. 3.3 Exponential decay of inductor current with time

(c) **Natural response of a RC circuit:** Network for the study of the natural response of a RC circuit is shown in Fig. 3.4. It is obvious from the figure that the same current $i_c(t)$ will flow through both the total equivalent resistance R_T and the capacitor C. Application of KVL to the closed loop gives $i_c(t) = C\left(\frac{dv_c(t)}{dt}\right)$.

$$v_R(t) + v_c(t) = 0. \tag{3.10}$$

But $v_R(t) = R_T i_c(t) = R_T C \frac{dv_C(t)}{dt}$, substituting this value of $v_R(t)$ in (3.10), one gets

$$R_T C \frac{dv_C(t)}{dt} + v_c(t) = 0. \tag{3.11}$$

Trying a solution $v_C(t) = K\,e^{st}$ in (3.11) gives

$$s = -\frac{1}{R_T C} \quad \text{and} \quad v_C^n(t) = Ke^{-\frac{1}{R_T C}t} = Ke^{-\frac{t}{\tau}}. \tag{3.12}$$

Here, the time constant τ is given by

$$\tau = R_T C. \tag{3.13}$$

To determine the value of K we need some boundary condition, let us try two different boundary conditions; (i) $v_c^n(t) = V_0\ at\ t = 0$, putting this condition in (3.12) one gets $V_0 = Ke^0\ or\ K = V_0$. Substituting this value of K in (3.12) gives

$$v_C^n(t) = V_0 e^{-\frac{t}{\tau}}. \tag{3.14}$$

Fig. 3.4 Network for the study of natural response of RC circuit

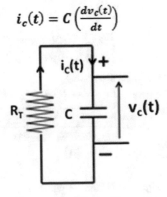

$$i_c(t) = C\left(\frac{dv_c(t)}{dt}\right)$$

When one compares (3.8) and (3.14), it is found that in both cases the natural responses i_L^n and v_C^n undergo exponential decay with time constants, respectively, $\frac{L}{R_T}$ and $R_T C$.

Natural response of current through the capacitor $i_C^n(t)$ may be easily found knowing that the currents through the capacitor and the resistor are of equal magnitude but of opposite signs. Hence,

$$i_C^n(t) = -i_R^n(t) = -\frac{1}{R_T}\left[v_C^n(t)\right] = -\frac{V_0}{R_T}e^{-\frac{t}{\tau}}. \tag{3.15}$$

It is left as an exercise to draw exponential decay curves for $v_C^n(t)$ and $i_C^n(t)$. Another form of boundary condition may be

(ii) $v_C^n(t = t(0^+)) = V_0$ *till* $t = t_0$, which means that v_C^n remains equal to V_0 from $t = 0$ to $t = t_0$. To find the value of K under this boundary condition, the value of capacitor voltage is substituted in (3.12) to get

$$v_C^n(t) = Ke^{-\frac{1}{R_T C}t} = Ke^{-\frac{t}{\tau}}; V_0 = Ke^{-\frac{t_0}{\tau}} \text{ and } K = V_0 e^{\frac{t_0}{\tau}}.$$

When the above value of K is substituted back in (3.12), one gets

$$v_C^n(t) = Ke^{-\frac{t}{\tau}} = V_0 e^{\frac{t_0}{\tau}} e^{-\frac{t}{\tau}} = V_0 e^{-\left(\frac{t-t_0}{\tau}\right)} \tag{3.16}$$

and

$$i_C^n(t) = -\frac{V_0}{R_T}e^{-\left(\frac{t-t_0}{\tau}\right)}. \tag{3.17}$$

Figure 3.5 shows the time variation of the capacitor voltage and current.

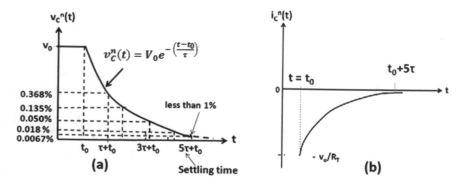

Fig. 3.5 Natural response functions for RC circuit **(a)** voltage across the capacitor; **(b)** current through the capacitor

First-order circuits with DC source

(d) Forced response of RC circuit

Let us consider the case when some step DC voltage is applied to a first-order circuit. As shown in Fig. 3.6 at $t = 0$ when the switch flips a step DC voltage of magnitude V_s is applied to the series combination of R and C. However, it may be noted that at time $t = (0^-)$ some voltage V_0 was already present across the capacitor. The differential equation describing the process at $t > 0$ may be written as

$$RC \frac{dv_c(t)}{dt} + v_C(t) = v_s. \tag{3.18}$$

The natural response function for this problem has already been obtained as

$$v_C^n = Ke^{-\frac{t}{\tau}}, \tag{3.19}$$

where time constant τ is given as RC.

The forced response v_C^f may be obtained by trying a solution $v_C^f = A$, since on the RHS of (3.18) is a constant v_s. Putting $v_C = A$ in (3.18) one gets

$$RC \frac{dA}{dt} + A = v_s \; or \; A = v_s \; that \; means \; that \; v_C^f = A = v_s.$$

Now both the natural and forced response functions are with us. The total response function is the sum of the two, and, therefore,

$$v_C = v_C^n + v_C^f = Ke^{-\frac{t}{\tau}} + v_s. \tag{3.20}$$

To determine the value of K we use the initial condition, $v_C(t = 0^-) = v_0$, putting this in (3.20) one gets

$v_0 = Ke^0 + v_s \; or \; K = (v_0 - v_s)$ putting this value of K back in (3.20), one obtains

Fig. 3.6 Step DC voltage applied to RC circuit

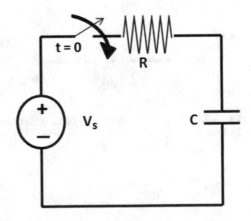

$$v_C = v_C^n + v_C^f = Ke^{-\frac{t}{\tau}} + v_s = (v_0 - v_s)e^{-\frac{t}{\tau}} + v_s$$

or

$$v_C = v_0 + v_s\left(1 - e^{-\frac{t}{\tau}}\right). \tag{3.21}$$

In case the switch is put on at time $t = t_0$ and earlier from $t = (0^-)$ to $t = t_0$ the voltage across the capacitor remains v_0, in that case (3.21) will become

$$v_C = v_0 + v_s\left(1 - e^{-\frac{(t-t_0)}{\tau}}\right). \tag{3.22}$$

Graphs showing the variation of capacitor voltage and capacitor current for the two initial conditions are shown in Figs. 3.7a and 3.7b.

Equations (3.22) may be written as

$$v_C \text{ at time } t = \{Initial\ value\ of\ v_c - Final\ value\ of\ v_c\} \times e^{-\left(\frac{t-t_0}{\tau}\right)} + Final\ value\ of\ v_c.$$

In fact, all state variables like voltages and currents may be represented by the same equation,

State variable at time t = {*initial value of state variable* − *Final value of state variable*}$xe^{-\left(\frac{t-t_0}{\tau}\right)}$ + *Final value of state variable*.

$$\tag{3.23}$$

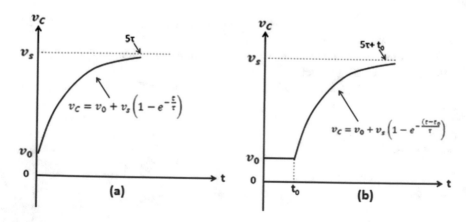

Fig. 3.7a Exponential capacitor charging curves for **a** initial condition $V_c = V_0$ at $(t = 0)$; **b** $V_c = V_0$ from $(t = 0^-)$ to $(t = t_0)$

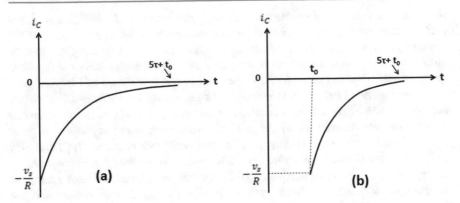

Fig. 3.7b Exponential rise of capacitor charging current curves for **a** initial condition $V_c = V_0$ at $(t = 0)$; **b** $V_c = V_0$ from $(t = 0^-)$ to $(t = t_0)$

(e) Forced response of RL circuit

It is possible to derive expressions for forced response of the RL combination in the same way as has been done for the RC case, however, it is possible to write the expression for inductor current at time t using the general formula given by (3.23) as follows

$$i_L(t) = (i_0 - i_s)e^{-\frac{t}{\tau}} + i_s. \tag{3.24}$$

Here, i_0 is the initial value of inductor current at $t = (0^-)$; i_s is the final value of inductor current (provided by the current source) and τ is the time constant which in the case of an inductor circuit is given by $\tau = L/R$.

Solved example (SE3.2) As shown in the figure, the switch which was on for a considerable period of time was made off at time $t = 0$, discuss the transient response of the circuit at $t > 0$ (Fig. SE3.2).

Fig. SE3.2 Circuit, for example (SE3.2)

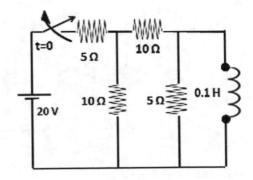

Solution *It is given in the problem that initially at time t < 0, the switch was closed and the circuit was in that condition for a considerable time. That means that the circuit has already attained a steady state by time t = (0⁻). As a matter of fact, in most cases, the circuit is taken to be in steady state at all times before t = 0, unless specified otherwise in the problem. The circuit at t < 0 is shown in Fig. SE3.2a i.*

Since at all times before t = 0, circuit is in steady state, the inductor in the circuit behaves as a short circuit. As a result 5Ω resistance in parallel to the short circuit behaves as zero resistance and the two 10Ω resistances come in parallel to each other. The 20 V battery, therefore, finds a total resistance of 5Ω + (10||10)Ω =10 Ω. The current through the battery is 20/10 = 2 A. When this 2 A current passes through the parallel combination of 10 Ω resistances, current through each parallel branch becomes 1 A. So the current through the inductor branch at time just before t = 0, $i_L(0^-) = 1\,A$.

Next, we calculate the equivalent resistance of the network looked back from the location of the inductor, after removing of inductor and replacing all sources of voltage by open circuit. The network in this condition looks like as shown in Fig. SE3.2a ii, **which reduces to the parallel combination of (20||5) Ω** $= \frac{20 \times 5}{20 + 5} = 4\Omega$. *Thus, the equivalent resistance is 4 Ω. Further in a RL circuit the time constant τ =L/R = 0.1/4.*

Since behaviour of the circuit is desired after the opening of the switch when no source of energy is connected to the circuit, the required response will be the natural response that is given by

$$i_L(t) = i_L(0^-)e^{-\frac{t}{\tau}} = 1e^{-\frac{t}{0.1/4}} = 1e^{-40t}. \tag{SE3.2.1}$$

After t > 0, current will exponentially decay from 1 A to zero. Though zero value of current may take long time, it will reach a value of 0.001A in about 5 τ, that is in 5 × 0.1/4 = 0.125 s.

One can also obtain expression for the voltage v_L(t) at t > 0 using the formula
$$v_L(t) = L\frac{di_L(t)}{dt} = 0.1[-40e^{-40t}]V = -4e^{-40t}\,V.$$

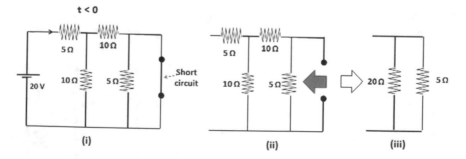

Fig. SE3.2a Circuit at t < 0

Fig. SE3.3 Network for exercise (SE3.3)

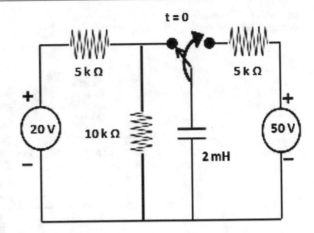

Solved example (SE3.3) For the circuit shown in figure obtain expression for capacitor voltage $v_c(t)$ at t > 0 (Fig. SE3.3).

Solution: *As shown in the figure, for t < 0, the capacitor is connected in parallel with a resistance of 10 k which is connected to another resistance of 5 k and to a battery of 20 V. Since the system is in steady state for t < 0, the capacitor is like an open circuit, the voltage $v_c(0^-)$ on the capacitor is same as across the resistance of 10k. The total resistance across the battery of 20 V is 10 k + 5 k = 15 k. Therefore, current through the combination of resistances is i_0 =20/15 mA . The voltage drop across 10 k resistance = $v_c(0^-)$ =10 k × i_0 =13.34 V.*

At t = 0, the capacitor got connected to 50 V battery via a 5 k resistance. After sufficient time more than 5τ, the system will again reach a steady state when capacitor will again behave as an open circuit and at that time the final voltage across capacitor will be the same as the voltage of battery, i.e. 50 V. Therefore, $v_c(\infty)$ = 50 V. So now we know the initial voltage $V_c(0^-)$=13.34 V, final voltage $v_c(\infty)$ = 50V across the capacitor. We now find the time constant τ of the system. For that we need the value of the resistance in series with the capacitor which is given as 5 kΩ. Hence, the time constant τ= RC = 5 k × 2 m H =10. Now using (3.23), we may write

$$v_c(t) = [initial\, value v_C(0^-) - final\, value v_C(\infty)] e^{-\frac{t}{\tau}} + final\, value v_C(\infty)$$

or

$$v_c(t) = \left\{ [13.34 - 50] e^{-\frac{t}{10}} + 50 \right\} V = \left[50 + (13.34 - 50) e^{-0.1t} \right] V.$$

Solved exercise (SE3.4): Obtain an expression for the output voltage $v_0(t)$ for t > 0 (Fig. SE3.4).

Solution: *As shown in the circuit diagram, for t < 0, there was no energy source in the op-amp circuit and, therefore, the initial output voltage was $v_0(0^-) = 0$. The time*

Fig. SE3.4 Network, for
example (SE3.4)

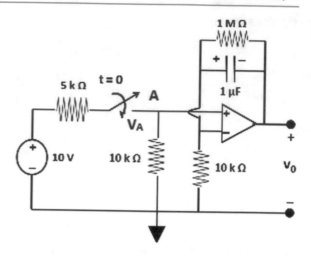

constant of the RC circuit is τ given by $\tau = (1M\Omega \times 1\mu F) =1$ s. After a time lapse of
5τ, i.e. $t = 5$ s, the circuit will again attain steady state and the final value of the
output voltage $v_0(\infty) = V_A$ (the voltage at the input of op-amp amplifier, point A)
$\times$ gain A_v of the op-amp amplifier.

Now, $V_A = 10$ V (10 k/10 k +5 k) = **6.67 V** and gain of non inverting amplifier A_v
= (1+1M/10 k) =1+10^3 ≅ 10^3. And so vo(∞) = 6.67x10^3. Now using the general
expression for any circuit observable given by (3.32), we have

$$v_0(t) = (v_0(0^-) - v_0(\infty))e^{-\frac{t}{\tau}} + v_0(\infty) = (0 - 6.67 \times 10^3)e^{-t} + 6.67 \times 10^3$$

or

$$v(t) = 6.67 \times 10^3(1 - e^{-t})V.$$

Solved example (SE3.5) Derive the first-order differential equation representing the
response of the output voltage $v_o(t)$ of the following op-amp circuit and show that
the circuit behaves as an integrator. Assume that the initial voltage across the
capacitor $v_C(0) = 0$ (Fig. SE3.5).

Solution: *Let us assume that a voltage source given by $A_0F(t)$ is applied at the
input of the circuit. Here, A_0 may be constant and $F(t)$ may be a function like $u(t)$ or
any other similar functions. The non-inverting input (A) of the op amp is at ground
potential, so the inverting terminal B will also be at ground (0 V) potential. Current
through the resistance R is given as $i_R = \frac{A_0F(t)}{R}$ and current through capacitor C,*
$i_c = C\frac{dv_C}{dt} = C\frac{d(v_0 - 0)}{dt} = C\frac{dv_0}{dt}$.
But

Fig. SE3.5 Network, for
example (SE3.5)

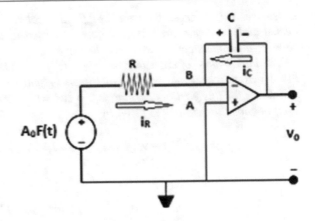

$$i_C = -i_R \ \ or \ \ C\frac{dv_0}{dt} = -\frac{A_0F(t)}{R} \quad Or \quad \frac{dv_0}{dt} = -\frac{A_0F(t)}{RC}. \qquad \text{(SE3.5.1)}$$

First-order differential equation (SE3.5.1) describes the output voltage response
of the given circuit.

If (SE3.5.1) is integrated, one gets

$$\int_0^t \frac{dv_0}{dt'}dt' = -\frac{1}{RC}\int_0^t A_0F(t)dt' \quad Or \quad v_0 = -\frac{1}{RC}\int_0^t A_0F(t)dt'. \qquad \text{(SE3.5.2)}$$

*Equation (SE3.5.2) tells that the output voltage v_0 is proportional to the integral
of the input signal.*

Solved example (SE3.6) Analyze the circuit given in Fig. SE3.6

Solution: *As is shown in the figure, there are two switches SW-1 which gets closed
at time t = 0 and the other SW-2 that flips at time t = 3 s, i.e. after 3 s of the closing
of SW-1. Further, initially at t < 0, there was no energy source in the circuit and at
t = 0 a step source 10 u(t) V gets connected to the capacitor in series with 1000 Ω*

Fig. SE3.6 Circuit, for
example (SE3.6)

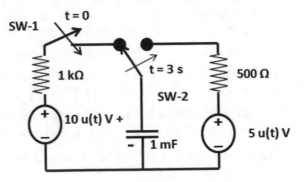

*resistance. This charges the capacitor for 3 seconds, then this source is discon-
nected and another source 5u(t) V gets connected to the capacitor and remains
connected till the steady state is reached at t = ∞.*

*The initial value of capacitor voltage $v_C(0^-) = 0$. The final value of v_C(final) if
SW-2 was not operated would have been v_C(final) = 10 V. Further, the time
constant τ for this part of charging is $\tau = RC =1000 \times 0.001 =1$ s. The response
$v_C(t)$ of capacitor voltage is given by*

$$v_C(t) = (v_c(0^-) - v_c(final))e^{-\frac{t}{\tau}} + v_c(final) = (0 - 10)e^{-t} + 10$$

or

$$v_C(t) = 10(1 - e^{-t}) \quad for \quad 3s < t > 0. \tag{SE3.6.1}$$

*We now calculate the value of capacitor voltage $v_C(3)$ at t = 3 s which will serve as
the initial value of capacitor voltage for the next part of charging when SW-2 is flipped.*

$$v_C(3) = 10(1 - e^{-3}) = 10(1 - 0.05) = 9.05V.$$

*At t =3 s, the voltage across the capacitor is 9.05 V, which serves as the initial
value of voltage for the next part. Since the circuit is allowed to remain connected
to 5 u(t) source till the steady state is reached, the final value of voltage at the
capacitor will be 5 V. It may be realized that after the closing of SW-2 the capacitor
will discharge from 9.05 V to 5 V via resistor of 500 Ω. The time constant for the
discharge will be RC = 500 x 0.001 =0.5 s. Therefore, from the general equation,*

For t > 3 s, $v_c(t > 3) = \left[(9.05 - 5)e^{-\frac{(t-3)}{0.5}} + 5\right] V = \{5 + 4.05e^{-2(t-3)}\}V.$

Self-assessment question: Draw a rough sketch of the capacitor voltage profile
from $t = 0$ to $t = 5$ s.

3.3 Second-Order Circuits

When a circuit contains two dynamic or energy storing elements, capacitor and
inductor along with resistances, it is called a second-order circuit since the circuit
response of such circuits may be represented by a second-order differential
equation.

(I) Natural response of a series RLC circuit (source free series RLC circuit)

Let us consider a source free series RLC circuit as shown in Fig. 3.8 and assume
that initial capacitor voltage be $v_C(0)$ is V_0 and initial inductor current be $i_L(0)$ is I_0.

Therefore, at $t = 0$, $v_C(0) = \frac{1}{C} \int_{-\infty}^{0} i dt = V_0$ and $i_L(0) = I_0$.

Fig. 3.8 Source free RLC circuit

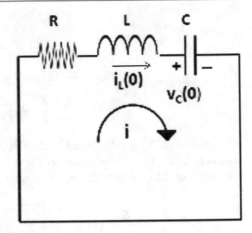

We now apply KVL to the closed loop of the circuit to get

$$Ri + L\frac{di}{dt} + \frac{1}{C}\int_{-\infty}^{0} i\,dt. \tag{3.25}$$

On differentiating the above equation with respect to time t,

$$R\frac{di}{dt} + L\frac{d^2i}{dt^2} + \frac{1}{C}i = 0 \tag{3.26}$$

$$or \quad \frac{d^2i}{dt^2} + \frac{R}{L}\frac{di}{dt} + \frac{1}{LC}i = 0. \tag{3.27}$$

Equation (3.27) is a typical second-order differential equation that describes the natural response of inductor current through a series RLC circuit.

Coming back to (3.25), we evaluate it at $t = 0$ time,

$$Ri(0) + L\frac{di(0)}{dt} + \frac{1}{C}\int_{-\infty}^{0} i(o)\,dt = 0 \Rightarrow RI_0 + L\frac{di(0)}{dt} + V_0 = 0.$$

Therefore,

$$\frac{di(0)}{dt} = -\frac{1}{L}(RI_0 + V_0). \tag{3.28}$$

It may be observed that (3.28) is a first-order differential equation that may have a solution of the type $i = Ae^{st}$. We try this solution for (3.27), and substitute this value of i in (3.27) to get

$$As^2 e^{st} + \frac{AR}{L} s e^{st} + \frac{A}{LC} e^{st} = 0$$

or

$$s^2 + \frac{R}{L}s + \frac{1}{LC} = 0. \tag{3.29}$$

Equation (3.29) is called the **Characteristic equation** because the roots of this equation decide the characteristics of current i through the series combination. The two roots of the characteristic equation may be written as

$$s_1 = -\frac{R}{L} + \sqrt{\left(\frac{R}{L}\right)^2 - 2 \cdot \frac{1}{LC}} = -\frac{R}{2L} + \sqrt{\left(\frac{R}{2L}\right)^2 - \frac{1}{LC}}$$

or

$$s_1 = -\frac{R}{2L} + \sqrt{\left(\frac{R}{2L}\right)^2 - \frac{1}{LC}} \tag{3.30}$$

and

$$s_2 = -\frac{R}{2L} - \sqrt{\left(\frac{R}{2L}\right)^2 - \frac{1}{LC}}. \tag{3.31}$$

It is customary to define $\alpha = \frac{R}{2L}$ and $\omega_0 = \frac{1}{\sqrt{LC}}$; substituting these values the two roots become

$$s_1 = -\alpha + \sqrt{\alpha^2 - \omega_0^2} \text{ and } s_2 = -\alpha - \sqrt{\alpha^2 - \omega_0^2}. \tag{3.32}$$

The two roots s_1 and s_2 are called **natural** frequencies that are measured in units of **neper per second** (Np/s). α is called the damping factor or neper frequency. $\omega_0 = \frac{1}{\sqrt{LC}}$ is the **undamped or resonant frequency**, measured in radians per second (rad/s).

Since (3.27) is a linear second-order differential equation with roots s_1 and s_2, therefore a linear combination of the two roots will also be a solution of the equation; therefore,

$$i(t) = A_1 e^{s_1 t} + A_2 e^{s_2 t}. \tag{3.33}$$

The constants A_1 and A_2 are determined from the boundary conditions, i.e. from the initial values of $i(0)$ and $di(0)/dt$.

The relative magnitudes of α and ω_o decide the four different types of solutions.

(a) **Overdamped case:** When $\alpha > \omega_0$, the solution is called overdamped.

This means that the two roots are given by

$$s_1 = -\alpha + \sqrt{\alpha^2 - \omega_0^2} = -\alpha + (quantity\ smaller\ than\ \alpha) = -ve\ real\ quantity$$
$$= -p(say)$$

$$(3.34)$$

and

$$s_2 = -\alpha - \sqrt{\alpha^2 - \omega_0^2} = -\alpha - (quantity\ smaller\ than\ \alpha) = -ve\ real\ quantity$$
$$= -q(say).$$

$$(3.35)$$

It is obvious that $|q| > |p|$
i.e. $C > \frac{4L}{R^2}$; both roots are real and negative; the response profile of current is

$$i(t) = A_1 e^{s_1 t} + A_2 e^{s_2 t} = A_1 e^{-pt} + A_2 e^{-qt}.$$

$$(3.36)$$

The values of constants A_1 and A_2 are determined from the boundary conditions of $i(0)$ and $di(0)/dt$. The current $i(t)$ will be the sum of two exponential decays. Figure 3.9 shows the time variation of the two roots and the total current.

Fig. 3.9 Time response of current $i(t)$ and the two roots

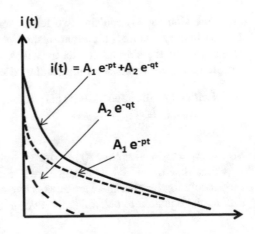

$i(t)$

$i(t) = A_1\ e^{-pt} + A_2\ e^{-qt}$

$A_2\ e^{-qt}$

$A_1\ e^{-pt}$

(b) **Underdamped case:** $\alpha < \omega_0$ In this case $\sqrt{\alpha^2 - \omega_0^2} = \sqrt{-(\omega_0^2 - \alpha^2)}$.

If now one defines

$$\omega_d = \sqrt{(\omega_0^2 - \alpha^2)} \text{ , which is a positive quantity} \tag{3.37}$$

ω_d is called the damping frequency. It may be noted that both ω_0 and ω_d are both parameters of the natural response of the series RLC circuit, ω_0 **is called the natural undamped frequency while** ω_d **is called natural damped frequency**. With these substitutions s_1 and s_2 become

$$s_1 = -\alpha + \sqrt{-1(\omega_0^2 - \alpha^2)} = -\alpha + j\omega_d$$

and $s_2 = -\alpha - \sqrt{-1(\omega_0^2 - \alpha^2)} = -\alpha - j\omega_d$.

Therefore,

$$i(t) = A_1 e^{-(\alpha - j\omega_d)t} + A_2 e^{-(\alpha + j\omega_d)t} = \left\{ A_1 e^{j\omega_d t} + A_2 e^{-j\omega_d t} \right\} e^{-\alpha t}. \tag{3.38}$$

But $e^{j\omega_d t} = \cos \omega_0 t + j \sin \omega_0 t$ and $e^{-j\omega_d t} = \cos \omega_0 t - j \sin \omega_0 t$ putting these values in (3.38) one gets

$$i(t) = \left\{ A_1 e^{j\omega_d t} + A_2 e^{-j\omega_d t} \right\} e^{-\alpha t} = \left\{ (A_1 + A_2) \cos \omega_0 t + j(A_1 - A_2) \sin \omega_0 t \right\} e^{-\alpha t}. \tag{3.39}$$

Constants $(A_1 + A_2)$ and $(A_1 - A_2)$ may be replaced, respectively, by, B_1 and B_2 to get

$$i(t) = \left\{ B_1 \cos\omega_0 t + B_2 \sin\omega_0 t \right\} e^{-\alpha t}. \tag{3.40}$$

Expression for $i(t)$ contains two terms, term $\{B_1 \cos \omega_0 t + B_2 \sin \omega_0 t\}$ exhibits the oscillatory nature of current response while term $e^{-\alpha t}$ indicates that the amplitudes of the sin and cos terms will keep decrease exponentially with time. Profile of current response in underdamped case is shown in Fig. 3.10.

(c) **Critically damped case:** This refers to the situation when $\alpha = \omega_0$, *i.e. when* $C = \frac{4L}{R^2}$.

So that $s_1 = s_2 = -\alpha$; Now $i(t) = A_1 e^{-\alpha t} + A_2 e^{-\alpha t} = (A_1 + A_2) e^{-\alpha t} = A e^{-\alpha t}$,
where $A = (A_1 + A_2)$. It is quite easy to realize that $i(t) = A e^{-\alpha t}$ cannot be a proper solution because there are always two boundary conditions in a second-order differential equation to determine the two unknown constants. Since in this special

Fig. 3.10 Current response of a series RLC circuit for the underdamped case

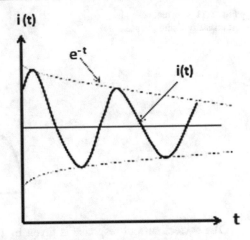

case the two roots have become identical, the current response has only one unknown constant A. It is, therefore, required to use a different method for finding the two solutions.

We go back to (3.25) $\frac{d^2 i}{dt^2} + \frac{R}{L}\frac{di}{dt} + \frac{1}{LC}i = 0$ and substitute $\alpha = \omega_0 = \frac{R}{2L}$ to get

$$\frac{d^2 i}{dt^2} + 2\alpha\frac{di}{dt} + \alpha^2 i = 0 \Rightarrow \frac{d}{dt}\left[\frac{di}{dt} + \alpha i\right] + \alpha\left[\frac{di}{dt} + \alpha i\right] = 0. \qquad (3.41)$$

Let us put $\frac{di}{dt} + \alpha i = f$ in (3.41) to get

$$\frac{df}{dt} + \alpha f = 0. \qquad (3.42)$$

Equation (3.42) is a first-order differential equation with solution $f = B_1 e^{-\alpha t} = \frac{di}{dt} + \alpha i$
or

$$\left(\frac{di}{dt} + \alpha i\right)e^{\alpha t} = B_1 \text{ which may be written as } \frac{d}{dt}(e^{\alpha t}i) = B_1. \qquad (3.43)$$

Integrating both sides of (3.43), we get

$$(e^{\alpha t}i) = B_1 t + B_2$$

or

$$i(t) = (B_1 t + B_2)e^{-\alpha t} = B_1 t e^{-\alpha t} + B_2 e^{-\alpha t}. \qquad (3.44)$$

Fig. 3.11 Current response
in critically damped case

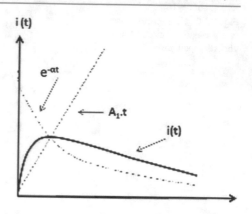

The correct current response is given by (3.44) which has two terms with two unknown constants B_1 and B_2. The first term contains $B_1 t e^{-\alpha t}$, that is constant B_1 multiplied by t times the negative exponential and the second term contains constant time negative exponential.

Term $(te^{-\alpha t})$ initially increases with the increase of t as exponent term $e^{-\alpha t}$ decreases very slowly at smaller value of t but for large values of t, $(e^{-\alpha t})$decays fast. As a result, the term $(te^{-\alpha t})$ shows a peak structure. The graph for both terms along with the current response function $i(t)$ for critically damped case is shown in Fig. 3.11.

(d) **Undamped response**:

If R is absent from the circuit, i.e. $R = 0$, $\alpha = \frac{R}{2L}$ *is also zero*.

That means $s_1 = +\sqrt{-\omega_0^2} = +j\omega_0 \, and \, s_2 = -j\omega_0$ and the solution for $i(t)$ will be

$$i(t) = A_1 e^{j\omega_0 t} + A_2 e^{-j\omega_0 t} = A_1\{\cos(\omega_0 t) + \sin(\omega_0 t)\} + A_2\{\cos(\omega_0 t) - \sin(\omega_0 t)\}$$
$$= (A_1 + A_2)\cos(\omega_0 t) + (A_1 - A_2)\sin(\omega_0 t)$$

or

$$i(t) = B_1 \cos(\omega_0 t) + B_2 \, \sin(\omega_0 t). \tag{SE3.45}$$

Here,

$$B_1 = (A_1 + A_2) \, and \, B_2 = (A_1 - A_2). \tag{SE3.46}$$

The current response given by (SE3.45) is an undamped sinusoidal response. As may be observed, there is no damping, i.e. no reduction in the amplitude of the signal with time. In other cases, where R is present, damping occurs because of the energy dissipation by the resistance. In absence of resistance, the energy stored in the circuit keeps changing from magnetic to electrostatic and oscillating between the inductor to capacitor and vice versa.

Self-assessment question: What will you infer about a circuit the response of which does not change with time?

(II)   Step voltage source response of a series RLC circuit

Figure 3.12 shows that at time $t = 0$ a step voltage of magnitude V_s is applied to the series combination. If $v_R(t), v_L(t),$ *and* $v(t)$, respectively, denote the voltage drops against R, L and c, then application of KVL yields

$$v_R(t) + v_L(t) + v(t) = V_s. \tag{3.47}$$

However, $v_R(t) = R\,i(t); \ v_L(t) = L\frac{di(t)}{dt}$ and $i(t) = C\frac{dv(t)}{dt}$ substituting these values in (3.47) yields

$$\frac{d^2v(t)}{dt^2} + \frac{R}{L}\frac{dv(t)}{dt} + \frac{1}{LC}v(t) = \frac{1}{LC}V_s. \tag{3.48}$$

The solution to (3.48) has two components:

(i)   A transient part that may be represented by $v_n(t)$ *or* $v_t(t)$. The transient component decreases with time and ultimately after some time dies out. Transient component is the solution of the source-free equation

$$\frac{d^2v(t)}{dt^2} + \frac{R}{L}\frac{dv(t)}{dt} + \frac{1}{LC}v(t) = 0. \tag{3.49}$$

Fig. 3.12 Step voltage applied to a series RLC circuit at $t = 0$

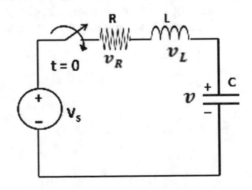

A comparison of (3.49) with (3.27) shows that the two equations are quite similar except for the fact that in (3.49) the variable is the capacitor voltage $v(t)$ while in (3.27) it is capacitor current $i(t)$. As such results obtained for transient current $i_n(t)$ in the last section may be extended for transient voltage response $v_n(t)$. Therefore, transient capacitor voltage response may also have four possibilities namely

(a) **Over damped, when both roots are real and negative**

$$v_n(t) = A_1 e^{-pt} + A_2 e^{-qt}$$ (3.49a)

(b) **Critically damped, when $\alpha = \omega_0$**

$$v_n(t) = (A_1 + A_2 t)e^{-\alpha t}$$ (3.49b)

(c) **Underdamped, when $\omega_0 > \alpha$, and roots complex**

$$v_n(t) = (B_1 cos(\omega_d t) + B_2 sin(\omega_d t))e^{-\alpha t}$$ (3.49c)

(d) **Undamped, when resistance R is not present**

$$v_n(t) = (B_1 cos(\omega_d t) + B_2 sin(\omega_d t))$$ (3.49d)

Here, $\alpha = \frac{R}{2L}$; $\omega_0 = \sqrt{\frac{1}{LC}}$; $\omega_d = \sqrt{\omega_0^2 - \alpha^2}$;

Further, constants appearing in these equations may be determined from the values of initial inductor current $I_L(0^-)$ and initial capacitor voltage $V_C(0^-)$.

(ii) The second component of the solution of (3.48) is the steady-state response and is often denoted by $v_s(\infty)$ or $v_{ss}(\infty)$. Steady-state response corresponds to the final value of the variable $v(t)$ when transient has died out and the system has stabilized. In the present case ultimately when the system will reach a steady state, the capacitor will get fully charged and the potential across it will be equal to the source voltage V_s. Again inductor will behave as short circuit and capacitor as an open circuit. No current will pass through the series combination when steady state is attained.

Therefore,

$$v_s(\infty) = V_s. \tag{3.50}$$

Hence, the complete solution of step source serial RLC circuit will be

$$v_{complete}(t) = v_n(t) + v_s(\infty). \tag{3.51}$$

In (3.51), $v_n(t)$ may be given by one of the four (3.49a)–(3.49d) and $v_s(\infty)$ by (3.50).

(iii) **Natural or source free response of a parallel RLC circuit**

A source free parallel RLC circuit is shown in Fig. 3.13. It is a property of any parallel network that the potential difference between the ends of all elements is the same but different amount of currents flow through different elements.

Let the voltage at the top nodes, A, B and C with respect to node D which is at ground potential be v and different currents, i_R, i_L and i may flow, respectively, through resistance R, inductance L and capacitance C. It is obvious that all currents and voltages in the circuit are time dependent, i.e. their value may change with time. Normally this is done by writing v as $v(t)$, etc. to indicate the time dependence of v. We are, however, not writing (t) in front of each variable to save space and ease of writing. Applying KCL at node A (or B or C), one may write

$$i_R + i_C + i = 0 \; Or \; \frac{v}{R} + \frac{i}{L}\int_{-\infty}^{0} v dt + C\frac{dv}{dt} = 0.$$

Fig. 3.13 Source free parallel RLC circuit

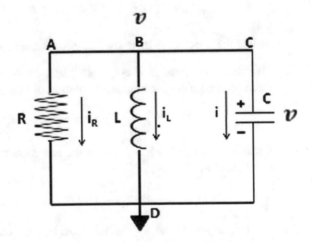

On differentiating the last equation with respect to t and dividing by C one gets

$$\frac{d^2v}{dt^2} + \frac{1}{RC}\frac{dv}{dt} + \frac{1}{LC}v = 0. \tag{3.52}$$

Equation (3.52) is a second-order equation. Shortcut method of obtaining the characteristic equation corresponding to (3.52) is to replace $\frac{d^2v}{dt^2}$ by s^2 and first derivative by s, so that the characteristic equations becomes

$$s^2 + \frac{1}{RC}s + \frac{1}{LC} = 0. \tag{3.53}$$

Roots of this characteristic equation may be given as

$$s_1 = -\frac{1}{2RC} + \sqrt{\left(\frac{1}{2RC}\right)^2 - \left(\sqrt{\frac{1}{LC}}\right)^2} \quad and$$

$$s_2 = -\frac{1}{2RC} - \sqrt{\left(\frac{1}{2RC}\right)^2 - \left(\sqrt{\frac{1}{LC}}\right)^2}. \tag{3.54}$$

Let us make the following substitutions $\beta = \frac{1}{2RC}$ and $\omega_0 = \sqrt{\frac{1}{LC}}$, and $\omega_d = \sqrt{\omega_0^2 - \beta^2}$ so that (3.54) becomes

$$s_1 = -\beta + \sqrt{(\beta)^2 - \omega_0^2} \text{ and } s_2 = -\beta - \sqrt{(\beta)^2 - (\omega_0)^2}.$$

Just like the series case there may be four possible solutions:

(a) **Overdamped case,** If $\beta > \omega_0$, s_1 and s_2 will be both real and negative numbers, and natural or transient solution of voltage against the capacitor is

$$v_n(t) = A_1 e^{-|s_1|t} + A_2 e^{-|s_2|t}. \tag{3.55a}$$

(b) **Critically damped,** when $\beta = \omega_0$, the two roots become equal and the solution becomes,

$$v_n(t) = (B_1 + B_2 t)e^{-\beta t}. \tag{3.55b}$$

(c) **Underdamped,** when $\beta < \omega_0$, the roots become complex with solution

$$v_n(t) = [B_1 \cos(\omega_d t) + B_2 \sin(\omega_d t)]e^{-\beta t}. \tag{3.55c}$$

(d) **Undamped,** when resistance is not present in the circuit, i.e. $R = \infty$ (in a parallel circuit the branch containing R should be removed or R be put equal to infinity), the roots are only imaginary and $\beta = 0$; *the solution is*;

$$v_n(t) = [B_1 \cos(\omega_d t) + B_2 \sin(\omega_d t)].$$

(e) **Evaluation of constants A_1, A_2, B_1, B_2, etc.**

Values of constants are determined from the initial conditions. In most of the problems initially at $t < 0$, the circuit is in steady state when inductance is like a short circuit and capacitor like an open circuit. From the analysis of the steady-state circuit, one can calculate the value of the current $I_L(0^-)$ through the branch of inductance and the voltage $V_C(0^-)$ across the capacitor. These calculated values are then used to find the constants. If the expression obtained is in the voltage form, for example, like (3.55), then one of the constant A_1 or A_2 (or B_1 or B_2) may be found directly by finding the value of $v_n(0)$, i.e. by putting $t = 0$ in the relevant expression and equating $v_n(0) = V_C(0)$. Suppose in this way the value of A_1 has been determined. To find a relation between the other constant A_2 and A_1, the value of $\frac{dv_n(0)}{dt}$ is determined by differentiating the relevant expression for $v_n(t)$ with respect to t and then putting $t = 0$. Now, one can correlate the value of $\frac{dv_n(0)}{dt}$ with the initial values in the following way.

In case of a parallel RLC combination, $\frac{v(t)}{R} + i_L(t) + C\frac{dv(t)}{dt} = 0$

or $\frac{dv(t)}{dt} = -\frac{1}{C}\left[\frac{v(t)}{R} + i_L(t)\right]$.

When the above equation is evaluated at $t = 0$, it gives

$$\frac{dv(0)}{dt} = -\frac{1}{C}\left[\frac{v(0)}{R} + i_L(0)\right] = -\frac{1}{C}\left[\frac{V_C(0)}{R} + I_L(0)\right],$$

$$\frac{dv_n(0)}{dt} = \frac{dv(0)}{dt} = -\frac{1}{C}\left[\frac{V_C(0)}{R} + I_L(0)\right].$$

All quantities on the right-hand side of the above equation are known, one can find out the value of the other constant also. The method will become still clearer in solved examples.

(IV) **Forced response of a parallel RLC circuit to a step current source**

Figure 3.14 shows a parallel RLC circuit connected to a constant current source of strength I_s A at time $t = 0$.

At $t > 0$, v be the potential at node A (or B or C) and $i_R, i,$ *and* i_C be the currents through resistance R, inductance L and capacitance C. Applying KCL at node A,

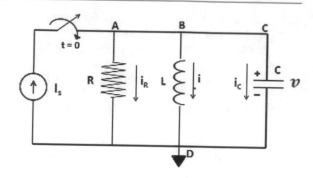

Fig. 3.14 Parallel RLC with step current source

$$i_R + i + i_C = I_s \quad Or \quad \frac{v}{R} + i + C\frac{dv}{dt} = I_s. \tag{3.56}$$

But $v = L\frac{di}{dt}$ substituting this value for v and dividing by C gives

$$\frac{d^2i}{dt^2} + \frac{1}{RC}\frac{di}{dt} + \frac{1}{LC}i = \frac{1}{LC}I_s. \tag{3.57}$$

The complete solution of (3.57) will contain two components: (I) steady-state component denoted by $i(\infty)$ and (II) the transient (or natural) component written as $i_n(t)$. The complete solution is

$$i_{complete}(t) = i(\infty) + i_n(t). \tag{3.58}$$

The steady-state component is equal to the source current that will pass through the inductor branch when steady state is reached. Therefore,

$$i(\infty) = I_s. \tag{3.59}$$

The transient component may have one of the following forms depending on the relative values of circuit components,

(a) **Overdamped case**, If $\beta > \omega_0$, s_1 and s_2 will be both real and negative numbers, and natural or transient solution of voltage against the capacitor is

$$i_n(t) = A_1 e^{-|s_1|t} + A_2 e^{-|s_2|t}. \tag{3.60a}$$

(b) **Critically damped**, when $\beta = \omega_0$, the two roots become equal and the solution becomes

$$i_n(t) = (B_1 + B_2t)e^{-\beta t}. \tag{3.60b}$$

(c) **Underdamped, when** $\beta < \omega_0$, the roots become complex with solution

$$i_n(t) = [B_1 \cos(\omega_d t) + B_2 \sin(\omega_d t)]e^{-\beta t}. \tag{3.60c}$$

(d) **Undamped,** when resistance is not present in the circuit, i.e. R $= \infty$ (in a parallel circuit the branch containing R should be removed or R be put equal to infinity), the roots are only imaginary and $\beta = 0$; *the solution is;*

$$i_n(t) = [B_1 \cos(\omega_d t) + B_2 \sin(\omega_d t)]. \tag{3.60d}$$

Adding $i(\infty) = I_S$ to the appropriate $i_n(t)$ provides the complete solution.

Self-assessment question: What will be the nature of the response if the network contains only resistances?

Solved example (SE3.7) Determine the values of $I_L(0)$, $V_c(0)$, $V_R(0)$, $I_R(0)$ and $I_R(\infty)$ for the given circuit (Fig. SE3.7).

Solution:

Initial conditions: at t < 0, circuit is in steady state so inductance is like a short circuit and capacitor in open circuit. The 5 A source is connected to the network. The total current delivered by the current source (5 A) passes through the resistance and inductance arm, therefore,

$I_l(0^-) = I_0(0)$
        $- 5A$ *(as current can not change suddenly through inductor) in the direction of arrow*

$V_C(0^-) = V_C(0) = 20\,\Omega \times 5\,A$
           $= 100V$ *(voltage cannot change suddenly across capacitor)*

Fig. SE3.7 Network, for example (SE3.7)

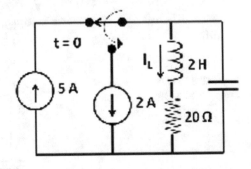

$I_R(0^-) = I_R(0) = 5A$ (*in the initial steady state inductance behaves as a short circuit*)   *in the direction of arrow*

$$V_R(0) = V_C(0) = 100V$$

$I_R(\infty) = -2A$, *in final steady state second current source of 2 A sends current in the opposite direction*

Self-assessment question: Why settling time is not very relevant for second-order systems?

Solved example (SE3.8) After a considerable period of time switch in circuit of Fig. SE3.8 got flipped at $t = 0$. Determine i(t) for t > 0.

Solution: *Initial conditions: at t < 0, the circuit was connected to 16 V battery, capacitor behaved as an open circuit and inductance as a short circuit. Total resistance across the battery is 8 Ω as inductor is like a short circuit and does not offer any resistance. Hence, current $I_L (0) = 16/8 = 2$ A.*

Since capacitor is not in circuit, one end of it is not connected, therefore, $V_L(0) = 0$ V.

*At and after $t \geq 0$, the network becomes a source free series network of RLC. The complete solution of the network will contain two parts; a component due to final steady state $i_{steady}(\infty)$ which will be given by $i_{steady}(\infty) = 0$, **as all initial energy that was stored will be dissipated by the resistance.***

The other component that is transient or natural component $i_n(t)$ may have one of the possible four types, which depends on the magnitudes of circuit elements. It may be observed that at t > 0, the circuit becomes a source free series RLC circuit. For a series circuit, the important parameters are α and ω_0. We calculate these parameters

$$\alpha = \frac{R}{2L} = \frac{8\Omega}{2\times 1H} = 4; \omega_0 = \sqrt{\frac{1}{LC}} = \sqrt{\frac{1}{1\times 1/7}} = \sqrt{7}; \text{Now } (\alpha^2 - \omega_0^2)^{1/2} = 3.$$

Since $\alpha > \omega_0$; the characteristic equation will have two real roots, both negative.

Fig. SE3.8 Network, for example (SE3.8)

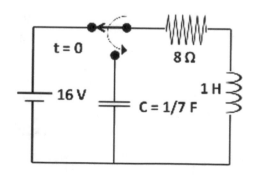

The two roots are $s_1 = -\alpha + \sqrt{(\alpha^2 - \omega_0^2)} = -4 + 3 = -1; s_2 = -\alpha - \sqrt{(\alpha^2 - \omega_0^2)} = -7$.

Therefore, this is the case of overdamped response, the transient or natural current response is given as

$$i_n(t) = A_1 e^{s_1 t} + A_2 e^{s_2 t} = A_1 e^{-t} + A_2 e^{-7t}. \qquad (SE3.8.0)$$

To find the values of constants A_1 and A_2, let us find the magnitude of current at $t = 0$;

i.e. $[i_n(0)] = A_1 + A_2$. But because current cannot change across inductance suddenly, this must be equal to inductor current determined under initial conditions as $I_L(0) = 2A$.

So,

$$A_1 + A_2 = 2. \qquad (SE3.8.1)$$

Since there are two unknown constants, one more relation is required to calculate the value of both constants. The second relation between A_1 and A_2 is obtained by differentiating $i_n(t)$ with time and evaluating the value of the derivative at $t = 0$.

$$\left[\frac{di_n(t)}{dt}\right]_{t=0} = [-A_1 - 7A_2]. \qquad (SE3.8.2)$$

Circuit at time $t > 0$ is shown in Fig. SE3.8a where voltage drops against each element are also shown. Applying KVL around the closed loop of the circuit, one gcts

$$v_c(t) + Ri_n(t) - L\frac{di_n(t)}{dt} = 0 \; Or \; \frac{di_n(t)}{dt} = \frac{v_c(t) + Ri_n(t)}{L}.$$

Evaluating the above expression at $t = 0$ gives

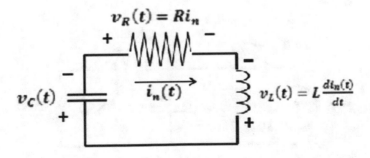

Fig. SE3.8a Network at $t > 0$

$$\left[\frac{di_n(t)}{dt}\right]_{t=0} = \frac{v_C(0) + Ri_n(0)}{L} = \frac{V_c(0) + 8 \times I_L(0)}{1} = 8 \times 2 = 16. \qquad (SE3.8.3)$$

Equations (SE3.8.2) and (SE3.8.3) give $A_1 = 5\ and\ A_2 = -3$, when these values are put in (SE3.8.0), one gets

$$i_{complete}(t) = \left(5e^{-t} - 3e^{-7t}\right)A.$$

Solved example (SE3.9) Circuit shown in the figure remained source free for a considerable period of time and then at $t = 0$ got suddenly connected to a 15 A current source. Analyze the circuit at $t = (0^-)$, $t = (0^+)$ and $t = (\infty)$. Determine the value of the rate of change of current at $t = (0^+)$ (Fig. SE3.9).

Solution: *Initially circuit was not connected to any source of energy and was in steady state. No energy was stored in the system and, therefore, all voltages and currents were zero at t = (0^-), i.e.*

$i_1(0^-) = 0, i_L(0^-) = 0, i_c(0^-) = 0, i_2(0^-) = 0,\ v_{R1}(0^-) = 0, v_{R1}(0^-) = 0, v_C(0^-) = 0;$
$v_{R2}(0^-) = 0.$

Since voltage across capacitor and current through inductor cannot change abruptly,
$v_c(0^+) = v_c(0^-)\ and\ i_L(0^+) = i_L(0^-).$ *Also, the voltage at node Y at time t will be equal to $v_c(t)$ and the current*

$i_{R1}(0^+) = \dfrac{v_c(0^+)}{R_1}$

   $= 0;$ *As already mentioned, because of the continuity of voltage across C;*

$v_C(0^+) = 0 = v_{R1}(0^+).$
Applying KCL at node X at t = (0^+), one may write

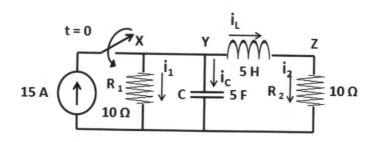

Fig. SE3.9 Network, for example (SE3.9)

$$i_{R1}(0^+) + i_C(0^+) + i_L(0^+) = 15\,A, \; since \; i_{R1}(0^+) \; \& \; i_L(0^+) \; are \; zero,$$

Therefore, $i_C(0^+) = 15A$.

Since $i_C(0^+) = C\dfrac{dv_c(0^+)}{dt}, \dfrac{dv_c(0^+)}{dt} = \dfrac{i_C(0^+)}{C} = \dfrac{15A}{5F} = 3V/s$.

Hence, $\dfrac{dv_c(0^+)}{dt} = 3V/s$. But the voltage drop across R_1 is the same as across C,
therefore,

$$\frac{dv_{R1}(0^+)}{dt} = \frac{dv_c(0^+)}{dt} = 3V/s \; and \; \frac{di_{R1}(0^+)}{dt} = \frac{\frac{dv_{R1}(0^+)}{dt}}{R_1} = \frac{3V/s}{10\Omega} = 0.3A/s.$$

So, $\dfrac{di_{R1}(0^+)}{dt} = \dfrac{di_1(0^+)}{dt} = 0.3A/s$.

At $t = \infty$ circuit will attain steady state, inductance will behave as a short circuit and capacitor as open circuit. This will leave the two resistances R_1 and R_2 in parallel connected to 15 A current source. Since the two resistances are of equal value, 8 A current will flow through each of them. The voltage across the capacitor at $t = \infty$ will be 10 Ω x 8 A =80 V.

Solved example (SE3.10) Network shown in the figure remained connected to a current source 2 $u(-t)$ A for a considerable period of time till at $t = 0$ it got connected to a step current source of 15 A. Discuss the initial conditions and derive the complete solution of the network (Fig. SE3.10).

Solution: *First about the current source 2 $u(-t)$, it is a step current source that provides current of 2 A for the period $t < 0$. For $t \geq 0$ source behaves as a short circuit and does not provide any current. Before closing of the switch, i.e. for $t < 0$, when there was steady state and capacitor was like an open circuit, current of 2 A was flowing in the series combination of $15\Omega + 20\Omega + 5\Omega$ resistances. The voltage across the capacitor at $t = (0^-)$, $V_c(0^-)$ is equal to the voltage drop across the 20 Ω resistance= $2 \times 20= 40$ V. At the same time current of 15 A was flowing through the inductance at $t = (0^-)$.*

Therefore, $V_C(0^-) = 40$ V, $I_L(0^-) = 15$ A.

At $t = 0$, when switch got flipped and current source 2 $u(-t)$ went out leaving a short circuit, the network turned into a parallel combination of LRC coupled to a step current source as shown in Fig. SE3.10a.

Because of the properties of the capacitor and inductor,

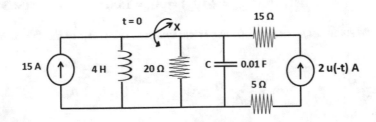

Fig. SE3.10 Network, for example (SE3.10)

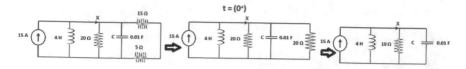

Fig. SE3.10a Network at $t = (0+)$

$$v_C(0^-) = v_C(0^+) = 40V; \text{ and } i_L(0^-) = i_L(0^+) = 15\,A. \qquad (SE3.10.1)$$

A look at the rightmost circuit of Fig. 3.10a, it may be seen that at $t = 0^+$, current through inductance is $i_L(0^+) = 15A$, as such no current is passing either through the resistance or the capacitance, i.e. $i_R(0^+) = 0 = i_C(0^+)$. But in the case of capacitor $i_C(0^+) = 0 = C\frac{dv_C(0^+)}{dt}$.

Hence,

$$\frac{dv_C(0^+)}{dt} = 0. \qquad (SE3.10.1a)$$

The complete response of the circuit will contain two terms, a steady-state term $I(\infty)$ or $V_C(\infty)$ and a transient or natural response term that may either be $i_n(t)$ or $v_n(t)$, depending on the type of analysis. It is worth noting that $i_n(t)$ refers to the inductor current, while $v_n(t)$ refers to the voltage across the capacitor.

In the analysis of a parallel RLC circuit, the nature of the transient is decided by the values of parameters $\beta = \frac{1}{2RC} = \frac{1}{2x10\Omega x0.01F} = 5; \omega_0 = \left(\frac{1}{LC}\right)^{1/2} = \left(\frac{1}{4Hx0.01F}\right)^{1/2} = 5.$

Since $\beta = \omega_0$, the two roots s_1 and s_2 are equal and response corresponds to critically damped class. The transient current or voltage response is given by

$$i_n(t) = (A_1 + A_2 t)e^{-\beta t} = (A_1 + A_2 t)e^{-5t}. \qquad (SE3.10.2)$$

To find the value of constants A_1 and A_2, we find the value of current $i_n(t)$ at $t = 0$ and equate it to the initial value of inductor current.

$$[i_n(t)]_{t=0} = I_L(0^-) = A_1 = 15A. \qquad (SE3.10.3)$$

To find A_2 we differentiate $i_n(t)$ w.r.t time and find the value of the derivative at $t = 0$.

$$\left[\frac{di_n(t)}{dt}\right]_{t=0} = \left[-5(A_1 + A_2 t)e^{-5t} + A_2 e^{-5t}\right]_{t=0} = (-5A_1 + A_2). \qquad (SE3.10.4)$$

But $L\frac{di_n(t)}{dt} = v_L(t) = v_C(t)$ *(in a parallel circuit voltage drop against each*
circuit element is same)

Therefore, $\left[L\frac{di_n(t)}{dt}\right]_{t=0} = [v_C(t)]_{t=0} = V_C(0^-) = 40V$

or, $\left[\frac{di_n(t)}{dt}\right]_{t=0} = \frac{40}{L} = \frac{40V}{4H} = (-5A_1 + A_2)$

or $-5 \times 15 + A_2 = 10$ Or $A_2 = 85$. Thus $A_1 = 15$ and $A_2 = 85$; *The transient current*
response is

$i_n(t) = (A_1 + A_2t)e^{-5t} = (15 + 85t)e^{-5t} A.$

At $t = \infty$, *steady state will reach when capacitance will work as an open circuit*
and inductance as short circuit; the inductance will short circuit all resistances in
parallel to it. As a result, total current of 15 A will pass through the short-circuited
arm of the inductance. Hence

$$I_L(\infty) = 15A.$$

The complete current response will be
$i_{complete}(t) = I_L(\infty) + i_n(t) = [15 + (15 + 85t)e^{-5t}]A.$

It is possible to write the transient circuit response in terms of $v_n(t)$, (the
capacitor voltage) also. Since the type or nature of the response does not depend on
whether it is in terms of current or voltage rather it is decided by the circuit
parameters. Therefore, the transient voltage response will also be of the type
$v_n(t) = (B_1 + B_2t)e^{-5t}$, *though the constants B_1 and B_2 may have values dif-*
ferent than those of A_1 and A_2

Again we evaluate $v_n(0^+) = B_1 = V_C(0^+) = 40V; \boldsymbol{B_1 = 40V}.$
To find B_2, we differentiate $v_n(t)$ and evaluate the derivative at $t = 0^+$
$\left[\frac{dv_n(t)}{dt}\right]_{t=0} = -5B_1 + B_2.$ *But from (SE3.10.1 a), it is known that* $\left[\frac{dv_n(t)}{dt}\right]_{t=0} = 0.$
Hence, $-5B_1 + B_2 = 0,$ *or* $B_2 = 5B_1 = 5\times40 = 200V.$
Thus, $v_n(t) = (B_1 + B_2t)e^{-5t} = (40 + 200t)e^{-5t}.$
Let us also calculate $v_c(\infty)$. *In the final steady state the inductor that will act as*
a short circuit will make the potential drop against resistance and capacitance
zero. Therefore,

$$v_c(\infty) = 0.$$

The complete response in terms of voltage may be given as
$\boldsymbol{v_{complete}(t)} = v_c(\infty) + \boldsymbol{v_n(t)} = 0 + (40 + 200t)e^{-5t} V.$

Solved example (SE.11) After remaining in off position for a considerable period
of time, switch (SW) of circuit (Fig. SE3.11) was flipped to on position at $t = 0$.
Calculate important circuit parameters including current through the capacitor
$i_C(0^+)$ and $\frac{dv(0^+)}{dt}$ at $t = (0+)$ and derive complete response for voltage v.

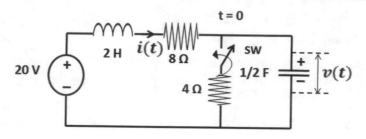

Fig. SE3.11 Network, for example (SE3.11)

Solution: *Since circuit remained in switch-off condition for a considerable period of time it has attained steady state at t =(0^-). Therefore, at t = (0^-), inductor is like a short circuit and capacitor an open circuit. Since capacitor is like an open circuit, no current is drawn from the source and* $i(0^-) = 0$. *Further, since there is no current, there is no voltage drop across 8Ω resistance and the voltage across the capacitor is equal to the source voltage, i.e.* $v(0^-) = 20V$. *The inductor current* $i_L(0^-) = 0$, *as* $i_L(0^-) = i(0^-)$.

So at = (0^-) : $i(0^-) = 0; i_L(0^-) = 0; v(0^-) = 20V$.

At t = 0, the switch is flipped on, however, currents and voltages associated with inductance and capacitance will not change immediately and will remain same at $t = 0^+$ *as they were at* $t = 0^-$. *Therefore,*

$$v(0^+) = v(0^-) = 20\, V \, and \, i_L(0^+) = i(0^-) = 0.$$

Status of the circuit at t > 0 is shown in Fig. SE3.11a. Let us concentrate at node X, Applying KCL at X one gets

$$i(0^+) = i_c(0^+) + i_R(0^+), \, but \, i(0^+) = 0, \, hence \, i_c(0^+) = -i_R(0^+).$$

Further, the voltage of node X is same as that of the capacitor, i.e. $V_X(0^+) = v(0^+) = 20V$.

Hence, $i_R(0^+) = \frac{20V}{4\Omega} = 5A; Therefore, \, i_c(0^+) = -i_R(0^+) = -5A$.

But current through capacitor is C times the rate of change of voltage across it, i.e.

$$C\frac{dv_C(0^+)}{dt} = i_C \, or \quad \frac{dv(0^+)}{dt} = \frac{dv_C(0^+)}{dt} = \frac{-5A}{C} = \frac{-5A}{0.5} = 10 \, V/s,$$

$$\frac{dv(0^+)}{dt} = 10V/s.$$

Having discussed the condition of circuit parameters at t = (0^-) and at t = (0^+) we now proceed to derive the characteristic equation for the system. It may be observed that the given circuit is neither a series circuit nor a pure parallel circuit, therefore, the characteristic equation for this circuit has to be derived from the circuit analysis. To obtain natural or transient response, the source of energy is removed and the circuit becomes one shown in Fig. SE3.11a. Applying KCL at node X,

$$i(t) = \frac{v(t)}{4} + C\frac{dv(t)}{dt} = \frac{v(t)}{4} + 0.5\frac{dv(t)}{dt}$$

or

$$i(t) = \frac{v(t)}{4} + 0.5\frac{dv(t)}{dt}. \qquad (SE3.11.1)$$

Applying KVL to the closed loop AXEFA in Fig. SE3.11a,

$$L\frac{di(t)}{dt} + 8i(t) + v(t) = 0, putting\ L = 2H\ and\ i = 5 + 0.5\frac{dv(t)}{dt}, one\ gets$$

$$2\left[\frac{1}{4}\frac{dv(t)}{dt} + 0.5\frac{d^2v(t)}{dt^2}\right] + 2v(t) + 4\frac{dv(t)}{dt} + v(t) = 0$$

or $\frac{d^2v(t)}{dt^2} + 4.5\frac{dv(t)}{dt} + 3v(t) = 0.$

To obtain the characteristic equation, replace $\frac{d^2v(t)}{dt^2}$ by s^2 and $\frac{dv(t)}{dt}$ with s. So the desired characteristic equation is,

$$s^2 + 4.5s + 3 = 0;$$

the two roots of this equation are given by

$$s_1 = -0.81\ and\ s_2 = -3.69.$$

Fig. SE3.11a Network for
natural response

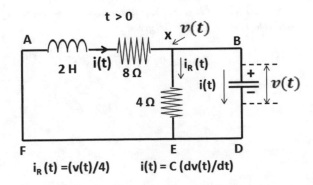

Since both roots are real and negative, this corresponds to overdamped case and the transient or natural response is

$$v_n(t) = Ae^{-0.81t} + Be^{-3.69t}.$$

After considerable period of time or at $t = \infty$, the circuit will again attain steady state. The inductance will then behave as a short circuit and capacitor as open circuit. The current drawn by the circuit will be $i(\infty) = \frac{20}{8+4} = 1.67\,A$ and the voltage drop against $4\,\Omega$ resistance $= 4x1.67 = 6.67\,V$.

The voltage across capacitor will be equal to voltage drop against $4\,\Omega$ resistance.

Hence, $v(\infty) = 6.67\,V$.

The complete response is $v_{complete}(t) = v(\infty) + v_n(t)$

$$v_{complete}(t) = 6.67V + Ae^{-0.81t} + Be^{-3.69t}. \qquad (SE3.11.2)$$

To find the value of constants A and B, we put $t = 0$ in (SE3.11.2) and equate it to $v(0) = 20V$.

So,

$$6.67 + A + B = 20; \ Or\ (A + B) = 13.33. \qquad (SE3.11.3)$$

Next, we differentiate (SE3.11.2) w.r.t to get $\frac{dv(t)}{dt}$ and then evaluate its value at $t = 0$ to get $\frac{dv(0)}{dt}$

$$\left[\frac{dv(t)}{dt}\right]_{t=0} = (-0.81A - 3.69B).$$

But

$$\left[\frac{dv(t)}{dt}\right]_{t=0} = \frac{dv(0^+)}{dt} = 10V/s = (-0.81A - 3.69B). \qquad (SE3.11.4)$$

Solving (SE3.11.3) and (SE3.11.4), one gets
$A = 20.53$ and $B = -7.2$ putting back these values of A and B in (SE3.11.2) the complete response may be given as

$$v_{complete}(t) = 6.67V + 20.53e^{-0.81t} - 7.2e^{-3.69t}.$$

3.4 Phasor Representation of Electrical Quantities

Many physical quantities like the steady-state voltages and currents, distance of the particle from mean position in simple harmonic motion, etc. vary with time like a sine function or cosine function. Such variables are called sinusoidal variables. Sinusoidal voltage and currents are often encountered in electric circuits. A sinusoidal quantity is characterized by two parameters, the amplitude or the maximum value denoted by putting subscript m to the variable and the cyclic frequency ω. Let there be two sinusoidal variables $P(t)$ and $Q(t)$ that are functions of time with amplitudes P_m and Q_m and cyclic frequency ω. These variables may be represented as

$$P(t) = P_m \sin \omega t \tag{3.61}$$

and

$$Q(t) = Q_m \sin(\omega t - \alpha)). \tag{3.62}$$

An important consideration about sinusoidal variables is that they may be represented by vertical projections of rotating vectors. For example, $P(t)$ and $Q(t)$ may be depicted by the vertical projections of vectors $\vec{P}$ and $\vec{Q}$ of lengths P_m and Q_m, rotating with cyclic frequency ω (rad/s) in anti-clock wise direction as shown in Fig. 3.15. The algebraic sum of these vectors, that will also be sinusoidal, may also be represented by a vector $\vec{R}$ of length equal to the vector sum of the lengths of vectors $\vec{P}$ and $\vec{Q}$.

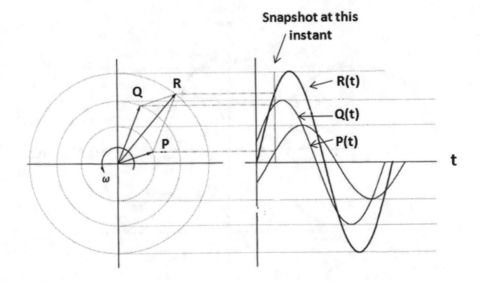

Fig. 3.15 Phasor representation of three sinusoidal variables $P(t)$, $Q(t)$ and $R(t)$

or

$$R(t) = P(t) + Q(t). \tag{3.63}$$

Rotating vectors $\vec{P}, \vec{Q}$ and $\vec{R}$ are shown in Fig. 3.15 on the left-hand side while their projections are shown on the right. Method of algebraic summing of vector lengths to obtain the resultant vector is also shown in the figure.

It is clear from the description of sinusoidal quantities given above that analysis of sinusoidal quantities can be easily done by the geometric study of rotating vectors. Snapshot of these vectors at a particular instant, i.e. rotating vectors frozen at a particular time make these studies still simpler. Rotating vectors depicting sinusoidal variables frozen at a particular instant of time are often referred as **phasors**.

3.4.1   Representation of a Sinusoidal Variable by a Phasor

Let us consider a sinusoidal voltage $v(t) = V_m \sin \omega t$, here, V_m is the amplitude or maximum value of voltage and, ω, the cyclic frequency, i.e. the rate in radians per second at which the vector corresponding to voltage $v(t)$ rotates in anti-clockwise direction. We start with the instant when the vector corresponding to voltage $v(t)$ is aligned with the X-axis, which is taken as the reference axis. With cyclic frequency ω the angle covered by the vector in time t will be, say, θ, which will be given by $\theta = \omega t$. Angle θ may also be written in terms of the linear frequency f, since

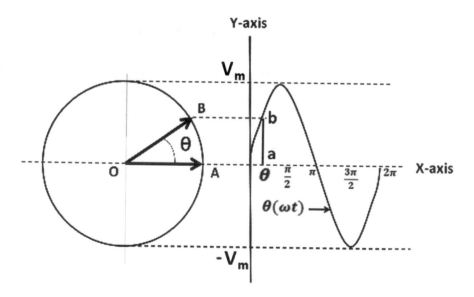

Fig. 3.16 Phasor representation of sinusoidal voltage

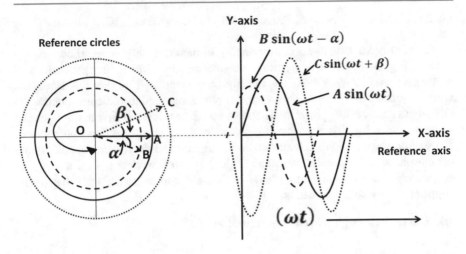

Fig. 3.17 Leading and lagging sinusoids

$\omega = 2\pi f$. Voltage vector freezed at time $t = 0$ is represented by the vector OA and shows the phasor corresponding to voltage $v(t)$ at $t = 0$ in Fig. 3.16. Same rotating vector is again freezed at $t = t$, after rotating by angle θ, and shows AB as the new position of phasor at time t.

Projection of the phasor on Y-axis gives the instantaneous value of the voltage (i.e. $V_m \sin\theta$). The projections on Y-axis are shown on the right-hand side of Fig. 3.16. As may be observed in the figure, instantaneous value of the Y-projection is 0, indicated by point 'a' at t or $\theta = 0$ and the projection has a value 'ab' at instant t. It is obvious that the projection of rotating voltage vector or freezed phasors will fall on the sinusoidal curve shown in the figure.

Figure 3.17 shows three vectors OA, OB and OC, of amplitudes A, B and C, respectively. All the three vectors are rotating in anti-clockwise direction with the same angular frequency ω (rad/s). Three concentric circles, called reference circles, with radii A, B and C may be drawn to follow the motion of the tips of these vectors. Figure 3.17 may be treated as a snapshot of these vectors at $t = 0$. If X-axis is taken as the reference axis, then vector OA is aligned with the reference axis, vector OB is below the reference axis, the angle that the initial position of OB makes with the reference axis is termed as the phase angle and in case of OB the phase angle is $(-\alpha)$. The negative sign comes because of the fact that one has to move in clockwise direction to reach OB at $t = 0$ while motion in anti-clockwise direction is assumed to be positive.

If the phasor corresponding to a sinusoid makes a negative phase angle with the reference axis, the phasor and the corresponding sinusoid are said to be lagging in phase with respect to a phasor which was aligned with reference axis at $t = 0$. In Fig. 3.17, phasor OB is lagging behind the phasor OA by phase angle α. On the other hand, phasor OC which makes a $+\beta$ angle with respect to the reference axis (as it is ahead in clockwise direction), is said to be ahead or leading phasor OA by phase angle β.

The sinusoidal waveforms corresponding to the three phasors are shown on the right in Fig. 3.17. These waveforms are obtained by plotting the projections of rotating vectors on Y-axis. As has already been mentioned, phasor OA is aligned with reference axis and, therefore, has zero phase angle. Sinusoid corresponding to OB which has negative phase angle shows up its first maxima first in the figure, at angle α before the first maxima of sinusoid of OA. Then, the first maximum of sinusoid corresponding to OA appears in the figure. And finally, the first maximum of sinusoid of OC appears at last, at an angle β after the maxima of OA. It may thus be inferred that first maxima of lagging sinusoid appear earlier than that of the aligned or the leading sinusoids.

(i) Current through a series combination of RLC

To further illustrate the concept of leading and lagging phasors, let us consider the series combination of R, L and C which is connected to a sinusoidal current source $I_0 \sin \omega t$ of maximum value I_0 and cyclic frequency ω, as shown in Fig. 3.18a. Let us further assume that the phasor corresponding to current is aligned to the reference axis X. Therefore, phasor representing current, of length I_0 is shown aligned to the X-axis in Fig. 3.18b. Let v_R, v_L and v_C denote, respectively, the voltage drops against resistance R, inductance L and capacitance C.

Then,

$$v_R = Ri = RI_0 \sin \omega t = V_R \sin \omega t, \ here V_R = RI_0. \tag{3.64}$$

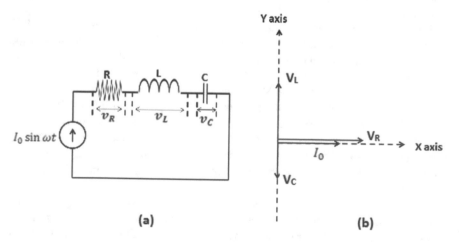

(a)   **(b)**

Fig. 3.18 **a** Series RLC circuit powered by sinusoidal current; **b** snapshot of phasors corresponding to current and voltage drops against R, L and C

Since potential drop against resistance is given by $V_R \sin \omega t$, the phasor corresponding to it will be of magnitude V_R and aligned to the reference axis. Thus both current phasor and voltage drop against resistance phasor are along the X-axis.

Voltage drop against inductance L is given as

$$v_L = L\frac{di}{dt} = L\frac{dI_0 \sin \omega t}{dt} = LI_0\omega \cos \omega t = V_L \sin(\omega t + 90°); \text{ here } V_L = LI_0\omega.$$

$$(3.65)$$

Equation (3.65) tells that the magnitude of phasor representing the voltage drop across the inductor is $V_L (= LI_0\omega)$ and that this phasor is leading the current phasor by 90°.

It is known that the current through capacitor is related to the differential of the voltage drop v_C across the capacitor by

$$c\frac{dv_c}{dt} = i, \text{ integrating both sides gives; } v_c = \frac{1}{C}\int idt = \frac{1}{C}\int I_0 \sin \omega t dt$$

$$= -\frac{I_0}{C\omega}\cos \omega t$$

or

$$v_c = V_C(-\cos \omega t) = V_C \sin(\omega t - 90°); \text{ where } V_C = \frac{I_0}{C\omega}. \qquad (3.66)$$

It may be inferred from (3.66) that the voltage drop against the capacitor and corresponding phasor of magnitude V_C lags behind the current by 90°.

Phasor diagram of current and three voltage drops is shown in Fig. 3.18b. Since phasors corresponding to current and the voltage drop against resistance are both aligned to X-axis, the V_R phasor has been shown slightly above the current phasor to keep the figure clean.

3.4.2 Representing a Phasor in Polar, Cartesian and Complex Number Forms

Phasors representing sinusoids may also be written in polar and Cartesian forms. In polar form, a phasor may be expressed as

$$Phasor\ OA = \vec{A} = A\angle 0°;$$

$$Phasor\ OB = \vec{B} = B\angle(-\alpha°)$$

$$and\quad Phasor\ OC = \vec{C} = C\angle\beta°.$$

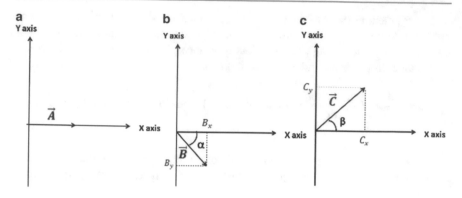

Fig. 3.19 Cartesian components of phasors A, B and C

As may be observed that, in polar form representation the magnitude of the vector followed by the phase angle that it makes with the reference axis is given. For example, if a sinusoid (or phasor) in polar form is written as $25\angle15°$, it will mean that the magnitude or the maximum value of the sinusoid is 25 (in appropriate units) and at $t = 0$, it makes an angle of $+15°$ with the reference axis.

Like any other vector, a phasor may be resolved in components in two mutually perpendicular directions. The two preferred directions are the reference axis or X-axis and the Y-axis. For example, Fig. 3.19 shows the x- and y-components of the three phasors. As expected phasor B has negative y-component. It is simple to write

$$A_X = A, \; A_y = 0; \quad B_x = B\cos\alpha, \quad B_y = -B\sin\alpha; \quad C_x = C\cos\beta,$$
$$C_y = C\sin\beta$$

and $\alpha = tan^{-1}\left(\frac{-B_y}{B_x}\right); \quad \beta = tan^{-1}\left(\frac{C_y}{C_x}\right).$

Thus, it is possible to write the x- and y- components of a phasor and if these components and phase angle are given then to write the phasor in polar form.

Phasors may also be represented as complex numbers. Complex algebra is one of the most valuable analytical tools for the study of phasors.

(i) **Complex number:**

A complex number Z may be defined as

$$z = x + jy,$$

where factor j is defined as $j^2 = -1$.

Fig. 3.20 Complex number z in complex plane

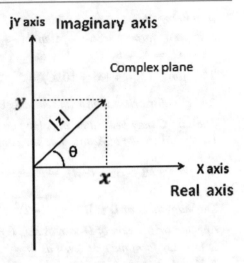

X is called the real part of z and y the imaginary part. The complex plain consists of a real axis, often taken along the X-axis and an imaginary or jy-axis, taken perpendicular to the real axis. The modulus of complex number z is given as

$$|z| = \sqrt{x^2 + y^2}.$$

Further, the angular orientation or angle θ is given as

$$\theta = tan^{-1}\left(\frac{y}{x}\right).$$

Also, $x = |z| \cos\theta$ *and* $y = |z| \sin\theta$.
Euler's theorem states that

$$e^{j\theta} = \cos\theta + j\sin\theta.$$

Therefore, $z = x + jy = |z| \cos\theta + |z|j\sin\theta = |z|e^{j\theta}$.
Figure 3.20 shows the complex number z mapped in complex plane.

Solved example (SE3.12) Two complex numbers A and B are given as $A = 4 + j5, B = 6 - j2$. Determine the values of A, B, $(A + B)$, $(A - B)$, $(A \cdot B)$, (A/B) in polar form.

Solution: *Let the sum of the two complex number be C,*

$$A = 4 + j5, |A| = \sqrt{16 + 25} = 6.40,$$
$$Angular\ orientation\ \theta_A = tan^{-1}\left(\frac{5}{4}\right) = 51.34°.$$

A in polar form is $6.40\angle51.34°$.

B in polar form: $(\sqrt{36+4})\angle tan^{-1}(-\frac{2}{6}) = 6.32\angle -18.43$.

Then, $C_x = A_x + B_x, C_y = A_y + B_y$.

Therefore, $C_x = 4 + 6 = 10;\ C_y = 5 - 2 = 3$ *and* $|C| = \sqrt{C_x^2 + C_y^2} = 10.44$.

And angular orientation with real axis $\theta = tan^{-1}\left(\frac{C_y}{C_x}\right) = tan^{-1}(0.3) = 16.7°$.

The sum C may be written in polar form as $C = 10.44 \angle16.7°$.

Let the difference of A and B be the complex number D. So,

$$D_x = A_x - B_x, D_y = A_y - B_y;\ Or \quad D_x = -2, D_y = 7.$$

The magnitude of D is $|D| = \sqrt{(-2)^2 + (7)^2} = 7.28$.

Angular orientation of D to real axis $\theta = tan^{-1}\left(-\frac{7}{2}\right) = -74°$.

D in polar form may be given as $D = 7.28\angle -74°$.

If M = A. B, then $M = (4 + j5)(6 - j2) = 24 + j30 - j8 - j^2 10 = 34 + j22$.

The magnitude of M $= \sqrt{34^2 + 22^2} = 40.49$.

Angular orientation $\theta = tan^{-1}\left(\frac{22}{34}\right) = 32.90°$.

M in polar form

$$40.49\angle32.90°. \tag{SE3.12.1}$$

Let $A/B = Q = \frac{4+j5}{(6-j2)} = \frac{(4+j5)(6+j2)}{(6-j2)(6+j2)} = \frac{14+j38}{40} = 0.35 + j0.95$.

In polar form, Q may be written as

$$Q = \sqrt{0.35^2 + 0.95^2}\, tan^{-1}\left(0\frac{0.95}{0.35}\right) = 1.01\angle69.77°. \tag{SE3.12.2}$$

Some important observations may be made here:

1. *A and B in polar notation have been calculated as*

$$A = 6.40\angle51.34° \textbf{ and } B = 6.32\angle -18.43$$

The multiplication $\textbf{A}.\textbf{B} = (6.40\angle51.34°)(6.32\angle -18.43)$

If one multiplies the magnitudes $|A|$ *and* $|B| = 6.40 \times 6.32 = 40.49$

And if one adds the angles of A and B $= 51.34 + (-18.43) = 32.90$

$$\textbf{A}.\textbf{B} = (6.40\angle51.34°)(6.32\angle -18.43) = 40.49\angle32.90. \tag{SE3.12.3}$$

If (SE3.12.1) and (SE3.12.3) are identical, from here it may be inferred that for multiplication of complex numbers it is easy to write the given numbers in polar form and multiply the modulus with each other to get the modulus of the resultant and add angles of multiplicands to get the angle of the resultant.

In short one may write $A.B = (|A|.|B|)\angle(\theta_A + \theta_B)$.

Similarly division of complex numbers is easy if they are in polar form. Dividing the modulus of the number in numerator by the modulus of the number in denominator gives the modulus of the dividend and subtracting the angle of the number in denominator from the angle of the numerator gives the angle of the final dividend.

OR $\frac{A}{B} = \frac{|A|}{|B|}(\angle(\theta_A - \theta_B))$.

2. *It may, however, be noted that for addition and subtraction of complex numbers it is better to first convert complex numbers in Cartesian form and then add or subtract the real and imaginary components separately to obtain the components of the final number.*

3.4.3 Representing Non-phasor Electrical Quantities by Complex Number

Since electrical quantities like sinusoidal current and sinusoidal voltage may be represented by phasors and phasors may be expressed as complex numbers, it is, therefore, possible to represent sinusoidal electric quantities by complex numbers. However, it is also possible to represent non-phasor electrical quantities like impedance, etc. as complex number.

Fig. 3.21 Graphic representation of the multiplication of a phasor by a complex number

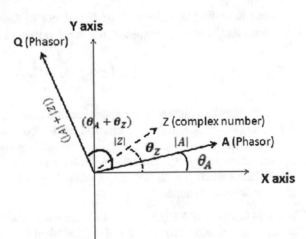

Let us consider the multiplication of a phasor $A = |A|\angle\theta_A$ and a complex number $Z = |Z|\angle\theta_Z$. Since both these quantities are in polar form, their multiplication may be done using the rule of multiplication of complex numbers. The resultant phasor Q may be written as

$$Q = (|A| + |Z|)\angle(\theta_A + \theta_Z). \tag{3.67}$$

Both (3.67) and Fig. 3.21 tell that by multiplication with a complex number, the magnitude of the phasor increases and that the angular orientation of the new phasor Q is obtained by rotating the old phasor A in anti-clockwise direction by angle θ_Z.

Carrying on this analogy further, it may be said that multiplication of any vector, phasor or complex number by j means a rotation of the quantity by 90° in anti-clockwise direction, while multiplication by $-j$ corresponds to a rotation of 90° in clockwise direction.

Let us now reconsider the case of flow of sinusoidal current through a series combination of RLC studied earlier. Figure 3.18 show phasors corresponding to different voltage drops in the circuit. It may be observed that phasor corresponding to voltage drop against inductance is 90° ahead of the phasor v_R, while phasor v_c is 90° behind. Since 90° rotation in anti-clockwise direction is equivalent to multiplication by j and 90° rotation in clockwise direction by $-j$, one may write

$$v_R = Ri, \ v_L = j\omega Li \ and \ v_C = -j\frac{1}{\omega C}i = \frac{1}{j\omega C}i.$$

If total voltage across the series circuit is v, then

$$v = Ri + j\omega Li + \frac{1}{j\omega C}i = \left(R + j\omega L + \frac{1}{j\omega C}\right)i.$$

Since the ratio $\frac{v}{i}$ is defined as the impedance of the circuit, the impedance Z of the series RLC circuit in complex number form may be given as

$$Z = \left(R + j\omega L + \frac{1}{j\omega C}\right).$$

It may thus be observed that impedance of a network may be written in complex form. Since admittance Y is reciprocal of impedance, it may also be given in complex notation. It may, however, be observed that complex values of impedance have meaning only when currents and voltages in the circuit have sinusoidal form. For DC case when $\omega = 0$, impedance offered by inductance becomes zero and that by a capacitance infinite.

Solved example (SE3.13) A series RLC circuit with R=100 Ω, L=0.5 H and C=0.01 F is fed by a sinusoidal current 10 sin(377t). Calculate impedance of each element in complex form, voltage drop against each element total and voltage drop in the circuit.

Solution:

$$Z_R = 100 = 100\Omega; \quad Z_L = j\omega L = jx377x0.5 = j188.5\Omega;$$

$$Z_C = \frac{1}{j\omega C} = -j\frac{1}{377x0.01} = -j(0.265)\Omega.$$

Total impedance

$$Z_{Total} = Z_R + Z_L + Z_C = 100 + j(188.5) - j(0.265)$$
$$= 100 + J(188.235) = 213.15\angle62.02.$$

Total voltage drop in the circuit $V = Z_{Total}.(10\ \sin(377t)) = (213.15\angle62.02)$
$(10\angle0)$

or Total voltage drop in the circuit $V = 2131.5\ \angle62.02$, $V = 2131.5\ \sin\ (377t +62.02)$.

Voltage drop against R $V_R = Ri = 100x10\ \sin(377\ t) = \boldsymbol{1000\ \sin\ (377\ t)}$.

Voltage drop against L, $V_Z = Z_L.i = j(188.5)\ 10\sin(377\ t) = \boldsymbol{(1885)\ \sin\ (377t +90)}$.

Voltage drop against C, $V_C = Z_C.i = -j(0.265)(10\sin(377t)) = 2.65\boldsymbol{sin}$
$(377t - 90)$.

3.5   Fourier Analysis

Basically Fourier analysis is based on one theorem or assumption that most of the functions may be decomposed into sum of trigonometric or exponential functions with definite frequencies and different coefficients. Fourier analysis is the study of how functions can be decomposed into characteristic trigonometric (essentially sin and cosine functions) or exponential functions. In general, electrical/electronic signals depict the time variation of voltage or current, i.e. signal $f(t)$ is a function of time. According to Fourier theorem, any signal $f(t)$ is the result of the superimposition of many sinusoidal signals of different frequencies and amplitudes. Fourier analysis provides a mathematical tool to find out the constituent sinusoids, their frequencies and amplitudes. Fourier analysis decomposes a given function $f(t)$ of time domain into corresponding function of frequency domain. A convenient example is that of a sound signal, in time domain it tells how the amplitude of the voltage changes with time; Fourier analysis of this sound signal will tell how the signal is distributed as a function of frequency. However, Fourier analysis has much wider applications in various fields including electrical/electronic systems, physics, mathematics, etc.

Broadly speaking functions can be of two types, periodic and those which are not periodic. Depending on the nature of the function, there may be two different types of Fourier expansions:

(a) **Fourier transform:** This applies to functions that are essentially non-periodic but are reasonably well behaved. Such functions can be written as a continuous integral of trigonometric or exponential function with a continuum of possible frequencies.
(b) **Fourier series:** A reasonably well-behaved periodic function may be written as discrete sum of trigonometric or exponential functions with specific frequencies.

The importance of Fourier analysis lies in two facts: (i) since most of the functions may be decomposed into trigonometric, i.e. into sinusoidal functions or exponential functions, essential properties of functions may be studied by analyzing the component sinusoidal or exponential functions. It is much simpler to analyze the component sinusoidal or exponential functions than to analyze the complicated function as a whole. (ii) Since most of the differential equations describing different phenomenon in nature are linear, solutions of component sinusoidal or exponential functions (obtained by Fourier analysis) may be added to get the final solution of the function. Thus, Fourier analysis provides a mathematical tool to study and obtain solutions of linear differential equations using the solutions of Fourier component sinusoidal or exponential functions.

In simple language, Fourier theorem says that all functions (except a very few) may be thought to have been built by the superposition of either sinusoids of discrete frequencies and of different amplitudes or of exponential functions having different exponents and amplitudes. And, therefore, the properties of parent functions may be deduced by the study of Fourier component sinusoids or exponential functions.

Since expansion of a given function into a trigonometric series is possible only if the function is periodic, that is a special case of functions, we first study periodic functions.

3.5.1   Expanding Periodic Function in Sinusoidal Series

Let us consider a function $f(x)$ that is periodic in an interval $0 \leq x \geq L$ as shown in Fig. 3.22. Fourier theorem states that $f(x)$ can be written as

$$f(x) = a_0 + \sum_{n=1}^{\infty} \left\{ a_n \cos\left(\frac{2\pi nx}{L}\right) + b_n \sin\left(\frac{2\pi nx}{L}\right) \right\} \tag{3.68}$$

or as

$$f(x) = \frac{a_0}{2} + \sum_{n=1}^{\infty} \left\{ a_n \cos\left(\frac{\pi nx}{L}\right) + b_n \sin\left(\frac{\pi nx}{L}\right) \right\}. \tag{3.68a}$$

Fig. 3.22 Function f
(t) periodic in an interval
$0 \le x \ge L$

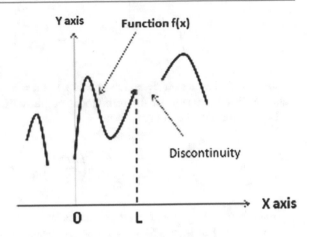

As shown in Fig. 3.22, L is the period of function $f(x)$. Two expressions for Fourier series given by (3.68) and (3.68a) are both in use by different authors; however, we shall adhere to the one given by (3.68).

The quantity with in summation sign in (3.68) has two terms, the periodicity of each of these terms depends on the value of n; if $n = 1$ the period will be L, if $n = 2$ period will be $L/2$ and so on. If odd terms are non-zero, then the maximum period will be L. Suppose coefficients a_n and b_n are zero for all odd terms, then the maximum period will be $L/2$. It is now required to determine the values of the coefficients a_n and b_n.

(a) **Determination of coefficients**

Since following integrals will be used for obtaining the values of coefficients, it is required to give the values of these integrals.

$$\int_0^L \sin\left(\frac{2\pi n x}{L}\right) \cos\left(\frac{2\pi m x}{L}\right) dx = 0, \qquad (3.69a)$$

$$\int_0^L \sin\left(\frac{2\pi n x}{L}\right) \sin\left(\frac{2\pi m x}{L} dx\right) = \frac{L}{2}\delta_{nm}, \qquad (3.69b)$$

$$\int\int_0^L \sin\left(\frac{2\pi nx}{L}\right)\cos\left(\frac{2\pi mx}{L}\,dx\right) = \frac{L}{2}\delta_{nm},\qquad (3.69c)$$

δ_{nm}, called Kronecker delta, has a value 1 if $n = m$, otherwise it is zero. Above equations are the results of the orthogonal property of sine and cosine terms.

Let us consider the integral,

$$\int_0^L f(x)dx = a_0\int_0^L dx + \sum_{n=1}^{\infty}a_n\int_0^L \sin\left(\frac{2\pi nx}{L}\right)dx + b_n\int_0^L \cos\left(\frac{2\pi nx}{L}\right)dx. \quad (3.70)$$

Since both $\sin\left(\frac{2\pi nx}{L}\right)$ and $\cos\left(\frac{2\pi nx}{L}\right)$ terms will undergo even number of oscillations in interval of length L, these terms will contribute zero. Equation (3.70) will then yield

$$\int_0^L f(x)dx = a_0\int_0^L dx = a_0 L.$$

And, therefore,

$$a_0 = \frac{1}{L}\int_0^L f(x)dx. \qquad (3.71)$$

To determine a_n, let us consider the integral

$$\int_0^L f(x)\cos\left(\frac{2\pi nx}{L}\right)dx.$$

And substitute the value of $f(x)$ in terms of the series expansion given by (3.68). Using the orthogonal property of integral of (3.69a) all sine terms will become zero and all cosine terms will also be zero except when n=m, in which case the integral will give the value $L/2$.

Hence,

$$\int_0^L f(x)\cos\left(\frac{2\pi nx}{L}\right)dx = a_m\frac{L}{2}$$

or

$$a_m = a_n = \frac{2}{L} \int_0^L f(x) \cos\left(\frac{2\pi nx}{L}\right) dx. \tag{3.72}$$

Similarly, it may be shown that

$$b_m = b_n = \frac{2}{L} \int_0^L f(x) \sin\left(\frac{2\pi nx}{L}\right) dx. \tag{3.73}$$

Solved example (SE3.14) Find the Fourier series for function f(x) given as

$$f(x) = \begin{cases} A, & 0 \le x \le L \\ 0, & L < x \le 2L \end{cases}$$

Solution: *Fourier series for a periodic function f(x) is given as*

$$f(x) = a_0 + \sum_{n=1}^{\infty} \left\{ a_n \cos\left(\frac{2\pi nx}{L}\right) + b_n \sin\left(\frac{2\pi nx}{L}\right) \right\},$$

where $a_0 = \frac{1}{L} \int_0^L f(x) dx$; $a_n = \frac{2}{L} \int_0^L f(x) \cos\left(\frac{2\pi nx}{L}\right) dx$ and $b_n = \frac{2}{L} \int_0^L f(x) \sin\left(\frac{2\pi nx}{L}\right) dx$.

Hence,

$$a_0 = \frac{1}{L} \int_0^L f(x) dx = \frac{1}{L} \int_0^L A dx = \frac{A}{L} x \Big|_0^L = A. \tag{SE3.14.1}$$

Also, $a_n = \frac{2}{L} \int_0^L f(x) \cos\left(\frac{2\pi nx}{L}\right) dx = \frac{2}{L} \int_0^L A\cos\left(\frac{2\pi nx}{L}\right) dx = \frac{2A}{L} \int_0^L \cos\left(\frac{2\pi nx}{L}\right) dx$

or $a_n = \frac{2A}{L} \int_0^L \cos\left(\frac{2\pi nx}{L}\right) dx = \frac{2A}{L} \left[\frac{L}{2\pi n} \sin\left(\frac{2\pi nx}{L}\right)\right]\Big|_0^L = \frac{A}{\pi n}[\sin 2\pi n - \sin 0] = 0$

or

$$a_n = 0. \tag{SE3.14.2}$$

Further, $b_n = \frac{2}{L} \int_0^L f(x) \sin\left(\frac{2\pi nx}{L}\right) dx = \frac{2}{L} \int_0^L A\sin\left(\frac{2\pi nx}{L}\right) dx = \frac{2A}{L} \int_0^L \sin\left(\frac{2\pi nx}{L}\right) dx$

or $b_n = \frac{2A}{L} \int_0^L \sin\left(\frac{2\pi nx}{L}\right) dx = \frac{2A}{L} \left[-\frac{L}{2\pi n} \cos\left(\frac{2\pi nx}{L}\right)\right]\Big|_0^L = \frac{A}{\pi n}[\cos 0 - \cos(2\pi n)]$

or $b_n = \frac{A}{\pi n}[\cos 0 - \cos(2\pi n)] = \frac{A}{\pi n}[1 - (-1)^n] = \frac{A}{\pi n}\left[1 + (-1)^{n+1}\right].$

It may be noticed that if n is even, i.e. n = 2k, where k = 1, 2, 3 ....

$$b_{2k} = \frac{A}{\pi(2k)}\left[1 + (-1)^{2k+1}\right] = 0.$$

And for the case when n is odd, i.e. n = 2k + 1, k = 1, 2, 3,........

$$b_{2k+1} = \frac{A}{\pi(2k+1)}\left[1 + (-1)^{2k+1+1}\right] = \frac{A}{\pi(2k+1)}\,[2],$$

$$b_{2k+1} = \frac{2A}{\pi(2k+1)}. \qquad\qquad (SE3.14.3)$$

Therefore, Fourier series expansion of the function is given as

$$f(x) = a_0 + \sum_{n=1}^{\infty}\left\{ a_n \cos\left(\frac{2\pi n x}{L}\right) + b_n \sin\left(\frac{2\pi n x}{L}\right)\right\}$$

$$= A + \sum_{k=1}^{\infty}\left\{\frac{2A}{\pi(2k+1)}\sin\left(\frac{2(2k+1)x\pi}{L}\right)\right\}$$

or

$$f(x) = A + \sum_{k=1}^{\infty}\left\{\frac{2A}{\pi(2k+1)}\sin\left(\frac{2(2k+1)\pi x}{L}\right)\right\}. \qquad (SE3.14.4)$$

Equation (SE3.14.4) may be written as

$$f(x) = A + \frac{2A}{\pi}\left[\frac{1}{3}\sin\frac{6\pi x}{L} + \frac{1}{5}\sin\frac{10 x\pi}{L} + \frac{1}{7}\sin\frac{14 x\pi}{L} + \frac{1}{9}\sin\frac{18 x\pi}{L} + \ldots\ldots\right].$$

The series within bracket is a converging series and approximation to f(x) will become better with the inclusion of more terms.

3.5.2 Expanding Periodic Function in Fourier Exponential Series

Since any function that may be written in terms of sines and cosines can also be expressed as an exponential, using Euler's theorem. As such it is always possible from Fourier analysis to express any periodic function as a sum of exponential functions, as given below

$$f(x) = \sum_{-\infty}^{+\infty} C_n e^{\left(\frac{i2\pi nx}{L}\right)}. \tag{3.74}$$

It may be observed that summation in (3.74) extends from $-\infty$ to $+\infty$. The coefficients C_n are given by

$$C_n = \frac{1}{L}\int_0^L f(x)e^{-\left(\frac{i2\pi nx}{L}\right)}dx. \tag{3.75}$$

Above relation may be obtained using the orthogonal property

$$\int_0^L e^{\left(\frac{i2\pi nx}{L}\right)}e^{-\left(\frac{i2\pi mx}{L}\right)}dx = L\delta_{mn}. \tag{3.76}$$

Solved example (SE3.15) Find Fourier exponential series for function $f(x)$ given as

$$f(x) = \begin{cases} A, & 0 \le x \le L \\ 0, & L < x \le 2L \end{cases}$$

Solution: *The exponential Fourier series expansion for function $f(x)$ is given as*

$$f(x) = \sum_{-\infty}^{+\infty} C_n e^{\left(\frac{i2\pi nx}{L}\right)}.$$

The coefficient $C_n = \frac{1}{L}\int_0^L f(x)e^{-\left(\frac{2\pi nx}{L}\right)}dx = \frac{1}{L}\int_0^L Ae^{-\left(\frac{2\pi nx}{L}\right)}dx = \frac{A}{L}\left[\frac{L}{2\pi in}e^{-\left(\frac{2\pi nx}{L}\right)}\right]\Big|_0^L$

or $C_n = \frac{A}{L}\left[\frac{L}{2\pi in}e^{-\left(\frac{2\pi nx}{L}\right)}\right]\Big|_0^L = \frac{A}{2\pi in}[e^{-2\pi in}-1] = \frac{A}{2\pi in}[(\cos 2\pi n - i\sin 2\pi n)-1]$

or $C_n = \frac{A}{2\pi in}[(\cos 2\pi n - i\sin 2\pi n)-1] = -\frac{Ai}{2\pi n}[(-1)^n - 1].$
If $n = 2k$, $k = 1, 2, 3........$

$$C_{2k} = -\frac{Ai}{2\pi 2k}\left[(-1)^{2k}-1\right] = 0.$$

If $n = 2k - 1$, $k=1,2,3 ....$

$$C_{2k-1} = -\frac{Ai}{2\pi(2k-1)}\left[(-1)^{2k-1}-1\right] = +\frac{Ai}{\pi(2k-1)},$$

$$f(x) = \sum_{-\infty}^{+\infty} C_n e^{\left(\frac{i2\pi nx}{L}\right)} = \sum_{-\infty}^{\infty} \frac{Ai}{\pi(2k-1)} e^{\left(\frac{i2\pi nx}{L}\right)}.$$

Solved example (SE3.16) Find the Fourier exponential series for the rectangular wave given by

$$f(x) = \begin{cases} 4 & 0 < x < 1 \\ -4 & 1 < x < 2 \\ f(x+2) = f(x) \end{cases}$$

Solution: *The given function is a rectangular wave of period x = 2*

$$C_n = \frac{1}{L} \int_0^L f(x) e^{-\left(\frac{i2\pi nx}{L}\right)} dx = \frac{1}{2}[\int_{-1}^0 (-4) e^{-\left(\frac{i2\pi nx}{2}\right)} dx + \int_0^1 (4) e^{-\left(\frac{i2\pi nx}{2}\right)} dx$$

or $C_n = \frac{1}{2}[\int_{-1}^0 (-4) e^{-(i\pi nx)} dx + \int_0^1 (-4) e^{-(i\pi nx)} dx] = \frac{1}{2}\left[\left\{\frac{4}{i\pi n} e^{-i\pi nx}\Big|_{-1}^0\right\} - \left\{\frac{4}{i\pi n} e^{-i\pi nx}\Big|_0^1\right\}\right]$

or $C_n = \frac{4}{i\pi n}(1 - (-1)^n)$
   or If n is even then $C_n = 0$.
   If n is odd, i.e. n=2k-1, k=1,2,3........
   Then $C_n = -\frac{8i}{(2k-1)\pi}$.
   And $f(x) = -\frac{8i}{\pi} \sum_{k=-\infty}^{\infty} \frac{1}{2k-1} e^{i(2k-1)\pi x}$.

3.5.3 Fourier Transform and Inverse Transform

So far Fourier expansion of only periodic functions has been considered. Is it possible to expand functions that are non-periodic? The answer is yes, if a little mathematical mischief is done. It is to assume that the period L of non-periodic function is infinite (∞), it is perfectly legitimate to assume that any function repeats itself after an infinite distance or time, as the case may be. With this assumption in place, it is possible to use all results obtained earlier for the case of periodic functions and then to extend them for the case $\lim_{L \to \infty}$ (*Result of periodic analysis*).

Let us start with expression for the expansion of a periodic function into exponential series, (3.74),

$$f(x) = \sum_{-\infty}^{+\infty} C_n e^{\left(\frac{i2\pi nx}{L}\right)}, \tag{3.77}$$

where

$$C_n = \frac{1}{L} \int_{-L/2}^{L/2} f(x) e^{-\left(\frac{i2\pi nx}{L}\right)} dx. \tag{3.78}$$

Let us make a substitution,

$$k_n = 2\pi n/L. \tag{3.79}$$

The difference between the two successive values of k_n may be obtained by differentiating (3.79)

$$dk_n = \frac{2\pi}{L} dn, \text{ but } dn = 1 \text{ as } n \text{ is an integer number.} \tag{3.80}$$

Hence,

$$dk_n = \frac{2\pi}{L}. \tag{3.81}$$

Since we will be interested in L going to the limit $L \rightarrow \infty$, dk_n *under the limit* $L \rightarrow \infty$ will be a very small quantity and, therefore, k_n may be treated as a continuous variable. Equation (3.77) may be written as

$$f(x) = \sum_{-\infty}^{+\infty} C_n e^{\left(\frac{i2\pi nx}{L}\right)} - \sum_{-\infty}^{+\infty} C_n e^{(ik_n x)}(dn). \tag{3.82}$$

RHS of (3.82) has been multiplied by (dn) as $dn = 1'$ but from (3.80) $dn = \frac{L}{2\pi} dk_n$, substituting this back in (3.82) one gets

$$f(x) = \sum_{-\infty}^{+\infty} C_n e^{(ik_n x)}(dn) = \frac{L}{2\pi} \sum_{-\infty}^{+\infty} C_n e^{(ik_n x)} dk_n.$$

Since dk_n is very small, this sum may be replaced by integration,

$$f(x) = \frac{L}{2\pi} \int_{n=-\infty}^{\infty} C_n e^{(ik_n x)} dk_n = \int_{n=-\infty}^{\infty} \left(C_n \frac{L}{2\pi}\right) e^{(ik_n x)} dk_n = \int_{-\infty}^{\infty} C(ik_n) e^{(ik_n x)} dk_n, \tag{3.83}$$

where

$$C(ik_n) \equiv \left(C_n \frac{L}{2\pi} \right).$$
(3.84)

Now from (3.78),

$C_n = \frac{1}{L} \int\limits_{-L/2}^{L/2} f(x) e^{-\left(\frac{i2\pi nx}{L} \right)} dx$, which in the limit $L \rightarrow \infty$ becomes

$C_n = \frac{1}{L} \int\limits_{-\infty}^{\infty} f(x) e^{-ik_n x} dx$, substituting this value of C_n in (3.84) gives

$C(ik_n) \equiv \left(C_n \frac{L}{2\pi} \right) = \frac{1}{2\pi} \int\limits_{-\infty}^{\infty} f(x) e^{-ik_n x} dx$, dropping the subscript n gives

$$C(ik) = \frac{1}{2\pi} \int\limits_{-\infty}^{\infty} f(x) e^{-ikx} dx.$$

Corresponding to (3.77) and (3.78), one may write

$$f(x) = \int\limits_{-\infty}^{+\infty} C(ik) e^{ikx} dk \ and \ C(ik) = \frac{1}{2\pi} \int\limits_{-\infty}^{\infty} f(x) e^{-ikx} dx$$
(3.85)

$C(ik) = \frac{1}{2\pi} \int\limits_{-\infty}^{\infty} f(x) e^{-ikx} dx$ **is called the Fourier transform of $f(x)$.**

And $f(x) = \int\limits_{-\infty}^{+\infty} C(ik) e^{ikx} dk$ **is called the reverse Fourier transform of C(k).**

Note: Depending on how $C(ik)$ is defined (3.85) may also be written as

$$f(x) = \frac{1}{\sqrt{2\pi}} \int\limits_{-\infty}^{+\infty} C(ik) e^{ikx} dk \ and \ C(ik) = \frac{1}{\sqrt{2\pi}} \int\limits_{-\infty}^{\infty} f(x) e^{-ikx} dx.$$

Different authors have used different notations to denote Fourier transform, in analogy with Laplace transform, Fourier transform of function $f(t)$ is often represented as $F(ik)$, which we are denoting by $C(ik)$. Further, like Laplace transformation, the operation of getting Fourier transform of a function is often represented by curly letter $\mathcal{F}$, i.e. one may write

$\mathcal{F}[f(x)] = C(ik) \ or \ F(ik) = \frac{1}{\sqrt{2\pi}} \int\limits_{-\infty}^{\infty} f(x) e^{-ikx} dx$ and $\mathcal{F}^{-1}[C(k)] or \mathcal{F}^{-1}[F(ik)] = f(x)$.

Since every function can be split up into even and odd parts $E(x)$ and $O(x)$, so that

$$f(x) = \frac{1}{2}\left[f(x) + f(-x)\right] + \frac{1}{2}\left[f(x) - f(-x)\right] = E(x) + O(x). \tag{3.86}$$

Taking Fourier transform of above equation gives

$$\mathcal{F}[f(x)]$$

$$= \int\limits_{-\infty}^{\infty} E(x)\cos(2\pi kx)\mathrm{d}x - iE(x)\sin(2\pi kx)\mathrm{d}x + O(x)\cos(2\pi kx)\mathrm{d}x - iO(x)\sin(2\pi kx)\mathrm{d}x.$$

$$\tag{3.87}$$

Since cosine is an even function and sine is an odd function, therefore, E(x) sin (2πkx) and O(x) cos (2πkx) terms will be overall odd terms and under integration give zero. The remaining terms may be written as

$$\mathcal{F}[f(x)] = \int\limits_{-\infty}^{\infty} E(x)\cos(2\pi kx)\mathrm{d}x - i \int\limits_{-\infty}^{\infty} O(x)\sin(2\pi kx)\mathrm{d}x. \tag{3.88}$$

Thus, Fourier transform of any function may be given in cosine and sine terms.

3.5.4 3.5.4 Properties of Fourier Transform

In the following important properties of Fourier transform are listed without giving proof for them.

- **Linearity:** $\mathcal{F}[ax(t) + by(t)] = a\mathcal{F}[x(t)] + b\mathcal{F}[y(t)]$
- **Time shift:** $\mathcal{F}[f(t \mp t_0)] = \mathcal{F}[f(t)]e^{\mp ikt_0}$
- **Frequency shift:** $\mathcal{F}^{-1}[C(k \mp \omega_0)] = f(t)e^{\mp i\omega_0 t}$
- **Time reversal:** $\mathcal{F}[f(-t)] = C(-k)$
- **Even and odd signals:** *If signal f(t) is an even (or odd) function of t (time), its Fourier transform C(ik) (also called frequency spectrum) is an even (or odd) function of k (often called frequency).*

  If $f(t) = f(-t)$, then $C(ik) = C(-ik)$ or if $f(t) = -f(-t)$, then $C(ik) = -C(-ik)$

- **Time and frequency scaling:** $\mathcal{F}[f(at)] = \frac{1}{a}C\left(\frac{k}{a}\right)$ *or* $\mathcal{F}[af(at)] = C\left(\frac{k}{a}\right)$
- **Complex conjugation:** If $\mathcal{F}[f(t)] = C(ik)$, *then* $\mathcal{F}[f^*(t)] = C^*(-ik)$
- **Symmetry:** if $\mathcal{F}[f(t)] = C(ik)$, *then* $\mathcal{F}[C(t)] = 2\pi f(-ik)$
- **Multiplication theorem:** $\int\limits_{-\infty}^{\infty} f(t)y^*(t) = \frac{1}{2\pi}\int\limits_{-\infty}^{\infty} C_f(ik)C_y^*(ik)dk$
- **Time derivative:** $\mathcal{F}\left[\frac{\mathrm{d}}{\mathrm{d}t}f(t)\right] = ikC(ik)$
- **Frequency derivative:** $\mathcal{F}[tf(t)] = i\frac{d}{dk}C(ik)$.

Table. 3.1 Fourier transform pairs

Function	$f(t)$	$C(ik) = F[f(t)]$
Fourier transform-1	1	$\delta(k)$
Fourier transform-cosine	$\cos(2\pi k_0 t)$	$\frac{1}{2}[\delta(k-k_0) + \delta(k+k_0)]$
Fourier transform-sine	$\sin(2\pi k_0 t)$	$\frac{1}{2}i[\delta(k+k_0) - \delta(k-k_0)]$
Fourier transform-exponential function	$e^{-2\pi k_0|t|}$	$\frac{1}{\pi}\frac{k_0}{(k^2+k_0^2)}$
Fourier transform-Gaussian	e^{-at^2}	$\sqrt{\frac{\pi}{a}}e^{-\frac{\pi^2 k^2}{a}}$
Fourier transform-delta function	$\delta(t-t_0)$	$e^{-2\pi ikt_0}$

Table 3.1 **gives some common Fourier transform pairs.**

3.5.5 Real, Imaginary, Even and Odd Functions and Fourier Transforms

There exist relationships between the nature of time-domain function $f(t)$ and its frequency-domain Fourier transforms $C(ik)$ or $F(\omega)$. The following chart provides a shortcut to find the nature of the frequency-domain Fourier transform if the nature of time-domain function is known or vice versa (FIg. 3.23).

Reading of the chart is simple, for example, if signal $f(t)$ is imaginary and odd, its Fourier transform will be real and odd.

3.5.6 Rectangular Pulse Function and Periodic Function

A rectangular signal may be a pulse signal or it may be a periodic signal as shown in (a) and (b) parts of Fig. 3.24. At the moment, we are interested in rectangular pulse signal or function of Fig. 3.24a.

A rectangular pulse function in short may be written as A rect $\left(\frac{t}{\tau}\right)$. The important parameters of the pulse function are A the amplitude and τ the period of the

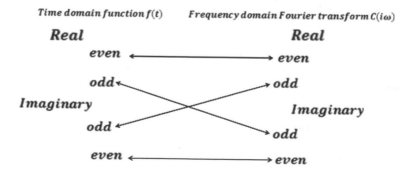

Fig. 3.23 Relation between the natures of time-domain signal and its frequency-domain Fourier transform

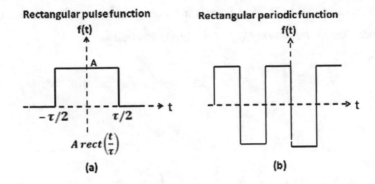

Rectangular pulse function

Rectangular periodic function

(a)

(b)

Fig. 3.24 **a** Rectangular pulse function **b** rectangular periodic function

Fig. 3.25 Function sin $c(z)$

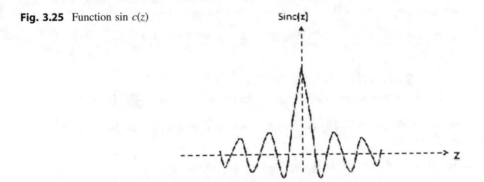

function. The function is symmetrical about the Y-axis and in mathematical terms may be written as

$$Arect\left(\frac{t}{\tau}\right) = \begin{cases} A & -\frac{\tau}{2} \leq t \leq \tau/2 \\ 0 & for\ all\ other\ values\ of\ t \end{cases}.$$

The Fourier transform of $f(t) = Arect\left(\frac{t}{\tau}\right)$ is given as

$\mathcal{F}\left[Arect\left(\frac{t}{\tau}\right)\right] = \frac{2A}{\omega}\sin\left(\frac{\omega\tau}{2}\right) = \frac{A}{\pi f}\sin(\pi f \tau)$, where $\omega = 2\pi f$.

Function $\frac{\sin(\pi f \tau)}{\pi f}$ is called the **sinc** function of (πf). sinc(z) is defined as **sinc(z) =** $\frac{\sin(z)}{z}$ and has the shape as shown in Fig. 3.25.

Solved example (SE3.17) Find Fourier transform of $f(x) = e^{-bx}u(x)$, where b>0.

Solution: *Fourier transform $C(k)$ of the given function is*

$$C(k) = \frac{1}{2\pi} \int_{-\infty}^{\infty} f(x)e^{-ikx}dx = \frac{1}{2\pi} \int_{-\infty}^{\infty} \left[e^{-bx}u(x)\right]e^{-ikx}dx$$

or

$$C(k) = \frac{1}{2\pi} \int_{-\infty}^{0} e^{-(b+ik)x}u(x) + \frac{1}{2\pi} \int_{0}^{\infty} e^{-(b+ik)x}u(x). \qquad (SE3.17.1)$$

First integral on the RHS is zero because function u(x) is zero for x<0 and u(x) =1 for x>0

Hence, $C(k) = \frac{1}{2\pi} \int_{0}^{\infty} e^{-(b+ik)x}u(x) = \frac{1}{2\pi} \int_{0}^{\infty} e^{-(b+ik)x} = \frac{1}{2\pi} \left\{ \frac{-1}{(b+ik)} e^{-(b+ik)x} \right\} \Big|_{0}^{\infty}$

or $C(k) = \frac{-1}{2\pi(b+ik)} [0-1] = \frac{1}{2\pi(b+ik)} = \frac{(b-ik)}{2\pi(b^2+k^2)}.$

The Fourier transform of the given function is $C(k) = \frac{(b-ik)}{2\pi(b^2+k^2)}.$

Solved example (SE3.18) Determine Fourier transform for the function shown in Fig. SE3.18.

Solution: *The function in the figure may be written as*

$$f(x) = A - \frac{T_1}{2} \le x \le \frac{T_1}{2}.$$

From the definition of Fourier transform (3.85),

$$C(k) = \frac{1}{2\pi} \int_{-\infty}^{\infty} f(x)e^{-ikx}dx = \frac{1}{2\pi} \int_{-T_1/2}^{T_1/2} Ae^{-ikx}dx = \frac{A}{2\pi} \int_{-T_1/2}^{T_1/2} e^{-ikx}dx$$

Fig. SE3.18 Function, for example (SE3.18)

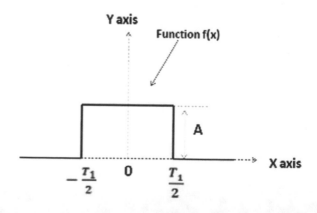

or

$$C(k) = \frac{A}{2\pi} \int\limits_{-T_1/2}^{T_1/2} e^{-ikx}\,dx = \frac{Ai}{2\pi k}\left[e^{-ikx}\right]\Big|_{-T_1/2}^{T_1/2}$$

$$= \frac{Ai}{2\pi k}\left[\left\{\cos\left(k\frac{T_1}{2}\right) - i\sin\left(k\frac{T_1}{2}\right)\right)\right\} - \left\{\cos\left(-k\frac{T_1}{2}\right) - i\sin\left(-k\frac{T_1}{2}\right)\right)\right\}\right]$$

$$= \frac{Ai}{2\pi k}\left[-2i\sin\frac{kT_1}{2}\right] = \frac{A}{\pi}\left[\frac{\sin\frac{kT_1}{2}}{k}\right].$$

The required Fourier transform $C(k) = \frac{A}{\pi}\left[\frac{\sin\frac{kT_1}{2}}{k}\right].$

Solved example (SE3.19)   Drive expression for the Fourier transform of the derivative of function $f(t)$.

Solution: *Let us consider*

$$\mathcal{F}\left[\frac{d}{dt}f(t)\right] = \int\limits_{-\infty}^{\infty}\left\{\frac{d}{dt}f(t)\right\}e^{-2i\pi kt}\,dt. \qquad (SE3.19.1)$$

Let us make the substitution, $du = \left\{\frac{d}{dt}f(t)\right\}$ *and* $v = e^{-2i\pi kt}$
so that $u = f(t)$ *and* $dv = -2i\pi k\, e^{-2i\pi kt}\,dt.$
So that (SE3.19.1) becomes $\mathcal{F}\left[\frac{d}{dt}f(t)\right] = \int\limits_{-\infty}^{\infty} v\,du.$

Now,

$$\int\limits_{-\infty}^{\infty} v\,du = (uv)\Big|_{-\infty}^{\infty} - \int\limits_{-\infty}^{\infty} u\,dv = \left[f(t)e^{-2i\pi kt}\right]\Big|_{-\infty}^{\infty} - \int\limits_{-\infty}^{\infty} f(t)\left(-2i\pi k e^{-2i\pi k}\,dt\right).$$

$$(SE3.19.2)$$

However, $\left[f(t)e^{-2i\pi kt}\right]\Big|_{-\infty}^{\infty} = 0$ *as it is the multiplication of a bounded function f
(t) and an oscillating function* $e^{-2i\pi kt}.$

Therefore, $\mathcal{F}\left[\frac{d}{dt}f(t)\right] = \int\limits_{-\infty}^{\infty} v\,du = -\int\limits_{-\infty}^{\infty} f(t)\left(-2i\pi k e^{-2i\pi k}\,dt\right)$

or $\mathcal{F}\left[\frac{d}{dt}f(t)\right] = 2i\pi k\left[\int\limits_{-\infty}^{\infty} f(t)\left(e^{-2i\pi k}\,dx\right)\right] = 2i\pi k\mathcal{F}[f(t)].$

Finally, $\mathcal{F}\left[\frac{d}{dt}f(t)\right] = 2i\pi k\mathcal{F}[f(t)].$

Solved example (SE3.20) Show that

$$\mathcal{F}[\cos(2\pi k_o t)f(t)] = \frac{1}{2}[C(k - k_0) + C(k + k_0)],$$

where $\mathcal{F}[f(t)] = C(ik)$.

Solution: $\mathcal{F}[\cos(2\pi k_o t)f(t)] = \int\limits_{-\infty}^{\infty} f(t)\cos(2\pi k_0 t)e^{-2\pi i k t}dt.$

But $(e^{i2\pi k_0 t} + e^{-i2\pi k_0 t}) = 2\cos(2\pi k_0 t);$

$$putting the value of \cos(2\pi k_0 t) = \frac{1}{2}(e^{i2\pi k_0 t} + e^{-i2\pi k_0 t})$$

$$\mathcal{F}[\cos(2\pi k_o t)f(t)] = \frac{1}{2}\int\limits_{-\infty}^{\infty} f(t)e^{2\pi i k_0 t}e^{-2\pi i k t}dt + \frac{1}{2}\int\limits_{-\infty}^{\infty} f(t)e^{-2\pi i k_0 t}e^{-2\pi i k t}dt$$

$$= \frac{1}{2}\int\limits_{-\infty}^{\infty} f(t)e^{-2\pi i(k - k_0)t}dt + \frac{1}{2}\int\limits_{-\infty}^{\infty} f(t)e^{-2\pi i(k_0 + k)t}dt$$

or $\mathcal{F}[\cos(2\pi k_o t)f(t)] = \frac{1}{2}(\mathcal{F}(k - k_0)) + \frac{1}{2}(\mathcal{F}(k + k_0)).$

Solved example (SE3.21) Derive Fourier transform for unit impulse $f(t) = \delta(t)$.

Solution: $\mathcal{F}[\delta(t)] = \int\limits_{-\infty}^{\infty} \delta(t)e^{-2i\pi k t}dt$

Since $\delta(t)$ has non-zero value only at $t = 0$, the integral on the RHS has to be evaluated only at $t = 0$, when $t = 0$ is put in the exponential term it gives 1, therefore,

$$\mathcal{F}[\delta(t)] = \left[\delta(t)e^{-2i\pi k t}\right]\big|_{t=0} = 1$$

or $\mathcal{F}[\delta(t)] = 1.$

Solved example (SE3.22) Find Fourier transform of $\sin k_0 t$

Solution: $\mathcal{F}\{\sin(k_0 t)\} = \int\limits_{-\infty}^{\infty} \sin(k_0 t)e^{-ikt}dt$

Fig. SE3.22 Fourier
transform of $\sin(k_0 t)$

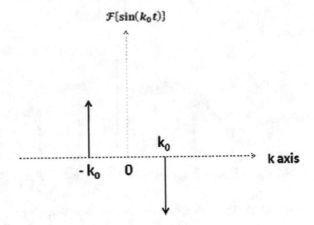

But $\sin(k_0 t) = \frac{1}{2i}\left[e^{ik_0 t} - e^{-ik_0 t}\right]$ *substituting this in the above equation,*

$$\mathcal{F}\{\sin(k_0 t)\} = \int\limits_{-\infty}^{\infty} \sin(k_0 t)e^{-ikt}\,dt = \frac{1}{2i}\int\limits_{-\infty}^{\infty} e^{ik_0 t}e^{-ikt}\,dt - \frac{1}{2i}\int\limits_{-\infty}^{\infty} e^{-ik_0 t}e^{-ikt}\,dt$$

or $\mathcal{F}\{\sin(k_0 t)\} = \frac{1}{2i}\int\limits_{-\infty}^{\infty} e^{-i(k-k_0)t}\,dt - \frac{1}{2i}\int\limits_{-\infty}^{\infty} e^{-i(k+k_0)t}\,dt$

$$= \frac{2\pi}{2i}\delta(k - k_0) - \frac{2\pi}{2i}\delta(k + k_0)$$

or $\mathcal{F}\{\sin(k_0 t)\} = -i\pi\delta(k - k_0) + i\pi\delta(k + k_0).$
   Figure SE3.22 shows Fourier transform of sin $(k_0 t)$.

Solved example (SE3.23) A periodic signal f(x) is given as $f(x) = \sum\limits_{n=-\infty}^{\infty} A_n e^{ink_0 x}$,
write function $f(t)$ as Fourier series and draw a rough sketch of $\mathcal{F}[f(x)]$.

Solution:

$$\mathcal{F}[f(x)] = \sum\limits_{n=-\infty}^{\infty} A_n \int\limits_{-\infty}^{\infty} e^{i(nk_0 - k)x}\,dx = 2\pi \sum\limits_{n=-\infty}^{\infty} A_n \delta(k - nk_0).$$

Graph for $\mathcal{F}[f(x)]$ *is shown in Fig. SE3.23.*

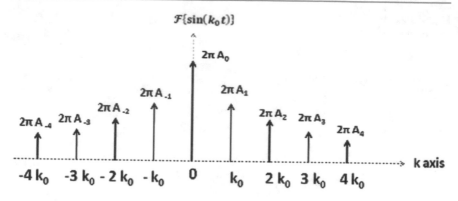

Fig. SE3.23 Distribution of Fourier transform for function $f(x)$

Solved example (SE3.24)   Obtain Fourier transform for signal $f(t) = e^{-(a|t|)} (a > 0)$ and draw rough sketches for the signal and its Fourier transform.

Solution:

$$\mathcal{F}\left[e^{-(a|t|)}\right] = \int_{-\infty}^{\infty} e^{-(a|t|)} e^{-j\omega t} dt = \int_{-\infty}^{0} e^{-a(-t)} e^{-j\omega t} dt + \int_{0}^{\infty} e^{-a(+t)} e^{-j\omega t} dt$$

$$= \frac{1}{(a - j\omega)} \left[e^{(a-j\omega)t}\right]\Big|_{-\infty}^{0} + \frac{1}{(a+j\omega)} \left[e^{-(a+j\omega)t}\right]\Big|_{0}^{\infty}$$

or $\mathcal{F}\left[e^{-(a|t|)}\right] = \frac{2a}{a^2 + \omega^2}$.

Sketches of signal $f(t)$ and its Fourier transform are shown in Fig. SE3.24.

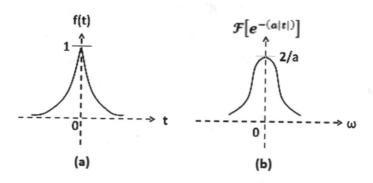

Fig. SE3.24 a Signal $f(t)$; **b** Fourier transform of $f(t)$

Problems

P3.1 Response of a first-order system is described by the differential equation,

$5\frac{dx}{dt} + 10x = 5cost - 15sint$.

Obtain the value of the system response $x(t)$ under the boundary condition $x(0) = 1$.

Answer: $[x(t) = e^{-2t} + 5cost - 15sint]$.

P3.2 Derive first-order differential equation describing the response of output voltage v_0 and show that the circuit works as a differentiator of the input signal (Fig. P3.2).

P3.3 Network given in Fig. P3.3 is in steady state for $t < 1$ s. Derive an expression for the inductor current i_L for $t > 1$ s.

Answer: $i_L(t) = 0.4e^{-200(t-1)} - 0.2]A$.

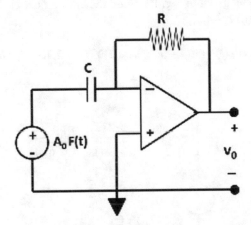

Fig. P3.2 Network for problem (P3.2)

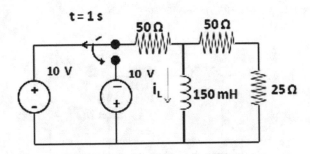

Fig. P3.3 Network for problem (P3.3)

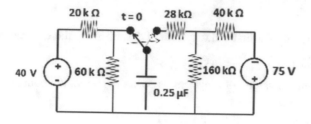

Fig. P3.4 Network for problem (P3.4)

P3.4   Drive expressions for the current $i(t)$ through 28 kΩ resistance and the voltage $v_C(t)$ across the capacitor (Fig. P3.4).

Answer: $[i(t) = -2.25e^{-100t}mA; v_C(t) = -60 + 90e^{-100t}V]$.

P3.5   For the circuit given in Fig. P3.5 find the inductor current $i_L(t)$ (i) for $4 > t > 0$ and for (ii) $t > 4$ s

Answer: $[(i)i_L(t) = 4(1 - e^{-2t})A; (ii)i_C(t) = 2.73 + 1.27e^{-1.47(t-4)}A]$.

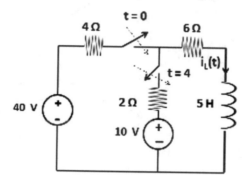

Fig. P3.5 Circuit for problem (P3.5)

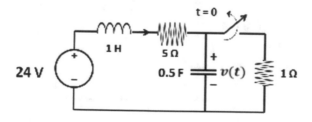

Fig. P3.6 Circuit for problem (P3.6)

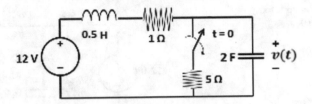

Fig. P3.7 Circuit for problem (P3.7)

P3.6 After a considerable period of time the switch of circuit of Fig. P3.6 got flipped to off at $t = 0$. Find the complete voltage response of the circuit

   Answer: $[v_{complete}(t) = 24 + 1.33e^{-4t} - 21.33e^{-t}$.

P3.7 For the circuit of Fig. P3.7 find the values of $v_c(t)$ and $i_c(t)$ for $t > 0$

   Answer: $\left[v_c(t) = 12 - (t+2)e^{-t}; i(t) = C\dfrac{dv_c(t)}{dt} = 2(t+1)e^{-t}\right].$

P3.8 After a considerable time the switch of Fig. P3.8 was flipped to connect the 20 A current source in the circuit. Find $i_L(0^-)$; $i_L(\infty)$; $v_C(0^+)$ and $v_R(0^+)$.

   Answer: $[i_L(0^-) = 10\ A; i_L(\infty) = -20A; v_C(0^+) = 400\ V$ and $v_R(0^+) = 400]$.

P3.9 The inductor current in a circuit is given as $\dfrac{d^2 i_L}{dt^2} + \dfrac{1}{2}\dfrac{di_L}{dt} + i_l = 0$. Determine the general response of the inductor current.

   Answer: $[i_L(t) = (A\cos(0.97t) + B\sin(0.97t))e^{-\frac{t}{4}}]$.

   ote since initial conditions are not given A and B are undermined

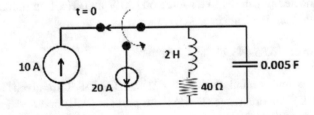

Fig. P3.8 Circuit for problem (P3.8)

Fig. P3.10 Network for
problem (P3.10)

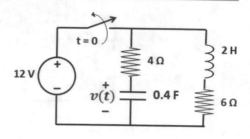

Fig. P3.11 Network for
problem (P3.11)

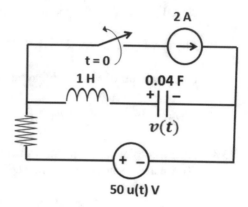

P3.10   For the circuit of Fig. P3.10 determine $i(0^+); v(0^+); i(\infty); \frac{di(0^+)}{dt}; \frac{dv(0^+)}{dt}$

Answer: $[i(0^+) = 2A; v(0^+) = 12V; i(\infty) = 0; v(\infty) = 0; \frac{di(0^+)}{dt} = -4\frac{A}{s}; \frac{dv(0^+)}{dt} = -5V/s].$

P3.11   For the circuit of Fig. P3.11 determine $v(t)$ for $t > 0$

Answer: $[\{50 + [(-62cos4t - 46.5sin4t)e^{-3t}]\}V].$

P3.12   Resistance of 20 Ω, Inductance of 0.04 H and Capacitor of 300 μF are
connected in parallel to a source of 100 V a 60 Hz. Determine the current
dawn from the source.

Answer: $[6.85\angle43.11°].$

P3.13   Two complex numbers $A = 4 - j2$ and $B = 3 + j1$ are given, find the sum
$(A+ B)$, the difference $(B - A)$, the product $(A.B)$ and division B/A for
these numbers.

Answer: $[(7 - j1), (-1 + j3), (14 - j2), (0.5 + j0.5)].$

P3.14   Use phasor technique to evaluate the expression and find its value at $t = 10$ ms.

$$i(t) = \frac{\mathrm{d}}{\mathrm{d}t}\{\cos(100t - 30°)\} + 150\cos(100t - 45°) + 500\sin(100t)$$

Answer: [204.088].

P3.15   A phasor is given as $(-2 + j4)$. Draw the phasor in jY-X plane and write it in polar form.

Answer: [4.47 $\angle116.56°$].

P3.16   Show that Fourier transform of Gaussian function $f(t) = e^{-at^2}$ is $\sqrt{\frac{\pi}{a}}e^{-\left(\frac{\pi^2 k}{a}\right)}$

P3.17   Show that $\mathcal{F}[tf(t)] = i\frac{d}{dk}C(ik)$, where $C(ik) = \mathcal{F}[f(t)]$

P3.18   State and proof the property of time reversal of Fourier transform.

P3.19   Find Fourier trigonometric series for the periodic saw-tooth signal defined as

$$f(t) = At \; for \; -\frac{L}{2} < t < +\frac{L}{2}, of \; period \; L$$

Answer: $\left[\frac{AL}{\pi}\left\{\sin\left(\frac{2\pi t}{L}\right) - \frac{1}{2}\sin\left(\frac{4\pi t}{L}\right) + \frac{1}{3}\sin\left(\frac{6\pi t}{L}\right) - \ldots\right\}\right]$.

Multiple-Choice Questions

Note: *In some of the multiple-choice questions more than one alternative are correct. Pick all correct alternatives for the complete answer.*

MC3.1 When the roots of characteristic equation are real and equal, the response will be

(a) Critically damped
(b) underdamped
(c) overdamped
(d) undamped

ANS: [(a)].

MC3.2 When the roots of characteristic equation are real and unequal, the response will be

(a) Critically damped
(b) underdamped
(c) overdamped
(d) undamped

   ANS: [(c)].

MC3.3 When the roots of characteristic equation are complex conjugate, the response will be

(a) Critically damped
(b) underdamped
(c) overdamped
(d) undamped

   ANS: [(b)].

MC3.4 After a considerable time the switch in Fig. MC3.4 got closed at $t = 0$. The correct equation(s) giving current in the circuit is(are)

$$10i + 2\frac{di}{dt} + 2\int_0^t idt = 50t > 0$$

(a) $10\,i = 0 \; t < 0$

(b) $10i - 2\frac{di}{dt} + 2\int_0^t idt = 0 \; t < 0$

(c) $10i + 2\frac{di}{dt} + 0.5\int_0^t idt = 50 \; t > 0$

Fig. MC3.4 Circuit for
question (MC3.4)

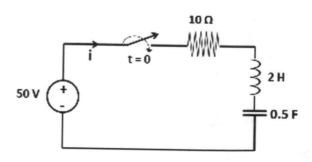

ANS: [(a), (b)].

MC3.5 The transient response of a second-order network is the sum of two real exponentials, the system is

(a) Critically damped
(b) underdamped
(c) overdamped
(d) undamped

ANS: [(c)].

MC3.6 Damping effect in complete response of a second-order circuit is essentially due to

(a) Inductance L onl
(b) Capacitance onl
(c) Resistance onl
(d) C and L both

ANS: [(c)].

MC3.7 If transient response of a second-order network does not decay with time, it means

(a) L = C
(b) L = 0
(c) C = 0
(d) R = 0

ANS: [(d)].

MC3.8 Characteristic equation for natural response of a RC system is

(a) $s + \frac{1}{RC} = 0$
(b) $s^2 + s + \frac{1}{RC} = 0$
(c) $s + RC = 0$
(d) $s^2 + s + RC = 0$

ANS: [(a)].

MC3.9 When roots of characteristic equation are complex, the response is

(a) Critically damped
(b) underdamped
(c) overdamped
(d) undamped

ANS: [(b)].

MC3.10 In RC circuit transient current is given by

(a) $\frac{V}{R}e^{(t/RC)}$
(b) $\frac{V}{R}e^{-(t/RC)}$
(c) $\frac{V}{R}+e^{(t/RC)}$
(d) $\frac{V}{R}-e^{(t/RC)}$

ANS: [(b)].

MC3.11 After a considerable period of time the switch in Fig. MC3.11 flipped to on at $t = 0$. The current through resistance at time $t = 0$ is

(a) 0 A
(b) 1 A
(c) 1 A
(d) 3 A

ANS: [(d)].

MC3.12 Transient current in circuit of MC3.11 is

(a) $1e^{-t}$

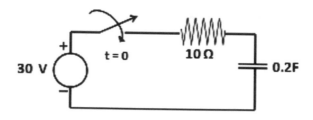

Fig. MC3.11 Circuit for (MC3.11)

(b) $2e^{-2t}$

(c) $3e^{-2t}$

(d) $3e^{-t}$

ANS: [(c)].

MC3.13 Time constants for RL and RC circuits are, respectively,

(a) RL, RC

(b) R/L , R/C

(c) L/R, C/R

(d) L/R, RC/1

ANS: [(d)].

MC3.14 1% settling time for a circuit is 1 second. Time constant of the circuit is

(a) 0.2 s

(b) 0.1 s

(c) 5 s

(d) 0.5 s

ANS: [(a)].

MC3.15 The phase angle in series RLC circuit is

(a) $tan^{-1}\frac{(Z_C - Z_L)}{R}$

(b) $tan^{-1}\frac{(Z_L - Z_C)}{R}$

(c) $tan^{-1}\frac{(Z_C - R)}{Z_L}$

(d) $tan^{-1}\frac{(R - Z_L)}{Z_C}$

ANS: [(b)].

MC3.16 The polar form of phasor $(4 + j\,3)$ is

(a) $25\angle 36.87°$

(b) $25\angle 53.13°$

(c) $5\angle 36.87°$

(d) $5\angle 53.13°$

ANS: [(c)].

MC3.17 The reference axis in case of phasor representation represents

(a) Time
(b) voltage
(c) current
(d) phase angle

ANS: [(a), (d)].

MC3.18 Product of two complex numbers $Z_1 = (1 + j)$ and $Z_2 = (1 - j)$ is

(a) $\sqrt{2}\angle 45°$
(b) $\sqrt{2}\angle -45°$
(c) $2\angle 45°$
(d) $2\angle 0°$

ANS: [(d)].

MC3.19 Which of the following is an even function of x?

(a) $\sin 2x + 3x$
(b) $x^3 + 4$
(c) $x^2 - 4x$
(d) x^2

ANS: [(d)].

MC3.20 Which of the following define a periodic function?

(a) $f(t+T) = f(t)$
(b) $f(t+T) = -f(t)$
(c) $f(t+T) = f(t) + T$
(d) $T = \pi$

ANS: [(a)].

MC3.21 A pulse of 1 V amplitude and 0.1 s duration is applied to an un-energized series RLC circuit having $R = 1\ \Omega$, $L = 2$ H and $C = 1/10$ F. Current through the circuit (in Ampere) at $t = 0$ and $t = \infty$ will, respectively, be

(a) $\infty; 0$
(b) $0; \infty$
(c) $0; 0$
(d) $\infty; \infty$

ANS: [(c)].

MC3.22 Fourier series for function $f(t) = \sin^2(t)$ is

(a) $0.5 + 0.5 sin2t$
(b) $0.5 - 0.5 sin2t$
(c) $0.5 + 0.5 cos2t$
(d) $0.5 - 0.5 cos2t$

ANS: [(d)].

MC3.23 Fourier series of a signal of time domain will correspond to

(a) Time domain
(b) both time and frequency domains
(c) Frequency domain only
(d) will depend on the function, it may be in time domain or frequency domain

ANS: [(c)].

MC3.24 If $G(f)$ represents the Fourier transform of signal $g(t)$ which is real and odd symmetric in time then $G(F)$ is

(a) complex
(b) imaginar
(c) real
(c) real and non-negative

ANS: [(b)].

MC3.25 The Fourier transform $C(i\omega)$ of time-domain signal $f(t)$ is imaginary and even symmetric in frequency domain, then $f(t)$ is

(a) Real and odd
(b) Real and even
(c) Imaginary and odd
(d) Imaginary and even

ANS: [(d)].

MC3.26 For a function g(t), it is given that $\int_{-\infty}^{\infty} g(t)e^{-i\omega t}dt = \omega e^{-2\omega^2}$ for any real

value of ω. If $y(t) = \int_{-\infty}^{t} g(\tau)d\tau$, then $\int_{-\infty}^{\infty} y(\tau)d\tau$ is,

(a) 0
(b) $-i$
(c) $-i/2$
(d) $i/2$

ANS: [(b)].

MC3.27 Which of the following figure corresponds to $f(t) = Rect(t - 1/2)$?

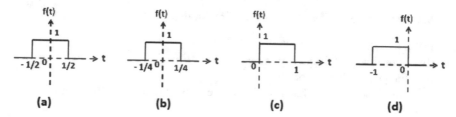

(a)   (b)   (c)   (d)

ANS: [(c)].

Short Answer Questions

SA3.1   Differential equation describing a second-order circuit is 5
$\frac{d^2 v(t)}{dt^2} + 20\frac{dv(t)}{dt} + \frac{7}{8}v(t) = 15$, obtain the characteristic equation and
exponential terms in transient.

SA3.2   What element causes damping in transient response of a circuit? Show
that by removing the damping component response becomes undamped.

SA3.3   Transient response of inductor current in a parallel RLC circuit is given
as
$$i_L^{transient}(t) = Ae^{-3t} + Be^{-7t}$$

If $I_L(0^+)$ and $V_C(0^+)$, the inductor current and capacitor voltage at $t = 0^+$ are known, how values of A and B can be calculated?

SA3.4   Discuss nature of transient response for series RLC circuit
SA3.5   Explain different types of transient responses of parallel RLC circuit
SA3.6   What is a phasor and how is it different from a vector?

SA3.7   what happens when an electrical quantity is multiplied by +J?

SA3.8   Multiplication of a phasor by a complex number amounts to the increase in the magnitude and a rotation of the phasor. Explain.

SA3.9   Which form of representation is better for taking product of phasors and why?

SA3.10   Show that Fourier transform of pulsed function $f(t) = A$ rect $\left(\frac{t}{\tau}\right)$ is $\frac{2A}{\omega}\sin(\omega\tau/2)$

SA3.11   If Fourier transform of $f(t) = A$ rect $\left(\frac{t}{\tau}\right)$ is $\frac{2A}{\omega}\sin(\omega\tau/2)$, use time shift property to obtain Fourier transform for function $f(t) = \text{rect}(t - 1/2)$

SA3.12   Fourier transform of a function $f(t)$ is $2\sin(f_1 t) + 5\sin(f_2 t) + 3\sin(f_3 t)$. Draw the function graph in frequency domain.

SA3.13   What is Fourier transform? Discuss its significance.

SA3.14   What is Fourier transform? Compare it with Laplace transform and discuss its significance

Sample Answer To SA3.14

What is Fourier transform? Compare it with Laplace transform and discuss its significance

Answer:

Jean-Baptiste Joseph Fourier a French mathematician and physicist in 1807 formulated the function known as Fourier transform. The Fourier transform of a signal f(t) is another function F(ω) defined as

$$F(\omega) = \mathcal{F}[f(t)] = \int_{-\infty}^{\infty} f(t)e^{-j\omega t}\,\mathrm{d}t \qquad (1)$$

Here, F(ω) is a function of real variable ω but in general the function value F(ω) is complex number. The signal f(t) is often termed as a function of time domain while F(ω) is called a function of frequency domain. Though f(t) tells how the amplitude of the signal is distributed with time, Fourier transform tells how the function amplitude is distributed with the constituent frequencies. Equation (1) may be written down in terms of sinusoids using Euler theorem as

$$F(\omega) = \int_{-\infty}^{\infty} f(t)\cos(\omega t)\,\mathrm{d}t - j\int_{-\infty}^{\infty} f(t)\sin(\omega t)\,\mathrm{d}t$$

Modulus |F(ω)| gives the amplitude spectrum of signal f(t), while the angle $\angle F(\omega)$ gives the phase spectrum of signal f(t).

There are some points of similarities between Laplace transform and Fourier transform.

The Laplace transform of signal f(t) is given as

$$\mathcal{L}[f(t)] = \int\limits_{0}^{\infty} f(t)e^{-st}dt \tag{2}$$

Comparison of (1) and (2) tells that Laplace transform integral is over 0 to ∞; the Fourier transform integration goes from −∞to∞. Further, in Laplace transform s can be any complex number in the region of convergence, while Fourier transform lies over jω, only over imaginary axis. If f(t) = 0 for t < 0, then Fourier transform is the Laplace transform evaluated on the imaginary axis. However, if f(t) ≠ 0 for t <0, then the Fourier and Laplace transforms may be very different.

According to Fourier theorem, any periodic function may be expended into either a trigonometric series or an exponential series. These series, called Fourier series, provide means to decompose a given periodic time-domain signal into corresponding frequency domain. Since Fourier series are converging series, it is sufficient to retain only few initial terms of the series to represent the signal. Since most of the physical processes can be written in terms of linear differential equations, the solutions of the trigonometric or exponential Fourier series may be added to get the general solution of the differential equation describing physical processes, without analyzing the complicated parent signal f(t). Thus, Fourier series go a long way in solving physical problems.

The problem arises if the function is non-periodic, in that case one cannot directly apply the results obtained for periodic signals. Fourier, however, showed that non-periodic signals may also be treated as periodic if it is assumed that there period is infinite, and therefore, results obtained for periodic signals may be borrowed for non-periodic signals in the limit of signal period L going to infinity. As expected, in the limit L → ∞, summation is replaced by integration. This resulted in expression given by equation (1) for Fourier transform for both periodic and non-periodic signals.

Long Answer Questions

LA3.1   What is meant by steady state and transient response of a circuit? In what way the complete response of a first-order circuit differs from that of a second-order circuit?. Discuss the complete response of a LR circuit.

LA3.2   In what respect a first-order circuit differs from a second-order circuit? Discuss how complete response of a series RLC circuit may be obtained.

LA3.3   With the help of a suitable example discuss how initial conditions are used to determine the unknown constants of complete response of second-order circuit.

LA3.4   Drive expression for complete response of a first-order RC circuit

LA3.5   What is meant by the time constant and how is it related to the 1%settling time? Derive expression for time constant of a RL circuit.

LA3.6   In what respect a phasor is different from a vector? Explain how the anti-clockwise rotation of a phasor may be converted into corresponding sinusoidal wave. Three sinusoids A, B and C are represented as

$$A = A_0 \sin(\omega t + 30°); B = B_0 \cos(\omega t) \ and \ C = C_0 \cos(\omega t + 30°)$$

Draw sketches for these sinusoids and indicate which is the most leading of the three.

LA3.7   The snapshot of a phasor of amplitude 5 V at time $t = 0$ makes an angle of 30° with the reference axis. Another phasor that also represents a sinusoidal voltage has amplitude of 7 V and makes an angle of 45° at $t = 0$. Write both phasors in polar and Cartesian components. Obtain expressions for the sum, differences, product and division of the two phasors. Do you think that sum, difference, products, etc. will change with the rotation of the phasor?

LA3.8   What is meant by Fourier analysis of a function? Give Fourier trigonometric expansion of a periodic function and determine the value of constants. How results of periodic functions may be extended to the case of non-periodic functions?

Electrical Properties of Materials

4

Abstract

Solid materials, on the basis of their electrical and thermal properties, may be classified as conductor, semiconductor and insulator. Electron band theory provides a basis to understand the difference between the three classes of solids. Fundamentals of band theory and further classification of semiconductors are detailed in this chapter. All modern electronic elements like junction diodes, bipolar junction transistors (BJTs) and field effect transistors (FETs), etc. use semiconductor material. It is, therefore, required to understand the physics underlying the working of these elements. Physics of semiconductors is discussed in sufficient details in the chapter.

4.1 Introduction

A material may possess several intensive properties that do not depend on the amount of the material. These quantitative properties are often used as a metric by which the advantages of one material over the other can be compared for material selection for a specific purpose. The material properties may be classified into various groups, like Acoustical properties, Atomic properties, Chemical properties, Electrical properties, Magnetic properties and many more. In the following, we shall talk about the electrical properties of materials that are important from the point of view of the physics of Electrical cum Electronic Engineering materials and devices.

© The Author(s), under exclusive license to Springer Nature Switzerland AG 2021   289
R. Prasad, *Analog and Digital Electronic Circuits*, Undergraduate Lecture
Notes in Physics, https://doi.org/10.1007/978-3-030-65129-9_4

4.2   Electrical Properties and Classification of Materials

The first and the most important electrical property of a material is its *'resistivity'*, (or specific resistance) generally denoted by Greek letter 'ρ' (rho). Resistivity is defined as the resistance offered by a unit cube (a block of 1 m × 1 m × 1 m) of the material between its opposite faces. The MKS unit of resistivity is 'ohm-metre' written as 'Ω-m' in short. The resistance R of a block of a material of length L and uniform area of cross section A may be written as (Fig. 4.1)

$$R(\Omega) = \rho(\Omega - m)\frac{L(m)}{A(m^2)}$$
$$\textbf{or}\quad \rho(\Omega - m) = R(\Omega)\frac{A(m^2)}{L(m)} \tag{4.1}$$

Though both the resistance and the resistivity are measures of the opposition to the flow of electric current through the material, resistivity is intrinsic property (does not depend on the amount of the material and has a fixed value for a given material at a given temperature) while resistance is an extrinsic property that depends also on the shape and size of the material.

The reciprocal of resistivity is called 'conductivity' or specific conductance and is represented by Greek letter σ (sigma). The SI units of conductivity are 'Siemens per metre' written as (S/m) in short.

$$\sigma\left(\frac{S}{m}\right) = \frac{1}{\rho(\Omega - m)}$$

Conductivity is a measure of the ease with which electric current can flow through the material. Both resistivity and conductivity of a given material depend on the temperature. The variation of resistivity with temperature T K (Kelvin) to a good approximation may be represented by the equation,

$$\rho_T = \rho_0(1 + KT)$$

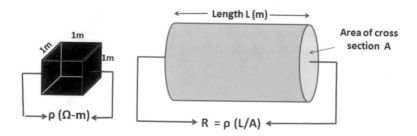

Fig. 4.1 Resistivity and resistance

Here, ρ_T and ρ_0 are, respectively, the magnitudes of the resistivities at absolute zero and absolute temperature T, while K is called the temperature coefficient of resistivity. The units of K are (absolute temperature K)$^{-1}$.

Table 4.1 lists the value of resistivity (ρ), conductivity (σ) and coefficient of temperature (K) of some materials important for electrical/electronic devices.

A close look at Table 4.1 reveals the following facts:

1. Materials/elements at the top of the table have very small value of resistivity of the order of 10^{-7} (Ω-m) and consequently large magnitude of conductivity $\sim 10^7$ (S/m). All these materials have positive small value of temperature coefficient. This means that these elements/materials offer little opposition to the flow of current and also their opposition (or resistance) to the flow of current increases with temperature. These materials/elements are called 'Conductors'.

2. At the bottom of the table, namely, diamond, which is an allotropic form of carbon and Teflon (a synthetic material) have very large values for resistivity, as large as 10^{25}(Ω-m). Obviously, these materials have negligibly small value for conductivity. Such materials are termed 'Insulator'. Since the resistivity of insulators is very large, it is difficult to record any small change in the value of the resistivity with temperature.

Table 4.1 Resistivity (ρ) and conductivity (σ) at 20 °C with temperature coefficient K (where ever available) for some elements/materials

Element/material	Resistivity (Ω-m) at 20 °C	Conductivity (S/m) at 20 °C	Temperature coefficient K (K)$^{-1}$
Gold	2.44×10^{-8}	4.10×10^7	3.40×10^{-3}
Silver	1.59×10^{-8}	6.30×10^7	3.80×10^{-3}
Copper	1.68×10^{-8}	5.96×10^7	4.00×10^{-3}
Iron	9.70×10^{-8}	1.00×10^7	5.01×10^{-3}
Platinum	1.06×10^{-7}	9.43×10^6	3.90×10^{-3}
Gallium	1.40×10^{-7}	7.10×10^6	4.00×10^{-3}
Carbon (amorphous)	5.0×10^{-4} to 8.0×10^{-4}	1.25×10^3 to 2.0×10^3	-0.5×10^{-3}
Carbon (graphite) Parallel to basal plane Perpendicular to basal plane	2.5×10^{-6} to 5.0×10^{-6} 3.0×10^{-3}	2.0×10^5 to 3.0×10^5 3.30×10^2	
Gallium Arsenide (GaAs)	1.0×10^{-3} to 1.0×10^8	1.0×10^{-8} to 1.0×10^3	
Germanium	4.60×10^{-1}	2.17	-48.0×10^{-3}
Silicon	6.41×10^2	1.56×10^{-3}	-75.0×10^{-3}
Diamond	1.00×10^{12}	1.00×10^{-13}	
Teflon	1.0×10^{23} to 1.0×10^{25}	1.0×10^{-25} to 1.0×10^{-23}	

3. Materials like Gallium arsenide (GaAs), an alloy of metals Gallium and Arsenic, Germanium and Silicon show moderate values both for resistivityand conductivity and strikingly have negative large values for temperature coefficient of resistivity. It means that the resistance of a given piece of Germanium or silicon will decrease sharply if the temperature of the specimen is increased. It is in sharp contrast to conductors, for which the resistivity increases with temperature and also insulators for which resistivity stays almost constant with temperature. These materials are called 'Semiconductors'. Moderate value of resistivity and sharp fall in the value of resistivity with the rise of temperature are two distinguishing features of semiconductors.

4. In Table 4.1, carbon stands out as a very peculiar element, in diamond configuration it is one of the best insulators, in graphite form it behaves like a conductor and also in crystalline form it exhibits different values for resistivity in different directions. Carbon, in graphite form, makes layered planner crystals, i.e. carbon atoms join together to form a layer and crystal is made by stacking of the piles of such planner layers. The resistivity of crystal has different values along the plane of the layer and perpendicular to it. Moreover, the cohesive forces between different layers are also relatively weak; therefore, these layers may easily slide over each other. This gives graphite the property being used as solid lubricant.

Materials for which resistivity depends on the orientation are termed as anisotropic materials. An anisotropic substance will have different properties in different directions. The general definition of resistivity given by (4.1) that is for isotropic substances does not hold for anisotropic substances.

To establish current in any material, it is essential to have an electric field. If $\vec{J}$ represents the current density at a particular location in the medium where the electric field is $\vec{E}$, then

$$\vec{j} = \sigma \vec{E} \quad or \quad \vec{E} = \rho \vec{J} \tag{4.2}$$

In (4.2), resistivity ρ and conductivity σ are tensors of rank -2 which may be represented by 3×3 matrices. Current density $\vec{j}$ and electric field $\vec{E}$ are three-dimensional vectors. The tensor form of the two equations contained in (4.2) may be written as

$$\begin{bmatrix} E_x \\ E_y \\ E_z \end{bmatrix} = \begin{bmatrix} \rho_{xx} & \rho_{xy} & \rho_{xz} \\ \rho_{yx} & \rho_{yy} & \rho_{yz} \\ \rho_{zx} & \rho_{zy} & \rho_{zz} \end{bmatrix} \begin{bmatrix} j_x \\ j_y \\ j_z \end{bmatrix} \tag{4.3}$$

$$\text{And} \quad \begin{bmatrix} j_x \\ j_y \\ j_z \end{bmatrix} = \begin{bmatrix} \sigma_{xx} & \sigma_{xy} & \sigma_{xz} \\ \sigma_{yx} & \sigma_{yy} & \sigma_{yz} \\ \sigma_{zx} & \sigma_{zy} & \sigma_{zz} \end{bmatrix} \begin{bmatrix} E_x \\ E_y \\ E_z \end{bmatrix} \tag{4.4}$$

Matrix equation (4.3) may be expanded as

$$E_x = \rho_{xx}j_x + \rho_{xy}j_y + \rho_{xz}j_z$$
$$E_y = \rho_{yx}j_x + \rho_{yy}j_y + \rho_{yz}j_z$$
$$E_z = \rho_{zx}j_x + \rho_{zy}j_y + \rho_{zz}j_z$$

If the direction of current in the material is taken as the X-axis, then $j_y = j_z = 0$ and

$$\rho_{xx} = \frac{E_x}{j_x}; \rho_{yx} = \frac{E_y}{j_x} \text{ and } \rho_{zx} = \frac{E_z}{j_x} \tag{4.5}$$

The resistivity and conductivity metrics given by (4.3) and (4.4) are reciprocal of each other; however, it does not mean that in general each corresponding element of the two metrics is reciprocal of each other, i.e. in general $\sigma_{xy} \neq \frac{1}{\rho_{xy}}$.

Self-assessment question (SAQ): What is meant by an intrinsic property of a material?

Self-assessment question (SAQ): A conducting rod of some metallic alloys is given, how one can ascertain whether it is isotropic or not?

4.3 Physics of Resistivity: Electron Band Theory of Solids

All elements in their solid form and materials listed in Table 4.1 are made of atoms, either of the same kind (elements) or of atoms of different kinds, each of which is electrically neutral. Further, most of the solids have crystalline structure: regular arrangement of atoms in a specific 3-D pattern. It then poses the question why different materials show different electrical properties? Why different materials have different values of resistivity? Answer to these questions lies in (i) size of the constituent atoms (distance of the most distant electrons from the centre of the atom) and (ii) relative separation between constituent atoms in the crystal. A single isolated atom of any element has a positively charged nucleus, the positive nuclear charge is equal to $(+Ze)$, where Z is the Atomic Number of the element in the periodic table and 'e' is the quantum of charge ($e = 1.6 \times 10^{-19}$ C). The positively charged nucleus is surrounded by Z number of electrons, each with negative charge of $-1e$. Now, these Z-electrons are revolving around the nucleus in different electron shells or orbitals, enveloping the nucleus with a negatively charged electron cloud. Thus if looked at from a distance, the atom as a whole appears electrically neutral; however, there are both positive and negative charges in the atom. Electrical properties of crystalline solids are essentially determined by the electrons which are in the outermost orbital, called the valence electrons and the relative separation between neighbouring atoms in the crystal.

According to the quantum mechanical model of the atom, electrons in an **isolated** atom are distributed around the nucleus of the atom with a probability density that varies with distance from the centre of the atom; the regions of larger probability density giving rise to orbitals. Further, electrons have discrete energies which means that they are distributed in different energy states. Since electrons have ½ $\hbar$ spin, they obey Fermi–Dirac statistics and the exclusion principle, according to which no two electrons in a given energy state can have same values for all their quantum numbers. As such the maximum number of electrons that may be accommodated in a given energy state is fixed. Discrete energy states are specified through the principal quantum number N and the relative orbital angular momentum ℓ of the state as N_ℓ. States with $\ell = 0, 1, 2 \ldots$ are, respectively, designated as s, $p, d \ldots$ etc. Thus, $1_s, 2_p, 3_s$ represents, respectively, states with $\ell = 0$ and principle quantum number 1, state with $\ell = 1\hbar$ and principle quantum no. 2, state with $\ell = 0$ and principle quantum number 3. The multiplicity or the maximum number of electrons that a given energy state may accommodate is given by $2(2\ell + 1)$. Thus in the s-state the maximum number of electrons may be $2(2 \times 0 + 1) = 2$, with one electron having its spin up and the other down. State-p may hold up to 6 electrons and state-d up to 10 and so on. Energy sequencing of these energy states is provided by quantum analysis and filling of electrons in states is done starting from the lowest energy state. In this way, electron configuration of isolated atoms of all elements of the periodic table may be built. For example, let us consider an isolated

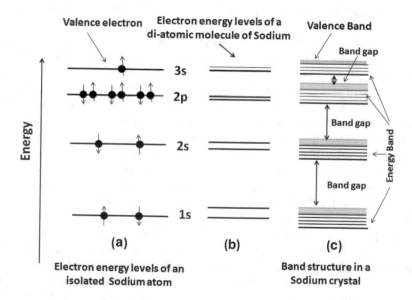

Fig. 4.2 **a** Electronic configuration of an isolated sodium atom. **b** Energy levels of a diatomic sodium molecule. **c** Electron energy bands in a crystal of sodium

atom of $^{23}_{11}Na$ that has 11 electrons. The electronic configuration of an isolated sodium atom at absolute zero is shown in Fig. 4.2a. Electrons occupying the states of highest principle quantum number (N) are called valence electrons. Valence electrons play a crucial role in chemical bonding and in conduction of current in materials. In case of sodium, the one electron in 3s state is valence electron. Further, the 3s state still has a vacancy, the maximum number of electrons that an s-state can have is 2, but in sodium it has only one electron. At 0 K temperature, in case of sodium, there are many empty electronic states at energies higher than the energy of 3s state.

Next, let us bring another atom of sodium near to the isolated sodium atom the electronic structure of which is shown in Fig. 4.2a. Now, there are two sodium atoms close to each other. The nuclei and the electron clouds of the two atoms will repel each other but the nucleus of one atom will attract the electron cloud of the other. Thus at a certain relative distance the diatomic system attains dynamic equilibrium forming a diatomic molecule, the electronic structure of which is shown in part (B) of Fig. 4.2. As may be observed in this figure, there are two energy levels, very near to each other, corresponding to 1s, 2s, 2p and 3s states of the diatomic molecule. Thus in case of a diatomic molecule each energy level of the isolated atom gets split into two components. Suppose now another, third atom of sodium is brought near to the diatomic molecule and forms a tri-atomic molecule. By analogy, each level of an isolated atom will split into three closely spaced components and so on. Thus adding of more and more atoms generates clusters of electron energy levels that are closely spaced in energy. These clusters or bundles of energy levels are called '**energy bands**'. Even in a small sodium crystal there are more than 10^7 sodium atoms and, therefore, energy bands of a small sodium crystal may contain up to 10^7 closely packed energy levels. Since discrete energy levels in a band are so close to each (separation between levels in a band may be of the order of 10^{-22} eV) and also because the separation between levels is of the order of the energy uncertainty in them that it becomes reasonable to assume that electrons may have almost continuous energies within the band.

Another important feature of band structure is the band gap. Generally, bands are separated from each other by a region of energy where there are no energy states for electrons of the parent atoms. Therefore, no electron of the parent material can have energies that lie in the band gap region. It is for this reason that these intra-band energy gaps are also called forbidden energy gaps. Before moving ahead, let us clarify one important fact. Suppose that there is a sodium crystal then electrons associated with sodium atoms cannot have energies lying in forbidden energy gaps, however, if some impurities are deliberately introduced in sodium crystal while growing the crystal, then electrons associated with impurity atoms may have energy levels that lie in forbidden energy gaps. That is why the word parent atoms are used to distinguish them from impurity atoms. It might also be observed in Fig. 4.2c that the magnitudes of forbidden energy gaps decrease as one moves to bands at higher energies and that at some energy bands originating from two successive states of the

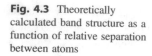 Theoretically calculated band structure as a function of relative separation between atoms

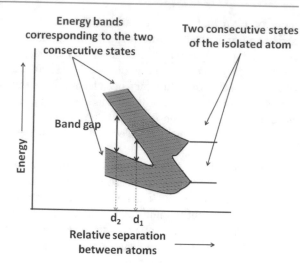

Fig. 4.3 Theoretically calculated band structure as a function of relative separation between atoms

parent isolated atom may overlap. As will become clear later that forbidden energy gap plays a crucial role in deciding the resistivity of the material.

So far we discussed only qualitatively the origin of band structure in crystalline solids. Though exact quantum mechanical calculation of band structure for any multi-body system is not possible, however, approximate quantum mechanical calculations show that the band gaps, in general, varies with the relative separation of atoms in the crystal. Figure 4.3 shows typical variation in the band gap with the relative separations d_1 and d_2 between atoms based on approximate quantum mechanical calculations. Theoretical calculations further show that separation between consecutive levels in energy bands may be as low as 10^{-22} eV. The thickness of a band depends on the degree of the overlap of the corresponding orbitals of individual atoms of the crystal. Since the degree of overlap for inner electron orbitals is less, the bands corresponding to them are comparatively narrower and are separated by large band gaps. On the other hand, orbitals corresponding to the outer electrons of individual atoms have considerable overlap and, therefore, the bands corresponding to outer electrons are thick and in some cases bands corresponding to two consecutive outer orbitals merge into each other without any band gap.

Band theory of crystalline solids is based on the following assumptions:

1. *Infinite-size system*: In order to have smooth and continuous band structure, it is required that the system must contain statistically large number of atoms. This assumption, however, does not pose any problem in case of crystalline materials as even a crystal of microscopic dimensions contains a very large number of atoms. Therefore, the band theory remains valid even for micro-sized transistors, etc.
2. *Homogeneous system*: Band structure is an intrinsic property of a homogeneous system, which assumes that the chemical makeup of the material must be uniform throughout the specimen.

3. *Non-interactivity*: Band structure evolves out of the '**single-electron states**'. Existence of single-electron states assumes that electrons travel in static potential without dynamically interacting with lattice vibrations, other electrons and photons.

Assumptions 2 and 3 are frequently violated in actual case and results in some special effects like band-bending, etc. In case of small size systems like quantum dot, small size molecule, etc., there is no continuous band structure. Further, strongly correlated materials like Mott insulators cannot be understood in terms of single-electron states, the band structure of such materials is not uniquely defined.

Self-assessment question (SAQ): Isolated atom of some elements behaves as an electric dipole while isolated atom of some other elements behaves like a neutral particle. What is the reason for this difference?

Self-assessment question (SAQ): Low energy bands of crystalline materials are generally narrower than the bands at higher energies, why?

Self-assessment question (SAQ): What is meant by the 'negative energy electron sea'?

Self-assessment question (SAQ): Suppose you are provided all necessary equipment to measure resistance and to establish current of desired magnitude in a conductor rod of 1 cm x 1 cm cross section and 20 cm length, how will you ascertain whether the material of the rod is isotropic or not?

Self-assessment question (SAQ): How can one physically justify the variation of band gap with relative separation of constituent atoms?

4.3.1   Valence and Conduction Bands

Low lying bands in the band structure of any crystalline material contain electrons that are nearer to their respective nuclei and are, therefore, strongly bound. The most loosely bound electrons in the system are those which reside in the band called '**valence band**' and at absolute zero temperature of the specimen, it is the band with highest energy that has electrons. Band next higher in energy than the valence band is called the '**conduction band**'. These two energy bands and the band gap between the valence and conduction bands play a very important role in deciding the electrical properties of the material. Since electrons of the low lying energy bands are tightly bound, they do not take part either in the chemical bonding of atoms or in conduction of electricity through the material and, therefore, it is customary to consider only the valence and conduction bands while discussing the electrical properties of the matter.

4.3.2   Fermi Level or Fermi Energy

Fermi energy or level is a hypothetical energy used to describe the top of the collection of electron energy levels at absolute zero temperature. Being Fermions obeying exclusion principle, electrons at absolute zero pack in to the lowest available energy states and build up a 'Fermi sea' of electron energy states. Fermi level is the surface of this Fermi sea at absolute zero, where no electron can have energy to rise above the Fermi level. Fermi energy plays an important role in characterization of materials regarding their thermal and electrical properties.

4.4   Conductors

On account of high degree of overlap in outer electron orbitals of individual atoms in some crystals, the conduction and the valence bands in such crystalline solids become so broad that they overlap; that means that there is no boundary between them. This occurs because in such materials the valence electrons in individual atoms are rather far away from their nucleus and are, therefore, loosely bound and also the relative separation between atoms in the crystal is large so that the band gap does not exist. Such materials are called conductors.

In any material including conductors at absolute zero temperature all electrons are bound, i.e. electrons in the valence band are also bound to the system. However, in conductors, the binding energy of valence band electrons is very small. When temperature of a conductor is raised above 0 K, the heat energy absorbed by the crystal sets its lattice into vibrations. A part of this lattice vibration energy is transferred to the valence electrons and if the transferred energy is more than the binding energy of the valence electrons, valence electrons become free and do not remain attached to any given atom. These electrons freely move in inter-atomic space of the crystal and are called **free electrons**. Since the conduction and the valence bands in a conductors overlap, the electrons are set free by the heat energy, shift from the valence to the conduction band. In conductors, a small rise of temperature shifts almost all valence electrons to the conduction band turning them into free electrons. Further increase of temperature of the conductor essentially does not increase the number density or concentration of free electrons but increases the average speed of their random motion.

Figure 4.4 shows the electron band structure of a conductor (a) at absolute zero temperature and (b) at a temperature higher than absolute zero. As is apparent, the low lying completely filled electron bands do not participate in heat and electric conduction through the material and, therefore, they are generally not shown in diagrams.

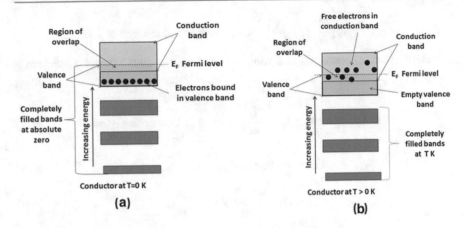

Fig. 4.4   a Band structure of a conductor at 0 K. **b** Band structure of a conductor at $T > 0$ K

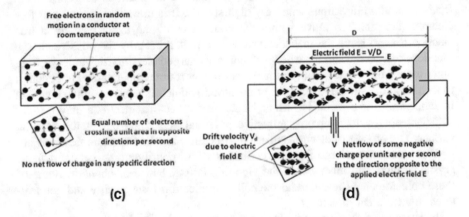

Fig. 4.4   c Free electrons in random motion with Fermi velocity, no net flow of charge in any specific direction. **d** Each free electron acquires an additional drift velocity V_d in the direction opposite to the applied electric field resulting in a net flow of negative charge per unit area per second in the direction opposite to the electric field E

At room temperature (≈ 300 K) almost all electrons that were in the valence band of a conductor get shifted to the conduction band as free electrons. Metals, most of which are conductors, therefore, at room temperature have sufficiently large number of free electrons per unit volume that may freely move in inter-atomic space and may be treated as a free electron (or fermions) gas. In a piece of conductor at room temperature, free electrons hop from one atom to the other and then to the other in most random way with an average speed v_F called Fermi velocity. Fermi velocity is related to the Fermi energy E_F through the following relations.

Fermi energy $E_F = \dfrac{\hbar^2}{2m_e}\left(\dfrac{3\pi^2 N}{V}\right)^{2/3}$, where N is the number of fermions,

each of mass m_e contained in a volume V and $\hbar$ the reduced Planck's constant

$$(4.6)$$

Fermi temperature $T_F = \dfrac{E_F}{k_B}$, here k_B is Boltzmann constant $\qquad (4.7)$

Fermi momentum $p_F = \sqrt{2m_e E_F}$ $\qquad\qquad (4.8)$

Fermi velocity $v_F = \dfrac{p_F}{m_e}$ $\qquad\qquad\qquad (4.9)$

In case of metals, Fermi energy is of the order of 2–10 electron volt (eV) and Fermi velocity of free electrons is of the order of 10^6 m/s. Fermi velocity is quite substantial almost 1% of the speed of light. Thus, free electrons in metals move freely in random directions with very high speeds. It is important to note that free electrons are 'free' to move within the boundaries of the metal piece, if a free electron tries to leave the metallic surface it is pulled back by the positive charge that develops on the surface when a negatively charged electron tries to leave it. In other words, the kinetic energy of free electrons at room temperature is not enough to overcome the attractive potential of the atomic nuclei. It is interesting to note that if temperature of a metal is further increased by heating, electrons may acquire sufficient energy to overcome attractive potential and come out of the metallic surface. The process of electron emission by heating is called **thermionic emission**.

Free electrons undergo frequent collisions with impurity atoms, crystal defects, crystal surface, and mostly with the vibrating crystal lattice or phonons. Some of these collisions are inelastic and the colliding electron loses energy and get scattered in a different directions.

In absence of any external electric field in a piece of conductor, the average number of free electrons crossing a unit area in any direction is same, and therefore, there is no net flow of charge and hence no current, as shown in Fig. 4.4c.

When a potential difference V is established between the two ends of the conductor, an electric field E $(=V/D$, D is the length of the conductor) gets set up in the conductor pushing each free electron in a direction opposite to the direction of E. The electric field, therefore, accelerates free electrons in the direction opposite to the electric field. This acceleration takes place only during the time interval between two successive collisions. For every conductor at a given temperature there is a characteristic mean time between two successive electron collisions which is generally denoted by Greek letter τ (Tau). Assuming that electron comes to rest at each collision, the velocity v_D acquired by the electron under the electric field E between two successive collisions may be obtained as

$$v_d = 0 + \frac{Ee}{m_e}\tau \qquad (4.10)$$

Thus, apart from the Fermi velocity which is randomly directed, each electron also acquires a velocity v_d, called the **drift velocity**. Drift velocity is very small $\approx 10^{-4}$ m/s. It is because electrons are continuously colliding with crystal lattice/ impurity/ defect, etc. and the mean time between collisions τ is very small. Typical value of τ is $\approx 10^{-14}$ s, which means that on an average an electron undergoes 10^{14} collisions per second.

Let us imagine an area of magnitude A (m^2) held normal to the direction of the electric field E. In unit time electrons held within the cylindrical volume $(A \cdot v_d)$ will all cross through this area. If 'n' (m^{-3}) is the number density of free electrons in the conductor, then current I will be given by

$$I = (A \cdot v_d)n\,e$$

And current density $j = \dfrac{I}{A} = e \cdot n \cdot v_d = nEe^2\tau/m_e = \left(\dfrac{ne^2\tau}{m_e}\right)E \qquad (4.11)$

$$\text{But} \quad j = \sigma E$$

$$\text{Hence,} \quad \boldsymbol{\sigma} = \left(\frac{ne^2\tau}{m_e}\right) \text{ and } \rho = \frac{m_e}{ne^2\tau} \qquad (4.12)$$

Equation (4.12) indicates that conductivity of a material depends on two factors (i) number density 'n' of free electrons in the material and (ii) the average time between two successive collisions τ.

Table 4.2 shows the dependence of conductivity (or resistivity) on the number density of free electrons in a material. It may be observed that in metals like silver, gold, copper and aluminium, conductivity decreases with the increase of the

Table 4.2 Number density of free electrons n, conductivity σ and resistivity ρ of some materials at room temperature

Material	Number density of free electrons 'n' (m^{-3}) at 300 K	Conductivity σ(S/m) at 300 K	Resistivity ρ (Ω-m) at 300 K
Silver	5.85×10^{28}	6.30×10^7	1.59×10^{-8}
Gold	5.90×10^{28}	4.10×10^7	2.44×10^{-8}
Copper	8.45×10^{28}	5.96×10^7	1.68×10^{-8}
Aluminium	18.06×10^{28}	3.50×10^7	2.82×10^{-8}
Gallium	15.30×10^{28}	7.10×10^6	1.40×10^{-7}
Silicon	1.45×10^{16}	1.56×10^{-3}	6.40×10^2
Germanium	2.40×10^{19}	2.17	4.6×10^1
Diamond	$<10^8$	$\approx 10^{-13}$	1×10^{12}

number density of free electrons. It happens because the probability of collisions increases as n^2, and the current density j increases as n; therefore, beyond a certain optimum value of 'n' any further increase in it decreases the mean time τ which in turn reduces the conductivity. Aluminium and Gallium that have much higher number densities of free electrons have lower conductivities. On the other hand when one compares the number density of free electrons in Aluminium and Germanium, the effect of increased number density is clearly reflected in the magnitudes of the conductivity of these two materials.

Assuming that the drift velocity for a given conductor has the value 10^{-4} m/s, an electron will take roughly ($10^4/3600=$) 2.78 h to reach the other end of a 1 m long rod of the material. As such it is apparent that actual current in a conductor does not flow by the motion of electrons from one end to the other end. Instead, the flow of net charge per unit time across an area held normal to the electric field constitutes the current flow in conductors (see Fig. 4.4b).

In metallic conductors, the number density of free electrons at room temperature is very large $\approx 10^{28}$ m^{-3}. On further increasing the temperature of the conductor the number density of free electrons does not change much (as all valence electrons are already free and electrons in the lower energy bands are tightly bound) but the speed of their random motion increases. As a result, free electron collision frequency increases reducing the mean time τ between two collisions. This reduction in τ reduces conductivity of the material. Therefore, the resistivity for conductors increases with the increase in temperature.

Self-assessment question (SAQ): When a current-carrying wire is cut, free electrons do not fall off from the wire. Explain.

Self-assessment question (SAQ): The mean random velocity of free electrons in conductors is quite large while their mean drift velocity is very small. Explain the reason for this difference.

Self-assessment question (SAQ): If the area of cross section of a current-carrying wire is reduced to half what will happen to the drift velocity of free electrons?

4.4.1 Metallic Bonding

Most of the metals have high boiling and melting points indicating that the chemical bonding between their atoms is quite strong. This strong chemical bonding arises from the strong force of attraction between the delocalised electrons cloud and the positive ionic cores of atoms. In the earlier example of the sodium crystal, it was mentioned that when another sodium atom is brought near to the isolated sodium atom, the valence electrons of the two sodium atoms each in 3s orbital of their respective atoms overlap making a diatomic molecular orbital. This essentially means that the two valence electrons, one from each atom, now are associated with

Fig. 4.5 A metallic crystal
with positive cores and
delocalized electron cloud

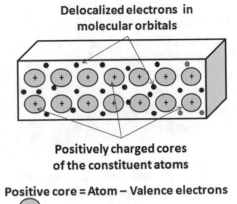

**Delocalized electrons in
molecular orbitals**

**Positively charged cores
of the constituent atoms**

Positive core = Atom − Valence electrons

both the atoms. On bringing more and more atoms close to each other still larger molecular orbitals corresponding to multi-atomic molecules of sodium are produced where the valence electrons are shared by all the atoms of the molecule. The molecular orbital of a group of large number of atoms must have the overlap of several orbitals as only two electrons can be accommodated in one orbital. Thus in metallic crystals valence electrons of the individual atom get detached from the parent atom and are shared by a group of atoms. These electrons are called **delocalised** electrons. Electrostatic force of attraction between the negatively charged cloud of delocalised electrons and the array of positively charged atomic cores gives rise to metallic bonding. The strength of metallic bond is expected to increase with the number of delocalised electrons and the positive charge on the atomic core. This is supported by the fact that in case of Magnesium metal where there are twice as many delocalised electrons as compared to the Sodium and also the positive charge on the atomic core is also twice as large that of sodium atom core, the bonding in Magnesium is stronger than that in Sodium. Figure 4.5 shows the structure of a typical metallic crystal with delocalized electron cloud.

4.4.2   Half Metals and Semimetals (Metalloids)

Valence electrons in any material may have their spins either up or down, i.e. in two mutually opposite orientations. It is, therefore, possible to build separate band structures for the two kinds of electrons. In some materials, it so happens that the valence band for electrons with one spin orientation is partially filled while there is a forbidden energy gap for electrons of the opposite spin orientation. As a result when some external voltage is applied, electrons with that orientation for which there are vacant states in the valence band contribute to current conduction while those with opposite orientation do not. Such materials are called **half metals** as electrons of only one specific orientation contribute to conduction. Chromium

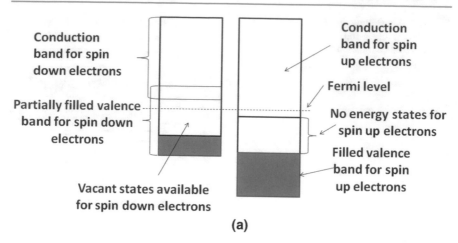

Conduction band for spin down electrons

Partially filled valence band for spin down electrons

Conduction band for spin up electrons

Fermi level

No energy states for spin up electrons

Filled valence band for spin up electrons

Vacant states available for spin down electrons

(a)

Fig. 4.6 a Typical band structure of a half metal

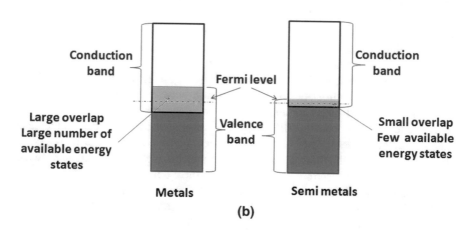

Conduction band

Fermi level

Conduction band

Large overlap Large number of available energy states

Valence band

Small overlap Few available energy states

Metals   Semi metals

(b)

Fig. 4.6 b Band structure of metals and semimetal

oxide, magnetite, lanthanum-strontium-magnetite (LSMO) are some half metals. All half metals are ferromagnetic but all ferromagnetic materials are not half metals (Fig. 4.6).

Semimetals are materials in which the bottom of the conduction band either just touches the top of the filled valence band or there is very little overlap between the two bands. As a result, the density of electron states near Fermi level is small in semimetals as compared to the case of metals where there is considerable overlap of valence and conduction bands. Because of the few available energy states near Fermi level, semimetals behave like semiconductors with negligible forbidden energy gap. Some metalloids are Arsenic, Antimony and Tellurium, etc. (Fig. 4.6b).

Self-assessment question (SAQ): In what respect delocalized electrons are different from free electrons in metals?

Self-assessment question (SAQ): What is the essential difference between semi-metal and half metal?

Solved example (SE-4.1) Calculate the number density of free electrons in copper at room temperature, given that the valency of copper is 1, density 8.96, Molecular mass 63.5 gm/mole and Avogadro number $A_N = 6.02 \times 10^{26}$ kmole^{-1}.

Solution: *Number density of free electrons 'n' in a material at room temperature may be calculated if it is assumed that all valence electrons of all atoms contained in 1m³ volume of the material are in conduction band. If z is the valency of the atom and V_a is the molar volume of the material, then*

$n = \frac{z \cdot A_N}{V_a}$ *But* $V_a = \frac{M_r \times 10^{-3}}{\rho}$ *where M_r and ρ are, respectively, molar mass of the molecule in grams and density of the material. Factor of 10^{-3} is included to convert the molar mass in kg.*

$$\text{Thus,} \quad n = \frac{z \rho A_N}{M_r \times 10^{-3}} \tag{4.13}$$

Substituting the values of given quantities in (4.13), one gets

$$n_{(copper)} = \frac{1 \times 8.96 \times 6.02 \times 10^{26}}{63.5 \times 10^{-3}} = 8.49 \times 10^{28} \text{ m}^{-3}$$

Solved example (SE-4.2) An aluminium wire of cross-sectional area of 4×10^{-6} m³ is carrying current of 5.0 A. Assuming that each Aluminium atom contributes one electron and that the density of aluminium is 2.7 and molar mass 26.98 gm/mole, calculate the free electron number density and their drift velocity.

Solution: *The number density of free electrons* $n = \frac{z \rho A_N}{M_r \times 10^{-3}} = \frac{1 \times 2.7 \times 6.02 \times 10^{26}}{26.98 \times 10^{-3}} =$ 6.02×10^{28} m^{-3}.

Also, the current $I = (A \cdot v_d) \, n \, e$

$$\text{or} \quad v_d = \frac{I}{A n e} = \frac{5.0}{4 \times 10^{-6} \times 6.02 \times 10^{28} \times 1.6 \times 10^{-19}} = 1.29 \times 10^{-4} \text{ m/s}$$

Solved example (SE-4.3) A metallic wire is drawn four times of its original length. What is the ratio of the new resistance of the wire to the old value of the resistance?

Solution: *Let R_1 and R_2, respectively, the new and old resistances and L_1 and L_2, respectively, the new and old lengths. If ρ is the resistivity of the metal, then:*
$L_1 = 4L_2$

$$R_1 = \frac{\rho(4L_2)}{A_1} \text{ and } R_2 = \frac{\rho L_2}{A_2} \text{ or } \frac{R_1}{R_2} = \frac{4A_2}{A_1} \qquad (4.14)$$

Here, A_2 and A_1 are, respectively, the new and old areas of cross section of the wire.

Since volume of the metal must be conserved, hence,
$A_1(4L_2) = A_2L_2$, therefore, $A_1 = A_2/4$ substituting this value of A_1 in (4.14) one gets

$$\frac{R_1}{R_2} = \frac{4A_2}{\frac{A_2}{4}} = 16$$

The resistance of the new piece of wire will be 16 times the original value.

4.5 Insulators

Those crystalline solids for which the forbidden energy gap between the valence band and the conduction band is either of the order of or larger than about 4 eV are termed as insulators. An insulator material contains negligible number of free electrons in conduction band at room temperature. The valence band is completely filled and the valence electrons in an insulator are so tightly bound with atoms that they do not jump to conduction band on absorbing heat energy.

That is why insulators are generally also bad conductors of heat. A typical band structure of crystalline insulating material is shown in Fig. 4.7, where it may be observed that the valence band is completely filled and the conduction band is totally empty with a large forbidden energy gap. Since electrons in low lying completely filled bands do not participate in either thermal conduction or electric

Fig. 4.7 Energy band picture of an insulator

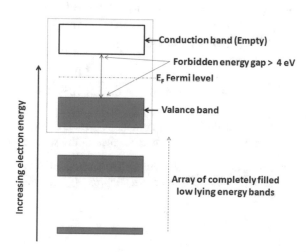

Table 4.3 Dielectric strength for some insulators

Material	Dielectric strength (kV/in.)
Mica	5000
Glass	2500
Teflon	1500
Paper	1250
Rubber	600
Porcelain	150
Air	50

conduction, it is customary to show only the valence and the conduction bands (enclosed in dotted rectangle in figure) while discussing the electrical properties of materials. As mentioned earlier, since there are no free electrons in an insulator even at higher temperatures, negligible current flows through a piece of insulator when some voltage is applied across it. However, on increasing the voltage to a very high value, called the **breakdown voltage**, the structure of the material breaks down and large currents may flow. An interesting example is lightening, normally air is insulator but it breaks down under the large voltage of lightening. The thickness of the material plays a role in breakdown. Specific dielectric strength is often listed in terms of kilo-volt per inch (kV/in.) and for some materials listed in Table 4.3.

In a crystal atoms of the constituent elements are held together via chemical bonding. Broadly speaking all elements in the periodic table can be divided into two categories: metals and non-metals. Atoms of metallic elements are held together by **metallic bonding**, in which **many** atoms share their valence electrons. Metallic atoms combine with non-metallic atoms forming **ionic bond** where actual transfer of electrons from one atom to the other takes place. For example, in NaCl, the valence electron of sodium leaves the sodium atom and shifts to the chlorine atom. This transfer of electron makes sodium atom a positive ion and the chlorine atom a negative ion, which are held together by the strong electrostatic force of attraction between the two ions. The bonding between two non-metallic atoms is mostly **covalent bonding** in which **few atoms** share their valence electrons. As such, bonding in insulators is either ionic or covalent, depending on the type of constituent atoms.

Glass is a very good insulator, and pure glass is also transparent to white light. It is because the valence electrons of the constituent atoms of glass are so tightly bound with their respective atoms that the energy contained in any component colour of white light is not enough to excite the atoms. Therefore, the white light passes through unabsorbed. However, if some impurities are introduced in glass lattice, the atoms of impurity element may absorb some specific component colour giving a colour to glass. Impurity of few chromium atoms in ruby crystal is responsible for the typical colour of the crystal.

Though it is customary to classify materials as insulators on the basis of their large forbidden energy gaps, however, band gap alone is not the criteria. Mobility of charge carries in the material also plays a very important role in classifying a material as semiconductor or insulator. For example, diamond that has a forbidden energy gap of 5.5 eV is a perfect insulator but synthesized materials like Aluminium nitride and Silicon nitride which have forbidden energy gaps of 6.3 eV and 5.0 eV, respectively, are treated as semiconductors because of the large mobility of charge carriers. As a matter of fact semiconductors with large band gaps are in some respects superior to normal semiconductors as they may be used at higher temperatures with little background noise.

Self-assessment question (SAQ): What is essentially responsible for the specific colours of gems?

Self-assessment question (SAQ): In your opinion in a crystal of some insulating material atoms are closely packed or far apart compared to a similar metallic crystal? Give reasons for your answer.

4.6 Semiconductors

Materials that have their electrical properties between conductors and insulators are termed as semiconductors. However, on account of vast applications of semiconductors considerable efforts have been made in recent times to synthesize tailor-made semiconductors for specific uses. Four parameters, namely, (i) the forbidden energy gap Eg, (ii) the magnitudes of the crystal wave number or crystal momentum vectors (k) at the bottom of the conduction band and the top of the valence band, (iii) number of available electron energy states around Fermi energy and (iv) the mobility of charge carriers play important role in characterization of semiconductors for different applications. It may, however, be mentioned that all these properties are interrelated and not totally independent.

Materials in the conduction and valence bands of which are separated by small forbidden energy gaps, generally <3 eV, with available electron density around Fermi level of the order of 10^{20} m^{-3}, are considered as common type of semiconductors. However, materials with larger energy gaps having charge carriers of high mobility are also very useful semiconductors for high-temperature applications.

Silicon and Germanium in their ultra-high pure form are the two naturally occurring semiconductors, and are, therefore, called **intrinsic** semiconductors. A look on the periodic table of elements will tell that C, Si, Ge, Sn and Pb are all members of the 14th group in periodic table and are all tetravalent, i.e. have four valence electrons. In crystalline form, they all have same 'diamond' structure. In spite of so many common properties they have substantially different electrical properties; Carbon in diamond form is one of the best insulators, Silicon is semiconductor, Germanium is both a semiconductor and semimetal, Tin (Sn) is

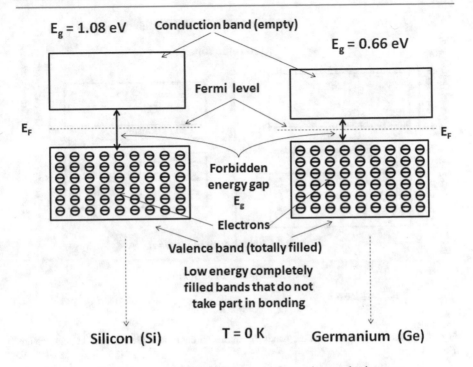

Fig. 4.8 Energy band systems for intrinsic Silicon and Germanium at absolute zero

semimetal or metalloid while Lead (Pb) is a metal. As we go down column 14th of the periodic table, the charge and the size of atoms increase and, therefore, the relative separation between two neighbouring atoms in the crystal structure changes from one element to the next affecting the electrical properties of elements.

Energy band structures for intrinsic (>99.9% pure) silicon and germanium crystals at absolute zero temperature are shown in Fig. 4.8. As may be observed, the band structure of Ge is on the right and that of Si on the left. The two band structures are independent of each other and have different energy scales as they are for two different crystals. The forbidden energy gaps for Si and Ge are, respectively, 1.08 eV and 0.66 eV. The Fermi levels (E_F) for the two systems lie in the middle of their respective forbidden energy gaps. The valence bands of both are completely filled with valence electrons while the conduction bands, under semi-classical limit, are totally empty at absolute zero temperature.

The conduction and valence bands of intrinsic silicon and germanium crystals at temperature $T > 0$ K are shown in Fig. 4.9. As may be observed in this figure, some valence electrons on gaining thermal energy overcame the forbidden energy gap and jump to the conduction band. These electrons in conduction band are now free, like that in a metal, and may contribute to the conduction of current. It may further be observed that at a given temperature $T > 0$ K, more electrons will jump from valence band to conduction band in Germanium as compared to Silicon since

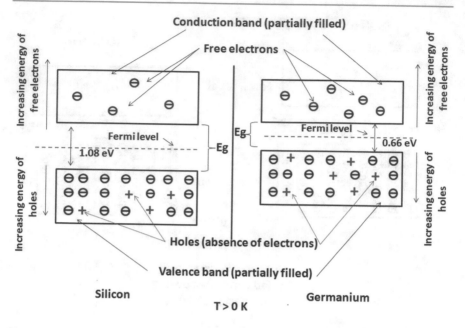

Fig. 4.9 Valence and conduction bands of intrinsicSilicon and Germanium at temperature $T > 0$ K

the forbidden energy gap for Ge is smaller than that of Si. The absence of electron in valence bands behaves as a positive charge and is called '**hole**'. It may, however, be realized that hole is a fictitious entity which has been assumed to explain the process of current flow in a semiconductor. The concept of hole originates from the quantum mechanical treatment of current flow in semiconductors. It so happens that Schrödinger's equation when applied to a semiconductor, under some approximations, separates out into two independent components one describing the motion of electrons and the other the motion of a positively charged particle. The absence of electron in the valence band is thus assumed to be the positive particle, hole, the motion of which is described by the second component of Schrodinger's equation.

In semiconductors, Fermi level assumes added importance, it may be treated as a reference of energy; energy of electrons in conduction band increases as one moves upwards from the Fermi level, while the energy of holes increases as one goes downwards from the Fermi level. That means that an electron at the top of the conduction band is most energetic while a hole at the bottom of valence band has largest energy. Further, the probability of finding an electron, say X units of energy above the Fermi level is same as the probability of finding a hole same X units below the Fermi level.

Free electrons in the conduction band undergo collisions with impurities (if any), sites of defects and crystal lattice (like that in case of metals) and lose/gain energy in these inelastic collisions. It often happens that on losing its energy the electron may fall back in the valence band and recombine with some holes in the valence band. In such recombination of an electron with a hole, energy either equal or slightly more than the forbidden energy gap, is released. At a constant temperature T, the two processes (i) generation of electron and hole pairs and (ii) their recombination proceed simultaneously such that ultimately thermal equilibrium is established. In the state of thermal equilibrium, the rate of generation of electron–hole pairs is equal to the rate of recombination of electron with hole. In an intrinsic semiconductor, the number densities of electrons (in conduction band) and holes (in valence band) are equal and depend on the temperature. When potential difference is applied across a semiconductor (which means that an electric field is established within the semiconductor) both electrons and holes develop drift velocities in opposite directions and establish current that is the sum of the electron current and the hole current. It is, however, to be noted that while free electrons can move within the boundaries of the crystal, the holes are constrained to move within the valence band only.

The basic difference in the electrical properties of conductors and semiconductors arises because of the very large difference in the number of charge or current carriers in the two. The number density of free electrons which are the current carriers in a metal is $\approx 10^{28}$ m^{-3}, while in semiconductors, the number of free electrons and holes is only around 10^{20} m^{-3}. When temperature of a conductor is further raised, there is no further increase in the number density of free electrons but the collision rate of free electrons increases considerably increasing the resistivity of metals with temperature. On the other hand, when temperature of a semiconductor is raised, more electron–hole pairs are created increasing the number density of charge carriers by large amount, collision rate of charge carriers also increases but relative increase in the number of charge carries is much more, wiping out the effect of increased collision rate. Hence, the resistivity of semiconductors decreases sharply with the increase of temperature.

Self-assessment question (SAQ): Why does the resistivity of a semiconductor decreases with the rise of temperature?

4.6.1 Covalent Bond Picture

The other equivalent way of understanding the phenomena of current flow in a semiconductor is to use the covalent bond picture of semiconductors. Both Si and Ge atoms have four valence electrons that take part in chemical bonding. If these four valence electrons are taken out, the remaining part of the atom is called the **core** of the atom. Since atom is neutral, the cores of both Si and Ge atoms have (+4e) charge, where e = 1.6×10^{-19} C is the unit of charge. In covalent bonding

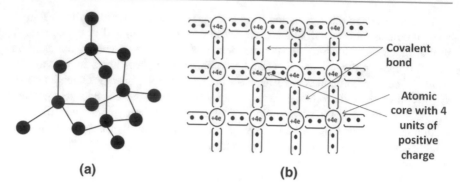

Fig. 4.10 **a** Diamond like structure of Si crystal. **b** Covalent bonds of Si in two dimensions

both in Si and Ge, four neighbouring atoms share two electrons, one from each atom to make four covalent bonds.

Crystals of both Ge and Si have 'diamond' structure which consists of two interpenetrating face centred cubic lattices which are displaced along the body diagonal by one-fourth of the distance (Fig. 4.10a). The four nearby atoms are held together by four covalent bonds, each covalent bond formed by the sharing of one electron from each atom as shown in the 2-D representation of Fig. 4.10b. The covalent bond energy in case of Ge is 0.66 eV and is 1.08 eV in case of Si, which corresponds to their respective forbidden energy gaps. At 0 K, all covalent bonds are intact and, therefore, there are no free charge carriers to conduct electric current. If temperature is raised to some value $T > 0$ K, electrons from some bonds break their covalent bonds and become free; they are now free to move around within the crystal boundary in the inter lattice space. The vacancy of electron in broken bond behaves like a positive charge and is called hole. Holes under the applied electric field drifts, but only from one bond to the next, never coming out in inter-atomic space. Thus in a semiconductor both conduction band free electrons and valence band holes bound in broken bonds contribute to the flow of current. This is shown in Fig. 4.11.

At $T > 0$ K, breaking of covalent bonds to create electron–hole pairs and the reverse process of annihilation of electron–hole pair to recreate covalent bonds goes on simultaneously. However, in a short time, thermal equilibrium is established when the rates of two opposite processes become equal and the average number of free electrons (and holes) per unit volume stabilizes to some value, say n_i^e which depends on the type of semiconductor and the temperature T. In intrinsic semi-conductor, the number densities of free electrons and holes are always equal, i.e. $n_i^e = n_i^p = n_i$ where n_i^p is the number density of holes.

$$(covalent\ bond \rightleftharpoons electron + hole)_{Thermal\ equilibrium}$$

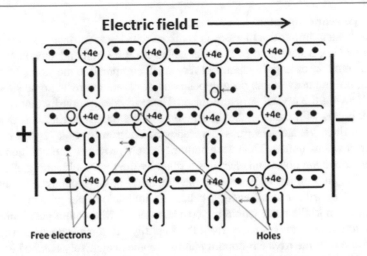

Fig. 4.11 In an intrinsic semiconductor at $T > 0$ K some covalent bonds break releasing the electrons which may move within the crystal boundary. Absence of electrons from broken bonds behaves as positively charged hole. Both electrons and holes contribute to current

Self-assessment question (SAQ): The electrons held in covalent bonds are bound electrons or delocalised electrons or free electrons? Give reasons for your Solution: wer.

4.6.2 Extrinsic or Doped Semiconductors

The electrical and thermal properties of materials depend, to a large extent, on the number density of free charge carriers in it. The number density of current carriers in semiconductors may be controlled (and hence its electrical properties) by introducing specific impurities deliberately in the crystal, in a controlled way, at the time of crystal growth. The process of introducing impurities in a crystal is called doping and the crystal so grown as doped semiconductor. Doping is highly technical process requiring extreme control on the crystal growing environment. It is in itself a complete science. Generally, impurities either of atoms that have five valence electrons or three valence electrons are added in a very minute amount (≈ 1 part in 10^7) at the time of crystal growth, and care is taken that basic crystal structure remains intact. Therefore, impurities of nearby elements in the periodic table, the atoms of which have nearly same size, are used for doping. Commonly, trivalent elements Boron (B), Aluminium (Al), Gallium (Ga), Indium (In) and pentavalent elements Phosphorus (P), Arsenic (As) and Antimony (Sb) are used to dope intrinsic Silicon and Germanium to develop the desired doped semiconductor.

(i) **P-type semiconductor**

If some trivalent impurities like Boron is mixed with some intrinsic semiconductors (like Si or Ge), a new kind of material called P-type semiconductor is produced. Since the number of impurity atoms is very small compared to the atoms of the bulk material, the impurity atoms occupy positions of parent atoms at some sites in the crystal, leaving the crystal structure intact. These trivalent impurity atoms develop covalent bonds with their neighbouring atoms of the bulk material. However, they have only three valence electrons, a vacancy of electron (or hole) is left in one of the four covalent bonds. Thus each impurity atom provides one additional hole, over and above the holes and electrons formed by the thermal breaking of covalent bonds. The additional hole around the impurity atom remains attached with impurity atom with a small energy, called the ionization energy, and may move from one atom to the next, within the covalent bonds, when atom gets ionized. On ionization, the hole moves away from the impurity atom and an electron replaces hole. As a result, the trivalent impurity atom becomes negatively charged ion. Thus on ionization of most of the impurity atoms, a P-type material has very large number of holes, essentially equal to the number of impurity atoms, plus some electron–hole pairs because of the thermal breakup of some covalent bonds, and negatively charged ions of impurity element. The number density of holes in a p-type material is much larger than that of free electrons created by thermal breaking of covalent bonds. Suppose there are N_A impurity atoms per unit volume all of which are ionized, then there will be N_A additional holes per unit volume,

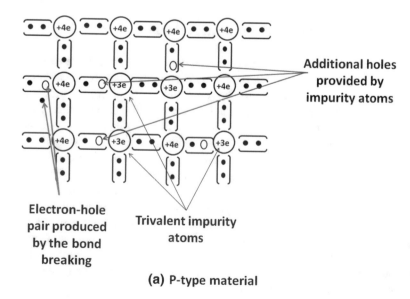

(a) P-type material

Fig. 4.12 a Covalent bond structure of a P-type material

each impurity atom providing one hole. While there will be only few say 100 electron–hole pairs per unit volume due to the thermal breaking of covalent bonds of parent material. As such the total number of holes is much larger (N_A + 100) than the number of electrons (100 only). Since positively charged holes are the majority carriers and electrons are minority carriers, the material is called P-type material. Figure 4.12a shows the 2-D structure of a P-type material. The trivalent impurity is called the acceptor impurity as it accepts electrons from the parent atoms.

(ii) N-type material

N-type semiconductor is formed when a pentavalent impurity is mixed with an intrinsic semiconductor. Pentavalent atoms surrounded by atoms of the parent bulk material get imbedded in the crystal lattice with four covalent bonds, one on each side, and the fifth electron of the impurity atom remains loosely attached to the impurity atom. At a temperature $T > 0$ K, the impurity atom may ionize and the loosely bound additional electron of impurity atom may become a free electron. Thus, on ionization each pentavalent impurity atom provides one additional free electron and becomes a positively charged ion. Since impurity concentration N_D may be of the order of 10^{10} or larger, large number of free electrons per unit volume becomes available for conduction. Then, there are few electron–hole pairs from the thermal breaking of covalent bonds of the intrinsic parent atoms. Therefore, in an N-type material the majority carriers are electron and minority carrier holes. Also, there are positively charged ions of the impurity element. Since impurity atoms in

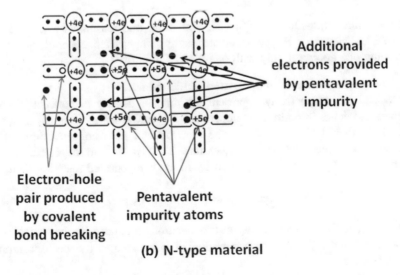

Additional electrons provided by pentavalent impurity

Electron-hole pair produced by covalent bond breaking

Pentavalent impurity atoms

(b) N-type material

Fig. 4.12 b Covalent bond structure of an N-type material

N-type material provide additional free electrons, the pentavalent impurity atoms are called donor atoms (Fig. 4.12b).

On the basis of the ionization energy, the impurities may be classed as **shallow** or **deep**. Those impurities the ionization energy for which is either equal to thermal energies ($\approx kT \approx 0.025$ eV) or less are called shallow, while those which have higher ionization energy are called deep. Deep impurities are used in semiconductors for high-temperature applications. If the ionization energy of the impurity is more than $3E_g$, where E_g is the magnitude of the forbidden energy gap of bulk material, it is unlikely to ionize. Such deep impurity atoms become sites in the crystal for the recombination of electron and holes and work as recombination **traps**.

Self-assessment question (SAQ): Is it true to say that an N-type crystal is overall negatively charged?

4.6.3 Compensated Semiconductors

If both, shallow donor and shallow acceptor impurities are simultaneously put in same intrinsic semiconductor, the resulting doped material is called **compensated semiconductor**. Let N_D and N_A, respectively, denote the donor and the acceptor concentrations when all impurity atoms have ionized, and if $N_D = N_A$, then the compensated semiconductor will behave as an intrinsic semiconductor. If $N_D > N_A$, the compensated semiconductor will behave as an N-type material and in the other case of $N_A > N_D$ as P-type material.

Self-assessment question (SAQ): Can you think one reason why it is easy to fabricate an N-type or P-type specimen using compensated material rather than fabricating one doping it only with one type of impurity?

(a) **Charge density in a compensated semiconductor**
Let N_A and N_D, respectively, represent the number density or concentrations of acceptor and donor impurities at temperature $T > 0$ K when all impurity atoms are ionized. Further, let n_e and n_p be the electron and hole concentrations in the compensated semiconductor at temperature T. If ρ (not to be confused with resistivity) denotes the free charge density in the compensated material, then

$$charge\ density\ \rho = \left(n_p - n_e + N_D^- - N_A^+\right) \tag{4.15a}$$

Here, N_D^- is the number density of free electrons produced by the ionization of donor atoms and N_A^+ is the concentration of holes created by the ionization of acceptor atoms.

Since there is no free charge density in an equilibrium material, equating (4.15a) to zero, one gets

$$n_p - n_e + N_D^- - N_A^+ = 0 \qquad (4.15b)$$

Also from mass action law, $n_p n_e = n_i^2$ combining this with (4.15b) yields

$$n_p^2 + \left(N_D^- - N_A^+\right)n_p - n_i^2 = 0$$

$$n_e^2 + \left(N_A^+ - N_D^-\right)n_e - n_i^2 = 0$$

Above quadratic equations may be solved to get

$$n_e = \left(\frac{N_D^- - N_A^+}{2}\right) + \sqrt{\left(\frac{N_D^- - N_A^+}{2}\right)^2 + n_i^2} \qquad (4.15c)$$

$$\text{And} \quad n_p = \left(\frac{N_A^+ - N_D^-}{2}\right) + \sqrt{\left(\frac{N_A^+ - N_D^-}{2}\right)^2 + n_i^2} \qquad (4.15d)$$

Equations (4.15c) and (4.15d) give the equilibrium concentration of electrons and holes. Now there are two possibilities: (i) $\left(N_D^- \sim N_A^+\right) \gg n_i$, i.e. the concentration of either electrons or of holes produced by doping is much larger than the charge concentration n_i of intrinsic material, then n_i^2 in (4.15c, d) may be neglected and the majority carrier concentration is simply the difference between the concentrations of the two types of impurities. (ii) However, if $\left(N_D^- \sim N_A^+\right) \approx n_i$, (4.15c) or (4.15d) depending on whether the majority carriers are electrons or holes be used to calculate the concentration n_{maj} of majority carriers. The concentration of minority carriers n_{mino} is, however, obtained using the fact that $n_{\text{maj}} \cdot n_{\text{mino}} = n_i^2$, so that $n_{\text{mino}} = \frac{n_i^2}{n_{\text{maj}}}$. For calculating the intrinsic concentration n_i formulation given in Sect. 4.6.2 may be used.

4.6.4 Mass Action Law

The law of mass action in chemistry defines the condition of chemical equilibrium in a chemical reaction. Similarly, in semiconductor physics law of mass action defines the condition for thermal equilibrium in a semiconductor. The law is applicable to both the intrinsic and extrinsic or doped semiconductors. The law says that 'the product of the electron concentration n_e and the hole concentration n_p in any semiconductor (either intrinsic or extrinsic) at temperature T K is constant and is equal to the product of electron and hole concentrations in the intrinsic semiconductor at same temperature T. In the mathematical form, the law of mass action may be written as

$$n_e n_p = n_i^e n_i^p = n_i^2 \qquad (4.16)$$

4.6.5   Non-degenerate and Degenerate Semiconductors

If only a small amount of dopant atoms are added to an intrinsic semiconductor so that the dopant atoms are far apart in the crystal structure and do not interact with each other, the semiconductor is called **non-degenerate**. The donor or acceptor levels are discrete, and the Fermi level for a non-degenerate semiconductor lies in forbidden energy gap. The general extrinsic semiconductors are non-degenerate.

However, if impurity concentration is high and impurity atoms are close to each other in the crystal, and they interact with each other, the donor or acceptor level broadens into a band and merges into either the conduction band for N-type material or into the valence band for P-type semiconductor. This type of semiconductor is called **degenerate** semiconductor (Fig. 4.13).

The Fermi level for degenerate semiconductor lies in either the conduction or the valence band.

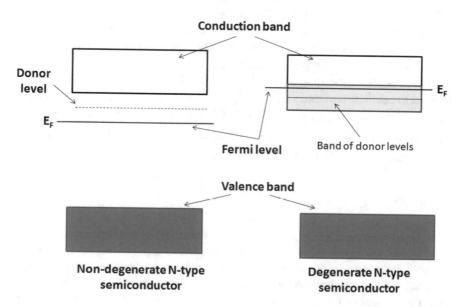

Fig. 4.13 Non-degenerate and degenerate N-type semiconductors

Self-assessment question SAQ: In a non-degenerate semiconductor, there are energy bands for electrons of bulk material but only single levels for the donor or acceptor atoms. Why?

(i) Electron and hole densities in non-degenerate semiconductor

The conductivity of intrinsic semiconductors is quite low. By controlled doping by a very small impurity the conductivity of semiconductor can be increased by many folds. The dopant element is selected in such a way that the energy levels of dopant fall in the forbidden energy gap of the intrinsic material. A shallow donor will have an energy level E_D in the forbidden band gap close to the bottom of the conduction band at energy E_C, while a shallow acceptor will have an energy level E_A in the forbidden energy gap close to the top of the valence band at energy E_V. It may be noted that though no energy levels of parent intrinsic material may be found in forbidden energy gap, however, it is quite possible that dopant atoms have their energy levels lying in the forbidden energy gap.

Phosphor is a frequent dopant for intrinsic silicon crystal and its dopant level E_D is 0.045 eV below the bottom of the conduction band of silicon. Similarly for boron, which is often used as a P-type dopant, the dopant level E_A is also around 0.045 eV just above the top of the valence band. Since both these dopants have low ionization energy, the thermal energy at room temperature is enough to completely ionize all dopant atoms, releasing specific charge carriers (either electrons or holes) per unit volume equal to dopant concentration. Thus,

$$n_e \simeq N_D \text{ and } n_p \cong N_A \qquad (4.17)$$

where n_e, n_p, N_D and N_A are, respectively, the free electron, hole, donor dopant and acceptor dopant concentrations (number per unit volume) at room temperature.

4.6.6 Effective Mass of Electron and Crystal Momentum

In the conduction band of a semiconductor, the so-called free electrons are really not so free; they are bound by the periodic potential of lattice nuclei. In order to take into account the effect of this periodic potential on the motion of electrons, an **effective mass** m_n^* is assigned to the electron. The energy–momentum relation for the conduction electron may be written as

$$E = \frac{p_c^2}{2m_n^*} \qquad (4.18)$$

Here, p_c is the '**crystal momentum**' which is analogous to particle momentum along a given crystal direction defined by the Miller index. For some crystals, like the silicon and gallium arsenide (GaAs) crystals, the maximum valence band energy occurs at crystal momentum $p_c = 0$. The minimum energy for conduction band electron in GaAs also occurs at $p_c = 0$. Hence, in case of GaAs, a transition across the forbidden energy gap requires only the absorption or emission of energy E_g. Because of this reason those semiconductors for which the maximum of valence band energy and the minimum of the conduction band energy that occur at the same value of the crystal momentum are called **direct semiconductors** and their band gaps as direct band gaps. On the other hand, in case of silicon, the maximum of valence band energy occurs at $p_c = 0$, but the minimum of the conduction band energy occurs in a direction where $p_c \neq 0$, hence in such crystals where the maximum energy of valence band and the minimum energy of conduction band occur for different values of crystal momentum p_c are called indirect semiconductors. In indirect semiconductors, an electronic transition across the band gap not only requires the exchange of the energy quantum E_g but also a change in the crystal momentum. The indirect semiconductors are both slow and inefficient for photon transition. It is for this reason that silicon is not good for making light-emitting diodes or semiconductor lasers.

The effective mass of electron may be easily obtained from second derivative of energy using the relation of (4.18) as

$$m_n^* = \frac{1}{\frac{d^2 E}{dp_c^2}} \qquad (4.18a)$$

Effective mass of electron or of hole may have different values depending on the material and the process for which it is required, for example, effective masses of electron and hole for calculating energy states and conduction for different materials are given below (Table 4.4).

The main advantage of using the concept of effective mass is that electron and hole may be treated as classical particles.

Table 4.4 Effective mass of electron and hole

Property	Symbol	Germanium	Silicon	Gallium arsenide
Forbidden energy gap	E_g (eV)	0.66	1.08	1.42
Effective mass for density of states calculations				
Electron	m_n/m_e	0.56	1.08	0.067
Hole	m_h/m_e	0.29	0.57 to 0.81	0.47
Effective mass for conductivity calculations				
Electron	-do-	0.12	0.26	0.067
Hole	-do-	0.21	0.36 to 0.38	0.34

4.6.7 Theoretical Calculation of Carrier Density in a Semiconductor

Charge carriers, electrons or holes, in a semiconductor reside either in the conduction band or the valence band. Both these bands have large number of energy states or levels. These levels are so close to each other in energy that it becomes impossible to talk about a specific level at a given energy E. Instead, one talks about the **level density**, the number of levels per unit volume within energy E and $(E + dE)$. The level density for allowed energy states in conduction band may be denoted by $N(E)$ and for holes in valence band by $P(E)$.

Let us first calculate theoretically the concentration or number density for free electrons (number of free electrons per unit volume) in a semiconductor. We chose a small energy interval dE around energy E in the conduction band and denote the concentration of electrons in this by $n_e(E)$. Now this density $n_e(E)$ is given by the product of (i) the density of allowed energy states per unit volume $N(E)$ at energy E and $(E + dE)$ and (ii) the probability $F(E)$ that this energy range is occupied by electrons. Therefore,

$$n_e(E) = N(E)F(E) \tag{4.19}$$

To obtain 'n_e' the total number of electrons per unit volume in full conduction band (4.19) must be integrated over energy from the bottom of the conduction band to the top of the conduction band. Hence, the electron concentration or number density in conduction band is given by

$$n_e = \int_{E_C}^{E_{top}} n(E)dE = \int_{E_C}^{E_{top}} N(E)F(E)dE \tag{4.20}$$

The factor $N(E)$ may be theoretically calculated using the tools of quantum statistics. Exact calculations for a semiconductor crystal are impossible as in an actual crystal the free electron faces periodic potentials due to atomic nuclei. However, approximate calculations, assuming that electron behaves as a free particle in a box, give the following expression for $N(E)$.

$$N(E) = 4\pi \left(\frac{2m_n^*}{h^2}\right)^{3/2} E^{1/2} \tag{4.21}$$

Here, m_n^* is the 'effective mass' of electron and h is Planck's constant.

The value of $F(E)$, the probability that electron occupies the state at energy E, may be obtained from the Fermi–Dirac distribution function as

$$F(E) = \frac{1}{1 + e^{\frac{(E - E_F)}{kT}}} \tag{4.22}$$

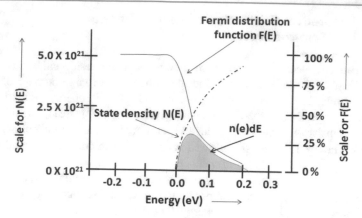

Fig. 4.14 Fermi distribution function and state density as functions of energy

where E_F, k and T are, respectively, Fermi energy, Boltzmann constant and T absolute temperature in Kelvin. Fermi level may be defined as the energy level the probability that an electron occupies is 0.5.

Figure 4.14 shows the variation of $N(E)$ and $F(E)$ as a function of energy. The shaded area of the curve gives the value of the integral $\int_{E_C=0}^{E_{top}=\infty} N(E)(F(E)dE$, the electron density in the conduction band. The lower limit of integration which is the bottom of conduction band E_C is taken as zero in the graph, and the upper limit that should be the top of conduction band is taken as infinity since $F(E)$ rapidly goes to zero for higher energies.

In the case where $(E - E_F) > 3kT$, 1 may be neglected in comparison to $e^{\frac{(E-E_F)}{kT}}$ in the denominator of (4.22) and so $F(E)$ may be written as

$$F(E) = e^{-(E-E_F)/kT} \tag{4.23}$$

Therefore,

$$n_e = \int_{E_C}^{E_{top}} n_e(E)dE = \int_{E_C}^{\infty} 4\pi \left(\frac{2m_n^*}{h^2}\right)^{3/2} E^{1/2} e^{-(E-E_F)/kT} dE \tag{4.23a}$$

$$\text{or} \quad n_e = 2\left(\frac{2\pi m_n^* kT}{h^2}\right)^{\frac{3}{2}} \left(e\left(-\left(\frac{E_C-E_F}{kT}\right)\right)\right)$$

$$\text{or} \quad n_e = N_C\left(e\left(-\left(\frac{E_C-E_F}{kT}\right)\right)\right) \tag{4.24}$$

$$\text{where} \quad N_C \equiv 2\left(\frac{2\pi m_n^* kT}{h^2}\right)^{\frac{3}{2}} \tag{4.25}$$

Similarly, the number density of holes in valence band may be written as

$$n_p = N_V\left(e^{\left(-\left(\frac{E_F - E_V}{kT}\right)\right)}\right) \tag{4.26}$$

$$\text{where} \quad N_V \equiv 2\left(\frac{2\pi m_p^* kT}{h^2}\right)^{\frac{3}{2}} \tag{4.27}$$

Self-assessment question (SAQ): 4.18: In (4.23a), the upper limit of integration which was top of conduction band has been replaced by infinity. What is the justification for this change of limit?

4.6.8 Positioning of Fermi Level

(a) **Intrinsic semiconductor** In an intrinsic semiconductor the free electron and hole densities are equal at all temperatures, i.e. $n_e = n_p = n_i$, therefore, on dividing (4.24) by (4.26) and putting E_F^i for the Fermi energy of intrinsic material one gets

$$\frac{n_e}{n_p} = 1 = \frac{N_C\left(e^{\left(-\left(\frac{E_C - E_F}{kT}\right)\right)}\right)}{N_V\left(e^{\left(-\left(\frac{E_F - E_V}{kT}\right)\right)}\right)} = \frac{N_C}{N_V}\, e^{\left(2E_F^i - E_V - E_C\right)}$$

Which on substituting the values of N_C and N_V gives

$$E_F^i = \frac{(E_C + E_V)}{2} + \frac{3kT}{4}\ln\left(\frac{m_p^*}{m_n^*}\right) \tag{4.28}$$

Since the value of Boltzmann constant $k = 8.617 \times 10^{-5}$ eV K^{-1} and $\ln\left(\frac{m_p}{m_e}\right)$ is also small, the value of the IInd term on RHS of (4.28) is negligible and may be neglected. Hence,

$$E_F^i \cong \frac{(E_C + E_V)}{2} \tag{4.29}$$

Equation (4.29) tells that the Fermi level for an intrinsic material lies in the middle of the conduction and valence bands and that it has weak dependence on temperature T.

Also for intrinsic materials, from mass action law, it follows that

$$n_e n_p = n_i^2 = N_C N_V e^{\left(\frac{E_V - E_C}{kT}\right)}$$

$$\text{or} \quad n_i = \sqrt{N_C N_V} e^{\left(-\frac{E_g}{2kT}\right)} \qquad (4.30)$$

The above relation is independent of Fermi energy and holds both for intrinsic and doped semiconductors. It signifies that an increase of carriers of one type is accompanied with the decrease of the carries of the other type so that their product remains constant at given temperature. The value of the product $n_e n_p$ depends on temperature and the band gap E_g.

(b) **Doped semiconductor** In an N-type semiconductor, the number density n_e of electron is essentially equal to the concentration N_D of the donor impurity when all impurity atoms have ionized. Therefore, for a completely ionized N-type semiconductor,

$$n_e = N_D = N_C \left(e^{\left(-\left(\frac{E_C - E_F}{kT}\right)\right)} \right)$$

$$\text{or} \quad E_F = E_C - kT \ln \frac{N_C}{N_D} \qquad (4.31)$$

The term $\left(kT \ln \frac{N_C}{N_D} \right)$ decreases with increasing value of N_D, the dopant concentration, and, therefore, the Fermi level shifts towards the bottom of the conduction band.

Similarly, for P-type semiconductor

$$n_p = N_A = N_V \left(e^{\left(-\left(\frac{E_F - E_V}{kT}\right)\right)} \right)$$

$$\text{And} \quad E_F = E_V + kT \ln \left(\frac{N_V}{N_A}\right) \qquad (4.32)$$

Equation (4.32) tells that with increasing value of N_A, the concentration of acceptor dopant, the Fermi level shift nearer to the valence band.

Self-assessment question (SAQ): Does the Fermi level in an intrinsic semiconductor is exactly at the middle of band gap? If not, then to which side, upwards or downwards?

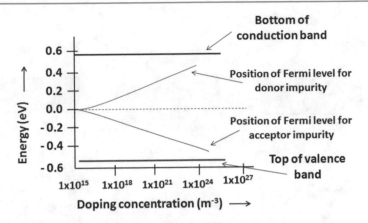

Fig. 4.15 Shift in the position of Fermi level in doped silicon semiconductor with doping concentration

From the above derivations, it may be concluded that the position of the Fermi level shifts from almost middle of the forbidden energy gap towards the bottom of the conduction band for n-type materials, the amount of shift increasing with the dopant concentration and for p-type materials the Fermi level shifts towards the top of the valence band, amount of shift increasing with the increase of dopant concentration. This is shown in Fig. 4.15.

4.6.9 Energy Band Diagram of Doped Semiconductor

Figure 4.16 shows the energy band diagrams of an intrinsic, N-type and P-type semiconductors. The Fermi level E_F^i for intrinsic semiconductor lies in the middle of the forbidden energy gap Eg.

However, the Fermi levels E_F^N and E_F^P, respectively, for the N-type and P-type semiconductors are not in the middle of the band gap but have shifted towards the conduction band for N-type and towards the valence band for P-type materials. Another important observation is the presence of E_D, the donor level just below ($<0.05\ eV$) the bottom of the conduction band. Donor level is an energy state of the pentavalent shallow donor impurity which gets completely ionized at room temperature leaving positively charged donor ions. Similarly, in case of the P-type material, the acceptor level E_A just above the top of the valence band, contains negatively charged ions of trivalent acceptor impurity. Trivalent acceptor atoms get ionized by accepting an extra electron from atoms of the bulk material to produce holes. Moreover, it may be observed in the above figure that the number of holes and electrons are equal for the intrinsic material while the concentration of electrons in conduction band is very much higher as compared to the concentration of holes in valence band for N-type material and the reverse is true for the P-type material.

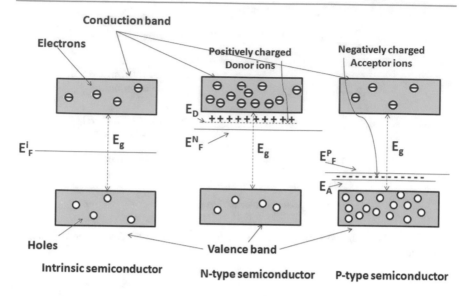

Fig. 4.16 Energy band diagram of intrinsic, N-type and P-type semiconductors

The electrons are, therefore, called majority carriers and holes minority carriers in N-type material, similarly holes are majority and electrons are minority carriers in P-type semiconductors.

Self-assessment question (SAQ): Where does the Fermi level for a heavily doped P-type semiconductor lies?

4.6.10 Compound Semiconductors

Many compounds of trivalent and pentavalent elements like Gallium arsenide (GaAs), Indium phosphate (InP), Gallium nitrate (GaN), etc. show semiconductor properties and are very suitable for optoelectronic devices. Most of them are direct semiconductors and, hence, have fast response to photon absorption/emission. Sulphur (S) and Selenium (Se) are often used as donor impurities while zinc (Zn) as acceptor impurity in compound semiconductors. A trivalent atom has three valence electrons while a pentavalent atom five; the two atoms form four covalent bonds sharing their total eight electrons as shown in Fig. 4.17. It is interesting to note that tetravalent atoms of Si and Ge may also be used as both donor and acceptor impurity. If a tetravalent atom replaces a trivalent atom then it works as acceptor and if it replaces a pentavalent atom it behaves as a donor. Such impurities are called 'amphoteric' dopant. Compound semiconductors are used to fabricate light-emitting diodes and semiconductor lasers.

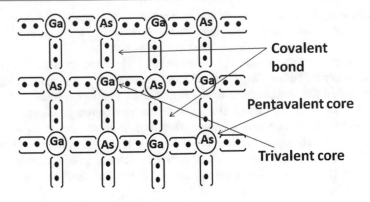

Fig. 4.17 Covalent bonding in tri and pentavalent atoms

Self-assessment question (SAQ): What is the advantage of a compound semiconductor?

4.6.11 Current Flow in Semiconductors

Both intrinsic and extrinsic semiconductors doped by shallow impurities have free electrons and holes at room temperature. Paul Drude, a German physicist, in the year 1900 proposed a theory to explain the current flow in metals. He assumed that free electrons in metals may be considered as an ideal gas of electrons to which kinetic theory of gases may be applied. The same concept of electron gas may also be extended to semiconductors. If so, a free electron in a semiconductor at temperature T may have $3/2kT$ of kinetic energy because of its three degrees of freedom and since $\frac{1}{2} kT$ of energy is associated with each degree of freedom. An electron of mass 9.1×10^{-31} kg and having kinetic energy of $3/2kT$ ($=6.21 \times 10^{-21}$ J) will move with a speed $v_{th} \approx 10^5$ ms^{-1}. However, in absence of any external electric field or any charge gradient, the motion of electrons will be random because of their frequent collisions with vibrating crystal lattice and scattering at sites of impurity atoms. Collision with vibrating lattice may be understood in the following way: at any temperature above absolute zero, the crystal lattice undergoes vibrations due to thermal energy. These thermal vibrations disturb the lattice periodic potential allowing the transfer of energy between lattice and the electron. The process is inelastic scattering where electron may lose or gain energy and change their direction of motion. Since lattice vibrations are also quantized, the quanta of energy associated with lattice vibrations are called **phonon** and hence, the collision between vibrating lattice and electron may also be described in terms of phonon–electron interaction. Shallow impurities are completely ionized at room temperature; therefore, sites of impurity atoms are sites of static charges, when an electron crosses the impurity atom it undergoes Coulomb scattering giving rise to the randomization of the direction of motion. Hence, in absence of any static charge gradient or external electric field, there is no net flow of charge in any direction when averaged over a considerably long period of time.

The current is seen as the net movement of electrons through the periodic potential of crystal lattice. Since in a semiconductor crystal the positive charges are all fixed within the atomic nuclei of the lattice, there can be no positive current in the same sense as for electrons. However, when an electron of the neighbouring atom is captured by its positive charge, resulting in an ionized atom and if this process successively repeats in one direction, it is regarded as current of positive charged holes. Although the motion is only of electrons within valence band, it appears as if holes are moving in the opposite direction and, therefore, the current is called hole current.

A piece of semiconductor at temperature $T K$ is said to be in thermal equilibrium when creation and recombination rates of electron–hole pairs are equal so that the average concentration of electrons n_e and of holes n_p remains constant. Further, at thermal equilibrium $n_e \cdot n_p = n_i^2$. In the state of thermal equilibrium though electrons move with considerable speeds v_{th}, their motion is random and so there is no net current in any direction. Current may, however, be made to flow in a piece of semiconductor (in thermal equilibrium) by many processes. Some of the important processes that may develop current flow are (i) application of external electric field (ii) accumulation of charges generating a spatial charge gradient in semiconductor (iii) injection of excess charges and by (iv) high field operation. Current flow due to these processes is briefly discussed below.

(a) Drift current

As already discussed in the case of metals, if an electric field $\mathcal{E}$ is established in the semiconductor in direction X by applying a potential difference V between the two ends of the crystal D distance apart $(\mathcal{E} = V/D)$, each free electron of charge $(-e)$ experiences a force $F = -(e\mathcal{E})$ in a direction opposite the direction of $\mathcal{E}$. The force F produces acceleration $a = F/m_n^*$, where m_n^* is the effective mass of electron in the material. If τ is the time between two successive collisions and if it is assumed that the electron comes to momentary rest after each collision, the velocity gained by the electron in X-direction opposite to the electric field, called **drift velocity** is given by $v_D^n = a \cdot \tau = -\frac{(e\,\mathcal{E}) \cdot \tau}{m_n^*}$. The proportionality factor between the drift velocity and the applied electric field is called **mobility** of electron and is denoted by μ_n, which is given by

$$v_D^n = -\frac{e \cdot \tau}{m_n^*}\mathcal{E} \text{ and } \mu_n = -\frac{e \cdot \tau}{m_n^*} \tag{4.33}$$

Negative sign on the right side shows that electron moves opposite to the direction of applied electric field.

$$\text{or} \quad \tau = \frac{v_D^n . m_n^*}{e.\mathcal{E}} \tag{4.33a}$$

$$\text{or} \quad v_d^n = -\mu_n \mathcal{E} \tag{4.34}$$

In a similar way, one may define the mobility of holes as

$$v_d^p = \mu_p \mathcal{E} \tag{4.35}$$

The electron mobility μ_n in (4.33) is directly proportional to the collision time τ, larger the magnitude of τ, lesser is collision frequency and more is electron mobility. Electrons may undergo two types of collisions: (i) with the vibrating lattice and (ii) with the impurity atoms. Higher the temperature, more will be the random velocity v_{th} of electrons and higher will be the frequency of vibration of the lattice; resulting in higher number of electron–lattice collisions per unit time. Theoretical calculations show that decrease in electron mobility due to lattice scattering goes as $T^{-3/2}$.

Impurity scattering occurs when electron travels past an ionized impurity atom and its path is deflected due to Coulomb scattering. Slower the electron, more time it will take to pass through the Coulomb field of the impurity ion and hence larger deflection in its path. The probability of impurity scattering depends on the total concentration of both positive and negative impurity atom concentrations. However, impurity scattering probability decreases with the rise of temperature. At higher temperature, the random speed of electron is high and, therefore, it takes lesser time to pass through the Coulomb field of the impurity ion, resulting in smaller probability of scattering. The mobility due to impurity scattering is theoretically estimated to vary as $T^{3/2}/N_{total}$, where N_{total} is the total concentration of both positive and negative impurity ions. This explains the experimentally observed fact that semiconductors have maximum mobility for low doping concentrations. Table 4.5 shows the variation in electron mobility in silicon for different dopants concentrations.

The drift velocity v_D^n gets superimposed along with the random thermal speed v_{th} on every electron under the influence of the electric field. This coordinated motion of electrons in the direction opposite to the applied electric field generates an electron drift current I_D^n. To calculate the magnitude of this drift current, let us

Table 4.5 Electron mobility in silicon at room temperature for different dopants and concentrations

Dopant concentration (atoms m^{-3})	Dopant/Mobility (m^2/V-s)		
	Boron	Arsenic	Phosphorous
10^{21}	46.2×10^{-3}	136×10^{-3}	136×10^{-3}
10^{23}	32.0×10^{-3}	72.0×10^{-3}	71.7×10^{-3}
10^{25}	7.1×10^{-3}	11.0×10^{-3}	12.1×10^{-3}

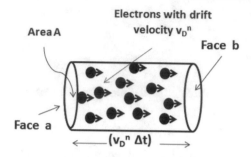

Fig. 4.18 All electrons contained in the volume of cylinder of length $(v_D^n \cdot \Delta t)$ and area of cross section A will cross face b in time Δt

consider an area A in the semiconductor and calculate how much charge will cross this area in time Δt. As shown in Fig. 4.18, all electrons contained in a cylinder of base area A and height $(v_D^n \cdot \Delta t)$ will cross the face b in time Δt. Therefore, the charge Q that will pass through area A in time Δt.

$Q = $ *charge of electron* x *concentration of electrons* $\times$ *volume of the cylinder*
$$- e \cdot n_e \cdot \left(A \cdot v_D^n \cdot \Delta t \right)$$

Therefore, electrons drift current through area $A = I_D^n = \frac{Q}{\Delta t} = -e \cdot n_e \cdot A \cdot v_D^n$.

And the electron current density $j_D^n = \frac{I_D^n}{A} = -e \cdot n_e \cdot v_D^n$ \hfill (4.36)

But $v_D^n = -\mu_n \mathcal{E}$ putting this in (4.36) gives

$$j_D^n = \frac{I_D^n}{A} = -e \cdot n_e \cdot v_D^n = e \cdot n_e \cdot \mu_n \mathcal{E}$$

Similarly, hole drift current density j_D^p may be given as

$$j_D^p = e \cdot n_p \cdot \mu_p \mathcal{E} \tag{4.37}$$

The total drift current density $j_D = j_D^n + j_D^p = \left(e \cdot n_e \cdot \mu_n + e \cdot n_p \cdot \mu_p \right) \mathcal{E}$

$$\text{or} \quad j_D = \sigma \, \mathcal{E} \tag{4.38}$$

where conductivity $\quad \sigma = \left(e \cdot n_e \cdot \mu_n + e \cdot n_p \cdot \mu_p \right)$ \hfill (4.39)

And resistivity

Fig. 4.18a Concentration dependence of electron and hole motilities for intrinsic silicon at room temperature

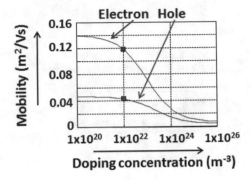

$$\rho = \frac{1}{\sigma} = \frac{1}{\left(e \cdot n_e \cdot \mu_n + e \cdot n_p \cdot \mu_p\right)} \qquad (4.40)$$

Since in doped semiconductors the concentration of majority carrier is much larger than that of the minority carriers, the contribution of the minority carrier to resistivity or conductivity may be neglected. Therefore, above equations reduce to

$$\sigma = e \cdot n_e \cdot \mu_n \text{ and } \rho = \frac{1}{e \cdot n_e \cdot \mu_n} \text{ for N-type materials} \qquad (4.41)$$

$$\text{And} \quad \sigma = e \cdot n_p \cdot \mu_p \text{ and } \rho = \frac{1}{e \cdot n_p \cdot \mu_p} \text{ for P-type materials} \qquad (4.42)$$

The time τ for free electrons between two successive collisions for intrinsic silicon at room temperature is of the order of 10^{-13} s while their thermal random speed v_{th} is $\approx 10^5$ ms^{-1}. The mean free path λ which is the mean distance travelled by an electron between two consecutive collisions is given by $= \tau \cdot v_{th} \approx 10^{-13} 10^5 = 10^{-8}$ m $= 0.01$ μm. Figure 4.18a shows the variation of electron and hole motilities as a function of impurity concentration for silicon at room temperature.

Self-assessment question (SAQ): Can you explain the motion of holes when some external electric field is applied to a semiconductor?

(b) Diffusion current

Diffusion is a universal process in which particles, charged or uncharged, move out from the region of high concentration to the region of low concentration so as to equalize concentrations throughout the system. Diffusion occurs naturally without any external force or field. In semiconductors, if electron or hole concentrations are not uniform, diffusion of charge carriers will take place from higher concentration side towards the lower concentration region and thus generating diffusion currents.

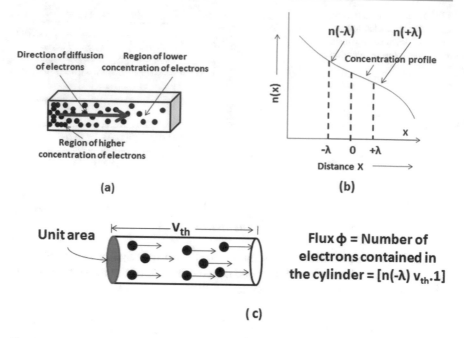

Fig. 4.19 a Diffusion of electrons from higher concentration to lower concentration region.
b One-dimensional concentration profile of electrons. **c** Flux is the number of electrons contained
in a cylinder of base unit and height v_{th}

Concentration gradient of charge carriers, electron and holes, in a semiconductor
may be produced in several ways; for example, by non-uniform doping or by
injecting additional charge carriers on one side by irradiating the crystal by
accelerated ions, etc.

Diffusion occurs at temperatures higher than absolute zero and, therefore, thermal energy of particles undergoing diffusion plays an important role in the process
of diffusion. Let us consider a semiconductor crystal which is non-uniformly doped
by, say, by N-type or donor impurity. Let the concentration of electrons be highest
on the extreme left and then it goes on reducing towards the right side as shown in
Fig. 4.19a. The electrons will diffuse from left side to as is indicated by arrow in the
figure.

Electrons either on the left or on the right will be moving in random directions
with average velocity v_{th} that depends on the temperature $T\,K$ of the specimen.
Further, electrons will be undergoing collisions with vibrating lattice and impurity
atoms. If τ is the mean time between two consecutive collisions, then one may
assign a mean free path λ, as the mean distance travelled by electron between two
consecutive collisions, such that

$$\lambda = v_{\text{th}} \cdot \tau \tag{4.43}$$

For simplicity, let us consider only one-dimensional diffusion along the X-axis. The electron concentration profile is shown in Fig. 4.19b. Let us consider two regions (i) one free path $(-\lambda)$ distance to the left from $X = 0$ and the other (ii) one free path $(+\lambda)$ distance on the right from $X = 0$. We now consider the flow of carrier (electrons) across the plane $X = 0$. If $n(-\lambda)$ denotes the concentration of electrons at $X = -\lambda$, then it is reasonable to assume that half of this number will move towards right and half towards left. Now, these $1/2n(-\lambda)$ will not encounter any collision till they all reach the point $X = 0$ (mean free path λ by definition is the distance in which no collision occurs). Therefore, the flux of electrons (number of electrons crossing a unit area per second) reaching $X = 0$ from the left side is given by

$$\Phi_{\text{left to right}}^{\text{elec}} = all\ electrons\ contained\ in\ cylinder\ of\ base\ unity\ and\ length\ v_{\text{th}}$$

$$\Phi_{\text{left to right}}^{\text{elec}} = \frac{1}{2} n(-\lambda) \cdot v_{\text{th}} \tag{4.44}$$

Similarly, flux of electrons reaching $X = 0$ from the right side

$$\Phi_{\text{left to right}}^{\text{elec}} = \frac{1}{2} n(+\lambda) \cdot v_{\text{th}} \tag{4.45}$$

It is obvious that $\Phi_{\text{left to right}}^{\text{elec}} > \Phi_{\text{right to left}}^{\text{elec}}$, since $n(-\lambda)$ is larger than $n(+\lambda)$. Net flux of charge carriers moving from left to right is, therefore,

$$\Phi_{\text{net}}^{\text{elec}} = \Phi_{\text{left to right}}^{\text{elec}} - \Phi_{\text{right to left}}^{\text{elec}} = \frac{1}{2} \cdot v_{\text{th}} [n(-\lambda) - n(+\lambda)] \tag{4.46}$$

It may be observed that the concentration of electrons reduces from $n(-\lambda)$ to $n(+\lambda)$ over a distance of 2λ. If the mean free path λ is small, the rate of decrement of concentration $(-\frac{dn}{dx})$ may be written as

$$-\frac{dn}{dx} = \frac{n(-\lambda) - n(+\lambda)}{2\lambda} \tag{4.47}$$

Replacing $\frac{[n(-\lambda)-n(+\lambda)]}{2}$ in (4.46) by $(-\lambda \frac{dn}{dx})$, one gets

$$\Phi_{\text{net}}^{\text{elec}} = -v_{\text{th}} \lambda \frac{dn}{dx} \tag{4.48}$$

The electron current density due to diffusion $j_n^{\text{Diff}} = -e\Phi_{\text{net}}^{\text{elec}} = -e \cdot \left(-v_{\text{th}} \lambda \frac{dn}{dx} \right)$

$$\text{or} \quad j_n^{\text{Diff}} = e \cdot v_{\text{th}} \lambda \frac{dn}{dx}$$

Putting $v_{th}\lambda = D_n$, where D_n is called the diffusion coefficient, one gets

$$j_n^{Diff} = e \cdot D_n \frac{dn}{dx} \qquad (4.49)$$

And similarly,

$$j_p^{Diff} = -e \cdot D_p \frac{dn}{dx} \qquad (4.50)$$

At first, it appears that diffusion coefficient D and the mobility of the carrier μ should have no relationship because mobility originates from drift, a process initiated by the application of an external electric field $\mathcal{E}$, while diffusion coefficient D is the outcome of non-uniform concentration of carriers. However, the two parameters, D and μ are related through the collision time τ, which in both cases have the same cause, collisions between charge carriers and vibrating lattice and impurity ions.

In case of one-dimensional motion of charge carrier electrons, their kinetic energy is equal to ½ kT, from the law of equipartition of energy. If m_n^* denotes the **effective mass of electron** in the semiconductor, then

$$\frac{1}{2}kT = \frac{1}{2}m_n^* v_{th}^2 \quad \text{or} \quad kT = m_n^* v_{th}^2$$

$$\text{Also,} \quad \lambda v_{th} = \frac{m_n^* v_{th}^2}{e} \frac{e\tau}{m_n^*} = \frac{kT}{e}\mu_n$$

$$\text{Therefore,} \quad D_n = \mu_n \frac{kT}{e} \text{ and } D_p = \mu_p \frac{kT}{e} \qquad (4.51)$$

Relations described by (4.51) are called Einstein's relations. Further, k stands for Boltzmann constant and 'e' electron charge $= 1.6 \times 10^{-19}$ C. The diffusion coefficients may, therefore, be calculated if the temperature and the mobility of the charge carrier are known.

The sum of drift and diffusion current densities for both type of charge carriers may, therefore, be written as

$$j_n^{total} = en_e\mu_n \mathcal{E} + eD_n \frac{dn}{dx} \text{ and } j_p^{total} = en_p\mu_p \mathcal{E} - eD_p \frac{dn}{dx} \qquad (4.52)$$

The total current density is the sum of the total electron and total hole current densities, i.e.

$$j_{total} = j_n^{total} + j_p^{total} \qquad (4.53)$$

Self-assessment question (SAQ): In what respect diffusion is related to thermal velocity?

(c) Recombination and generation current

At thermal equilibrium in a semiconductor the condition $n_i^2 = n_e n_p$ holds good and the rate of generation of electron–hole pairs by thermal dissociation of covalent bonds is balanced by the rate of their recombination, as such there is no net generation or recombination current. However, if additional charge carriers are injected in the system by some external means breaking thermal equilibrium, the system will become unstable and the extra charge carriers will slowly die out by recombination giving rise to recombination current.

Additional charge carriers may be introduced in the semiconductor crystal by several methods, the two most important are: (i) irradiating the crystal by photons (or light) of energy higher than the forbidden energy gap and (ii) by irradiating the crystal with ionizing radiations like charged particles. In both these processes, equal number of additional electron–hole pairs is created. Suppose a semiconductor crystal at temperature T is in thermal equilibrium and has n_e, n_p densities of electrons and holes, respectively. If the crystal is of N-type, then $n_e \gg n_p$ and electrons will be majority carriers; on the other hand, if it is a P-type crystal, then $n_p \gg n_e$. Now if some additional Δn and Δp electron–hole pairs are introduced in the system by some methods such that the concentration of additional carriers introduced in the system is much smaller than the concentration of the majority carrier but it is comparable to the concentration of minority carriers then the condition is referred as **low-level injection**. For example, for an N-type crystal, low-level injection occurs when $n_e > > \Delta n$ and $n_p \approx \Delta p$. However, if the concentration of injected carriers is comparable or higher than the concentration of majority carriers, the injection is termed as **high-level** injection.

In the case of direct semiconductors, where crystal k-value is same for the top of the valence band and the bottom of the conduction band, direct absorption/emission of photon during excitation/de-excitation is possible in a single step. For such systems and for low-level injection, it can be shown that the net recombination rate U is proportional to the excess minority concentration (Δp for N-type crystal), i.e.

$$U \cong \beta n_e \Delta p \qquad (4.54)$$

Here, β is proportionality constant, n_e is the concentration of majority carriers in thermal equilibrium and Δp is the excess minority concentration.

$$But \quad \frac{dp}{dt} = -U = -\beta n_e \Delta p$$

$$Or \quad p(t) = p(t = 0) + G\tau_p e^{-t/\tau_p}$$

Here, $p(t = 0)$ is initial hole concentration, $\tau_p = \frac{1}{\beta n_e}$ and G a constant.

The above expression tells that the concentration of excess carriers decay exponentially.

For an indirect semiconductor and also for high-level injection the treatment is more complex and involved.

4.6.12 Operation of Semiconductor Under High Field

When a semiconductor is subjected to high electric field $\mathcal{E}$ by applying a large voltage across it, two changes occur.

(i) With the increase of electric field $\mathcal{E}$, the drift velocity v_D increases and when it approaches v_{th}, the collision time τ becomes shorter and the mobility of carriers does not remain constant. Also, the drift velocities of majority and minority carriers attain a saturation value given by

$$v_n^s = \frac{v_s}{\left[1 + \frac{7\times10^5(Vm^{-1})}{\mathcal{E}(Vm^{-1})}\right]^{1/2}} \text{ and } v_p^s = \frac{v_s}{\left[1 + \frac{2\times10^2(Vm^{-1})}{\mathcal{E}(Vm^{-1})}\right]} \tag{4.55}$$

Here, v_s is a constant that has the value 10^5 m/s for silicon at room temperature.

(ii) If the applied electric field is very high, the conduction electrons may acquire very high speed and may be able to create further large number of electrons on collision with other lattice atoms. These liberated electron–hole pairs also acquire high velocities under the applied high electric field, ionizing more atoms of the lattice. This process is called **avalanche** and very high currents, order of magnitude larger, may be produced by this process.

Self-assessment question (SAQ): Under high field operation large avalanche current flows; however, what is the source of the large number of electrons required for large current? Are they free electrons of the conduction band? If not then they come from where?

4.6.13 Hall Effect

Hall Effect measurements provide a practical method of determining the nature (or charge of majority) of carriers as well as their mobility in a semiconductor. The experiment of measuring Hall Effect is based on the fact that free charged carriers moving in a specific direction when subjected to a magnetic field, normal to their direction of motion, experience a magnetic force and get accumulated after getting deflected by the magnetic field. This accumulation of charged particles gives rise to an induced electric field that opposes further deflection of particles. The magnitude

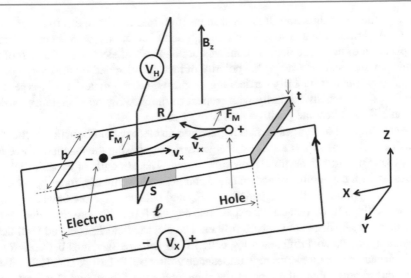

Fig. 4.20 Experimental setup for observing Hall Effect

of this induced electric field increases as more and more particles undergo deflection. However, ultimately a steady state is reached when the magnetic deflection force is counter-balanced by the electric force due to induced electric field. There is no further deflection of charge carriers and consequently no further rise of the electric field. The direction and the magnitude of this induced electric field provide the polarity of majority charge carriers and their mobility.

Experimental setup for observing Hall Effect is shown in Fig. 4.20. A voltage V_x is applied to the two terminals of a rectangular piece of some doped semiconductors of dimensions ℓ, b *and* t. The applied voltage V_x establishes an electric field $E_x\left(=\frac{V_x}{\ell}\right)$ in X-direction that generates a **drift current I_D** to flow in positive X-direction. This drift current will be the sum of both electron current and hole current. The hole current will consist of holes moving with **drift velocity v_x** in positive X-direction while electrons in electron current will be moving with **drift velocity $(-v_x)$** in negative X-direction. Now if a magnetic field B_Z is applied to the semiconductor in Z-direction, the moving charge carriers, both holes and electrons will experience a force F_M, the magnitude of which will be given by Lorentz expression as $F_M = |e(B_z \ X \ v_x)|$. However, the magnetic field is in Z-direction which is normal to the direction of velocity v_x, therefore, the cross product in the bracket will yield

$$F_M = eB_Z v_x \qquad (4.56)$$

The direction of the force F_M is given by right-hand rule as indicated in the figure. As shown in the figure, the magnetic force F_M acts on both electrons and holes in the same direction. It is because electron and hole have electric charges of opposite sign and are also moving in opposite directions with same velocity. Application of magnetic field accumulates both holes and electrons on face R of the semiconductor slab. However, the polarity of face R is decided by the charge of the majority carriers. If majority carriers are electrons, i.e. the material is N-type, face R will acquire negative polarity with respect to the face S, and if majority carriers are hole, R will become positive with respect to S.

Accumulation of charges on face R generates an electric field E_y in the Y-direction which exerts a force $F_E = e \cdot E_y$ on charge carriers opposing the magnetic force F_M, thus reducing further accumulation of charges. The strength of electric field E_y increases with the number of accumulated charges and eventually, force F_E becomes equal in magnitude to the magnetic force F_M and balances it. Now onwards charge carriers in the drift current do not experience any net force, F_M gets cancelled by F_E, and therefore, E_y does not increase further. The electric field E_y is called **Hall field**. Eventually, a potential difference V_H, called **Hall voltage**, develops between R and S which can be measured by a high impedance voltmetre. Polarity of the Hall voltage gives information about the type of the semiconductor. If terminal R is positive it means that the majority carriers are holes and if it is negative they are electrons.

Let us assume that the semiconductor is of P-type so that terminal R is positive and terminal S negative.

In equilibrium, $F_M = F_E$

$$\text{or} \quad e \cdot B_Z \cdot v_x = e \cdot E_y$$

$$\text{or} \quad E_y = B_Z \cdot v_x \tag{4.57}$$

where v_x is the drift velocity of holes. Further, J_D, the drift current density may be obtained from I_D and the area of cross section of semiconductor (b.t) as $j_D = \frac{I_D}{b \cdot t}$.
But from (4.36) and (4.37), we have

$$j_D^p = \frac{I_D^p}{A} = e \cdot n_p \cdot v_D^p = e \cdot n_p \cdot \mu_p \mathcal{E} \tag{4.58}$$

That for the present case gives.
$v_x = \frac{j_D}{e \cdot n_p}$, putting back in (4.57) one gets

$$E_y = B_Z \cdot v_x = \frac{j_D}{e \cdot n_p} B_Z \tag{4.59}$$

The Hall Coefficient R_H is defined as

$$R_H = \frac{Hall\ Field\ E_y}{j_D B_z} = \frac{1}{e n_p} \tag{4.60}$$

Table 4.6 Some important constants

Constant	MKS system	Electron volt
Velocity of light c	2.9979×10^8 m s^{-1}	–
Unit of charge e	1.6×10^{-19} C	–
Mass of electron	9.1091×10^{-31} kg	$511 \times 10^3/c^2$ eV
Planck's constant h	6.62608×10^{-34} J s	4.1357×10^{-15} eV s
Boltzmann's constant k	1.3807×10^{-23} j K^{-1}	8.6173×10^{-5} eV K^{-1}
Avogadro's number N_A	6.0221×10^{26} kilomol^{-1}	–
Energy	1.6×10^{-19} J	1.0 eV

Also the mobility μ_p of majority carrier holes is given as

$$\mu_p = \frac{j_D}{e \cdot n_p E_y} = \frac{j_D R_H}{E_y} \tag{4.61}$$

Hall field E_y may be experimentally obtained as $E_y = \frac{measured\ value\ of\ V_H}{the\ breath\ b}$.
And Hall coefficient

$$R_H = \frac{\frac{V_H}{b}}{\frac{I_D}{t \cdot b}} = \frac{V_H \cdot t}{I_D B_z} \tag{4.62}$$

Thus, the Hall coefficient may be experimentally determined and from it the concentration n_p and mobility μ_p of majority carrier may be obtained. Above formulations will also hold if majority carrier is electron.

Self-assessment question (SAQ): What will happen if magnetic field is applied along the X-direction (Table 4.6)?

Solved example (SE-4.4): Calculate the forbidden energy gap for a semiconductor which is transparent to light of wavelength longer than 0.85 μm.

Solution: *A photon of light of wavelength 0.85 μm is absorbed by the semiconductor, why? It is absorbed because it shifts an electron from the valence band to the conduction band. It means that the forbidden energy gap is equal to the energy of the photon of wavelength 0.85 μm. Now the energy of photon is given by*

$$E = hv = h\frac{c}{\lambda} = \frac{1.24}{\lambda(in\ \mu m)} eV = \frac{1.24}{0.85} eV = 1.46\ eV$$

The forbidden energy gap for the semiconductor is 1.46 eV.
It is worth remembering that energy of a photon in eV can be easily calculated if its wavelength is given in μm, using the relation: Photon energy $hv(in\ eV) = \frac{1.24}{\lambda(\mu m)}$.

Solved example (SE-4.5) Forbidden energy gap E_g for GaAs semiconductor is 1.4 eV at 500 °C. If the Fermi level is 0.20 eV below the bottom of the conduction band, identify the type of the material and calculate the equilibrium majority carrier concentration given that the effective masses of electron is 0.067 m_e.

Solution: *It is given that $E_g = 1.4$ eV and $E_C - E_F = 0.20$ eV, hence $E_F - E_V = 1.20$ eV. Since the Fermi level is not in the middle of E_g and is near to the bottom of conduction band, the material is N-type.*

It follows from (4.24) that

$$n_e = 2\left(\frac{2\pi m_n kT}{h^2}\right)^{\frac{3}{2}}\left(e^{\left(-\left(\frac{E_C - E_F}{kT}\right)\right)}\right)$$ *Substitute the given values of parameters,*

$$n_e = 2\left(\frac{2\pi \times 0.067 \times 9.1 \times 10^{-31} \times 1.38 \times 10^{-23} \times 500}{(6.63 \times 10^{-34})^2}\right)^{3/2}\left(e^{-\left(\frac{0.2 \times 1.6 \times 10^{-19}}{1.38 \times 10^{-23} \times 500}\right)}\right)$$

$$= 9.29 \times 10^{23} \times e^{-4.64} = 9.29 \times 10^{23} \times 9.7 \times 10^{-3} = 9.01 \times 10^{21} \text{ m}^{-3}$$

The majority carrier density is 9.01×10^{21} m^{-3}.

Solved example (SE-4.6) A semiconductor crystal at temperature 300 K has forbidden energy gap of 1.40 eV, effective electron mass of 0.1m_e and effective hole mass of 0.5m_e. Calculate (a) the energy shift of the Fermi level from the middle of the forbidden energy gap (b) the charge carrier concentration of intrinsic material.

Solution: *(i) From (4.28), we have*

$$E_F^i = \frac{(E_C + E_V)}{2} + \frac{3kT}{4}\ln\left(\frac{m_p}{m_e}\right)$$

$$or \quad E_F^i - \frac{(E_C + E_V)}{2} = \frac{3kT}{4}\ln\left(\frac{m_p}{m_e}\right)$$

Therefore, the desired shift of Fermi level from the middle of band gap $E_F^i - \frac{(E_C + E_V)}{2}$

$$= \frac{3kT}{4}\ln\left(\frac{m_p}{m_e}\right) = \frac{3 \times 8.61 \times 10^{-5} \times 300}{4}\ln\frac{0.5m_e}{0.1m_e} = 19.37 \times 10^{-3} \times 1.609$$

$$= 0.031\text{eV}$$

(b) *From (4.30), we have*

$$n_i = \sqrt{N_C N_V} e^{\left(-\frac{E_g}{2kT}\right)} = \left[\left(2\left(\frac{2\pi m_n kT}{h^2}\right)^{\frac{3}{2}}\right)\left(2\left(\frac{2\pi m_p kT}{h^2}\right)^{\frac{3}{2}}\right)\right]^{1/2} e^{\left(-\frac{E_g}{2kT}\right)}$$

$$2\left(\frac{2\pi kT m_e}{h^2}\right)^{3/2}\left(\frac{m_n m_p}{m_e m_e}\right)^{\frac{3}{4}} e^{\left(-\frac{E_g}{2kT}\right)}$$

$$= 2\left(\frac{2\pi \times 1.38 \times 10^{-23} \times 300 \times 9.1 \times 10^{-31}}{(6.62 \times 10^{-34})^2}\right)^{3/2} (.1 \times .5)^{3/4} e^{-\left(\frac{1.4 \times 1.6 \times 10^{-19}}{2 \times 1.38 \times 10^{-23} \times 300}\right)}$$

$$2(1.26 \times 10^{25})(0.106)e^{-27.0} = 0.267 \times 10^{25} \times 1.88 \times 10^{-12}$$
$$= 5.0 \times 10^{12} \text{ m}^{-3}$$

The carrier density in intrinsic crystal is 5×10^{12} *m*$^{-3}$.

Solved example (SE-4.7) Electron carrier density in an intrinsic semiconductor at room temperature is 5.0×10^{12} m^{-3}. The crystal is doped with shallow impurity atoms so that the Fermi level shifted towards the conduction band by 0.35 eV from the middle of the forbidden energy gap. Determine the nature of impurity and its concentration.

Solution: *Since the Fermi level has shifted towards the conduction band, the impurity is donor impurity.*

$$\text{Also,} \quad N_D \cong n_e = n_i e^{-\left(\frac{E_i - E_F}{kT}\right)} = 5 \times 10^{12} e^{-\left(\frac{-0.35}{0.0258}\right)}$$
$$= 5 \times 10^{12} \times e^{13.56} = 5 \times 10^{12} \times 7.74 \times 10^5$$

The impurity concentration is $N_D = 3.87 \times 10^{18}$ *m*$^{-3}$.

Solved example (SE-4.8) Forbidden energy gap for GaAs compound semiconductor is 1.42 eV at 300 K temperature. If introduction of some impurity shifts Fermi level from the middle of band gap by 0.22 eV towards the valence band, find the number density of majority and minority carriers. Take effective masses of electron and holes, respectively, as 0.91×10^{-31} kg and 4.55×10^{-31} kg. Also, calculate the concentration of charge carriers in intrinsic material.

Solution: *The energy band diagram of the system is shown in Fig. SE-4.8. The concentrations of electron and holes are given as*

$$n_e = N_C e^{-\left(\frac{E_C - E_F}{kT}\right)} \quad \text{and} \quad n_p = N_V e^{-\left(\frac{E_F - E_V}{kT}\right)}$$

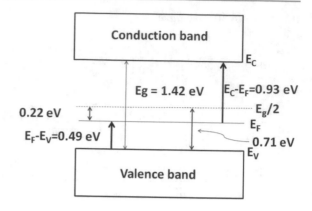

Fig. SE-4.8 Energy band diagram of the given system

Therefore,

$$n_e = 2\left(\frac{2\pi \times 0.91 \times 10^{-31} \times 1.38 \times 10^{-23} \times 300}{(6.62 \times 10^{-34})^2}\right)^{3/2} e^{-\left(\frac{0.93 \times 1.6 \times 10^{-19}}{1.38 \times 10^{-23} \times 300}\right)}$$

$$2 \times 3.97 \times 10^{23} \times 2.2 \times 10^{-16} = 1.74 \times 10^8$$

The electron concentration is $1.74 \times 10^8 \ \mathrm{m}^{-3}$.

The hole (majority carriers) concentration $n_p = N_V e^{-\left(\frac{E_F - E_V}{kT}\right)}$

$$= 2\left(\frac{2\pi \times 4.55 \times 10^{-31} \times 1.38 \times 10^{-23} \times 300}{(6.62 \times 10^{-34})^2}\right)^{3/2} e^{-\left(\frac{0.49 \times 1.6 \times 10^{-19}}{1.38 \times 10^{-23} \times 300}\right)}$$

$$= 2 \times 4.4 \times 10^{24} \times 5.66 \times 10^{-9} = \mathbf{4.98 \times 10^{16} \ m^{-3}}$$

Concentration of intrinsic carriers' $n_i = \sqrt{n_e n_p} = \sqrt{1.74 \times 10^8 \times 4.98 \times 10^{16}}$
$= 2.9410^{12} \ \mathrm{m}^{-3}$.

Solved example (SE-4.9) E_g for a semiconductor is 0.71 eV at 300 K and it is doped with donor impurity of concentration 3×10^{12} and acceptor impurity of concentration $1 \times 10^{12} \ \mathrm{cm}^{-3}$. Calculate the number density of electrons and holes in the compensated semiconductor. Take effective masses of electron and hole, respectively, as 0.91×10^{-31} kg and 4.55×10^{-31} kg.

Solution: *Let us first calculate* $n_i = \sqrt{N_C N_V} e^{-\left(\frac{E_g}{2kT}\right)}$

$$n_i = \left[2\left(\frac{2\pi m_n x k T}{(h)^2}\right)^{3/2} 2\left(\frac{2\pi m_n x k T}{(h)^2}\right)^{3/2}\right]^{1/2} e^{-\left(\frac{E_g}{2kT}\right)}$$

$$or \quad n_i = \left[\frac{2\left(\frac{2\pi \times 0.91 \times 10^{-31} \times 1.38x\,10^{-23} \times 300}{(6.62 \times 10^{-34})^2} \right)^{3/2}}{2\left(\frac{2\pi \times 4.55 \times 10^{-31} \times 1.38 \times 10^{-23} \times 300}{(6.62 \times 10^{-34})^2} \right)^{3/2}} \right]^{1/2} e^{-\left(\frac{0.71}{2 \times 0.0258} \right)}$$

$$= \left[2(54.01 \times 10^{14})^{\frac{3}{2}} 2(270.07 \times 10^{14})^{3/2} \right]^{1/2} e^{-13.76}$$

$$= \left[(7.93 \times 10^{23})(8.87 \times 10^{24}) \right]^{1/2} 1.06 \times 10^{-6}$$

$$or \quad n_i = 26.52 \times 10^{23} \times 1.06 \times 10^{-6} = \mathbf{2.81 \times 10^{18}}$$

Since impurity concentration is given in cm^{-3}, it should be converted into m^{-3} by multiplying by 10^6.

$$N_D = 3 \times 10^{12} cm^{-3} = 3 \times 10^{18}\ m^{-3} and\ N_A = 1 \times 10^{12} cm^{-3} = 1 \times 10^{18}\ m^{-3}$$

To calculate equilibrium concentrations of electrons and holes, following equations may be used.

$$n_e = \left(\frac{N_D^- - N_A^+}{2} \right) + \sqrt{\left(\frac{N_D^- - N_A^+}{2} \right)^2 + n_i^2} \qquad (4.15c)$$

$$And \quad n_p = \left(\frac{N_A^+ - N_D^-}{2} \right) + \sqrt{\left(\frac{N_A^+ - N_D^-}{2} \right)^2 + n_i^2} \qquad (4.15d)$$

$$Therefore, \quad n_e = \left(\frac{3 \times 10^{18} - 1 \times 10^{18}}{2} \right) + \sqrt{(1 \times 10^{18})^2 + (2.81 \times 10^{18})^2}$$

$$or \quad n_e = 1 \times 10^{18} + \sqrt{8.8961} \times 10^{18} = \mathbf{3.98 \times 10^{18}\ m^{-3}}$$

$$And \quad n_p = -1 \times 10^{18} + \sqrt{8.8961} \times 10^{18} = \mathbf{1.98 \times 10^{18}\ m^{-3}}$$

Solved example (SE-4.10) The intrinsic concentration of charge carriers in a semiconductor is $2.0 \times 10^{18}\ m^{-3}$. If electron and hole nobilities are, respectively, 0.38 and 0.12 $m^2\ V^{-1}\ s^{-1}$, what is the resistivity and conductivity of the material?

Solution: *Conductivity* $\sigma = n_i e(\mu_e + \mu_p) = 2.0 x 10^{18} x 1.6 x 10^{-19}(0.38 + 0.12) = 0.16 \Omega^{-1} m^{-1}$

$$or \quad \sigma = 0.16\Omega^{-1}m^{-1}$$

$$And \; resistivity \quad \rho = \frac{1}{\sigma} = \frac{1}{0.16} = 6.25 \; \Omega - m$$

Solved example (SE-4.11) The Hall coefficient for a certain specimen at room temperature is $7.0 \times 10^{-5} \, m^3C^{-1}$ and its conductivity is $150 \, \Omega^{-1} \, m^{-1}$. What are the values of the concentration and mobility of majority carriers?

Solution: *Hall coefficient* $R_H = 1/n.e$

$$Hence, \quad n = \frac{1}{e \cdot R_H} = \frac{1}{1.6 \times 10^{-19} \times 7.0 \times 10^{-5}} = 8.92 \times 10^{22}$$

$$Also \quad \mu = \frac{\sigma}{n \cdot e} = \sigma \cdot R_H = 150 \times 7.0 \times 10^{-5} = 1.05 \times 10^{-2} \, m^2V^{-1}s^{-1}$$

Solved example (SE-4.12) A Hall voltage of 30 μV developed across the breath of a rectangular piece of a semiconductor, when current of 20 mA is made to pass lengthwise through the semiconductor slab and a magnetic field of $0.5 Wb \, m^{-2}$ is applied perpendicular to the largest face of the slab. Calculate the magnitude of Hall coefficient given that the thickness of the slab is 1 mm.

Solution: *From (4.62), we have Hall coefficient* $R_H = \frac{V_H \cdot t}{I_D B_z}$.
Substituting the values of different parameters given in the question one gets

$$R_H = \frac{V_H \cdot t}{I_D B_z} = \frac{30 \times 10^{-6} \times 1 \times 10^{-3}}{20 \times 10^{-3} \times 0.5} = 3.0 \times 10^{-6} \; per \; Coulomb \; per \; meter \; cube$$

Solved example (SE-4.13) Find the resistance of a semiconductor slab of size 10 cm x 2 mm x 1 mm, given that intrinsic carrier concentration is $2.0 \times 10^{19} \, m^{-3}$ and mobility of electrons and holes are, respectively, 0.40 and 0.20 $m^2 \, V^{-1} \, s^{-1}$.

Solution: *Using the relation, Conductivity* $\sigma = n_i e (\mu_e + \mu_p)$.
And substituting the given values, one gets

$$\sigma = n_i e (\mu_e + \mu_p) = 2 \times 10^{19} \times 1.6 \times 10^{-19}(0.4 + 0.2) = 1.92 \, \Omega^{-1} \, m^{-1}$$

Therefore, resistivity $\rho = \frac{1}{1.92} = 0.52$ *Ω-m.*
Resistance of the slab $R = \rho \frac{L}{A} = 0.52 \times \frac{10 \times 10^{-2}}{2 \times 10^{-3} \times 1 \times 10^{-3}} = 2.5 \times 10^4 \, \Omega$.

Solved example (SE-4.14) In semiconductor silicon the drift velocity of electron remains unchanged from 1×10^5 m/s when field is changed from 30 kV/cm to

150 kV/cm. Calculate the mean scattering time for both fields and explain why the drift velocity does not change. Take the effective electron mass to be $0.26\, m_e$.

Solution: *From (4.33a), we have* $\tau = \frac{v_D^n \cdot m_n}{e \cdot \mathcal{E}}$.

Let us calculate the mean scattering time τ at $\mathcal{E} = 30\, \text{kV/cm} = 30 \times 10^3 / 1 \times 10^{-2}\, \text{V/m} = 3 \times 10^6\, \text{V/m}$

$$\tau\left(at\, 3 \times 10^6\, \frac{V}{m}\right) = \frac{1 \times 10^5 \times 0.26 \times 9.1 \times 10^{-31}}{1.6 \times 10^{-19} \times 3 \times 10^6} = 4.929 \times 10^{-14}\, s$$

$$Also\, at \quad \tau\left(at\, 15 \times 10^6\, \frac{V}{m}\right) = \frac{1x10^5 \times 0.26 \times 9.1 \times 10^{-31}}{1.6 \times 10^{-19} \times 15 \times 10^6} = 9.85 \times 10^{-15}\, s$$

The drift velocity constant between 3 MV/m and 15 MV/m fields as the scattering time becomes shorter, allowing the electron to accelerate for shorter times.

Solved example (SE-4.15) If the mobility of electron in some semiconductors at 300 K is 1100 $\text{cm}^2\text{V}^{-1}\,\text{s}^{-1}$ and the effective mass is 0.26 times the rest mass of electron, calculate the mean free path of electron.

Solution: *The mobility μ is given in CGS system; let us convert it in MKS system.*

$$\mu = 1100\, \text{cm}^2\, per\, Volt\, per\, second = 0.11 \text{m}^2\text{V}^{-1}\text{s}^{-1}$$

Now from (4.33), we have $\mu_n = -\frac{e \cdot \tau}{m_n}$.

Leaving the negative sign that only implies that the direction of motion of electron is opposite to the direction of electric field, we have

$$\tau = \frac{\mu_n m_n}{e} = \frac{0.11 \times 0.26 \times 9.1 \times 10^{-31}}{1.6 \times 10^{-19}} = 1.627 \times 10^{-13}\text{s}$$

Also, the thermal velocity v_{th} at 300 K may be obtained from the law of equipartion of energy as

$$\frac{1}{2}m_n v_{\text{th}}^2 = \frac{3}{2}kT \ or \ v_{\text{th}} = \sqrt{\frac{3 \times 1.38 \times 10^{-23} \times 300}{0.26 \times 9.1 \times 10^{-31}}} = 2.291 \times 10^5\, \text{m/s}$$

The mean free path $\lambda = v_{\text{th}} \cdot \tau = 2.291 \times 10^5 \times 1.627 \times 10^{-13} = 3.727 \times 10^{-8}$ **m**.

Solved example(SE-4.16): The electron concentration in a non-uniformly doped semiconductor is given as $n(x) = \left(1 \times 10^{23}\right)e^{-\left(\frac{x}{3\times10^{-4}}\right)}$ for $x > 0$. If Diffusion coefficient $D_n = 30 \times 10^{-4}\, \text{m}^2$ s, compute the drift current density at $x = 3 \times 10^{-4}$ m.

Solution: *From EQ. (4.49), we have*

$$j_n^{\text{Diff}} = e. \quad D_n \frac{dn}{dx}. \tag{4.49}$$

In order to use the above expression for diffusion current density, we first calculate (dn/dx) at $x = 3 \times 10^{-3}$ m.

$$\text{Now,} \quad n(x) = \left(1 \times 10^{23}\right) e^{-\left(\frac{x}{3 \times 10^{-4}}\right)}$$

$$\text{Therefore,} \quad -\frac{dn}{dx} = \left(1 \times 10^{23}\right) \left(-\frac{1}{3 \times 10^{-4}}\right) e^{-\left(\frac{x}{3 \times 10^{-4}}\right)}$$

$$\text{And} \quad \left(\frac{dn}{dx}\right)_{x=3 \times 10^{-4}} = 3.33 \times 10^{26} \left(e^{-1} = 0.367\right) = 1.22 \times 10^{26}$$

Therefore,

$$j_n^{\text{Diff}} = e \cdot D_n \frac{dn}{dx} = 1.6 \times 10^{-19} \times 30 \times 10^{-4} \times 1.22 \times 10^{26} = 5.86 \times 10^4 \text{ Am}^2$$

Problems

P4.1 With the following data calculate number density of free electrons and the frequency of collisions of free electrons at 300 K in a gold rod of length 1 m and area of cross section 0.1 m². Given data: Valency of Gold 1; Molar mass of Gold 196.96 gm/mole; density of Gold 19.32; conductivity of Gold 4.10×10^7 S/m; mass of electron 9.1×10^{-31} kg; charge of electron 1.6×10^{-19} C.

Answer: [n = 5.90×10^{28} m^{-3}; 4.0×10^{13} collisions per second].

P4.2 A wire of diameter 0.02 m carries current of 100 A. If the number density of free electrons in the material of the wire is 1×10^{28} m^{-3}, calculate the drift velocity of electrons in the wire.

Answer: [3.9×10^{-4} m/s].

P4.3 For a semiconductor of forbidden energy gap of 0.7 eV, effective electron and hole masses, respectively, of 0.1 m_e and 0.6 m_e (m_e = 9.1 X 10^{-31} kg) determine the position of Fermi level at temperature 300 K.

Answer: 0.384 eV that me Solution: 0.034 eV towards the conduction band from the middle of band gap.

P4.4 The intrinsic carrier density in a germanium crystal was 2.1 × 10^{19} m^{-3}. It was doped with Boron of concentration 4.5 × 10^{23} atoms m^{-3}. If electron and hole mobility are, respectively, 0.4 and 0.2 m^2 V^{-1} s^{-1} calculate the conductivity of the intrinsic and doped semiconductor.

Answer: [2.01 and 1.44 × 10^4 Ω^{-1} m^{-1}].

P4.5 The forbidden energy gap for germanium is 0.67 eV. What is the intrinsic carrier concentration at 300 K if the effective electron and hole masses are, respectively, 0.12 and 0.28 times the electron mass?

Answer: [2.34 × 10^{18} m^{-3}].

P4.6 Determine the Hall coefficient for N-type semiconductor for which electron density is 3 × 10^{22} and electrical conductivity 100 Ω^{-1} m^{-1}. Also compute the value of electron mobility.

Answer: [R_H = 2.08 × 10^{-4} m^{-3}C^{-1}; μ = 2.08 × 10^{-2} m^{2}V^{-1} s−1].

P4.7 Calculate the resistance of a germanium road of length 10 cm, radius 1 mm, given that the intrinsic carrier density at room temperature is 2.37 × 10^{19} m^{-3} and electron and hole mobility are, respectively, 0.38 and 0.18 m^{2}V^{-1} s^{-1}.

Answer: [14989 Ω].

P4.8 Calculate the value of the thermal energy (kT) at room temperature of 300 K both in joules and eV.

Answer: [4.1421 × 10^{-21} J; 0.02588 eV].

P4.9 The conductivity of an N-type silicon semiconductor at room temperature (300 K) is 0.3(Ω cm)$^{-1}$. Calculate the position of the Fermi level in the material.

Answer: [0.27 eV below the bottom of conduction band].

P4.10 A piece of a semiconductor is doped with $N_D = 3 \times 10^{21}$ m^{-3} and $N_A = 2 \times 10^{21}$ m^{-3}. If Eg for the semiconductor is 1.0 eV and it is at room temperature, calculate the number density of intrinsic charge carriers and hence find the concentration of majority and minority carriers in the doped material. The effective mass of electron and hole in the material are, respectively, 0.12 and 0.26 times the electron mass.

Answer: $[n_i = 7.07 \times 10^{16}$ m^{-3}, *Since* $n_i \ll N_D$, majority carriers are electrons and their concentration ne = $(N_D - N_A) =$ (3–2) × 1021 m^{-3} = 1 × 10^{21} m^{-3}.

Minority concentration n$_p$ = n$_i$^{2}/n$_e$. = 4.99 × 10^{12}].

Multiple-Choice Questions

Note: Some of the questions may have more than one correct alternative; all correct alternatives must be ticked for complete answer in such cases.

MC4.1 Which of the following quantity(ies) of charged carriers remains nearly constant over a range of the applied voltage across the semiconductor rod?

  (a) Thermal speed
  (b) Drift velocity
  (c) collision frequency
  (d) minority concentration.

ANS: [(a), (b) and (d)].

MC4.2 With the increase in the dopant concentration the bulk resistance of the semiconductor

  (a) Decreases
  (b) remains constant
  (c) increases
  (d) may increase or decrease depending on the material.

ANS: [(a)].

MC4.3 If n_i, n_e and n_p, respectively, denote the number density of charge carriers in intrinsic material of electrons and holes in a P-type semiconductor, then

  (a) $n_i > n_p$
  (b) $n_i > n_e$
  (c) $n_p < n_e$

(d) $n_e > n_i$.

ANS: [(b)].

MC4.4 The resistance of a piece of a conductor increases with temperature because

(a) Charge carrier concentration decreases with the rise of temperature
(b) charge carrier collision time increases with temperature
(c) charge carrier collision time decreases with temperature
(d) charge carrier mean free path increases with temperature.

ANS: [(c)].

MC4.5 Which of the following have covalent bonding?

(a) GaAs
(b) Diamond
(c) NaCl
(d) Cu.

ANS: [(a), (b)].

MC4.6 Electron energy bands in crystalline material develop because of

(a) nuclear-nuclear interactions
(b) nuclear-electron interactions
(c) core electron interactions
(d) valence electrons interactions.

ANS: [(d)].

MC4.7 Forbidden energy gap in metals is of the order of

(a) > 10 eV
(b) > 3 eV
(c) < 3 eV
(d) 0 eV.

ANS: [(d)].

MC4.8 Materials in which valence electrons of only one spin orientation contribute to current conduction are called

(a) Metalloids
(b) semiconductors
(c) metals

   (d)   half metals

ANS: [(d)].

MC4.9 Semiconductors with negligible band gap are called

   (a)   Intrinsic semiconductors
   (b)   semimetals
   (c)   non-metals
   (d)   half metals

ANS: [(b)].

MC4.10 Diffusion current is caused by

   (a)   Potential gradient
   (b)   concentration gradient
   (c)   temperature gradient
   (d)   none of these.

ANS: [(b)].

MC4.11 Which of the following is/ are better for photonic applications?

   (a)   Metalloids
   (b)   half metals
   (c)   indirect semiconductors
   (d)   direct semiconductors.

ANS: [(d)].

MC4.12 Avalanche current in semiconductors flows at

   (a)   Very high temperature
   (b)   very high acceptor concentration
   (c)   very high donor concentration
   (d)   very large drift voltage.

ANS: [(d)].

MC4.13 In a P-type non-degenerate semiconductor donor level is

   (a)   In a band just above the top of valence band
   (b)   in a band just above the Fermi level
   (c)   is a discrete level just above the valence band

(d) is a discrete level just above the Fermi level.

ANS: [(c)].

MC4.14 A compensated semiconductor is doped with, respectively, N_D and N_A shallow donor and acceptor concentrations. The free charge density in semiconductor is

(a) $(n_p - n_e + N_A - N_D)$
(b) $(n_e - n_p + N_A - N_D)$
(c) $(n_p - n_e + N_D - N_A)$
(d) $(n_e - n_p + N_D - N_A)$.

ANS: [(a)].

MC4.15 If m_e, m_n^* and m_p^*, respectively, represent the rest mass of electron, effective mass of electron and effective mass of hole, then

(a) $m_e > m_n^* < m_p^*$
(b) $m_e > m_p^* < m_n^*$
(c) $m_e < m_n^* < m_p^*$
(d) $m_e = m_n^* = m_p^*$

ANS: [(a)].

MC4.16 The common factor that affects both the mobility μ and the diffusion coefficient D is

(a) Concentration n_p of holes
(b) concentration n_e of electrons
(c) collision time τ
(d) electric field $\mathcal{E}$

ANS: [(c)].

MC4.17 If in Hall affect experiment n, e, j, μ and $\mathcal{E}_H$ respectively, stand for majority carrier concentration, unit charge, current density, majority carrier mobility and the Hall electric field, then

(a) $R_H = n.e$
(b) $R_H = \frac{1}{n.e}$
(c) $R_H = \frac{\mathcal{E}_H.\mu}{j}$
(d) $R_H = \frac{j}{\mathcal{E}_H.\mu}$

ANS: [(b), (c)].

MC4.18 Tick the correct statements.

(a) Phonon is a quanta of vibrational energy
(b) diffusion of charge carriers in semiconductorsoccurs because of their electrical charges
(c) diffusion of carriers will still occur at absolute zero temperature
(d) thermal velocity v_{th} plays important role in diffusion.

ANS: [(a), (d)].

MC4.19 Which of the following atoms may be used as amphoteric dopant for compound semiconductors of tri and pentavalent elements?

(a) Si
(b) Ge
(c) Sn
(d) Au

ANS: [(a), (b)].

MC4.20 Let Δn and Δp denote the number density of additional electrons and holes injected in a P-type semiconductor having carrier densities n_p and n_e at temperature $T\ K$. For low-level injection

(a) $\Delta p > n_p$ and $\Delta n > n_e$
(b) $\Delta p < n_p$ and $\Delta n > n_e$
(c) $\Delta p > n_p$ and $\Delta n \approx n_e$
(d) $\Delta p < n_p$ and $\Delta n \approx n_e$.

ANS: [(d)].

Short Answer Questions

SA4.1   Give a brief description of the classification of crystalline materials on the basis of their electrical properties.
SA4.2   How electron energy bands originate in crystalline materials?
SA4.3   What are metals, half metals and semimetals? In what respect they are different from each other?
SA4.4   What is the difference between free electrons and delocalized electrons?
SA4.5   Explain why the resistance of a piece of conductor increases with rise of temperature while it decreases for the case of semiconductor.

SA4.6   Define mobility of a charge carrier and qualitatively explain why it decreases with the increase in the dopant concentration.

SA4.7   What is meant by the direct and indirect band gaps?

SA4.8   State the mass action law and give one of its applications.

SA4.9   Draw a rough sketch in 2-D for covalent bonds between tri and pentavalent elements constituting compound semiconductors.

SA4.10   Discuss in short the origin of Hall Effect.

SA4.11   What is meant by thermal equilibrium in the case of semiconductors?

SA4.12   Briefly outline the flow of drift current in a doped semiconductor.

SA4.13   What is thermal velocity of charge carriers in semiconductor? Estimate its magnitude at room temperature.

SA4.14   Under what conditions a piece of semiconductor material will behave as a conductor?

SA4.15   Discuss in short the process of diffusion in one dimension in semiconductors.

Sample answer to short answer question SA4.11

Thermal equilibrium in semiconductors

Equilibrium of a system essentially means that the system is in a stable state. For example, a chemical reaction attains equilibrium when rates of forward and backward reactions become equal. Similarly, in a semiconductor at any temperature $T > 0$ K, both creation of new electron–hole pairs by thermal breaking of covalent bonds and their reformation by the recombination of electron with hole go on simultaneously. In the state of thermal equilibrium, the rates of these two opposite processes become equal so that the average number densities of electron n_e and holes n_p over a period of time become constant. Another outcome of thermal equilibrium in semiconductors is that there is no flow of current in the system and the Fermi level of the system has no gradient. Any gradient in Fermi level will result in current flow. Further, the free charge density in thermal equilibrium is zero. For example, in case of a compensated semiconductor doped, respectively, with N_D and N_A concentrations of shallow donor and acceptor impurities such that all impurity atoms are ionized, the density of free positive charge will be $(n_p + N_A^+)$, where N_A^+ is the number density of holes created by the ionization of acceptor atoms, and the density of free negative charge will be $-(n_e + N_D^-)$, where N_D^- is the density of (negative) electrons provided by the ionization of N_D donor atoms. Therefore, the total density of free charges in the system is

$$\rho(density\ of\ free\ charges) = n_p - n_e + N_A^+ - N_D^-$$

If the semiconductor system is in thermal equilibrium, then ρ must be equal to zero.

Long Answer Questions

LA4.1 Discuss the electron band theory for crystalline solids, giving its assumptions and their validity. What are delocalized and free electrons? Distinguish between metals and semimetals.

LA4.2 What is covalent bonding? Discuss covalent bond structure of intrinsic and of doped semiconductors. What is meant by electron gas in semiconductors? Define thermal velocity and obtain its value at room temperature.

LA4.3 What may be the possible reasons for current flow in a semiconductor? Discuss drift current and obtain expressions for mobility of charge carriers. Why does the mobility decreases with the increase of dopant concentration?

LA4.4 Under what circumstances diffusion current may flow in a semiconductor? Derive expression for drift current in one-dimensional diffusion. What parameter correlates mobility with diffusion and why?

LA4.5 What is Hall Effect? With the help of a suitable diagram explain the experimental setup for the measurement of Hall coefficient.

LA4.6 What do you understand from 'level density' and 'occupation number' with reference to the electron band theory? Obtain expression for electron number density in conduction band for a semiconductor crystal.

p-n Junction Diode: A Basic Non-linear Device

5

Abstract

A p-n junction diode is formed when two sides of an intrinsic semiconductor crystal are doped by n-type and p-type materials. The transfer characteristics of a junction diode are non-linear. The non-linear property of junction diode is used in making devices like signal detector, etc. Junction diodes have many other applications. Physics of the working of a junction diode and large number of its applications are presented in this chapter.

5.1 Introduction

p-n junction, generally called diode, is the basic non-linear device which has been extensively used in electronics as a rectifier, detector, current limiter and voltage regulator, etc. As light-emitting diode (LED), it is replacing conventional light producing bulbs, fluorescent tubes, etc. Non-linear response of diodes has been used to detect electronic signals. Some details of the physics of operation of p-n junction diode along with some important applications are presented in this chapter.

5.2 p-n Junction in Thermal Equilibrium

A junction diode is fabricated by developing a p-type and an n-type region **in a single crystal** either by epitaxial growth, implantation or by diffusion. It is a two-terminal device. In order to understand the working of a diode, it is convenient to start with two separate crystals, one p-type and the other n-type and later to join them, removing the surfaces of the joint face so that it becomes a single crystal. P-type and N-type semiconductors are shown in Fig. 5.1. The P-type material has

© The Author(s), under exclusive license to Springer Nature Switzerland AG 2021
R. Prasad, *Analog and Digital Electronic Circuits*, Undergraduate Lecture
Notes in Physics, https://doi.org/10.1007/978-3-030-65129-9_5

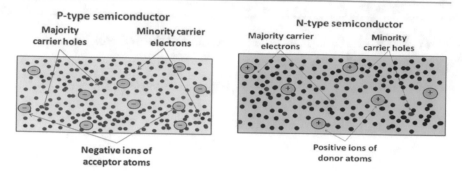

Fig. 5.1 P- and N-type semiconductors

holes as majority carriers, electrons as minority carriers and negatively charged ions of acceptor atoms which are bound with the crystal lattice. On the other hand, in N-type material electrons are majority carrier, holes minority carrier and immobile positive ions of donor atoms bound with the lattice. When the P-type or the N-type crystal is in thermal equilibrium, the total negative and positive charges in the crystal are equal in magnitude and the crystal is electrically neutral. The corresponding energy band diagram of the two types of material is shown in Fig. 5.2. It may be noted that Fermi levels E_F^P and E_F^N, respectively, for the P-type and the N-type isolated crystals are at different energies.

Self-assessment question (SAQ): Two separate crystals are doped one with donor impurity and the other with acceptor impurity in equal amounts. Fermi levels in these two crystals will be at same energy or at different energies?

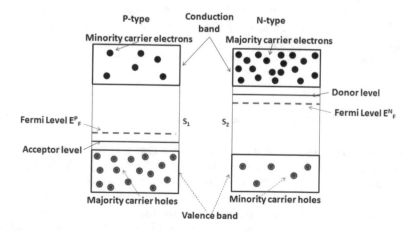

Fig. 5.2 Energy band diagram of P-type and N-type crystals.

The two crystals are now brought near to each other such that their surfaces S_1 and S_2 touch each other. Normally one will expect that holes which are in majority on the P-side and electrons which are in majority on the N-side will diffuse to the other side. But this will not happen as the interfaces S_1 and S_2 of two crystals will not allow the transfer of charges. However, if the interface surfaces S_1 and S_2 are removed, so that the two crystals become a single crystal and charges may move freely throughout, then following processes will occur.

Self-assessment question (SAQ): Free electrons and holes in a semiconductor at room temperature move with large speeds but they do not come out of the crystal, why?

(i)   Holes will diffuse from P-side to the N-side and electrons will diffuse from N-side to P-side on account of the difference in their concentrations on the two sides.

(ii) Before diffusion of charge carriers both the P and the N sides were electrically neutral as they were in thermal equilibrium and also because atoms of both the parent bulk material and of the dopants were neutral, hence there was no excess charge on either side.

(iii) However, the diffusion of positively charged holes from the P-side leaves negatively charged ions of acceptor atoms on the P-side and similarly, diffusion of electrons from the N-side leaves positively charged donor atoms. These negative and positive stationary charges that develop, respectively, on P and N-sides as a result of diffusion of majority carriers are referred as **uncovered** charges.

(iv) With the generation of uncovered stationary charges on the two sides of the interface or the metallurgical junction, an electric field ϵ, directed from the N-side towards the P-side, develops across the p-n junction. The strength of the electric field increases with the diffusion of majority carriers. However, the direction of electric field ϵ is such that it opposes further diffusion of majority carriers.

(v)   With the establishment of the electric field ϵ a drift current made up of minority carriers also comes in to play. The minority carriers drift under the influence of the electric field ϵ, holes from N-side to P-side and electrons from P-side drift to N-side.

(vi) Diffusing and drifting free charges across the junction recombine and those left are swept towards respective layers of uncovered charges. As such in a small region around the junction consisting of negatively charged totally and partially uncovered acceptor ions on the P-side and positively charged totally and partially uncovered donor ions on the N-side is a region where no free charges may stay. Since the region is depleted of all mobile charges this region is called **depletion region (or depletion layer) or the space charge region.**

(vii) On the P-side from the metallurgical boundary there develops a strip of totally uncovered negative ions and then narrow transition region of partially uncovered negative ions. Beyond the transition region is the bulk

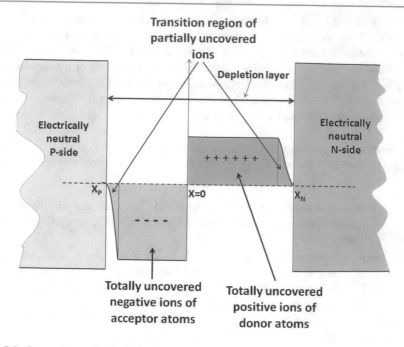

Fig. 5.3 Space charge distribution of an abrupt p-n junction in thermal equilibrium.

P-type semiconductor which is electrically neutral. Similarly, on the N-side, there is a layer of totally uncovered positive ions which is followed by a thin transition layer of partially uncovered positive ions which merges into electrically neutral bulk N-type material. Thin layers of partially uncovered ions on both sides of the junction are called **transition regions**. The region lying between the beginning of transition region on P-side till the end of the transition region on the N-side is the region of space charge distribution as shown in Fig. 5.3.

The processes listed above occur within less than 10^{-12} s of the formation of metallurgical junction and during this time the system passes through unsteady state of no thermal equilibrium. However, soon thermal equilibrium is established leading to the steady state where there are no currents in the system.

Self-assessment question (SAQ): At the instant a p-n junction is formed both majority and minority currents flow through the junction, before the establishment of thermal equilibrium. What causes these currents and why do these currents stop at thermal equilibrium?

5.2.1 Extension of Depletion Layer on Two Sides of the Junction

Since the width of transition regions on the two sides of the junction is very small as compared to the widths of the respective totally uncovered charges, it is assumed that charge distributions are rectangular up to the distance X_p on the P-side and X_N on the N-side of the junction. Further, to maintain the overall electric neutrality of the system in thermal equilibrium, the total amount of uncovered negative charge on the P-side of the junction must be equal to the total amount of positive uncovered charge on the N-side. Let A be the area of cross section of the crystal and N_A, N_D, respectively, the concentrations of acceptor and donor impurities and if it is assumed that impurities were shallow so that all impurity atoms have ionized at the crystal temperature, then

Total uncovered negative charge on P-side = A. N_A. X_P
Total uncovered positive charge on N-side = A. N_D.X_N
On equating the charges on the two sides, one gets

$$N_A X_P = N_D X_N \text{ or } \frac{N_A}{N_D} = \frac{X_N}{X_P} \tag{5.1a}$$

Equation (5.1a) suggests that extension of depletion layer on a given side of the junction is inversely proportional to the dopant concentration on that side. If $N_A > N_D$ then $X_P < X_N$.

5.2.2 Position of Fermi Level for a p-n Junction in Thermal Equilibrium

As explained above as soon as the p-n junction is formed, when the system is still not in thermal equilibrium, both the drift and the diffusion currents start to flow across the junction. However, soon the system attains thermal equilibrium and flow of all types of currents comes to an end. No current may flow in thermal equilibrium. From (4.52), the total hole current which is the sum of the drift and diffusion currents of holes is given by

$$j_p^{total} = e n_p \mu_p \varepsilon - e D_p \frac{dn_p}{dx} \tag{5.1}$$

Also from (4.51) $D_p = \mu_p \frac{kT}{e}$ substituting this value of D_p back in (5.1), one gets

$$j_p^{total} = e n_p \mu_p \varepsilon - \mu_p kT \frac{dn_p}{dx} \tag{5.2}$$

In (5.2), n_p is the concentration of holes, 'e' the quantum of charge (1.6×10^{-19} C) that each hole has, μ_p the hole mobility and ϵ electric field responsible for drift of holes. Under electric field ϵ each hole experiences a force F_p in the direction of the electric field (in +X-direction) which is given as

$$F_p = e\epsilon = -\frac{dE_V}{dx} \tag{5.3}$$

Here, E_v is the potential energy at the top of valence band. Further, the concentration of holes n_p is given by

$$n_p = N_V e^{-\left(\frac{E_F - E_V}{kT}\right)}$$

Therefore,

$$\frac{dn_p}{dx} = \frac{n_p}{kT}\left(\frac{dE_V}{dx} - \frac{dE_F}{dx}\right) \tag{5.4}$$

Substituting the values of $e\epsilon$ from (5.3) and of $\frac{dn_p}{dx}$ from (5.4) in (5.2) one gets

$$j_p^{total} = n_p\mu_p\left[-\frac{dE_V}{dx}\right] - \mu_p kT\left[\frac{n_p}{kT}\left(\frac{dE_V}{dx} - \frac{dE_F}{dx}\right)\right] = n_p\mu_p\left[\frac{dE_F}{dx}\right]$$

$$\text{Or} \quad j_p^{total} = n_p\mu_p\left[\frac{dE_F}{dx}\right] \tag{5.5}$$

However, in thermal equilibrium $j_p^{total} = 0$, and, therefore, from (5.5)

$$\left[\frac{dE_F}{dx}\right]_{thermalequilibrium} = 0 \tag{5.6}$$

Equation (5.6) tells that **in thermal equilibrium the Fermi level is a straight line or has the same energy throughout the crystal from the P-side to the N-side across the p-n junction.** This may be compared to the case when two crystals were separate and not fused into each other, as shown in Fig. 5.2 the Fermi levels E_F^P and E_F^N were at different energies. Once the two crystals are fused to become a single crystal and thermal equilibrium is established, Fermi level of the system becomes a straight line, pulling down the band structure of the N-side with respect to the P-side as shown in Fig. 5.4.

5.2.3 Built-In Potential V_{bi}

With the uncovering of immobile negative and positive charges, respectively, on the P-and N-sides of the junction, a potential difference, called **built-in potential,** develops between the bulk P- and N- sides of the junction. It is denoted by V_{bi} and

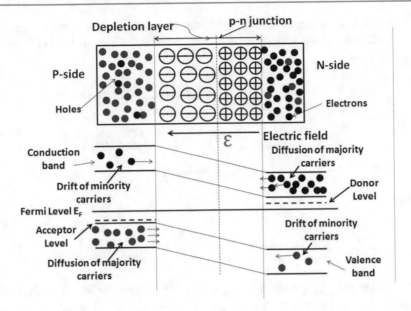

Fig. 5.4 Band structure of a p-n junction in thermal equilibrium

is equal to the difference in the potentials of the N-side V_N and the potential V_P of the P-side, i.e.

$$V_{bi} = V_N - V_P \tag{5.7}$$

Since at thermal equilibrium no current may flow in the system and, therefore, equating the total hole current density j_p^{total} given by (5.2) to zero, one gets

$$j_p^{total} = en_p\mu_p\varepsilon - eD_p\frac{dn_p}{dx} = 0$$

$$\text{Or} \quad en_p\mu_p\varepsilon = eD_p\frac{dn_p}{dx} \tag{5.8}$$

$$\text{Or} \quad \frac{\mu_p}{D_p}\varepsilon(x) = \frac{1}{n_p(x)}\frac{dn_p(x)}{dx}$$

Since presently we are discussing only one-dimensional problem, the dependence of the electric field ε and the hole density n_P on distance x from the metallurgical junction has been included in (5.8). Further, using the relation $D_p = \mu_p\frac{kT}{e}$, (5.8) may be written as

$$\frac{e}{kT}\varepsilon(x) = \frac{1}{n_p(x)}\frac{dn_p(x)}{dx} \quad \text{or} \quad \varepsilon(x) = \frac{kT}{e}\left(\frac{1}{n_p(x)}\frac{dn_p(x)}{dx}\right) \tag{5.9}$$

It is well known that an electric field may be derived as negative gradient of a scalar potential. Let the scalar potential V be the cause of electric field $\epsilon(x)$ so that one may write

$$-\frac{dV}{dx} = \epsilon(x) = \frac{kT}{e}\left(\frac{1}{n_p(x)}\frac{dn_p(x)}{dx}\right)$$

$$\text{Or}\quad \int_{-X_P}^{X_n} Vdx = -\frac{kT}{e}\int_{(n_P)_{at\,x=-X_P}}^{(n_P)_{at\,x=X_N}}\left(\frac{dn_p(x)}{n_p(x)}dx\right)\tag{5.10}$$

Or

$$V_{bi} = V_N - V_P = \int_{-X_P}^{X_n} Vdx = \frac{kT}{e}\ln\left(\frac{(n_P)_{at\,x=-X_P}}{(n_P)_{at\,x=X_N}}\right)\tag{5.11}$$

In (5.11), $(n_P)_{at\,x=-X_P}$ and $(n_P)_{at\,x=X_N}$ stand, respectively, for the concentration of holes at $x = -X_P$, that is at the bulk P-side and the concentration of holes at $x = X_N$, that is at the bulk N-side. If it is assumed that the P- and N-sides are doped with shallow dopants of concentration N_A at P-side and N_D on N-side and that all dopant atoms are fully ionized at temperature T, then

$$(n_P)_{at\,x=-X_P=N_A}\tag{5.12}$$

However, holes are minority carriers on N-side. The concentration of majority carrier electrons is N_D. Therefore, the concentration of minority carrier holes on N-side is given by

$$(n_P)_{at\,x=X_N} = \frac{n_i^2}{N_D}\tag{5.13}$$

Substituting the values of $(n_P)_{at\,x=-X_P}$ and $(n_P)_{at\,x=X_N}$ from (5.12) and (5.13) back in (5.11), one gets

$$V_{bi} = \frac{kT}{e}\ln\left(\frac{N_A N_D}{n_i^2}\right)\tag{5.14}$$

5.3 Highly Doped Abrupt p-n Junction in Thermal Equilibrium

Heavily or highly doped abrupt junctions are often used for the detection of ionizing radiations. Generally, in such junctions, the P-side is heavily doped to about 10^{25} acceptor atoms per cubic metre, while the N-side is lightly doped to around

10^{20} donor atoms per metre cube. Heavily doped P-side is often designated as P^+-type and the lightly doped N-region simply as N-side. The space charge distribution of a highly doped p+ -n abrupt junction in thermal equilibrium is shown in Fig. 5.5. Because of heavy doping on P-side, the width of the uncovered negative charge x_P is very narrow and extends to a larger depth. The total width W_d of the depletion layer is given by

$$W_d = X_P + X_N \qquad (5.15)$$

As shown in Fig. 5.5, there are two charge distributions in the depletion region, one of negative charge from $x = -X_P$ to $x = 0$ that is in the range $-X_P < x < 0$ and of positive charge for the range $0 < x < X_N$. If N_D and N_A are, respectively, the dopant concentrations on the N and P-sides and if it is assumed that dopants are shallow so that all dopant atoms have ionized at temperature T, then charge density $\rho_{(-)}$ of the negative charge in the range $-X_P < x < 0$ may be given by

$$\rho_{(-)} = -eN_A \text{ in the range } -X_P < x < 0 \qquad (5.16)$$

And similarly, the density of positive charge $\rho_{(+)}$ is given by

$$\rho_{(+)} = eN_D \text{ in the range } 0 < x < X_N \qquad (5.17)$$

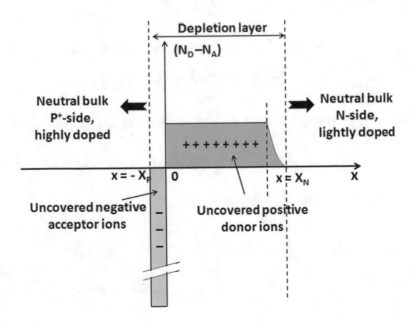

Fig. 5.5 Space charge distribution of a heavily doped (P^+) abrupt junction in thermal equilibrium

It is known that any static distribution of charge of density ρ may be assigned a scalar potential, V, such that

$$\nabla^2 V = -\frac{\rho}{\varepsilon_m} \text{ (Poisson's equation, } \varepsilon_m \text{ being the permitivity of the medium)} \quad (5.18)$$

And

$$E(\text{electric field}) = -\nabla V \quad (5.19)$$

For one-dimensional distribution (in X-direction only) the above (5.18) and (5.19) reduce to

$$\frac{d^2 V(x)}{dx^2} = -\frac{\rho(x)}{\varepsilon_m} \quad (5.20)$$

And

$$\text{Electric field } E = -\frac{dV(x)}{dx} \quad (5.21)$$

Substituting the values of $\rho_{(-)} = -eN_A$ in the range $-X_P < x < 0$ and $\rho_{(+)} = eN_D$ in range $0 < x < X_N$ in (5.20), one gets

$$\frac{d^2 V(x)}{dx^2} = -\frac{\rho(x)}{\varepsilon_m} = -\frac{(-eN_A)}{\varepsilon_m} = \frac{eN_A}{\varepsilon_m} \quad -X_P < x < 0$$

$$\text{Or } \frac{dV(x)}{dx} = \frac{eN_A}{\varepsilon_m}(x + x_P) \quad -X_P < x < 0 \quad (5.22)$$

$$\text{And } E^-(x) = -\frac{dV(x)}{dx} = -\frac{eN_A}{\varepsilon_m}(x + x_P) \quad -X_P < x < 0$$

Similarly, for the other regions, the electric field E(x) may be given as

$$E^+(x) = -\frac{dV(x)}{dx} = \frac{eN_D}{\varepsilon_m}(x - x_N) \quad 0 < x < x_N \quad (5.22a)$$

The built-in potential V_{bi} is then given as

$$V_{bi} = -\int_{-x_P}^{x_N} E(x)dx = -\int_{-x_P}^{0} E^-(x)dx - \int_{0}^{x_N} E^+(x)dx \quad (5.23)$$

$$\text{Or } V_{bi} = \frac{eN_A x_P^2}{2\varepsilon_m} + \frac{eN_D x_N^2}{2\varepsilon_m}$$

However, from the overall neutrality of charges in thermal equilibrium and from (5.1a), one has

$$N_A x_P = N_D x_N \tag{5.24}$$

Using (5.23) and (5.24), it is easy to get

$$N_A x_P = N_D x_N x_P = \sqrt{\frac{2\varepsilon_m}{e} \frac{N_D}{N_A(N_A + N_D)}} V_{bi} \tag{5.25}$$

And

$$x_N = \sqrt{\frac{2\varepsilon_m}{e} \frac{N_A}{N_D(N_A + N_D)}} V_{bi} \tag{5.26}$$

Therefore, the width of the depletion layer W_d may be given as

$$W_d = \sqrt{\frac{2\varepsilon_m}{e} \frac{(N_A + N_D)}{(N_D N_A)}} V_{bi} \tag{5.27}$$

In case $N_A \gg N_D$, as is the case of P^+ doping,

$$W_d \cong \sqrt{\frac{2\varepsilon_m}{e} \frac{V_{bi}}{N_D}} \tag{5.28}$$

Equation (5.28) tells that in highly doped abrupt junction, the width of depletion layer almost entirely depends on the concentration of the lighter dopant.

As shown in Fig. 5.6, a semiconductor crystal doped with N_A and N_D shallow impurities on the two sides and in thermal equilibrium may be divided into three sections: (i) the bulk P-side, having holes as majority and electrons as minority carriers. The resistance of this part is small as it has large number of mobile charge carriers. (ii) The bulk N-side with electrons as majority and holes as minority carriers. The resistance of this part is also low. (iii) The central part is called the depletion layer. There are no mobile charges in this part and, therefore, depletion layer has very high resistance and behaves as an insulator. Further, there is a **potential barrier** or potential drop equal to V_{bi} across the depletion layer, N-side being at higher potential than the P-side. When some voltage say, V_{ex} is applied externally to the crystal by connecting the external source to the P and N terminals of the crystal, the total external voltage V_{ex} essentially drops across the depletion layer as the bulk P and N-parts have negligible resistances. Depending on the polarity of the applied voltage V_{ex}, it may add to V_{bi} or may reduce it. Thus, the effective value of the potential barrier across the depletion layer may be increased or decreased by properly selecting the polarity of the external voltage. Another

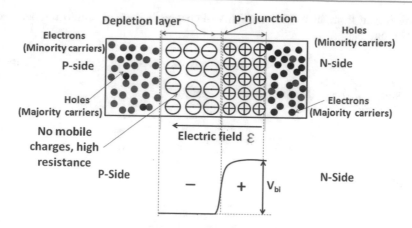

Fig. 5.6 p-n junction in thermal equilibrium and the built-in voltage V_{bi}

important feature of depletion layer is that on its side in contact with bulk P-type and extending up to the metallurgic junction, there is a layer of uncovered negative ions of acceptor atoms. Similarly, on the other side, there is a layer of positive uncovered donor ions. The total negative charge $(-Q)$ is, however, equal to the total positive charge $(+Q)$. The widths of the negative and the positive charge distributions are in inverse proportion to their respective concentrations. Thus, depletion layer is like a charged parallel plate capacitor of thickness W_d, where W_d is the total width of the depletion layer. According to (5.27), the width of depletion layer W_d is given by $W_d = \sqrt{\frac{2\varepsilon_m}{e} \frac{(N_A + N_D)}{(N_D N_A)} V_{bi}}$, and for given values of N_D and N_A, it is proportional to the square root of the built-in potential V_{bi}. Since the effective value of V_{bi} may be enhanced or reduced through the externally applied voltage V_{ex}, it is possible to change the width of the depletion layer which in turn may change the capacitance of the depletion layer parallel plate capacitor. Thus, depletion layer may be used as a voltage-controlled **varicap** (variable capacitance) the capacitance of which may be controlled by some external voltage. The capacitance C_{dep} of the depletion layer, assuming it to be a parallel plate capacitor, is given by

$$C_{dep} = \frac{\varepsilon_m . A}{W_d} = \sqrt{\frac{e\varepsilon_m (N_A N_D)}{2V_{eff}(N_A + N_D)}} \qquad (5.29)$$

Here,

$$V_{eff} = V_{bi} \pm V_{ex} \text{ and } A \text{ is the area of the junction.} \qquad (5.30)$$

The effective resistance of the depletion region may also be varied by changing the magnitude of the effective potential barrier across it. Therefore, the depletion layer may also be used as a voltage-controlled variable resistance.

Self-assessment question (SAQ): What is the cause that develops a potential barrier at depletion layer?

Self-assessment question (SAQ): Let there be a silicon crystal of length L which contains a p-n junction and an external voltage V_{ex} be applied across it. The potential gradient because of this external potential will be uniform and will be given by V_{ex}/L. True or falls? Give reasons for your answer.

5.3.1 p-i-n Junction

A normal p-n junction diode has low working voltage and also does not operate properly at high frequencies. However, when an additional layer of intrinsic material is sandwiched between the highly doped P- and N-sides, a p-i-n diode is made. This additional intrinsic material layer effectively increases the width of the depletion layer and provides additional thickness of matter for the absorption of radiations, when diode is used for detecting ionizing radiations. The increased thickness of depletion region reduces the capacitance of depletion layer making its switching action fast. The p-i-n devices may be used for fast switching in high-frequency applications. They are also used in photodetectors and high voltage power electronics applications.

Solved example (SE-5.1) Assuming that intrinsic carrier concentration n_i at room temperature for silicon is 1×10^{16} m^{-3}, and its relative permittivity is 11.0, calculate the value of built-in potential and total width of depletion layer for a crystal which is doped with acceptor impurity of 1×10^{22} m^{-3} and donor impurity of 6×10^{22} m^{-3}. The absolute permittivity of vacuum is 8.85×10^{-12} F/m.

Solution: *Let us first calculate the built-in potential V_{bi} using the (5.14)*

$$V_{bi} = \frac{kT}{e} \ln\left(\frac{N_A N_D}{n_i^2}\right) = \frac{1.38 \times 10^{-23} \times 300}{1.6 \times 10^{-19}} \ln\left(\frac{1 \times 10^{22} \times 6 \times 10^{22}}{10^{16^2}}\right)$$
$$= 2.5875 \times 10^{-2} \ln\left(6 \times 10^{12}\right)$$

Or

$$V_{bi} = 2.5875x10^{-2} \times 29.422 = \textbf{\textit{0.761V}}$$

The total width of depletion layer is given by (5.27) as
Or

$$W_d = \sqrt{\frac{2\varepsilon_{0x\varepsilon_r}}{e} \frac{(N_A + N_D)}{(N_D N_A)} V_{bi}} = \sqrt{\left(\frac{2 \times 8.85 \times 10^{-12} \times 11}{1.6 \times 10^{-19}}\right)\left(\frac{1 \times 10^{22} + 6 \times 10^{22}}{1 \times 10^{22} \times 6 \times 10^{22}}\right) \times 0.761}$$

$$W_d = 3.28 \times 10^{-8} \text{ m}$$

Solved example (SE-5.2) A p-n junction of area 1×10^{-8} m^2 is doped with N$_A$ = 10^{24} m^{-3}, N$_D$ = 10^{22} m^{-3}. What type of junction is it? Calculate the junction capacitance if the relative permittivity of the junction material is 11 and the carrier density of intrinsic material is 10^{16} m^{-3}.

Solution: *Since the doping level of acceptor impurity is much higher than that of donor impurity, the junction is an abrupt p$^+$-n junction. For calculating the capacitance, we have to calculate V_{bi} and W_d. Let us first calculate V_{bi}. Though the temperature is not explicitly given in this problem, it is normally taken as 300 K.*

$$V_{bi} = \frac{kT}{e} \ln \left(\frac{N_A N_D}{n_i^2} \right) = \frac{1.38 \times 10^{-23} \times 300}{1.6 \times 10^{-19}} \ln \left(\frac{1 \times 10^{24} \times 10^{22}}{10^{162}} \right) = 2.5875 \times 10^{-2} \ln 10^{14}$$

Or $V_{bi} = 2.5875 \times 10^{-2} \times 32.23 = $ **0.834 V**

The width of depletion layer is given by (5.27) as
$W_d = \sqrt{\frac{2\varepsilon_m}{e} \frac{(N_A + N_D)}{(N_D N_A)}} V_{bi}$, *however, if $N_A \gg N_D$, it reduces to*

$$W_d = \sqrt{\frac{2\varepsilon_m}{e} \frac{1}{(N_D)} V_{bi}} = \sqrt{\left(\frac{2 \times 11 \times 8.85 \times 10^{-12}}{1.6 \times 10^{-19}} \right) \frac{0.834}{10^{24}}} = \mathbf{3.185 \times 10^{-8}} \text{ m}$$

Assuming depletion layer to be a parallel plate capacitor, the capacity is given by (5.29) as

$$C_{dep} = \frac{\varepsilon_m . A}{W_d} = \frac{11.0 \times 8.85 \times 10^{-12} \times 1 \times 10^{-8}}{3.185 \times 10^{-8}} = 30.56 \times 10^{-12} F = \mathbf{30.56pF}$$

5.4 Biased p-n Junction in Thermal Equilibrium

A p-n junction is represented by the symbol shown in Fig. 5.7. The arrowhead in the figure stands for the P-side, while the vertical line for the N-side. The arrowhead is also called anode and the vertical line as cathode.

Most electronic devices require appropriate potentials at various terminals of the device for the proper working of the device. The arrangement of power sources to supply required potentials is called biasing. A p-n junction has two terminals, i.e. two Ohmic connections one on the P-side and the other on the N-side. A power source may be connected between these two terminals in two different configurations which will be discussed in the following.

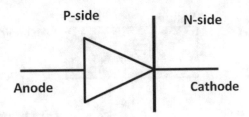

Fig. 5.7 Symbol for p-n junction

An unbiased p-n junction in thermal equilibrium shown in Fig. 5.8a has the following properties: (i) there is no current through the junction. Before the establishment of thermal equilibrium, the majority carriers, holes from the p-side and electron from the n-side were crossing the junction to the other side due to the concentration difference, constituting diffusion current. This resulted in developing a potential difference V_{bi}, called built-in potential, across the depletion layer. The built-in potential difference opposed the diffusion of majority carriers and ultimately stopped the flow of diffusion current. The built-in voltage V_{bi} in thermal equilibrium at 300 K temperature is about 0.7 V for Si and 0.3 V for Ge. The diffusion current depends on the number density or concentration of shallow acceptor and donor impurities and is not much affected by the temperature. (ii) Before thermal equilibrium, the developing built-in potential difference encouraged the flow of minority carriers, electrons on the p-side and holes on the n-side, to cross the junction constituting the drift current. At thermal equilibrium, the diffusion and drift currents mutually balanced each other such that no net current flows through the junction. It may be remembered that drift current due to minority carriers depends strongly on temperature since minority carriers both on p-side and on n-side are created by the breaking of covalent bonds by thermal energy. At higher ambient temperature more bonds get broken increasing the number density of minority carriers which in turn increases minority carrier drift current. (iii) A small region of a few micrometre width gets developed around the junction where no mobile charges can stay. This mobile charge-free region, called **depletion layer,** has very large resistance as compared to the bulk p- and n-type end sections which because of large number of mobile charges, have negligible resistance. Any external voltage applied to the system via Ohmic terminals at P- and N-sides essentially drops across the depletion layer and thus either enhances or decreases the built-in potential V_{bi}. The width of the depletion layer depends both on the concentrations of the donor and acceptor impurities as well as the effective potential difference across the p-n junction. With this background let us analyze the behaviour of the p-n junction under forward and reverse bias conditions.

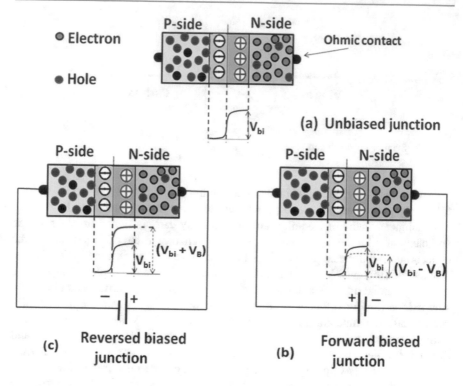

Fig. 5.8 Reverse and forward biasing connections of a p-n junction.

5.4.1 Forward Bias

When the positive terminal of the external power source (a cell in the present case) is connected to the P-side and the negative terminal to the N-side, as shown in Fig. 5.8b, the arrangement is called **forward biased**. In forward bias, the externally applied voltage by the power source denoted by V_B and called bias voltage, opposes the built-in voltage V_{bi}. With the decrease in the magnitude of the effective potential drop across the depletion layer, the width of the depletion layer reduces and ultimately when V_B becomes equal to V_{bi} and effective potential difference becomes zero, the depletion layer vanishes, for $V_B > V_{bi}$, an electric field is established at the junction pointing from P-side to N-side. Notice that the direction of this new electric field is opposite to the direction of the earlier electric field in the unbiased junction. Now there is no depletion layer and also there is an electric field from P-side towards N-side. This new electric field makes the majority carrier holes from the P-side and majority carrier electrons from N-side to drift to the other side. So, in forward bias above a critical value of the bias voltage V_B, a drift current consisting of majority careers of the two sides flow across the junction and the device as shown in Fig. 5.9a. The circuit diagram for forward biasing is shown in Fig. 5.9b.

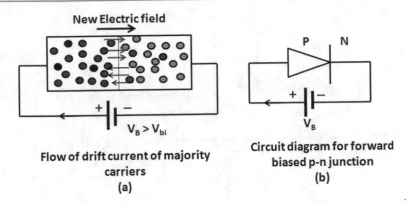

New Electric field

Flow of drift current of majority
carriers
(a)

Circuit diagram for forward
biased p-n junction
(b)

Fig. 5.9 a Flow of drift current of majority carriers; **b** circuit diagram of a forward-biased p-n junction

The process of current flow in forward biasing may be understood in the following way. When the positive terminal of the external source is connected to the P-side, electrons in the valence band partially bound to covalent bonds on the P-side move towards the positive terminal. This amounts to the motion of holes towards the junction. Excess holes that reached the depletion layer take electrons from the negatively charged ions of the acceptor atom and thus reduce the thickness of the depletion layer on the P-side. Similarly, on the N-side, electrons are repelled towards the junction by the negative voltage applied by the external source. Excess numbers of electrons reaching the depletion layer combine with the positively charged donor atoms and neutralize them reducing the thickness of the layer of positive ions. Finally, when V_B increases beyond a critical value an electric field gets established across the junction that makes the majority carriers to drift to the other side constituting a large drift current.

I–V characteristic of a forward-biased p-n junction is shown in the first quadrant of Fig. 5.10. As may be observed in this figure, when forward bias voltage $+V_B$ is initially increased no current flows till a critical value of the applied bias voltage called the **cut-in voltage**. The cut-in voltage for Si is $\approx 0.5V$ while for Ge it is $\approx 0.2V$. Beyond the cut-in voltage the forward current I_F increases sharply with a small increment in the bias voltage. I–V characteristic for forward-biased p-n junction is non-linear, i.e. it is not a straight line. It may be shown that the current density j_F for forward-biased p-n junction may be given by

$$J_F = en_i^2 \left[\frac{D_n}{\lambda_n N_A} + \frac{D_p}{\lambda_p N_D} \right] \left(e^{\left(\frac{eV_B}{kT} \right)} - 1 \right)$$

Here, symbols have their usual meaning.

However, Shockley derived the following equation for the current I_F, which is generally in milliamp given as

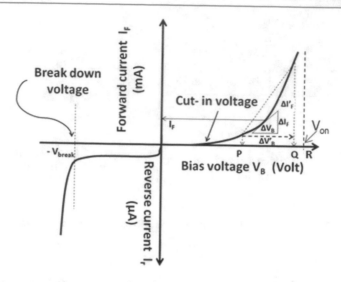

Reverse bias V-I characteristic   Forward bias V-I characteristic

Fig. 5.10 I–V characteristics of a p-n junction

$$I_F = I_s \left(e^{\left(\frac{V_B}{nV_T} \right)} - 1 \right) \tag{5.31}$$

Here, I_s is the saturation current which may be of the order of 10^{-9} to 10^{-18} A, n is a constant which is taken equal to 1 for germanium based diodes and varies between 1 and 2 for Si-based devices, and V_T is temperature equivalent voltage (= kT) which is around 26 mV at room temperature of around 300 K.

Since the I-V characteristic of a forward-biased junction is non-linear, its slope changes from one point on the curve to another point. Therefore, the AC or dynamic resistance r_{ac} of the forward-biasedjunction at a particular value of the current is given by the slope of the curve at that point, as shown in the figure by

$$r_{ac} = \frac{\Delta V_B}{\Delta I_F} \tag{5.32}$$

However, the average resistance r_{ac} over the selected range of operation PQ of the p-n junction is simply,

$$r_{av} = \frac{\Delta V_B'}{\Delta I_F'} \tag{5.32}$$

As the forward bias voltage is increased beyond the cut-in voltage (V_{cut-in}), the forward current increases almost exponentially till the forward bias voltage reaches the value V_{on} shown by point R in Fig. 5.10. Any attempt to further increase the forward bias voltage beyond V_{on} only increases the forward current keeping the forward voltage constant at V_{on}. As such it is not possible to increase the forward bias beyond V_{on}, and the forward bias p-n junction starts behaving like a battery of voltage V_{on}. Forcing the bias voltage beyond V_{on} may result in damaging the diode because of excessive current flow and heating. The value of V_{on} for silicon-based diode is 0.7 V and for germanium diode it is 0.3 V.

5.4.2   Reverse Bias

A reverse-biased p-n junction and its circuit diagram are shown in Fig. 5.11. As shown in this figure, the reverse biasing is achieved by connecting the positive terminal of the power source to the N-side of the junction and the negative terminal to the P-side.

The source voltage ($-V_B$) in reverse bias adds up to the already existing built-in potential difference V_{bi}. With this increase in the effective value of V_{bi}, the width of depletion layer increases (the width of depletion layer is proportional to the square root of effective V_{bi}). As a result, any flow of majority carriers from either side through the depletion layer becomes impossible. However, enhanced potential barrier across the extended depletion layer forces minority carriers from both sides to drift through the depletion layer. Thus, a small drift current due to minority carriers, called **reverse saturation current**, flows in the circuit. The reverse saturation current I_r is small ($\approx microamp$) since the number density of minority carriers is very small. But the reverse bias current is very sensitive to temperature since the concentration of minority carriers depends strongly on temperature. For every 10 °C rise of temperature I_r almost doubles itself. As will be discussed later,

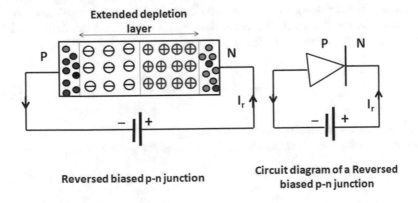

Reversed biased p-n junction

Circuit diagram of a Reversed biased p-n junction

Fig. 5.11 Reverse-biased p-n junction and its circuit diagram

the temperature dependence of reverse bias current requires special arraignments to be made for bias stabilization in case of junction transistors.

Physical process that increases the width of depletion layer under reverse bias may be explained in the following way. Since negative terminal of the power source is connected to the P-side, electrons are injected by the negative source terminal. Holes (majority carriers) are attracted towards the negative end combine with the injected electrons. This reduces the number density of holes, to replenish this decrease more acceptor impurity atoms near the depletion layer take up electrons creating holes and itself becoming negatively charged ions. Thus, new layers of uncovered negative immobile ions develop on the P-side. Similarly, On the N-side, majority carrier electrons move towards the positive terminal of the source, some of which are taken up by the positive terminal for the current flow in the external circuit. This reduction of electrons is compensated by the uncovering of more positively charged donor atoms near the depletion layer extending the width of existing positive ion layer. Thus, in reverse bias, new layers of negative and positive ions are formed that extend the thickness of depletion layer.

The reverse bias I−V characteristic is shown in third quadrant of Fig. 5.10. As may be observed in this figure, the reverse current initially rises exponentially with the reverse bias voltage $(-V_B)$ and soon stabilizes to an almost constant value till the breakdown voltage $(-V_{break})$, where it suddenly increases to a very large value. Since the magnitude of the reverse bias current I_r remains almost constant over a wide range of operating voltage (almost $V_B = 0$ to $V_b = -V_{break}$) it is also called reverse saturation current. The magnitude of reverse saturation current is generally of the order of few μA and is limited because of the small and fixed concentrations of minority carriers at a given temperature. However, with the rise of temperature, the concentration of minority carriers increases almost exponentially and hence I_r also increases rapidly with temperature.

(a) **Breakdown in reverse bias**

When the reverse bias voltage $(-V_B)$ exceeds a critical value, suddenly the reverse current I_r increases to a large value with in a small increase of the bias voltage. This is referred as breakdown of the depletion layer or simply the breakdown of the junction. The breakdown occurs when the electric field across the depletion layer exceeds a critical value ε_{crit} which is given as

$$\varepsilon_{crit} = \sqrt{\frac{2eN(V_{bi} + V_{break})}{\varepsilon_m}} \tag{5.33}$$

Here, N is the concentration of dopants.

The following two mechanisms are responsible for breakdown.

(i)   Zener breakdown

This kind of breakdown is dominant when **both sides of junction are very heavily doped so that the thickness of depletion layer is very small**. The high electric field across the depletion layer established by the large voltage (sum of the built-in potential V_{bi} and the applied V_{break}) causes an offset in the bands such that it becomes possible for carriers to tunnel through the narrow depletion layer.

Figure 5.12 a shows an unbiased p-n junction where a bound electron in the valence band of P-side cannot jump to the conduction band on the N-side. However, in the case of reverse-biased p-n junction at Zener breakdown, the large potential difference across the thin depletion layer pushes down the band structure of N-side so much that the conduction band of N-side almost align in energy with the valence band on P-side or goes down below it. It is now possible for bound electrons in valence band of P-side to **quantum mechanically tunnel** through the thin depletion layer and reach the conduction band of N-side. The large number of bound electrons on tunnelling becomes free and constitutes large reverse current. The reverse bias V_{break} at which Zener breakdown will occur is given by

$$V_{break} = \frac{\varepsilon_m \varepsilon_{crit}^2}{2eN} - V_{bi} \qquad (5.34)$$

Typical values of $\varepsilon_{crit} \approx 10^8 \text{V/m}$ and $V_{break} < 5V$ for Zener breakdown. Moreover, it is possible to control the value of breakdown voltage by controlling the dopant level. Specially designed Zener diodes that have predetermined value of breakdown voltage are commonly used for voltage reference and for voltage regulation.

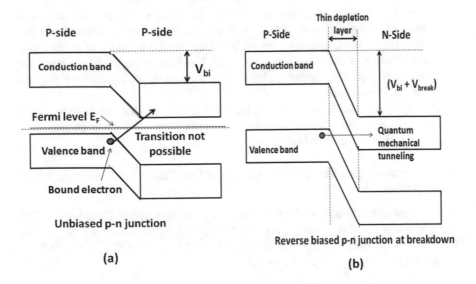

Fig. 5.12 Band structure of **a** Unbiased p-n junction ; **b** reversed biased p-n junction at Zener breakdown

Fig. 5.13 Zener diode in
reverse bias breakdown mode

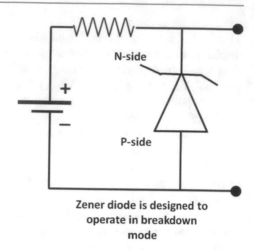

Zener diode is designed to
operate in breakdown
mode

Such diodes are called Zener diodes and are designed to operate in reverse bias breakdown mode as shown in Fig. 5.13. Zener diode is represented by a tilted cathode to distinguish it from normal diode, as shown in this figure.

(ii) Avalanche breakdown
The other mode of breakdown is termed **avalanche breakdown** and dominates for junctions both sides of which are lightly doped. In lightly doped junctions the depletion layer is thick and under large reverse bias, is subjected to large electric field. Minority carrier electrons drifting across the depletion layer from the P-side gain sufficient energy under the influence of large electric field and knockout more electrons from the immobile positive and negative dopant ions in depletion layer.

Newly knocked-out electrons also gain high energy and initiate secondary ionization of immobile ions present in depletion layer and thus librating more electrons. This cascade process goes on building like the building of a snow ava-lanche by snow-bowling effect and is shown in Fig. 5.14. Thus, a copious supply of electrons is generated in depletion layer drift towards the N-side constituting the large breakdown current.

Both types of breakdowns are reversible which means that the junction is not damaged permanently when breakdown takes place, provided the temperature is not allowed to exceed the specified value.

Self-assessment question (SAQ): What prohibits Zener breakdown in lightly doped p-n junctions?

Solved example (SE-5.3) A diode that may be described by the Shockley equation has current of 1.0 mA when a potential of 0.5 V is applied across it at 100 °C. Calculate the current through the diode at the potential of 0.8 V, given that the parameter n for diode has the value 2.

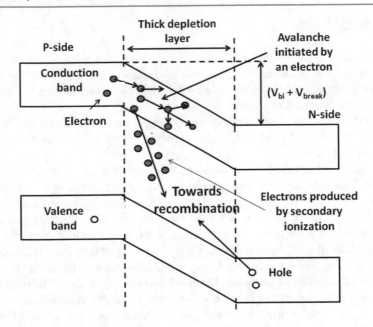

Fig. 5.14 Avalanche breakdown

Solution: *[Some tips: A simple formula to calculate V_T is: $V_T = \dfrac{\text{Temperature in Kelvin}}{11600}$ V*

Also at room temperature of 300 K, Shockley equation may be written as

$$I = I_s\left(e^{\frac{40V_B}{n}} - 1\right) \text{ A.}]$$

According to Shockley equation $I_F = I_s\left(e^{\left(\frac{V_B}{nV_T}\right)} - 1\right)$. Let us first calculate the

value of I_s, for which we have to calculate the value of V_T at 100 °C. Now,

$V_T = \frac{kT}{e} = \frac{(1.38x10^{-23})(373.15)}{1.6x10^{-19}} = 0.032$ V. *Substituting this value of V_T in Shockley equation, one gets*

$$1x10^{-3} = I_s\left(e^{\left(\frac{0.5}{2x0.032}\right)} - 1\right) = I_s\left(e^{7.81} - 1\right) = I_s(244.14 - 1) = 243.14I_s$$

Therefore, $I_s = 4.1x10^{-6}A$

Current at 0.8 V will be given by $I = 4.1x10^{-6}\left(e^{\left(\frac{0.8}{2x0.032}\right)} - 1\right) = 1.597x10^{-3}$ A

5.5   Ideal Diode

An ideal diode is an imaginary two-terminal device which is characterized by an I
−V characteristic shown in Fig. 5.15. The diode acts as a closed switch or a short
circuit if current flows in the direction as indicated by the arrowhead and as an off
switch when current flows in the opposite direction.

A simple two-port system may be made by applying an input between the anode
of the diode and the ground and taking the output between the cathode and ground
as shown in Fig. 5.16.

If a sinusoidal signal of single frequency f_1 is applied at the input of this two-port
network, one gets the waveform shown on the output side. Let us analyze the
changes brought by the diode device in the original signal. Firstly, in the original
input signal, the average value of signal was zero. However, the average value of
the output signal is not zero but V_{ave}. Further, it can be shown that the output signal
is the resultant of a DC value (equal to V_{ave}) plus waves of frequency f_1, $2f_1$, $3f_1$......
(higher harmonics of f_1) with diminishing amplitudes. As such it may be concluded
that **a non-linear device generates harmonics of the input signal. Generation of
harmonics of the input frequency is an essential property of non-linear devices.**
If the output signal is filtered through suitable frequency filter, signal of desired
frequency may be sorted out.

The three main differences between a p-n junction diode and an ideal diode are
(i) there are nothing like cut-in voltage V_{cut-in} and the 'on' voltage V_{on} in case of an
ideal diode while a real junction diode has a cut-in voltage of about 0.5 V and V_{on}
of 0.7 V for Si diode and about 0.2 V and 0.3 V, respectively, for a Ge diode.
(ii) Similarly there is no possibility of very large current in reverse-biased ideal
diode, while in the real diode large current flows in reverse bias if voltage increases
beyond breakdown voltage V_{break}. (iii) In an ideal diode, no current flows when it is
reverse biased. On the other hand, small reverse saturation current of the order of
few micro-ampere flows in reverse-biased diode till the breakdown voltage.
However, the reverse saturation current in a real diode is at least three orders of
magnitudes smaller than the forward current and may, therefore, be neglected in
most situations.

Fig. 5.15 Ideal diode and its
I−V characteristic

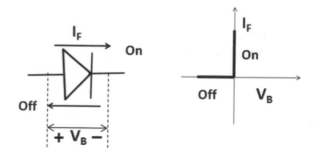

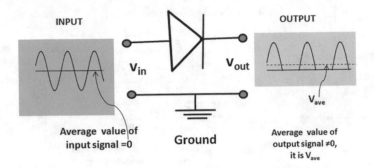

Fig. 5.16 Simple two-port network using ideal diode

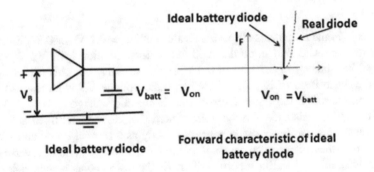

Fig. 5.17 Ideal battery diode and its transfer characteristic

5.5.1 Transfer Characteristic of a Real Diode

Transfer characteristics of a device are graph showing relationships of output quantities with corresponding input quantities. Voltage transfer characteristic of a p-n junction diode is shown in Fig. 5.18b.

Let us start from point A on transfer characteristic where the diode is reverse biased. Since a reverse-biased diode is like an 'open switch', the signal present at the input passes to the output. The state of 'off' for diode continues for the whole duration till the diode is in reverse bias, i.e. up to point o on the graph. From point o, if one moves to the right, diode becomes forward biased, however, it remains off till point B that corresponds to the 'on' voltage V_{on}. For silicon-based diode the V_{on} is about 0,7 V, while for germanium one it is 0.3 V. Therefore, between O and B diode is non-conducting and is like an open switch, hence output is equal to the input. It may thus be observed that from A to B whatever signal is present in the input will pass over to output. This is represented by straight line AB of slope 1. At point B and beyond it on the right side, diode becomes a closed switch, like a short circuit, and, therefore, a constant voltage, 0.7 V in case of Si and 0.3 V in case of

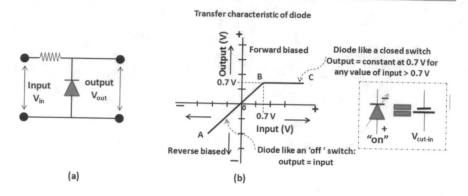

Fig. 5.18 a, b Transfer characteristic of a diode

Ge, appears in the output for all positive values of $V_{in} > 0.7$ V (or 0.3 V in case of Ge). It may be said that once the diode is in forward bias with $V_{in} \geq V_{on}$, it behaves like a battery of 0.7 V or of 0.3 V, as the case may be. This is indicated on extreme right in inset of Fig. 5.18 b. When input voltage V_{in} is increased beyond V_{on}, current through the diode increases such that the voltage $(V_{in} - V_{on})$ drops across the series resistance and the voltage across diode remains fixed at V_{on}.

In forward bias, a Zener diode behaves exactly as a normal diode and after the diode is 'on', the voltage drop across it remains fixed at V_{on}. Also in reverse bias a Zener diode behaves just like a normal diode till the reverse bias voltage is less than the breakdown voltage V_{break}; it is like an 'off' switch and whatever signal is present in the input is passed on to the output. If reverse bias is increased beyond V_{break}, Zener breaks down and now voltage across Zener becomes constant equal to the breakdown voltage V_{break}. It may, therefore, be said that a Zener diode behaves like a battery of V_{on} volts for forward bias larger than 'on' voltage and also it behaves as a battery of V_{break} volt for reverse bias voltages larger than the breakdown voltage.

Transfer characteristic of a Zener diode is shown in Fig. 5.19b. In this case, lines BC in forward bias and AD in reverse bias show that output voltage remains constant once the diode is on and also when diode is under breakdown.

From the above discussion another important observation may be made, and it is that inclusion of a series resistance is a must whenever a normal diode or a Zener diode is used to avoid any damage to the diode. Any excess voltage drops across this series resistance and the diode is not subjected to unbearable large potentials.

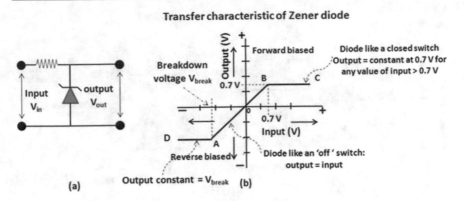

Fig. 5.19 Transfer characteristic of a Zener diode

5.6 Some Applications of Diode

A junction diode has many applications. It is used in communication systems as limiters, gates, clippers and mixers; in computers as logic gates, clamps, clippers; in radios as mixers, signal detector, in automatic volume control circuits; in radar circuits for parameter amplification, phase detection, automatic gain control and in televisions for signal clamping, gain control, phase locking, etc. Diodes are also used for converting alternating current into direct current, the process called rectification. Rectified output is required for assembling power supplies that are extensively used in electronic/electrical systems. Some important diode applications are discussed here.

5.6.1 Half-Wave Rectifier

In most countries of the World electric power is supplied in alternating sinusoidal waveform. It is because of the ease of transmitting alternating power (AC) from the place of generation to other places where it is used. However, most of our electrical and electronic devices run on direct current (DC) power. DC power supplies are fabricated to run such systems and devices.

A sinusoidal AC wave is shown on extreme left of Fig. 5.20 which is made of positive and negative half cycles. The first step in converting AC to DC is to remove the negative half cycles or to convert negative half cycles into positive half cycles. This process is called **rectification**. If only positive half cycles of the sinusoidal wave are used and negative half cycles are rejected, the process is termed as **half-wave rectification** and if negative half cycles are converted into positive half cycles and in this way both the negative and the positive half cycles of the input sinusoidal wave are used for conversation to DC, the process is called **full-wave**

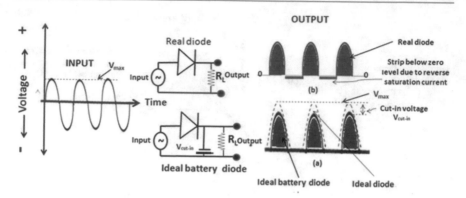

Fig. 5.20 Half-wave rectification

rectification. Rectified wave has only positive half cycles, is unipolar (i.e. the polarity remains positive) but is still not d.c. since the magnitude of the voltage is still increasing and decreasing with time. The variation of the magnitude of voltage with time is controlled in the second step, called **smoothing** or filtering after which the output wave becomes a straight line on the voltage–time graph indicating that the output voltage remains constant with time, i.e. the output is now DC. In good power supplies a third section, called **stabilizer**, is also added so that the output voltage does not change even when there are some fluctuations in the parameters of the input AC wave. Thus a DC power supply consists of (i) a rectifier (ii) a filter or smother and (iii) a stabilizer.

A single diode half-wave rectifier circuits are shown in the middle part of Fig. 5.20. The sinusoidal AC is applied at the input ports of the rectifier circuit while the half-wave rectified output is delivered at the output port across the load resistance R_L. Since diode acts like an 'on' switch when current flows in the direction of the arrowhead, there is output across the load resistance R_L only during the positive half cycles of the applied AC wave. During the negative half cycles of the AC input the diode works as an 'off' switch and, therefore, no current flows through the load resistance R_L, hence no output. We have so far dealt with three types of diodes, namely an Ideal diode, an ideal battery diode and a real diode. Let us first analyze the output that one will get if an ideal diode is in use.

The I–V characteristic of an ideal diode is shown in Fig. 5.15. It is, therefore, obvious that the ideal diode will start giving output signal as soon as the positive half cycle of the input wave starts and will reach the peak value when the positive half of input attains maximum value and will then decrease to zero with the input AC. The output wave of the ideal diode is shown by the dotted curve in Fig. 5.20.

Next, let us consider an Ideal battery diode, where a battery of voltage $-V_{on}$ is connected in parallel to the load resistance in the output to simulate the effect of the on voltage. The output of this diode will start slightly late, till the voltage of the input positive half-wave reaches the value equal to the cut-in voltage. Also the peak of the positive half cycle at the output will be at a lower voltage $(V_{max} - V_{cut\text{-}in})$ as compared to the ideal diode. The output of the ideal battery diode is shown by shaded curve in Fig. 5.20. The difference between the peak voltages of the ideal and ideal battery diodes is equal to $V_{cut\text{-}in}$, as indicated in the figure.

Finally, we consider the real diode. The output in case of the real diode will start a little after the start of the positive cycle in the input AC signal, as the diode will conduct only beyond the cut-in voltage. This is exactly same as in the case of the ideal battery diode. Further, the maximum of the output positive wave will occur at the voltage $(V_{max} - V_{cut\text{-}in})$ just like the ideal battery diode. However, there will be one major difference between the outputs of the ideal battery diode and the real diode. It will be due to the reverse saturation current. Both, the ideal and the ideal battery diodes assume that the diode is 'off' during the negative half cycle of the input AC. But in case of the real diode a small and constant current, called reverse saturation current, flows in output during the negative half cycles of the input AC wave. As such the output wave pattern in case of a real diode stands on a pedestal of a thin strip below the zero voltage line, as shown in output (b) in Fig. 5.20.

The output waveforms shown in Fig. 5.20 are called half-wave rectified output.

Self-assessment question (SAQ): What is the likely cause of cut-in voltage?

Self-assessment question (SAQ): A half-wave rectified wave is not pure DC, explain?

Self-assessment question (SAQ): A diode may generate waves of higher frequencies. Is it a correct statement? Explain

(a) **Performance analysis of half-wave rectifier** Let a sinusoidal wave given by

$V = V_{max}\sin\theta$ be applied to the input port of a diode through the secondary of a transformer as shown in Fig. 5.21. As already discussed the diode will be in forward bias for the positive half cycles of the input wave and will be in reverse bias for the negative half cycles. During the positive half cycles, the current through the diode will be given by $I_{diode}^{+} = \frac{V}{R_s + R_f + R_L}$ where R_s, R_f and R_L are, respectively, the resistance of the secondary of input transformer, the resistance of the diode in forward bias and the load resistance in the circuit. In the case $R_L \gg (R_s + R_f)$ so that the sum may be neglected in comparison to the load resistance R_L, the diode current becomes $I_{diode} = \frac{V}{R_L}$ and the output voltage across the load resistance becomes

$$V_{out}^{+} = \frac{V}{R_L} \cdot R_L = V = V_{max}\sin\theta \text{ between } \theta = 0 \text{ to } \pi \qquad (5.35)$$

During the negative half cycle, the current through the diode will be given as $I_{diode}^{-} = \frac{V}{R_s + R_r + R_L}$, here, R_r is the resistance of the diode under reverse bias. Since R_r and R_L both very large, I_{diode}^{-} will be so small that it may be neglected. It is already

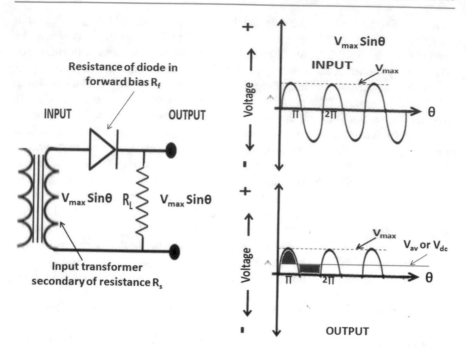

Fig. 5.21 Performance analysis of half-wave rectifier

known that reverse saturation current is few orders of magnitudes smaller than the forward current and thus be neglected in comparison to it. It may thus be concluded that the output of the diode across the load resistance R_L will contain only positive halves of the input sinusoidal wave as shown in Fig. 5.19. That means,

$$V_{out}^- = 0 \text{ between } \theta = \pi \text{ to } 2\pi \tag{5.36}$$

Therefore, the total output wave

$$V_{out} = V_{out}^+ + V_{out}^- \tag{5.37}$$

The purpose of a rectifier is to convert AC into DC. However, the output wave obtained from a half-wave rectifier is unidirectional pulsating wave, not a totally DC wave. It contains both DC and AC components. The performance of a rectifier is judged through a set of parameters called **performance factors** which are given in the following;

i. **Average or DC voltage**

The first performance factor is V_{dc} or V_{av} the DC component of the wave. It is just the average value of the output voltage over one full cycle, i.e. from $\theta = 0$ to $\theta = 2\pi$

$$V_{dc}^{half} = V_{av}^{half} = \frac{1}{2\pi} \int\limits_0^{2\pi} V_{out} = \frac{1}{2\pi} \left[\int\limits_0^{\pi} V_{out}^+ d\theta + \int\limits_{\pi}^{2\pi} V_{out}^- d\theta \right]$$

$$= \frac{1}{2\pi} \left[\int\limits_0^{\pi} V_{max} \sin\theta d\theta + \int\limits_{\pi}^{2\pi} 0 d\theta \right] = \frac{V_{max}}{2\pi} [-\cos\theta]_0^{\pi} = \frac{V_{max}}{\pi} = 0.318 V_{max}$$

$$\text{Or } V_{dc}^{half} = V_{dc}^{half} = \frac{V_{max}}{\pi} = 0.318 V_{max}$$

$$(5.38)$$

The half-wave rectified output voltage varies as $V_{max} \sin\theta$ during the positive half period and remains zero for the next half period. The average value of the output voltage corresponds to the value of the output voltage such that if it is uniformly distributed over the full cycle, the area under the average value should be same as the area of the positive half cycle. As shown in Fig. 5.19, the area of the shaded part of positive half cycle is distributed over the negative half cycle, where in original rectified wave there was no output voltage.

Equation (5.38) shows that the DC component in the output of a half-wave rectifier is only 31.8 %, which is obviously not very good.

ii. **RMS voltage**
   The rms value of the output voltage over one full cycle gives the effective value of the output voltage so far as heating effect is concerned. Further, it shows the joint effect of both the DC and AC components. It is defined as

$$V_{rms} = \sqrt{\frac{1}{2\pi} \int\limits_0^{2\pi} (V_{out})^2 d\theta} = \sqrt{\frac{1}{2\pi} \int\limits_0^{2\pi} (V_{out}^+ + V_{out}^-)^2 d\theta} = \sqrt{\frac{1}{2\pi} \int\limits_0^{\pi} (V_{max} \sin\theta)^2 d\theta}$$

$$\text{Or } V_{rms}^{half} = \sqrt{\frac{V_{max}^2}{2\pi} \int\limits_0^{\pi} \sin^2\theta d\theta} = \sqrt{\frac{V_{max}^2}{2\pi} \int\limits_0^{\pi} \left(\frac{1 - \cos 2\theta}{2} \right) d\theta} = \sqrt{\frac{V_{max}^2}{4\pi} \left[\theta - \frac{\sin 2\theta}{2} \right]_0^{\pi}}$$

$$\text{Or } V_{rms}^{half} = \sqrt{\frac{V_{max}^2}{4\pi} [\pi]} = \frac{V_{max}}{2} = 0.5 V_{max}$$

$$(5.39)$$

iii. **Peak factor**
   Peak factor is defined as the ratio of the maximum value to the rms value of a time varying quantity. The peak factor for half-wave rectified output is given by,

$$\text{Peak factor} = \frac{V_{max}}{V_{rms}} = \frac{V_{max}}{\frac{V_{max}}{2}} = 2 \tag{5.40}$$

iv. **Ripple factor**

The ratio of the rms value of AC component to the DC component is defined as the Ripple factor and is denoted by γ. It is a measure of the relative strengths of the AC and DC components in the output wave.

The output wave from a half-wave rectifier contains both the DC and AC components. The V_{rms}, calculated earlier, is the rms value jointly due to the AC and DC components. If V_{ac} and V_{dc}, respectively, denote the rms values of AC and DC voltages in the output, respectively, then

$$V_{rms} = \sqrt{V_{dc}^2 + V_{ac}^2} \text{ Or } V_{ac} = \sqrt{V_{rms}^2 - V_{dc}^2}$$

Or Ripple factor $\gamma = \frac{V_{ac}}{V_{dc}} = \frac{\sqrt{V_{rms}^2 - V_{dc}^2}}{V_{dc}} = \sqrt{\frac{V_{rms}^2}{V_{dc}^2} - 1} = \sqrt{\frac{\left(\frac{V_{max}}{2}\right)^2}{\left(\frac{V_{max}}{\pi}\right)^2} - 1}$

Therefore,

$$\text{Ripple factor } \gamma^{half} = \sqrt{\frac{\pi^2}{4} - 1} = \sqrt{1.4674} = 1.211 \tag{5.41}$$

It is apparent from the above equation that in the half-wave rectified output the AC component is 1.21 times the DC component.

v. **Efficiency of rectifier**

The ratio of DC output power to the AC input power of a rectifier is called its efficiency and is denoted by η.

In a half-wave rectifier, the peak load current I_{max} may be written as

$I_{max} = \frac{V_{max}}{R_s + R_f + R_L} \cong \frac{V_{max}}{R_L}$ if $R_L \gg (R_f + R_s)$

Now, $I_{dc} = \frac{I_{max}}{\pi} = \frac{V_{max}}{\pi R_L}$

Also, $I_{rms} = \frac{I_{max}}{2} = \frac{\frac{V_{max}}{R_L}}{2} = \frac{V_{max}}{2R_L}$

The DC output power $P_{output}^{dc} = I_{dc}^2 R_L = \left(\frac{V_{max}}{\pi R_L}\right)^2 R_L$

The AC input power $P_{input}^{ac} = I_{rms}^2 R_L = \left(\frac{V_{max}}{2R_L}\right)^2 R_L$

Therefore, **Efficiency of rectifier**

$$\eta^{hal;f} = \frac{P_{output}^{dc}}{P_{input}^{ac}} = \frac{\left(\frac{V_{max}}{\pi R_L}\right)^2 R_L}{\left(\frac{V_{max}}{2R_L}\right)^2 R_L} = \frac{4}{\pi^2} = \textbf{\textit{0.405}} \qquad (5.42)$$

This shows that only about 40.5% of the input AC power is converted into DC power in the output of a half-wave rectifier. This is not a good percentage.

vi. **Form factor**

The ratio of the rms value to the average value of a time-varying signal is called the form factor.

$$\text{Form factor for a half} - \text{wave rectifier} = \frac{V_{rms}^{half}}{V_{av}^{half}} = \frac{\frac{V_{max}}{2}}{\frac{V_{max}}{\pi}} = \frac{\pi}{2} = 1.57 \qquad (5.43)$$

vii. **Peak inverse voltage (PIV)**

It is defined as the maximum voltage that may get applied to the diode under the reverse bias condition without damaging it. In case of a half-wave rectifier$(PIV)^{half}$ is equal to V_{max}.

viii. **Transformer utilization factor (TUF)**

All rectifiers use an input transformer that supplies sinusoidal AC power to the rectifier circuit. Generally, the input (or primary) winding of this transformer is connected to the AC power line that delivers AC power either at 220 volt or at 110 volt. However, most of the solid-state rectifier circuits require sinusoidal AC at much lower voltage of less than 50 V. Therefore, the transformer steps down the AC from 220 or 110 V to the desired lower voltage. Let us denote by $V_{max} \sin\theta$ the sinusoidal wave with peak value V_{max} that is delivered by the secondary of the transformer and applied to the rectifier circuit.

Transformers are characterized by their **volt-ampere (VA)rating**. The **effective VA rating** of a transformer is the average value of the VA ratings of its primary and secondary windings. VA rating of a transformer tells about the maximum ac power that may be delivered by the transformer. The cost of a transformer depends on its VA rating and increases with it. It is important to realize that any circuit that is connected to the secondary of a given transformer should in principle draw AC power from the transformer not more than its rated VA value and also not very much less than the rated VA, because if it draws much less power than the VA rating of the transformer, it will not be cost effective. A cheaper transformer of lower VA rating might have been used for saving on the transformer cost. To judge the effective utilization of the VA rating of a transformer by the rectifying circuit, a quantity called transformer utilization factor (TUF) for rectifiers is defined as follows:

$$TUF = \frac{\text{dc power in the output of the rectifier}}{\text{Effective VA rating of the transformer}} \tag{5.43a}$$

Knowledge of the TUF helps in choosing the appropriate transformer for a rectifier.

ix. **TUF for a half-wave rectifier**:
   The average or DC voltage and DC current for a half-wave rectifier are, respectively, given by

$V_{av}^{half} = \frac{V_{max}}{\pi}$, and $I_{av}^{half} = \frac{I_{max}}{\pi} = \frac{V_{max}}{R_L \pi}$, therefore, the DC output power P_{dc}^{half} is

$$P_{dc}^{half} = V_{av}^{half} I_{av}^{half} = \frac{V_{max}^2}{R_L \pi^2} \tag{5.43b}$$

Next one should find the effective VA rating of the transformer. Since the signal at the output of the transformer, i.e. at its secondary is a sinusoidal wave of maximum amplitude V_{max}, the rms value of transformer secondary output voltage will be $\left(\frac{V_{max}}{\sqrt{2}}\right)$. It may be noted that the rms value of voltage for a sinusoidal wave and also for the full-wave rectified output wave is $\left(\frac{V_{max}}{\sqrt{2}}\right)$. Now, let us find the rms value of the current that passes through the secondary when the half-wave rectifier circuit is connected to it. It is obvious that the current passing through the secondary of the transformer will be same as the current that passes through the load resistance R_L. Therefore, the rms value of the current through transformer secondary will be same as the rms value of the current passing through the load resistance R_L. It means that the

$$I_{rms}^{secondary} = I_{rms}^{half} = \frac{V_{rms}^{half}}{R_L} = \frac{V_{max}}{2R_L} \tag{5.43c}$$

Therefore, the effective VA rating of the transformer $= \left(\frac{V_{max}}{\sqrt{2}}\right)\left(\frac{V_{max}}{2R_L}\right)$

$$\text{Effective VA rating of the transformer} = \frac{V_{max}^2}{2\sqrt{2}R_L} \tag{5.43d}$$

Hence,

$$TUF \text{ off or half} - \text{wave rectifier} \frac{\frac{V_{max}^2}{R_L \pi^2}}{\frac{V_{max}^2}{2\sqrt{2}R_L}} \frac{2\sqrt{2}}{\pi^2} = 0.286 \qquad (5.43e)$$

TUF for a half-wave rectifier is only 0.286. It means that the VA rating of the input transformer required for half-wave rectifier is approximately (1/0.286 =) 3.5 times the DC power output. For example, if the DC load at the output is, say, of 100 watt, then the VA rating of the required transformer must be 3.5 × 100 = 350 watt. It is obviously very poor utilization of transformer rating. It may be noted that larger the magnitude of TUF, smaller will be the value of VA rating of the required transformer.

Self-assessment question (SAQ): What is the physical significance of rms value of an AC quantity? If rms value of a signal is equal to its average value what is the nature of the signal?

5.6.2   Full-Wave Rectifier

In a half-wave rectifier, almost half of the input power during the negative half cycles of the input AC goes waste. To avoid this loss of power, efforts were made to utilize both the positive as well as the negative halves of the input sinusoidal wave for rectification. The resulting circuits which deliver unidirectional pulsating DC in their output for both the positive as well as for the negative half cycles of the input sinusoidal wave are termed as full-wave rectifiers. Two types of full-wave rectifier circuits are in use: (i) those which require input transformer the secondary winding of which is split equally in two halves with a common centre tapped connection. Only two power diodes are required in such circuits. (ii) The one which uses a normal transformer but requires four power diodes. This type of circuit is also called bridge rectifier.

(a)   **Full-wave rectifier using input transformer with centre tapped secondary**
      Figure 5.22a shows the circuit diagram of a full-wave rectifier that uses input transformer with secondary winding split into two equal parts. The center tapped terminal CT of the secondary is always at 0 V while the other terminals A and B assume positive and negative polarities alternately after each half cycle of the input sinusoidal wave shown in part (b) of the figure. Let us consider the instant when terminal A is positive and B is negative. In this condition diode D_1 will be forward biased and current will flow through D_1 in the direction of arrow along the path $AD_1EF(CT)$ through the load resistance R_L. Since current will enter the load resistance R_L at terminal E and will leave it at terminal F, E will be at higher potential than F. The voltage drop across the load resistance will appear as the output signal. After positive half cycle of the input sinusoidal wave when

terminal A was positive, the polarity will reverse, terminal B will become positive with respect to terminal A but the potential of the centre tapped CT will remain zero. Now, in the negative half cycle, diode D_2 will be forward biased and current through it will flow along the path $BD_2EF(CT)$. It may be noted that current through load R_L still flows from E to F keeping the terminal E at higher potential than F. In this way, for both the negative and the positive cycles of the input wave, the output across the load resistance will be positive from E to F. The full-wave rectified output is shown in part (c) of the figure. Alternate positive half cycles in the rectified output correspond to the conduction of diode D_1 and diode D_2, respectively. It may be observed that in this arrangement centre tapped output of the transformer secondary plays the key role of keeping diodes D_1 and D_2 forward biased, respectively, in the positive and negative half cycles. Further, both the negative and the positive half cycles of the input sine wave are now rectified to deliver pulsating unidirectional DC at the output.

There are two basic issues with this method of full-wave rectification: (i) it requires a special input transformer, the secondary of which must be split into two equal halves. If it is required to get the maximum voltage in the rectified output as V_{max}, then the secondary of the transformer must produce maximum voltage of $2V_{max}$ between terminals A and B so that the potential difference between the centre tapping (CT) and the two ends A and B may be V_{max} each. This requires double the number of turns in the secondary as compared to a normal transformer that delivers maximum output voltage of V_{max}. This almost doubles the cost of the special transformer. (ii) The second issue required the PIV rating of the diodes used for rectification. PIV or the peak inverse voltage is the maximum voltage that may be applied to a diode without damaging it when it is in reverse bias (not conducting). Let us investigate the PIV requirement for diodes D_1 and D_2 used in the circuit shown in Fig. 5.22. Let us consider the instant when terminal A is at $+V_{max}$ voltage with respect to the centre tapping

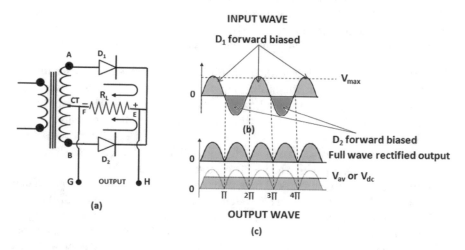

Fig. 5.22 a Circuit diagram of a full-wave rectifier using input transformer of centre tapped secondary; **b** the input wave; **c** the full-wave rectified output

(CT). At this instant terminal B will be at $-V_{max}$ volt. Now the anode of diode D_2 (not conducting) which is reversed biased is at potential $-V_{max}$ while its cathode is connected through diode D_1(which is in forward bias and is like a closed switch) to terminal A at $+V_{max}$ potential. Thus, the maximum potential drop across the reversed biased diode D_2 is $[V_{max} - (-V_{max})] = 2V$max. Similarly, for the next swing of the input cycle, diode D_1 will become reversed biased and the maximum voltage across it will be $2V_{max}$. Therefore, the diodes D_1 and D_2 should have their PIV rating at least $2V_{max}$, i.e. these diodes must not go bad when they are subjected to 2Vmax voltage under reverse-biased condition. Generally, the cost of diode increases with its PIV rating.

(b) **Full-wave rectifier using bridge circuit**

The full wave rectifier discussed above requires a special transformer with secondary winding almost double in size of a normal transformer. Since copper or aluminium wire is used for secondary winding, special transformers required for such full-wave rectifiers are costly. In a full-wave bridge rectifier, four power diodes are used along with an ordinary transformer. Since diodes are relatively cheaper as compared to the cost of special transformers, therefore, bridge rectifier circuits are more often employed for full-wave rectification. The circuit diagram of a bridge rectifier is shown in Fig. 5.23.

Figure 5.23a shows the circuit diagram of a full-wave bridge rectifier. Let us assume that end A of the input transformer secondary is positive and end J is negative. In this condition, diodes D_2 and D_4 will be forward biased while diodes D_1 and D_3 will be reversed biased. Arrows show the direction of the current flow in the circuit in this condition. As shown in the figure, current

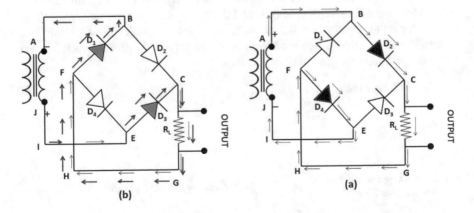

Fig. 5.23 Bridge rectifier **a** diodes D_2 and D_4 forward biased; **b** diodes D_3 and D_1 forward biased

enters the load resistance R_L at C and leaves it at G, therefore, the end C of the load resistance will be at higher potential than end G.

After positive half cycle, the end A of the input transformer secondary will become negative and end J will become positive. Therefore, for the negative half cycle of the input wave, diodes D_3 and D_1 will become forward biased- while D_2 and D_4 will be reversed biased. The forward-biased diodes will conduct and current will flow through the circuit, the direction of flow of current for the negative half cycle is shown by arrows in part (b) of the figure. The important point to note is that the direction of flow of current through the load resistance R_L remains same for both the positive as well as for the negative swings of the input sinusoidal wave. As a result, the voltage drop across R_L is always from C to G, i.e. unidirectional. The rectified output of the bridge circuit is same as the output of the full-wave rectifier that uses a centre tapped secondary and is shown in Fig. 5.23c.

There are two distinct advantages of bridge rectifier over the other rectifier that uses special input transformer. Firstly, there is no need of a special transformer that has its secondary split into two equal halves. Also, the diodes used in bridge circuit do not require high PIV rating; rather one may use diodes with PIV rating less than V_{max}, as a pair of diodes connected in series gets reversed biased together and, therefore, each diode is subjected to only half of the total voltage drop against the pair. The only negative point of the bridge rectifier is the use of four diodes, however, it is not a big disadvantage as the cost of low PIV diodes is not very much.

(c) **Performance analysis of a full-wave rectifier**

The output waveform of a full-wave rectifier is shown in Fig. 5.24. Assuming that the sinusoidal wave at the rectifier input is given by $V = V_{max} sin\theta$ and that the forward resistance of diodes R_f and the resistance of the secondary windings of the input transformer R_s are both negligible in comparison to the load resistance R_L, the voltage waveform at the output of the full-wave rectifier over one cycle ($\theta = 0$ to 2π) may be written

$$V_{out}^{onecycle} = V_{out}^{+} + V_{out}^{-} \quad \text{And} \quad V_{out}^{+} = V_{out}^{-} = V_{max} \sin \theta \qquad (5.44)$$

Fig. 5.24 Output waveforms of a full-wave rectifier

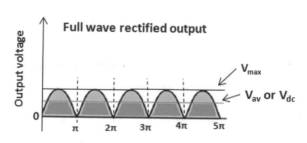

Here, V_{out}^{+} refers to the rectified output corresponding to the positive half cycle of the input sinusoidal wave between $\theta = 0$ to $\theta = \pi$ and similarly V_{out}^{-} refers to the rectified output wave corresponding to the negative half cycle of input sinusoidal wave between $\theta = \pi$ to 2π. It may be remarked that in case of the half-wave rectification V_{out}^{-} was zero.

i. **Average or DC value of voltage for full-wave rectifier** It follows from the definition of the average or DC value that

$$V_{av}^{\text{full}} = V_{dc}^{\text{full}} = \frac{1}{2\pi} \int_{0}^{2\pi} V \sin\theta d\theta = \frac{1}{2\pi}\left[\int_{0}^{\pi} V_{\text{out}}^{+} d\theta + \int_{0}^{2\pi} V_{\text{out}}^{-} d\theta \right]$$

$$\text{Or } V_{av}^{\text{full}} = V_{av}^{\text{full}} = \frac{1}{2\pi}\left[2\int_{0}^{\pi} V_{\text{out}}^{+} d\theta \right] = \frac{1}{\pi}\int_{0}^{\pi} V_{max} \sin\theta d\theta = \frac{2V_{max}}{\pi} \qquad (5.45)$$

$$\text{Or } V_{av}^{\text{full}} = V_{av}^{\text{full}} = \frac{2V_{max}}{\pi}$$

It may be observed that $V_{av}^{\text{full}} = 2V_{av}^{\text{half}}$

ii. **rms voltage for a full-wave rectifier** rms voltage is defined as

$$V_{rms}^{\text{full}} = \sqrt{\frac{1}{2\pi}\left[\int_{0}^{\pi}(V_{\text{out}}^{+})^2 d\theta + \int_{\pi}^{2\pi}(V_{\text{out}}^{-})^2 d\theta \right]} = \sqrt{\frac{1}{\pi}\left[\int_{0}^{\pi} V_{max} \sin\theta)^2 d\theta \right]} \qquad (5.46)$$

$$\text{Or } V_{rms}^{\text{full}} = \sqrt{\frac{V_{max}^2}{2}} = \frac{V_{max}}{\sqrt{2}}$$

Therefore, $V_{rms}^{\text{full}} = \sqrt{2}V_{rms}^{\text{half}}$

iii. **Peak, average and rms currents for full-wave rectified wave.** Under the assumption that $R_L \gg R_S + R_f$, the maximum current through the circuit will flow when the voltage is maximum, i.e.

$$I_{max}^{\text{full}} = \frac{V_{max}}{R_L};$$

Also, the average value of current $I_{av}^{\text{full}} = I_{dc}^{\text{full}} = \dfrac{V_{av}^{\text{full}}}{R_L} = \dfrac{2V_{max}}{\pi R_L} \qquad (5.47)$

Similarly, rms current $I_{rms}^{\text{full}} = \dfrac{V_{rms}^{\text{full}}}{R_L} = \dfrac{V_{max}}{\sqrt{2}R_L} \qquad (5.48)$

iv. **Peak factor for full-wave rectifier** Peak factor is defined as the ratio of the maximum value of the voltage to the rms value of voltage. Therefore,

$$\text{\textit{Peak factor for full wave rectification}} = \frac{V_{max}}{V_{max}/\sqrt{2}} = \sqrt{2} = \textit{1.414} \quad (5.49)$$

v. **Form factor for full-wave rectification** The ratio of the rms value to the average value of a time-varying quantity is called the form factor.

$$\textbf{Form factor for full-wave rectification} = \frac{V_{rms}^{full}}{V_{av}^{full}} = \frac{V_{max}/\sqrt{2}}{(2V_{max})/\pi} = \frac{\pi}{2\sqrt{2}} = 1.11$$

$$(5.50)$$

vi. **Ripple factor for full-wave rectifier** The unidirectional but pulsating output voltage of a full-wave rectifier contains both the DC and AC components. The DC component is, however, given by V_{av}^{full}. On the other hand, the square of the rms value $\left(V_{rms}^{full}\right)^2$ gives the sum of the squares of DC and AC components, i.e.

$$\left(V_{rms}^{full}\right)^2 = \left(V_{dc}^{full}\right)^2 + \left(V_{ac}^{full}\right)^2, \text{therefore, } V_{ac}^{full} = \sqrt{\left(V_{rms}^{full}\right)^2 - \left(V_{dc}^{full}\right)^2}$$

$$\text{Ripple factor for full wave rectifier } \gamma^{full} = \frac{V_{ac}^{full}}{V_{dc}^{full}} = \sqrt{\frac{\left(V_{rms}^{full}\right)^2}{\left(V_{dc}^{full}\right)^2} - 1}$$

$$\text{Or } \gamma^{full} = \sqrt{\left(\frac{V_{rms}^{full}}{V_{dc}^{full}}\right)^2 - 1} = \sqrt{\left(\frac{V_{max}/\sqrt{2}}{2V_{max}/\pi}\right)^2 - 1} = \sqrt{\left(\frac{\pi}{2\sqrt{2}}\right)^2 - 1} = \sqrt{\frac{\pi^2}{8} - 1} = 0.483$$

$$\text{Or } 2\,\gamma^{full} = \textbf{0.483} \quad (5.51)$$

It may be observed that $\gamma^{full}(=0.483)$ is much smaller than γ^{half} that was 1.21. A smaller value of ripple factor means smaller relative strength of AC component in the rectified wave.

vii. **Rectification efficiency of full-wave rectifier** The ratio of the DC power in the output to the AC power is called the rectification efficiency. Now AC power $P_{ac}^{full} = (R_L + R_f)I_{rms}^2$ and the DC power in the output $P_{dc}^{full} = R_L I_{dc}^{full}$, therefore,

$$\textbf{Rectification efficiency for full-wave rectifier } \eta^{full} = \frac{P_{dc}^{full}}{P_{ac}^{full}} = \frac{R_L\left(\frac{2V_{max}}{\pi R_L}\right)^2}{(R_f + R_f)\left(\frac{V_{max}}{\sqrt{2}R_L}\right)^2}$$

Or

$$\eta^{full} = \frac{8R_L}{(R_f + R_f)\pi^2} = 0.810\frac{R_L}{(R_f + R_f)} \qquad (5.52)$$

If $R_L \gg R_f$, then $\eta^{full} = 0.810$ or 81.0 %. This is almost twice of η^{half}.

viii.   **PIV of diodes used in full-wave rectifiers** The PIV of diodes used in bridge full-wave rectifiers may be V_{max}, but it should be $2V_{max}$ for the other rectifier circuit that uses special input transformer with secondary winding split into two equal parts with centre tapping connection.

ix.   **TUF of full-wave rectifiers**

(a) *Centre tapped transformer:* Finding the effective VA rating in this case is a little more involved because of the two reasons, firstly the transformer's secondary winding is split into two equal halves and secondly the current flows alternatively, through the two halves. For the positive half cycle, it flows through the upper half and during the negative half cycle through the lower half part of the secondary.

The DC output power in this case is given as

$$P_{dc}^{Full} = V_{av}I_{av} = \left(\frac{2V_{max}}{\pi}\right)\left(\frac{2I_{max}}{\pi}\right) = \left(\frac{4V_{max}I_{max}}{\pi^2}\right) \qquad (5.53)$$

Next, let us calculate the effective VA rating of the transformer. Since the voltage through each part of the secondary winding is sinusoidal with peak value V_{max}, therefore, the rms value of voltage will be $(V_{max}/\sqrt{2})$. However, the current through each half of the secondary flows only for the half cycle the rms value of current will be $(I_{max}/2)$. The VA rating of each half of the secondary is, therefore, $[V_{max}.I_{max}/2\sqrt{2}]$. Hence, the total VA rating of the complete secondary of the transformer is

$$VA_{Sec} = 2\left[V_{max}.I_{max}/2\sqrt{2}\right] = \left[V_{max}.\frac{I_{max}}{\sqrt{2}}\right] = 0.707V_{max}I_{max} \qquad (5.54)$$

The VA rating of the transformer primary $= VA_{pri} = (V_{max}I_{max})/2$
$$= 0.5V_{max}I_{max} \qquad (5.55)$$

Effective VA rating of the trsansformer $= \dfrac{\left[VA_{Sec} + VA_{pri}\right]}{2}$
$$= 0.603V_{max}I_{max} \qquad (5.56)$$

Therefore, the TUF of centre tapped full wave rectifier

$$(TUF)^{CT\,full} = \frac{\left(\frac{4V_{max}I_{max}}{\pi^2}\right)}{0.603V_{max}I_{max}} = \frac{4}{0.603x\pi^2} = \mathbf{0.672} \tag{5.57}$$

(b) **Bridge full-wave rectifier**

The DC power output of the bridge rectifier is also same as that of the centre tapped rectifier,

$$P_{dc}^{bridge} = P_{dc}^{Full} = \left(\frac{4V_{max}I_{max}}{\pi^2}\right) \tag{5.58}$$

Let us now consider the VA rating of the transformer. Here, we need not consider the VA ratings of primary and secondary separately. It is because in this case current flows for the complete cycle through the secondary. Since the voltage and currents at transformer secondary are both sinusoidal, their rms values will be $\frac{V_{max}}{\sqrt{2}}$ and $\frac{I_{max}}{\sqrt{2}}$, respectively. Hence,

$$(TUF)^{bridge,full} = \frac{\left(\frac{4V_{max}I_{max}}{\pi^2}\right)}{\left(\frac{V_{max}}{\sqrt{2}}\right)\left(\frac{I_{max}}{\sqrt{2}}\right)} = \frac{8}{\pi^2} = 0.810 \tag{5.59}$$

The TUF for bridge full-wave rectifier is 0.810, which means the VI rating of the required transformer will be (1/0.810 =) 1.23 times the DC power of the output. Suppose it is desired that the DC power at the load is say 100 watt, then the VI rating of the transformer required to feed input to the bridge circuit may be 100 x 1.23 =123 watt. This tells that the transformer utilization in a bridge rectifier is very good. Following Table 5.1 summarizes the importance of TUF for different rectifiers.

The performance parameters for single phase different rectifier circuits are tabulated in Table 5.2. As may be observed in this table, full-wave bridge rectifier is the most efficient rectifier that has minimum ripple. It may further be observed that even in bridge full-wave rectifier the ripple component is almost 48%, and

Table. 5.1 Transformer utilization factors for rectifiers

Type of rectifier	Transformer utilization factor (TUF)	VA rating of required transformer for 100 W load
Full-wave bridge	0.810	123 VA
Full-wave centre tapped	0.672	149 VA
Half wave	0.286	350 VA

Table. 5.2 Performance parameters of half- and full-wave rectifiers

Performance factor	Half-wave rectifier	Full-wave rectifier
Average or DC value	V_{max}/π ; I_{max}/π	$2V_{max}/\pi$; $2I_{max}/\pi$
rms value	$V_{max}/2$; $I_{max}/2$	$V_{max}/\sqrt{2}$; $I\,max/\sqrt{2}$
Peak factor	2	$\sqrt{2} = 1.414$
Form factor	$\frac{\pi}{2} = 1.57$	$\frac{\pi}{2\sqrt{2}} = 1.11$
Ripple factor	1.21	0.48
Rectification efficiency	$\frac{4}{\pi^2} = 0.405; 40.5\%$	$\frac{8}{\pi^2} = 0.810; 81\%$
Transformer utilization factor (TUF)	0.286	Centre tapped Sec 0.672 Bridge circuit 0.810

therefore, it is required to further treat the rectifier output to remove the AC ripple component. It is for this reason ripple removing circuits, also called **smoothing circuits** or **filter circuits,** are used in DC power supplies.

Self-assessment question (SAQ): A rectifier has 50% TUF and another one 75%, which is better and why?

Self-assessment question (SAQ): What should be the value for peak factor for an ideal rectifier?

5.6.3 Three-Phase Rectifiers

In the foregoing, we discussed three different types of single-phase rectifiers, out of these the full-wave bridge rectifier circuit appears the best as it has high rectification efficiency, low ripple factor and highest value of TUF. In spite of all these properties, a single-phase full-wave rectifier is not very efficient for delivering high output DC power. Three-phase rectifiers are, therefore, frequently used for high DC power requirements. A single-phase supply has two delivery wires, namely phase and neutral or ground on the other hand a three-phase supply is delivered through four wires, three for the three different phases that differ by 120° phase angle from one to the other. The fourth wire is ground or neutral. The following Table 5.3 summarizes the relative merits of three-phase rectifiers.

Since the topic of three-phase rectifiers is beyond the scope of the current presentation, we end the discussion on rectifiers by giving circuit diagrams for three-phase half-wave and full-wave rectifiers along with a table summarizing their performance parameters. Fig. 5.25.

Table. 5.3 Single- and three-phase rectifier comparison

Single-phase rectifiers	Three-phase rectifiers
Two wire input supply	Four wire input supply
Power delivery low	High power delivery
High ripple factor	Ripple factor low
Relatively low rectification efficiency	High rectification efficiency
Less TUF	High TUF
Low PIV	High PIV
Additional filters are required	No additional filters are required

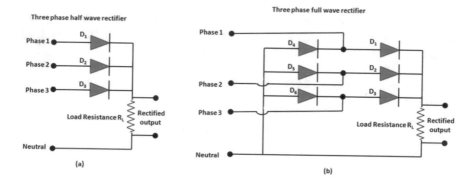

Fig. 5.25 Circuit diagrams **a** three-phase half-wave rectifier; **b** three-phase full-wave rectifier.

Table. 5.4 Performance parameters for three-phase half-wave and full-wave rectifiers

Performance parameter	Three-phase half-wave rectifier	Three-phase full-wave rectifier
Number of diodes required	3	6
Conduction period of each diode	120°	120°
Average current	$I_{dc}/3$	$I_{dc}/3$
Average voltage	$0.82 V_{max}$	$0.955 V_{max}$
PIV	$\sqrt{3}\ V_{max}$	$\sqrt{3}\ V_{max}$
Ripple factor	0.17	0.04
Rectification efficiency	97%	99.8%
Transformer utilization factor TUF	–	0.954

5.6.4 Ripple Filters or Smoothing Circuits

Most electronic devices, like television sets, computers, etc., require stabilized DC power to run them properly. DC power supplies are fabricated to deliver stabilized DC power. A DC power supply may be divided into three fundamental blocks (i) the rectifier (ii) the filter and (ii) the stabilizer, as shown in the block diagram of Fig. 5.26. As may be observed in this figure, the input sinusoidal wave is first rectified. The rectified output wave of both single-phase half-wave and full-wave rectifiers contains fundamental and higher harmonics of the input sinusoidal wave superimposed over a constant DC component. These AC components are termed as ripples and are undesirable. Generally, four types of filter circuits are employed to reduce ripple component. The choice depends on the requirements of the following devices. For example, if it is required to draw heavy currents from the rectifier circuit then inductor input filter is better while if low current and higher output voltage are required, it is better to use a capacitive filter.

(a) **Resistance-Capacitance (RC) Filter**

The simplest ripple filter is a **resistance-capacitor (R-C)** combination. In case of rectifiers, the load resistance and an additional capacitor in parallel to the load resistance make the R-C combination (see Fig. 5.27). The reactance (opposition offered by capacitance) of capacitor of capacitance C for a signal of frequency f is given as $X_C = \frac{1}{2\pi f C}$. For a DC signal, the reactance is infinite and for higher frequencies it is low. Therefore, if capacitor is connected in the output of a rectifier parallel to the load resistance R_L, AC components present in

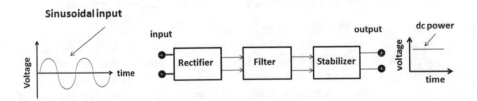

Fig. 5.26 Block diagram of a stabilized power supply

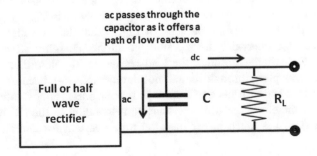

Fig. 5.27 The R-C filter connected to a full or half-wave rectifier

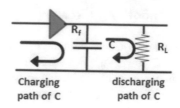

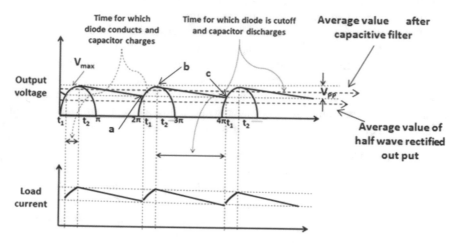

Fig. 5.28 a Working of a capacitive filter with a half-wave rectifier

the rectified output will pass through the capacitor while the DC component will move to the output as the capacitor will offer large reactance to it.

i. R-C filter with half-wave rectifier

Figure 5.28a explains the working of a R-C filter on the half-wave rectified output. As shown in this figure, during rising part of the rectified half wave, the capacitor C charges to the maximum voltage V_{max}. The charging of the capacitor is fast as charging takes place through the series combination of diode's forward resistance R_f and the secondary resistance R_s both of which are low and hence the time constant $((R_f+R_s) .C)$ is small. Once the capacitor C is charged to V_{max}, the diode gets reversed biased (the charged capacitor works like a battery of voltage V_{max}) because the voltage at the anode of the diode becomes less than V_{max} while the voltage at the cathode is V_{max} (because of the charged capacitor). As may be seen in Fig. 5.28a, the capacitor will charge during the time interval t_1 to t_2 . With no further charging after time t_2, the capacitor C discharges through the load resistance R_L. Note that discharge of capacitor takes place through R_L and not through (R_f+R_s), because the diode is in reverse bias and resistance of a diode in reverse bias is almost infinite. But R_L $>>(R_f+R_s)$, therefore, the time constant corresponding to the discharge of the capacitor (CR_L) is much larger than the time constant for charging (CR_f) so

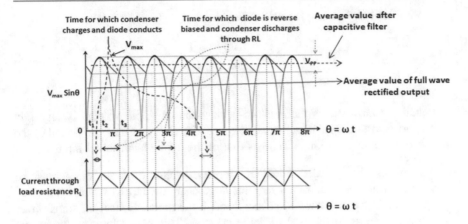

Fig. 5.28b b Working of a capacitive filter with full-wave rectifier

capacitor loses charge slowly, keeping the diode in reverse bias till the rising part of the next rectified wave.

With the inclusion of capacitor C, the diode conducts for time t_1 to t_2 (point a to b) in each positive half cycle. And the capacitor discharges from point b to point c in each cycle. This results in two things: (i) the ripple size which before the inclusion of capacitor was as much as the size of the rectified positive halves, with the inclusion of the capacitor has reduced to a very low value the peak-to-peak value V_{PP} as shown in Fig. 5.28a. (ii) The average value of the output without capacitor was was V_{max}/π, which has become much larger as is shown in the figure. The lower section of Fig. 5.28a shows the variation of current through the load resistance R_L. Further, it may be observed in this figure that the frequency of the left over ripple is one per cycle, which means the same as the frequency f of the input sinusoidal wave.

It may be observed that with the proper choice of R_L and C the capacitor charging time may be reduced to a low value and the capacitor discharge time may be assumed to be equal almost to the time period T of the input sinusoidal wave. Further, if average current through the load is I_L, then the charge lost by the capacitor during its discharge ΔQ may be given by

$$\Delta Q = I_L T \tag{5.60}$$

However, if V_{PP} represents the peak−to−peak value of the ripple voltage in the output,

Then,

$$V_{PP} = \frac{\Delta Q}{C} \tag{5.61}$$

From (5.60) and (5.61), it follows that

$$V_{PP} = \frac{\Delta Q}{C} = \frac{I_L T}{C} \tag{5.62}$$

Equation (5.62) shows that peak-to-peak value of ripple voltage is inversely proportional to the value of the capacitor. It also tells that the time period of the ripple left after filtration is same as that of the applied sinusoidal input wave.

This means that inorder to reduce ripple capacitor of large value must be selected. But a large capacitor requires large charging current which is supplied by the input transformer and passes through the diode. Hence, the transformer and the diode must be rated for large currents. This puts a limit on the capacitance which may be used.

Self-assessment question (SAQ): Suppose a given rectifier has to feed a 100 Ω load and also 10000 Ω load, what type of filter circuits will be required for the two cases?

Solved example (SE-5.4) (i) A 220 V(rms) 50 Hz sinusoidal wave is rectified using a half-wave rectifier and a load resistance of 50 Ω. Calculate the maximum voltage, DC voltage, peak-to-peak ripple voltage, percentage of ripple and ripple frequency. (ii) If a 10000 μF capacitor is used to filter the half-wave rectified output, determine the ripple parameters still left in the output.

Solution:

(i) The rms value of sine wave is 220 V, hence $V_{max} = \sqrt{2}V_{rms} = 311V$
  Also, $V_{av} = V_{dc} = \frac{V_{max}}{\pi} = \frac{311}{\pi} = 99.0V$
  Further, $(V_{PP})_{ripple} = V_{max} - 0 = 311V$
  Hence, Percentage ripple $= \frac{V_{PP}}{V_{dc}} x100 = \frac{311}{99} x100 = 314.14\%$
  And The ripple frequency $f_r = 50$ Hz

(ii) After filtration the average or $Vdc \cong V_{max} = 311V$
  The load current $I_L = \frac{V_{dc}}{R_L} = \frac{311}{50} = 6.22A$
  Also, The time period T of sinusoidal wave $T = \frac{1}{50}s = 0.02\ s$
  Therefore,   The   charge   lost   by   capacitor   during   discharge
  $\Delta Q = I_L \cdot T = 6.22x0.02$

$$= 0.1244\ \text{Coulomb}$$

Now, The peak-to-peak value of residual ripple $= (V_{PP})_{ripp} = \frac{\Delta Q}{C} = \frac{0.1244}{10000 \times 10^{-6}} =$ 12.44 V

Percent ripple $= \frac{12.44}{311} \times 100 = 3.85\%$

It may be noted that after filteration by capacitor the percent ripple has gone down from 314.14% to 3.85%.

ii. **R-C filter with full-wave rectifier** The charging and discharging of the filter capacitor in case of the full-wave rectified output is shown in Fig. 5.29b.

In case of full-wave rectifier, main difference with the half-wave rectifier lies in the times of charging and discharging of the capacitor. In the rectified output of the full-wave rectifier, two positive pulses occur per time period T (while in case of half-wave rectified output one positive pulse appears per cycle or in time T). Therefore, the capacitor charges and discharges approximately for half of the time as compared to the half-wave case. Since the capacitor discharges approximately for T/2 time, the peak-to-peak amplitude of ripple $(V_{PP})_{ripple}$ is almost half as compared to the half-wave case and also the repletion frequency f_r is twice of the frequency of input sinusoidal wave. It is simple to show that in case of full-wave rectified wave subjected to capacitor C in parallel to the load resistance R_L the $(V_{PP})_{ripple}$ is given by

$$(V_{PP})_{ripple} = \frac{I_L T}{2C}$$

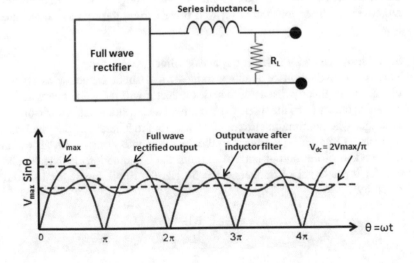

Fig. 5.29 Circuit diagram and the input and output wave shapes for a series inductor filter coupled to a full-wave rectifier

It may be concluded that the ripple component after shunt capacitive filter depends on three factors: (i) type of rectified output; half or full wave (ii) magnitude of the capacitance C and (iii) the value of the load resistance R_L ($I_L \cong V_{max}/R_L$). Ripple component will decrease with increasing C, decreasing T and decreasing R_L. Shunt Capacitive filter is very convenient as it has small size and is not costly. Further, it gives DC output almost equal to the peak input voltage V_{max} but does not work well if R_L is small, i.e. heavy current is drawn by the load.

(b) **Series inductor filter**

Series inductor filter is better suited for those systems that draw large currents. In this case, inductance or choke of large inductance is connected in series with the rectified output as shown in the top of Fig. 5.29.

In series inductor filter, the property of an inductor of opposing rapid changes in current through it is used. The inductor of inductance L offers reactance $X_L = 2\pi f L$ to a signal of frequency f. f is zero, i.e. for a DC current X_L is zero, while X_L increases with the frequency of the signal. Therefore, AC components of currents will face opposition, depending on their frequencies, and will be attenuated while passing through the inductance. Fig. 5.29 shows the result of series L filter on full-wave rectified output. Two facts may be observed in this figure, (c) output DC component is equal to V_{dc} (not V_{max}) and the magnitude of ripple (Vpp) depends very much on the magnitude of L. It may be shown that the ripple factor for series inductor filter is given by

$$\gamma^{seriesL} = \frac{R_L}{3\sqrt{2}\omega L}.$$

Series inductance filter is bulky and not very convenient for transportation, moreover, the filter does not work well if the current through the inductance is low.

(c) **Series inductance and shunt capacitor filter**

This is a combination of a series inductor and shunt capacitor as shown in Fig. 5.30. As already discussed a series inductor will reduce ripple when large current is drawn from the filter. On the other hand, a shunt capacitor works well as a filter when minimum current is drawn from it. Therefore, the combination of L and C provides a filter for the output ripple component of which is independent of the output current through the combination. It may be shown that for input signal of frequency f = 50 Hz, the ripple factor for LC filter is given by

$$\gamma^{LC} = \frac{1.194}{LC}$$

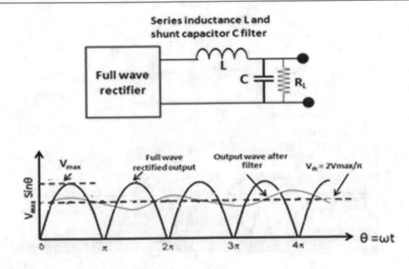

Fig. 5.30 Circuit diagram and waveforms of a LC filter

(d) **Pi-section or CLC filter**

The CLC filter is a combination of a shunt capacitor filter and a LC filter as shown in Fig. 5.31. Since it appears like the Greek letter $\prod$ (pi), therefore, it is called the pi-section filter. It is very versatile and is frequently used for filtration. The output of this filter is almost ripple free because filtration of AC components is done at three successive stages. First, the input capacitor bypasses AC components, next the inductance attenuates the AC component which is finally removed by the last capacitor. Thus, the output is almost DC. The ripple factor for pi-section filter is given by

$$\gamma^{pi} = \frac{\sqrt{2}}{8\omega^3 c_1 c_2 L R_L}$$

The main advantage of pi-filter is high output voltage and low ripple factor; however, the PIV requirement of diodes is high. A big disadvantage of the filter is poor regulation, which means that output voltage depends very much on the current drawn by the load.

Self-assessment question (SAQ): What is the frequency of ripples in full-wave rectifier output?

Self-assessment question (SAQ): Why the output voltage is not near about the peak voltage V_{max} in series inductance filter?

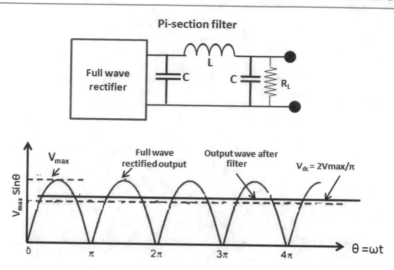

Fig. 5.31 Pi-section filter with output waveforms

(e) **Zener diode stabilization**

There are two important properties associated with a Zener diode. (i) Under reverse bias a Zener diode breaks down at a fixed voltage and (ii) the breakdown voltage does not vary or change if the amount of current drawn from the Zener diode is varied within specified limits. For example, in Fig. 5.32a, the voltage across the reversed biased Zener diode remains fixed at V_{break} when current through the Zener varies between I_{min} and I_{max}. These properties of a Zener diode make it a versatile voltage regulator.

Figure 5.32b shows a Zener diode regulated power supply. As may be identified in this figure, there is a full-wave bridge rectifier, a pi-section filter which is followed by Zener diode regulator. Zener is connected with a series resistance R_s to the positive terminal of the pi-section filter and the anode to the negative terminal of filter circuit. As such the Zener diode is under reverse bias. Let us assume that the output voltage of filter circuit is 9 V and the breakdown voltage of Zener diode is 5.0 V. Now, the current through the Zener will adjust in such a way that the difference of 9−5 = 4V will drop across the series resistance R_s so that the voltage drop against Zener diode remains 5.0 V. Next, let us connect some load in the output which draws some additional current. With this change in current distributions, the current through the Zener diode will again adjust such that the output voltage across the load as well as across the Zener diode remains 5.0 V. Thus, the voltage drop against the Zener diode always remains fixed to its breakdown voltage and in this way Zener diode provides a stabilized voltage power supply, in this case of 5.0 V.

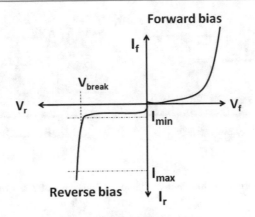

I-V graph for Zener diode

Fig. 5.32a V-I graph for a Zener diode

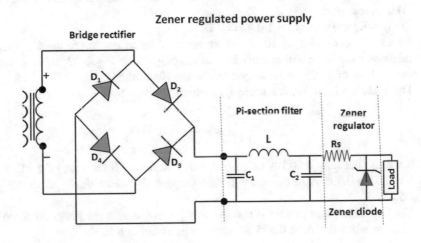

Fig. 5.32b Zener regulated power supply

It is possible to fabricate Zener diodes of any desired breakdown voltage from few volts to few hundred volts by controlling the doping concentrations at the time of growing the semiconductor crystal. Further, Zener diodes are specified by two parameters, the breakdown voltage and the wattage.

Let us assume that the wattage of the Zener used in the power supply is of 2 Watt and 5.0 V breakdown voltage. The maximum current that may be drawn from this diode I_{max} is given by

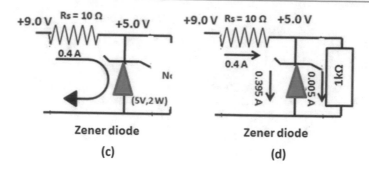

Fig. 5.32c, d Current distribution through different components of Zener regulator

$$I_{max} = \frac{2W}{5V} = 0.4A$$

Therefore, up to a maximum of 0.4 A current may pass through the diode without damaging it. Now, (9–5) = 4 V of voltage must drop against the resistance R_s. This means that

(9−5) = R_s x 0.4 or R_s = 4/0.4 = 10 Ω.

So a series resistance of 10 Ω will be required to be put in the circuit. These calculations were done without any load in the output (see Fig. 5.32c). Let us now connect a load of 1 kΩ in the output across the Zener diode.

The 1 kΩ load will draw current I_{load} which is given by

$$I_{Load} = \frac{5\,V}{1\,k\Omega} = \frac{5}{1000} = 0.005A$$

It means that now 0.005 A of current will flow through the load of 1 kΩ and (0.4–0.005) = 0.395 A of current will flow through the Zener diode, as shown in Fig. 5.32d.

The important point to note is that the voltage drop across the Zener diode does not change when 0.4 A or 0.395 A current passes through it.

Some Formulations For Solving Numerical Problems

1. Most electronic devices run on 24 V or at lesser DC voltages. Stabilized voltage DC power supplies are used to supply power to electronic devices. However, in most of the countries AC power at 110 V, 60 Hz or 220 V, 50 Hz is distributed for public use. In order to run low voltage DC power supplies from high voltage AC lines of public distribution system, step down transformers are used to reduce the line voltage. These step down transformers have primary windings of larger number of turns and secondary windings of smaller number of turns. The turn ratio of secondary to primary is same as their voltage ratio, therefore,

Output voltage at secondry (V_s) = Voltage at the input of primary$(V_p)\dfrac{\text{number of turn sin secondary}(n_s)}{\text{number of turn sin primary}(n_p)}$

$$\text{Or }\quad \frac{V_s}{V_p} = \frac{n_s}{n_p}$$

2. While specifying the line voltage as 220 V, 50 Hz, 220 V is the rms value of the sinusoidal line voltage. rms value for a sinusoidal function is related to the peak nor maximum value V_{max} by the relation: V_{max} or $V_{peak} = \sqrt{2}V_{rms} = 1.414V_{rms}$

3. In case of a full-wave rectified output which is filtered by a shunt capacitor of capacitance C (in Farad) and load resistance R_L(in ohm), the rectified DC output voltage $V_{dc}^{\text{full wave rectifier}} = \left(1 - \frac{1}{2fR_LC}\right)V_{max}$ Volt

4. The peak-to-peak ripple voltage $V_{r(p-p)}^{\text{fullwave}} = \left(\frac{1}{2fR_L}\right)I_{dc}$ Volt

5. The percent ripple factor $r_{\text{capacitor filter}}^{\text{fullwave}} = \dfrac{V_{r(p-p)}^{\text{fullwave}}}{V_{dc}^{\text{fullwave rectifier}}} \text{x} 100$

Solved example (SE-5.5) A 110 V (rms), 60 Hz AC supply energizes the primary winding of a transformer of turn ration 5:1. The secondary of the transformer is connected to a bridge rectifier circuit that uses silicon diodes. Calculate the effective DC value of the output voltage.

Solution: *Peak value of AC input at the primary of transformer* $= \sqrt{2}V_{rms}$
$V_{primary}^{peak} = \sqrt{2}\text{x}110 = 155.56V; \; V_{secondary}^{peak} = \frac{155.56}{5} = 31.11$ V

Peak voltage across diode $V_{Diode}^{peak} = 31.11 - 2(0.7) = 29.71V$

$$V_{out}^{dc} = 0.636\text{x}V_{Diode}^{peak} = 0.636\text{x}29.71 = 18.89V$$

The average or DC output $= 18.89V$

Solved example (SE-5.6) In Solved Example 5.5 if a load of 100 Ω and capacitor of 1000 μF is connected in the output, calculate peak-to-peak value of ripple voltage and the output DC voltage after filtration.

Solution: *In the last example, the DC voltage of rectified output after diode was calculated as 29.71 V. Now, calculate the average or DC current through the load*
$I_{dc} = \frac{29.71}{100} = 0.2972A$.
   The frequency of ripple in full-wave rectifier is 2x frequency of the input AC= *2x60=120 Hz.*
   The admittance G offered by the capacitance of 1000 μF at a frequency of 120 Hz is given by $G = \frac{1}{1000\text{x}10^{-6}\text{x}120} = 8.33$.

Therefore, peak-to-peak voltage of ripple = $GxI_{dc} = 8.33x0.2972 = 2.47V$
Average DC voltage in the output after filtration = $29.71-(2.47/2) = 28.47$ V
The peak-to-peak ripple voltage = 2.47 V
And DC output voltage after filtration = 28.47 V

5.7 Some Other Applications of Diodes

5.7.1 Voltage Multiplier

Combinations of diodes and capacitors may be used to build voltage multiplier circuits. When sinusoidal AC is applied at the input of a N-stage voltage multiplier, rectified unidirectional pulsating DC with maximum voltage N times the maximum input voltage is delivered at the output. Cockcroft and Walton used this voltage multiplying technique to produce high DC voltage for accelerating charged nuclear particles to higher energies in their particle accelerator.

Figure 5.33a, b shows voltage doubler and tripler circuits. If a sinusoidal AC of maximum voltage V_{max} is applied at the input of the doubler circuit (Fig. 5.33a) and for the cycle when terminal A is positive, diode D_1 being forward bias, conducts charging the capacitor C_1 to voltage V_{max}. In the next half cycle when terminal B assumes positive potential, diode D_2 becomes forward bias and charges capacitor C_2 to potential V_{max}. Since capacitors C_1 and C_2 are in series, the potential difference across them adds up producing voltage $2 V_{max}$ across the load. Of course, it is assumed that negligible voltage drops across the load.

It may be noted that a series combination of a diode and capacitor makes a basic unit for voltage multiplication. Three diodes in series with three condensers make a voltage tripler as shown in part (b) of the figure.

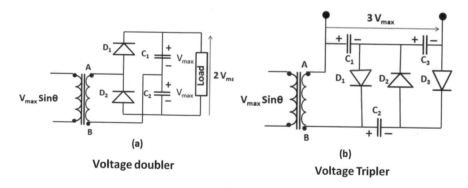

Fig. 5.33 a voltage Doubler circuit; **b** voltage Tripler circuit

Fig. 5.34 **a** OR gate; **b** AND
gate using D-D logic

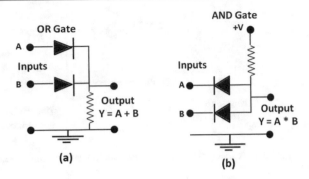

(a)

(b)

5.7.2   Diodes as Logic Gates

In digital electronics, Boolean algebra is implemented through logic gates like AND, NAND, NOT and OR gates, etc. Diodes are often used to construct gates for D-D logic. For example, Fig. 5.34a, b demonstrates the principle of operation for AND and OR logic gates using diodes.

Part (a) of the figure represents an OR gate based on diodes. If logic state- 1 is represented by +5 V (say) and logic state-0 by zero volt, then for logic states A = 0 and for B = 0 and also for A = B = 0, the diodes will not conduct and hence output Y will be 0, that verifies the element, 0 + 0 = 0 of the truth table of OR gate. Also if one of the two inputs is in logic state 1 or both the inputs A = 1 = B, then at least one or both diodes will be forward biased and conducts so that the output Y will be in logic state 1, verifying the element 1 + 0 = 0 + 1 = 1 + 1 = 1 of the truth table.

Figure 5.34b represents logic gate AND. If it is assumed that V=5 Volt, the output Y will remain zero (Y = 0) for A = 0, or B = 0 or A = B = 0. The output Y will be non-zero only when both A and B are in state −1. This demonstrates that the output Y confirms to the truth table of logic gate AND.

5.7.3   Envelop Detector

For transmission of audio signals to long distances, radio stations use amplitude modulation (AM). In AM transmission, the amplitude of a high-frequency radio wave, called carrier, is modulated according to the audio wave. This amplitude modulated carrier wave may travel long distances without much attenuation and can be received at faraway places.

Amplitude modulated signal is transmitted by the radio station and is received by listeners having radio sets. In a radio receiver, modulated carrier waves are demodulated using diode peak detector as shown in Fig. 5.35.

Diode envelope detector circuit is shown in Fig. 5.35. The capacitor in the output of the circuit offers a low reactance path to high-frequency carrier waves while low-frequency audio signal is passed to the output.

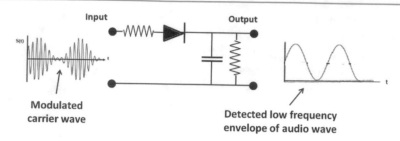

Fig. 5.35 A diode circuit may extract audio signal from the amplitude modulated carrier wave

5.7.4   Limiting or Clipping Circuits

Limiting/clipping of a signal means to restrict the height or amplitude of the signal at some preset value, without altering any other characteristics like frequency of the signal. Diodes may be used to make circuits that may clip some part of the signal at a preset voltage level. Diode clipping circuits may be divided into two classes: (a) Series clipping circuits and (b) shunt or parallel clipping circuits. Each of these two categories may be further subdivided into (a.1) series negative clipping circuit, (a.2) series positive clipping circuit, (b.1) parallel negative clipping circuit and (b.2) parallel positive clipping circuit.

(a)   **Series Negative Clipper**

The circuit of a series diode clipper is shown in the central part of Fig. 5.36. When a sinusoidal signal is applied at the input of the circuit, diode remains forward biased for only positive swings of the signal and gets reversed biased for negative swings. Since a forward-biased diode is like a 'closed' switch, the input signal is passed over to the output and appears across the resistance R, only for each positive swing of the input signal. During negative swings when diode is reversed biased and is like an 'open' switch, the output gets cut off from the input, no current flows through resistance R and no signal appears in the output. Thus, output will appear only for positive swings and the negative part of the input signal will be clipped in the output. In the discussion so far it has been assumed that the diode behaves as an ideal diode. However, in case of

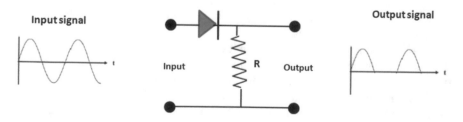

Fig. 5.36 Circuit diagram of a series negative clipper with waveforms of input and output signals

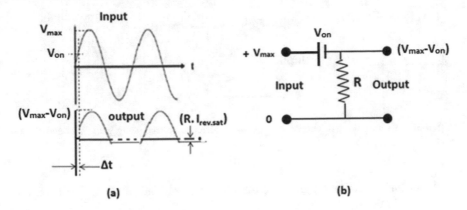

Fig. 5.37 Input and output signals for a real diode

a real diode, even in case of forward bias the diode will start conducting only when input voltage exceeds the cut-in voltage V_{cut-in}, therefore, the output signal will not start at the same instant when the positive swing starts in the input, it will be delayed by the time (Δt) the signal at input takes to reach the value V_{cut-in}. Moreover, the maximum of the output signal will occur at a voltage ($V_{max}-V_{cut-in}$), which is less than the maximum amplitude of the input signal by the amount of cut-in voltage. Similarly, in case of a real diode, during negative swing small reverse saturation current ($I_{rev.sat}$) will flow through resistance R resulting in a ($-R.I_{rev.sat}$) voltage drop in the output. This is shown in Fig. 5.37a. It may be recalled that while discussing the transfer characteristic of a real diode, it was mentioned that a real diode in forward bias starts conducting after the bias voltage reaches the cut-in voltage V_{cut-in} and that a conducting diode is like a battery of Voltage V_{on}. Figure 5.37b shows the situation when input voltage has reached the value $+V_{max}$, the diode is conducting and may be replaced by a battery of voltage V_{on}. The output voltage at this instant will be ($V_{max} - V_{on}$), as shown in part (a) of this figure. It may be noted that V_{on} has the value 0.7 V and 0.3 V, respectively, for silicon and germanium diodes and may not be neglected in comparison of V_{max}, but (Δt) and the potential drop ($-R.I_{rev.sat}$) are certainly so small in comparison to the time period of the input signal and its maximum potential V_{max}, that they may be neglected.

i. **Series negative clipper with positive reference voltage**
A series clipper with positive reference voltage is shown in Fig. 5.38. Operation of the circuit is simple, the diode remains forward biased only for the part of the positive swing of the input when the input voltage is equal or greater than the reference voltage V_R. For rest of the positive swing and for whole of the negative swing, the diode remains cut off and no signal appears at output. The

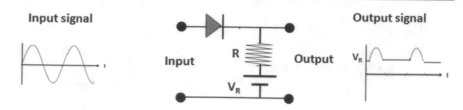

Fig. 5.38 A series negative clipper with positive reference voltage along with input and output signals

input and the output signals of the clipper, assuming diode to be ideal, are shown in the figure.

Self-assessment question (SAQ): Draw output signal for a series negative clipper with positive reference voltage in case of a real diode when (i) $V_R > V_{\text{cut-in}}$ (ii) $V_R < V_{\text{cut-in}}$.

ii. **Series negative clipper with negative reference voltage**

The negative reference voltage increases the duration for which the diode remains forward biased and conducts producing the output. Now, the diode remains forward biased not only for the positive swing of the input but also for the part of negative swing when the negative swing voltage is less in magnitude to the magnitude of the reference voltage. Therefore, whole of the positive swing and a part of the negative swing of the input signal for which diode is forward biased are passed to the output and appear across the resistance R. The output signal for an ideal diode is shown in the figure. Fig. 5.39.

Self-assessment question (SAQ): Draw output signal for a series negative clipper with negative reference voltage in case of a real diode when (i)$| V_R |>| V_{\text{cut-in}} |$ (ii) $| V_R | < |V_{\text{cut-in}}|$.

iii. **Series positive clipper**

Circuit diagram along with input and output signal waveforms for a series positive clipper, assuming diode to be an ideal diode, are shown in Fig. 5.40a.

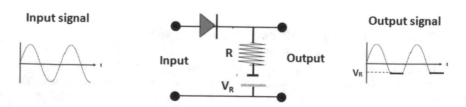

Fig. 5.39 Series negative clipper with negative reference voltage and its input-output signals

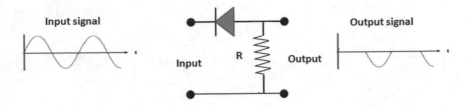

Fig. 5.40a circuit diagram and input, output signals of a series positive clipper

As is obvious, the diode becomes forward biased for the negative swings of the input signal when the upper terminal of the input port is negative with respect to the lower terminal which always remains at ground potential. Forward-biased diode behaves as a 'closed' switch and whatever signal is present at the input (i.e. the negative swing) is passed to the output and appears against resistance R. During positive swings, diode remains reverse biased, behaves like an 'open' switch isolating output from the input, hence no signal appears in the output.

Series positive clipper with positive reference voltage is shown in the central portion of Fig. 5.40b and the input, output signals, respectively, on the left and to the right of the circuit diagram. Because of the positive reference voltage V_R, the diode remains forward biased for complete negative swing of the input signal and also for the part of positive swing when input voltage is less than or equal to V_R, the reference voltage. Therefore, whole of the negative swing and the part of positive swing below V_R appears in the output.

Figure 5.40c shows the circuit diagram of a series positive clipper with negative reference voltage and the output signal corresponding to the sinusoidal input signal. The analysis of the circuit is straight forward; diode becomes forward biased only for the part of the negative cycle when $|V_{input}| \geq |V_R|$, otherwise it remains reversed biased. The corresponding waveform of the signal is shown in the figure. It may be kept in mind that in the present analysis the diode is assumed to be an ideal diode. Output waveforms for a real diode when diode may be replaced by a battery of V_{on} volt are shown in Fig. 5.40b'.

Self-assessment question (SAQ): Draw waveform of the output signal obtained from a series positive clipper with negative bias (reference voltage) of −5V, when a square wave of peak-to-peak value of 20 V is applied at the input. Treat diode to be a real diode with cut-in voltage of 0.7 V.

(b) **Shunt clippers** In a shunt clipper, diode is connected in parallel to the applied signal. It may be recalled that in series clippers diodes are connected in series to the input signal and signal in the output is obtained only when diode is forward biased and conducting. On the other hand, in shunt clippers diodes are connected in parallel to the applied signal and signal in the output is obtained only when diode is reversed biased, non-conducting and like an 'open' switch. In

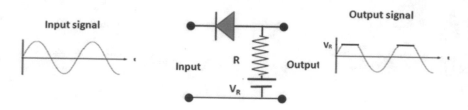

Fig. 5.40b Circuit diagram and input, output signal for a series positive clipper with positive reference voltage

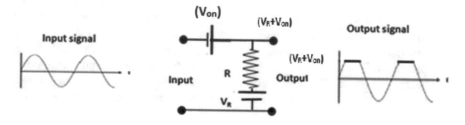

Fig. 5.40b' Equivalent circuit when diode conducts in positive swing

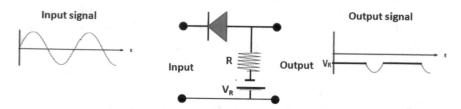

Fig. 5.40c c Circuit diagram and input, output signals for a series positive clipper with negative reference voltage

shunt clippers, there is no output when diode is forward biased, conducting and is like a 'closed' switch. Shunt clippers may be of following types: (i) shunt negative clipper, (ii) shunt negative clipper with positive reference voltage, (iii) shunt negative clipper with negative reference voltage, (iv) shunt positive clipper, (v) shunt positive clipper with positive reference voltage and (vi) shunt positive clipper with negative reference voltage.

i. Shunt negative clipper

Figure 5.41a shows the circuit and the input, output signals of a shunt negative clipper. During the positive half cycle of the input square wave when terminal A is positive with respect to B, assuming diode to be ideal it remains reversed bias, like an 'open' switch. Therefore, the input signal, i.e. the positive part of the

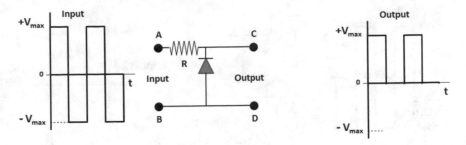

Fig. 5.41a Shunt negative clipper with input, output signals

square wave passes to the output. However, during negative half cycle when terminal A is negative with respect to B which is at zero potential, the diode becomes forward biased, works like a closed switch and grounds the negative half pulse during each cycle. The resulting output wave is shown on the right side in Fig. 5.41a.

However, in case of real diode, the diode did not conduct for whole of the positive swing and also for a part of the negative swing, till the negative potential at terminal A reaches the value $(-V_{on})$. Therefore, whole of the positive swing and a part of negative swing appear in the output. This is represented in Fig. 5.41a'. It may be noted in this figure that the potential across a conducting diode remains fixed at V_{on} as shown in the figure.

ii. **Shunt negative clipper with positive bias or reference voltage**
As shown in Fig. 5.41b, the positive bias makes diode to remain forward biased and conducting for whole of the negative half of the input wave plus for the duration of the positive half till the positive cycle voltage reaches the value $(V_R - V_{on})$. Once the input voltage in positive swing increases beyond

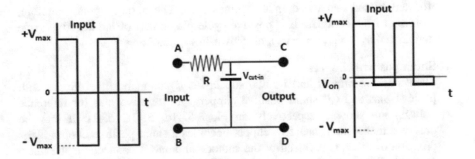

Fig. 5.41a' Equivalent circuit and the output signal for shunt negative clipper in the case of real diode

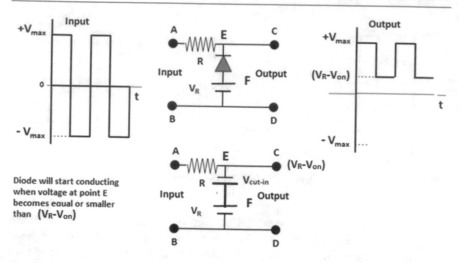

Fig. 5.41b Circuit and wave shapes of input, output signals of shunt negative clipper with positive reference voltage

(V_R-V_{On}), diode becomes reversed biased and non-conducting, like an 'open' switch. As such only part of the input signal that lies between V_{max} and (V_R-V_{on}) goes to the output.

iii. Shunt negative clipper with negative bias

A simple way to analyze a shunt clipper is to find out the duration of the input signal for which the diode remains reversed biased. It is because of the fact that output signal will appear only during the period when diode is reversed biased and like an 'open' switch. It may be observed in Fig. 5.41c that the diode remains in reverse biased for whole of the positive half of the input signal plus during the time in negative swing when the magnitude of(negative) input voltage is less than the magnitude (negative) bias voltage V_R. Once the magnitude of negative input voltage increases beyond $-V_R$, the diode becomes forward biased and the output becomes zero. The output signal, therefore, consists of the complete half positive cycle plus a part of the negative cycle terminated at $(-V_{input} = -V_R)$, as shown in Fig. 5.41c.

(b) Shunt positive clippers

Shunt positive clipper and its two subclasses (i) shunt positive clipper with positive bias and (ii) shunt positive clipper with negative bias (or reference voltage) are shown, respectively, in Figs. 5.41d, 5.41e, 5.41f. It may be observed that a shunt negative clipper becomes a shunt positive clipper if the direction of diode is reversed or the cathode and anode connections are inter-changed. The circuits are easy to analyze if it is kept in mind that in shunt diode circuits output appears only for the duration of time of the input signal for which diode is non-conduction or reverse biased. Also, it is important to

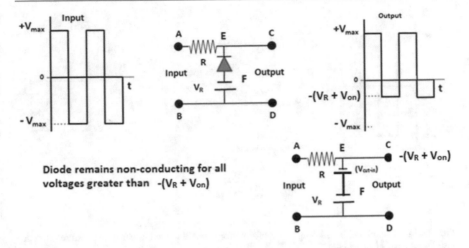

Fig. 5.41c Circuit and input, output waveforms of shunt negative clipper with negative reference voltage

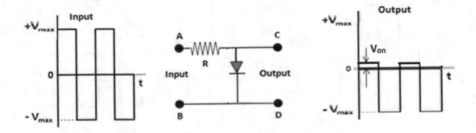

Fig. 5.41d Shunt positive clipper and its input, output signals

remember that terminal B or D is always at ground potential zero and the potential as well as polarity of terminal A changes with the positive and negative swings of applied signal. In a positive clipper without any bias, diode remains non-conducting for the duration the positive voltage increases from zero to V_{Cut-in} . Hence, the part of the input signal in the interval 0 V to V_{on} V is passed to output. From V_{on} to V_{max}, and from V_{max} to V_{on} (during the falling part of the positive swing of input signal) diode becomes forward biased and starts conducting, therefore, no signal appears in the output. However, when the potential at terminal A becomes less than zero, i.e. negative, diode becomes reverse biased and whatever signal is present at the input is passed to the output. The output waveform corresponding to the square wave input obtained using a real diode is shown on the right side of the figure. Important point to observe is that portion of the positive swing of input signal below 'on' voltage

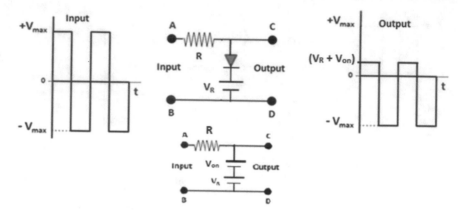

Fig. 5.41e Circuit and input, output waveforms for shunt positive clipper with positive reference voltage in case of real diode

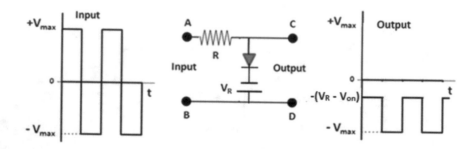

Fig. 5.41f Circuit and input, output waveforms for shunt positive clipper with negative reference voltage

of the diode also appears in the output. However, if diode is taken as an ideal diode then output will start from zero base line.

(i) Shunt positive clipper with positive bias

The circuit diagram, input and output waveforms for a positive clipper with positive bias or reference voltage are shown in Fig. 5.41e.

In unbiased shunt positive clipper, during the positive swing diode remains non-conducting up to V_{on}, however, when a positive reference voltage V_R ($V_R > V_{on}$) is included the diode remains non-conducting up to the voltage ($V_{on} + V_R$) of the positive swing. Hence, the input signal up to the voltage ($V_R + V$) of the positive swing is passed to the output. Further, diode remains non-conducting under reverse bias during the negative swings and hence complete negative swings also appear in the output.

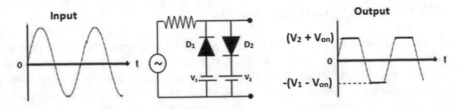

Fig. 5.42a Positive–Negative clipper with biasing

iii. **Shunt positive clipper with negative bias**
   Without any biasing, diode remains non-conducting from $(+V_{on})$ volt to $(-V_{max})$ volt of the input signal. With the inclusion of negative biasing of V_R volt, diode will remain non-conducting for $(-V_R + V_{on} -V_{max})$ volt of the input signal. This means that only the part of the negative swing of the input signal that lies between $\{-(V_R-V_{on})\}$ and up to $(-V_{max})$ will pass on to the output. The output waveform obtained from a shunt positive clipper with negative basing of V_R volts is shown on the right side in Fig. 5.41f.

(f) **Positive–Negative clipper with two different biasing**
   It is possible to clip both the positive and negative swings of the input signal at two different (or same) voltage levels using a combination of shunt positive and negative clippers with appropriate biases as shown in Fig. 5.42a.
   Zener diodes may also be used for making positive–negative clipper as shown in Fig. 5.42b. It is assumed that both Zeners are of the same type (either silicon based or germanium based) so that their 'on' voltages are same.

Self-assessment question (SAQ): Replacing conducting diodes by a battery of voltage V_{cut-in} and Zener under breakdown by a battery of voltage V_{break}, obtain upper and lower cut values for the positive–negative limiter of Fig. 5.42b.

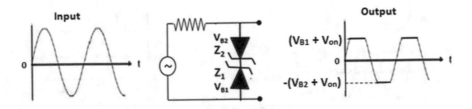

Fig. 5.42b Positive–Negative clipper using Zener diodes

5.7.5 Clamper Circuits Using Diode

Electronic device that shifts the DC level of any time-varying signal to a preset value is called a clamper or level shifter. A clamper either pulls up or pushes down on the voltage axis the entire input signal without altering its other parameters like peak-to-peak value, frequency, amplitude, phase, etc. Level shifters or clampers may be broadly classified into two types: **positive clamper** which shifts the input signal up on the voltage axis or puts the input signal on a predetermined DC pedestal such that **the minimums of the signal touch the preset DC level.** For example, the positive clamper in Fig. 5.43 has lifted the entire input signal and put it on the preset DC level of V = X (volt).

A **negative** clipper on the other hand pushes the entire input signal such that **the maximums of the signal touch a preset DC level** and the entire signal remains below this preset DC level. As an example in Fig. 5.43, the negative clamper has pushed the entire input signal below the DC level V = −Y (volt). It may be observed in this figure that no other parameter of the input signal has been affected by the clamping circuit. It may also be mentioned that the value of the preset DC level in both positive and negative clampers may be positive or negative; i.e. a positive clamper may clamp a signal at a negative DC level and similarly, a negative clamper may clamp the signal to a positive DC level. The point of distinction is if the entire signal is clamped above the preset DC level (whether positive or negative) the clamping is called positive and if the entire signal is clamped below the preset DC level (no matter whether the DC level is positive or negative) the camping is negative.

Both positive and negative clampers are made using a diode and capacitor. The direction of diodes is opposite in the two cases. Figure 5.44a, b, respectively, shows the unbiased positive and unbiased negative clampers. When an AC signal v_{in} is applied to the input of the positive clamper (Fig. 5.44a), the diode conducts during the negative swing of the input and charges the capacitor via the path FBAG,

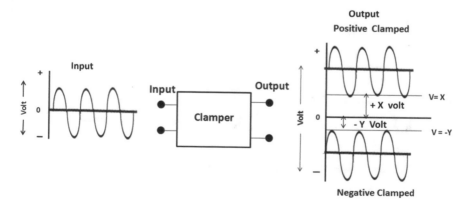

Fig. 5.43 Illustration of typical positive and negative clampers

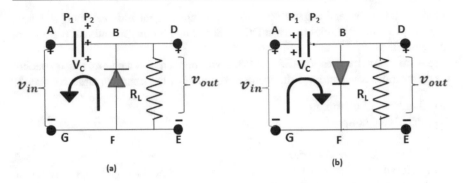

Fig. 5.44 a positive clamper b negative clamper

indicated by the arrow in the figure. Thus plate P2 of the capacitor acquires positive charge and plate P1 negative. Fig. 5.44a.

The time constant τ_{cha} for capacitor charging is essentially given by $\tau_{cha} \approx R_f \cdot C$, where R_f is the forward resistance (or dynamic resistance) of the diode and C is the magnitude of the capacitance. Since R_f is very small, the charging time of the capacitor is very low and within first few negative cycles of the input signal the capacitor gets charged to a voltage V_C, which is very near to the maximum positive voltage V_{max} of the input signal. During positive swings of the input signal when diode is reversed biased and is like an 'open' switch, the capacitor tries to discharge through the load resistance R_L. Since R_L has large value of the order of kilo ohms, the discharge time constant $\tau_{dis} \approx R_L C$ is at least ten orders of magnitude larger than τ_{cha} and, therefore, the capacitor does not lose any appreciable amount of charge during positive swings. As such after few cycles the charge and voltage on the capacitor become constant. The plate P2 of the capacitor which is positively charged and is at positive potential V_C makes the diode reversed biased even during the duration of the negative swings of the input signal. As such after initial few cycles diode remains reversed biased both for the positive and the negative swings of the input signal. Since a reversed biased diode is like an 'open' switch, the input signal superimposed over a DC level is passed over to the output and appears across the load resistance R_L. The magnitude of the DC pedestal depends on V_c, the voltage drop across the capacitor. It is worth noting that fast charging of the capacitor during negative swings and essentially negligible discharge during positive swings of the input wave keep diode reversed biased and put the input signal on a DC pedestal decided by the potential at plate P2 of the capacitor.

In case of a negative clamper, in the initial few cycles of the input signal diode conducts during positive swings charging capacitor via the path BFGAB as shown in the figure by the arrow. Because of relatively very small time constant for charging and considerably large time constant for discharge, the potential across the capacitor becomes constant with plate P2 as negative. Also, the diode remains cut off for both positive and negative swings of the input signal and, therefore, the input wave appears in the output below a DC level decided by the potential at plate P2 of the capacitor.

Self-assessment question (SAQ): What will happen if in a clamper the capacitor discharge time constant is of the same magnitude as the charging time constant?

(a) **Analysis of clampers**
(b) **Positive Clampers**

i. Unbiased positive clamper:

The input signal, a positive clamper and the output positively clamped signal are shown in Fig. 5.45i, ii, iii, respectively.

The equivalent circuit at the instant when capacitor is charging is shown in part (iv) of the figure, where the conducting diode is replaced by a battery of V_{on} volt. Here, V_{on} is the voltage at which diode starts conducting when it is forward biased during negative swing of the input signal which is a rectangular wave in the present case. As the first step of analysis, we calculate the potential V_c across the diode. For this let us apply Kirchhoff's voltage law (KVL) to the closed-loop ABFGA, starting from A and keeping in mind the convention that while tracing the closed path a voltage drop is taken positive if one goes from positive terminal to negative terminal otherwise it is taken as negative.

$$-V_C - V_{on} - v_{in} = 0$$
$$Or \quad V_C = -V_{on} - v_{in} \tag{5.63}$$

The maximum value of V_C occurs when diode conducts in the negative swing and reaches the value-V_{max} .After a few cycles the value of V_C stabilizes at this maximum value. Therefore,

on substituting $(-v_{in}) = -(-V_{max})$, one gets
$$V_C = -V_{on} + V_{max} \tag{5.64}$$

In order to obtain a relationship between the instantaneous values of input signal and the output signal, we again apply KVL in the closed loop ABDEGA and get,

$$-V_C + v_{out} - v_{in} = 0$$
$$Or \quad v_{out} = V_c + v_{in} = (-V_{on} + V_{max}) + v_{in} \tag{5.65}$$

Equation (5.65) provides the required relation between the instant values of output voltage corresponding to the input value, for example, let us calculate the instantaneous values of output for three values of input.

$$v_{in} = +V_{max}; v_{out} = (-V_{on} + V_{max}) + V_{max} = 2V_{max} - V_{on} \tag{5.66a}$$

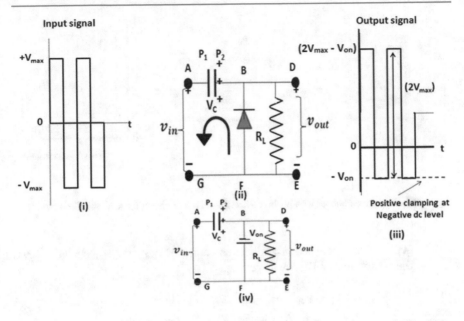

Fig. 5.45 Input signal, positive clamper, output signal and equivalent circuit for analysis of an unbiased positive clamper

$$v_{in} = 0; v_{out} = (-V_{on} + V_{max}) + 0 = V_{max} - V_{on} \qquad (5.66b)$$

$$v_{in} = -V_{max}; v_{out} = (-V_{on} + V_{max}) - V_{max} = -V_{on} \qquad (5.66c)$$

It may be verified that the output signal shown in part (iii) of the figure confirms to the conditions specified by the set of equations (5.66). An unbiased positive clamper, therefore, puts the entire input signal at a negative DC pedestal of $-V_{on}$.

ii. **Positive clamper with positive reference voltage or bias**

Figure 5.46a shows the circuit of a positive clamper with positive bias of V_0 volt. Equivalent circuit for analysis is shown in part (b) of this figure. Writing KVL for the closed-loop ABFGA, one gets

$$-V_C - V_{on} + V_0 - v_{in} = 0$$
$$\text{Or } V_C = (V_0 - V_{on}) - v_{in}$$

For maximum value of V_C, we substitute $(-v_{in}) = V_{max}$ to get,

$$V_C = (V_0 - V_{on}) + V_{max} \qquad (5.67)$$

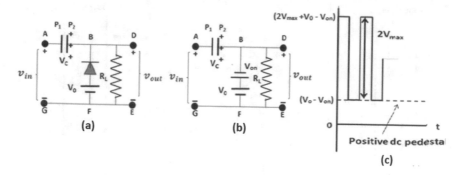

Fig. 5.46 **A** Positive clamper with positive bias; **B** equivalent circuit for analysis; **C** output signal

To get a relation between instantaneous values of v_{out} and v_{in}, we write KVL for the closed-loop ADEGA to get

$$- V_C + v_{out} - v_{in} = 0$$
$$\text{Or} \quad v_{out} = V_C + v_{in} = (V_0 - V_{on}) + V_{max} + v_{in} \tag{5.68}$$

$$\text{If } v_{in} = + V_{max}; v_{out} = (2V_{max} + V_0 - V_{on}) \tag{5.69a}$$

$$\text{If } v_{in} = 0; v_{out} = (V_{max} + V_0 - V_{on}) \tag{5.69b}$$

$$\text{If } v_{in} = - V_{max}; v_{out} = (V_0 - V_{on}) \tag{5.69c}$$

Equations (5.69) tells that the positive clamper has put the entire input signal on a positive DC level equal to $(V_0 - V_{on})$ as shown in Fig.5.46c.

iii. **Positive clamper with negative bias**

Figure 5.47b shows a positive clamper with a negative bias of V_0 volts. The analysis of the circuit is left as an exercise. It may be shown that the output signal will have the waveform as shown in Fig. 5.47c. The positive level shifter has put the input signal on a negative DC level of $-(V_0 + V_{on})$.

(c) **Negative clampers**

i. **Negative clamper with positive bias**

Figure 5.48a shows a negative clamper with a positive reference voltage of V_0 volts.

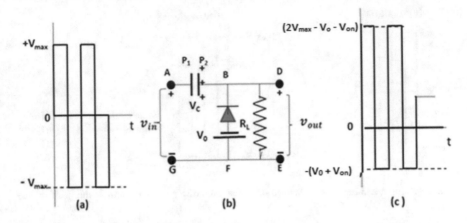

Fig. 5.47 **a** Input signal; **b** positive clamper with negative bias of V_0 volt; **c** output waveform

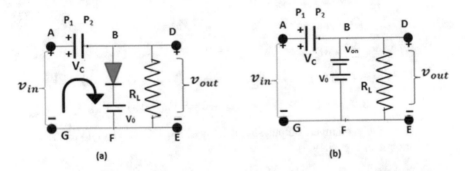

Fig. 5.48 **a** Negative clamper with positive reference voltage V_0; **b** Equivalent circuit for analysis

As already discussed, diode conducts during the positive swing of the input signal for the duration the voltage drop across diode exceeds its cut-in voltage and remains forward biased. Conduction of diode charges the capacitor to the voltage V_C via the path indicated by the arrow putting a positive charge on plate P1 of the condenser. The equivalent circuit for the purpose of analysis is shown in Fig. 5.48b. In order to obtain the value of potential V_C, one may apply KVL to the closed-loop ABFGA to get

$$+ V_C + V_{on} + V_0 - v_{in} = 0$$
$$\text{Or} \quad V_C = -(V_{on} + V_0) + v_{in} = -(V_{on} + V_0) + V_{max} \tag{5.70}$$

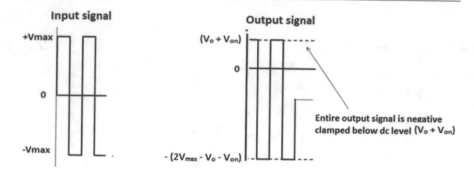

Fig. 5.49 Input and negatively clamped output signals of a negative clamper with positive bias

To get the relation between the instantaneous values of v_{out} and v_{in}, we apply KVL to the closed-loop ADEG to get

$$+ V_c + v_{out} - v_{in} = 0$$
$$Or \quad v_{out} = -V_C + v_{in} = (V_{on} + V_0) - V_{max} + v_{in} \tag{5.71}$$

When $v_{in} = V_{max}$; $v_{out} = (V_{on} + V_0) - V_{max} + V_{max} = (V_{on} + V_0)$

If $v_{in} = -V_{max}$; $v_{out} = (V_{on} + V_0) - V_{max} - V_{max} = (V_{on} + V_0) - 2V_{max}$

If $v_{in} = 0$; $v_{out} = (V_{on} + V_0) - V_{max} - 0 = (V_{on} + V_0) - V_{max}$

Negatively clamped output signal at positive DC level of $(V_0 + V_{cut-in})$ is shown in Fig. 5.49.

ii. **Negative clamper with negative bias**

It is simple to obtain the input-output relationship in case of negative clamper with negative bias by replacing the value of bias voltage from $+V_0$ to $-V_0$ in the result obtained for the previous case of negative clamper with positive bias. Equation (5.71) may, therefore, be (Fig. 5.50) rewritten as

$$v_{out} = -V_C + v_{in} = (V_{on} - V_0) - V_{max} + v_{in} \tag{5.72}$$

So that when

$$v_{in} = V_{max}; \ v_{out} = -(V_0 - V_{on}) \tag{5.73a}$$

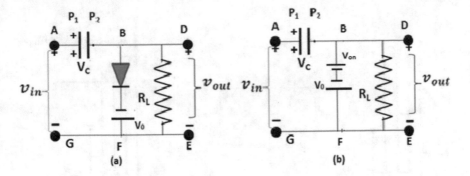

Fig. 5.50 **a** Circuit diagram of a negative clamper with negative bias; **b** Equivalent circuit for analysis

When

$$v_{in} = -V_{max}; v_{out} = -(V_0 - V_{on}) - 2V_{max} \qquad (5.73b)$$

The output of the negative clamper with negative bias of V_0 volt is shown in Fig. 5.51b.

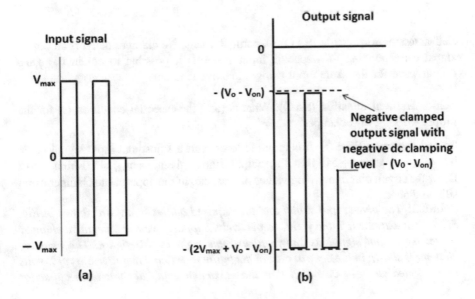

Fig. 5.51 **a** Input signal; **b** output signal of a negative clamper with negative reference voltage $(-V_0)$.

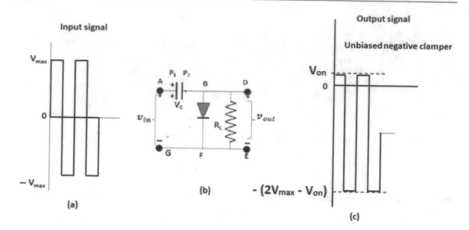

Fig. 5.52 **a** Input signal; **b** Circuit of an unbiased negative clamper; **c** Negative clamped output signal

iii. **Unbiased negative clamper**

The characteristic of unbiased negative clamper may be obtained by setting the bias voltage V_0 to zero in the analysis of either negative clamper with positive or that with negative bias. The result will be that the output signal will hang down from the DC level $+V_{on}$ volt, as shown in Fig. 5.52b.

Self-assessment question (SAQ): A particular negative clamper delivers an output signal corresponding to a specific input wave, is it possible to obtain the same output wave for the same input using a positive clamper?

Self-assessment question (SAQ): What is (are) the essential condition (s) for the proper working of a clamper?

Solved example (SE-5.7) It is required to convert a sinusoidal input signal given by $v_{input} = 10\sin\omega t + 5V$ into the output signal given as $v_{out} = 10\sin\omega t - 5V$. Draw the circuit diagram giving values of bias, etc. of the appropriate clamper using silicon diode.

Solution: *The given input signal and the required output signal are shown in Fig. SE-5.7a at extreme left and right, respectively. Suppose, we use a negative clamper to get the output signal as shown at extreme right in this figure. The negative clamper is shown in the central part of the figure to which a negative bias of X volts is connected. It is to be noted that the maximum value at which the capacitor*

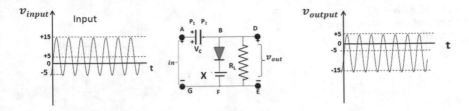

Fig. SE-5.7a Input signal, negative biased negative clamper and the output signal

voltage V_C will stabilize will occur when $v_{input} = +15V$. Figure SE-5.7b shows the equivalent circuit of clamper for analysis. To obtain the value of CC, we apply KVL to the closed-loop ABFGA to get,

$$+V_C + 0.7 + X - v_{input} = 0; \text{ Hence } V_C = v_{input} - (X + 0.7)$$
$$\text{Putting } v_{input} = +15 \text{ V, one gets}; V_C = 15 \text{ V} - (X + 0.7) \text{ V} \qquad (SE - 5.7.1)$$

In order to obtain a relation between the instantaneous values of output signal and input signal, the KVL is applied in the closed-loop ADEGA to get

$$+V_C + v_{output} - v_{input} = 0 \text{ Or } v_{output} = v_{input} - V_C$$

Substituting the value of VC from above, one gets

$$v_{output} = v_{input} - [15V - (X + 0.7)] \qquad (SE - 5.7.2)$$

Equation (SE-5.7.2) provides the required relation. In the output signal, it may be observed that the minimum value of the output signal is -15V, which should occur corresponding to the minimum value of $-5V$ of the input signal. Substituting these value in (SE-5.7.2), one gets

$$-15 = -5 - [15V - (X + 0.7)] \text{ or } (X + 0.7) = 5$$

Hence, $X = 5 - 0.7 = 4.3V$
Therefore a negative clamper with positive bias of 4.3 V will produce the desired output signal from the given input signal.

Solved example (SE-5.8) Is it possible to solve the above problem (SE-5.7) using a positive clamper? If yes obtain the parameters of the positive clamper that performs the same function as has been performed by negative clamper.

Solution: *Let us assume that a positive clamper with positive bias of Y volt produces the desired output signal corresponding to the input signal of the last solved*

Fig. SE-5.7b NEquivalent
circuit of negative clamper for
analysis

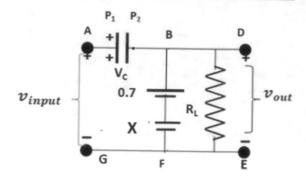

example. The circuit diagram of the positive clamper and its equivalent circuit for analysis are given in Fig. SE-5.8.1.

Writing the KVL for the loop ABFGA, one gets $-V_C - 0.7 + Y - v_{in} = 0$

Or $V_C = (Y - 0.7) - v_{in}$, *the maximum value at which Vc stabilizes is obtained when* $v_{in} = -5V$

Therefore, the stabilized value of

$$V_C = (Y - 0.7) - (-5) = (Y + 4.3) \qquad (SE - 5.8.1)$$

Applying KVL to loop ADEGA, $-V_C + v_{out} - v_{in} = 0$

or $v_{out} = V_C + v_{in} = (Y + 4.3) + v_{in}$

Now if v_{out} *is taken as +5 which should correspond to the value of* $v_{in} = +15$, *putting these values in the above equation, one gets*

$$+5 = (Y + 4.3) + 15 \, or \, (Y + 4.3) = -10 \, or \, Y = -14.3V$$

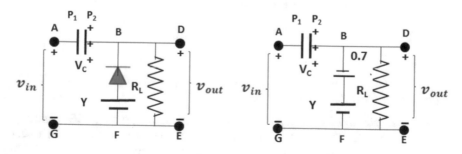

Fig. SE-5.8.1 Positive clamper with positive bias and corresponding equivalent circuit for analysis.

Fig. SE-5.8.2 Positive
clamper that will perform the
same function as performed
by the negative clamper.

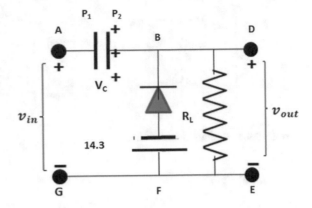

We obtain a negative sign for Y which means that the positive clamper with
negative bias of 14.3 V (shown in Fig. SE-5.8.2) will produce the same output as is
produced by the negative clamper with positive bias of 4.3 V.

Solved example (SE-5.9) *Design a clipper using germanium diodes that may clip
a rectangular wave of peak-to-peak value of 20V and average value of zero volt at
+7 V and −5 V.*

Solution: *The clipper circuit given below clips the positive swing of the signal at
(V_2+V_{cut-in}) v and the negative swing at $−(V_1− V_{cut-in})$ volt. For a germanium diode
V_{cut-in} is 0.3 V. It is required that in the positive direction the input should be
clipped at +7V (Fig. SE-5.9.1)*
   Therefore, (V_2+V_{cut-in}) = +7 Or $V_2 = 7 − V_{cut-in} = 7−0.3 = 6.7$ V
   Similarly, $−(V_1−V_{cut-in})$ =-5 Or $V1 = 5+V_{cut-in} = 5 + 0.3 = 5.3$ V
   The desired clipper is shown in Fig. SE-5.9.2.

Solved example (SE-5.10) A triangular wave shown as Fig. a is applied to the
biased series clipper as shown in Fig. b. Proof that the output signal from the clipper
will have the shape as given in Fig. c.

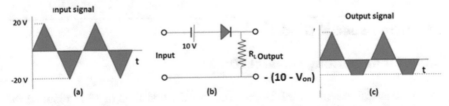

Solution: *The output from the clipper will be obtained only when diode is forward
biased, conducting and is like a closed switch. This will happen when diode is
forward biased and bias voltage is larger than the 'on' voltage V_{on} of the diode. No
output will appear if diode is reversed biased and is like an open switch. If there
was no biasing, the diode will conduct during the positive swing when positive
voltage exceeds V_{on}. However, in the present case, a positive bias of 10 V is given*

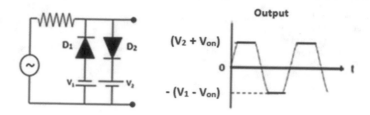

Fig. SE-5.9.1 Positive–negative clipper

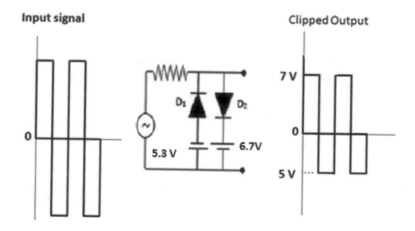

Fig. SE-5.9.2 Positive–negative clipper that clips at +7 and −5 volts

to the diode, hence it will conduct for entire positive swing and also during the negative swing of the input signal till voltage say $(-X$ volt), such that $[10 - X = V_{on}]$, which gives $X = (10 - V_{on})$. Therefore, part of the input signal that lies between $[-(10 - V_{on})]$ and +20 volt will appear in the output, as shown in Fig. c.

5.8   Some Special Diodes

5.8.1   Light-Emitting Diode (LED)

It was the British scientist H. J. Round who for the first time in 1907 reported the emission of visible light from some solid-state devices on application of voltage. However, it was Nick Holonyak Jr., who in 1962 developed the first practical visible-spectrum light-emitting diode. Holonyak Jr. is often called the father of LED. Schematically a LED is represented as Fig. 5.53

A LED is essentially a p-n$^+$ junction diode made up of a highly doped n-side and lightly doped p-side. As a result, the depletion layer extends mostly on the p-side.

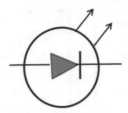

Fig. 5.53 Symbol for light-emitting diode (LED)

When such a junction is forward biased, electrons, the majority carriers on n-side are injected into the p-region where electrons are minority carriers. Hence, this is called minority injection. Since p-side is lightly doped, very few holes cross from the p-side to the n-side. Injected electrons recombine with holes on the p-side-emitting incoherent visible light. This process is called injection electroluminescence. The visible light photons should be allowed to escape from the device without being reabsorbed. The process of recombination of injected electrons and holes may be classified into two types: (i) Direct recombination (ii) Indirect recombination.

(i) Direct recombination takes place in materials that have direct band gap. In such materials, minimum energy of the conduction band lies directly above the maximum energy of the valence band in momentum–energy space, as shown in Fig(5.54a). Therefore, in such materials, electrons from the conduction band may directly combine with holes in the valence band spontaneously emitting photons of visible light. GaAs is a typical example of direct band gap material.

(ii) Indirect recombination occurs in those cases where the energy minimum of the conduction band and energy maximum of valence band are displaced from each other in energy–momentum space. As a result, direct recombination of electron and holes becomes less probable. In such materials, additional shallow donor impurities or dopants are added. These donor states capture injected free electrons locally, providing the necessary momentum shift for recombination. GaP is an example of the material that has indirect energy gap. Often nitrogen (N) is used as the shallow additional dopant in this case. The mechanism of indirect recombination is shown in Fig. 5.54b, c. The main difference between direct and indirect recombination lies in the fact that in case of indirect recombination, apart from photons of visible light, phonons (energy quantum of surface vibrations) are also emitted to compensate for momentum difference.

The wavelength of emitted radiations is essentially decided by the forbidden energy gap E_g of the semiconductor material. If ν is the frequency of the emitted radiation, then

$h\nu \cong E_g$ But $= \frac{c}{\lambda}$; Therefore, $\lambda = \frac{hc}{E_g}$

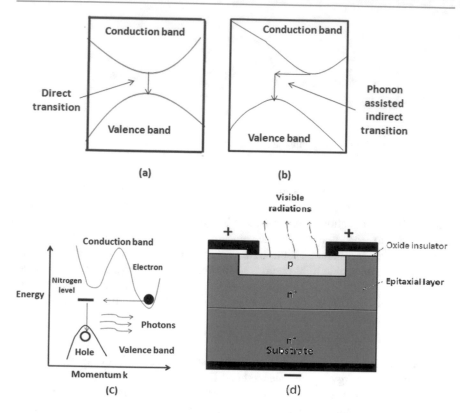

Fig. 5.54 **a** direct recombination; **b** indirect recombination; **c** Assisted indirect recombination in Energy–Momentum space; **d** Layout of a typical LED

Here λ, h and c are, respectively, the wavelength of emitted light, Planck's constant and the velocity of light. Structure of LED plays a crucial role in efficient emission of visible light. It is necessary for the emission of light that most of the recombination events occur very near to the surface of LED so that they have negligible chance of absorption. One way to achieve it is to make p-layer so thin that almost entire p-thickness is covered by depletion layer. Other methods include increasing the doping concentration of substrate so that additional charge carrier electrons move to the surface and recombination occurs predominantly at the surface.

A list of some semiconductor materials and the colour of light emitted by them is given below.

1. Aluminium gallium indium phosphide (AlGaInP) : High brightness Orange-red, Orange, Yellow and green
2. Aluminium gallium arsenide (AlGaAs): Red

3. Aluminium gallium phosphide (AlGaP): Green
4. Silicon carbide(SiC): Blue

5.8.2 Photodiode

Photodiodes are p-n junction diodes made in such a way that the area of their depletion layer may be exposed to visible light or any desired radiations. They are generally kept under reverse bias and when not exposed to light a very small reverse saturation current, called dark current, produced due to the thermal breakdown of covalent bonds, flows through them. However, when light or any other ionizing radiation of sufficient energy falls on depletion layer, it creates electron–hole pairs, the number density of which is proportional to the flux and the intensity or the power of the incident light. As a result, a considerably large reverse current flows through the circuit. This current may be used to activate other devices. The current through the diode stops when the source of radiations is withdrawn. However, in a normal p-n junction diode, the thickness of depletion layer is small and it has large capacitance which increases the switching time of the diode. As such normal junction diodes are slow in their switching response. To improve the switching time p-i-n type, diodes are often used for making fast photo diodes. Additional layer of intrinsic material adds to the depletion layer, reducing the capacitance of the device and makes it fast. In order to increases the radiation detection efficiency and to increase the gain of the device, photodiodes are some times operated under large reverse bias voltage. Large value of reverse voltage produces a stronger electric field through the depletion region. Initial electron–hole pairs produced by the incident radiations acquire large kinetic energy under the high electric field and while moving create secondary electron hole pairs. Therefore, this internal multiplication of electron–hole pairs results in constituting considerably large reverse current. Since additional secondary electron–hole pairs are created by avalanche process, diodes are called avalanche diodes. In a photodiode, reverse saturation current does not depend much on the reverse bias voltage but it strongly depends on the power of the illuminating radiations. Photodiodes may be used in two different modes, namely, in photovoltaic mode, and in photo conductive mode. In photovoltaic mode, no bias voltage is applied to the diode and the dark current is very small. But diode response is slow because of the thin depletion layer which results in large value of capacitance. When radiations from outside fall on diode and create electron–hole pairs in depletion region, the created free charge carriers (electrons and holes) are swept to their respective sides by the built-in potential difference at depletion layer. This sweeping of charge carriers sends current in the external circuit which may be detected by a sensitive ammetre or some other instruments indicating the presence of radiations. Further, the magnitude of current is proportional to the power of the incident radiations. In conduction mode, photodiode is reverse biased which widens the thickness of depletion layer, reducing its capacitance. Therefore, in conduction mode diode response time becomes small,

450

5 p n Junction Diode. A Basic Non-linear Device

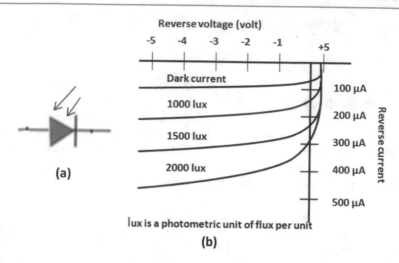

Fig. 5.55 **a** Symbol; **b** typical I-V characteristics of a photodiode

and it becomes a fast switching device. The magnitude of reverse current essentially depends on the power of incident radiations.

A photodiode is characterized in terms of its three performance parameters: (i) **Responsivity**, which is defined as the ratio the photocurrent generated to the power of the incident radiations. (ii) **Quantum efficiency** is the ratio of number of electron–hole pairs generated per incident radiation photon. (III) **Transit or response time** is the time taken by the charge carriers (generated by incident radiations) in crossing the p-n junction. Symbol of a photodiode and a typical reverse biase I-V characteristic of a photodiode are shown in Fig (5.55).

5.8.3 Laser Diode

LED emit visible light of different colours but the light emitted by them is not monoenergetic, which means that all emitted photons do not have the same frequency. LED emits light in a frequency band of $(v \pm \Delta v)$. Further, the light emitted by LED is incoherent, which means that all emitted photons are not in same phase. A laser diode emits light photons which have exactly the same frequency (and hence energy) and they are all in same phase, i.e. photons are monoenergetic and coherent.

Laser diode has nip structure, i.e. there are three layers of semiconducting materials, n-type, intrinsic and p-type. The laser action takes place in the middle intrinsic material which is sandwiched between the n- and p-type materials. The diode is kept forward biased and the forward current initiates the laser action in the middle layer of intrinsic material.

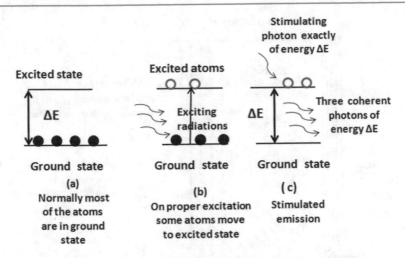

Fig. 5.56 **a** Ground and excited states of atom; **b** Excitation of atoms; **c** stimulated emission of photon

LASER stands for 'Light Amplification by Stimulated Emission of Radiation' and is a method of producing intense beam of monoenergetic and coherent photons. The laser action involves three distinct processes named (i) stimulated emission (ii) optical pumping and (iii) population inversion.

(a) **Stimulated emission** According to quantum theory, atoms (of any material) may have many discrete energy states. The state with lowest energy is called the ground state and generally atoms live in their ground state, as shown in Fig. 5.56a. However, ground state atoms may absorb energy from some incident radiations and may go to one of their excited states (see Fig. 5.56b). In general, the mean lives of excited states are very short of the order of 10^{-9} s, and, therefore, the excited atoms undergo **spontaneous** decay back to the ground state or to some intermediate excited states. In spontaneous decay, excess excitation energy is emitted in the form of photons, but these photons are not coherent. It has, however, been observed that if atoms in an excited state of energy ΔE are hit by a photon of exactly the same energy ΔE, all atoms in the excited state (for example, two in Fig. 5.56c plus the photon that has stimulated are all instantly emitted as monoenergetic and coherent photons. This way of producing monoenergetic coherent photons is called stimulated emission.

(b) **Population inversion** It is evident that to produce an intense beam of monoenergetic and coherent photons using stimulated emission, there must be large number of atoms in the excited state. However, it is generally not possible, because of the law that says that the population of atoms in excited state cannot be more than the population of atoms in the ground state. In technical language, it is said that for the production of intense coherent photon beams

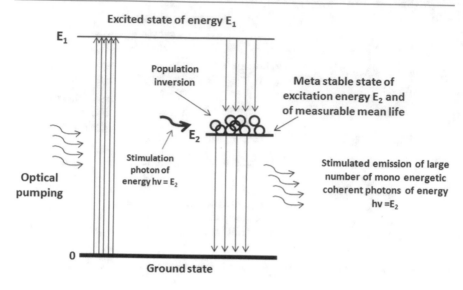

Fig. 5.57 Representation of LASER action

using stimulation emission, it is essential to have population inversion, i.e. there must be more atoms in the excited state than in the ground state.

(c) **Optical pumping** Optical pumping is the process of producing population inversion. In this method, one chooses an atomic system that has several excited states but one of which is metastable, i.e. it has long mean life of the order of a few milliseconds or more. In such a system, the ground state atoms are continuously excited to some higher states by bombarding them with photons of required energy, for example, in Fig. 5.57, ground state atoms are excited to the state of excitation energy E_1. Now, most of the atoms of excited state E_1 decay to the intermediate metastable state which has a measurable lifetime. It means that atoms reaching the metastable state stay in this state for some time and do not decay to the ground state immediately. Continuously exciting of ground-state atoms to the excited state E_1 using photons is called optical pumping. Using optical pumping, it is possible to achieve population inversion of the metastable state at energy E_2. Once population inversion of metastable state is achieved, it is possible to de-excite all the atoms in this metastable state by hitting this state with a photon of energy E_2, through stimulated emission of monoenergetic coherent photons of energy E_2. In a laser diode, optical pumping and stimulated emission is achieved by photons generated in the intrinsic material layer when diode is forward biased and current flows through it.

Laser diodes work at relatively higher forward currents, about 80% of the maximum value prescribed for the diode. This high current is required to initiate laser action. Considerable amount of heat is generated because of the high current which is further enhanced by the large number of photons produced in stimulated emission. By the time a laser diode is working properly it is at the brink of disaster, a little reduction in current stops laser action and a little higher current will damage the diode. It is, therefore, essential to have adequate heat dissipation when laser diode is working.

5.8.4 Schottky Diode

A Schottky diode, also called hot-carrier diode, hot electron diode, barrier diode or surface barrier diode, is a semiconductor–metal junction diode with very small forward voltage drop or cut-in voltage and very fast switching time. In normal silicon and most of the other diodes the cut-in voltage or 'On' voltage ranges from 0.6 V to 1.5 V, while in Schottky diodes the forward voltage drop ranges from 0.15 V to 0.45 V. This smaller value of voltage drop results in less wastage of input power across the diode, production of smaller amount of heat and makes the diode very fast providing higher switching speed and better system efficiency. In order to understand the working of a Schottky diode, one must understand the properties of a metal–semiconductor junction.

Metal–Semiconductor junction Let us consider a metal and N-type semiconductor junction the energy band diagram of which is shown in Fig. 5.58. In this figure, on the metal side, there are overlapping valence and conduction bands filled with free electrons up to certain energies. On the semiconductor side, there are separated conduction and valence bands, valence band having holes (not shown in the figure) and conduction band filled with free electrons up to a certain energy level. E_0, in this figure, denotes the energy required to pull out electron from the material to the vacuum or air, in other words the work function. It may be noted that work function of free electrons in the metal is larger than the work function of free electrons in semiconductor. This means that smaller amount of energy is required to pull out free electron from the semiconductor than from the metal. However, E_0 for both, the metal and the semiconductor are measured from the Fermi level which has the same value throughout the material. Smaller value of workfunction for semiconductor means that the potential energy of free electrons on semiconductor side is more than the potential energy of free electrons on metal side. As a result, some free electrons from semiconductor side move to the metal side. The transfer of electrons from semiconductor side to metal side results in the formation of a thin depletion layer at the junction. The depletion layer has negative ions on the metal side and positive ions on the semiconductor side. A potential barrier is thus generated across the junction with built-in potential that depends on the extent of doping of the semiconductor and the metal used. In case the doping of the semiconductor is low or moderate, the built-in potential is sufficient to restrict the flow of electrons across the junction and the metal–semiconductor junction behaves as an ordinary p-n junction but with very small of the order of 0.2 V potential barrier. Such a junction is called **rectifying junction.** On the other hand, if the doping of semiconductor is

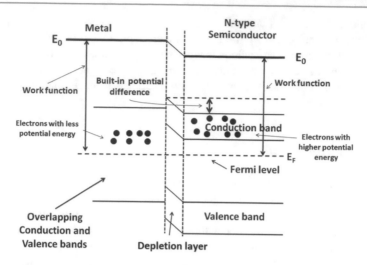

Fig. 5.58 Energy band diagram of a metal–N-type semiconductor junction

heavy, the potential barrier is negligibly small and the junction behaves as an
Ohmic junction. Ohmic junction serves as a contact terminal. Ohmic junctions are
made at the ends of a semiconductor to create contact terminals through which bias
voltage may be applied.

In the case of rectifying metal–semiconductor junction, the depletion layer
extends mostly on the semiconductor side and is quite thin. Also the positive and
negative charges contained in depletion layer are small, and therefore, the capacitance of depletion layer is also very small. As a result, the switching time of the
junction, the time it takes in going from 'on' state to the 'off' state is very less. The
Schottky diodes, therefore, are very fast and may be used to rectify high frequency,
of the order of megahertz (MHz) without any distortion. This property of Schottky
diodes is used in switch-mode power supplies as well as in other applications in
radio frequency range.

Figure 5.59a shows the symbol that is used to denote Schottky diode. The
forward bias configuration of Schottky diode is shown Fig. 5.59b. The total forward
current I_T is the sum of three currents such as (i) the diffusion current, due to
diffusion of electrons across the junction when potential barrier at the junction is
reduced because of the forward bias, (ii) tunnelling current, which comes into play
because of the very thin depletion layer; some electrons may tunnel through the
potential barrier and (iii) thermionic current due to the temperature of the junction.

The V-I characteristic of Schottky diode is shown in Fig. 5.59c. It may be
observed that on account of the low cut-in voltage, large forward currents may flow
at small value of forward bias. However, the big disadvantage of Schottky diode lies
in the considerably large reverse saturation current, which also generates larger
noise.

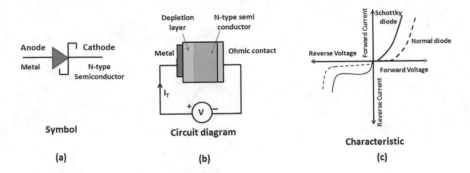

Fig. 5.59 a Symbol for Schottky diode; **b** Circuit diagram for current flow in forward bias; **c** I–V characteristics of Schottky and normal diode.

Problems

P5.1   Assuming that intrinsic carrier concentration n_i at room temperature for silicon is 1×10^{16} m^{-3}, and its relative permittivity is 11.0, calculate the value of built-in potential and partial widths of depletion layer on two sides of the junction for a crystal which is doped with acceptor impurity of 1×10^{22} m^{-3} and donor impurity of 6×10^{22} m^{-3}. The absolute permittivity of vacuum is 8.85×10^{-12} F/m.

Answer: [0.761 V; 2.81×10^{-8} m ; 0.47×10^{-8} m].

P5.2   A germanium diode that obeys Shockley equation has a reverse saturation current of 0.3 µA at room temperature. Calculate current through the diode at bias voltage of 0.15 V.

Answer: [120.73 µA].

P5.3   Calculate reverse saturation current for a silicon diode (n = 2) if current of 1.17 A flows on applying 0.6 V at 25° C.

Answer: [10 µA].

P5.4   A DC power supply of bridge rectifier uses silicon diodes and is fed by a 120 V(rms) power at 60 Hz through a transformer of turn ratio 5:1. A load of 240 Ω along with a shunt capacitor of 470 µF is connected in the output for filtration. Determine the DC voltage across the load and peak-to-peak value of ripple in the output.

Answer: [31.3 V; peak-peak ripple 2.4 V].

Fig. P5.6 Circuit for
problem (P5.6)

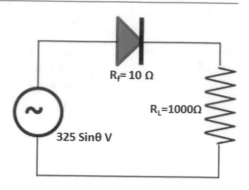

$R_f = 10\ \Omega$

$R_L = 1000\Omega$

325 Sinθ V

P5.5 A bridge rectifier is energized by 220V (rms) 50 Hz AC through a
transformer of turn ration 10:1. Determine the rms value of the rectified
output, average value of rectified output, percent rectification efficiency and
the PIV of diodes. Assume diodes are ideal.

Answer: [rms output = 31.11 V; DC output 19.81 V; %rectification effe:
81.08%; PIV 62.44 V].

P5.6 Figure shows a half-wave rectifier that uses a silicon diode of forward
resistance 100Ω and a load resistance of 1000 Ω. A sinusoidal signal of 325
V amplitude is applied at the input. Calculate (i) maximum output current
I_{max} (ii) DC output current I_{dc} (iii) rms output voltage and (iv) percent
rectification efficiency η. Fig. P5.6

Answer: [(i) 147.7 mA (ii) 47.01 mA (iii) 147.7 V (iv) 36.84%.

P5.7 A sinusoidal signal of maximum voltage of V_m has DC value of zero volt. It
is required to clamp it with a positive clamper that uses silicon diode in such
a way that its DC value becomes 5 volt. Design the clamper circuit.

Answer: [Positive clamper with negative bias of $\{-(V_m-5.7)\}$ volt.

P5.8 In problem 5.7, if it is required to use a negative clamper with silicon diode,
determine the magnitude and sign of bias.

Answer: [negative clamper with positive bias of $(V_m-5.7)$ V].

P5.9 It is required to convert the input signal given in fig(a) into an output signal
given in Fig. b. Describe with necessary details the circuit based on silicon
diode that may perform the desired action. Figure P5.9

Answer: [A shunt negative clipper with positive bias of 9.7 V].

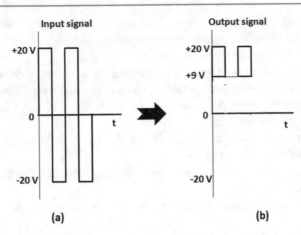

Fig. P5.9 Circuit for problem (P5.9)

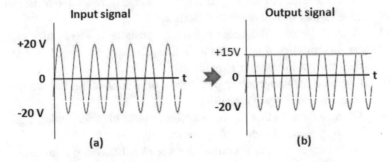

Fig. P5.10 Circuit for problem (P5.10)

P5.10 Describe with necessary details a diode circuit that may convert the given input signal (Fig. a) into the output signal given in Fig. b using an ideal diode. Figure P5.10

Answer: [A series positive clipper with positive reference voltage of 15 V].

Short Answer Questions

SA5.1 Describe the motion of electrons and holes in p-n junction at thermal equilibrium.

SA5.2 Discuss the development of built-in potential in a p-n junction at thermal equilibrium.

SA5.3 What mechanism(s) is (are) responsible for current(s) in a p-n junction?

SA5.4   Which part of a p-n junction has maximum resistance and why?

SA5.5   Write Shockley equation for current in a p-n junction and explain the meaning of each symbol used.

SA5.6   Without derivation write an expression for the width of depletion layer in terms of doping concentrations and built-in voltage of a p-n junction.

SA5.7   Name the two types of breakdown mechanisms of a p-n junction and briefly discuss them.

SA5.8   List some important applications of diodes. What is the special outcome of non-linearity of diodes?

SA5.9   What is the physical significance of Transformer Utilization Factor of a rectifier circuit?

SA5.10 Compare the performance parameters of a half-wave and a full-wave rectifier.

SA5.11 Name different kinds of filter circuits and select the one which is best suited for a load that draws high current, give reasons for your selection.

SA5.12 Why the value of Peak-to-peak ripple voltage in full-wave rectification is smaller that of half-wave rectification?

SA5.13 What is the main difference between a positive and a negative clamper?

SA5.14 'A forward-biased p-n junction is like a battery if forward bias is more than the cut-in voltage' explain.

SA5.15 Compare the performance parameters of a three-phase full-wave and a single-phase full-wave rectifier.

SA5.16 What is the difference between a clipper and a clamper circuit, explain by drawing rough sketches of signal outputs.

SA5.17 What causes the cut-in voltage for a forward-biased p-n junction?

SA5.18 Name different types of clipper and clamper circuits and draw circuit diagram of one of them.

SA5.19 It is required to put a + X volt DC pedestal on a signal using a silicon diode positive clamper, draw the circuit that may perform the desired task.

SA5.20 Explain how Zener diode works as a voltage stabilizer.

SA5.21 What are the differences between the light emitted by a LED and a laser diode Laser diode?

SA5.22 What is the advantage of using the pin type photodiodes?

SA5.23 Why Schottky diodes are fast and efficient?

SA5.24 What is Fermi energy? How is it located in metals, semiconductors and insulators?

Sample Answer to 5.24

What is Fermi energy? How is it located in metals, semiconductors and insulators?

Answer:

Fermi energy is a quantum mechanical concept. In quantum mechanics, particles in a system can have only discrete energies. When one considers an assembly of non-interacting Fermions like electrons, then there are electrons that occupy lowest energy state and there are some electrons that have highest energy state. The difference between the energy of the highest occupied energy state and the lowest occupied energy state at absolute zero temperature (0 K) is called Fermi energy and the level with highest energy occupied by electrons is called the Fermi level. In solids particularly, Fermi level plays a crucial role. For example, in metals there is large number of electrons that are very loosely bound to atomic cores, however, at absolute zero temperature they are all bound. These electrons in metal at absolute zero K are not stationary and have different energies. The energy possessed by electrons of highest energy is the Fermi energy. In other words, it may be said that at absolute zero Fermi level is the top of the collection of electron energy levels.

At temperatures higher than 0 K, a certain fraction of electrons exists above the Fermi level. The energy distribution function for electrons at any temperature T is given by Fermi Function f(E) given as

$$f(E) = \frac{1}{e^{\frac{(E-E_F)}{kT}} + 1}$$ *, here E_F is Fermi energy.*

Figure 11. Sample shows the position of Fermi Level in metals, intrinsic semiconductors and insulator. It may be recalled that in semiconductors the Fermi Level lies almost in the middle of the forbidden energy gap but no electron of the semiconductor possesses this energy. In doped semiconductors, Fermi level shifts from the middle of band gap towards conduction band in case of N-type material and towards valence band in P-type material.

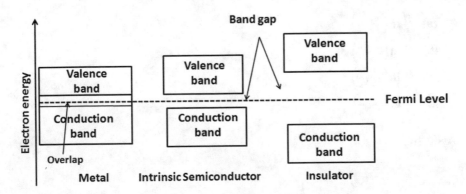

Fig. 11. Sample Fermi Level in metals, intrinsic semiconductors and Insulators

Fig. MC5.1 Circuit for
(MC5.1)

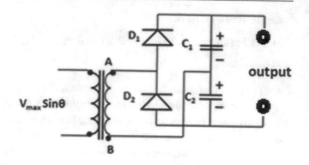

Fermi level may also be defined as the energy level such that the probability of finding an electron X units of energy above it is same as finding a hole X unit of energy below it.

Multiple-Choice Questions

Note: Some of the multiple-choice questions may have more than one correct alternative; all correct alternatives must be marked for a complete answer in such cases.

MC5.1 The circuit shown in Fig. MC5.1 is a

   (a) Full-wave rectifier
   (b) Positive–negative clipper
   (c) Positive clamper
   (d) voltage doubler

ANS: [(d)].

MC5.2 The built-in voltage in a p-n junction is given by

   (a) $\frac{kT}{e} \ln\left(\frac{N_A N_D}{n_i^2}\right)$

   (b) $\frac{eT}{k} \ln\left(\frac{N_A N_D}{n_i^2}\right)$

   (c) $\frac{kT}{e} \ln\left(\frac{n_i^2}{N_A N_D}\right)$

   (d) $\frac{eT}{k} \ln\left(\frac{n_i^2}{N_A N_D}\right)$

ANS: [(a)].

MC5.3 The ratio of the rms value of rectified pulsating DC obtained from a full-wave rectifier and from a half-wave rectifier is:

(a) 2

(b) $\sqrt{2}$

(c) $\frac{1}{2}$

(d) $\frac{1}{\sqrt{2}}$

ANS: [(b)].

MC5.4 Shockley expression for diode current is:

(a) $I_o\left(e^{\left(\frac{V}{nV_T}\right)} - 1\right)$

(b) $I_o\left(e^{\left(\frac{V}{nV_T}\right)} + 1\right)$

(c) $I_o\left(e^{\left(n\frac{V}{V_T}\right)} - 1\right)$

(d) $I_o\left(e^{\left(\frac{nV_T}{V}\right)} - 1\right)$

ANS: [(a)].

MC5.5 If $\left(I_{dc}^{3ph}\right)_{half}$ and $\left(I_{dc}^{3ph}\right)_{full}$, respectively, represent the DC current of three phase half-wave and full-wave rectifiers, then

(a) $\left(I_{dc}^{3ph}\right)_{full} > \left(I_{dc}^{3ph}\right)_{half}$

(b) $\left(I_{dc}^{3ph}\right)_{full} = \left(I_{dc}^{3ph}\right)_{half}$

(c) $\left(I_{dc}^{3ph}\right)_{full} < \left(I_{dc}^{3ph}\right)_{half}$

(d) no definite relation between the two.

ANS: [(b)].

MC5.6 If $E_F^p, E_F^n and E_g$, respectively, denote, the Fermi energy on p-side, Fermi energy on n-side and the forbidden energy gap for a silicon diode, then $|(E_F^p - E_F^n)|$ is equal to:

   (a) (k) E_g
   (b) $E_g/2$
   (c) 0
   (d) $1/E_g$

ANS: [(c)].

MC5.7 A Zener diode behaves like a battery for

   (a) Forward bias < cut-in voltage
   (b) (b) Forward bias > Cut-in voltage
   (c) reverse bias < breakdown voltage
   (d) Reverse bias > breakdown voltage.

ANS: [(b), (d)].

MC5.8 Same AC signal is applied at the input of each of the following circuits, outputs of which circuits will be identical Fig. MC5.8

ANS: [(i) & (ii); (iii) & (iv)].

MC5.9 A 20 V peak voltage sinusoidal AC signal is applied at the input of the following circuit, the maximum voltage of the output signal will be Fig. MC5.9

   (a) 25 V
   (b) 20 V
   (c) 15 V
   (d) 10 V

ANS: [(c)].

MC5.10 A 20 V peak voltage sinusoidal AC signal is applied at the input of the following circuit, the maximum voltage of the output signal will be Fig. MC5.10

   (a) 10 V
   (b) 20 V

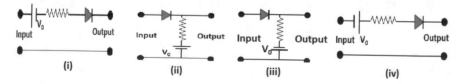

Fig. MC5.8 Circuits for MC5.8

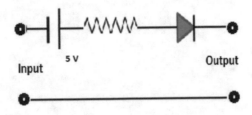

Fig. MC5.9 Circuit for MC5.9

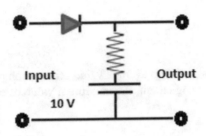

Fig. MC5.10 Circuit for MC5.10

    (c) 30
    (d) 40

ANS: [(a)]

MC5.11 A sinusoidal signal with average value zero and amplitude 30 V is applied at the input of the following circuit, the maximum value of the output voltage will be Fig. MC5.11

    (a) 25 V
    (b) 30 V
    (c) 35 V

Fig. MC5.11 Circuit for MC5.11

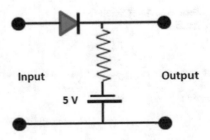

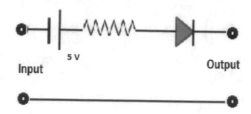

Fig. MC5.12   Circuit for MC5.12

(d)   40 V

ANS: [(c)].

MC5.12 A sinusoidal signal with average value zero and amplitude 30 V is applied at the input of the following circuit, the maximum value of the output voltage will be Fig. MC5.12

(a)   25 V
(b)   30 V
(c)   35 V
(d)   40 V

ANS: [(c)].

MC5.13 Which of the following circuit(s) change the DC level of the input signal

(a)   Full-wave rectifier
(b)   Half-wave rectifier
(c)   Clamper
(d)   symmetrical positive–negative clipper

ANS: [(a), (b) and (c)].

Fig. MC5.14   Circuit for
MC5.14

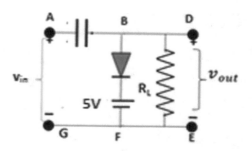

MC5.14 A square wave of peak voltage 25 V is applied at the input of the following ideal diode circuit. The maximum voltage in the output will be Fig. MC5.14

(a) 0 V
(b) 5 V
(c) 10 V
(d) 15 V

ANS: [(b)].

MC5.15 Which of the following provides incoherent radiations in a band of frequencies

(a) LED
(b) Zener diode
(c) photodiode
(d) Laser diode

ANS: [(a)].

MC5.16 The wattage of the transformer feeding AC power to a bridge rectifier that delivers 200 W DC power should be of the order of

(a) 550 W
(b) 450 W
(c) 350 W
(d) 250 W

ANS: [(d)].

MC5.17 Which of the following provides coherent monoenergetic beam of photons?

(a) LED
(b) Zener diode
(c) photodiode
(d) Laser diode

ANS: [(d)].

MC5.18 Which of the following(s) is (are) generally used in reverse-biased mode?

(a) Laser diode
(b) photodiode

(c) Zener diode

(d) Schottky diode

ANS: [(b), (c)].

MC5.19 Diode with least value of forward voltage drop and depletion capacitance is

(a) Laser diode

(b) photodiode

(c) Zener diode

(d) Schottky diode

ANS: [(d)].

MC5.20 Phonon assisted transitions occur in

(a) Photodiode

(b) Schottky diode

(c) direct gap semiconductor

(d) indirect gap semiconductor

ANS: [(d)].

MC5.21 A metal–semiconductor junction where semiconductor is highly doped behaves like

(a) Rectifying junction

(b)   Ohmic junction

(c) LED

(d) Laser diode

ANS: [(b)].

MC5.22 A metal–semiconductor junction where semiconductor is lightly doped behaves like

(a) Rectifying junction

(b)   Ohmic junction

(c) LED

(d) Laser diode

ANS: [(a)].

Long Answer Questions

LA5.1 Discuss the mechanism of generation of the built-in voltage in a p-n junction and derive an expression for it. At a given temperature the mobility of holes is less than that of free electrons, why?

LA5.2 Define Fermi level and show that in a p-n junction at thermal equilibrium it is continuous throughout the p- and the n- sides.

LA5.3 Discuss the region of space charge in a p-n junction diode at thermal equilibrium, and show that the extension of depletion layer is inversely proportional to the dopant concentration. What is the special characteristic of an abrupt junction?

LA5.4 What is depletion layer and why it has high resistance? Derive an expression for the capacitance of depletion layer in terms of the dopant concentrations on the two sides. Is it possible to manipulate this capacitance? If yes, how?

LA5.5 Draw a neat circuit diagram of a Zener regulated full-wave rectified power supply using a shunt capacitor filter. What is ripple factor? Estimate its value for the abovementioned power supply?

LA5.6 Define the performance parameters and obtain their values for a half-wave and full-wave rectifier using bridge circuit. What is the physical significance of transformer utilization factor?

LA5.7 What are the differences between a clipper and a clamper and between positive and negative clamping? Discuss the working of a positive clamper with V_0 volts positive bias.

LA5.8 It is required to clip a sinusoidal wave of amplitude V_m, at $+x$ volt and at $-y$ volt using silicon diodes. Design the circuit discuss its working and specify the voltage of biasing batteries.

LA5.9 With the help of suitable diagrams explain the forward and reverse bias characteristics of a p-n junction. What is meant by breakdown under reverse biasing and what are the possible reasons?

LA5.10 Discuss the working of a Schottky diode explaining the properties of a metal–semiconductor junction. Why this diode may be used at high frequencies while normal p-n junction diode cannot be?

LA5.11 What is the configuration of a Laser diode and how does it work? What is meant by stimulated emission and population inversion?

LA5.12 Draw the transfer characteristics for a p-n junction Zener diode and discuss the conditions under which it may be treated as a battery both in the forward and the reverse bias conditions. What is meant by PIV of a diode?

LA5.13 With the help of a circuit diagram explain the working of a single phase full-wave rectifier using a centre tapped transformer. Compare it with a bridge circuit.

LA5.14 Write a detailed note on the applications of diodes, explaining one of them in detail.

LA5.15 Give a list of special diodes, indicating the special property each has. Discuss the working of one of them in detail.

Transistor Bipolar Junction (BJT) and Field-Effect (FET) Transistor

6

Abstract

Transistors is one of the most versatile solid-state device. A bipolar junction transistor is formed when two p-n junctions of specific widths are developed back-to-back on a single intrinsic crystal. In a BJT both the minority and the majority carriers play an important role, however, the operation of a FET is governed by carriers of only one kind. Fabrication details, working, simple small-signal equivalents and important applications of both type of transistors are discussed in the following.

6.1 Introduction

Transistors are three-terminal devices that are used in most integrated circuits (IC's). There are two types of transistors: (i) bipolar junction transistors (BJTs) in which both the majority as well as the minority currents play role in its operation, (ii) FETs or field-effect transistors in which only one type of current, i.e. either the hole current or the electron current plays a role in the operation of the device. Generally, transistors are extremely small, highly efficient, incredibly reliable, quite robust, stable and usually inexpensive. In this chapter, we shall discuss both the BJT and the FET.

6.2 Types and General Construction of BJT

A BJT may be considered as a combination of two semiconductor diodes held back-to-back and developed on a single semiconductor crystal. However, the doping levels and widths of the two diodes are different from each other. There may

© The Author(s), under exclusive license to Springer Nature Switzerland AG 2021   457
R. Prasad, *Analog and Digital Electronic Circuits*, Undergraduate Lecture
Notes in Physics, https://doi.org/10.1007/978-3-030-65129-9_6

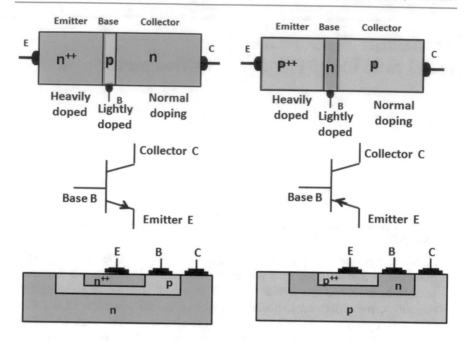

Fig. 6.1 Physical structure, symbol and IC implementation diagrams of npn and pnp transistors

be two different arrangements: **npn** and **pnp**. Ohmic contacts and connecting leads are provided on each of the three sections, which are called **Emitter**, **Base** and **Collector** and denoted by **E, B and C.**

Simplified physical structure, symbols used to denote in circuit diagrams and an implementation version on an IC for npn and pnp transistors are shown in Fig. 6.1.

The emitter section of the transistor which is n^{++} type in npn transistor and p^{++} type in pnp transistor, is **heavily doped**. Therefore, the majority concentration in the emitter is considerably high. The middle section, called base, is **quite thin or narrow and lightly doped,** so that the majority carrier concentration in the base region is quite low as compared to the emitter section. The collector section, n-type in npn transistor and p-type in pnp transistor, is doped to normal concentration and is designed to withstand large currents, therefore has a large area. The arrow on the emitter arm in their symbols distinguishes between the npn and pnp transistors. The arrow on emitter arm in **npn points outwards** while in **pnp it points inwards**.

6.3 Working of a BJT

The operation of a BJT basically depends on two factors: (i) heavy doping of emitter section (ii) very narrow or thin and lightly doped base region. The collector section is moderately doped but has a large surface area. It is because when in operation considerable heat is generated in the collector section of the BJT and a large surface area helps in rapid dissipation of this heat. If there is not adequate heat loss, the device may burn. It is to be realized that developing two pn junctions on a single semiconductor crystal does not create a BJT; for proper transistor action, the doping levels and surface area of the three sections must be properly manipulated so that the device works as a transistor. As already shown, there may be two classes of BJT, the npn- and the pnp-types. However, the normal operation of both these types is almost identical, except that the role played by majority carrier electrons in npn-type is played by holes in pnp-type transistor. Further, in most applications, an npn-type BJT is preferred over the pnp-type because, at a given temperature, the mobility of electrons in semiconductors is larger than that of holes; therefore, npn-based devices are faster as compared to those based on pnp. In the following the working of an npn transistor will be discussed in detail and as expected the same description will hold good for pnp transistors if the role played by electrons is replaced by that of holes. An important point to remember is that directions of conventional currents and DC voltages in a pnp transistor circuit are opposite to that in npn circuits.

In normal operation, both in npn and pnp transistors, the emitter-base junction (EB junction) is kept forward biased and the collector–base junction (CB junction) reverse biased.

Figure 6.2 shows the biasing of (a) npn transistor and (b) of a pnp transistor. Battery V_1 provides forward bias to the emitter–base junction while battery V_2 keeps the collector–base junction reverse biased. Let us now consider the operation of the npn transistor. Since emitter–base (EB) junction is forward biased, the majority carrier electrons from the emitter are pushed in the base region and the majority carrier holes from the base flow to the emitter. Thus, the emitter-base junction behaves like an ordinary semiconductor diode. However, the doping of emitter is much heavier than that of base, therefore, the electron flow from the emitter to the base is much heavier than the flow of holes from the base. Some electrons injected from the emitter side recombine with holes in the base but most of these electrons reach the other end of the base region since the thickness of the base is deliberately kept small. The base–collector junction is reverse biased and, therefore, all most all electrons coming from emitter (with a slight reduction in number because of recombination with holes in base region) are attracted by the positive potential at the far end of the collector region. Thus a current i_C almost equal to the emitter current i_E flows through the reverse-biased collector–base (CB) junction.

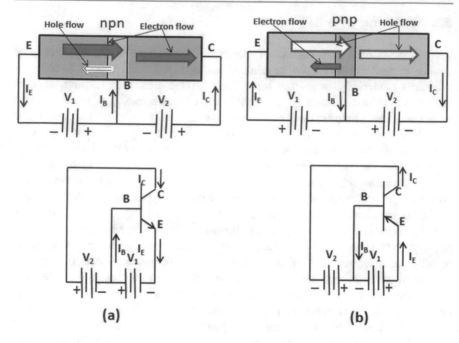

Fig. 6.2 Biasing of **a** of npn **b** of pnp transistors

It may be observed that though the forward-biased emitter–base junction works as an ordinary diode, the reverse-biased collector–base junction does not behave as an ordinary reverse-biased diode. In a normal reverse-biased diode, a very small-reverse saturation current (of the order of few microamps) can flow. However, in a transistor, the emitter–base junction works as a diode but the reversed-biased collector–base junction can have currents much larger than the reverse saturation current. This has been made possible by heavy doping of emitter; very light doping of base and making base quite narrow.

Flow of charge carriers in an npn transistor under normal operation, when EB junction is forward biased and CB junction is reversed bias, is shown in Fig. 6.3. As shown in this figure, a heavy current, predominantly of electrons flow from emitter to base and a very weak hole current from base to emitter. The narrow base region is p-type and the majority carriers there are holes. We know that in a semiconductor the concentration of majority (and also minority) carrier at a given temperature is fixed. In the base region, the concentration of holes decreases below the fixed value because of the recombination of holes with electrons coming from the emitter and also because some holes are flowing out of the base toward the emitter under forward bias. To make up for this reduction in the hole concentration in the base region, more covalent bonds break producing both holes and electrons. However, there was no depletion of electron concentration in the base region, and therefore,

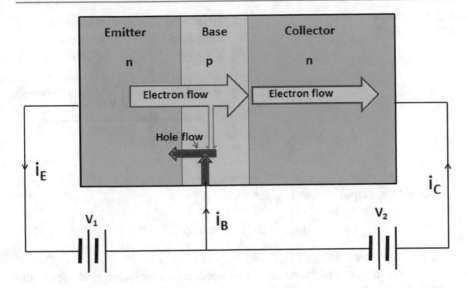

Fig. 6.3 Flow of charge carriers in an npn transistor in normal operation

additional electrons created by covalent bond breaking move out of the base constituting the base current i_B.

The magnitude of the base current i_B depends strongly on the magnitude of the forward bias voltage v_{BE}. If v_{BE} is less than the cut-in voltage ($V_{cut-in} \approx 0.5$ V for Si) of the forward-biased BE junction (which works as a diode) the diode will not get 'on' and no forward current will flow, i.e. $i_E = 0$. Once v_{BE} increases beyond V_{cut-in}, i_E increases rapidly which in turn increases i_B. The base current i_B also depends on the potential difference v_{CE} between collector and emitter but not very strongly.

A forward-biased junction that behaves like a 'closed' switch has a very small resistance. The emitter-base junction, therefore, may be represented by a small resistance 'r'. On the other hand, a reverse-biased junction that is like an 'open switch' offers a much larger resistance and, therefore, the collector–base junction may be represented by a large resistance 'R'. As such, a BJT may be represented by a series combination of a small resistance r and a large resistance R as shown in Fig. 6.4. Heavy doping of emitter and light doping of base coupled with a narrow width of base makes it possible that nearly same current $i_c \approx i_E$ flows through this series combination of resistances. Let us assume that current i_E is changed by a small amount Δi_E, as a result the change in power Δp_{in} in the emitter–base section, which is also called the input section, is given by

$$\Delta p_{in} = r.(\Delta i_E)^2 \tag{6.1}$$

Fig. 6.4 Resistance
equivalent of a BJT

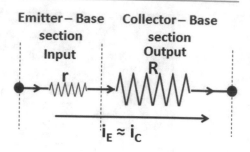

And the corresponding change in the power Δp_{out} of the collector–base section, also called the output section is given by

$$\Delta p_{out} = R.(\Delta i_C)^2 \approx R(\Delta i_E)^2 \tag{6.2}$$

It may be observed from (6.21) and (6.2) that $\Delta p_{out}/\Delta p_{in} \approx R/r \gg 1$. Thus, a small change in power in the emitter section produces a much larger change in the power of the output section. That is to say that the transistor action of the BJT amplifies the input power by a large factor.

The input section of a transistor is a low resistance system and changes in current or voltage in this section may be done with small power consumption. It may be observed that by changing emitter current i_E or the base current i_B the collector current Ic gets changed. In principle, current i_E or i_B may be changed by changing the effective value of resistance r of the input section. Hence any change in the magnitude of the emitter current or base current may be interpreted as the change in the effective resistance r of the emitter section. Similarly, any change in the magnitude of current i_C in the output section may be interpreted as if the resistance R of this section has been changed. It is because from Ohm's law we know that a rise in current in a circuit means a decrease in circuit resistance and vice-versa. Hence, it may be said that by changing the effective resistance r of the input section it is possible to change the effective resistance R of the output section. The transistor action may, therefore, be defined as the control on the output resistance by controlling the input resistance. The term transistor stands for 'trans-resistance' meaning that the changes in effective input resistance are amplified and transferred to the output section. It is important to remember that the transistor does not produce or add power to the system, it simply controls it. Power delivered in the output of a transistor comes from the batteries used in providing bias to the two junctions.

Both the emitter–base and the collector–base junctions have resistances, respectively, r and R. When current i_E (or i_C) pass through these junctions heating of these junctions takes place. Heat produced per unit time at emitter–base junction is $(r.i_E^2)$ J and at collector–base junction is $(R.i_C^2 \approx R.i_C^2)$ J. Since r is small the rate of heat generation at the emitter is much smaller as compared to that at the collector. The area of the collector is, therefore, kept large for easy heat dissipation.

To further facilitate heat loss, metallic caps called heat sinks, are wrapped around transistors.

A BJT may be compared to an electronic tap that is able to control a large flow of electrons (or holes) with only a small variation in base current i_B. The collector current i_C remains zero until there is flow of i_B in the circuit. A bipolar junction transistor is thus a current operated device.

A BJT has three ports or connecting leads. The normal operation of BJT requires that all the three leads are connected to voltage sources. If only two leads are connected to voltage sources and the third lead is left floating, then BJT does not behave as a transistor, it behaves as two diodes joint back to back.

6.4 Discrete BJT, Packaging, Type and Testing

With the advancements in integrated circuit (IC) technology large number of circuit elements including a number of BJT's are now fabricated in a single IC chip. However, single BJTs, termed as discrete BJT, are also available in various packaging by different manufacturers. Each BJT has three leads, respectively, for emitter (E), base (B) and collector (C). The manufacturer provides the manual or data sheets that specify all details like which of the lead corresponds to E, B, C, etc.

It is always better to consult the manufacturer's datasheet/manual before using a device, however, some common types of BJT packaging are shown in Fig. 6.5. Figure (i) shows the package where the body of BJT looks like a half-cylinder. Holding the device in hand such that the flat face of the BJT body faces the observer, the first lead from the right is emitter (E), the second base (B) and the third collector (C). Similarly, in case of metal-caped BJT package (Fig. iii), a tip of metal protrudes out as shown in the figure. The lead nearest to the protruding tip is emitter (E), the lead farthest from E is collector and the lead between C and E is base (B).

In order to find whether the given BJT is npn-type or pnp-type, one needs an Ohm-metre that may measure the resistance between two terminals or a multi-metre. A multi-metre can perform multiple functions including that of an ohm-metre and the desired function can be selected through a knob on the metre. While using a multi-meter, one should select resistance measurement setting. The Ohm or multi-metre has two leads, one red marked as '+' and the other black marked 'ground'.

Suppose it is required to find the type of the given BJT, as the first step the three leads E, B and C of the transistor are identified using the manual or datasheet. Once that is done, the positive lead of the ohm-meter is touched with base B and the negative lead with emitter E, as shown in Fig. 6.6.

Now there are two possibilities: the meter will either show (i) a very low resistance or (ii) a high resistance. If it is case (i) of very low resistance, then the transistor is npn-type and if it is case (ii) of high resistance, then a pnp transistor.

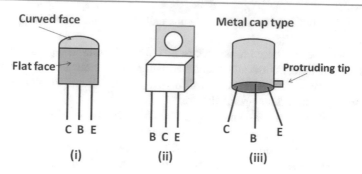

Fig. 6.5 Different packaging of discrete BJT

This may be confirmed further by connecting negative meter lead to collector terminal.

Self-assessment question SAQ: Explain why in case (i) of low resistance, the BJT will be an npn-type?

Self-assessment question SAQ: What will be the resistance reading of meter (low or high) when negative metre lead is connected to collector terminal C? Give a reason for your answer.

To check if the given transistor is properly working or not, one may use the fact that if only two terminals of a BJT are connected to the voltage sources then it works as a combination of two diodes and not as a transistor. Therefore, with the help of a metre one can check the working of two individual diodes: both diodes must separately show that on forward biasing they have low resistance and on reverse biasing a high resistance.

Fig. 6.6 Identifying BJT type

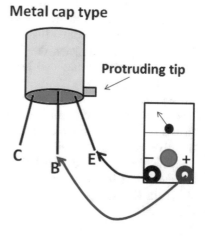

Self-assessment question SAQ: A pnp transistor is given and its E, B and C terminals are known, state the steps to be taken to check if it is alright or not?

6.5   Current–Voltage Characteristics of a BJT

Figure 6.7 shows the nomenclature and signs of different potential differences and terminal currents in (a) npn and (b) in a pnp transistor. As may be observed in these figures, the subscripts of potential difference v_{xy}, for example, indicate that terminal x is at a higher potential than terminal y. It is for this reason that subscripts of a given voltage v in case of pnp transistor are interchanged as compared to the npn transistor. There are three terminal currents and three voltages between the three terminals of a BJT. One may apply KCL to the three terminal currents to get

$i_B + i_C = i_E$, this equation holds good for both the npn and pnp types of BJTs.

$$(6.3)$$

One may also apply KVL to the three voltage drops; however, while applying KVL one must use the proper sign convention for the voltage drops. Let us assume that voltage drop is taken positive if it is traversed in the direction of higher to lower voltage.

For the npn transistor, we start from point C and moves in the clockwise direction towards E and so on to get

$$v_{CE} + (-v_{BE}) + v_{BC} = 0 \quad \text{Or} \quad v_{BC} = v_{BE} \; v_{CE} \quad \text{(for npn)} \qquad (6.4)$$

Fig. 6.7 Nomenclature and signs of voltages and terminal currents in **a** npn and **b** pnp transistors.

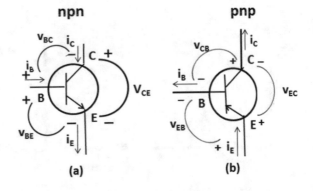

(a)   (b)

And

$$(-v_{EC}) + v_{EB} + (-v_{CB}) = 0 \text{ Or } v_{CB} = v_{EB} - v_{EC}(\text{for pnp}) \qquad (6.5)$$

Equations (6.4) and (6.5) are not different as any $v_{XY} = -v_{YX}$; *for example* $v_{CB} = -v_{BC}$ and so on.

In a BJT the collector current i_C (which is $\approx i_E$) starts flowing when the forward bias voltage v_{BE} of EB junction exceeds the cut-in voltage V_{cut-in} which for silicon devices is around 0.5 V, and the collector current increases exponentially with v_{BE} reaching a saturation value at around $v_{BE} = 0.7V$, which is called the **'on' voltage**. When attempt is made to increase v_{BE} beyond 'on' value, current i_C increases rapidly without any appreciable increase in v_{BE}. Further, at a constant value of current i_C, v_{BE} decreases by about 2 mV with each 1 °C rise in temperature. The current–voltage characteristics of a BJT are shown in Fig. 6.8.

Another important i-v characteristic of a BJT is the graph between the base current i_B and the base-emitter voltage v_{BE} that is shown in Fig. 6.9. In normal operation when the collector-base junction is reverse biased and emitter–base junction is forward biased, i_B follows i_C, since both of them depend on i_E. The emitter current i_E increases with the increase in the forward bias voltage v_{BE}; with the increase of i_E the recombination rate of charge carriers in the base region increases and also the magnitude of the majority carrier current flowing from the base to the emitter (of holes in npn and of electrons in case of pnp transistors) also increases. This results in the increase of the base current, as shown in the figure.

Self-assessment question SAQ: It is said that unwanted electronic signals called noise are generated whenever electrons and holes combine (or recombination takes place). Identify the locations where there are chances of noise generation in a BJT.

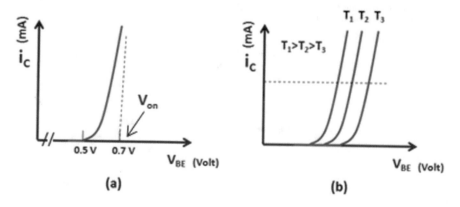

Fig. 6.8 Current–voltage characteristic of a BJT and its temperature dependence

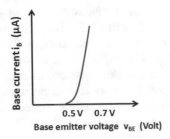

Fig. 6.9 Base current versus base–emitter voltage characteristic of a BJT

6.6 Modes of Operation of a BJT

A BJT may operate in three different modes depending on the biasing condition of the two junctions, the emitter–base (EB) junction and the collector–base (CB) junction.

1. **Cut-off mode**:When both, the EB and the CB junctions are reverse biased.
2. **Active mode**:When EB junction is forward biased beyond cut-in voltage and CB junction is reverse biased.
3. **Saturation mode**: Both, the EB and CB junctions are forward biased.

BJT operates in active mode when used in amplifier circuits. Both saturation and cut-off modes are used for switching applications of BJT.

6.7 BJT Configurations and Parameters

A BJT because of its transistor action has the property of amplifying power and /or current, therefore, it is frequently used for signal amplification. Transistors as an amplifier may be classified by:

(a) Circuit configuration,
(b) Signal level,
(c) Class of operation,
(d) Type of transistor used.

Classification based on circuit configuration will be discussed in the following. BJT has three leads or terminals E, B and C, but an amplifier circuit needs two terminals where the input signal may be applied and two terminals from where the output signal may be drawn. Therefore, it is obvious that in a BJT amplifier one of its three terminals must be common to both the input and the output. Depending on which terminal remains common to input and output both, there may be three different arrangements or configurations for BJT amplifiers. These three configurations are: **(A) Common base or CB configuration. (B) Common Emitter or CE configuration. (C) Common collector or CC configuration, which is also referred to as Emitter follower**. Since the common terminal is generally grounded, therefore, these configurations may also be called as grounded base, grounded emitter and grounded collector configuration, respectively. The most commonly used configuration is the common emitter configuration.

6.7.1 Common Base Configuration

In this configuration, the base terminal is grounded and remains common to both the input and the output of the transistor. Emitter–base section is the input where the signal to be amplified is connected and the output signal is drawn across the collector and base as shown in Fig. 6.10 Common base current gain α is defined as

$$Common\ base\ current\ gain\ \alpha \equiv \frac{i_C}{i_E} < 1. \tag{6.6}$$

$$The\ common\ base\ voltage\ gain\ A_V = \frac{R_L \cdot i_C}{R_{in} \cdot i_E} = \left(\frac{R_L}{R_{in}}\right)\alpha \tag{6.7}$$

The voltage gain A_V is given by the multiplication of the ratio between the load resistance R_L and the input resistance R_{in} with α, the common base current gain. Though α is less than one but R_L/R_{in} is generally much larger than one, therefore, the voltage gain is quite large. The amplitude of the output signal is larger than that of the input signal and there is no phase difference between the input and the output signals. Also, applying KCL to the npn transistor one gets

$$i_E = i_B + i_C \tag{6.8}$$

As may be seen from Fig. 6.10, the input impedance of the amplifier is low (Small R_{in}) while the output impedance is quite large (large R_L). This configuration is often used for a single-stage amplifier as in microphones, etc., because of its large voltage gain and almost unit current gain ($\alpha \approx 1$).

The input characteristics of a BJT in common base configuration are shown in Fig. 6.11. As may be observed in this figure, when v_{CB} is zero and the emitter–base junction is forward biased, emitter current starts flowing at cut-in voltage V_{cut-in} which for silicon devices is around 0.5 V and for germanium devices is around 0.2 V. On further increasing the emitter–base voltage v_{BE}, the emitter current increases rapidly and attains a saturation value at $v_{BE} = 0.7$ V (the V_{on}) for silicon and

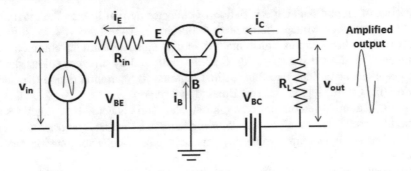

Fig. 6.10 Common base BJT amplifier

at $v_{BE} = 0.3$ V for germanium devices. After that it is not possible to increase v_{BE} anymore, it remains fixed at V_{on}.

On increasing the v_{CB} from 0 to say 5 V, the collector–base junction becomes reverse biased which increases the width of the depletion layer. Since the base is lightly doped, the depletion layer extends almost completely in the base region. Extension of the depletion layer in the base region reduces the effective thickness of the base and, therefore, the cut-in voltage becomes lower. Emitter current starts flowing at a lower value of v_{BE}. On further increasing the reverse bias of BC junction to say 15 V, the width of the depletion layer further increases reducing further the effective thickness of base, which in turn further reduces the cut-in voltage as shown in the figure.

Fig. 6.11 Input characteristics of a BJT in common base configuration

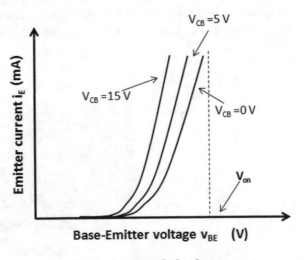

Input characteristics in common base configuration

Series of graphs showing a variation of collector current i_C with the magnitude of the reverse bias voltage between the collector and the base v_{CB} for different values of emitter currents constitute the output characteristics in common base configuration and are shown in Fig. 6.12. As indicated by different colour shades, the graph can be divided into four distinct regions: (i) Saturation region (ii) Active region (iii) Cut-off region and (iv) Break down region.

In the saturation region, both the emitter–base and collector–base junctions are forward biased and, therefore, the collector current i_C suddenly and almost vertically raises when the output voltage v_{CB} makes the base–collector junction forward biased.

When the emitter–base voltage is below the cut-in voltage or the BE junction is reversed bias and the CB junction is also reversed bias, no emitter current and no collector current flow in the circuit. This region is called the cut-off region and the BJT does not work as an amplifier in this region. The transistor behaves as an 'off' switch.

For emitter current $i_E > 0$ and when CB junction is reverse biased (below the breakdown voltage), the BJT works as a voltage amplifier. This region of operation is called the active region. It may be observed in the figure that the magnitude of the reverse bias voltage does not have any appreciable effect on the magnitude of collector current i_C for a fixed value of emitter current i_E. The different branches of characteristics for different values of emitter currents i_E are almost horizontal in the active region.

Early Effect

In a BJT, there are two junctions (EB junction and CB junction) and two depletion layers one at each junction. The width of the depletion layer at the EB junction does not very much depend on the magnitude of the forward bias voltage applied at the EB junction. However, the width of the depletion layer substantially increases with the increase in the magnitude of the reverse bias voltage at the collector–base

Fig. 6.12 Output characteristics of a BJT in common base configuration

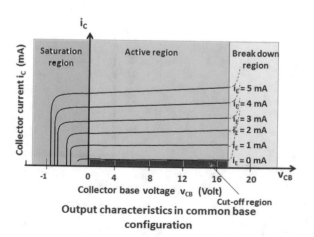

Output characteristics in common base
configuration

junction. Moreover, the depletion layers at both junctions extend more in to the base region on account of its low concentration of doping. With the increase of the reverse bias voltage at collector–base junction, the width of depletion layer increases narrowing the effective thickness of base region. This dependence of the effective thickness of base region on the output voltage (v_{CB}) is called early effect. On applying such a large reverse bias voltage that the effective thickness of base region becomes zero and the two depletion layers touch each other, the whole of the base is filled with depletion layers that are insulator and no current passes through the device. The phenomenon is called **punch through**.

Self-assessment Question SAQ: Though the depletion layer has charged ions, it behaves like an insulator; explain.

Transistor parameters in common base configuration

(a) **Dynamic input resistance**: Dynamic input resistance r_i is defined as the ratio of the change in input voltage or emitter–base voltage (Δv_{BE}) to the corresponding change in emitter current (Δi_E) at a constant output voltage v_{CB}.

$$r_0 = \frac{\Delta v_{BE}}{\Delta i_E}, at\ v_{CB}\ constant \tag{6.9}$$

The input resistance or impedance of a **CB** BJT amplifier is very low.

(b) **Dynamic Output Resistance**
   The dynamic output resistance r_0 is defined as the ratio of change in output voltage or collector– base voltage (Δv_{CB}) to the corresponding change in output or collector current (Δi_C) at constant emitter current.

$$r_0 = \frac{\Delta v_{CB}}{\Delta i_c}, at\ constant\ i_E \tag{6.10}$$

The output resistance or impedance of a CB amplifier is very high.

(c) **The current gain**
   The current gain α of a BJT in common base configuration is defined as the ratio of the output or collector current i_C to the input or emitter current i_E.

$$\alpha = \frac{i_C}{i_E} \tag{6.11}$$

The current gain in CB configuration is less than 1 and typically of the order of 0.98

Self-assessment question SAQ: Draw a circuit diagram for a common base amplifier using pnp-transistor.

6.7.2 Common Emitter Configuration

Common emitter configuration is the most widely used configuration because in this configuration both the voltage and the current gains have moderate values and the power gain has a high value. Since its input impedance is moderate and output impedance is high, several stages of amplifiers may be easily cascaded with each other to get large voltage, current and power gains. Further, as has already been said, an npn transistor is faster than an pnp transistor, we shall discuss the common emitter amplifier using npn transistor in some details. The discussion may, however, be extended to include the CE-pnp transistor amplifier with suitably altered battery polarities and current directions.

Figure 6.13 shows the common emitter amplifiers using (a) npn and (b) pnp BJTs. In any configuration the amplifier circuit may be in active mode; when EB junction is forward biased and CB junction is in reverse bias, in saturation mode; when both the EB and CB junctions are forward biased and in cut-off mode if both junctions are reverse biased. A common emitter BJT amplifier has medium values for both the current and voltage gains, high value for power gain, moderate input impedance and high output impedance.

(a) Operation of an npn transistor in active mode
    As has already been discussed in active mode, electrons from the emitter region are injected in the base region due to the forward biasing of EB junction and most of the injected electrons reach the edge of the CB junction before they recombine with holes in the narrow base region. Electrons reaching the edge of CB junction are swept by the reverse bias at this junction. Some holes that are majority carriers in the base region are also injected in the emitter region from the base. If i_E denotes the emitter current, i_{En} the current due to the electrons injected from E to B; i_{Ep} current due to holes injected from B to E; i_C the collector current; i_{CBO} the reverse saturation current through reversed

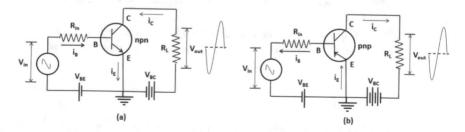

Fig. 6.13 Common emitter **a** npn **b** pnp BJT amplifiers

biased CB junction when the emitter is open; i_B the total base current; i_{B1} current due to the injection of holes from B to E; and i_{B2} the current due to the recombination of electrons and holes in base; then;

$$i_E = \left(i_{En} + i_{Ep}\right); i_B = \left(i_{B1} + i_{B2}\right) \quad \text{and} \quad i_C = \left(i_{Cn} + i_{CBO}\right) \tag{6.12}$$

The current distribution in an npn transistor operating in active mode is shown in Fig. 6.14. Some parameters of this configuration are defined as specified below.

$$\textbf{Emitter injection efficiency } (\boldsymbol{\gamma}) = \frac{i_{En}}{\left(i_{En} + i_{Ep}\right)} \quad \textit{less than 1} \tag{6.13}$$

$$\textbf{Base transport factor } (\boldsymbol{\alpha_T}) = \frac{i_{Cn}}{i_{En}} \quad \textit{less than 1} \tag{6.14}$$

$$\textbf{Common base current gain } (\boldsymbol{\alpha}) = \frac{i_{Cn}}{i_E} = \gamma\alpha_T \quad \textit{less than 1} \tag{6.15}$$

$$\textbf{Common emitter current gain } (\boldsymbol{\beta}) = \frac{i_C}{i_B} \quad \text{much greater than 1} \tag{6.16}$$

Emitter injection effeciency γ, base transport factor α_T and common – base current gain α may have values of the order of 0.9 or so.

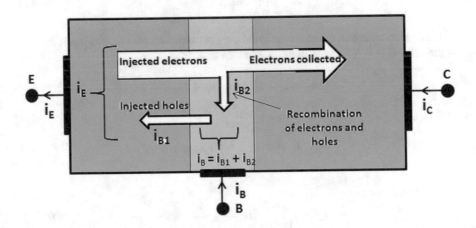

Fig. 6.14 Distribution of currents in an npn transistor operating in active mode

Terminal Currents

Both i_B and i_C flows into the transistor and emitter current i_E flows out, hence applying KCL,

$$i_E = i_C + i_B \tag{6.17}$$

Neglecting the reverse saturation current i_{CBo}, the collector current may be written as

$$i_c \approx i_{Cn} = I_s e^{\frac{v_{BE}}{V_T}} \tag{6.18}$$

And

$$i_B = \frac{i_C}{\beta} \tag{6.19}$$

Also,

$$Emitter\ current\ i_E = i_C + i_B = \frac{\beta+1}{\beta} i_C = \frac{i_C}{\alpha} \tag{6.20}$$

It follows from (6.15) that

$$\alpha = \frac{\beta}{\beta+1} \ and\ \beta = \frac{\alpha}{1-\alpha} \tag{6.21}$$

Further, in common emitter configuration the output current i_C and input current i_B are related through the following equations:

$$i_E = i_C + i_B; i_C = \alpha i_E + i_{CBO}$$

or

$$i_C = \alpha(i_C + i_B) + i_{CBO}$$

or

$$i_C = \frac{\alpha}{(1-\alpha)}(i_B) + \frac{1}{(1-\alpha)} i_{CBO} \tag{6.22}$$

In case $i_B = 0$ (open base), (6.22) reduces to

$$i_C = \frac{1}{(1-\alpha)} i_{CBO} \tag{6.23}$$

Fig. 6.15 Currents i_{CBO} and i_{CEO} when base is open

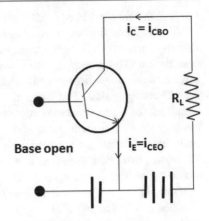

However, when $i_B = 0$ (open base), $i_C = i_E$ and from Fig. 6.15, $i_E = i_{CEO}$. With these substitutions in (6.23) one gets

$$i_C = i_E = i_{CEO} = \frac{1}{(1 - \alpha)} i_{CBO}$$

or

$$i_{CEO} = \frac{1}{(1 - \alpha)} i_{CBO} \tag{6.24}$$

But from (6.21), $\frac{1}{(1-\alpha)} = (\beta + 1)$ substituting this in (6.24) one gets

$$i_{CEO} = (\beta + 1) i_{CBO} \tag{6.25}$$

Since β is much larger than 1, $(\beta + 1) = \beta$, which reduces (6.25) to

$$i_{CEO} = \beta i_{CBO} \tag{6.26}$$

Input–Output Characteristics in Common Emitter Configuration

In common emitter configuration, i_B and v_{BE} are input current and voltage while i_C and v_{CE} are output current and voltage. The input characteristics in CE configuration are a series of curves between i_B and v_{BE} for different fixed values of output voltage v_{CE}. Input characteristics for two values of v_{CE} are shown in Fig. 6.16a. It is easy to understand the nature of these curves in terms of the variation in emitter current i_E with the emitter-base voltage v_{BE}. The emitter current of the forward-biased EB junction starts showing up when v_{BE} exceeds the cut- in voltage (0.5 V for Si and 0.2 V for Ge) and then rises almost exponentially with v_{BE} till the voltage v_{BE} reaches V_{on} the 'on' voltage (0.7 V for Si and 0.3 V for Ge). The voltage v_{BE} cannot be increased beyond V_{on}. The base current i_B is proportional to i_E ($i_B = i_E/\beta$) and therefore, it follows the variations of i_E as shown in Fig. 6.16a.

Further, the forward biased EB junction injects majority carriers from the emitter into the base and the base current i_B has two components i_{B1} due to the injection of majority carriers of the base into the emitter and i_{B2} because of recombination in the base region. The recombination component i_{B2} depends on the magnitude of the reverse bias voltage between collector and base or between collector and emitter in the CE configuration. The larger the reverse bias voltage v_{CE} smaller will be the magnitude of i_{B2} as chances of recombination in the base region will get reduced as the injected carriers will get quickly swept by the larger reverse bias voltage. As such, base current i_B decreases with the increase of v_{CE} if v_{BE} is kept constant. Therefore, to obtain the same i_B at a higher v_{CE}, the forward bias v_{BE} must be increased so that more carriers from the emitter are injected in the base. For example, in Fig 6.16a the same current indicated by point A at lower v_{EB} may be obtained at a higher value of v_{CE} at a higher value of voltage v_{CE} as is indicated by point B. This explains why the curve corresponding to higher value of v_{CE} lies on the higher v_{BE} side. The input resistance r_i^{CE} in common emitter configuration is given by

$$r_i^{CE} = \frac{\Delta v_{BE}}{\Delta i_B} \tag{6.27}$$

The input resistance in CE configuration is medium, neither very high nor very low.

The output characteristics for CE configuration are shown in Fig. 6.16b. The total area of the characteristic curves may be divided into four parts; the cut-off region, the saturation region, the active region and the breakdown region. Different branches of the output characteristics corresponding to increasing values of base

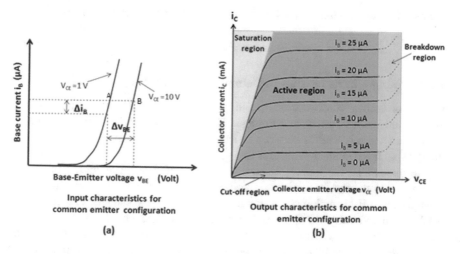

Input characteristics for common emitter configuration

(a)

Output characteristics for common emitter configuration

(b)

Fig. 6.16 Input and output characteristics of a BJT in common emitter configuration

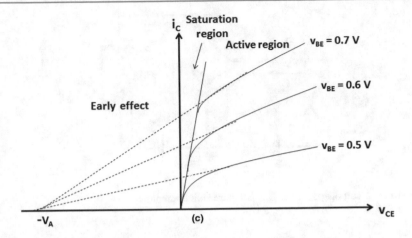

Fig. 6.16c Input and output characteristics of a BJT in common emitter configuration

current i_B are nearly horizontal and almost parallel to each other, indicating that v_{CE} does not very much influence the collector current i_C if i_b is kept constant. Though it is not quite visible in Fig. 6.16b but different branches of characteristics have slopes which increase with the magnitude of the reverse bias voltage v_{CE}, due to the Early effect. As the reverse bias voltage increases, the width of the depletion layer at the BC junction increases reducing the effective thickness of the base region. Reduction in the effective thickness of base reduces the chances of recombination in the base region and hence i_C increases. This is called Early effect.

Early effect is more explicitly shown in Fig. 6.16c where i_C versus V_{CE} curves are drawn for fixed values of v_{EB}. Extending back the linear portions of characteristic curves in the active region are found to meet at a point $-V_A$ as shown in Fig. 6.16c. This happens because of the Early effect. The value of V_A typically lies between 100 V and -50 V. It can be shown that the linear dependence of collector current i_C may be given as

$$i_C = I_s e^{\left(\frac{v_{BE}}{V_T}\right)}\left(1 + \frac{v_{CE}}{V_A}\right) \tag{6.28}$$

The output resistance in CE configuration r_0^{CE} maybe given as

$$r_0^{CE} = \frac{\Delta v_{CE}}{\Delta i_C}, \text{ at constant } v_{BE} \tag{6.29}$$

Phase Difference Between the Input and Output Signals

Let us consider the phase relationship between the input and output signals. The circuit diagram of an npn common emitter amplifier is redrawn in Fig. 6.17. Let us

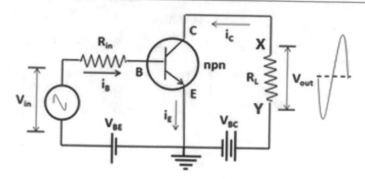

Fig. 6.17 Circuit diagram of an npn common emitter amplifier

assume that initially when no signal is applied to the input some base current i_B^{ini} is flowing in the circuit corresponding to which current i_C^{ini} flows in the output section. Therefore, in the initial state the potential at point X, denoted by v_X^{ini}, will be given by

$$v_X^{ini} = V_{BC} - R_L\left(i_C^{ini}\right) \tag{6.30}$$

And the potential at point Y, V_Y will always remain fixed at the battery potential V_{BC}.

Now suppose an AC signal is applied at the input as shown in the figure. During the positive swing of the input signal the base current i_B increases, say, to the value i_B^f and since the output current $i_C = \beta i_B$, the output current i_C also increases. The current i_C flows through the load resistance R_L from Y to X. The potential v_Y at point Y remains fixed at the battery voltage V_{BC} but the potential of point X, v_x^f now decreases to the value,

$$v_x^f = V_{BC} - R_L\left(i_C^f\right) \tag{6.31}$$

Since $i_c^f > i_C^{ini}$, $hence$, $v_x^f < v_x^{ini}$. Thus, it may be observed that corresponding to the increase in the input signal, the output signal decreases, i.e. there is a phase difference of 180° between the input and the output signals. The phase relationship between the input and output signals of a common emitter amplifier is shown in Fig. 6.18.

6.7.3 Common Collector Configuration

The circuit diagram of a common collector BJT amplifier is shown in Fig. 6.19. In this configuration, the input signal is applied between the base and collector and the

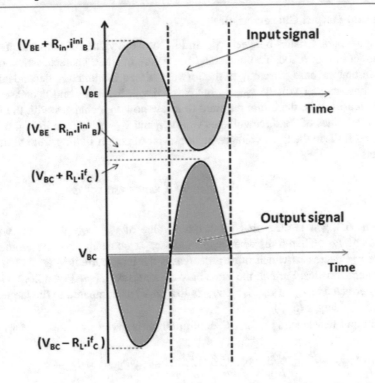

Fig. 6.18 Phase relationship between the input and output signals of a common emitter BJT amplifier

Fig. 6.19 Common collector BJT amplifier

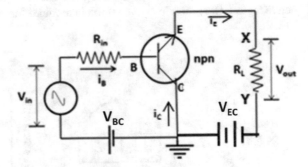

output is drawn across the load resistance R_L connected between emitter and collector, as shown in the figure. Base current i_B and potential difference v_{BC} are, respectively, the input current and voltage, while emitter current i_E and the emitter–collector voltage v_{EC} are the output current and voltage, respectively.

Input and Output Characteristics

The input characteristics between v_{BC} and i_B for fixed values of output voltage v_{EC} are shown in Fig. 6.20c. To understand the nature of these characteristics, one has to recall that the base current i_B in the forward-biased EB junction depends directly on the base-emitter voltage v_{BE}, i_B increases if v_{BE} increases and decreases when v_{BE} decreases. It is, therefore, required to know how v_{BE} changes with the change in v_{BC}. Polarities of the three voltages v_{BE}, v_{EC} and v_{BC} are shown in Fig. 6.20a, b. Applying KVL to the three voltages in clockwise direction starting from terminal E, one gets

$$-v_{EC} + v_{BC} + v_{BE} = 0 \text{ Or } v_{BE} = v_{EC} - v_{BC} \tag{6.32}$$

From (6.32), it is clear that for a fixed value of v_{EC}, v_{BE} decreases with the increase of v_{BC}. Therefore, when v_{BC} increases at constant v_{EC}, the emitter-base voltage v_{BE} decreases which in turn decreases the base current i_B as shown in the input characteristics. Further, the same base current, say y, (see Fig. 6.20c) will flow at a lower value of v_{BC}, (say X_1) if v_{EC} is low (3 V) as compared to the larger value X_2, if v_{EC} is larger (5 V).

The input resistance r_i^{CC} for CC configuration is given by

$$r_i^{CC} = \frac{\Delta v_{BC}}{\Delta i_B} = \frac{few \ volts}{few \ micro \ amp} \approx 10^6 \Omega ..... \text{very high}$$

The output characteristics, series of graphs between i_E and v_{EC} for different values of i_B are shown in Fig. 6.21. The area of the characteristic curves may be divided into four regions: saturation region, cut-off region, breakdown region and active region. Operation in active region is required for amplification purpose, while

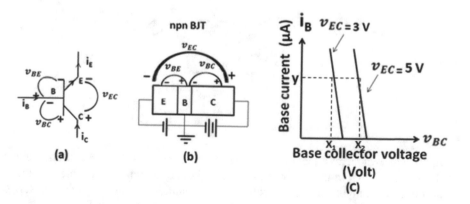

Fig. 6.20 Designated terminals voltages with their polarities **a** for npn BJT symbol; **b** in the corresponding block diagram; **c** input characteristics in common collector configuration

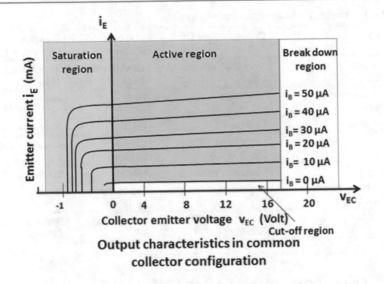

Fig. 6.21 Output characteristics of a BJT in common collector configuration

operation in saturation and cut-off regions is used when the BJT is required to perform switching action.

The output characteristics in the CC configuration very much resemble output characteristics of CB configuration. It is expected also because, the emitter current and collector currents are nearly equal, i_E is proportional to i_B and $v_{EC} = v_{BE} + v_{BC}$. So that for a given value of base current v_{BE} is fixed and v_{EC} is equal to v_{BC} plus some constant. As a result, the variation of v_{EC} follows the variation of v_{BC}.

The output resistance r_o^{CC} is given by

$$r_o^{CC} = \frac{\Delta v_{EC}}{\Delta i_E} \approx \frac{few\ volt}{few\ mili\ amp} \approx few\ kilo\ ohm$$

Thus the input impedance (r_i^{CC}) in CC configuration is very high and the output impedance r_o^{CC} is low. The current gain in CC configuration ($\Delta i_E/\Delta i_B$) is quite large but the voltage gain ($R_L i_E / r_i^{CC} \cdot i_B$) is generally slightly less than 1.

Because of its high input impedance with low output impedance, and near unit voltage gain, BJT amplifiers in CC configuration are used as a **buffer** between two devices to isolate one another without loading. They are also used for impedance matching between two devices for maximum power transfer.

Since in CC configuration output current i_E follows the input current i_B and the voltage gain is nearly unity, the input and output signals are in-phase; the CC amplifier is also called emitter follower.

The following table summarizes the properties of different configurations (Table 6.1).

Table 6.1 Comparative study of three configurations

Property	Common Base CB	Common emitter CE	Common Collector CC
Input resistance or impedance	Very low, less than 100 Ω	Low, less than 1 kΩ	Very high, up to 1MΩ
Output resistance or impedance	Very high	High	low
Current gain	Less than 1	High	Very high
Voltage gain	Larger than CC but less than CE	Very High, highest	Lowest, less than 1
Power gain	Moderate, medium	Highest	Medium
Phase relation between input and output	In phase	Out of phase by 180°	In phase
Application	In high-frequency switching	In amplifiers	Buffer and emitter follower

6.7.4   Class of Operation of Amplifiers

Amplifiers may be classified according to different criteria, for example, CB, CE and CC classification of amplifiers is based on the criteria as to which terminal is common between the input and the output of the amplifier. Another important criterion on the basis of which amplifiers may be classified is the location of the operating point on load line. If the Q-point is in the middle of the active region, i.e. at or near the centre of the load line and the transistor conducts for both the negative and positive swings of the input signal the amplifier is called **class A amplifier**. However, if the Q-point is towards the lower end of the load line, near the cut-off region such that transistor remains operative only for the positive swing of the input signal the amplifier is classified as **class B amplifier**. Similarly, if the operating point allows the transistor to remain operative for the whole of the positive and also for a part of the negative swing, it is class AB amplifier and so on. Classification based on conduction duration of the amplifier is summarized in the following table (Table 6.1a).

Table 6.1a Classification of amplifiers on the basis of the duration of conduction

Class of amplifier	Duration of conduction	Angle of conduction
CLASS A	Over the full 360° of the cycle	2π
CLASS B	Over the half of the cycle	π
CLASS AB	Over slightly more than half cycle,	$<2\pi$
CLASS C	Over less than half cycle	$<\pi$
CLASS D and above	Not applicable	Not applicable

6.8 BJT Biasing Using Single Battery V$_{CC}$

So far, two batteries were used to provide proper voltages to the three terminals of a BJT. However, it is both possible and convenient to use a single battery to provide the required potentials to different sections of a junction transistor. This single battery and the potential provided by it is often denoted by V$_{CC}$.

A single-stage CE amplifier using npn transistor and a single battery V$_{CC}$ is shown in Fig. 6.22. Appropriate potential to the collector is provided by battery V$_{CC}$ in series with the collector resistance R$_C$. The potential divider arrangement of resistances R$_1$ and R$_2$ provides the potential $V_{BE} = \frac{V_{CC}R_2}{(R_1+R_2)}$ to the base of the transistor. Further the collector potential V_{EC} is given by $V_{EC} = (V_{CC} - I_C R_C)$. In this way desired DC potentials to the three terminals of the BJT may be provided by suitably selecting the values of V$_{CC}$, R$_C$, R$_1$ and R$_2$. The same circuit will also hold in case of the pnp transistor with reversed connections of V$_{CC}$ and the direction of flow of currents. The AC signal may be applied at the input through a capacitor that isolates the DC level. Similarly, the AC output is delivered through the isolating or coupling capacitor in the output.

6.8.1 DC Load Line

Figure 6.23 shows the transfer or output characteristics of an npn common emitter amplifier. The X-axis in this graph shows the collector voltage with respect to emitter, V$_{CE}$ and the Y-axis the collector current I$_C$. Considering DC analysis in which coupling capacitors are treated as open circuits, the maximum value that V$_{CE}$ can have, occurs when I$_C$ is zero and is limited by the maximum voltage delivered by the battery V$_{CC}$. The point A on the X-axis corresponding to V$_{CE}$ = V$_{CC}$. Similarly, the maximum collector current $(I_C)^{max}$ that may flow through the transistor is given by (V$_{CC}$/R$_C$). The point B corresponding to this maximum collector

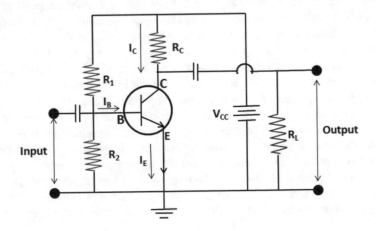

Fig. 6.22 A single-stage CE amplifier using npn transistor

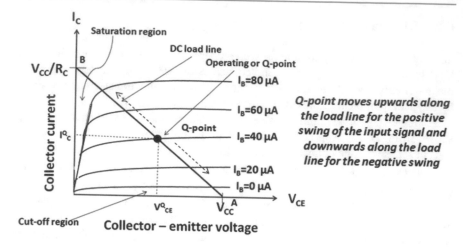

Fig. 6.23 DC load line across the output (or transfer) characteristics of a BJT amplifier using npn transistor in Common emitter configuration

current is shown on the Y-axis by V_{cc}/R_C. A straight line AB joining the maximum current point with the maximum voltage point is called the DC load line. DC load line represents the equation $V_{CE} = V_{CC} - R_C I_C$ and is the locus of points that give the allowed values of current I_C and voltage V_{CE}. The load line cuts the output curves corresponding to different values of input current I_B. The desired value of the input current I_B can be selected independently by choosing the values of resistances R_1 and R_2. Suppose by adjusting the resistances R_1 and R_2 the potential V_{BE} is so adjusted that $I_B = 40\ \mu A$ flows in the circuit of Fig. 6.23. Under these conditions the point of intersection of the characteristic curve of $I_B = 40\ \mu A$ with the load line gives the operating or Quiescent or simply Q-point. In CE configuration input signal is applied between the base and emitter (which is at ground potential) and it essentially changes the magnitude of the base current I_B. During the positive swing of the applied input signal I_B increases and the operating point shifts upward along the AC load line, and during the negative swing of the input signal the operating point moves downwards along the load line. As has already been mentioned, it is possible by adjusting the values of V_{CC}, R_C, R_1 and R_2 to select the position of the Q-point on the output characteristics. Selecting the initial position of the Q-point when no signal is present at the input is very important. If the Q-point is very close to point B, the upper end of the load line, then it is possible that the positive swing of the input signal may drive the operating point to the saturation region where transistor does not work as an amplifier. Similarly, choosing the Q-point near the other end A may take the system to the cut-off region during the negative swing of the input signal. Therefore, for normal operation of BJT as an amplifier which may amplify both the positive and the negative swings of the input signal, it is required that the Q-point may be selected in the middle of the transfer characteristics.

AC Load Line

The DC load line is required to fix the Q-point of the amplifier to any desired value. Further, while drawing DC load line it is assumed that coupling capacitors are like open switches and components beyond them like R_L and the resistance R_S of the input source etc are not present. Once the Q-point is fixed and an AC signal is applied at the input, the operating point moves in accordance with the applied AC signal. The motion of the operating point during the presence of the input AC signal takes place along the AC load line. While drawing AC load line, coupling capacitors are treated as closed switches or as short circuit. Therefore, load resistance R_L and input source resistance R_S etc if present in the circuit need to be considered while drawing AC load line. The AC load line cuts the Y-axis (on which I_C is plotted) at AC-saturation current point given by

$$I_C^{acsatu} = \frac{V_{CE}^Q}{(R_C||R_L)} + I_C^Q$$

And the X-axis (on which V_{CE} is plotted) at AC cut-off point given by,

$$V_{CE}^{accut-off} = V_{CE}^Q + I_C^Q.(R_C||R_L)$$

Where V_{CE}^Q and I_C^Q are, respectively, the values of collector-emitter voltage and collector current at Q-point. Further, for AC analysis coupling capacitors are shorted, the load resistance R_L comes in parallel with the collector resistance R_C. AC load line along with graphical representation of variations in input current i_B, collector current i_C and collector emitter voltage v_{CE} are shown in Fig. 6.23a. It may be noted that I_c is generally of the order of few mille-amp, V_{CE} of the order of few volts and i_B is only a few ten to few hundred of micro-amp.

Self-assessment question SAQ: Are DC and AC load lines very different from each other? What is the reason of this difference? Which of the two load lines is likely to have higher slope?

DC and AC Values of Current Gain β

For a given transistor the graph between collector current I_C and the base current I_B is non-linear as shown in Fig. 6.23b. The DC value of the current gain β_{dc} at the Q-point is given as,

$$\beta_{dc} = \frac{I_C^Q}{I_B^Q}$$

However, β_{ac} is defined through the expression,

$$\beta_{ac} = \frac{\Delta I_C}{\Delta I_B}$$

Fig. 6.23a AC load line and variations of collector current, base current and collector-emitter voltage according to applied AC signal. Note that I_B and I_C have different scales

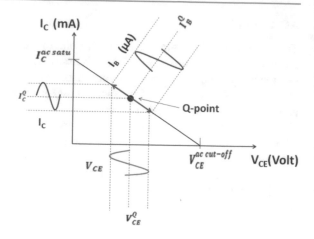

Therefore, β_{ac} may have a value different than β_{dc} because of the non-linearity of the graph. While working out problems one may use the value of β_{ac} for AC analysis and β_{dc} for DC analysis. However, if two separate values are not provided, the only value available may be used for both DC and AC analysis.

6.8.2 Stability of Q-Point

The Q-point on the load line decides the initial conditions of the amplifier circuit before some signal is applied at the input. The applied input signal moves the Q-point along the load line. It is, therefore, required that the Q-point should remain stable when there is no signal and that it should not shift of its own without the application of input signal. In practice, however, it is found that in the case of BJT, the operating point may shift (without the application of any input signal) on account of (a) heating of the transistor (b) replacement of the transistor.

(a) **Heating of the transistor**

The base-collector junction is reverse biased and presents large resistance to the collector current I_C, as a result considerable heat is generated at the collector of the transistor. Now, the collector current I_C is made up of two components, βI_B and $(\beta+1) I_{CBO}$, i.e.;

$$I_C = \beta I_B + (\beta+1)I_{CBO} \qquad (6.33a)$$

Here I_{CBO} is the reverse saturation (or leakage) current at the BC junction due to the flow of minority carriers. The minority current I_{CBO} is very sensitive to temperature; in general it doubles itself for every 10 °C rise of temperature. As a result the Q-point starts shifting upwards on the load line after some time of operation when the temperature of the transistor rises. If no arrangement is made to keep the Q-point stable even when temperature rises, the operating

Fig. 6.23b Illustration of the difference between DC and AC values of gain β

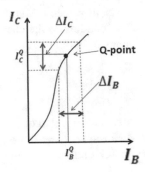

point may land up in saturation region without the application of any signal and thus the amplifier action of the transistor may get lost. This running away of the Q-point without any input signal due to temperature is termed as **thermal runaway**.

(b) **Replacement of Transistor**

Another cause of voluntary shifting of the Q-point is the change in the magnitude of current gain β of the transistor. With use transistor in an amplifier may worn out and has to be replaced. When a transistor in a given circuit is replaced by a new transistor of same make and specifications it is quite likely that the magnitude of the current gain β for the new transistor is different from the value for the old transistor. As a matter of fact no two transistors have exactly the same value of β. Manufacturers only specify the range of the β-values for a given lot of transistors produced by them. As such, the Q-point is bound to change with the replacement of the transistor, which is undesirable.

(c) **Change in V_{BE}**

The operating point is very sensitive to the value of base current I_B which in turn depends on the base emitter potential difference V_{BE}. If there is even a slight change in V_{BE}, may be due to fluctuations in the power supply feeding voltage to the base or due to some instabilities in circuit, the operating point may change which is undesirable. The operating point should shift only when some signal is applied at the input, otherwise in absence of the input signal it should remain fixed.

(d) **Stability Factor**

As the name suggests, stability factor is a parameter that judges the capability of the bias arrangement as to how far it can hold the Q-point stable against (i) temperature rise which changes I_{CBO}, (ii) change in β and (iii) change in I_B or in V_{BE}. As such one may introduce three stability factors S, S_β and S_{VBE}, one each for the above mentioned three causes. The most important amongst them is the stability factor S corresponding to the change in I_{CBO}.

Stability factor S is defined as the ratio of the change in collector current I_C corresponding to the change in the collector leakage current I_{CBO} for the constant values of current gain β and I_B.

$$S \equiv \left(\frac{dI_C}{dI_{CBO}}\right) at\ constant\ \beta, I_B \tag{6.33b}$$

A smaller value of S indicates better stabilization of the Q-point.

However, $I_C = \beta I_B + (\beta + 1)I_{CBO}$

Differentiating the above equation w.r.t I_C, one gets,

$$1 = \frac{\beta(dI_B)}{(dI_C)} + \frac{(\beta + 1)(dI_{CBO})}{(dI_C)} \tag{6.33c}$$

But $\frac{(dI_{CBO})}{(dI_C)} = \frac{1}{S}$; substituting in (6.33c)

$$\frac{(\beta + 1)}{S} = 1 - \frac{\beta(dI_B)}{(dI_C)}$$

Or

$$S = \frac{(\beta + 1)}{\left[1 - \frac{\beta(dI_B)}{(dI_C)}\right]} \tag{6.33d}$$

Equation (6.33d) gives a general expression for stability factor S that is applicable for any bias scheme. As already mentioned the stability factor S essentially measures the stability of the Q-point against variation in I_{CBO}.

Two other, but less important, stability factors may also be defined, one S_β for stability against variation in β (keeping I_{CBO} and V_{BE} constant) and the third $S_{V_{BE}}$ for stability against variation in the base emitter voltage V_{BE}, for constant I_{CBO} and β.

$$S_{V_{BE}} \equiv \frac{\partial I_C}{\partial V_{BE}} = \frac{\Delta I_C}{\Delta V_{BE}} \tag{6.33e}$$

$$S_\beta \equiv \frac{\partial I_C}{\partial \beta} = \frac{\Delta I_C}{\Delta \beta} \tag{6.33f}$$

A small value of stability factor indicates good bias stability and a large value a poor bias stability.

6.8.3 Different Schemes of Biasing and Their Stabilities

Biasing using a single battery may be done in following three different ways:
(a) Fixed biasing (b) Collector–base biasing and (c) Voltage divider biasing
We shall discuss these biasing schemes in details in the following.

(a) **Fixed or base biasing**: Circuit diagram of fixed or base biasing is shown in Fig. 6.24. The collector and the base are independently biased using two resistances, R_C and R_B , connecting one end of both to the positive terminal of the battery V_{CC} and the other end, respectively, to the collector and base. The coordinates (V_{CE}, I_C) of the operating point are given by the following Equations,

Applying KVL to emitter – collector loop, one gets

$$V_{CE} = V_{CC} - R_C I_C \tag{6.34}$$

Similarly, applying KVL to base emitter loop,

$$V_{CC} - R_B I_B - V_{BE} = 0 \, or \, I_B = \frac{(V_{CC} - V_{BE})}{R_B}$$

$$But \, I_C = \beta I_B + (\beta + 1) I_{CBO} = \beta \left(\frac{V_{CC} - V_{BE}}{R_B} \right) + (\beta + 1) I_{CBO} \tag{6.35}$$

Substituting the above value of I_C in (6.34) one gets,

Fig. 6.24 Fixed bias arrangement

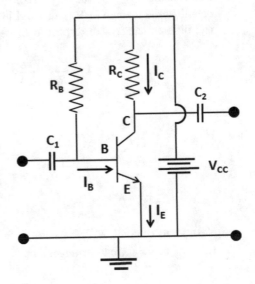

$$V_{CE} = V_{CC} - R_C \beta \left(\frac{V_{CC} - V_{BE}}{R_B} \right) + (\beta + 1) I_{CBO} \qquad (6.36)$$

Equations (6.35) and (6.36) define the Q-point.

There is no protection against thermal run away of the operating point in this type of biasing and, therefore, this type of biasing is not used frequently.

The stability factor S^{Fix} for fixed biased configuration may be calculated from (6.35), differentiating it with respect to I_C keeping β and V_{BE} fixed, which yields;

$$1 = 0 + (\beta + 1) \frac{dI_{CBO}}{dI_C} = \frac{(\beta + 1)}{S}$$

Therefore,

$$S^{Fix} = (\beta + 1) \qquad (6.36a)$$

Equation (6.36a) tells that stability factor for fixed bias arrangement is very large as the current gain β has a large value in the range of 20 to 100 or more. Therefore, as already discussed, this biasing is not effective in stabilizing the operating point.

To calculate S_{VBE}^{Fix}, the stability factor that measures the Q-point stability with regard to the change in V_{BE} in fixed bias arrangement, we go back to (6.35) and differentiate it with respect to I_C treating β and I_{CBO} as constants, to get

$$1 = \frac{-\beta \, dV_{BE}}{R_B \; dI_C} = -\frac{\beta}{R_B} \frac{1}{S_{VBE}^{Fix}}$$

Or

$$S_{VBE}^{Fix} = -\frac{\beta}{R_B} \qquad (6.36b)$$

Next, let us calculate S_{β}^{Fix}, the stability factor corresponding to variations in the value of β, for fix bias arrangement.

$S_\beta^{Fix} = \frac{\partial I_C}{\partial \beta}$, to calculate it we go back to (6.35) and differentiate it with respect to I_C treating I_{CBO} and V_{BE} constant, to get

$$1 = I_B \frac{\partial \beta}{\partial I_C} + I_{CBO} \frac{d\beta}{dI_C} = \frac{(I_B + I_{CBO})}{S_\beta^{Fix}}$$

Or

$$S_\beta^{Fix} = \frac{\partial I_C}{\partial B} = I_B + I_{CBO} \approx I_B \quad (\text{since} \quad I_{CBO} \text{ is very small}) \qquad (6.36c)$$

It may thus be concluded that for fixed bias arrangement , stability factor against temperature is not good as $(\beta + 1)$ is a large value, against variation in V_{BE} could be managed to a small value by choosing a large value for R_B and the stability factor against variation in β will depend on the magnitude of the base current.

(b) **Collector–base biasing** Figure 6.25 shows the arrangement for collector–base biasing. As may be observed in this figure, the collector is connected to the battery V_{CC} via a resistance R_C and the base is connected to V_{CC} through the series combination of R_C and R_B.

It may be noted that current $(I_B + I_C)$ flows through the resistance R_C, while only current I_B flows through the resistance R_B. Applying KVL to the collector– emitter and base–emitter loops, one gets

Fig. 6.25 Collector–base biasing

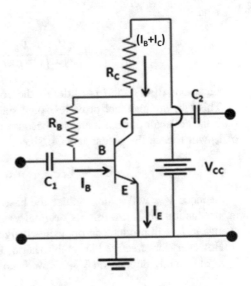

$$V_{CC} = (I_C + I_B)R_C + V_{CE} \qquad (6.37)$$

$$V_{CE} = R_B I_B + V_{BE} \qquad (6.38a)$$

Hence,

$$V_{CC} = (I_C + I_B)|R_C + R_B I_B + V_{BE} \qquad (6.38b)$$

Or

$$I_B = \frac{V_{cc} - V_{BE} - R_C I_C}{(R_B + R_C)} \qquad (6.39a)$$

Since

$$I_C = \beta I_B + (\beta + 1)I_{CBO} \qquad (6.39b)$$

$$I_B = \frac{(V_{CC} - V_{BE})}{[R_B + (1 + \beta)R_C]} \qquad (6.40)$$

Therefore,

$$V_{CE} = V_{CC} - R_C(\beta + 1)\left[\frac{V_{CC} - V_{BE}}{R_C(\beta + 1) + R_B}\right] \qquad (6.41)$$

And

$$I_C = \beta \frac{(V_{CC} - V_{BE})}{[R_B + (1 + \beta)R_C]} + (\beta + 1)I_{CBo} \qquad (6.42a)$$

Equations (6.41) and (6.42) define the coordinates of the Q-point.

This bias arrangement provides some stabilization of the operating point. Suppose the collector current increases because of the rise of temperature, the increase of I_C will reduce the potential V_{BE}. Since,

$$V_{BE} = V_{CC} - [R_C(I_C + I_B) + R_B I_B]$$

Reduced V_{BE} will in turn reduce the base current I_B. Reduction in I_B will result in a reduction of I_C towards its original value. Thus it may be observed that any change in I_C will produce an opposite change in I_B, stabilizing the operating point.

However, for effective bias stabilization, it is required that the collector current I_C must strongly depend on V_{BE}. Now from (6.42a), we have,

$$I_C = \beta \frac{(V_{CC} - V_{BE})}{[R_B + (1 + \beta)R_C]} + (\beta + 1)I_{CBo} \qquad (6.42b)$$

Science I_{CBO} is very small, as a first approximation the second term in (6.42a) may be neglected. If the denominator of the first term of (6.42b) is large, the effect of V_{BE} on I_C will be less. Therefore, for effective bias stabilization, the denominator of (6.41) should be as small as possible. This may be achieved if $R_B \ll \beta R_C$, so that R_B may be neglected in comparison to βR_C. Then the denominator of the above equation reduces to $(1+\beta)$ R_C, and in that case collector current is given by;

$$I_C = \beta \frac{(V_{CC} - V_{BE})}{[(1 + \beta)R_C]} \qquad (6.42c)$$

That may provide some stabilization of Q-point.

The stability factor S^{CB} for this biasing may be obtained using the general equation (6.33b) for S, which says

$S = \frac{(\beta + 1)}{\left[1 - \frac{\beta (dI_B)}{(dI_C)}\right]}$. Now the value of $\frac{(dI_B)}{(dI_C)}$ required to calculate S may be obtained

from (6.39) that gives

$$I_B = \frac{V_{cc} - V_{BE} - R_C I_C}{(R_B + R_C)} - \frac{V_{cc} - V_{BE}}{(R_B + R_C)} - \frac{R_c}{(R_B + R_C)} I_c$$

Therefore,

$$\frac{dI_B}{dI_C} = 0 - \frac{R_c}{(R_B + R_C)}$$

Hence, stability factor for collector–base biasing $S^{CB} = \frac{(\beta + 1)}{\left[1 - \frac{\beta (dI_B)}{(dI_C)}\right]} =$

$\frac{(\beta + 1)}{\left[1 - \beta \left(-\frac{R_C}{R_B + R_C}\right)\right]}$

Or

$$S^{CB} = \frac{(\beta + 1)}{\left[1 + \frac{\beta R_C}{R_B + R_C}\right]} = \frac{(\beta + 1)(R_B + R_C)}{(R_B + (\beta + 1)R_C)} \qquad (6.43a)$$

It may be noted that the stability factor for collector–base biasing S^{CB} is much smaller (hence better stability) than for the fixed bias for which $S^{Fix} = (\beta + 1)$.

To calculate the stability factor S^{CB}_{VBE} against variation in V_{BE} for collector–base biasing, we rewrite (6.38b) as follows:

$$V_{CC} = (I_C + I_B)R_C + R_B I_B + V_{BE}$$

Or

$$V_{CC} - I_C R_C - (R_C + R_B)I_B - V_{BE} = 0 \qquad (6.43b)$$

Also, it is known that the following equation holds for all configurations:

$$I_C = \beta I_B + (\beta + 1)I_{CBO}$$

Since I_{CBO} is much smaller and further in present calculations we are considering it to be constant, therefore one may write
$I_C = \beta I_B$ or $I_B = \frac{I_C}{\beta}$, substituting this value of I_B in (6.43b) one gets

$$V_{CC} - I_C R_C - (R_C + R_B)\frac{I_C}{\beta} - V_{BE} = 0$$

Or

$$I_c = \frac{\beta V_{CC}}{[(\beta + 1)R_C + R_B]} - \frac{\beta V_{BE}}{[(\beta + 1)R_C + R_B]} \qquad (6.43c)$$

Differentiating (6.43c) with respect to I_C, treating β as constant, one gets

$$1 = 0 - \frac{\beta}{[(\beta + 1)R_C + R_B]}\frac{\partial V_{BE}}{\partial I_C} = -\frac{\beta}{[(\beta + 1)R_C + R_B]}\frac{1}{S_\beta^{CB}}$$

Hence,

$$S_{V_{CE}}^{CB} = -\frac{\beta}{[(\beta + 1)R_C + R_B]} \qquad (6.43d)$$

To calculate $S_{\beta,}^{CB}$ we use (6.42b), which is
$I_C = \beta\frac{(V_{CC} - V_{BE})}{[(1 + \beta)R_C]}$ We differentiate it partially with respect to β, treating V_{BE} constant to get

$$S_\beta^{CB} = \frac{(V_{CC} - V_{BE})(R_C + R_B)}{[R_B + (1 + \beta)R_C]^2} \qquad (6.43e)$$

(c) **Voltage divider or self-biasing**: Figure 6.26a shows the potential divider (voltage divider/ self-bias/ emitter bias) biasing arrangement and Fig. 6.26ab its DC equivalent. It may be observed in this figure that potential V_B at the base is produced by a potential divider arrangement of two resistances R_1 and R_2 connected to the battery V_{CC}. Thus,

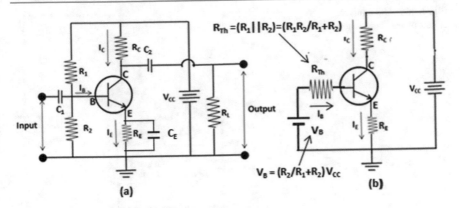

Fig. 6.26 **a** Potential divider biasing; **b** DC equivalent of (**a**)

$$V_B = \frac{R_2}{R_1 + R_2} V_{CC} \tag{6.44a}$$

Also, the series resistance R_{Th} connected to the base in the DC equivalent circuit is equal to the parallel combination of resistances R_1 and R_2, i.e.

$$R_{Th} = (R_1 \| R_2) = \frac{R_1 R_2}{R_1 + R_2} \tag{6.44b}$$

Equations (6.44a) and (6.44b) tells that the magnitudes of V_B and R_{Th} got fixed if V_{CC}, R_1 and R_2 are fixed. The capacitances C_1, C_2 and C_E in Fig. 6.26a are such that the impedance offered by each of them at the signal frequency f is negligible, and therefore, they may be treated as short circuit for the applied signal. Capacitance C_E allows the AC component of the signal at the emitter to pass to the ground while blocking the DC component that is required for proper biasing.

To obtain the coordinates of the Q-point, we apply KVL to the base–emitter loop to get

$$V_B - R_{Th}I_B + V_{BE} - (\beta + 1)I_B R_E = 0$$

Or

$$I_B = \frac{V_B + V_{BE}}{R_{Th} + (\beta + 1)R_E} \tag{6.45}$$

Applying KVL to collector–emitter loop, one gets

$$V_{CE} = V_{CC} - R_C I_C - R_E I_E = V_{CC} - R_C I_C - R_E (I_C + I_B) \tag{6.46}$$

And,

Fig. 6.26c DC equivalent circuit of self-bias scheme

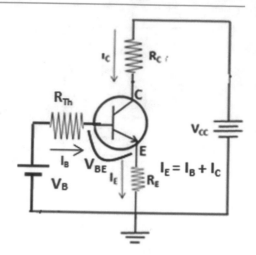

$$I_C \approx \beta I_B = \frac{\beta(V_B - V_{BE})}{R_{Th} + (\beta + 1)R_E} \qquad (6.47)$$

A look to (6.47) reveals that for any reason (like heating of the transistor or by the change in the value of β) if I_C increases then V_{BE} must decrease, as all other quantities in the equation are constant for fixed values of V_{CC}, R_1 and R_2. Any decrease in V_{BE} from (6.45) will reduce I_B which in turn will reduce I_C restoring it towards its initial value. Similarly, if I_c decreases, say because of the replacement of the transistor, then from (6.47) V_{BE} must increase. An increase in V_{BE} will increase I_B which in turn will increase I_C restoring it towards its initial value. Thus any change in I_C induces a change in I_B in the opposite sense; if I_C increases I_B decreases and if I_C decreases I_B increases. Such a process where the change in some output quantity produces an opposite effect on the input quantity is called **negative feedback**. Stabilization of the Q-point both in collector–base biasing and potential divider biasing is achieved by negative feedback.

The effectiveness of stabilization depends on how strongly the collector current depends on V_{BE}. In (6.47) V_{BE} will influence I_C by a large factor if the denominator $(R_{Th} + (\beta + 1)R_E)$ is small. Therefore, for better stabilization $R_{Th} \ll (\beta+1)\ R_E$, so that (6.47) reduces to

$$I_C = \beta I_B = \frac{\beta(V_B - V_{BE})}{R_{Th} + (\beta + 1)R_E} \approx \frac{\beta(V_B - V_{BE})}{(\beta + 1)R_E} \approx \frac{(V_B - V_{BE})}{R_E} \qquad (6.48)$$

Though negative feedback is good for stabilization of the operating point, it inherently reduces the gain of the amplifier.

To obtain an expression for the stability factor S^{Self} for this biasing, we write the KVL for the base–emitter loop once gain as

$$V_B - R_{Th}I_B - R_E(I_B + I_C) = 0$$

Or

$$I_B = \frac{V_B}{R_{Th} + R_E} - \frac{R_E}{R_{Th} + R_E} I_C$$

Differentiating the above equation with respect to I_C,
$\frac{dI_B}{dI_C} = 0 - \frac{R_E}{RT_h + R_E}$ Substituting this value in general expression for stability factor
(6.33b) $[S = \frac{(\beta+1)}{\left[1 - \frac{\beta(dI_B)}{(dI_C)}\right]}]$ one gets

$$S^{Self} = \frac{(\beta+1)}{\left[1 + \frac{\beta R_E}{RT_h + R_E}\right]} = \frac{(\beta+1)}{\left[1 + \frac{\beta}{1 + \frac{R_{Th}}{R_E}}\right]} \tag{6.49a}$$

It may be noted that if $\frac{R_{Th}}{R_E}$ is made very small so that it may be neglected in comparison to 1 then S^{Self} may approach 1, the limiting and best possible value for stability factor. It is because of this reason that collector base or self-biasing is used in most amplifier circuits.

To calculate $S^{self}_{V_{BE}}$, let us write the KVL for the base–emitter loop, as shown in Fig (6.26c),

$$-V_B + R_{Th}I_B + V_{BE} + R_E(I_C + I_B) = 0$$

Or

$$V_{BE} = V_B - (R_{Th} + R_E)I_B - R_E I_C \tag{6.49b}$$

But $I_B \approx \frac{I_C}{\beta}$ substituting this value of I_B in (6.49b), one gets

$$V_{BE} = V_B - (R_{Th} + R_E)\frac{I_C}{\beta} - R_E I_C$$

Or

$$V_{BE} = V_B - \frac{[R_{Th} + (\beta+1)R_E]}{\beta}I_C \tag{6.49c}$$

Differentiating the above equation with respect to I_C and treating β, V_B, R_{Th} and R_E as constants, one gets

$$\frac{dV_{BE}}{dI_C} = \frac{1}{S^{Self}_{V_{BE}}} = -\frac{[R_{Th} + (\beta+1)R_E]}{\beta}$$

Therefore,

$$S_{V_{BE}}^{Self} = -\frac{\beta}{[R_{Th} + (\beta + 1)R_E]} \tag{6.49d}$$

Finally, to calculate S_β^{Self} let us rewrite (6.48) as

$I_C = \frac{\beta[V_B - V_{BE}]}{[R_{Th} + (\beta+1)R_E]}$ Differentiating it with respect to β and treating V_B, V_{BE}, R_E and R_{Th} constant, one gets

$$\frac{dI_C}{d\beta} = S_\beta^{Self} = \frac{(V_B - V_{BE})(R_{Th} + R_E)}{[R_{Th} + (\beta + 1)R_E]^2} \tag{6.49e}$$

Self-assessment question SAQ: In fixed bias arrangement which of the three stability factors is likely to override the other two?

Solved example (SE6.1): Biasing arrangement for an npn BJT amplifier in CE configuration is shown in Fig. SE6.1. Obtain the parameters of the Q-point. Also, discuss the effect of the change of β from 200 to 150 on the operating point.

Solution: It is to be noted that an amplifier operates in the forward active region of output characteristics and, therefore, the base–emitter junction is forward biased. It means that the forward-biased BE junction is 'on' and the base–emitter voltage V_{BE} is 0.7 V for silicon device and 0.3 V for germanium device. Further, if the type of the device is not specified then it is normal to assume it to be a silicon device. In this problem the type of device is not specified, we take it to be silicon and assume $V_{BE} = 0.7$ V.

Applying KVL to the base–emitter *loop*, one gets

$$300x10^3 x I_B \ V + 0.7 \ V = 10 \ V,$$

$$I_B = \frac{9.3}{3x10^5}A = 3.1x10^{-5}A$$

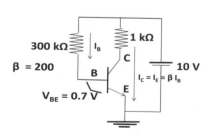

Fig. SE6.1 Circuit for problem (SE6.1)

Now,

$$I_E = I_C = \beta I_B = 200x3.1x10^{-5}A = 6.2x10^{-3}A$$

The potential drop against 1 kΩ *collector resistance* $1x10^3x6.2x10^{-3}V = 6.2V$
Hence, the voltage $V_{CE} = 10\ V - 6.2\ V = 3.8\ V$
The parameters specifying the operating point are

(h) $V_{CE} = 3.8\ V$; (ii) $I_B = 3.1\ x\ 10^{-5}\ A = 31\ \mu$; and (iii) $I_C = I_E = 6.2\ mA$.

Next, let us investigate the effect on the coordinates of the Q-point when β is changed from an initial value of 200 to 150.
With the change of β, the I_B will not change it will remain the same (31 μA) but $I_C = I_E$ will change and the new value *of emitter current will $I_E' = 150\ x\ 31\mu A = 4.65\ x\ 10^{-3}\ A$. Also the new value of $V_{CE} = 10 - 4.65 = 5.35\ V$. The new coordinates of the Q-point are: $V'_{CE} = 5.35\ V$; $I'_B = 31\ \mu A$; $I_C' = I'_E = 4.65\ mA$;*
It may be observed that this type of biasing (fixed or emitter biasing) is not at all good to keep the shifting of the operating point to a minimum against the change of β.

Solved example (SE6.2) Calculate the value of the base resistance R_B so that a base current of 31μA flows through the circuit of Fig. 6.16. Determine the coordinates of the Q- point and investigate the shift in the operating point when β is changed from 200 to 150.

Solution *Applying KVL to the 10 V battery-base–emitter loop, one gets*
$1x10^3x(I_B + I_C) + R_BI_B + 0.7V = 10V\ (SE6.2)$
Putting $I_B = 31x10^{-6}\ A$ and $I_C = \beta\ I_B = 200\ x\ 31x\ 10^{-6}\ A$ in (SE6.2.1), one gets

$$1x10^3x201x31x10^{-6} + R_Bx31x10^{-6} + 0.7 = 10$$

So,

$$R_B = 99 \times 10^3\Omega;\ I_B = 31\ \mu A = 31 \times 10^{-6}A$$

Also,

$$I_C = I_E = 200xI_B = 6.2 \times 10^{-3}A = 6.2\ mA$$

Fig. SE6.2 Circuit for
problem (SE6.2)

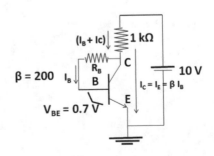

Fig. SE6.3.1 *Circuit for example (SE6.3)*

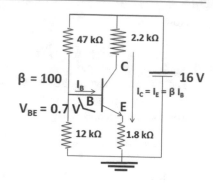

Hence,

$$V_{CE} = 10\ V - 1000 \times (31\ \mu A + 6.2\ mA) = 3.769\ V$$

The coordinates of the Q-point are: $V_{CE} = 3.769\ V$; $I_C = 6.2\ mA$ and $I_B = 31\ \mu A$

If these coordinates are compared with coordinates of Q-point in the last example (SE6.1), it may be observed that I_B and I_C are the same but V_{CE}, in this case, is 3.769V, while it was 3.8 V in the last case.

Next, let us see how Q-point parameters change when β changes from 200 t0 150. Change of β will change $I_C = I_E$ from 6.2 mA to (150 X 31X10^{-6} =) 4.65 mA (same change as in last example) and new value of $V'_{CE} = [10 - 1x\ 10^3 (150\ x\ 31\ x\ 10^{-6} + 31x10^{-6})] = 5.319\ V.$

Solved example (SE6.3) Calculate the parameters of the operating point and the collector current that may drive the circuit to saturation region, for the self-biased amplifier shown in the Fig. SE6.3.1.

Solution *Given self-biased circuit can be reduced to the circuit shown in Fig. SE6.3.2.*

Using (6.43a) and (6.44), one may calculate the values of V_B and R_B as given below.

$$V_B = \frac{12x10^3\Omega}{(12+47)x10^3\Omega}\ 16V = 3.25V; R_B = \frac{12x47}{59}x10^3\Omega = 9.56x10^3\Omega$$

$$(SE6.3.1)$$

Fig. SE6.3.2 *Reduced circuit*

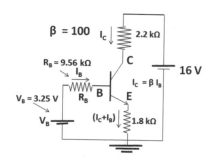

Applying KVL to base–emitter circuit,

$R_B I_B + 0.7 + 1.8 x 10^3 [I_B + I_C] = V_B$ *(SE6.3.2)*

Substituting the values of R_B, V_B from (SE6.3.1) and putting $I_C = \beta\ IB = 100\ I_B$ in above equation, one gets $I_B = 13.33 \mu A$

Therefore,

$$I_C = 100\ x\ I_B = 1.33\ mA;\ And\ I_E = (I_B + I_C) = 1.34\ mA.$$

To calculate VCE, we apply KVL to the collector–emitter loop to get

$$2.2 x 10^3 x 1.33 x 10^{-3} V + V_{CE} + 1.8 x 10^3 x 1.34 x 10^{-3} V = 16 V$$

Or

$$V_{CE} = (16 - 5.34)\ V = 10.66\ V$$

The parameters of the Q-point are: $V_{CE} = 10.66 V; I_C = 1.34 mA; I_B = 13.33 \mu A$

When the amplifier circuit goes into saturation region, both junctions of the BJT are forward biased and are like closed or "on" switches and offer no resistance. In case of saturation, the total voltage of the battery of 16 V drops across the combination of resistances of 2.2 kΩ plus 1.8 kΩ. Therefore, a current of (16/4 kΩ) = 4 mA should flow through the transistor. On the other hand, if the circuit is in the cut-off region then both junctions will be reverse biased, behave like open switches and no current be drawn from the 16 V battery. Therefore V_{CE} will become 16 V.

Self-assessment question SAQ: Which stability factor becomes important when a burnt out transistor has to be replaced in a circuit? Give reasons for your answer.

6.9 BJT Modelling and Equivalent Circuit: Small-Signal Model

Both circuit designers and manufacturing industry require simple models that may describe transistor performance in terms of some parameters. There are two types of models; the one that is suitable when the current and voltage of the input signal is small, they vary within about 10% of the current and voltage values at the Q-point. This ensures that the operating point stays within the linear region. Small-signal models are, therefore, also called linear models. The other models, termed non-linear models are better suited when the applied signal is large. In the present presentation, we shall discuss the small-signal equivalent circuits.

There are two prominent methods used for substituting BJT in case of small signals: (i) the use of hybrid or h-parameter of the transistor and the values of circuit components, (ii) use of some specific transistor parameter like the resistance r, the transconductance g_m or the current amplification β and the values of the circuit components. So far both the industry and circuit designers were mostly using the first method and, therefore, hybrid analysis got overemphasised in textbooks. Hybrid parameter analysis is still a favoured technique but the fact that h-parameters are not so readily and easily available, their values change considerably from one transistor to the other of the same batch and manufacturer, and that they produce accurate results only under specific operation conditions make h-parameter approach less attractive. On the other hand, the second approach which uses transistor β and component values are gaining favour on account of the easy availability of required parameter values, simple and straightforward procedure and generally quite accurate results. In the present text, small-signal resistance model (written as r-model), transconductance g_m model (also called hybrid-pi model), and hybrid model will be discussed.

6.9.1   Small-Signal r-Parameter Transistor Model

An r-parameter model for BJT is perhaps the most simple and straightforward model. The basic circuit of the r-model is shown in Fig. 6.27a. As may be observed in this figure, the BJT is replaced by three resistances and a dependent current source. In general, because of the very small value of r'_B it may be neglected and replaced by a short circuit. On the other hand, on account of the large value of collector resistance r'_c of the order of few hundred kilo ohms, it may be replaced by an open circuit. With these simplifications, the modified simple r-model equivalent

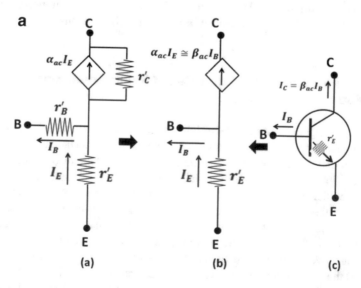

Fig. 6.27a **a** r-parameter equivalent of BJT. **b** Simplified r-parameter equivalent. **c** Relation of transistor symbol

is shown in part (b) of the figure. Part (c) of the figure depicts the relationship between r-parameter model and transistor symbol.

The AC operation of a transistor according to r-model may be interpreted in the following way. A resistance r'_E appears between the emitter and the base terminals, which is equal to the resistance that will be 'seen' looking into the emitter of a forward-biased junction. The collector effectively acts as a dependent current source of current $i_C = \alpha_{ac} i_E \cong \beta_{ac} i_B$.

The emitter resistance r'_E may be calculated using the expression

$$r'_E = \frac{25mV}{i_E} \tag{6.50a}$$

Above expression holds good for an abrupt p-n junction at 20°C, and resistance r'_E increases a little with temperature as well as in the case of a graded junction.

(a) **Input resistance at the base** In r-parameter model r'_E represents the resistance that will appear at the emitter when looking into the emitter from the input side. The value of the equivalent input resistance at base R^{Base}_{in} may be calculated as discussed in the following,

The input resistance looking in at the base $R^{Base}_{in} = \frac{v_{in}}{i} = \frac{v_B}{i_B}$

But base voltage $v_B = i_E r'_E$

Also, $i_C = \beta_{ac} i_B \cong i_E$; therefore, $i_B - \frac{i_E}{\beta_{ac}}$,

Hence,

$$R^{Base}_{in} = \frac{v_B}{i_B} = \frac{i_E r'_E}{\frac{i_E}{\beta_{ac}}} = \beta_{ac} r'_E \tag{6.50b}$$

Equation (6.50b) tells that resistance r'_E which is present in the emitter arm (between emitter and ground) is equivalent to a resistance $R^{Base}_{in} = \beta_{ac} r'_E$ in base arm (between base and the ground). The reverse is also true, that is a resistance R present in base arm is equivalent to a resistance $R_{emt} = (R/\beta_{ac})$ in emitter arm. These conversions are frequently required in numerical problems.

(b) **Analysis of a BJT amplifier in common emitter configuration using r-parameter model**

Let us carry out the analysis of a BJT amplifier in common emitter configuration shown in Fig. 6.27b using r-parameter model. As shown in figure two values, one AC and the other DC for the current gain beta are provided in the problem.

DC analysis of the circuit: This is the first step in which all capacitors are treated as open terminals and the circuit reduces to what is shown in Fig. 6.27c.

Fig. 6.27b Circuit diagram
of the given npn common
emitter amplifier

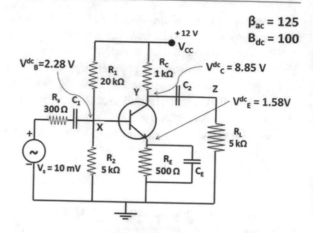

$\beta_{ac} = 125$
$B_{dc} = 100$

Fig. 6.27c Reduced circuit
for DC analysis

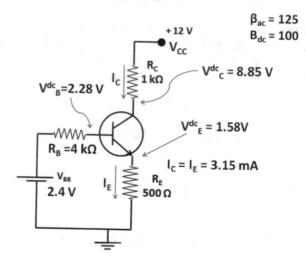

$\beta_{ac} = 125$
$B_{dc} = 100$

The voltage source $V_{BB} = \frac{R_2}{(R_2 + R_1)} V_{CC} = \frac{5k\Omega}{25k\Omega} x12\ V = 2.4\ V$; Resistance $R_B = R_2 \| R_1 = 4\ k\Omega$

$$I_E \cong I_C = \frac{V_{BB} - 0.7}{500 + R_B/\beta_{dc}} = \frac{1.7}{500 + 40} = 3.15 mA$$

$$V_E^{dc} = 3.15 \times 10^{-3} \times 500 = 1.58\ V; \quad V_B^{dc} = V_E^{dc} + 0.7\ V = 2.28\ V$$

$$V_C^{dc} = V_{CC} - I_C R_C = 12 - 3.15 x 10^{-3} x 1 x 10^3 = 8.85 V$$

$$V_{CE}^{dc} = V_C^{dc} - V_E^{dc} = 8.85 - 1.58 = 7.27V$$

Having computed the DC values of current and voltages at various points in the
network we now carry out AC analysis.

AC analysis: The main difference between the AC and the DC analysis lies in the fact that while in DC analysis both coupling capacitors C_1 and C_2 and the bypass capacitor C_E are taken as open circuits, in AC analysis these capacitors are taken as short circuits. As a result, the resistance R_S of the input source and load resistance R_L become parts of the AC circuit, while the capacitor C_E short circuits emitter resistance R_E which gets removed from the AC equivalent. Further, all DC sources are replaced by ground for AC analysis. The AC equivalent circuit to some extent depends on the model of the transistor that is used for analysis. For example, the present analysis is being done using r-parameter model that considers an inherent emitter resistance r_E' to be present in the emitter arm that is not short-circuited by the capacitor C_E.

The reactance X_C of a capacitor is given by the expression $X_C = 1/j\omega C$, which is treated as zero or negligible as compared to the value of resistance R_E in AC analysis, and therefore, C_E is treated as a short circuit. Detailed calculations have shown that this assumption of treating X_C as zero or negligible holds good only if X_C is at least 1/10 of R_E. This may help in selecting the proper value of capacitor C_E if the value of R_E and the frequency ω of the input signal are given. Figure 6.28 shows the AC equivalent circuits at different stages: (a) the DC voltage sources grounded and capacitors replaced by short circuits. Dotted lines indicate the changes that have taken place in the circuit. The r-parameter emitter resistance r_E' is included in the circuit in emitter arm as shown in the figure.

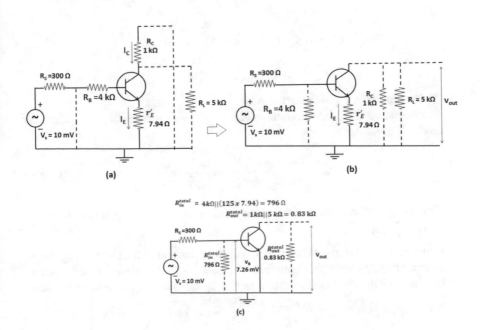

Fig. 6.28 AC equivalent circuits of the amplifier

The value of r_E' is calculated using the expression $r_E' = \frac{25mV}{I_E} = \frac{25mV}{3.15mA}$ that gives the value 7.94 Ω. The value of emitter current $I_E = 3.15$ mA is used from DC analysis.

To calculate the total input resistance R_{in}^{total}, looking to the base, the emitter arm resistance r_E' has been transferred to the base side by multiplying r_E' by 125, the AC current gain β_{ac}. With this transferring of emitter arm resistance to the base arm, there are two resistances $R_B = 4k\Omega$ and ($125x$ $r_E'=$) 0.993 kΩ in parallel. The value of R_{in}^{total} comes out to be 0.796 kΩ = 796Ω

The amplitude of the signal v_B that is applied at the base is given by

$$v_B = \frac{R_{in}^{total}}{\left(R_{in}^{total} + R_s\right)} . v_s = \frac{796}{796 + 300} 10mV = 7.26mV$$

It may be observed that though the source of input signal delivers a signal of 10 mV, the signal actually applied at base is only 7.26 mV. This attenuation of the signal occurs because of the division of the input signal against the series combination of source resistance R_s and the total input resistance R_{in}^{total}. Only that part of the input signal that drops against R_{in}^{total} is actually applied at the base.

Similarly, in the output, the collector resistance R_C and the load resistance R_L makes a parallel combination of the resultant equivalent resistance R_{out}^{total} is given by

$$R_{out}^{total} = R_c\ R_L = 1k\Omega\ 5\ k\Omega = 0.83\ k\Omega$$

(c) Voltage gain in r-parameter model

AC voltage gain A_v for common emitter configuration according to r-parameter model is given by

$$A_v = \frac{v_{out}}{v_{in}} = \frac{\alpha_{ac}i_ER_C}{i_Er_E'} \approx \frac{i_ER_C}{i_Er_E'} = \frac{R_C}{r_E'} \tag{6.50c}$$

For the given amplifier circuit, the gain from base to collector $A_v = \frac{R_C}{R_{in}^{totao}}$ $= \frac{1x10^3}{796} = 1.26$

Expression (6.50c) gives the voltage gain from base to collector (see Fig. 6.29), in order to find the overall gain from the source of input signal to the collector one has to consider the attenuation of the signal between source and the base.

The overall gain $A_V^{overall} = \frac{R_{out}^{total}}{R_{in}^{total}} = \frac{0.83x10^3}{796} = 1.04$

The overall gain decreases because of the lower value of R_{out}^{total} than that of R_C (= 1 kΩ).

Fig. 6.29 r-model equivalent circuit for calculating voltage gain

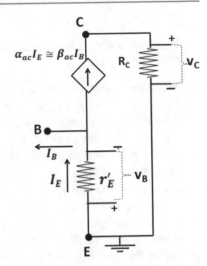

If it is assumed that the source V_s in Fig. 6.27b delivers a square wave of peak-to-peak value of 10 mV then the DC level, the amplitude and the phase of signals at the source, at base (point X), at collector (point Y) and across the load resistance R_L (point Z) are shown in Fig. 6.30. It may be noted that the DC level of the signal before the coupling capacitor is zero volt, at base the DC level is +2.28 V (as calculated in DC analysis), at collector it is +8.85V and at point Z after the capacitor is again 0 V. Further, the signals at source and at base are in-phase but amplified signals at points Y and Z are 180^0 out of phase with the input signal.

(d) **Effect of the bypass capacitor C_E on the voltage gain in r-model analysis**
 If it is assumed that capacitor C_E is not present in amplifier circuit then the total resistance in the emitter arm of r-equivalent circuit will be $R_E^{total} = r'_E + R_E = 7.94 + 500 = 507.94\Omega$

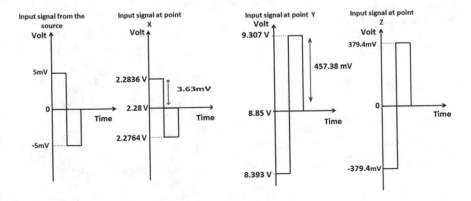

Fig. 6.30 Signal amplitude and DC level at different points of Fig. 6.27a

And the voltage gain $A_V^{overall} = \frac{R_{out}^{total}}{r_E' + R_E} = \frac{0.83 \times 10^3}{507.93} = 1.63$

It may be observed that the inclusion of any extra resistance in the emitter arm reduces the gain.

(e) **Stability of voltage gain in r-model**

The value of voltage gain in r-model very much depends on the value of r_E' which in turn depends on temperature. Change in temperature and also the replacement of the transistor by a new one of the same specification, are likely to change the voltage gain. In order to make the voltage gain less dependent on the emitter resistance r_E', the method called **swamping** is used. Swamping in fact is a compromise between having a bypass capacitor across R_E or having no bypass capacitor at all. In a swamped amplifier, R_E is partially bypassed so that a reasonable gain can be achieved and at the same time effect of r_E' is considerably reduced. Total external emitter resistance R_E is made of two resistances R_{E1} and R_{E2}, and the resistance R_{E2} is by passed while R_{E1} remains un-bypassed. The voltage gain is then given by

$A_v^{Swam} = \frac{R_{out}^{total}}{r_E' + R_{E1}}$ and if R_{E1} is $>> r_E'$, then gain becomes almost independent of r_E'. The circuit diagram of the swamped r-model reduced amplifier circuit and corresponding BJT circuit are shown in Fig. 6.31.

(f) **Current gain in r-model** Current gain A_i is given by,

$A_i = \frac{I_C}{I_S}$, here I_C is the collector current and I_S the current drawn from source V_s, which may be calculated as

$$I_S = \frac{v_S}{R_S + R_{in}^{total}}$$

Fig. 6.31 Circuit diagram of a swamped amplifier that uses a partially bypassed emitter resistance to minimize the effect of r_E' on the gain to achieve gain stability

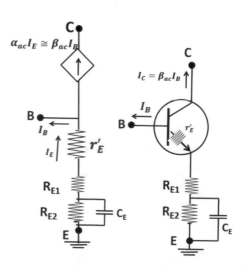

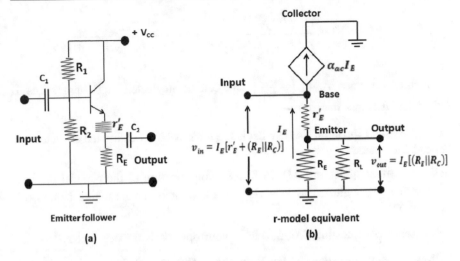

Fig. 6.32 **a** Emitter follower; **b** r-model equivalent of emitter follower

For the present example of amplifier, $I_s = \frac{0.707 \times 10 mV}{(300 + 796)\Omega} = 6.45 \mu A$

Therefore, the current gain $A_i = \frac{I_C}{I_s} = \frac{3.15 mA}{6.45 \mu A} = 488$

The power gain $A_p = A_v A_i$ and can be easily calculated.

(g) r- parameter analysis of a Common collector amplifier or Emitter follower

A common collector amplifier where the input signal is applied at base and the output is taken from the emitter, is called an emitter follower. The voltage gain of an emitter follower (CC amplifier) $\approx 1 (or 0.98)$ and the output signal is in phase with the input signal. But the current gain of emitter follower is quite large. Hence, emitter followers are used to delivering signals at the output that are an exact replica of the original signal (no phase change) with almost the same amplitude but with considerably higher current capacity. Emitter followers are often used as the first stage of a cascade amplifier. Their purpose is to match the output impedance of the signal source with the input impedance of the cascade amplifier. As already mentioned emitter follower increases, the current capacity of the input signal and thus reduces the chance of signal attenuation when applied at some device that has low input impedance.

The circuit diagram of an npn BJT emitter follower and its r-model equivalent are shown in Fig. 6.32. Although the positive terminal of the power supply V_{cc} in Fig (6.32 a) is connected at the collector, however, from the point of AC analysis the collector is connected to the **AC ground**. An AC ground is the terminal that is connected to the ground when AC analysis of the network is done. As all AC sources are grounded while doing AC analysis, the $+V_{CC}$ end of collector is also treated as AC ground.

As indicated in the r-model equivalent circuit, the input signal $v_{in} = I_E$ $[r'_E + (R_E || R_L)]$ and the output signal $v_{out} = I_E(R_E || R_L) = I_E \left(\frac{R_E R_L}{R_E + R_L} \right)$. The voltage gain $A_v^{E.F}$ of the emitter follower is, therefore, given by

$$A_v^{E.F} = \frac{v_{out}}{v_{in}} = \frac{I_E\left(\frac{R_E R_L}{R_E + R_L}\right)}{I_E\left[r_E' + \left(\left(\frac{R_E R_L}{R_E + R_L}\right)\right)\right]} = \frac{R_E R_L}{[(R_E + R_L)r_E' + R_E R_L]}$$

In the case when $r_E' \ll \left(\frac{R_E R_L}{R_E + R_L}\right)$ and can be neglected, $A_v^{E.F}$ approaches 1; Otherwise, it is less than 1.

Self-assessment question SAQ: What is the logic behind treating the collector as a dependable source?

Self-assessment question SAQ: In further simplification of the r-model the base resistance r_B' is neglected as it is very small, what is the reasoning behind this assumption?

Self-assessment question SAQ: What in your opinion is the cause of r_E'?

Self-assessment question SAQ: Why the resistance r_C' has a value that is considerably larger than that of r_B' and r_E'?

6.9.2 Small-Signal Transconductance or Hybrid-pi Model for CE Configuration

A small-signal model, as already mentioned is a linear model for a BJT which is essentially a non-linear device. For example, Fig. 6.33 shows the linear operation of the forward-biased base emitter junction in an npn transistor. Quantities represented by upper case letters are DC values and those with lower case represent AC values.

The small-signal equivalent may or may not have time dependence. Time dependence comes into play when effects produced by junction capacitances are taken into account.

Fig. 6.33 Operation of the base–emitter junction in linear part of the input characteristic of an npn transistor

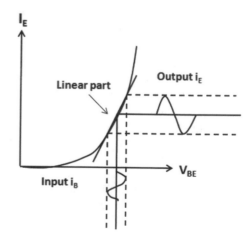

Fig. 6.34 BJT CE amplifier
with AC input

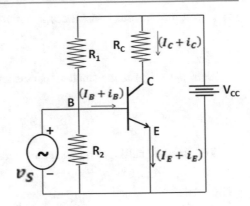

Before describing the hybrid-pi or transconductance model let us discuss some parameters that are required to understand the model.

(a) **Common Emitter–Active Region**

Figure 6.34 shows a CE amplifier. As may be seen in the figure, both the base current and the collector current have a DC component denoted by capital letters and an AC component denoted by small letters. The AC input signal v_i produces the AC base current i_B which in turn produces the AC collector current i_c. The transconductance g_m has two components; the AC component called the small-signal transconductance and denoted by g_m^s. The other component called large-signal transconductance is denoted by g_m^L. The small-signal transconductance is defined as

$$g_m^s \equiv \left[\frac{d(I_C + i_c)}{d(V_{BE} + v_{BE})}\right]_{Qpoint} = \frac{change\ in\ total\ collector\ current}{change\ in\ total\ base - emitter\ voltage} = \frac{i_C}{v_{BE}} = \frac{i_C}{v_i}$$

Hence,

$$i_C = g_m^s v_i \qquad\qquad (6.51a)$$

Also, the value of large-signal transconductance at Q-point is given as

$$g_m^L \equiv \left[\frac{d(I_C)}{d(V_{BE})}\right]_{Qpoint} = \frac{d(I_s e^{\frac{v_{BE}}{V_T}})}{d(V_{BE})} = \frac{I_s e^{\frac{v_{BE}}{V_T}}}{V_T} = \frac{I_C}{V_T} \qquad (6.51b)$$

It may be recalled that the numerical value of V_T at room temperature may be taken as 26 mV or 25 mV.

(b) **Input and output resistances** In active region, one may write

$$i_C = \beta_0 i_B \tag{6.52a}$$

Here $\beta_0 (or \beta_{ac})$ is the small-signal current gain given by

$$\beta_0 = \frac{i_C}{i_B} \tag{6.52b}$$

The small-signal input resistance r_π^S is defined as

$$r_\pi^S \equiv \frac{v_i}{i_B} = \frac{i_C}{g_m^S i_B} = \frac{\beta_0}{g_m^S} \tag{6.52c}$$

The small-signal output resistance r_0^S is related to the output conductance g_0^S as

$$\frac{1}{r_0^S} = g_0^S \equiv \frac{\Delta(I_C + i_C)}{\Delta(V_{CE} + v_{CE})} = \frac{i_C}{v_{CE}} \tag{6.53a}$$

Or

$$i_C = g_0^S v_{CE} \tag{6.53b}$$

Further, from (6.28) one has

$$I_C = I_{se}^{\left(\frac{v_{BE}}{V_T}\right)} \left(1 + \frac{v_{CE}}{V_A}\right)$$

And, therefore, the large-signal output transconductance g_0^L maybe given as

$$g_0^L = \left[\frac{\partial(I_C)}{\partial(V_{CE})}\right]_{Q-point} = I_{se}^{\left(\frac{v_{BE}}{V_T}\right)} \left(0 + \frac{1}{V_T}\right) = \frac{I_C}{V_T} \tag{6.54}$$

Hence, the large-signal output resistance r_0^L is

$$r_0^L = \frac{1}{g_0^L} = \frac{V_T}{I_C} \tag{6.55}$$

Since both, the small-signal parameters and the large-signal parameters are evaluated at Q-point they may be set equal to each other, i.e.

$$r_\pi^S = r_\pi^L = r_\pi; g_m^S = g_m^L = g_m; r_0^S = r_0^L = r_0 \text{ etc.} \tag{6.56}$$

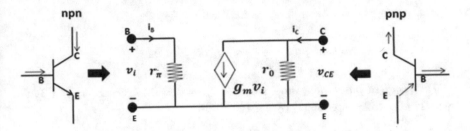

Fig. 6.35 Transconductance small-signal equivalent of BJT in CE configuration

Now in simple small-signal BJT model the input section (the base–emitter junction is replaced by an input resistance $r_\pi^s = r_\pi$ and the output section (collector–emitter) junction by a dependent current source of strength $g_m^s v_i = g_m v_i$ in parallel with the output resistance $r_0^s = r_0$, as shown in Fig. 6.35. The no-load voltage gain may be given as

$$Gain_{noload} = \frac{v_{out}}{v_{in}} = -g_m r_0 \qquad (6.57)$$

It is worth noting that the same circuit diagram for transconductance model holds good for both the npn and pnp transistors.

Self-assessment question SAQ: Why it is required to have an equivalent circuit of a transistor?

Self-assessment question SAQ: Point out the similarities and the point of difference between an r-model and the hybrid-pi model equivalent circuits.

Solved example (SE6.4) Find the small-signal input resistance, output resistance, transconductance, amplifier gain without any load for a BJT CE-amplifier, given that DC current is 2 mA, Early voltage is −50 V and small-signal forward current gain β_0 at room temperature is 100.

Solution: *The small-signal transconductance g_m is given by*

$$g_m = \frac{I_C}{V_T} = \frac{2 \times 10^{-3}A}{26 \times 10^{-3}V} = \frac{1}{13} \; Sieman \; (ohm^{-1})$$

Also, the input resistance $r_\pi = \frac{\beta_0}{g_m} = 100x13 = 1300\Omega = 1.3 \; k\Omega$

And,

$$r_0 = \frac{V_A}{I_C} = \frac{50}{2x10^{-3}} = 25x10^3 = 25k\Omega$$

But, $g_m v_i r_0 = -v_{CE}$, *therefore, no load gain* $= \frac{v_{CE}}{v_i}$

$$No\ load\ gain = \frac{v_{out}}{v_{in}} = -g_m r_0 = -1.92 \times 10^3$$

The required parameters have values: $g_m = \frac{1}{13} = 0.77$ *Sieman;* $r_\pi = 1.3\ k\Omega$; $r_0 = 25\ k\Omega$, *no load gain* $= -1.92 \times 10^3$

(c) **Further extension of small-signal transconductance model**

The transconductance model can be further extended to include the effects of (i) the change in depletion width with v_{CE} and (ii) the change in minority carrier concentration in base region due to the change of the thickness of the base region with v_{CE}. Let us first consider how the effect of the change in depletion width at the BC junction can be included in the model.

i. **Collector–base resistance in small-signal transconductance model**

It may be recalled that the thickness of the depletion layer at the base–collector junction increases with the increase of the reverse bias voltage between collector and emitter v_{CE}. Further, the extended depletion layer mostly lies in the base region on account of the low carrier concentration at base. The concentration of total minority charge carriers stored in the base region decreases with the increase of v_{CE}, as shown in Fig. 6.36. This reduction in minority charge concentration decreases the recombination rate of electron–hole pairs. The base current i_B has two components: the recombination component i_{B2} and the drift current i_{B1}. The effects originating from the variation in V_{CE} have been included in small-signal model in term of the collector–base resistance r_μ defined as

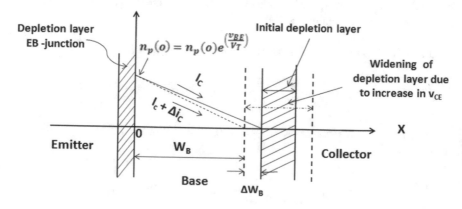

Fig. 6.36 Effect of the increase n collector–emitter voltage on the depletion layer and concentration of charge carriers.

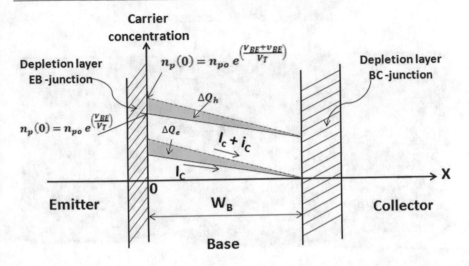

Fig. 6.37 Effect of the change in base–emitter voltage on charge carrier concentration

$$r_\mu = \frac{\Delta v_{CE}}{\Delta i_{B2}} = \frac{\Delta v_{CE}}{\Delta i_C}\frac{\Delta i_C}{\Delta i_{B2}} = \frac{\Delta i_C}{\Delta i_{B2}} r_0 \approx \beta_0 r_0 \qquad (6.58)$$

In practice, it has been observed that $r_\mu \geq 10\beta_0 r_0$ for npn transistor and around 2 to 5 $\beta_0 r_0$ for pnp transistors.

ii. Base charging capacitance in small-signal BJT modelling

Change in base–emitter voltage (Δv_{BE}) changes the concentration of minority carriers ($\Delta Q_e) = q_e$ in the base region. Concentration change of minority carriers induces a corresponding change in the concentration of majority carriers $\Delta Q_h = q_h$ (see Fig (6.37)). In equilibrium, $q_e = q_h$.
But

$$q_h = C_b v_i \qquad (6.59a)$$

Here C_b is the **base-charging capacitance** that is given by

$$C_b = \frac{q_h}{v_i} = \frac{\tau_F i_C}{v_i} = \tau_F g_m = \tau_F \frac{I_C}{V_T} \qquad (6.59b)$$

The base transit time τ_F is given by

$$\tau_F = \frac{W_B^2}{2D_n}$$

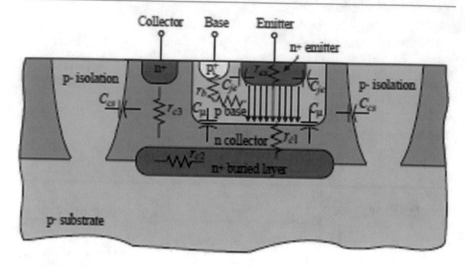

Fig. 6.38 Typical cross section of an npn BJT showing locations of different resistances and capacitances considered in small-signal models.

Inclusion of base charging capacitor in small-signal model makes it time dependent.

iii. **Unavoidable or parasitic elements of the small-signal model for BJT**

Figure 6.38 shows the detailed cross section of an npn BJT. Various resistances and capacitances that need to be included in the small-signal time-dependent model are shown in this figure. The resistances shown in the figure are all bulk Ohmic and important of them are, r_b, r_c and r_{ex}. It is worth noting that r_b is a function of I_C. Capacitance C_{je} represents the capacitance of the forward-biased base–emitter junction while C_μ the capacitance of the reversed-biased base–collector junction.

The complete small-signal model for an npn BJT is shown in Fig. 6.39.

Self-assessment question SAQ: what is meant by the parasitic elements of a transistor? How do they come into the picture?

Solved example (SE6.5) Draw the complete small-signal transconductance model circuit diagram using following parameters:

$I_c = 2$ mA; $V_{CB} = 3$ V; $V_{cs} = 5$ V; $C_\mu = 5.6$ fF; $C_{je} = 20$ fF; $C_{cs} = 10.5$ fF; $\beta_0 = 100$;

$\tau_F = 10$ps; $V_A = 50$ Vr$_b = 300$ Ω; $r_c = 50$ Ω; $r_\mu = 5\Omega$; $r_{ex} = 10$ $\beta_0 r_0$.

Solution: Let us calculate the values of the following quantities from the given data.

$$g_m = \frac{I_C}{V_T} = \frac{2 \times 10^{-3} A}{26 \times 10^{-3} V} = \frac{1}{13} \left(\Omega^{-1} \right) = 0.077 \left(\Omega^{-1} \right) = 77 \ \frac{mA}{V}$$

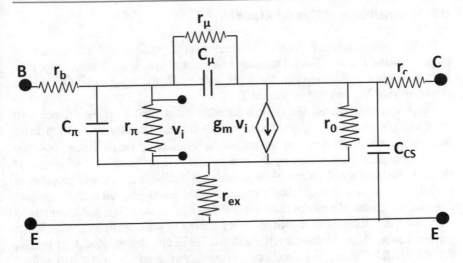

Fig. 6.39 Detailed small-signal model for an npn BJT

$$C_b = \tau_F g_m = (10ps)\left(77\frac{mA}{V}\right) = 0.77pF;$$

$$C_\pi = C_b + C_{je} = 0.77\ pF + 20\ fF = (0.77 + 0.2)\ pF = 0.79\ pF$$

$$r_\pi = \frac{\beta_0}{g_m} = 100x13\Omega = 1.3\ k\Omega;\ r_0 = \frac{V_A}{I_C} = \frac{50\ V}{2x10^{-3}A} = 25\ k\Omega$$

$$r_\mu = 10\beta_0 r_0 = 10x100x25\ k\Omega = 25\ M\Omega$$

The circuit diagram of the small-signal equivalent is given in Fig. SE6.5.

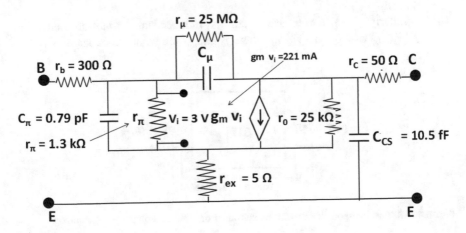

Fig. SE6.5 Circuit diagram of small-signal equivalent of (SE6.2)

6.9.3 Small-Signal Hybrid Model

A BJT is a two-port device with one terminal of input common with one terminal of output as shown in Fig. 6.40. There are two pairs of quantities that are important from the operational viewpoint, the input voltage V_i, the input current I_i and the output voltage V_o, and output current I_0.

For a circuit designer, the only information that matters is how these four parameters, $V_i, I_i,$ and V_0, I_0 are related to each other. Generally, any two of these four parameters may be treated as independent variables, which means that they may be assigned some values by the designer, and if the functional relationships between the known and unknown parameters are available then it will be possible to obtain the value of the other two variables, called dependent variables. Therefore, from the viewpoint of a circuit designer and the manufacturer, it is not important to know how and why these functional relationships got established or what is the physics behind those functional relationships. For circuit designers and, of course, for device fabricators, the functional relationships between input and output parameters are important, otherwise for them, the rectangular block representing the BJT in Fig. 6.40 is just like a black box. The hybrid model of transistor provides these functional relationships. This is the reason why this hybrid model approach is also called the black-box approach.

There are four variables $(V_i, I_i, V_0 \, and \, I_0)$ any two of them may be treated as independent variables; the remaining two will become dependent variables. The two independent variables out of these four may be selected in six different ways and, therefore, the functional relationships between dependent and independent variables may be represented by twelve different equations as indicated below.

(i)   Let V_i and I_i be the independent variables, then the two functional rela-
      tionships between the dependent and the independent variables may be
      written as

      $V_0 = f_1(V_i, I_i) \, and \, I_0 = f_2(V_i, I_i)$ Where f_1 and f_2 are some functions of
      V_i & I_i.

(ii) Similarly, if $V_0 \, and \, I_0$ are taken as independent variables, then $V_i =
      f_3(V_0, I_0) \, and \, I_i = f_4(V_0, I_0)$ where f_3 and f_4 are some functions of V_0 & I_0.

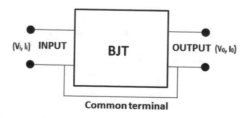

Fig. 6.40 BJT as two-port device with one terminal of input and output ports common

(iii)   In case I_i *and* I_0 are taken as independent variables, then
$V_i = f_5(I_i, I_o)$ *and* $V_0 = f_6(I_i, I_0)$ Where f_5 *and* f_6 are some functions of
I_i & I_0

In this way, three more sets of two equations each may be written corresponding
to the choice of independent variables as $(V_i, V_0), (V_0, I_i)$ and (I_0, V_i). Any of these
choices of independent variables is good and the only condition is that functions
corresponding to the choice are real. In actual practice, it has been observed that for
a BJT the choice of I_i *and* V_0, i.e. of input current and output voltage as independent variables is best suited as the corresponding functions are simple and
generally have real values. With this choice, one may write the following relations:

$$V_i = F_1(I_i, V_0) \tag{6.60a}$$

And

$$I_0 = F_2(I_i, V_0) \tag{6.60b}$$

Here F_1 and F_2 are two functions of I_i and V_0. Equations (6.60a) and (6.60b)
tells that changes in the values of I_i *and* V_0 will produce a change in the value of V_i
and I_0. Therefore, the total change δV_i in the magnitude of V_i may be written as

$$\delta V_i = \frac{\partial V_i}{\partial I_i} \delta_{I_i} + \frac{\partial V_i}{\partial V_0} \delta_{V_0} \tag{6.61a}$$

And similarly,

$$\delta I_0 = \frac{\partial I_0}{\partial I_i} \delta_{I_i} + \frac{\partial I_0}{\partial V_0} \delta_{V_0} \tag{6.61b}$$

Equations (6.61a) and (6.61b) holds only if partial derivatives $\frac{\partial V_i}{\partial I_i}, \frac{\partial V_i}{\partial V_0}, \frac{\partial I_0}{\partial I_i}$ and $\frac{\partial I_0}{\partial V_0}$
remain constant, respectively, over the corresponding changes δ_{I_i} and δ_{V_0} in I_i and
V_0. This is possible if one considers only small signals.
V_i, I_i, V_0 *and* I_0 in (6.61a) and (6.61b) are the values of total voltage and of total
current at the input and the output ports. Therefore, $\delta V_i, \delta I_i, \delta V_0$ *and* δI_0 may
represent the AC components of voltages and currents or the applied small signals.
That is if AC signals are represented as; $\delta V_i = v_i, \delta I_i = i_i, \delta V_0 = v_0,$ *and* $\delta I_o = i_o,$ *then* the above two equations may now be written as

$$v_i = h_{11}i_i + h_{12}v_0 \tag{6.62a}$$

And similarly,

$$i_0 = h_{21}i_i + h_{22}v_0 \tag{6.62b}$$

Where $\frac{\partial V_i}{\partial I_i} = h_{11}, \frac{\partial V_i}{\partial V_0} = h_{12}, \frac{\partial I_0}{\partial I_i} = h_{21}$ and $\frac{\partial I_0}{\partial V_0} = h_{22}$

It may be noted, that h_{11} represents the ratio of the change in voltage to change in current and therefore, has the units of resistance 'Ohm' (or Ω). Likewise, h_{12} and h_{21} are ratios of two voltages and of two currents, respectively, and hence are dimensionless. The coefficient h_{22} ($=\frac{\partial I_0}{\partial V_0}$) is the ratio of current to voltage and, hence, represents an admittance with units Siemens (or Mho, or Ω^{-1}). The coefficients appearing in Equations (6.62a) and (6.62b) have different dimensions and units; they are called hybrid parameters and the model based on these parameters the hybrid model.

The four coefficients or generalized (hybrid) h- parameters are defined by the following expressions:

$h_{11} = \frac{v_i}{i_i}$, *when* $v_o = 0$, is called input impedance with output short-circuited.

$h_{12} = \frac{v_i}{v_0}$, *when* $i_i = 0$, is called the reverse voltage gain or reciprocal of voltage gain when the input terminal is open.

$h_{21} = \frac{i_0}{i_i}$, *when* $v_o = 0$, is called the forward current gain when output is short-circuited.

$h_{22} = \frac{i_0}{v_0}$, *when* $i_i = 0$, is called the output admittance or reciprocal of output impedance when input terminals are open.

Since a BJT amplifier may be connected in three different configurations, the h-parameters are also designated accordingly, as given below:

1. Common emitter (CE) configuration

$$v_{BE} = h_{ie}i_B + h_{re}v_{CE} \tag{6.63a}$$

$$i_e = h_{fe}i_B + h_{oe}v_{CE} \tag{6.63b}$$

2. Common base (CB) configuration

$$v_{EB} = h_{i_b}i_E + h_{rb}v_{CB} \tag{6.64a}$$

$$i_C = h_{fb}i_E + h_{ob}v_{CB} \tag{6.64b}$$

3. Common collector (CC) configuration

$$v_{BC} = h_{ic}i_B + h_{rc}v_{EC} \tag{6.65a}$$

$$i_e = h_{fc}i_B + h_{oc}v_{EC} \tag{6.65b}$$

Hybrid equivalent circuits for the common emitter (CE), common base (CB) and common collector configurations are shown in Fig. 6.41.

Hybrid parameters for transistors are generally provided by the manufacturer either in their manuals or datasheets. Typical values of h-parameters are given in Table 6.2.

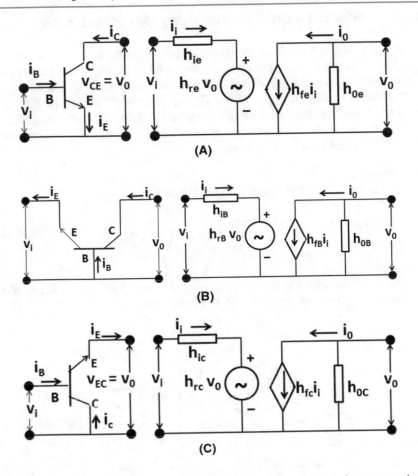

Fig. 6.41 Hybrid equivalent circuits for CE, CB and CC configurations are shown, respectively, in figures (A), (B) and (C)

Table 6.2 Typical values of hybrid parameters

Hybrid parameter	Common emitter (CE) configuration	Common base (CB) configuration	Common collector (CC) configuration
hi	1100 Ω	20 Ω	1100 Ω
h_f	50	-0.98	-50
h_r	2.5×10^{-4}	3.0×10^{-4}	1
h0	25×10^{-6} Siemens	0.5×10^{-6} Siemens	25×10^{-6} Siemens

6.9.4   Analysis of a BJT Amplifier Using Hybrid Parameters

A BJT amplifier can be made by taking a BJT, properly biasing it so that the operating point remains in the middle of the forward active region and connecting a signal source in the input and a load in the output, as shown in Fig. 6.42.

The hybrid equivalent circuit of the CE amplifier is shown in Fig. 6.43. Figure 6.44 shows how the output section may be simplified by calculating R_{eq}, the equivalent impedance of the parallel combination of $1/h_{oe}$ and Z_L, and the output current i_L is equal to $-i_0 = -h_{fe}i_i = -h_{fe}i_B$.

(a) **Current gain**

The current gain of the amplifier stage is defined as

$$A_i \equiv \frac{i_L}{i_i} = \frac{-i_0}{i_i} = \frac{i_L}{i_B} = -\frac{-h_{fe}}{(1+h_{oe}Z_L)} \tag{6.66a}$$

(b) **Input impedance**

The input impedance Z_i is defined as the impedance looking into the amplifier through the input terminals (A, A) in Fig (6.29d),

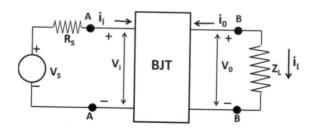

Fig. 6.42   A BJT CE-amplifier circuit

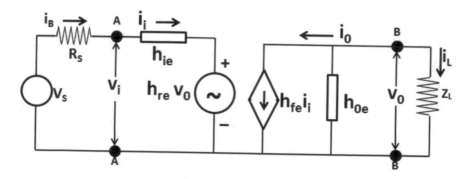

Fig. 6.43   A BJT CE-amplifier hybrid equivalent circuit

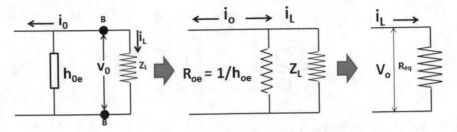

$$R_{eq} = Z_L/(1+h_{oe}Z_L); \quad I_L = -i_o = -h_{fe}i_i = -h_{fe} i_B;$$
$$V_o = R_{eq} i_L = -(Z_L h_{fe} i_B)/(1+h_{oe} Z_L)$$

Fig. 6.44 Simplification of output section

$$Z_i = \frac{v_i}{i_i} = \frac{h_{ie}i_B + h_{re}v_0}{i_B} = h_{ie} + \frac{h_{re}(Z_L i_L)}{i_B} = h_{ie} + \frac{h_{re}v_0}{i_B}$$

Or

$$Z_i = h_{ie} + \frac{h_{re}}{i_B}\left(-\frac{Z_L h_{fe} i_B}{1+h_{oe}Z_L}\right) = h_{ie} - \frac{h_{re}h_{fe}Z_L}{1+h_{oe}Z_L}$$

Or

$$Z_i = h_{ie} - \frac{h_{re}h_{fe}}{\frac{1}{z_l} + h_{oe}} = h_{ie} - \frac{h_{re}h_{fe}}{Y_L + h_{oe}} \tag{6.66b}$$

(c) **Output admittance**

Output admittance Y_o is defined as the ratio of the output current to the output voltage when input source is short-circuited, i.e.,

$$Y_0 \equiv \left(\frac{i_o}{v_0}\right)_{(v_s=0)}$$

To calculate Y_o, we apply KVL to the input loop to obtain

$$(R_s + h_{ie})i_i + h_{re}v_0 - v_s = 0$$

If the source v_s is short-circuited then $v_s = 0$, and the above equation reduces to

$$\left(\frac{i_i}{v_0}\right)_{(v_s=0)} = -\frac{h_{re}}{(R_s + h_{ie})} \tag{6.66c}$$

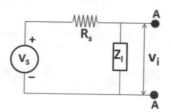

Fig. 6.45 Part of the input section

If one looks at the output loop, it is found that the output current i_o is made of two components: a component ($h_{fe}\, i_i$)passing through the dependent source and the other component ($h_{oe}\, v_o$) passing through the admittance h_{oe}. Therefore,

$$i_o = h_{fe} i_i + h_{oe} v_o$$

Or

$$\frac{i_o}{v_o} = h_{oe} + h_{fe} \frac{i_i}{v_o}$$

Substituting the value of $\left(\frac{i_i}{v_o}\right)_{(v_s=0)}$ in the above equation from (6.66c), one gets

$$Y_0 \equiv \left(\frac{i_o}{v_o}\right)_{(v_s=0)} = h_{oe} - h_{fe} \frac{h_{re}}{(R_s + h_{ie})} \qquad (6.66\text{d})$$

(d) **Voltage gain**

Voltage gain A_V is defined as the ratio of output voltage to the input voltage,

$$A_v = \frac{v_o}{v_i} = \frac{Z_L i_L}{Z_i i_i} = \frac{Z_L A_i}{Z_i} \qquad (6.66\text{e})$$

Where Z_i is the input impedance and A_i the current gain.

(e) **Voltage gain** A_{vs} taking into account the source resistance R_s

$$A_{vs} = \frac{v_o}{v_s} = \frac{v_o}{v_i} \frac{v_i}{v_s} = A_v \frac{Z_i}{Z_i + R_s} \text{(see Fig.(6.45))} \qquad (6.66\text{f})$$

Solved example (SE6.6) Draw the hybrid equivalent circuit for an npn BJT amplifier in common emitter configuration for which the hybrid parameters are $h_{ie}=1k\Omega$; $h_{re}=0.5 \times 10^{-4}$; $h_{oe}=100\times10^{-6}$ mho, $Z_L =1$ kΩ, Rs $= 1$ kΩ. Determine the input and output impedances and current and voltage gains.

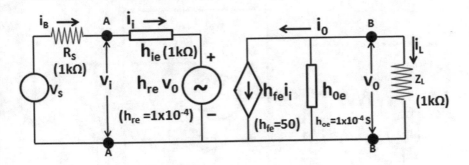

Fig. SE6.6 Hybrid equivalent of npn transistor in CE configuration

Solution: The hybrid equivalent of the amplifier is shown in Fig. SE6.6.

The current gain is given by $A_i \equiv \frac{i_L}{i_i} = -\frac{-h_{fe}}{(1 + h_{oe}Z_L)} = -\frac{50}{1 + 1 \times 10^{-4} \times 1 \times 10^3} = -45.45$

Input impedance $Z_i = h_{ie} - \frac{h_{re}h_{fe}}{Y_L + h_{oe}} = 1 \times 10^3 - \frac{1 \times 10^{-4} \times 50}{1 \times 10^{-3} + 1 \times 10^{-4}} = 995.46\Omega$

Output impedance $Z_0 = \dfrac{1}{\left(h_{oe} - h_{fe} \frac{h_{re}}{(R_s + h_{ie})}\right)} = \dfrac{1}{975 \times 10^{-7}} = 10.26k\Omega$

$$\text{Voltage gain } A_v = \frac{Z_L A_i}{Z_i} = \frac{1 \times 10^3(-45.45)}{995.46} - -45.66$$

6.10 General Approach to the Analysis of BJT Amplifier

A single-stage BJT amplifier using npn transistor in CE configuration is shown in Fig. 6.46.

As may be seen in this figure, there are three capacitors C_1, C_2 and C_E. Capacitors C_1 and C_2 are called coupling capacitors and block DC levels, allowing only AC input signal to reach the amplifier and only AC output signal to reach the next stage of amplification or to the load R_L. The values of these capacitors are so chosen that they offer negligible reactance to the frequencies of AC input and output signals and behave as short-circuits for these AC frequencies. On the other hand, they offer infinite resistance to the DC. The capacitance C_E is such that it offers a very low reactance path for AC and thus effectively removes the emitter resistance R_E for AC signals. This ensures that the Q-point (DC) of the amplifier does not shift due to AC signals.

The analysis of an amplifier stage may be done in steps: (a) in the first step one may draw the DC equivalent of the given amplifier using an appropriate large-signal or DC transistor model to find the operating or Q-point. (b) In the second step, the AC equivalent circuit may be drawn on the basis of some small-signal model and then to find the amplifier parameters, input impedance,

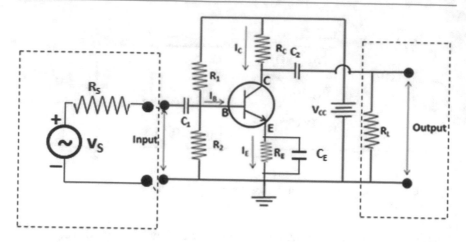

Fig. 6.46 Single-stage BJT amplifier using npn transistor in CE configuration

output impedance, current and voltage gains, etc. from the expressions of the model. The DC and AC equivalents may be drawn using the following rules.

DC analysis

- Draw DC equivalent by replacing all capacitors with open circuits and inductors with short circuits.
- Use an appropriate DC or large-signal transistor model to find the operating or Q-point, i.e. the values of collector current I_C and the collector–emitter voltage V_{CE} at the operating point.

AC Analysis

- Draw AC equivalent circuit using a small-signal model by replacing all capacitors with short circuits, inductors by open circuits, DC voltage sources by ground connections and DC current sources by open circuits.
- Replace transistor with its small-signal model
- Using small-signal AC equivalent determine the characteristic parameters of the amplifier
- Finally, combine the results of the DC and ACAC analysis to obtain total voltages and currents in the network.

The following example has been included to illustrate the methodology of amplifier analysis.

Figure 6.47 shows a typical single-stage BJT amplifier using npn transistor in CE configuration. It is given that β for the transistor is 100 and $V_A = 50$ V.

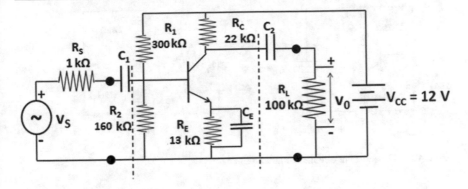

Fig. 6.47 Typical single stage BJT amplifier using npn transistor in CE configuration

DC equivalent As the first step of analysis, we draw the DC equivalent of the amplifier by removing capacitors by open circuits. The DC equivalent is shown in Fig. 6.47a, which may be further reduced to the circuit shown in Fig. 6.48, where

$$V_B = \frac{R_2}{R_1 + R_2} V_{CC} = \frac{160 \times 10^3 \times 12}{460 \times 10^3} = 4.17V \ and \ R_B = 160k\Omega \| 300k\Omega = 104.38k\Omega$$

DC equivalent circuit will be used to find the coordinates of the Q-point. Let us first find the value of the base current I_B. Since a BJT amplifier operates in forward active region, with base–emitter junction forward biased and in 'on' condition, the base–emitter voltage V_{BE} may be taken equal to the V_{on} which has the value 0.7 V for silicon device and 0.3 V for germanium device. As the type of the transistor is not mentioned explicitly, it is a convention to take it silicon based. Therefore, V_{BE} =0.7 V. Further, the collector current is β times the base current and β is given as 100, hence I_C =100 I_B; and $I_E = I_C + I_B$ = (100+1) I_B. Applying KVL to base–emitter loop, one gets

$$R_B I_B + R_E (101 I_B) + 0.7 = 4.17V$$

Or

$$104.38 \times 10^3 \times I_B + 13 \times 10^3 (101) I_B = 4.17 - 0.7 = 3.47V$$

Or

$$I_B = 2.45 \times 10^{-6} A = 2.45 \mu A \tag{6.67}$$

Hence,

$$I_C = 2.45 \times 10^{-6} \times 100 A = 0.245 \times 10^{-3} A = 0.245 mA$$

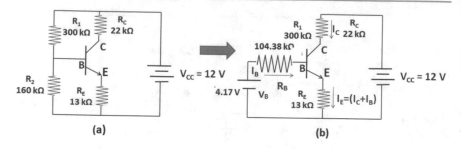

Fig. 6.48 DC equivalent circuits of the amplifier

And

$$I_E = I_C + I_B = (0.245 + 0.00245)mA = 0.247mA$$

To find the value of V_{CE}, the voltage between collector and emitter, we apply KVL to the Collector emitter section to get

$$R_c I_C + V_{CE} + R_E I_E = V_{cc}$$

Or

$$22x10^3 x0.245x10^{-3} + V_{CE} + 13x10^3 x0.247x10^{-3} = 12$$

Hence,

$$V_{CE} = 3.39V$$

The coordinates of the Q-point are: $V_{CE} = 3.39V$; $I_C = 0.245mA$; $I_B = 2.45\mu A$

AC analysis using transconductance or hybrid-pi model

Having found the coordinates of the Q-point from the DC analysis let us carry out AC analysis using the hybrid-pi model for small signals. The ACAC equivalent circuit of the amplifier using hybrid-pi model is shown in Fig. 6.49.

As may be observed in this figure, there are three important parameters that need to be calculated from the data obtained from the DCDC analysis and the given values of parameters.

The transconductance $g_m^s = \dfrac{(I_C)_Q}{V_T} = \dfrac{0.245 \times 10^{-3}A}{26mV} = 9.42 \times 10^{-3} \; Siemens$

$$(6.67a)$$

Also,

$$r_\pi^s = \frac{\beta}{g_m^s} = \frac{100}{9.42x10^{-3}} = 10.62x10^3\Omega \qquad (6.67b)$$

And

$$r_o^s = \frac{V_A}{(I_C)_Q} = \frac{50}{0.245x10^{-3}} = 204x10^3\Omega \qquad (6.67c)$$

Here

$(I_C)_Q$ and $(V_{CE})_Q$ represent, respectively, the values of collector current and the voltage drop between collector and emitter at Q–point.

The circuit shown in Fig. 6.49 may be further reduced by introducing resistances R_{in} and R_{out}, that is given by (see Fig. 6.33);

$$R_{in} = parallel\ combination\ of\ R_B\ and\ r_\pi^s = 9.68x10^3\Omega$$

And

$$R_{out} = parallel\ combination\ of\ r_o^s, R_C\ and\ R_L = 16.57x10^3\Omega$$

The amplifier voltage gain $A_V = \frac{v_0}{v_i} = -g_m^s R_{out} = -9.42x10^{-3}x16.57x10^3 = -156$

The overall voltage gain or source to output gain $A_v^S = \frac{v_0}{v_s} = \frac{v_0}{v_i}\frac{v_i}{v_s} = A_v\frac{R_B}{R_B+R_S} = -156x\frac{104.38}{104.38+1} = -154.5$

Or

$$A_v^S = -154.5$$

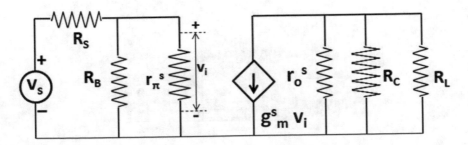

Fig. 6.49 AC equivalent circuit using transconductance small-signal model

Fig. 6.50 Reduced AC
equivalent circuit

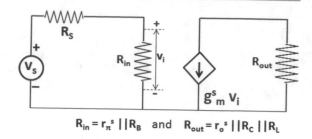

$$R_{in} = r_\pi^s \, || \, R_B \quad \text{and} \quad R_{out} = r_o^s \, || \, R_C \, || \, R_L$$

6.11 Ebers–Moll Model for BJT

So far we discussed three small-signal or linear models for a bipolar junction transistor and also worked out some examples to show their application. We now discuss in brief a large-signal or non-linear model for a BJT for the sake of completeness.

The npn bipolar junction transistor can be considered essentially as two pn junctions placed back-to-back, with the base p-type region being common to both diodes. This can be viewed as two diodes having a common third terminal. However, the two diodes are not in isolation, but are interdependent. This means that the total current flowing in each diode is influenced by the conditions prevailing in the other. In isolation, the two junctions would be characterized by the normal Diode Equation with a suitable notation used to differentiate between the two junctions. However, when the two junctions are combined to form a transistor, the base region is shared internally by both diodes even though there is no external connection to it. As seen previously, in the forward active mode, α_F times the emitter current reaches the collector. This means that α_F of the diode current passing through the base–emitter junction contributes to the current flowing through the base–collector junction. Typically, α_F has a value of between 0.98 and 0.99. This is shown as the forward component of current as it applies to the normal forward active mode of operation of the device. Note this current is shown as a conventional current in Fig. 6.51. It is equally possible to reverse the biases on the junctions to operate the transistor in the "reverse active mode". In this case, α_R times the

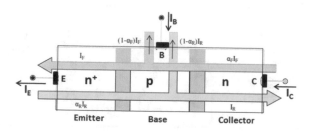

Fig. 6.51 Currents through the combined junctions

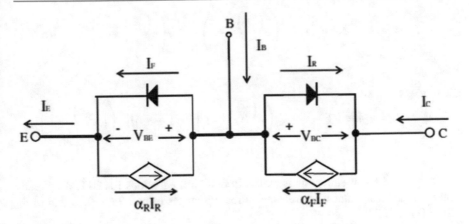

Fig. 6.52 Ebers–Moll equivalent of an npn BJT

collector current will contribute to the emitter current. For the doping ratios normally used the transistor will be much less efficient in the reverse mode and α_R would typically be in the range 0.1–0.5.

Ebers–Moll model is an attempt to create an electrical model of the device having two diodes whose currents are determined by normal diode law but their interdependence is modelled through transfer ratios α_F and α_R. Two dependent current sources are used to represent the interaction of the junctions as shown in Fig. 6.52.

Diode currents are given as

$$I_F = I_{ES}\left(e^{\left(\frac{V_{BE}}{V_T}\right)} - 1\right) \tag{6.68a}$$

And

$$I_R = I_{CS}\left(e^{\left(\frac{V_{BC}}{V_T}\right)} - 1\right) \tag{6.68b}$$

Where I_{ES} and I_{CS} are saturation current that are generally in nA.
Applying KCL to the circuit gives

$$I_E = [I_F - \alpha_R I_R] = [I_{ES}\left(e^{\left(\frac{V_{BE}}{V_T}\right)} - 1\right) - \alpha_R I_{CS}\left(e^{\left(\frac{V_{BC}}{V_T}\right)} - 1\right)] \tag{6.69a}$$

$$I_C = [\alpha_F I_F - I_R] = [\alpha_F I_{ES}\left(e^{\left(\frac{V_{BE}}{V_T}\right)} - 1\right) - I_{CS}\left(e^{\left(\frac{V_{BC}}{V_T}\right)} - 1\right)] \qquad (6.69b)$$

And

$$I_B = [I_E - I_C] = [(1 - \alpha_F)I_{ES}\left(e^{\left(\frac{V_{BE}}{V_T}\right)} - 1\right) + (1 - \alpha_R)I_{CS}\left(e^{\left(\frac{V_{BC}}{V_T}\right)} - 1\right)]$$

$$(6.69c)$$

The above set of equations is called Ebers–Moll equations for BJT. As already said, the typical values of transfer ratios α_F ranges from 0.98 to 0.99 and of α_R from 0.1 to 0.5.

6.11.1 Modes of Operation

Ebers–Moll model is a good steady-state (DC) model for a BJT which has the capability of specifying the conduction state for any mode of operation of the transistor.

(a) **Forward active mode** In this mode the base–emitter junction is forward biased, V_{BE} is +ve and $e^{\left(\frac{V_{BE}}{V_T}\right)} \gg 1$. The base-collector junction is reverse biased, V_{BC} is –Ve and $e^{\left(\frac{V_{BC}}{V_T}\right)} \ll 1$.

Then from the model,

$$I_E \approx \left[I_{ES}\left(e^{\left(\frac{V_{BE}}{V_T}\right)}\right)\right] \quad \text{Relatively large} \qquad (6.70)$$

$$I_C \approx \left[\alpha_F I_{ES}\left(e^{\left(\frac{V_{BE}}{V_T}\right)}\right)\right] \quad \text{Relatively large} \qquad (6.71)$$

And

$$I_B \approx \left[(1 - \alpha_F)I_{ES}\left(e^{\left(\frac{V_{BE}}{V_T}\right)}\right)\right] \approx (1 - \alpha_F)I_E. \quad \text{Relatively small} \qquad (6.72)$$

When there is no bias on three terminals of the BJT, the emitter, the base and the collector sections have nearly fixed concentrations of minority carriers that is uniform throughout the given section. For example, in an npn transistor, the

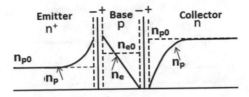

Fig. 6.53 Minority carrier charge distribution profile for forward active mode in npn transistor

concentration of holes in emitter and collector regions and of electrons in the base region in the initial unbiased state may be represented, respectively, by n_{p0} and n_{e0}. However, on biasing the transistor, the minority carrier concentration in each of the three sections change with distance and attains a new equilibrium value. It is interesting to study how the minority charge carrier concentration changes with distance in each section. The variation of minority carrier concentrations with distance for forward active operation mode in an npn transistor is shown in Fig. 6.53.

(b) **Reverse active mode** In this mode, base-emitter junction is reverse biased, V_{BE} is −ve and $e^{\left(\frac{V_{BE}}{V_T}\right)} \ll 1$. Base collector junction is forward biased, V_{BC} is +ve, and $e^{\left(\frac{V_{BC}}{V_T}\right)} \gg 1$. Hence from model equations,

$$I_E \approx -\alpha_R \left[I_{CS} \left(e^{\left(\frac{V_{BC}}{V_T}\right)} \right) \right] \quad \text{Moderately high} \tag{6.73}$$

$$I_C \approx -\left[I_{CS} \left(e^{\left(\frac{V_{BC}}{V_T}\right)} \right) \right] \quad \text{Moderate} \tag{6.74}$$

And

$$I_B \approx \left[(1 - \alpha_F) I_{CS} \left(e^{\left(\frac{V_{BC}}{V_T}\right)} \right) \right] \quad \text{may be as high as } 0.5|I_C| \tag{6.75}$$

This mode is not suitable for amplification and BJT in reverse active mode is often used for steering currents in logic switching circuits etc. Minority carrier charge profile for this mode is shown in Fig. 6.54.

(c) **Cut-off mode** In cut-off mode base-emitter junction is not biased, $V_{BE}=0$ V, and $\left(e^{\frac{V_{BE}}{V_T}} \right) = 1$; Also base-collector junction is reverse biased, V_{BC} is negative and $\left(e^{\frac{V_{BC}}{V_T}} \right) \ll 1$. Therefore, from model equations,

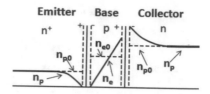

Fig. 6.54 Minority carrier charge profile for reverse active mode of an npn transistor

$I_E \approx \alpha_R I_{CS}$ Reverse saturation or leakage current $\approx$ nA
$I_C \approx I_{CS}$ Reverse saturation or leakage current $\approx$ nA
And $I_B \approx -(1 - \alpha_R) I_{CS}$ Reverse saturation or leakage current $\approx$ nA
From the above analysis, it is evident that under cut-off mode BJT behaves like an open switch as
shown in Fig. 6.55. Further, the minority charge carrier profile for cut-off mode is shown in Fig. 6.56.

(d) Saturation mode In this mode, BE and BC junctions are both forward biased, $V_{BE} \approx 0.8$ V, $V_{BC} \approx 0.7$ V and both $\left(e^{\frac{V_{BE}}{V_T}}\right)$ and $\left(e^{\frac{V_{BC}}{V_T}}\right)$ are $\gg 1$. Hence

$$I_E \approx I_{ESe}^{\left(\frac{V_{BE}}{V_T}\right)} - \alpha_R I_{CSe}^{\left(\frac{V_{BC}}{V_T}\right)} \tag{6.76}$$

Since only nA current flows between emitter and collector, the device is like an open switch

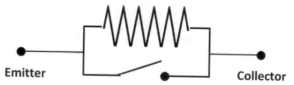

Fig. 6.55 In cut-off mode BJT is like an open switch

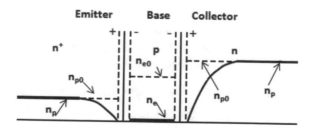

Fig. 6.56 Minority charge profile or cut-off mode

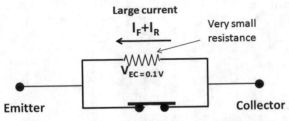

Fig. 6.57 In saturation mode, a BJT is like an 'on' switch

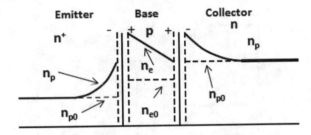

Fig. 6.58 Charge profile of minority carriers in saturation mode

$$I_C \approx \alpha_F I_{ES} e^{\left(\frac{V_{BE}}{V_T}\right)} - I_{CS} e^{\left(\frac{V_{BC}}{V_T}\right)} \qquad (6.77a)$$

And

$$I_B \approx (1 - \alpha_F) I_{ES} e^{\left(\frac{V_{BE}}{V_T}\right)} + (1 - \alpha_R) I_{CS} e^{\left(\frac{V_{BC}}{V_T}\right)} \qquad (6.77b)$$

In this case sufficient current flows through the device as both junctions are forward biased, also, the potential drop across the device that is from emitter to collector V_{EC} is only around $0.8 - 0.7 = 0.1$ V. Therefore, the transistor is like a closed or 'on' switch as shown in Fig. 6.57.

The charge profile of different sections of the BJT in saturation mode is shown in Fig. 6.58.

6.12 Summary of BJT Amplifiers

In electronics, one often encounters very small and weak signals which need amplification and strengthening before further processing. Since a single BJT is inherently an amplifier, most literature on bipolar junction transistors is devoted to

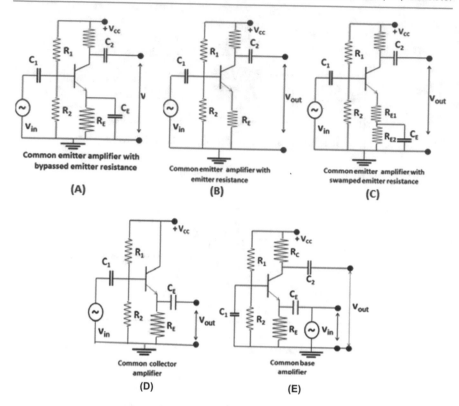

Fig. A, B, C, D and E Circuit diagrams for different configurations

their amplifier aspect. A BJT may be used in three different configurations each of which has its own characteristic properties as amplifier. Amplifier using an npn transistor in common emitter configuration is mostly used on account of its fast recovery time and moderate values for both current and voltage gains. In this writeup also only npn-BJT amplifier in CE configuration has been discussed in detail. Space and time requirements do not permit to go into a detailed discussion of other configurations. However, in the following, a summary of the essential features of BJT amplifiers in CC and CE configurations is presented. Symbols used here have their usual meaning. Further,

$$r_0 = \frac{V_A}{I_C}; g_m = \frac{I_C}{V_T}; r_\pi = \frac{\beta}{g_m}; r_E = \frac{25mV}{I_E}$$

6.12.1   Common Emitter

(a)   **With bypassed emitter resistance**

$$No-load\ voltage\ gain\ A_v = -\frac{\beta}{r_\pi}[R_C||r_0] \approx -\frac{\beta}{r_\pi}R_C = -\frac{R_C}{r_e}$$

$$Input\ resistance\ R_{in} = R_B||r_\pi;$$

$$Output\ resistance\ R_{out} = r_0$$

$$Lower\ cut-off\ frequency\ f_{cut}^l = \frac{1}{2\pi R_i C_1} + \frac{1}{2\pi R'_E C_E}$$

$$where\ R'_E = R_E||\left(r_E + \frac{R_B}{\beta}\right)$$

(b)   **With emitter resistance**

$$No-load\ voltage\ gain\ A_v = -\frac{R_C}{R_E + r_E} \approx -\frac{R_C}{R_E}$$

$$R_{in} = R_B||[\beta\{r_E + R_E\}];\quad R_0 = r_e\left(1 + \frac{R_E}{r_E}\right)$$

$$f_{cut}^l = \frac{1}{2\pi R_i C_1}$$

(c)   **Swamped emitter resistance**

$$No-load\ voltage\ gain\ A_v = -\frac{R_C}{R_{E1} + r_E} \approx -\frac{R_C}{R_{E1}}$$

$$R_{in} = R_B||[\beta\{r_E + R_{E1}\}];\ R_0 = r_e\left(1 + \frac{R_{E1} + R_{E2}}{r_E}\right)$$

$$f_{cut}^l = \frac{1}{2\pi R_i C_1} + \frac{1}{R''_E C_E}$$

$$Where\ R''_E = R_{E2}||\left\{R_{E1} + r_E + \frac{R_B}{\beta}\right\}$$

6.12.2 Common Collector

Emitter follower

$$No - load\ voltage\ gain\ A_v = -\frac{R_C}{R_{E1} + r_E} \approx -\frac{R_C}{R_{E1}}$$

$$R_{in} = R_B || [\beta \{r_E + R_{E1}\}];\ R_0 = r_E \left(1 + \frac{R_{E1} + R_{E2}}{r_e}\right)$$

$$f_{cut}^l = \frac{1}{2\pi R_i C_1} + \frac{1}{R_E'' C_E}$$

6.12.3 Common Base

Buffercircuit with current gain ≈ 1 Insert Fig(E) here

$$No\text{-}load\ voltage\ gain\ A_v = \frac{R_C(g_m r_0 + 1)}{R_C + r_0} \approx g_m R_C$$

$$Short\text{-}circuit\ current\ gain\ A_i = \frac{r_\pi + \beta r_0}{r_\pi + (\beta + 1) r_0} \approx 1$$

$$R_{in} = \frac{r_E(r_0 + R_C)}{r_0 + r_E + \frac{R_C}{\beta + 1}} \approx r_E \approx \frac{1}{g_m}$$

$$R_{out} = R_C || ([1 + g_m(r_\pi || R_B)] r_0 + r_\pi || R_B) \approx R_C || r_0$$

Note: In case bias is provided by a current source and R_B is not present then take $R_B = \infty$. In case a load R_L is present in the circuit, R_C should be replaced by ($R_C ||$ R_L) in all expressions.

Self-assessment question SAQ: What is the main difference between the hybrid-π and Ebers–Moll model? Are they comparable to each other?

Self-assessment question SAQ: What is the main limitation of hybrid model that hybrid-π model does not have?

Solved example (SE6.7) An npn silicon transistor is used to design an amplifier in CE configuration with the following data. Determine the Q-point and draw the DC load line. Hence show that the amplifier is in forward active region. What can be the maximum peak-to-peak value of the undistorted output signal?

Given data: $V_{CC} = 15$ V; $\beta = 150$; $V_{BE} = 0.7$ V; $R_E = 2.7$ kΩ; $R_C = 4.7$ kΩ; $R_1 = 47$ kΩ; $R_2 = 10$ kΩ; $R_L = 47$ kΩ; $R_s = 100$ Ω;

Solution: *Let us first calculate the values of V_B, and R_B*

$$V_B = \frac{R_2}{R_1+R_2}V_{CC} = \frac{10k\Omega}{(47+10)k\Omega}15V = 2.63V; R_B = R_1\|R_2 = 8.25k\Omega$$

Next by applying KVL to base–emitter section, we calculate I_B

$$R_B I_B + V_{BE} + R_E I_E = V_B; or 8.25x10^3 I_B + 0.7 + 2.7x10^3 x(151)I_B = 2.63V$$

$$I_B = 4.64\mu A$$

$I_C = 150x4.64\mu A = 0.696mA$ and $I_E = 0.706mA$.
To find the value of V_{CE} at Q-point we apply KVL to the collector–emitter loop
to get.

$$R_C I_C + (V_{CE})_Q + R_E I_E = V_{CC} or$$

$$(V_{CE})_Q = 15 - \left[(4.7x10^3 x0.696x10^{-3}) + (2.7x10^3 x0.706x10^{-3})\right] = 9.83V$$

It may be observed that $I_B > 0$ and $(V_{CE})_Q > 2$ V; therefore, the Q-point is in the
forward active region. If I_B is Zero then Q-point may be in cut-off region. Also, If
$I_C \approx V_{CC}/(R_E+R_C)$, then the Q-point may be in the saturation region.
Maximum value of IC will when $V_{CE}=0$, and $(I_C)max = V_{CC}/(R_E+R_C)=15V/(4.7$
$+2.7)k\Omega=2.02$ mA.
The DC load line is obtained by joining the $(I_C)_{max} =2.02$ mA point on Y axis
(representing collector current) with the point $V_{CE}=V_{CC}=15$ V on the X-axis representing V_{CE}, as shown in Fig. SE6.7.

Fig. SE6.7 DC load line of
example (SE6.7)

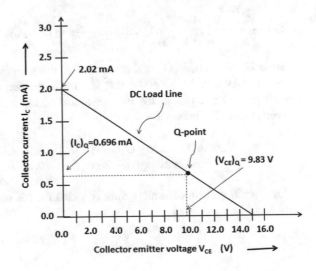

The coordinates of the Q-point are: $I_C = 0.696mA$; $V_{CE} = 9.83V$; $I_B = 4.64\mu A$

As may be seen in Fig. SE6.7, the Q-point is not in the middle of the load line but is towards the lower edge (cut-off region). The maximum negative swing of the output signal towards the cut-off point can be $(15-9.83) = 5.17$ V, hence the maximum swing towards both directions $= 2 \times 5.17 = 10.34$ V. If the peak-to-peak value of the output signal is larger than this value, the negative part of the signal will be clipped.

6.13 Gain in dB, Low-Pass and High-Pass Filters and Frequency Response

6.13.1 Gain in dB

Voltage or current gain of an amplifier is the ratio of the output voltage (or current) to the input voltage (or current) and has no units, being a ratio. However, gains are often represented in a special unit called decibel 'dB' in short, which is defined as

$$Voltage\ gain\ A_v\ in\ dB \equiv 20\ \log\left(\frac{V_{out}}{V_{in}}\right) \qquad (6.78a)$$

It means that an amplifier that amplifies the input signal 10 times will have a gain of 20 dB that amplifies to 100 times will have a gain of 40 dB and so on. One advantage of using gain in dB units is that the overall gain of a multistage amplifier (in dB) may be obtained by **adding** the gains (in dB) of individual stages. This may be clear from the following example. Suppose an amplifier has n-stages, where input voltage is V_{in}. Let output voltages of the first to n^{th} stage be $V_1, V_2, V_3, \ldots V_n$. The overall gain G may be written as

$$G = \frac{V_n}{V_{in}} = \frac{V_1}{V_{in}} \cdot \frac{V_2}{V_1} \cdot \frac{V_3}{V_2} \cdots \cdots \cdots \frac{V_n}{V_{n-1}} = G_1 G_2 G_3 \cdots \cdots \cdots G_n \quad (6.78b)$$

Here $G_1, G_2, G_3, \ldots G_n$ are, respectively, the gains of the I, II, III....and n^{th} stages. It may thus be observed that if gain is defined in a normal way then the overall gain G is equal to the multiplication of the individual gains. Now let us convert it into dB units, then

$$20\log G = 20\log(G_1 G_2 G_3 \cdots \cdots \cdots G_n)$$
$$= 20\log G_1 + 20\log G_2 + 20\log G_3 \cdots \cdots 20\log G_n$$

Or Overall gain in dB unit = Sum of gains (in dB unit) of individual stages.

Table 6.3 Voltage or current ratios to dB

Voltage or current ratio V_{out}/V_{in} or I_{out}/I_{in}	dB $20\log$ (V_{out}/V_{in}) Or $20\log$ (I_{out}/I_{in})	Voltage or current ratio $V_{out}/$ V_{in} or $I_{out}I_{in}$	dB $20\log$ (V_{out}/V_{in}) Or $20\log$ (I_{out}/I_{in})	Voltage or current ratio V_{out}/V_{in} or I_{out}/I_{in}	dB $20\log$ (V_{out}/V_{in}) Or $20\log$ (I_{out}/I_{in})
100 000	100	1 000	60	2	6
31 623	90	316.23	50	1.414	3
10 000	80	100	40	1.122	1
3 162	70	10	20	1	0
0.891	−1	0.1	−20	0.001	−60
0.707	−3	0.03162	−30	0.0003162	−70
0.5	−6	0.01	−40	0.0001	−80
0.316 2	−10	0.003162	−50	0.000 01	−100

Commonly Encountered dB Values

0 dB The reference level to which all +dB and −dB figures refer.

−3 dB Generally, used to specify the limits of bandwidth in amplifiers indicating the point. (frequency of the input signal) where the output voltage falls to $0.707(= 1/\sqrt{2})$ the maximum (mid-band) value and the output power falls to half of its maximum value.

−6 dB Voltage gain falls to half of its maximum value.

−20 dB Signal voltage amplitude divided by 10

−40 dB Signal voltage amplitude divided by 100

The following table provides dB values corresponding to some important output and input voltage (or current) ratios (Table 6.3).

6.13.2   High-Pass and Low-pass Filters

Filters are electronic devices that may allow electronic signals of a given frequency range to go ahead and do not allow signals of all other frequencies to pass through. Broadly speaking, filters may be classified into two types, the low- pass filter that allows **signals lower in frequency than a given frequency** f_{cut}^{ch} to pass to the filter

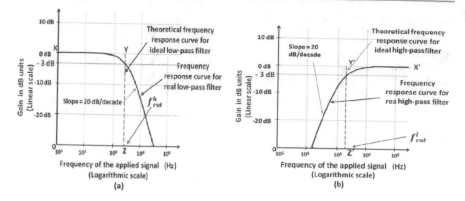

Fig. 6.59 Bode graphs (frequency vs. gain) for **a** low-frequency pass and **b** high-frequency pass filters.

output and inhabits signals of frequencies higher than f^h_{cut} to reach its output. The other class of filter is the high-pass filter, which let signals **higher in frequency than a given frequency** f^l_{cut} to reach its output but does not allow signals of lower frequencies than f^l_{cut} to reach the output. A combination of a high-pass and a low-pass filter makes a bandpass filter that allows signals lying in a frequency band (f^l_{cut} to f^h_{cut}) to pass to the output.

Graph showing the variation of the gain (ratio of the output to input voltage) of a circuit with the frequency of the input signal is called **Bode plot** or **frequency response characteristic**. In order to plot a large range of frequencies on a single graph one uses semi-logarithmic graph paper for drawing Bode plots. Frequencies are plotted on the X-axis that has a logarithmic scale while the gain is plotted on the Y-axis that has a linear scale. Bode curves for the low-pass and high-pass filters are shown in Fig. 6.59. Theoretically, an ideal low-pass filter should pass, without any attenuation in gain, signals up to some frequency f^h_{cut} and all other signals with frequency $f > f^h_{cut}$ must not appear in the output of the filter circuit, therefore, should be attenuated or grounded. Similarly, an ideal high-pass filter must pass all signals with frequencies $f > f^l_{cut}$ to the output but attenuate signals of frequencies $f < f^l_{cut}$ so that they may not appear at the filter output. The frequency f^l_{cut} or f^h_{cut} is called the cut-off frequency. Ideal frequency filters have a sharp cut at the cut-off frequency f_{cut}. However, the real filters do not have sharp cuts at cut-off frequency, but instead, the gain of the filter circuit starts falling a little before the cut-off frequency f_{cut} and gain becomes -3dB of the peak value at $f = f_{cut}$. The gain continues to fall at the rate of -20 dB per decade (gain falls by -20 dB for every change in frequency by a factor of 10, which corresponds to each cycle of the semi-log graph). A high-pass filter followed by a low-pass filter makes a bandpass filter allowing frequencies between f^l_{cut} and f^h_{cut} to reach the final output.

Simple filters may be assembled using the combination of resistance and capacitance. Such filters that have no source of energy are called passive filters.

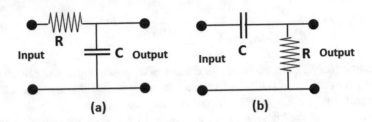

Fig. 6.60 Passive **a** low-pass **b** high-pass filters

Passive low-pass and high-pass filters are shown in Fig. 6.60. If v_{in} represents the voltage of the applied signal at the input of a low-pass filter (fig. a), then the voltage $v^L{}_o$ of the output signal will be given by

$$v_o^L = \frac{\text{impedance offered by the capacitance}}{\text{impedance offered by resistance} + \text{impedance offered by capacitance}}(v_{in})$$

Or $v_o^L = \frac{-jX_C}{R + (-jX_C)}(v_{in})$ where $X_C = \frac{1}{2\pi fC}$ is the reactance offered by the capacitance.

Therefore,

$$\text{the gain of the low} - \text{pass filter } G^L = \left|\frac{v_o^L}{v_{in}}\right| = \frac{X_C}{R - jX_C} = \frac{X_C}{\sqrt{R^2 + X_c^2}} \quad (6.79a)$$

It may be noted that in the denominator of (6.79a) the total impedance Z is obtained by taking the under the root of the sum of squares of resistance R and reactance X_C.

At low frequencies and for a small value of capacitance C, X_C may have a large value ($X_C \gg R$) and the gain G^L may be of the order of 1 (only slightly less than 1). Thus the capacitance will offer high reactance to low frequencies and, therefore, low frequencies will appear at the output, while for higher frequencies, X_C will become small, the capacitor will work like a short circuit and high frequencies will be grounded. Hence, the circuit shown in Fig (6.60a) works as a low-pass filter allowing low frequencies to appear at the output across the capacitor and high frequencies getting grounded. Let us now calculate the gain of the low-pass filer at the cut-off frequency $f_{cut} = (1/2\pi RC)$,

$$G_{fc}^L = \frac{\left(\frac{1}{2\pi(1/2\pi RC)C}\right)}{\sqrt{R^2 + \left(\frac{1}{2\pi(1/2\pi RC)C}\right)^2}} = \frac{R}{\sqrt{2}R} = \frac{1}{\sqrt{2}} = 0.707 = -3dB \quad (6.79b)$$

It may thus be observed that the gain of the low-pass filter falls to -3db from its original value at cut-off frequency. The cut-off frequency f_c is also called the corner frequency or the -3 dB frequency. Let us calculate the gain at a frequency $f = 10\ f_{cut}$.

$$G_{10fc}^{L} = \frac{R}{\sqrt{R^2 + (10R)^2}} = \frac{1}{\sqrt{101}} = \frac{1}{0.0998} \approx 0.1 = -20dB \qquad (6.79C)$$

Equation (6.79C) tells that with the change of frequency by a factor of 10 (called decade) gain falls by −20 dB. In technical terms, it is said that the gain of a filter falls or rolls-off at a rate of −20 dB per decade.

The analysis of the high-pass filter is similar; in this case, the output is obtained across the resistance R. The reactance of the capacitor is large at low frequencies, therefore, the series capacitor acts like an open switch and does not allow low frequencies to reach the output. However, at high frequencies, the capacitor behaves like a closed switch and the high-frequency signals appear in the output across the resistance. In this case also, the cut-off frequency f_{cut} is given by,

$f_{cut} = \frac{1}{2\pi RC}$; Where the gain falls to −3 dB. Further, the gain rolls off at the rate of −20 dB/decade.

In case of a low-pass filter, the phase of the output signal lags behind the input signal and the phase shift depends on the frequency of the input signal. For a signal that has a frequency equal to the cut-off frequency, the gain falls by −3dB and phase shifts by −45⁰. On the other hand, in case of high-pass filter, the phase of the signal at the output is ahead of the phase of the input signal, the amount of the phase shift depends on the frequency. For example, the phase shift of a signal of cut-off frequency is +45⁰.

6.13.3   Frequency Response of a Single-Stage BJT Amplifier

In this section, we shall study, as a representative case, the frequency response of an npn single-stage BJT amplifier in CE configuration. Figure 6.61 shows a typical frequency response characteristic (or Bode plot) for a single-stage BJT amplifier.

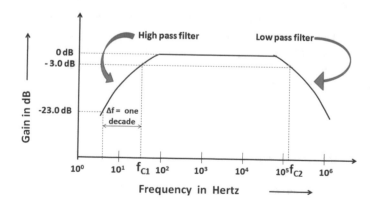

Fig. 6.61 Typical frequency response characteristic of a BJT

The curve can be divided into three distinct regions of frequencies. The central part of the curve lying between the **lower cut-off frequency f_{C1}** and the **upper cut-off frequency f_{C2}** is called the **mid-frequency region** and in this region the gain of the amplifier is constant over the whole range of frequencies. Gains in other frequency regions are normalized with respect to the gain in the mid-frequency region, which is assigned a value of 1 which is equivalent to (0 dB).

The frequency region below f_{C1} is called low-frequency region and the one having frequencies higher than f_{C2} the high-frequency region. As may be seen in the figure, the gain of the amplifier falls below the mid-frequency value both for the low and the high-frequency parts. Looking to the low-frequency part of the curve it appears as if the amplifier circuit behaves like a high-pass filter and for the high-frequency part as a low-pass filter.

Mid-frequency region Figure 6.26 shows the circuit diagram of a single-stage BJT amplifier using npn transistor in common emitter configuration. As may be seen in this figure, there are three capacitors C_1, C_2 and C_E that are connected externally. Apart from that three more capacitances C_{BE} of the depletion layer between base and emitter (also denoted by C_π), C_{BC} (also denoted by C_μ) of the base–collector junction and C_{EC} between collector and emitter, that are internal to a BJT, are also present as shown in Fig. 6.39. In the mid-frequency range all these six capacitances, three external and three internal have no effect on the gain of the amplifier. Capacitors like C_1 and C_2 which are in series in the circuit offer negligible reactance to the input signal if its frequency lies in the mid-frequency range and similarly, capacitances that are connected in parallel in the circuit, like C_E, offer infinite reactance at mid frequencies and may be treated as open circuits. As such the gain of the amplifier is totally independent of the values of the capacitances. It is possible to calculate mid-frequency gain using any small-signal model like the transconductance (or hybrid-pi) model or the hybrid model. Expressions for mid-frequency gain in the transconductance model is given in expression (6.57) and for hybrid model by (6.66e). The gains calculated in solved example (SE6.6), etc. are also mid-frequency gains, though it has not been explicitly mentioned there.

Low-frequency region: The three external capacitors including coupling capacitors C_1, C_2 and the emitter capacitor C_E are responsible for the decrease of amplifier gain in low-frequency region. The three internal capacitances have such values that they produce no affect on the amplifier gain in low-frequency region. Corresponding to the three external capacitances there are three different cut-off (or corner) frequencies, f_{c1}, f_{c2} and f_{c3}, where gains falls by -3 db of the mid-frequency value. In order to understand and simplify the treatment, it is reasonable to assume that the three capacitances do not interfere with each other and each of them may be treated separately and independently.

The cut-off frequency both for low-pass and high-pass filters is given by $f_{cut-off} = \frac{1}{2\pi RC}$ where R is the magnitude of the resistance in the circuit and C the magnitude of the capacitance. Therefore, at low frequencies, one will have three cut-off frequencies each with its own value of resistance R.

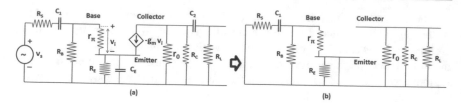

Fig. 6.62 a AC equivalent of BJT amplifier **b** local equivalent

$$f^1_{cut} = \frac{1}{2\pi R_1 C_1} ; f^2_{cut} = \frac{1}{2\pi R_2 C_2} \; and \, f^3_{cut} = \frac{1}{2\pi R_3 C_E}$$

If it is assumed that the magnitudes of the capacitances C_1, C_2 and C_E are known, then in order to know the magnitude of each cut-off frequency, it is required to find the value of the corresponding resistance. To do that one uses some small-signal model AC equivalent circuit. One may use the hybrid model but in the present discussion we use the AC equivalent circuit of transconductance model shown in Figure 6.35 where symbols have their usual meaning. Since we are interested in finding out the value of the cut-off frequency corresponding to one capacitor at a time, we short the remaining two capacitances and also short the AC and DC voltage sources. We first consider the input coupling capacitance C_1, keeps it in the circuit and short circuit the capacitances C_2 and C_E. As a result, the emitter resistance R_E is now shorted and the resistance r_π gets connected from base to ground. Further, since AC voltage sources are grounded, v_s is removed and replaced by the short circuit. Also as v_s is zero (grounded) so v_i is also zero and $g_m v_i$ is also zero. This means that no current passes through the dependent current source on the output side and the branch is like an open circuit. The modified circuit, called local equivalent in case when only C_1 is present, is shown in Fig 6.62b. From Fig. 6.62b, it is clear that the frequency f^1_{cut} does not depend on the output section and will be decided by the input section only.

As may be seen in the figure, total resistance R_1 across the two plates of capacitor C_1 is given by

$$R_1 = R_s + (R_B || r_\pi)$$

And

$$f^1_{cut} = \frac{1}{2\pi R_1 C_1} = \frac{1}{2\pi \left(R_s + \frac{R_B r_\pi}{R_B + R_\pi} \right) C_1} \tag{6.80a}$$

Once the amplifier circuit is given, the values of resistances R_s, R_B and of capacitor C_1 are known. Hence it is possible to calculate the value of f^1_{cut}.

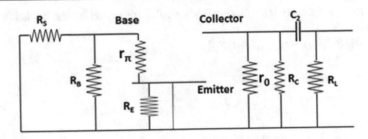

Fig. 6.63 Local equivalent for coupling capacitance C_2

To calculate f_{cut}^2

The cut-off frequency f_{cut}^2 due the capacitor C_2 will be decided by the resistances r_o, R_C and R_L. The total resistance R_2 across the two plates of the capacitor C_2 is given by

$$R_2 = (r_0 || R_C) + R_L = \left(\frac{r_0 R_C}{r_0 + R_C} \right) + R_L$$

Therefore,

$$f_{cut}^2 = \frac{1}{2\pi \left[\left(\frac{r_0 R_C}{r_0 + R_C} \right) + R_L \right] C_2} \qquad (6.80b)$$

To calculate f_{cut}^3 To calculate the cut-off frequency corresponding to the capacitor C_E, we short-circuit C_1 and C_2. The local equivalent circuit is shown in Fig. 6.64.

Again, the cut-off frequency f_{cut}^3 will be decided by the resistances R_s, R_B and R_E only. One important point to be noted in Fig. 6.64 is that the resistance R_E which was connected between the emitter and ground has been shifted to be between the base and ground. As is conventional, when an element is shifted from emitter to base its magnitude is multiplied by the factor $(\beta+1)$ while on shifting from base to emitter it is divided by $(\beta+1)$.

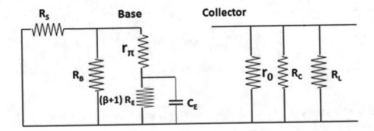

Fig. 6.64 Local equivalent circuit for calculating cut-off frequency for capacitor C_E

The total resistance R_3 connected across the two ends of the capacitor C_E is given by

$$R_3 = [\{(R_S\|R_B) + r_\pi\}\|(\beta+1)R_E]$$

And the

$$f_{cut}^3 = \frac{1}{2\pi R_3 C_E} \tag{6.80c}$$

Out of the three cut-off frequencies f_{cut}^1, f_{cut}^2 and f_{cut}^3 one will be largest. The largest of these frequencies will be the **lower cut-off frequency** represented by f_{C1} in Fig. 6.61 of the amplifier's Bode plot. It may be once again emphasized that the present analysis which is based on the assumption that each of the three capacitors C_1, C_2 and C_E independently influence the gain of the amplifier, will hold good if the three cut-off frequencies f_{cut}^1, f_{cut}^2 and f_{cut}^3 are quite different from each other.

High-frequency region High-frequency region includes those frequencies where the reactance offered by the three external capacitances is negligible and, therefore, they are like short circuits. The net result of this assumption is that in the local equivalent circuit based on transconductance model, the emitter resistance R_E is short-circuited and does not appear. Hence the emitter gets directly connected to the ground. Further, the three internal capacitances, C_π, C_μ and C_{CE} and stray capacitances C_{wi} and C_{wo} due to wirings both in the input and the output sections, also come into play.

The local high-frequency equivalent circuit of single-stage BJT amplifier based on transconductance model is shown in Fig. 6.65a. However, the capacitance C_μ between the base and the collector complicates further reduction of the network. To overcome this difficulty, one may use the Miller method of breaking the capacitor C_μ into two components $C_{m\mu i}$ and $C_{m\mu o}$, the former going to the input side and the later to the output side. These components have the values

$$C_{m\mu i} = C_\mu(1 + |A_v|) \ and \ C_{m\mu o} = C_\mu\left(\frac{1 + |A_v|}{|A_v|}\right) \tag{6.80d}$$

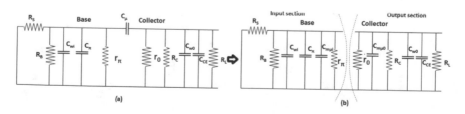

(a)

(b)

Fig. 6.65 Local equivalent circuits at high frequency **a** with C_μ **b** C_μ bifurcated into $C_{\mu i}$ and $C_{\mu o}$ using Miller concept

Where A_v is the mid-frequency gain of the amplifier. The final reduced network of the local equivalent is shown in Fig. 6.65b. As may be observed in Fig. 6.65b, there are two independent parts of the network, the input section and the output section, each of which has capacitances and resistances. Hence there may be two cut-off frequencies f_i^h and f_o^h, respectively, to the input and the output sections. These cut-off frequencies are given by

$f_i^h = \frac{1}{2\pi C_{in} R_{in}}$ and $f_h^o = \frac{1}{2\pi C_{out} R_{out}}$ where C_{in}, C_{out}, R_{in} and R_{out} are, respectively, the equivalent capacitance and resistance of the input and the output sections. It may be observed that all capacitances and resistances in a section are in parallel. Hence

$$C_{in} = C_{wi} + C_{m\mu i} + C_\pi \text{ and } C_{out} = C_{wo} + C_{n\mu o} + C_{CE}$$

Also,

$$R_{in} = R_s \| R_B \| r_\pi \text{ and } R_{out} = r_0 \| R_C \| R_L$$

When values of these resistances and capacitances are provided, it is possible to calculate the cut-off (or -3 dB) frequencies f_i^h and f_o^h. In general, the calculated values of the two frequencies will be different. The one which has a smaller magnitude will be the higher cut-off frequency of the amplifier, which is denoted by f_{C2} in Fig. 6.62.

Self-assessment question SAQ: What is the advantage of using dB scale for the gain of an amplifier?

Self-assessment question SAQ: At some frequency f_x gain of an amplifier falls to one fifth of its mid-frequency value. What will be the gain value at f_x in dB?

Solved example (SE6.8) Analyse the following emitter follower circuit using hybrid-pi model to obtain (i) parameters of the Q-point (ii) values of transconductance model parameters (iii) total input and output resistances and (iv) cut-off frequencies. **Neglect parasitic elements. The BJT is silicon-based one and has β = 200, VA =- 150**.

Solution For DC analysis, it is simple to get: $V_{BB} = \frac{22}{22+18} \times 9V = 4.95V$

And $R_B = 22 \text{ k} \| 18 \text{ k} = 9.90 \text{ k}\Omega$.

Applying KVL to base –emitter junction, one gets, $I_E R_E + V_{BE} + R_B I_B = 4.95$

Putting $V_{BE} = 0.7$ V (for Si device) and $I_B = I_E/(\beta+1)$ with $R_B = 9.9$ k, $R_E = 1k$ one gets,

$I_E = 4.05$ mA, $I_B = 20.3$ μA; $Ic \approx I_E$

Similarly by applying KVL to collector–emitter section

$V_{CE} + R_E I_E = 9.0V$ that gives $V_{CE} = 5.0V$

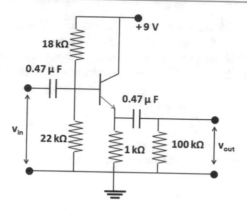

Fig. SE6.8 Circuit for example (SE6.8)

Therefore, the Q-point parameters are: $I_C = 4.05 \, mA$, $V_{CE} = 5.0 \, V$, $I_B = 20.3 \, \mu A$
According to transconductance model $g_m = \frac{I_C}{V_T} = \frac{4.05 \times 10^{-3} A}{26 \times 10^{-3} V} = 0.156 \, Siemen$

$$r_\pi = \frac{\beta}{g_m} = \frac{200}{0.156}\Omega = 1.28 k\Omega; r_0 = \frac{V_A}{I_C} = \frac{150}{4.05 \times 10^{-3}} = 37.04 k\Omega$$

The input resistance R_{in} is given by; $R_{in} = R_B||[r_\pi + (1+\beta)(r_0||R_E||R_L)]$
$= [9.9||\{1.3 + 194\}]k\Omega$
Or

$$R_{in} = 9.9 k\Omega$$

Also

$$R_{Out} = r_0||\frac{r_\pi}{\beta} = 6.4\Omega$$

The cu- off frequencies: $f_{1cut} = \frac{1}{2\pi(R_i)c_1} = \frac{1}{2\pi \times 9.9 \times 10^3 \times 0.47 \times 10^6} = 34.2 Hz$

$$f_{2cut} = \frac{1}{2\pi(R_0 + R_L)c_2} = \frac{1}{2\pi(6.4 + 100 \times 10^3) \times 0.47 \times 10^{-6}} = 3.39 Hz$$

6.13.4   BJT as a Switch

We have seen that a BJT works as an amplifier when the operating point lies in the forward active region, which is also called the linear region. However, a transistor may work as an 'on' switch when it is in saturation and as an 'off' switch when in cut-off region. Since the state of operation of the transistor can be easily controlled by a small change in base current, it is possible to turn 'on' or 'off' a transistor switch with very little effort or power consumption. BJT switches are often used to activate solenoid-operated electromagnetic relays. However, the proper choice of the transistor is important so that it may provide the required current to activate the relay. The maximum current rating and other technical details of the transistor are provided in manufacturer's datasheet or manual. It is a common practice to keep a 10% margin of safety while choosing the transistor. Choice of a proper value and wattage of the collector resistance R_C is also important. Since in switching action the transistor operates either in saturation or in cut-off regions, the collector current I_C is either of the order of (V_{cc}/R_C) or zero. Therefore, for reliable operation, it is required to have (V_{cc}/R_C) larger than the activation current of the relay. Once I_C is finalized, and the voltage V_{cc} of the power supply is known it is easy to determine the specifications of the collector resistance. The value of the base current that will switch the operating point to the saturation or full-on condition is obtained by dividing the collector current I_C by current gain β provided in datasheet. If base–emitter circuit is powered by some other device, say by a photodiode as shown in Fig. 6.66 then base resistance R_B is adjusted such that the required base current flows through the circuit when diode is operating.

Self-assessment question SAQ: Why in most of BJT devices employ an npn BJT in CE configuration?

Self-assessment question SAQ: What in your opinion is the biggest (i) advantage and (ii) the biggest disadvantage of using a BJT in electronic devices.

Self-assessment question SAQ: BJT is a current-activated or voltage-activated element?

6.14 Field-Effect Transistor (FET)

In a BJT current is constituted by the flow of both the electrons and the holes. In field-effect transistors, however, current is constituted by the flow of only one kind of charge carriers, either electrons or holes. Moreover, in a FET the current flow is controlled by an electric field other than the one that is responsible for the flow of charge carriers. The field-effect transistors are of two types: (i) junction field-effect transistor (JFET) and (ii) metal–oxide–semiconductor field-effect transistor (MOSFET) or simply insulated gate field-effect transistor (IGFET).

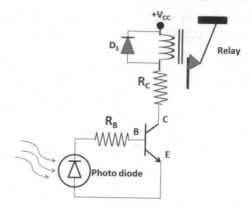

Fig. 6.66 BJT switch to operate an electromagnetic relay

6.14.1 Junction Field-Effect Transistor (JFET)

There may be two classes of JFET, one in which electrons constitute the current: is called the **n-channel FET** and the other in which holes constitute the current the **p-channel FET**. Like BJT, FET is also a three-terminal device with **Source S**, **Drain D** and **Gate G** as the three terminals. In circuit diagrams, the n-channel and the p-channel FET's are represented by the symbols shown in Fig. (6.67).

In order to understand the operation of an FET, we take the example of a rod of some doped semiconducting material, say of n-type, of length L , cross-sectional area A and connect a variable source of voltage V across it, as shown in Fig. 6.68a.

Since the given semiconductor rod will have a fixed value of resistance R, the current– voltage (I-V) graph of the rod, according to Ohm's law will be a straight line as indicated by dotted line OB in Fig (6.68b). However, if after reaching a certain voltage V_1 the cross-sectional area A of the rod is reduced (the rod is made narrow), the resistance of the rod will increase to the value $R_1 > R$ and the current will become less than what it would have been if the rod was not made narrow. If this process of reducing the value of A with the increase of V is continued in such a

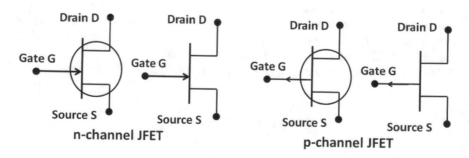

Fig. 6.67 Symbols for n-channel and p-channel JFET

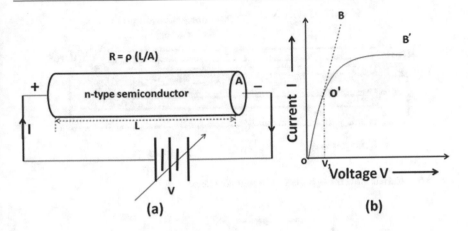

Fig. 6.68 Current flow through an n-type semiconductor rod

way that at each step reduction in the value of A counterbalances the increase in current due to the increase of voltage V, then current will get stabilized at some value and the resultant (I-V) plot will become a horizontal straight line as shown by the O'B' in Figure 6.68b. In a JFET, this control on the magnitude of current by reducing the effective area of cross section of the conducting channel is achieved via an electric field (or a voltage). Hence, these voltage-activated devices are called field-effect transistors. It may be recalled that in a BJT current is controlled essentially by the base current, and therefore, BJTs are current-activated devices.

We study the constructional details of a JFET in steps. Let us take a rod or block of some lightly doped semiconductor, say of n-type silicon and develop metallic contacts at its two ends, name one of the contacts **Source** (denoted by S) and other **Drain** (denoted by D). Let us connect a voltage source V_{DS} across the source and drain such that (conventional) current enters the semiconductor block through the drain and leaves it through the source terminal. This will establish a current I through the semiconductor which will be constituted by the flow of free electrons from S to D as the material is n-type (see Fig. 6.69). If instead a block of p-type semiconductor is taken, then current will be established by the flow of holes. Coming back to the n-type material, with the flow of current a voltage gradient will get established in the semiconducting material between D and S. Semiconductor material towards D will be at higher potential and the potential will go on decreasing as one moves towards source S. It is important to note that if the flow of current through the semiconducting material is stopped, the voltage gradient will also vanish. The cylindrical part of the semiconductor between D and S through which current I flows is called **channel**, which in case of the n-type material is termed as **n-channel**.

A schematic diagram of an n-channel JFET is shown in Fig. 6.70. As may be seen in this figure, a block of an n-type semiconductor that has conducting ports at two ends forms the n-channel. The terminal at one end is called source and is

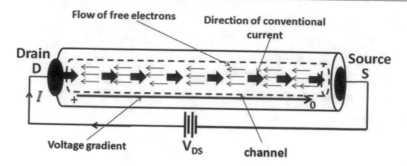

Fig. 6.69 Current flow through the n-type semiconductor road

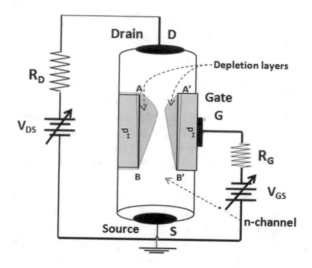

Fig. 6.70 Schematic diagram of an n-channel JFET

denoted by S. The terminal at the opposite end is called drain and is denoted by D. A variable voltage source V_{DS} may be connected between S and D along with a series resistance R_D. As may be seen in the figure the drain terminal D is connected to the positive terminal of the battery while the source terminal S is grounded. Since D is positive and S is at ground potential, a potential gradient gets established in the n-type material from D to S (also called the n-channel), with D side at higher potential than the S side. Two heavily doped p^{++} sections are developed, opposite to each other across the central part of the n-type block by diffusion. These two p^{++} sections are **electrically connected with each other** (internal connection is not shown in the figure) and serve as Gate (denoted by G). With the diffusion of two p-type sections, two p-n junctions are formed at the two edges (AB and A'B') of the p++ material (of gate) which are in contact with the n-type material. In case the gate terminal G is at ground potential and a current is passing through the n-channel from D to S, the two p-n junctions at the edges of p^{++} materials get reverse biased. It is because the p^{++} sides of these junctions are at ground potential and the n-sides are at positive potentials due to the voltage gradient produced in the n-channel by the flow of current. Moreover, the magnitude of the reverse bias voltage across the p-n junction will not be uniform all along the junction length; reverse bias voltage will be larger towards the drain and smaller towards the source. As a result of this non-uniform reverse biasing of p-n junction's depletion layers of non-uniform thickness develop across these junctions. Since reverse bias voltage is larger towards the drain end, the depletion layers are thicker towards D and narrower towards S. Another important characteristic of these depletion layers is that they extend mostly in the n-channel region since it is lightly doped as compared to the p$^+$ $^+$ region. As a matter of fact, the main reason behind heavy doping of gate material is to develop a depletion layer that extends essentially into the n-channel region. Since depletion layers behave as insulators, the volume of the n-channel occupied by them does not allow any current to pass through it, which results in the narrowing of the conducting part of n-channel. The widths of depletion layers increase further when the potential difference V_{DS} is increased keeping gate G at ground potential. It is because the increase of V_{DS}, in turn, increases the reverse bias potential across n-p junctions.

(a) **Drain or output characteristics**

   Family of curves showing the variation of drain current I_D with drain-source voltage V_{DS} for different values of the gate voltage V_{GS} constitutes drain characteristics of a JFET. A circuit arrangement shown in Fig. 6.71 may be used to study these characteristics. To start with we study, the variation of drain current I_D when drain voltage V_{DS} is increased keeping gate voltage $V_{GS} = 0$. When V_{DS} is positive and increased from zero to say V_1, the positive drain potential attracts free electrons (present in the n-type semiconductor as majority

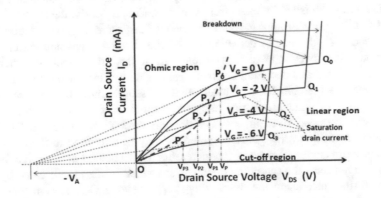

Fig. 6.71 Drain characteristics of an n-channel JFET

carriers) constituting the drain current I_D, the conventional current flowing from D to S. If it is assumed that the bulk resistance of the n-channel is say, R_1, then $I_{D1} = (V_1/R_1)$ Amp. With the flow of this current, a potential gradient gets established in the n-channel from D to S, with D-side at higher potential and the potential continuously decreasing from D to S. Consequently, the two n-p junctions at edges of the P^{++}-type gate material will get reverse biased, since their P^{++} sides will be at ground potential while the n-sides will be at positive potential (due to the potential gradient). Further, the reverse bias voltage across n-p junctions along their lengths will **not be uniform**; it will be higher near the drain end and will decrease towards the source end. As a result, depletion layers of non-uniform thickness will develop on the two opposite edges of the p++ type materials. These depletion layers which will extend into n-channel will make the channel narrow increasing its resistance from R_0 to R_1, where R_0 was the bulk resistance of the n-channel when V_{DS} was zero. On further increasing V_{DS} to say, V_2 ($V_2 > V_1$) the width of depletion layers will further increase, further reducing the conduction volume of the n-channel and in turn increasing the resistance of the channel from R_1 to R_2, where $R_2 > R_1$ but the new value of drain current $I_{D2} = (V_2/R_2)$ will still be larger than I_{D1}. Thus increasing the drain – source voltage V_{DS}, pushes much more electrons in the current that overrides the effect of the increased resistance. However, the increased value of channel resistance reduces the rate of rise of current. This explains the rise of I_D from point o to point P_0 on the graph. At point P_0 which corresponds to $V_{DS} = V_p$ on the graph (Fig. 6.71) the thick edges of the two depletion layers near the drain end touch each other which is referred to as **pinch-off** and the voltage V_P **the pinch-off voltage**. Often V_P is also termed as the **threshold voltage.** It is interesting to note that at pinch- off and beyond pinch-off, up to point Q_0, the drain current does not become zero, rather it becomes almost constant or increases very slowly with the increase of the voltage V_{DS}. The reason is simple, at pinch-off when channel is totally blocked, the flow of current I_D stops momentarily; with this momentary stoppage of current the potential gradient that was established by the current vanishes. Consequently, the reverse biasing at n-p junctions disappear. With no reverse bias, there are no depletion layers anymore. In absence of depletion layers, the channel opens and current flows through it. However, as soon as current starts to flow, the potential gradient gets re-established and depletion layers appear again blocking the channel. It is easy to imagine that blocking of channel, the disappearance of depletion layers, reopening of channel, all this happens in quick succession and, therefore, a current averaged over time keeps flowing through the channel. The slow increase of this average current with V_{DS} may be attributed to the pushing of some extra charge carriers by the increasing V_{DS}. The part of the curve from P_0 to Q_0 represents the saturation drain current which is often represented by I_{DSS}. On further increasing the drain voltage V_{DS} beyond Q_0, the semiconducting material undergoes an avalanche break down. As a result of breakdown large current which increases almost vertically, flows through the circuit. The series resistance R_D is kept in the circuit to protect the JFET from getting damaged if a large current flows through it.

When gate potential V_{GS} is given a negative value, say -2 V with respect to source (ground), the reverse bias of the two n-p junctions increases, and the widths of two depletion layers get enlarged by a fixed value as compared to the width at $V_{GS} = 0$. This results in pinch-off taking place at a drain potential V_{P1} which is smaller than V_P for $V_{GS} = 0$. Also, with enhanced reverse bias, the avalanche breakdown also sets in at a lower value of drain potential V_{DS}, shortening the voltage span of the drain saturation current. This trend follows at still larger negative values of $V_{GS} = -4$ V, -6 V, etc. However, at some negative value of V_{GS}, say at $V_{GS} = -X$, the drain current I_D becomes zero and the device is said to have gone in cut-off. This happens at that value of the negative gate potential where depletion layers are so wide that they touch each other and block the channel. This is a permanent blocking of the channel due to high reverse bias provided by the negative potential of the gate and has nothing to do with the flow of current. The channel blockage stays even when there is no internal potential gradient. This is the major difference in channel blocking at pinch-off and at cut-off. The potential –X that makes the JFET cut-off may have values from -8 V to -12 Volts, depending on the doping and the type of the semiconductor.

The area of drain characteristics may be divided into four regions: **Ohmic region, linear or active region, breakdown region** and **cut-off region**. If pinch-off points P_0, P_1, P_2, P_3, etc. are joined one gets a cure shown by dotted line, area to the left of this dotted curve is called the Ohmic region where the JFET behaves like a resistance the magnitude of which may be controlled by the gate voltage, as at different V_{GS} it has different values of resistance R_0, R_1, R_2, etc. The drain current I_D^{Ohm} in Ohmic region is given by

$$I_D^{Ohm} = \left(\frac{I_{DSS}}{V_P^2}\right)[2(V_{GS} - V_P)V_{DS} - V_{DS}^2] \tag{6.81a}$$

The region where drain current saturates is referred to as linear or active region which may be used for amplification of signal applied at the gate. Drain current in this region may be given by the following Shockley equation modified to include correction for its variation with V_{DS}

$$I_D^{active} = I_{DSS}\left(1 - \frac{V_{GS}}{V_P}\right)^2 (1 + \lambda V_{DS}) \tag{6.81b}$$

Here, λ, called 'channel length modulation parameter' is the reciprocal of V_A, i.e. $\lambda = 1/V_A$. The values of I_{DSS} and V_P are provided by the manufacturer in their datasheets.

The region where avalanche breakdown occurs is called breakdown region and care should be taken to avoid this region. The region of gate voltage beyond which no current flows through JFET is called the cut-off region and may have application when the device is used as a switch.

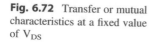

Fig. 6.72 Transfer or mutual characteristics at a fixed value of V_{DS}

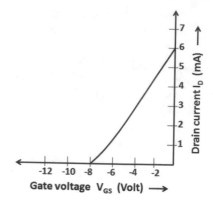

The drain or output characteristics of a p-channel JFT are similar to that of a n-channel JFT except that the potentials at drain and source are reversed, the current is constituted by the flow of majority carrier holes, and the controlling gate voltage is positive.

It is interesting to observe that the drain characteristic of a JFET resembles the output characteristics for a BJT in common emitter configuration given in Fig. 6.16c. The role played by the base–emitter voltage V_{BE} in case of BJT is played by V_{GS} in case of JFET. Further, the saturation drain currents when extended backwards meet at a point where $V = -V_A$, which is similar to the case of BJT. These plots also show how drain current may be controlled by the gate potential.

The only current that flows through the gate is the reverse saturation current constituted by the flow of minority carriers through the reverse-biased n-p junctions, this current is only of the order of a few hundred nanoamperes (10^{-9} A) or less and in most cases neglected. Since the input section of a JFET (gate-source) is reverse biased, the input impedance of JFET is quite high in comparison to a BJT where the input section (base–emitter) is forward biased and has low input impedance. As will be discussed later, the input impedance of a MOSFET is still larger because of the insulating layer that totally insulates the gate.

(b) **Transfer or mutual characteristics of JFET**

Transfer characteristics of a JFET show the variation of drain current I_D with gate voltage V_{GS} at a fixed drain-source voltage V_{DS}. Fig. 6.72 shows the variation of drain current with the gate voltage V_{GS} for some fixed value of drain-source voltage V_{DS}, say 5 V. As may be observed in this figure, the drain current decreases as the gate potential is changed from 0 to –2 V, to –4V .... and finally I_D becomes zero at V_{GS} = –8 V, which is the pinch-off voltage V_P for this setting. The mutual characteristic is almost linear and therefore, one may define the transconductance g_m as

$$g_m \equiv \left[\frac{\Delta I_D}{\Delta V_{GS}}\right]_{constant V_{DS}} \qquad (6.81c)$$

Fig. 6.73 Typical amplification characteristic of a JFET

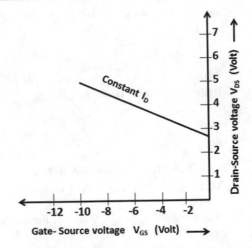

(c) **Amplification characteristic**

Amplification characteristic shows the variation of V_{DS} when V_{GS} is changed keeping the drain current constant. A typical amplification characteristic is shown in Fig. 6.73. Amplification factor μ is defined as

$$\mu \equiv - \left[\frac{\Delta V_{DS}}{\Delta V_{GS}}\right]_{constant I_D} \tag{6.81d}$$

(d) Relation between drain resistance r_d, amplification μ and conductance g_m

In the situation when both AC and DC signals are present in the system, one may define the instantaneous value of a parameter as the sum of the instantaneous DC value and the instantaneous AC value. For example, one may write the instantaneous total value of the drain voltage v_{DS}, then

$$v_{DS} = V_D + v_{ds} \tag{6.81e}$$

Where V_D is the pure DC value and v_{ds} is the instantaneous pure AC component. Further, i_D, the instantaneous total drain current will depend both on the instantaneous total values of drain voltage v_{DS} and gate voltage v_{GS}, which in mathematical terms may be written as

$$i_D = f(v_{DS}, v_{GS}) \tag{6.81.f}$$

One may use Taylor's expansion to obtain the following value for the small change Δi_D in i_D:

$$\Delta i_D = \left(\frac{\partial i_D}{\partial v_{DS}}\right)\Delta v_{DS} + \left(\frac{\partial i_D}{\partial v_{GS}}\right)\Delta v_{GS} \tag{6.81g}$$

The drain resistance r_d and the transfer conductance g_m are defined as

$$r_d \equiv \left[\frac{\partial v_{DS}}{\partial i_D}\right]_{constant\ v_{GS}} \quad and \quad g_m = \left[\frac{\partial i_D}{\partial v_{GS}}\right]_{constant\ v_{DS}} \tag{6.81h}$$

Substituting these values in (6.81g),

$$\Delta i_D = \frac{1}{r_d}\Delta v_{DS} + g_m \Delta v_{GS} \tag{6.81i}$$

Now if (6.81e) is differentiated then
$\Delta v_{DS} = \Delta V_D + \Delta v_{ds} = \Delta v_{ds}$ since $\Delta V_D = 0$ as V_D is pure DC value which is constant.
This shows that Δv_{DS} , which is the change in the DC value of the variable is equal to the change in its AC value Δv_{ds} . However, v_{ds} is itself a variable, therefore
$\Delta v_{DS} = v_{ds}$ and similarly $\Delta v_{GS} = v_{gs}$ substituting these values in (6.81), one gets

$$\Delta i_D = \frac{1}{r_d} v_{ds} + g_m v_{gs} \tag{6.82}$$

(e) **Small-signal hybrid-pi model equivalent for JFET**
Small signal transconductance model of a JFET is shown in Fig. (6.74). Values for the two parameters g_m and r_o may be calculated as follows:

$$g_m = \frac{\partial i_D}{\partial v_{GS}} = -\frac{2I_{DSS}}{V_P}\left(1 - \frac{V_{GS}}{V_P}\right)(1 + \lambda v_{DS})$$

And

$$r_o = \frac{\partial v_{DS}}{\partial i_D} = \frac{1}{I_{DSS}\left(1 - \frac{V_{GS}}{V_P}\right)^2 \lambda}$$

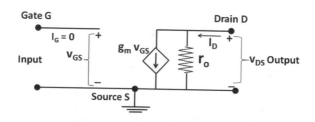

Fig. 6.74 Small-signal hybrid-pi model equivalent of FET

6.14.2   Metal–Semiconductor Field-Effect Transistor (MESFET)

MESFET is a modified form of JFET where a layer of metal aluminium is deposited over the FET structure to make a metal–semiconductor Schottky barrier junction. Generally, it is made of GaAs semiconductor because of the high electron mobility in it. Structural details of an n-channel MESFET are shown in Fig. 6.75a. On an insulating undoped Gallium Arsenide (GaAs) substrate an epitaxial layer of lightly doped n-type GaAs semiconductor is developed. Two Ohmic contacts at the two ends serve as Drain and Source. The middle portion of the top surface is covered will an Aluminium metal layer to develop a metal--semiconductor junction. In the central part of aluminium layer, a conducting contact is made for gate terminal.

Essential features of a metal–semiconductor junction (also called Schottky junction) are shown in part (b) of the figure. In case of an isolated n-type semi-conductor, the Fermi level lies just below the bottom of the conduction band, while in case of an isolated metal the Fermi level is just above the top of the conduction band. However, when a metal–semiconductor junction is formed, the Fermi levels of the two sides align themselves pulling the conduction band of metal at a lower energy than that of the n-type semiconductor. As a result some free electrons which have higher energies in the conduction band of the n-type semiconductor move to metal side leaving positive ions behind. This transfer of electrons from n-type semiconductor to metal creates a potential barrier at the metal–semiconductor junction, making n-type semiconductor positive with respect to the metal. This is called the Schottky barrier. Electrons moving out from the semiconductor leave behind uncovered positive ions that constitute a depletion layer at the junction. Since the n-type semiconductor is lightly doped, the depletion layer extends mostly in the n-type semiconductor.

When drain D is given a positive potential and source S and gate G are kept at ground potential, a drain current flows through the n-channel, which establishes a potential gradient from D to S. The potential gradient in the channel results in non-uniform reverse biasing of the metal–semiconductor junction as in the case of JFET. Any increase in reverse bias by applying negative potential at gate G with

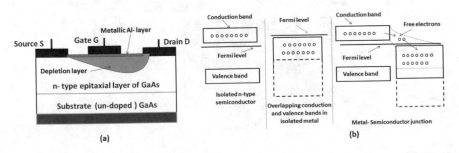

Fig. 6.75 **a** Structure of MESFET; **b** band structures of isolated n-type semiconductor, isolated metal and metal–semiconductor junction

respect to source S enhances the thickness of depletion layer, chocking the current channel. The operation and characteristics of MESFET are similar to that of a JFET. The main advantage of MESFET is its small size and low parasitic capacitances that make its response fast. It is for this reason that these devices are mostly used in microwave systems.

Self-assessment question SAQ: List two advantages of a metal–semiconductor junction over a conventional p-n junction?

6.14.3 Metal–Oxide–Semiconductor Field-Effect Transistor (MOSFET)

Both in the case of JFET and MESFET, the gate terminal is in contact with the conducting channel and some very small reverse bias saturation current ($\approx 10^{-9}$ A) flows through the gate which is reverse biased. In contrast to that in case of a metal–oxide–semiconductor field-effect transistor, the gate is totally insulated from the conducting channel by a thin SiO_2 (glass) layer. This makes the input impedance of a MOSFET almost infinite or in the order of several tens of megaohms.

Structural details of a typical MOSFET may be understood through Fig. (6.76). Initially, a near insulating block of say, p-type semiconductor is taken as substrate and in the second stage, two lightly doped n-sections are developed at two ends by diffusion that serve as drain and source. Next, the top surface of the block is covered with an insulating thin layer of SiO_2 (glass). Since SiO_2 layer readily interacts with sodium ions which are present in all environments, a layer of Si_3N_4, silicon nitrate

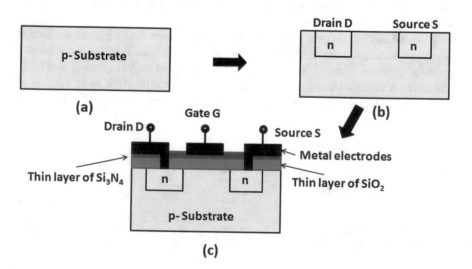

Fig. 6.76 a A near insulator block of p-type substrate. **b** Two lightly doped n-type blocks developed by diffusion. **c** Final structure of a MOSFET

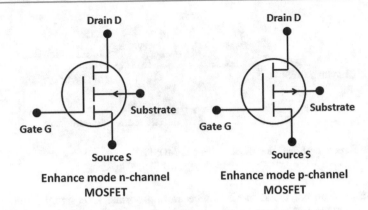

Fig. 6.77 Symbols for enhanced-mode n- and p-channel MOSFETs

is laid over the oxide layer. Nitrate layer does not allow sodium ions to pass through and in this way avoids the contamination of SiO_2. Holes are then drilled through the oxide and nitrate layers above the source and drain n-type regions to develop Ohmic contacts. Finally, the metallic (mostly aluminium) coating is done over the top surface covering the region between the drain and the source. Terminals for drain D, source S and gate G are then taken out as shown in the figure. It may be noted that there is no physical penetration of the nitrite and oxide layers in the gate region and the gate is in no way electrically connected to the channel below the gate.

Working Working of MOSFET may be easily understood if it is realized that the top metallic coating of the gate region and the semiconductor substrate under it forms a parallel plate capacitor with dielectric layer of oxide and nitrite sandwiched between them. When some positive charge (potential) is given to the gate, it induces a negative charge of an equal amount in the semiconductor layer immediately below the gate. On increasing the positive potential at the gate correspondingly larger amount of negative charge gets induced in the semiconductor below the gate. This induced negative charge in the semiconductor layer beneath the metallic gate region (between source and drain) makes the (otherwise almost insulator p-substrate) semiconductor an n-type conducting channel. If now drain D is given a positive potential and source S is connected to ground, a (conventional) current may flow from drain to source. The actual current, however, is constituted by the flow of electrons from source to drain. It is important to note here that the conductivity of the induced n-channel increases or enhances with the increase in positive gate potential. The MOSFET in which the gate potential enhances the current flow, or where no current can flow without the gate potential or those MOSFETS which in absence of gate voltage behave as an open switch are called EMOSFET or enhance-mode metal–oxide field-effect transistors. It is simple to realize that the gate potential has great influence on the resistance of the conducting channel and

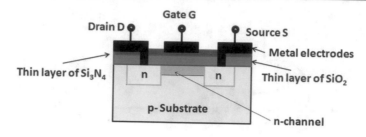

Fig. 6.78 Structure of depletion-mode n-channel MOSFET

that in case of an enhanced-mode device an appropriate gate signal is required to switch it on. The device just discussed is an n-channel enhancement MOSFET. To have a p-channel enhanced-mode device, one may start with a near insulator n-type substrate, diffuse two lightly doped p-type blocks at the two ends and apply a negative potential at the gate so that a p-type channel develops because of the induced positive charge. Symbols for the n- and p- channel enhanced-mode MOSFETs are shown in Fig. 6.77. Mark the central broken line in these symbols that is characteristic of an enhanced-mode device.

Figure 6.78 shows the structure of a depletion-mode n-channel MOSFET. As may be observed in the figure an n-channel is already developed just under the gate by doping a small strip of the substrate with a sufficient concentration of n-type impurities that may neutralize the p-type impurities and make the strip n-type. This structure is such that a conducting channel is already present and if a potential difference is applied between D and S, drain current will start flowing even in absence of any voltage at gate. If now some negative potential is given to the gate, the positive charges induced in the n-type pre-existing channel will neutralize some of the electrons and will thus deplete the conducting channel of charge carriers, reduce the drain current which will amount to the increase of channel resistance. Since in this case application of negative voltage at the gate controls the drain current by depleting the charge carriers, this is called the depletion-mode n-channel MOSFET. Similarly, there can be a depletion-mode p-channel MOSFET. The symbols for depletion-mode MOSFETS are shown in Fig. 6.79. It may be noticed that a continuous central line indicates the depletion mode.

It is interesting to note that the MOSFET structure shown in Fig. 6.78 may operate both in the depletion as well as in enhancement modes. If the gate potential is negative it operates in depletion mode and when gate potential is positive the device operates in enhance mode. Family of curves describing the variation of drain current I_D with drain voltage V_{DS} with respect to source is called transfer characteristics of the MOSFET. Typical transfer characteristics for the MOSFET that may operate both in enhanced and depletion modes are shown in Fig. 6.80.

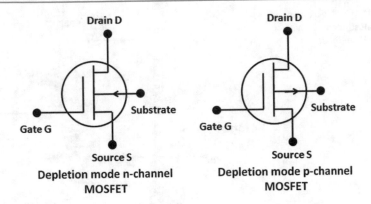

Depletion mode n-channel MOSFET

Depletion mode p-channel MOSFET

Fig. 6.79 Symbols for depletion-mode MOSFETs

Fig. 6.80 Transfer characteristics of an n-channel MOSFET

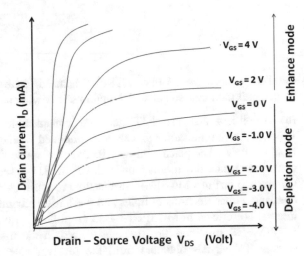

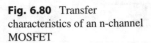

6.14.4   MOSFET Amplifier

MOSFET amplifiers are often used in the early stages of amplification. It is because of two reasons:

(a) MOSFETs have extremely high input impedance hence it does not load the source of input signal. Generally, small signals are produced by weak sources that are not capable to deliver sufficient current. In order to amplify the signal from such a weak source if the source is connected to an amplification device of low input impedance, the device will draw current from the source which the source will not be able to provide. As a result, the signal will get distorted and the amplified signal will not be an exact amplified replica of the actual input signal. If instead of a high

Fig. 6.81 Circuit diagram of
an enhanced-mode n-channel
MOSFET amplifier

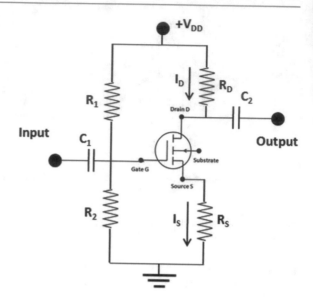

input impedance device like MOSFET is used, there will be no loading of the
source and the input signal will remain as it is without any distortion.

(b) Amplifiers that use BJT generate considerable internal electronic noise.
Unwanted electronic noise signals are generated whenever recombination of electron and hole takes place, and in a BJT recombination occurs at several stages of
operation, hence the internal noise level in BJT-based amplifiers is quite high.
Contrary to that in a MOSFET current flow is constituted by charge carriers of only
one kind, either electrons or holes and there is no recombination at any stage hence
internal noise level in MOSFET-based amplifiers is quite low.

MOSFET amplifiers are frequently used in the first stage of amplifying TV
signals. TV signals obtained from the receiver antenna are very small and of
negligible current capacity. When fed to a MOSFET amplifier of high input
impedance there is no distortion of the signal and also negligible internal noise is
generated. Once amplified, signals with good signal to noise level are available
from the MOSFET amplifier, they may be further amplified using BJT amplifiers
that have high gain.

The circuit diagram of a typical enhanced-mode n-channel MOSFET amplifier is
shown in Fig. 6.81. The gate voltage $V_G = (R_2/R_1+R_2)V_{DD}$ may be so adjusted by
choosing the values of resistances R_1 and R_2 that the operating point remains in the
middle of the linear (or saturation) region of the transfer characteristics. The output
is obtained through the coupling capacitor C_2 which isolates DC level. Input to the
circuit is also provided through a coupling capacitor C_1 to allow only the AC signal.
As is evident, the amplified output signal is 180^0 out of phase with the input signal.

6.14.5   MOSFET as Switch

Cut-off and saturation regions of MOSFET characteristics are used to make MOSFET switches. The main advantage of the MOSFET switch is its speed of operation which is much higher than BJT switches. Figure 6.82 shows switch circuits for n-channel and p-channel enhancement types MOSFET switches. As may be seen in the figure, gate voltage V_i initiates switching action. Let us consider n-channel MOSFET (Fig (a) If $V_i = 0$, the MOSFET is cut-off, switch is off, no current passes through the device, $I_{SD} = 0$ and hence there is no drop of voltage across the drain resistance R_D. The output voltage is equal to V_S. When V_i becomes positive (logic state 1), the transistor goes to saturation and heavy current passes through the transistor. MOSFET behaves like a closed switch offering negligible resistance and output gets directly connected to ground, the voltage V_0 goes to 0 V. It may be observed that when gate G is in state 0, output is high and when gate is in state 1, output is LOW. Operation of p-channel switch is just the opposite. When gate is in state 0, switch is off, current $I_{DS} = 0$, hence, $V_0 = 0$. When gate is in state 1 switch is on, source voltage reaches output and $V_0 = V_s$.

It is left to the reader to draw switch circuits for depletion-type MOSFETs.

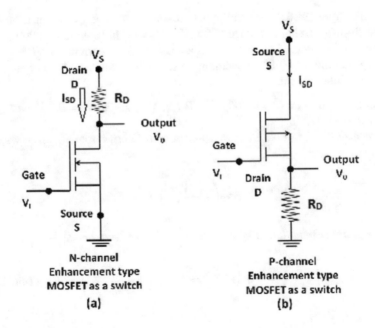

Fig. 6.82 MOSFET as Switch

Precautions in using MOSFET MOSFET are very sensitive to static charges and to electrostatic discharge. Electrostatic discharge occurs mostly when insulators rub each other liberating static charges due to friction. Momentary static voltages as large as 3000 V or more may be produced in this way. Care must be taken to avoid such happenings particularly during repair of MOSFET-based instruments.

Self-assessment question SAQ: In normal operations, a JFET works in depletion or enhance mode?

Self-assessment question SAQ: What is the special characteristic of a MESFET?

Self-assessment question SAQ: Under the identical condition of biasing minimum gate current will flow in (i) eMOSFET (ii) dMOSFET (iii) MESFET (iv) JFET. Give reason(s) for your answer.

Self-assessment question SAQ: Insulating depletion layers completely block the conducting channel at pinch-off but considerably large current keeps flowing at and beyond pinch-off in a field-effect transistor. Explain.

Self-assessment question SAQ: At a large value of reverse bias the drain current completely stops flowing, however, at pinch-off, it keeps flowing. In what respect pinch-off is different from the cut-off.

Solved example (SE6.9): When the gate-to-source voltage (VGS) of a MOSFET with pinch-off (threshold) voltage 400 mV , working in saturation is 900 mV, the drain current is observed to be 1 mA. Calculate the drain current for VGS of 1.4 V, assuming that the MOSFET is operating in saturation. Neglect all other effects like channel modulation, etc.

Solution; *It is given that $V_P = 400\ mV = 0.4\ V$; $V_{GS} = 900\ mV = 0.9\ V$; and $I_{DS} = 1$ mA $= 1 \times 10^{-3}$ A.*

Since MOSFET is operating in saturation or active region, therefore, from (6.81b)

$$I_D^{active} = I_{DSS}\left(1 - \frac{V_{GS}}{V_P}\right)^2 (1 + \lambda V_{DS})$$

Since channel modulation effect is neglected, λ is zero, hence;

$$I_D^{active} = I_{DSS}\left(1 - \frac{V_{GS}}{V_P}\right)^2 \ Or \ 1x10^{-3} = I_{DSS}\left(\left(1 - \frac{0.9}{0.4}\right)^2\right) = I_{DSS}(1.5625)$$

or

$$I_{DSS} = 6.4x10^{-4}A$$

Therefore,

$$I_D^{active} at(V_{GS} = 1.4V) = I_{DSS}\left(1 - \frac{V_{GS}}{V_P}\right)^2 = 6.4x10^{-4}x\left(1 - \frac{1.4}{0.4}\right)^2 A = 4.0mA.$$

The required value of I_{DS} is 4.0 mA.

Solved example (SE6.10): Source block of a silicon ($n_i = 10^{10}$ per cm^3) n-channel MOSFET has the area 1 sq.μm and depth of 1μm. If the dopant density in the source block is 10^{19} /cm^3, calculate the number density of holes in the Source block and the total number of holes in the source volume

Solution: *Volume of the source block V = {1 x 10^{-12} } x 1 x 10^{-6} m^3= 10^{-12} cm^3, Concentration of dopant electrons n_e= 10^{19} per cm^3. Therefore, the density of holes $n_h = n_i^2/n_e = (10^{10})^2/10^{19} = 10$ per cm^3.*
    The number density of holes in source block is 10 holes per cubic cm.
    The number of holes in the block of volume 10^{-12} cm^3 = 10 x 10^{-12} =10^{-11} = almost negligible or zero.

Problems

P6.1   For the amplifier circuit given in Fig. P6.1 find the coordinates of the operating point and using hybrid-pi model for small signals (transconductance model) to determine (i) g_m (ii) r_π, (iii) r_o and voltage gain of the amplifier. The forward current gain β of the silicon BJT is 150 and V_A is 100 V.

   Answer: [Q-point: (I_C= 1.47 mA, V_{CE}= 3.46 V, I_B= 9.8 μA); (i) g_m= 56.5 x 10^{-3} S, (ii) r_π= 2.65 kΩ (iii) r_o= 68.02 kΩ (iv) Voltage gain=- 219.78].

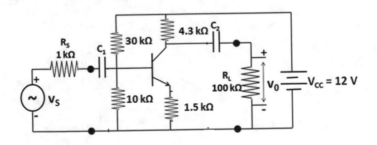

Fig. P6.1 Circuit for problem (P6.1)

P6.2   Hybrid parameters for an npn silicon transistor in CE configuration at room temperature are given below, draw the hybrid equivalent circuit and calculate input impedance and voltage gain if load resistance is 100 kΩ. $h_{ie} = 1.1$ kΩ; $h_{fe} = 50$; $h_{re} = 2.5 \times 10^{-4}$; $h_{oe} = 25 \times 10^{-6}$ S; $V_A = 50$ V and $V_T = 26$ mV.

Answer: [Input impedance $Z_i = 743$ Ω; Current gain $A_i = -14.3$; Voltage gain $A_V = -1.92 \times 10^3$].

P6.3   Calculate the magnitude of the emitter resistance r_E for the circuit shown in Fig. P6.3.

Answer: [25 Ω].

P6.4   An npn BJT common emitter amplifier with swamping emitter resistance has the following values of circuit elements. Draw the circuit diagram of the amplifier and carry out r-parameter model analysis to get the value of emitter resistance r'_E, voltage gain from base to collector, and overall gain. Given values of circuit elements: $V_s = 10$ mV, $R_s = 600$ Ω, $R_1 = 47$ kΩ,

Fig. P6.3 Circuit for problem (P6.3)

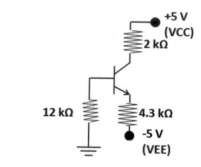

Fig. P6.6 Circuit for problem (P6.6)

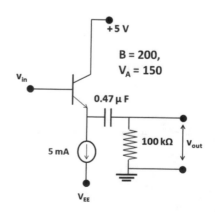

R_2= 10 kΩ; V_{CC}=+10 V; R_C = 4.7 kΩ; R_{E1}=470 Ω; R_{E2}=470 Ω; C_1 = C_2 = 10 μF; C_E = 100 μF.

Answer: [r_E' = 23.6Ω, *Voltagegainfrombasetocollector* = *9.09*, Overall gain = 8.45].

P6.5 Use the r-parameter model to calculate the voltage gain of an emitter follower that has an emitter resistance R_E of 1 kΩ, a load resistance R_L of 5 kΩ, with emitter current of 1.0 mA.

Answer: [0.97]

P6.6 Calculate Q-point parameters and transconductance model small-signal parameters for the following amplifier that is biased with a current source.

Answer: [I_c= 5 mA; I_B = 25 μA; V_{CE} = 4.3 V; g = 192 x 10^{-3} Siemen; r_0 = 30 kΩ; rπ= 1.04 kΩ; R_{in}= 4.6 MΩ; R_{out} = 5.2 Ω].

P6.7 An npn BJT amplifier in common emitter configuration has following values of components. Calculate bias point (Q- point), R_{in}, R_{out} and the lower cut-off frequency of the amplifier. r_π =5kΩ; r_0 = 100 kΩ; r_e =25 Ω; R_E=500 Ω; R_C =2 kΩ, V_{CC} = 15 V; R_1 = 33 kΩ; R_2 =5.9 kΩ; C_1= C_2 = 300 nF and β = 200.

Answer: [Q-point {I_C = 3 mA; V_{CE} =7.5 V, I_B = 15 μA}; R_{in} = 5.1 kΩ; R_0 = 525 Ω; Lower cut-off frequency = 104 Hz; Higher cut-off frequency =1010 Hz].

Fig. P6.8 Circuit for problem (P6.8)

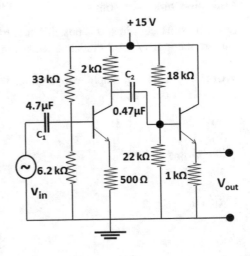

P6.8   Find the bias point for the two transistors and the overall gain of the circuit given in Fig. (P6.8). It is given that $\beta = 200$, $r_\pi = 5$ kΩ and $r_0 = 100$

Answer Hint: Parameters of the Q-point (or biasing) may be found by DC analysis of the given network. However, for DC analysis all capacitors are replaced by open circuits. When capacitors C_1 and C_2 are removed, the network breaks into two independent circuits. The circuit that contains transistor T_1 is a common emitter amplifier and that with T_2 an emitter follower. DC analysis of the two circuits will give $R_{B1} = 5.22$ kΩ, $V_{BB1} = 2.37$ V, $I_{C1} \approx I_{E1} = 3.17$ mA, $I_{B1} = 16 \mu A$, $V_{CE1} = 7.1$ V, $R_{i1} \approx R_{B1} = 5.22$ kΩ and for transistor T_2, $R_{B2} = 9.9$ kΩ, $V_{BB2} = 8.25$ V, $I_{C2} \approx I_{E2} = 7.2$ mA, $I_{B2} = 36 \mu A$, $V_{CE2} = 7.8$ V, $R_{i2} \approx R_{B2} = 9.9$ kΩ. Further, the input resistance R_{i2} *of the emitter follower circuit (T_2 transistor) works as the load for the circuit of transistor T_1.*
The voltage gain of amplifier circuit T_1 *is given by*

$$A_{v1} = -\frac{R_{C1} \parallel R_{i2}}{R_{E1}} = -\frac{2k\Omega \parallel 9.9\ k\Omega}{500} = -\frac{1.66\ k\Omega}{500} = -3.33$$

Since the gain of an emitter follower is ≈ 1, the overall gain $\approx -3.33 \times 1 = -3.33$ Ω. The negative sign in front of the gain tells that there is phase change of $180°$.

P6.9   Drain terminal D of an n-channel MOSFET is connected to the gate G so that $V_{GS} = V_{DS}$. If the pinch-off voltage of the device is 1 V and the drain current I_D is 1 mA for V_{GS} of 2 V, calculate the drain current for V_{GS} of 3 V.

   Answer: [4 mA].

Multiple-Choice Questions

Note: In some questions it is possible that more than one alternative is correct. All correct alternatives must be ticked for complete answer and full marks.

MC6.1 For a BJT in amplifier operation, the ratio of collector current to emitter current is denoted by

(a) Alpha
(b) Beta
(c) Delta
(d) Gamma

ANS: (a).

MC6.2 When BJTs are used in digital circuits they operate mostly in

   (a) Linear region
   (b) saturation or cut-off region
   (c) active region
   (d) non-linear region

ANS: (b).

MC6.3 In amplifier application of an npn transistor the base current is constituted by

   (a) Donor ions
   (b) Acceptor ions
   (c) Hole
   (d) Electrons

ANS: (d).

MC6.4 Tick the correct relation(s).

   (a) $I_B = I_C + I_E$
   (b) $I_B = I_C - I_E$
   (c) $I_E = I_C + I_B$ Cell
   (d) $I_E = I_C - I_B$ Cell

ANS: (c).

MC6.5 Tick the correct relation(s)

   (a) $I_C = \alpha I_E + I_B$
   (b) $I_C = \alpha I_E + \beta I_B$
   (c) $I_C = \alpha I_E + I_{CEO}$
   (d) $I_C = \alpha I_E + I_{CBO}$

ANS: (d).

MC6.6 Output impedance of a bipolar junction transistor is high

   (a) only in common base configuration
   (b) only in common emitter configuration
   (c) only in common collector configuration
   (d) in all configurations

ANS: (d).

MC6.7 Collector current of 10 mA flows through a BJT of $\beta = 100$. The emitter current of the circuit is

(a) 100.1 mA
(b) 100 mA
(c) 10.1 mA
(d) 1.01 mA

ANS: (c).

MC6.8 With the rise of temperature base-emitter resistance of a BJT

(a) decreases
(b) remains constant
(c) increases
(d) may increase or decrease depending on the material

ANS: (a).

MC6.9 which of the following has highest input impedance

(a) BJT configuration
(b) BJT configuration
(c) JFET
(d) MOSFET

ANS: 9 [(a) and (b)].

MC6.10 Select the voltage activated devices

(a) MOSFET
(b) FET
(c) dependent current source
(d) BJT

ANS: (d).

MC6.11 Tick the current-activated devices

(a) MOSFET
(b) FET
(c) dependent current source
(d) BJT

ANS: [(c) and (d)].

MC6.12 Depletion layers in a BJT extends mostly in

(a) Emitter region
(b) Base region
(c) Collector region
(d) emitter and collector regions

ANS: (b).

MC6.13 Best and the worst bias stability is exhibited, respectively, by

(a) fixed bias; collector–base
(b) bias
(c) self-bias; fixed bias
(d) collector–base

ANS: (c).

MC6.14 Stability factor S^{Self} against thermal runaway in self-bias is given by

(a) $\dfrac{(\beta+1)}{\left[1+\frac{\beta R_E}{RT_h + R_E}\right]}$

(b) $\dfrac{(\beta+1)}{\left[\frac{\beta R_E}{RT_h + R_E}\right]}$

(c) $\dfrac{1}{\left[1+\frac{\beta R_E}{RT_h + R_E}\right]}$

(d) $\dfrac{(\beta+1)}{\left[1-\frac{\beta R_E}{RT_h + R_E}\right]}$

ANS: (a).

MC6.15 According to hybrid -pi small-signal model of the BJT the no-load gain and the transconductance are given, respectively, by
(a) $-g_m r_\pi; I_C/V_T (h)$ (b) $-g_m r_0; I_C/V_A$ (c) $-g_m r_0; I_C/V_{T_T} (d)$ $-g_m r_\pi; I_C/V_A$

ANS: (c).

MC6.16

(a) $R_{in}^{base} = \beta_{ac} r_E'$

(b) $r_E' = \frac{\beta_{ac}}{R_{in}^{base}}$

(c) $R_{in}^{base} = \frac{r_E'}{\beta_{ac}}$

(d) $R_{in}^{base} = \frac{r_E'}{\alpha}$

ANS: (a).

MC6.17 A given configuration of a single BJT amplifier has these values of hybrid parameters $h_i = 1100\Omega$, $h_f = -50$, $h_r = 1$, and $h_0 = 25 \times 10^{-6}$ Siemen. The configuration is

(a) common base
(b) common emitter
(c) common collector
(d) class AB

ANS: (c).

MC6.18 As per Ebers–Moll model of BJT the approximate value of base current I_B in cut-off mode is given as

(a) $(1 - \alpha_F)I_E(b) - (1 - \alpha_R)I_{CS}$
(b) I_B
(c) $(1 - \alpha_R)I_{CS}$
(d) $-(1 - \alpha_F)I_E$

ANS: (b).

MC6.19 At (-6dB) the voltage gain of an amplifier falls to $1/X$ of its maximum value. X is

(a) 100
(b) 10
(c) 2
(d) $\sqrt{2}$

ANS: (c).

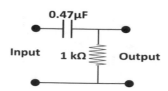

Fig. MC6.20 Circuit for (MC6.20)

MC6.20   Circuit shown in the figure will pass all signals with frequency f

    (a) f < 34 kHz
    (b) f < 340 kHz
    (c) f < 34 kHz
    (d) f < 34 kHz

ANS: (c).

MC6.21   Higher and lower cut-off frequencies for a BJT amplifier are decided, respectively, by

    (a) parasitic plus stray wiring capacitances; coupling plus emitter bypass capacitance
    (b) coupling plus emitter bypass capacitance, parasitic plus stray wiring capacitances
    (c) stray wiring plus emitter bypass capacitance; parasitic plus coupling capacitances
    (d) only coupling capacitances; only parasitic capacitances

ANS: (c).

MC6.22   In small-signal hybrid-pi model of FET, the output resistance r_0 is given as

    (a) $I_{DSS}\left(1 - \frac{V_{GS}}{V_P}\right)^2 \lambda$

    (b) $I_{DSS}\left(1 - \frac{V_{GS}}{V_P}\right)^2 \frac{1}{\lambda}$

    (c) $\dfrac{1}{I_{DSS}\left(1 - \frac{V_{GS}}{V_P}\right)^2 \lambda}$

    (d) $\dfrac{1}{I_{DSS}\left(1 - \frac{V_{GS}}{V_P}\right)^2}$

ANS: (a).

MC6.23   In a metal–semiconductor junction if E_M^{con} and E_S^{con}, respectively, denote the maximum energies of the conduction bands of metal and of semiconductor, then

    (a) $E_M^{con} > E_S^{con}$
    (b) $E_M^{con} = E_S^{con}$
    (c) $E_M^{con} < E_S^{con}$
    (d) E_M^{con} maybe > or < E_S^{con}

ANS: (c).

MC6.24 Symbols for depletion-mode n-channel and enhance mode p-channel MOSFETS are, respectively,

  (a) (i) , (iv)
  (b) (ii), (iii)
  (c) (i), (ii)
  (d) (iii), (iv)

ANS: (a).

MC6.25 Which of the following has least noise, high input impedance and high sensitivity to static charges?

  (a) BJT
  (b) JFET
  (c) MESFET
  (d) MOSFET

ANS: (d).

MC6.26 A n-channel MOSFET is better than p-channel type because

  (a) it is faster compatible (c) (d) it has less noise
  (b) it is TTL compatible
  (c) it has better immunity
  (d) it has less noise

MC6.27 Which of the following effects can be caused by a rise in the temperature?

  (a) increase in collector current in BJT
  (b) decrease in collector current in BJT
  (c) increase in I_D in MOSFET
  (d) decrease in I_{DS} in MOSFET

ANS: (a), (c).

Short Answer Questions

SA6.1   Draw small-signal transconductance model equivalent circuit for a MOSFET and discuss the dependence of g_m on V_{GS}.

SA6.2   With the help of a suitable diagram discuss the working of a MESFET.

SA6.3   What are depletion and enhance modes of a MOSFET? Give two advantages of MOSFET over BJT.

SA6.4   A nearly constant drain current I_D flows between the Pinch-off point up to the breakdown point in a JFET. Explain the process of current flow.

SA6.5   What are Ohmic, saturation and cut-off regions of JFET?

SA6.6   Draw a rough sketch of the transfer characteristics of a MOSFET.

SA6.7   Write the expression (without derivation) for drain current I_D in terms of the threshold potential V_P and gate voltage V_{GS}. Ignore channel width modulation effects.

SA6.8   Draw a rough sketch of Bode plot for an npn common emitter single-stage amplifier and show -3dB points.

SA6.9   What factors are responsible for the fall of the voltage gain of a BJT amplifier at low frequencies and at high frequencies?

SA6.10 Write a brief note on parasitic elements of a BJT

SA6.11 Describe in short Ebers–Moll model of BJT.

SA6.12 Define hybrid (model) parameters for BJT and give their order of magnitudes for CE configuration.

SA6.13 Give a brief account of r-parameter model of BJT

SA6.14 What is meant by bias stabilization? Why is it important?

SA6.15 What is transistor action? How is it achieved in a BJT?

SA6.16 A BJT is an inherent amplifier, elaborate on this.

SA6.17 What is Early effect? Does it occur only for BJT?

Sample Answer to Short Answer Question SA6.3

What are depletion and enhanced modes of a MOSFET? Give two advantages of MOSFET over BJT.

Answer

Figure 1 a shows an enhanced-mode n-channel MOSFET. As may be seen in the In this figure there is no conducting channel already present between the Drain D and the Source S. As such if any potential difference is applied between D and S, keeping gate G at ground potential, no current will flow through the device between the drain and the source. As such, in absence of any gate voltage the MOSFET behaves as an OFF switch. Since the Gate region behaves like a parallel plate capacitor, if a negative potential is applied at the gate, the induced positive charge below the gate surface in the P-substrate neutralizes some holes and develops an n-channel between the Drain and Source. The electron concentration (number of free electrons

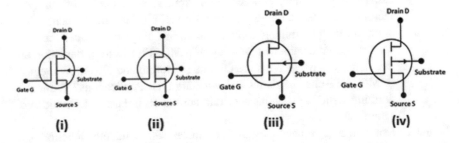

(i)   **(ii)**   **(iii)**   **(iv)**

Fig. 1 a Enhanced-mode n-channel MOSFET **b** Depletion-mode n-channel MOSFET

per unit volume) in this induced n-channel will depend on the amount of the positive charge developed under the gate area, which in turn will depend on the magnitude of the negative potential given to the gate terminal. If a potential difference is maintained between the Drain and the Source, currents proportional to the negative gate voltage will flow through the device. Since, for a fixed potential difference between the Drain and the Source, the magnitude of the current increases with the increase of the negative Gate voltage, the device is called an Enhancement mode device. In a similar way, there may be a p-channel enhanced-mode MOSFET.

On the other hand in a depletion-mode n-channel device, an n-channel is developed by doping the region of substrate semiconductor lying below the Gate with an n-type dopant and thus an n-type conducting channel already exists under the Gate. It means that when Gate potential is zero, some current will flow between Drain and Source if a potential difference is applied across them. As such, the depletion-mode MOSFET behaves as a CLOSED switch when there is no voltage at the Gate. When a negative voltage is applied at the gate, the induced positive charge in the pre-existing n-type channel neutralizes some of the free electrons and reduces the electron concentration, which in turn reduces or depletes the current through the device. It is for this reason that this is called a depletion-mode device. One may also explain the working of a p-channel depletion-mode MOSFET.

Advantages of MOSFET

1. Since there is no electrical connection between the Gate terminal and the conducting channel in a MOSFET, the input impedance of the device is very large. On the other hand, the input impedance of a BJT is considerably low. As a result MOSFET amplifier may be used as the first stage of amplification without loading the signal source.
2. Electronic noise is generated whenever recombination of charge carriers takes place in a device. Since current in a MOSFET is constituted by carriers of only one kind, either electrons or holes, no recombination takes place in a MOSFET and hence noise component is very low in MOSFET.

Long Answer Questions

LA6.1   Discuss in detail the working of a BJT. List the different biasing schemes using a single battery and derive an expression for the bias stability factor against temperature variation for self-biasing arrangement.

LA6.2   With the help of suitable figures and circuit diagrams discuss the constructional details and working of a MOSFET both in depletion and in enhanced modes. What are Ohmic, saturation and breakdown regions in MOSFET characteristics and what are their applications? Without deriving write expressions for drain current I_D in cases of Ohmic and saturation regions.

LA6.3   Draw an energy band diagram of a metal–semiconductor junction and use it to explain the working of a metal–semiconductor field-effect transistor (MESFET). Why MSFET made with GaAS are frequently used in microwave equipment?

LA6.4   What are FETs and in what respect they are superior to BJT? What are the special features of a MOSFET? Draw circuit diagrams of a depletion-mode MOSFET single-stage amplifier and its hybrid-pi model equivalent and discuss their output characteristics.

LA6.5   Draw the circuit diagram of an npn BJT single-stage amplifier with self-bias and discuss the nature of its frequency response characteristics. Using small-signal equivalents derive expressions for the lower and upper cut-off frequencies.

LA6.6   Why bias stabilization is required in circuits that use BJT? Drive expressions for the three stability factors in cases of fixed and self-biases.

LA6.7   With the help of a circuit diagram of a BJT amplifier explain the difference between the DC and the AC load lines. What is Q-point and what are its defining parameters. Discuss how the selection of Q-point decides the class of amplifiers. Derive an expression for the no-load voltage gain of a single-stage BJT amplifier.

LA6.8   What is the need for transistor modelling? Give a detailed account of the r-parameter small-signal model and use it to analyse the working of emitter follower?

LA6.9   Discuss with necessary details the hybrid model of transistor. Why it is termed as the black-box approach? Derive expressions for hybrid parameters in CE configuration.

LA6.10   Describe Ebers–Moll model of transistor. Why it is called the non-linear model? Discuss in light of this model the four regions of operation of a BJT. Also, discuss the variation of minority concentration in base for these four operations.

LA6.11   Discuss in detail the transconductance or hybrid-pi model of transistor. What is meant by parasitic parameters? Draw the complete hybrid-pi model equivalent of a single-stage BJT amplifier including the parasitic elements.

LA6.12   On what different parameters BJT amplifiers may be classified? Analyse the working of a common collector BJT amplifier using hybrid-pi model. What is its major application and why?

LA6.13   Take a suitable example with the help of which explain the method of DC and AC analysis of a BJT circuit.

LA6.14   Draw circuit diagrams of npn BJT single-stage amplifiers in CC, CE and CB configurations using self-bias. Give a detailed analysis of the amplifier in any one configuration driving expressions for voltage and current gains. Which configuration has maximum current gain?

Feedback in Amplifiers

7

Abstract

Signal amplifier is an electronic device that amplifies the signal applied at the input and delivers an amplified signal at the output. When a part of the output signal is taken and mixed with the input signal, the process is called feedback. Feedback may be of two types; positive feedback in which the feedback part of the signal is in-phase with the input signal, and therefore, the overall gain of the amplifier increases. However, if the feedback signal is out of phase, the overall gain decreases and the process is termed as negative feedback. Both positive and negative feedbacks are frequently used; former to convert an amplifier into Oscillator and later to stabilize the amplifier operation. Details of the two types of feedbacks, methods of achieving them, their effect on the performance of two port networks, etc. are discussed in this chapter.

7.1 Introduction

Most of the electronic and electrical circuits/devices have an input port and an output port, a signal is applied at the input, the circuit elements/device structure performs operations on the input signal and delivers the desired signal in the output. If a part of the output signal is taken out by a suitable network and re-applied to the input signal, the process is called **feedback**. The network that takes out the part of the output signal and connects it to the input is called the **feedback network**. The feedback network is generally passive, i.e. a combination of passive circuit elements like resistance, capacitor, inductance, etc. In some special cases, however, the feedback network may contain active elements like battery, transistors, etc. In this chapter, we shall discuss the effects of feedback on linear amplifiers.

© The Author(s), under exclusive license to Springer Nature Switzerland AG 2021   583
R. Prasad, *Analog and Digital Electronic Circuits*, Undergraduate Lecture
Notes in Physics, https://doi.org/10.1007/978-3-030-65129-9_7

The block diagram of a feedback system associated with a linear amplifier (now on we shall simply call it amplifier) is shown in Fig. 7.1a. It is assumed here that the amplifier is an ideal amplifier and is a unidirectional device that transmits signals only from input to output, i.e. in forward direction only. The signal source provides the input signal X_s which may either be a voltage or a current signal. In absence of the feedback network, the input signal will reach directly at the input of the amplifier, and an A times amplified signal of magnitude (AX_s) will appear at the amplifier output from where it goes to the load. Here A is the **open-loop gain** of the amplifier. However, with feedback system in place, the situation gets changed. The **feedback system** may be resolved into three components: (i) **Sampling system,** which takes out a part β of the amplifier output and applies it at the feedback network. If the amplifier output is voltage, the sampled signal will also be voltage and if the amplifier output is current, the feedback component will also be current. (ii) The **feedback network** is generally made of passive circuit elements which in principle may transmit a signal applied to it in both the forward and the backward directions, but it is **assumed** that the **feedback network is unidirectional** and transmits signals in backward direction only. The multiplicative factor β is called **feedback factor** and its magnitude is decided by the structure of the feedback network. The signal that appears at the back-end of the feedback network is denoted by $X_f = \beta X_o$. (iii) The third component of feedback network is the **mixer**. A mixer, generally represented by a summation sign Σ, mixes the feedback component with the input signal X_s before it reaches the input of the amplifier. Signal X_e which reaches the amplifier input after the mixing of the feedback signal X_f is called an **error signal**. Mixing may be of two types: (a) the feedback signal X_f is in-phase

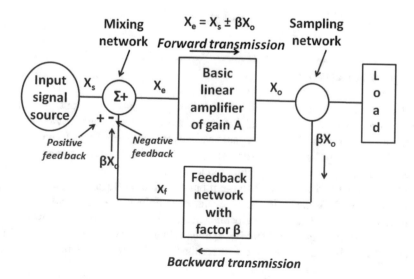

Fig. 7.1a Basic lay out of a feedback system applied to a linear amplifier

with the source signal X_s and adds to the input signal X_s, so that the signal reaching at amplifier input X_e is the sum of X_s and X_f. This is called **positive feedback**.

In positive feedback $X_e = X_s + X_f = X_s + \beta X_0 = X_s + \beta(AX_e)$

Or

$$X_s = (1 - \beta A)X_e \tag{7.1}$$

Also, amplifier output X_o after incorporating feedback network is given by

$$X_o = AX_e \tag{7.2}$$

Therefore, the overall gain or **closed-loop gain** of the amplifier after feedback, denoted by A_f is given by

$$A_f = \frac{X_0}{X_s} = \frac{AX_e}{(1 - \beta A)X_e} = \frac{A}{1 - \beta A} \tag{7.3}$$

Following points may be noted regarding (7.3);

(a) In case of positive feedback $|\beta A|$ cannot exceed 1.
(b) βA may be totally real or it may be a combination of real and imaginary parts.
(c) In case when $\beta A = 1$ and the angle of the real and imaginary part is zero (< 0), A_f becomes ∞ and the system starts behaving like an oscillator. If the above condition $(\beta A = 1; \angle 0^0)$ is fulfilled at only one frequency, say, ω, then the amplifier with positive feedback becomes an oscillator that generates sinusoidal waves of frequency ω.
(d) The condition $(\beta A = 1; \angle 0^0)$ is called Barkhausen criteria for sustained oscillations. We shall discuss details of oscillators, etc. when we take up positive feedback in more details.

7.1.1 Negative Feedback in Amplifiers

If the source signal X_s and the feedback signal X_f is 180° out of phase, on mixing the error signal X_e will be given by

$$X_e = X_s - X_f = X_s - \beta X_0 = X_s - \beta(AX_e)$$

Or

$$X_s = (1 + \beta A)X_e$$

And

$$A_f = \frac{X_0}{X_s} = \frac{AX_e}{(1+\beta A)X_e} = \frac{A}{(1+\beta A)} \tag{7.4}$$

Comparison of (7.3) with (7.4) tells that in case of negative feedback the denominator $(1 + BA)$ is greater than 1, and therefore, overall gain A_f (also called closed-loop gain) is less than the open-loop gain A. So in negative feedback the closed-loop gain decreases, while in positive feedback the closed-loop gain increases over the magnitude of the open-loop gain.

Summing up, it may be said that there may be two distinct effects of feedback: either the part of the output that is fed back via the feedback network (i) **decreases the magnitude of the input signal**, in that case, feedback is called **negative or degenerative** or (ii) it may **strengthen the input signal**, then the **feedback** is termed **positive or regenerative**. Obviously, in case of **negative feedback** the feedback component is **out of phase with the input signal by 180°**; while in case of **positive feedback** it is **in-phase** with the input signal.

Since the amplifier is a two port device (the input and the output ports), the basic feedback block diagram of Fig. 7.1a in case of amplifiers may be shown by Fig. 7.1b.

And for negative feedback the closed-loop gain A_f is given by (7.4). In case when $\beta A \gg 1$

$$A_f = \frac{A}{A\beta} = \frac{1}{\beta} \tag{7.5}$$

Equation (7.5) tells that the **closed-loop gain** of the amplifier with **high negative feedback depends only on the feedback factor β** which is the property of the feedback network. In such a situation $(\beta A \gg 1)$ closed-loop gain A_f becomes independent of amplifier parameters. This is important in view of the fact that any change in amplifier parameters now does not affect the closed-loop gain, giving high degree of stability to the operation of the negative feedback amplifier. It is known that amplifier parameters may change because of (i) the ageing of its

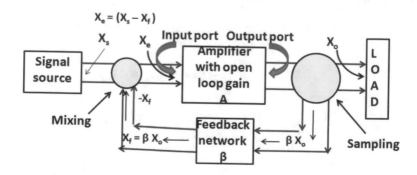

Fig. 7.1b Block diagram of an amplifier with feedback

components and or when some worn out components are replaced (ii) due to change in the ambient temperature. Of course, this additional stability has been achieved at the cost of reduction in closed-loop gain of the amplifier.

$$\text{Further, in case when } \beta A >> 1, \quad A_f = \frac{1}{\beta} = \frac{X_o}{X_S} \text{ or } X_S = \beta X_o \qquad (7.6)$$

But $X_e = X_s - X_f$ and from (7.6) $X_s = \beta X_o$ while $X_f = \beta X_0$

Therefore, in the limit $\beta A \gg 1$, X_e *approches zero*. This means that in this limiting case no signal is applied at the input of the amplifier irrespective of what happens to the internal parameters of the amplifier. That is why the closed-loop gain A_f becomes constant equal to $1/\beta$.

Self-assessment question: What happens to the inherent (open-loop) gain of the amplifier when negative feedback is employed? Does it change?

7.2 Classification of Amplifiers

Before going into further details of negative feedback, let us briefly review the characterization of amplifiers. Basic amplifiers may be classified into four types.

7.2.1 Voltage–Voltage Amplifier or Voltage Amplifier

These are amplifiers the input and the output of which are both voltage signals. These amplifiers are also called voltage-voltage transducer and in short VVT. The equivalent circuit for VVT is as shown in Fig. 7.1c.

The upper diagram in Fig. 7.1c shows the actual or practical VVT circuit where the input source has impedance Z_s which for satisfactory operation of the amplifier must be much smaller than the input impedance Z_i and similarly the output impedance Z_0 must be much smaller than the load resistance R_L. The circuit enclosed in the dotted rectangle is the equivalent circuit of the amplifier. V_s, Z_i, A_v, Z_o, Z_L and v_o, respectively, denote the strength of the voltage signal provided by the source, the input impedance, open-loop gain, output impedance, load impedance and voltage across the load of the amplifier. Open-loop gain A_v of the amplifier in case of VVT is a dimensionless number or fraction.

Self-assessment question: What will happen if $Z_i = Z_s$ and $Z_o = Z_L$?

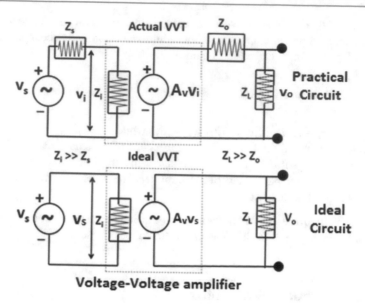

Fig. 7.1c Equivalent circuit of a real and an ideal voltage amplifier (VVT)

7.2.2 Voltage–Current or Transconductance Amplifier (VCT)

As the name suggests, the input of a voltage–current amplifier is a voltage signal while it delivers a current signal in the output. Therefore, in the equivalent circuit, the input section contains a voltage source while the output section has a dependent current source, the current of which depends on the input voltage. The practical and the ideal equivalent circuits for a VCT are shown in Fig. 7.1d.

Further, the CVT is also called the **transconductance amplifier** as the units of amplification factor A_g output current to input voltage (i/v) are those of conductance, i.e. Siemen, reciprocal of ohm. Another symbol G_m has also been used to represent gain of a Voltage-current amplifier. It needs to be remembered that gains of amplifiers may have units and not necessarily dimensionless quantities. The output current, in this case, is designated by i_o.

7.2.3 Current–Current Amplifier (CCT)

The input and the output of a current–current amplifier operate with current signals. The practical and the ideal equivalent circuits for a CCT are shown in Fig. 7.1e. It is self-explanatory why the practical equivalent turns into the ideal equivalent if Z_s >> Z_i and Z_0 >> Z_L. Also, the current amplification factor A_c (dimensionless) is to be taken when Z_L is short-circuited.

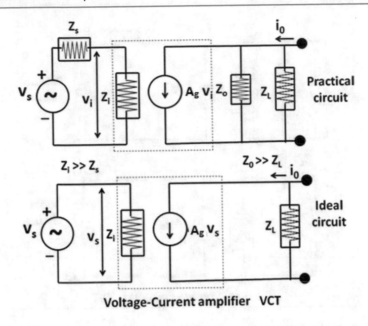

Voltage-Current amplifier VCT

Fig. 7.1d Equivalent circuits for a practical and an ideal voltage-current amplifier (ACT)

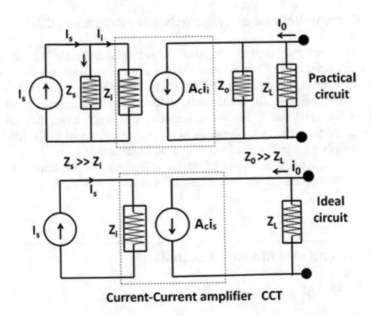

Current-Current amplifier CCT

Fig. 7.1e Practical and ideal equivalent circuits for a current-current amplifier (CCT)

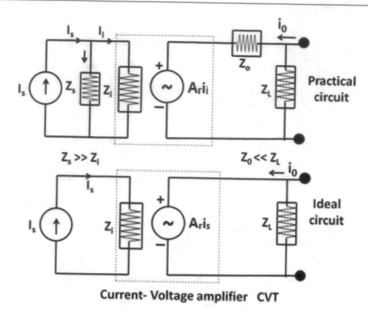

Current- Voltage amplifier CVT

Fig. 7.1f Practical and ideal equivalent circuits for transresistance amplifier or CVT

7.2.4   Current–Voltage or Transresistance Amplifier (CVT)

Figure 7.1f shows the practical and ideal equivalent circuits for current–voltage amplifier the input of which is a current signal and the output is a voltage signal.

Since the gain A_r (also represented by R_m) of the amplifier has units of resistance, therefore, it is also called the **transresistance amplifier**. A practical amplifier approaches the ideal one if the source impedance is much larger than the input impedance and the output impedance is negligible as compared to the load impedance. Symbols used in Fig. 7.1f have their usual meanings.

Before moving on to the topic of negative feedback in amplifiers, it will be worthwhile to understand the processes of sampling and mixing in feedback scenario.

7.3   Sampling and Mixing of Signals

7.3.1   Sampling

Sampling and mixing are two complementary processes in any type of feedback. In the processes of sampling, a part of the signal from the output of the amplifier is taken. It is obvious that if the amplifier is voltage-voltage (VVT) or Current–voltage (CVT) type, the output signal will be a voltage signal. **A part of the output voltage** signal can be taken by the sampling network only through a parallel connection to the output. Therefore, this type of sampling is termed as **Voltage sampling or Shunt sampling**. In the other case, when the amplifier delivers a current pulse in its output as in the case of VCT and CCT, the feedback sampling signal will also be a current signal. **A part of the current signal can be taken by the feedback network only through a series connection**. The sampling of a current pulse is, therefore, carried out using a series connection. The technical name for this sampling is **Current sampling or Series sampling**.

In short, the sampling may be either **Voltage sampling** (Shunt connection) or **Current sampling** (Series connection) as shown in Fig. 7.1a(i), (ii).

7.3.2 Mixing

The part of the output signal sampled by the feedback network is mixed with the signal at the input of the amplifier. Since a voltage can be mixed only with a voltage and a current only to a current; therefore, the feedback network converts the sampled feedback signal to the current signal or to the voltage signal, if required. For example, in the case of a VCT, the output signal is a current signal and the input signal is a voltage signal, therefore, the feedback network samples a current signal of magnitude βi_o (i_o being the output current) from the output and converts it into a voltage signal v_f which is mixed with the voltage signal v_s at the input. Similarly, in the case of a CVT, the feedback network samples a voltage signal βv_o (v_o being the output voltage) from the output and converts it into a current signal i_f before mixing it with the input current signal i_s. Obviously, no such conversion will be required in case of CCT or VVT amplifiers (Fig. 7.2a).

A very important aspect of the process of mixing is that **voltage signals can be added or subtracted only in series connection as shown in** Fig. 7.2b(i). Therefore, mixing of the voltage signal is done using **Series mixing**. It is not possible to add or subtract voltages in parallel connections. In case of negative feedback, the

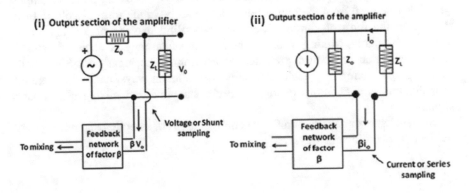

Fig. 7.2a Block diagram of (i) Voltage or Shunt sampling (ii) Current or Series sampling

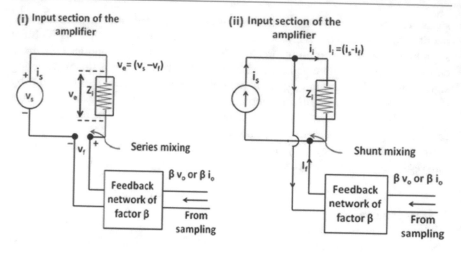

Fig. 7.2b Circuits for (i) Series mixing (ii) Shunt mixing

polarity of the feedback signal v_f should be such that the net voltage (error voltage) $v_e = (v_s - v_f)$. On the other hand, **current signals can be added or subtracted only in parallel connections,** and hence **shunt mixing** is employed for mixing of current signals at the input of the amplifier as shown in Fig. 7.2b(ii). Hence, in all there may be two types of mixing methods: the shunt mixing and the series mixing.

7.4 Sampling and Mixing Topologies (Configurations)

A feedback network takes a part of the output signal from the output port of the device and mixes it with the source signal at the input port. As already discussed, in technical terms the process of taking a part of the output signal is called **SAMPLING** and reapplying it at the input is called **MIXING**. The fraction of the output signal that is taken from the output port may either be in the form of voltage or it may be in the form of current. Therefore, **Sampling** may be of **Voltage** or of **Current**. Similarly, **Mixing** of the feedback component may be achieved using a **series** connection or a **shunt** (parallel) connection at the input port. Since both sampling and mixing may each be done in two different ways, hence there are all four different configurations or topologies for sampling and mixing of feedback signals.

Nomenclature or naming of different configurations of sampling and mixing may be done in two different ways:

(A) Writing first the **sampling method** and next the **type of mixing method**, for example.

(a) Voltage sampling-series mixing topology (in short: Voltage-series)
(b) Voltage sampling-shunt mixing topology (in short: Voltage-shunt)
(c) Current sampling-series mixing topology (in short: Current-series)
(d) Current sampling-shunt mixing topology (in short: Current-shunt).

(B) Writing first the name of the **mixing method** followed by giving the name of the **type of connection** used for sampling. In this method, the topologies given from (a) to (d) above may also be written as.

(a') Series mixing-shunt sampling topology (in short: Series-shunt).

(b') Shunt mixing-shunt sampling topology (in short: Shunt-shunt).

(c') Series mixing-series sampling topology (in short: Series-series).

(d') Shunt mixing-series sampling topology (in short: Shunt-series).

It may be noted that (a) Voltage sampling-series mixing topology and (a') Series mixing-shunt sampling topology refers to the same topology. Similarly, topologies given at (b), (c) and (d) also refers, respectively, to topologies named at (b'), (c') and (d').

Self-assessment question: Which feedback topology will be used to implement feedback in a transconductance amplifier?

7.4.1 Effects of Negative Feedback on Amplifier Properties

Inclusion of negative feedback in amplifiers make the following effects in operational properties of the amplifier.

7.4.2 Reduction in Overall Gain

Equation 7.3 demonstrates that the closed-loop gain (or the overall gain) of the amplifier circuit with negative feedback is given by.

$A_f = \frac{X_o}{X_s} = \frac{A}{(1+\beta A)}$ and since $|(1+\beta A)| > 1$;

$A_f < A$, therefore, the overall gain or closed-loop gain of the amplifier with negative feedback circuit decreases as compared to the open-loop gain of the amplifier.

7.4.3 Desensitization of Overall Amplifier Gain

The open-loop or inherent gain A of the amplifier may change mainly because of two reasons; (a) replacement of (worn out) circuit elements and (b) change in the temperature of the surroundings. The ratio (dA/A), the relative change in the value of A, called the sensitivity of gain A is a measure of the effect of the gain change. It is interesting to calculate the magnitude of the relative change in the closed-loop gain (overall gain after including the feedback network) (dA_f/A_f) with negative feedback and to compare it with (dA/A).

Now, from (7.3) $A_f = \frac{A}{(1 + \beta A)}$

Therefore, $dA_f = d\left[\frac{A}{(1+\beta A)}\right] = \frac{1}{(1+\beta A)} dA + \left\{\frac{-\beta A}{(1+\beta A)^2}\right\} dA = \frac{dA}{(1+\beta A)^2}$

Dividing both sides of the above equation by A_f, one gets;

$$\frac{dA_f}{A_f} = \frac{\frac{dA}{(1+\beta A)^2}}{\frac{A}{(1+\beta A)}} = \left[\frac{1}{(1+\beta A)}\right] \frac{dA}{A} \tag{7.7}$$

Since $|(1 + \beta A)| > 1$, in case of negative feedback, therefore,

$$\frac{dA_f}{A_f} < \frac{dA}{A} \tag{7.8}$$

Thus, inclusion of negative feedback network in amplifier circuit reduces the sensitivity of overall gain to circuit parameters or desensitizes the closed-loop gain. In the limiting case when $\beta A \gg 1$; the closed-loop gain A_f becomes totally independent of the amplifier circuit.

7.4.4 Increase in the Bandwidth of the Amplifier

In our discussion on amplifiers, the bandwidth of the amplifier was defined as the frequency range between the lower and upper cut-off (or corner) frequencies (where the gain falls by 3db value of the mid-frequency gain). It was also pointed out there that corresponding to several possible internal capacitances in parallel with each other in the high frequency equivalent circuit of the amplifier, one may find more than one corner frequencies. Each corner frequency is called a pole. Let us consider an amplifier that has only one upper corner frequency or pole denoted by f_{hi}. If A_M denotes the amplifier gain at mid-frequency (or mid-band gain) and f_{hi} the upper corner frequency then the gain $A(F)$ of the amplifier at some other frequency F is given by the expression

$$A(F) = \frac{A_M}{1 + \frac{F}{f_{hi}}} \tag{7.9}$$

Applying a negative feedback of (frequency independent) feedback factor β, the closed-loop gain $A^f(F)$ of the amplifier at any frequency F will become

$$A^f(F) = \frac{A(F)}{1+\beta A(F)} = \frac{\frac{A_M}{1+\frac{F}{f_{hi}}}}{\left[1+\beta\left(\frac{A_M}{1+\frac{F}{f_{hi}}}\right)\right]} = \frac{\frac{A_M}{1+A_M\beta}}{1+\frac{F}{(1+A_M\beta)f_{hi}}} \tag{7.10}$$

A comparison of (7.9) with (7.10), tells that with negative feedback the mid-frequency gain A_M has changed (decreased, as expected) to

$$A_M^f = \frac{A_M}{1+A_M\beta}, \tag{7.11}$$

And the higher corner frequency f_{hi} has become

$$f_{hi}^f = (1+A_m\beta)f_{hi} \tag{7.12}$$

Similarly, it may be shown that with negative feedback the lower corner frequency f_{lo} becomes

$$f_{lo}^f = \frac{f_{lo}}{(1+A_m\beta)} \tag{7.13}$$

Equations (7.11), (7.12) and (7.13) tell that with the inclusion of negative feedback of feedback factor β: (I) the mid-frequency gain of the amplifier decreases by the factor $(1+A_m\beta)$, (II) the magnitude of the higher corner frequency increases by the factor $(1+A_m\beta)$ and (III) the magnitude of the lower corner frequency is reduced by the factor $r(1+A_m\beta)$. It is, however, interesting to note that the multiplication of bandwidth with gain remains constant, since a decrease in the mid-frequency gain is compensated by the increase in the bandwidth. This behaviour is called **gain-bandwidth trade-off**.

Figure 7.3 shows the frequency response characteristic of an amplifier without any feedback (upper curve), where the mid-frequency gain A_M has the value 10^5 and the bandwidth (f_{lo} to f_{hi}) has the value $(1 \times 10^4 - 1 \times 10^{-4}) \approx 10^4 Hz$. The bandwidth X gain without feedback $= \approx 10^4 x 10^5 = 10^9$. When negative feedback is employed, the closed-loop gain (with feedback) A_M^f falls to 10^2 and the bandwidth increases to $(1x10^7 - 1x10^{-7}) \approx 10^7$. The magnitude of bandwidth X gain with negative feedback $= 10^2 \times 10^7 = 10^9$, constant.

Self-assessment question: Derive an expression for f_{lo}^f the lower corner frequency with negative feedback.

Self-assessment question: Calculate the value of feedback factor β for the case represented by Fig. 7.3.

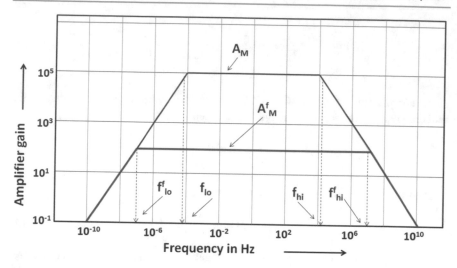

Fig. 7.3 Frequency response characteristic of an amplifier with and without negative feedback

7.4.5 Reduction in Amplifier Noise

Every electronic circuit/device generates noise. Noise consists of undesired random signals that are generated because of many different reasons in a circuit. A big cause of noise generation is the current flow through resistive elements in the circuit; such noise is called Johnson noise. In case of amplifiers, considerably lesser noise is produced in voltage amplifiers as compared to the power amplifiers. It is because in voltage amplifiers small currents flow through circuit elements, while in power amplifiers larger currents flow through circuit elements generating large noise. The open-loop gain of voltage amplifiers may have a large value of gain of the order of a few hundred or so, but their output signal does not have enough power. On the other hand, the open-loop gain of the power amplifier may be of the order of 1, but the power of their output signal is large. In order to produce highly amplified and high power signals, a few low power voltage amplifiers are cascaded in series and the output of the last voltage amplifier is fed to a power amplifier which delivers a high voltage high power signal. In all this process a substantial amount of noise is introduced in the output because of the power amplifier.

An important parameter for noise characterization of electronic circuits is the signal to noise ratio, denoted by SNR and defined as

$$SNR = \frac{Magnitude\ of\ actual\ signal}{Magnitude\ of\ noise\ signal} = \frac{V_s}{V_n} \tag{7.14}$$

Here V_s and V_n are, respectively, the amplitudes of the signal voltage and the noise voltage.

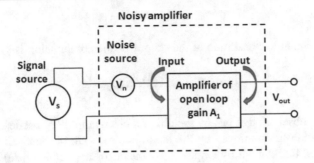

Fig. 7.4a Block diagram of a noisy amplifier. It is assumed that all noise in the amplifier is produced by a noise source connected at the input

Figures 7.4a and 7.4b shows a noisy amplifier of open-loop gain A_1. It is assumed for simplicity that all the noise of the amplifier is produced by a noise source of noise voltage V_n that is connected at the input of the amplifier. An external source that generates signals of voltage V_s is now connected to the amplifier. The output voltage of the amplifier will be given by

$$V_{out} = A_1(V_s + V_n) = A_1 V_s + A_1 V_n \qquad (7.15)$$

It may be observed from (7.15) that the signal to noise ratio at the output of the noisy amplifier still remains $SNR_{out} = \frac{A_1 V_s}{A_1 V_n} = \frac{V_s}{V_n}$, the same as it was at the input of the amplifier.

Next, let us couple a clean amplifier (like a voltage amplifier that has negligible noise) of open-loop gain A_2 with a noisy amplifier (may be a power amplifier) of open-loop gain A_1 in series (see and look for the noise in the output of this series combination. While analyzing this network, one must remember that the net gain of two amplifiers in series is given by the multiplication of the gains of individual amplifiers. The output of this series combination will be given by

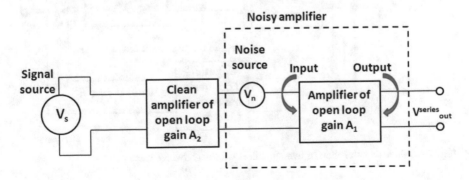

Fig. 7.4b Series combination of a clean and a noisy amplifier

$$V_{out}^{series} = (A_2 A_1 V_s + A_1 V_n)$$

And the signal to noise ratio at the output of the combination is given by

$$SNR_{out}^{series} = \frac{A_2 A_1 V_s}{A_1 V_n} = A_2 \frac{V_s}{V_n} \tag{7.16}$$

It may be observed that just by including a clean amplifier in series, the SNR of the combination has improved by the factor A_2, the gain of the clean amplifier. It may be noted that though by employing a clean (or negligible noise) amplifier in series the signal to noise ratio has improved but instabilities in the operation of the series combination have considerably increased. It is because now variations in the values of circuit elements of the clean amplifier as well as those of the noisy amplifier and also variations in ambient temperature will affect the performance of the circuit. Thus by incorporating a clean amplifier, the SNR of the combination has improved but the system has become more unstable with respect to the parameters of the two amplifiers and the temperature.

Finally, let us investigate the effect of negative feedback on the SNR of the series combination of the two amplifiers. As shown in Fig. 7.4c, a negative feedback with feedback factor β is included with the series combination of the two amplifiers.

It may be recalled that the overall gain of an amplifier of open-loop gain A becomes $\frac{A}{1+\beta A}$ when negative feedback of factor β is included in the circuit. Therefore, the output voltage V_{out}^f for the circuit of Fig. 7.4c will be given by

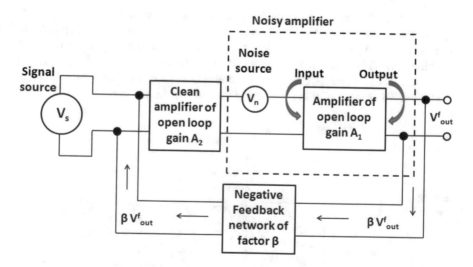

Fig. 7.4c Series combination of two amplifiers with negative feedback

$$V_{out}^f = \frac{A_2 A_1}{1 + \beta A_2 A_1} V_s + \frac{A_1}{1 + \beta A_2 A_1} V_n$$

And the signal to noise ratio of the combination with negative feedback SNR_{series}^f is given by

$$SNR_{series}^f = A_2 \frac{V_s}{V_n} \qquad (7.17)$$

It is worth noting that the signal to noise ratio with negative feedback SNR_{series}^f has the same magnitude as for the series combination (without feedback) SNR_{out}^{series} but the added advantage of negative feedback is that now the network is not much dependent on the variations in the parameters of the two amplifiers and temperature. It is more stable.

7.4.6 Reduction in Non-Linear Distortion:

In practice, the output v_o of an amplifier is related to its input signal v_i through an equation of the type given below.

$$v_0 = V_{offset} + A v_i + K_1 v_i^2 + K_2 v_i^3 + K_3 v_i^4 + \ldots \qquad (7.A)$$

In the above (7.A), V_{offset} may be treated like a noise signal which is generated by the amplifier circuit. The offset component varies with temperature like noise. $A v_i$ is the linear component of the output and all other terms ($K_1 v_i^2 + K_2 v_i^3 + K_3 v_i^4 + \ldots$) put together constitute the non-linear component. In a good linear amplifier, offset is low, and coefficients K_1, K_2, ..., etc. of the non-linear terms are also small. Now, if a negative feedback of factor β is employed, then the actual signal that reaches the input of the amplifier is the error signal given by $v_e = (v_i - \beta v_o)$. Substituting v_e in place of v_i in the above equation, one gets

$$v_o = V_{offset} + A(v_i - \beta v_o) + K_1(v_i - \beta v_o)^2 + K_2(v_i - \beta v_o)^3 + K_3(v_i - \beta v_o)^4$$

Or $(1 + \beta A)v_o = V_{offset} + A v_i + K_1(v_i - \beta v_o)^2 + K_2(v_i - \beta v_o)^3 + K_3(v_i - \beta v_o)^4$

And $v_o = \frac{V_{offset}}{(1 + \beta A)} + \frac{A}{(1 + \beta A)} v_i + \frac{1}{(1 + \beta A)}\left[K_1(v_i - \beta v_o)^2 + K_2(v_i - \beta v_o)^3 + K_3(v_i - \beta v_o)^4 .. \right]$

$$(7.B)$$

Equation (7.B) tells that with the inclusion of negative feedback the offset and the coefficients of all non-linear terms have reduced by the factor $(1 + \beta A)$. Thus, negative feedback reduces the offset, i.e. noise and also the contribution of non-linear components. It is, however, also true that the gain with negative feedback also get reduced over the open-loop gain by the factor $(1 + \beta A)$, a price to be paid for the reduction in noise and non-linear components.

A linear amplifier amplifies input signals of all magnitudes by the same amount, the gain A of the amplifier. Theoretically transfers characteristic (output Vs input graph) of a linear device is a straight line. It may, however, be remembered that the transfer characteristic remains linear only within a given range of the input signal. In the amplifying action of a transistor, it is known that, if the size of the input signal increases beyond a certain value, the operating point of the amplifying device, the transistor, goes into saturation or cut-off. Both saturation and cut-off regions make the transfer characteristic non-linear. Figure 7.5 shows typical transfers characteristic of an amplifier. As may be seen in the figure, origin O represents the Q-point and in absence of negative feedback the dynamic range of input signal is given by AA'. If the signal goes beyond A or beyond A' non-linear distortions appear in transfer characteristics. However, when negative feedback is employed in the amplifier circuit, the gain decreases, transfer characteristic becomes less slant, the dynamic range of input signal increases to BB' and relative non-linear distortion also decreases.

Typical transfers characteristics of two different amplifiers are shown in Fig. 7.6a, b. Dotted lines in two figures show the ideal or theoretical response curve. In figure (a), the solid line represents the actual transfer characteristic which has a distortion in the middle that might be due to the non-linear response of the device for the small range of the input parameters. In figure (b), the actual response without fee back is initially linear but at higher input signals it becomes non-linear. When negative feedback is incorporated in these circuits, both types of non-linear distortions get reduced.

Since non-linearity originates from the variation in the amplification factor (or gain) A of the amplifier, the open loop non-linear distortion D is defined as.

$D = \frac{\Delta A}{A}$, where ΔA is the fluctuation in the magnitude of the gain A. If ΔA is infinitesimally small, it may be replaced by the differential dA and the open loop distortion D may be written as

Fig. 7.5 Typical transfers characteristics of an amplifier

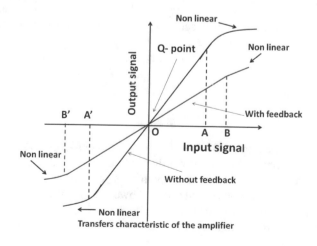

Transfers characteristic of the amplifier

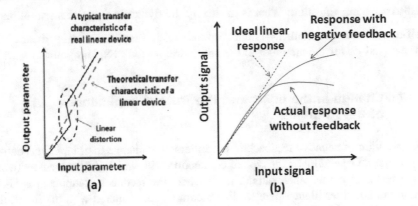

Fig. 7.6 Transfer characteristics of real and ideal linear device

$$D = \frac{dA}{A} \tag{7.18}$$

Similarly, one may define the distortion with negative feedback by $D^f = \frac{dA_f}{A_f}$ where A_f is the amplifier gain when negative feedback of factor β is integrated into the amplifier. Now it is required to find a relation between D and D^f to investigate the effect of negative feedback on distortion.

Again,

$$A_f = \frac{A}{(1 + \beta A)} \tag{7.19}$$

Differentiating both sides, one gets

$$dA_f = (1 + \beta A)^{-1} dA - \beta A dA (1 + \beta A)^{-2} = \frac{dA}{(1 + \beta A)^2} = \frac{dA}{(1 + \beta A)} \frac{1}{(1 + \beta A)} \tag{7.20}$$

But from (7.19) $\frac{1}{(1 + \beta A)} = \frac{A_f}{A}$, substituting this value in (7.20), one gets

$$dA_f = \frac{dA}{(1 + \beta A)} \frac{A_f}{A} \quad or \quad D^f = \frac{dA_f}{A_f} = \frac{1}{(1 + \beta A)} \frac{dA}{A}$$

$$Or \quad D^f = \frac{D}{(1 + \beta A)}. \tag{7.21}$$

The negative feedback, therefore, reduces the non-linear distortion by the factor $(1 + \beta A)$.

Self-assessment question: What is meant by the dynamic range of an amplifier?

Self-assessment question: Does the response of an amplifier remain linear if the input signal drives the amplifier in the saturation region? Explain your answer.

7.4.7 Change in the Input and the Output Impedance of the Amplifier

Incorporation of negative feedback in amplifiers may change their input and output impedances. Depending on the type of connection with which the feedback network is connected at the amplifier output, from where the feedback component is taken and at the amplifier input where feedback component is mixed with the input, the output and input impedances may get changed (increased or decreased) by a factor $(1 + \beta A)$. In order to understand these changes in impedances, it is required to know different configurations or topologies of sampling and mixing of negative feedback component.

(a) **Voltage amplifier with negative feedback**.
The input signal of a voltage amplifier is a voltage pulse and the amplifier also delivers a voltage signal at its output. It employs **Voltage sample-series mixing topology (or Series- shunt topology)**; i.e. the negative feedback network samples voltage from the output (using shunt connection) and feedback a voltage signal $v_f = \beta v_o^f$ using a series connection at the input, as shown in Fig. 7.7.

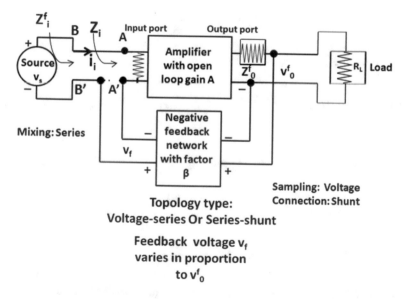

Fig. 7.7 Block diagram of a voltage amplifier using voltage-series topology for negative feedback

As shown in the figure, the fed back voltage signal v_f is proportional to the output signal v_o^f and is given by

$$v_f = \beta v_o^f \qquad (7.22)$$

We now investigate the effect of negative feedback series-shunt topology on the input and the output impedances of the amplifier. Even without doing detailed analysis, it is possible to predict the nature of change in these impedances. Since the feedback network is connected in parallel to the output of the amplifier, the output impedance of the amplifier with feedback Z_o^f will decrease compared to the output impedance Z_o without feedback as a result of the inclusion of feedback network. Similarly, the feedback network is connected in series at the input, the input impedance Z_i^f of the amplifier with feedback will increase as compared to the input impedance Z_i without feedback.

(i) *Change in the input impedance* It is important to understand that the open-loop input impedance Z_i is the magnitude of impedance across points AA'; while the input impedance with feedback Z_i^f appears across points BB' in Fig. 7.7. If a current i_i flows in the input circuit of the amplifier, then

$$Z_i = \frac{v_i}{i_i} \qquad (7.23)$$

And

$$Z_i^f = \frac{v_s}{i_i} \qquad (7.24)$$

But $v_i = v_s - v_f$

$$\text{Or} \quad v_s = v_i + v_f = v_i + \beta v_o^f = v_i + \beta A_v v_i = v_i(1 + \beta A_v) \qquad (7.25)$$

Substituting the value of v_s from (7.25) in (7.24), one gets

$$Z_i^f = \frac{v_s}{i_i} = \frac{v_i(1 + \beta A_v)}{i_i} = Z_i(1 + \beta A_v) \text{ [Since from (7.23) } Z_i = \frac{v_i}{i_i}] \qquad (7.26)$$

The input impedance with feedback becomes $(1 + \beta A)$ times the open-loop input impedance Z_i. Also, since $(1 + \beta A_v) > 1$; $Z_i^f > Z_i$

Therefore, the input impedance of a voltage amplifier with **series mixing of negative feedback** increases.

(ii) *Change in output impedance* Next we investigate the **effect of shunt sampling** on the output impedance of the amplifier. Let us denote the open-loop (without feedback) output impedance of the amplifier as Z_o and with negative feedback with **shunt sampling** as Z_o^f.

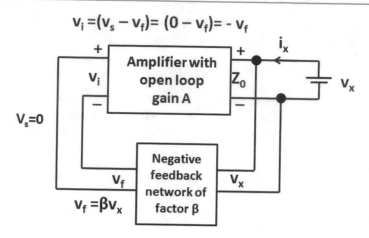

Fig. 7.7a Equivalent network for calculating output impedance

To calculate output impedance, it is required to draw the equivalent circuit of the amplifier network by replacing the input signal source with a short circuit (putting $v_s = 0$) and removing the load by an open circuit. Further, a voltage (or a battery) source of strength v_x is connected across the output port so that it sends a current say, i_x, into the amplifier when feedback network is connected in the circuit as shown in Figs. 7.7a and 7.7b.

As may be observed in the equivalent circuit, the voltage across the output of the amplifier is v_x which is also applied at the input side of the feedback network. The output from the feedback network that goes to the input of the amplifier is $v_f = \beta v_x$

Also, the error signal v_i at the input of the amplifier is given by

$$v_i = (v_s - v_f) = (0 - \beta v_x) = -\beta v_x \tag{7.27}$$

Further, the signal v_o^f at the output of the amplifier due to the voltage v_i will be

$$v_o^f = A_v v_i = (-A_v \beta v_x) \tag{7.28}$$

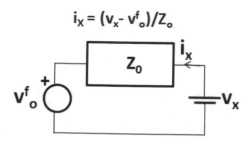

Fig. 7.7b Simplified equivalent at the output of the amplifier

Figure 7.5b shows the simplified equivalent network at the output of the amplifier where two voltage sources v_o^f and v_x are connected across the open-loop impedance Z_0 of the amplifier.

Therefore, the current through the impedance Z_0 is given by

$$i_x = \frac{(v_x - v_o^f)}{Z_o} = \frac{v_x - (-A_v\beta v_x)}{Z_o} = \frac{v_x}{Z_o}(1 + \beta A_v)$$

And

$$Z_o^f = \frac{v_x}{i_x} = \frac{Z_o}{(1 + \beta A_v)} \tag{7.29}$$

As expected, (7.29) shows that the output impedance of the amplifier with negative feedback using shunt sampling got reduced by the factor $(1 + \beta A_v)$ from the open-loop impedance.

Summing up, it may be said that a voltage amplifier with negative feedback has high input impedance (because of the series mixing) and low output impedance (on account of shunt sampling).

An important observation is that the magnitude of input impedance with negative feedback $Z_i^f = \frac{v_s}{i_i} = \frac{v_i(1 + \beta A_v)}{i_i} = Z_i(1 + \beta A_v)$ depends only on the input impedance without feedback, Z_i, feedback factor β and the open-loop gain A_v; it does not depend on any parameter of the output section. Hence, irrespective of the method of sampling at the output, Z_i^f will increase by the factor $(1 + \beta A_v)$ if there is series mixing at the input. The same is true for the output impedance Z_o^f with feedback, it also does not depend on any input parameter, hence irrespective of the method of mixing at the input, Z_o^f will decrease by the factor $(1 + \beta A_v)$ when there is voltage (or shunt) sampling at the output.

An equivalent circuit for an ideal voltage amplifier, that has a negligible value of source impedance Z_s, is shown in Fig. 7.7c. The network included in the dotted rectangle replaces the amplifier circuit with the feedback mechanism. It may be noted that since the input impedance Z_i is in parallel to the source of input voltage, it should have a high value so that maximum voltage appears at the input side of the equivalent network. Similarly, the output impedance Z_o^f is in series with the load, it should have a low value so that all of the output voltage appears against the load.

In any type of amplifier, the quantity βA has to be a dimensionless quantity; therefore, the units of the feedback factor β are reciprocal of the units of gain A. In case of voltage amplifier gain, A_v is a ratio having no units, therefore, the feedback factor β is a fraction without any units. It may further be observed that the output section of the equivalent circuit of the voltage amplifier (Fig. 7.7c) resembles a Thevenin equivalent, hence, a series-shunt or a voltage amplifier is called a Thevenin source.

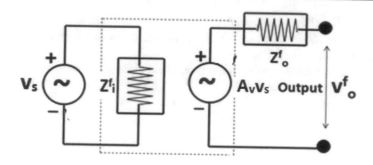

Fig. 7.7c Simple equivalent circuit of a voltage amplifier with negative feedback

(b) Current amplifier

An amplifier, in which both the input and the output signals are current signals, is called a current amplifier. It is also referred to as a current controlled current source or CCT (current–current transducer). The block diagram of the current amplifier with negative feedback is given in Fig. 7.8.

(i) *Input impedance with negative feedback for CCT*. Let us calculate the input impedance of the current amplifier with negative feedback and denote it by Z_i^f while the input impedance without feedback is denoted by Z_i.

As shown in Fig. 7.8, current I_i is given by

$$i_i = i_s - i_f \quad \text{or} \quad i_s = i_i + i_f \tag{7.30}$$

And

$$i_f = \beta i_o^f = \beta(A_c i_i) = \beta A_c i_i \tag{7.31}$$

Now

$$Z_i^f = \frac{v_i}{i_s} \text{ and } Z_i = \frac{v_i}{i_i} \tag{7.32}$$

Therefore, from (7.32) $\frac{Z_i^f}{Z_i} = \frac{i_i}{i_s} = \frac{i_i}{i_i + i_f} = \frac{i_i}{i_i + \beta A_c i_i} = \frac{1}{(1 + \beta A_c)}$

Or

$$Z_i^f = \frac{Z_i}{(1 + \beta A_c)} \tag{7.33}$$

Therefore, the input impedance with current mixing or shunt mixing decreases by the factor $(1 + \beta A_c)$.

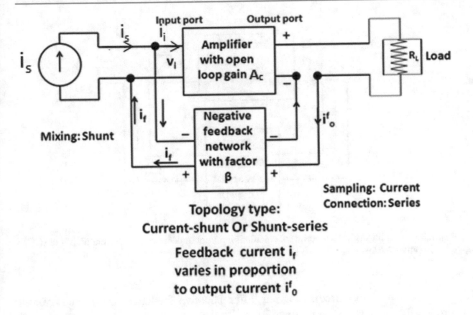

Fig. 7.8 Block diagram of a current amplifier with negative feedback using shunt mixing-series sampling topology

Once again, the new value of input impedance with negative feedback and shunt mixing does not depend on any parameter of output section, and will remain the same whether sampling is of voltage or of current.

(ii) *Output impedance with negative feedback for CCT*:
To determine the output impedance of a CCT with negative feedback, the input current source is replaced by an open circuit, the load is removed and at its place a voltage source of value V_x is connected so that the equivalent circuit looks like the one shown in Fig. 7.8a.

As shown in the figure, current I_x passes through the input of the feedback network and current signal of magnitude $i_f = (\beta\, I_x)$ is fed back to the input of the amplifier. Current i_i at the input of the amplifier is given by

$$i_i = i_s - i_f = 0 - \beta i_x = -\beta i_x \tag{7.34}$$

The current i_o^f at the output of the amplifier is given by

$$i_o^f = -\beta A_c i_x \tag{7.35}$$

The negative sign in (7.35) means that current i_o^f flows in a direction opposite to that of I_x, the direction of flow for which is taken as positive.

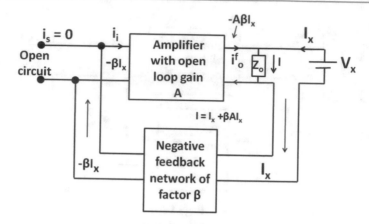

Fig. 7.8a Simplified block diagram for calculating output impedance of a current amplifier (CCT) with negative feedback

As shown in the circuit diagram, after negative feedback, the total current I through the impedance Z_o is given by

$$I = I_X + A_c\beta I_X = (1 + \beta A_c)I_X$$

And

$$Z_o I = V_X \, or \, Z_o[(1 + \beta A_c)I_X] = V_X$$

But

$$Z_o^f = \frac{V_X}{I_X} = Z_o[(1 + \beta A_c)] \tag{7.36}$$

Therefore, the output impedance of CCT with negative feedback increases by the factor $[(1 + \beta A_c)]$ and it does not depend on the type of mixing at the input end.

It may be concluded from the foregoing that (i) whenever there is a series connection of the feedback network with the input or the output port of the amplifier, the corresponding impedance with feedback will increase by the factor $(1 + \beta A_c)$. (ii) Whenever the feedback network is connected in parallel (or shunt) to the input or the output port of the amplifier, the corresponding impedance with feedback will decrease by the factor $(1 + \beta A_c)$.

It has also been observed from these derivations that the input and the output impedances with feedback do not depend on each other. However, in practice, it is not true, particularly in case of the feedback network which consists of passive elements that are inherently bidirectional. As a matter of fact, this independence of input and output quantities is the outcome of the assumption that both the amplifier and the feedback network are unidirectional, an idealized concept.

Fig. 7.8b Equivalent circuit of a current-current amplifier that uses shunt-series topology for negative feedback

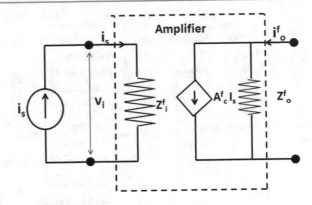

The equivalent circuit of a Shunt-series (or Current amplifier) is shown in Fig. 7.8b A CCT amplifier with negative feedback is also called a Norton current source.

Self-assessment question: Which feedback **topology is used** with transresistance amplifier?

Self-assessment question: Does β for a current amplifier with **negative feedback has units**?

(c) **Transconductance amplifier**:
The output signal in a transconductance amplifier is a current pulse, while the input signal is a voltage pulse. It may also be called a voltage controlled current source. This type of amplifier employs **Current sampling-series mixing** (series-series) topology for negative feedback. Since series connections are employed at both the input and the output ports for coupling the feedback network, the transconductance amplifier with feedback has a high input and high output impedances.
Figure 7.9 shows the block diagram of a transconductance amplifier.

An equivalent circuit for an ideal series-series feedback amplifier is shown in Fig. 7.9a.

In Fig. 7.9a, A_g is transconductance gain also denoted by the symbol G_m, and has units of Siemen (reciprocal of ohm). Since units of β are reciprocal of A_g, they are ohms. It may also be noticed that the input impedance and output impedance (both without feedback) are included in the basic amplifier circuit. A simplified equivalent circuit of the network of Fig. 7.9a is shown in Fig. 7.9b, where all quantities refer to their values after feedback.

It is easy to calculate the circuit parameters using Fig. 7.9a.

It is already known that closed-loop gain or overall gain with feedback

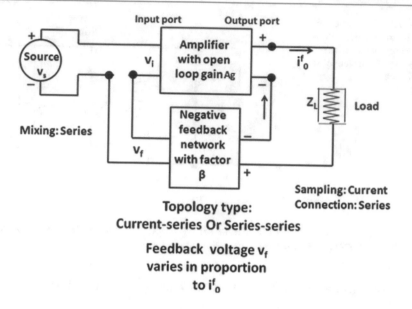

Fig. 7.9 Block diagram of a transconductance amplifies with negative feedback using current sampling-series mixing (Series-series) topology

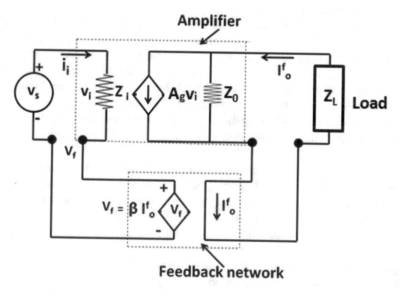

Fig. 7.9a Equivalent circuit of an amplifier that uses series-series topology

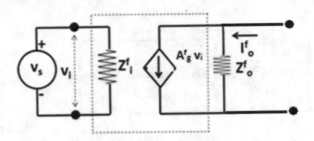

Fig. 7.9b Equivalent network of a transconductance amplifier after inclusion of negative feedback

$$A_g^f (\text{or } G_m) = \frac{i_o^f}{v_s} = \frac{A_g}{(1 + \beta A_g)} \quad (7.37)$$

It may be noticed that in (7.37) A_g and β have reciprocal units and units of A_g are Siemen. Therefore, units for feedback factor β in case of series-series topology are ohms.

The input impedance with feedback $Z_i^f = \frac{v_s}{(i_i)} = \frac{v_s}{(v_i/Z_i)} = Z_i \frac{v_s}{v_i} = Z_i \frac{(v_i + \beta v_o^f)}{v_i}$

Or

$$Z_i^f = Z_i \frac{(v_i + \beta v_o^f)}{v_i} = Z_i \frac{(v_i + \beta A_g v_i)}{v_i} = Z_i(1 + \beta A_g) \quad (7.38)$$

It is again observed that series connection at the input increases the input impedance (with feedback) by the factor $(1 + \beta A_g)$

To calculate output impedance with feedback, the voltage source v_s at the input is replaced by a short circuit, and the load at the output is replaced by a test voltage source V_x that sends a current I_x through the circuit. The modified circuit is shown in Fig. 7.9c.

As shown in Fig. 7.9c, the feedback voltage V_f is

$$v_f = \beta I_x \quad (7.39)$$

Also, $v_i = v_s - v_f$ (because of negative feedback).
And hence, $v_i = 0 - v_f = -\beta I_x$.
Therefore, the current i through the dependent source that is given by Av_i may be written as

$$i = -A_g \beta I_x \quad (7.40)$$

Negative sign in (7.40) shows that the actual direction of the current 'i' is opposite to the designated direction. Now currents I_x and i both pass through the impedance Z_0, and hence the voltage drop V against Z_0 is

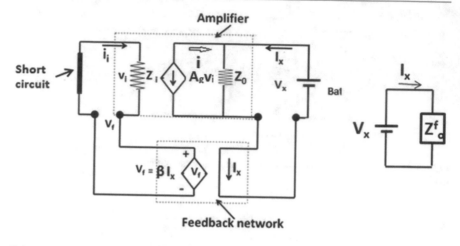

Fig. 7.9c Circuit for calculating output impedance with feedback

$$V = Z_o(|i| + I_x) = Z_o(A_g\beta I_x + I_x) = Z_o(1 + A_g\beta)I_x$$

But

$$V = V_x \quad \text{and} \quad Z_o^f = \frac{V_x}{I_x} = Z_o(1 + A_g\beta) \tag{7.41}$$

As expected, the output impedance with feedback is $(1 + A_g\beta)$ times the output impedance without feedback because of the series connection of the feedback loop at the output section.

Equivalent circuit given in Fig. 7.9b suggests that the transconductance amplifier with feedback works like a Norton's source.

(d) Transresistance amplifier:

The input signal in a transresistance amplifier is a current pulse, while the output signal is a voltage pulse, and therefore, it is also called a current controlled voltage source. On account of the shunt-shunt topology that is employed for feedback in such amplifiers, both the input and the output impedances of the amplifier with feedback have lower values than the corresponding values without feedback. Since the ratio of the output quantity (voltage) to the input quantity (current) has a unit of resistance, therefore, it is called a transresistance amplifier (CVT) and the gain A_r of this amplifier has units of ohm. The block diagram of the current–voltage amplifier is given in Fig. 7.10.

That uses voltage sampling-shunt mixing (shunt-shunt) topology.

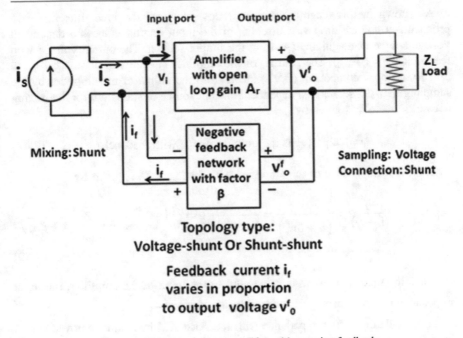

Fig. 7.10 Block diagram of a transresistance amplifier with negative feedback

The equivalent circuit of the (VCT) amplifier is given in Fig. 7.10a, where input and output impedances without feedback Z_i and Z_o are included within the basic amplifier circuit.

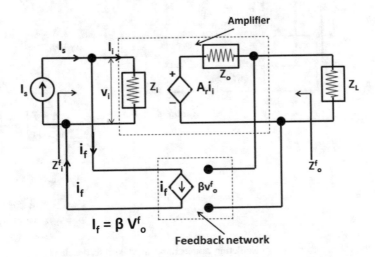

Fig. 7.10a Equivalent circuit for transresistance amplifier with negative feedback using shunt-shunt topology

As shown in this diagram, current i_i flows through the input impedance Z_i producing a voltage drop v_i. Current i_i in the input section induces a dependent voltage source of magnitude $(A_r\, i_i)$ at the output section. The output voltage with feedback $v^f_o = (\,A_r\,i_i)$. The gain A_r of the amplifier has units of ohm.

The feedback component $(\beta\, v^f_o)$ taken from the output of the amplifier by the sampling network serves as a dependent current source of strength i_f for the mixing network, where i_f is given by

$$i_f = \beta v^f_o \quad (v^f_o \text{ is the output voltage with feedback}) \tag{7.42}$$

The input impedance of the amplifier with feedback Z^f_i is given by

$$Z^f_i = \frac{v_i}{i_s} = \frac{v_i}{(i_i + i_f)} = \frac{v_i}{\left(i_i + \beta v^f_0\right)} = \frac{v_i}{(i_i + \beta A_r i_i)} = \frac{v_i}{i_i}\left(\frac{1}{1 + \beta A_r}\right) = Z_0\left(\frac{1}{1 + \beta A_r}\right)$$

$$\tag{7.43}$$

Since the mixing network is in parallel to the input of the amplifier, the input impedance with feedback gets reduced by the factor $\left(\frac{1}{1 + \beta A_r}\right)$.

For calculating output impedance with feedback Z^f_o, the load is removed and in its place, a test voltage source V_x is connected, while the input current source is replaced by an open circuit as shown in Fig. 7.10b . Now $I_s = 0$, therefore

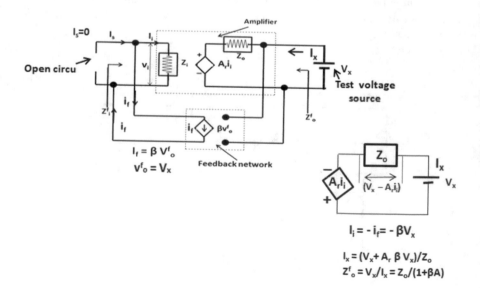

Fig. 7.10b Circuit diagram for calculating output impedance with feedback

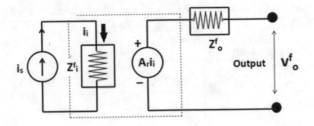

Fig. 7.10c Equivalent circuit of an ideal transresistance amplifier

$$i_i = 0 - i_f = -\beta v_0^f = -\beta V_x \tag{7.44}$$

And the magnitude of the dependent voltage source on the output side: $(A_r i_i)$ becomes

$$A_r i_i = -\beta A_r V_x \tag{7.45}$$

Hence, the total voltage across the impedance Z_0 is $[V_x - (-\beta A V_x)]$
Therefore,

$$\text{the current } I_X = \frac{[V_x - (-\beta A_r V_x)]}{Z_O} = \frac{V_X}{Z_O}(1 + \beta A_r) \tag{7.46}$$

But

$$Z_o^f = \frac{V_X}{I_x} = \frac{Z_o}{(1 + \beta A_r)} \tag{7.47}$$

The output impedance with negative feedback using Shunt-Shunt topology in case of transresistance amplifier reduces by the factor $(1 + \beta A_r)$ on account of the shunt connection of the sampling network at the output port. Finally, Fig. 7.10c shows the equivalent circuit for an ideal transresistance feedback amplifier (see Fig. 7.10c).

Self-assessment question: What are the units for the feedback factor β in case of the four topologies?

Self-assessment question: A transresistance amplifier is like a Thevenin or Norton source?

Table 7.1 Summary of relations for four types of feedback amplifiers or topologies

Quantity parameter	Voltage Amplifier (series-shunt)	Current amplifier (shunt-series)	Transconductance amplifier (series-series)	Transresistance amplifier (shunt-shunt)
Input source (X_s)	V_s Volt	I_s Amp	V_s Volt	I_s Amp
Input signal (x_i)	v_i	i_i	v_i	i_i
Output signal (x_o)	v^f_o	i^f_o	i^f_o	v^f_o
Feedback component (x_f)	v_f	i_f	v_f	i_f
Gain A	$A^f_v = v^f_o/v_i$	$A^f_c = i^f_o/i_i$	$A^f_g = i^f_o/v_{i\ \text{Siemen}}$	$A^f_r = v^f_o/i_{i\ \text{ohm}}$
Input impedance	$Z^f_i = Z_i(1 + \beta A_v)$	$Z^f_i = Z_i/(1 + \beta A_c)$	$Z^f_i = Z_i(1 + \beta A_g)$	$Z^f_i = Z_i/(1 + \beta A_r)$
Feedback factor β	$v_f/v^f_{o\ \text{no units}}$	$I_f/i^f_{o\ \text{no units}}$	$v_f/i^f_{o\ \text{ohm}}$	$I_f/v^f_{o\ \text{siemen}}$
Output impedance	$Z^f_o = Z_o/(1 + \beta A_v)$	$Z^f_o = Z_o(1 + \beta A_c)$	$Z^f_o = Z_o(1 + + \beta A_g)$	$Z^f_o = Z_o/(1 + \beta A_r)$
Nature of source	Thevenin	**Norton**	**Thevenin**	**Norton**

7.5 Problem-Solving Technique for Feedback Amplifiers

Solving problems based on feedback in amplifiers requires strict discipline, otherwise, it is quite likely that one may commit errors. Feedback amplifier problems may be solved using several different approaches like, first reducing the given network into a network without feedback retaining the loading of the input and the output sections by the feedback network. The non-feedback network may be further simplified using an appropriate small signal ac equivalent circuit (for example, hybrid equivalent or r-equivalent, etc.). The ac equivalent may then be resolved, making use of the laws of series and parallel combinations, KVL & KCL, Thevenin and Norton theorems, etc. However, such analysis becomes cumbersome on account of the fact that different ac equivalents are required for different feedback topologies and also several steps of simplifications complicate the analysis.

The other and a more powerful method is based on resolving the given feedback amplifier network into two circuit blocks: the A-circuit and the β-circuit. Both, the A-circuit and the β- circuit have different standard structures each for the four different feedback topologies. Using the method of A- and β-circuits for analysis requires first to identify the type of feedback topology that has been used in a given problem. Once the feedback topology is known, the next step is to reduce the given feedback amplifier circuit into A-circuit and β-circuit specific for the identified topology. The β-circuit is then used to obtain the feedback factor (β) and the values of some impedance that are parts of the A-circuit. With values of impedances obtained from the analysis of the β-circuit being switched in, the A-circuit may be analyzed to obtain other parameters like the input and output impedances, etc. Often, in case of simpler networks, it is possible to analyze the β-circuit and the A-circuits just by inspection. Since the structure of A- and β-circuits depends on the feedback topology; identification feedback topology and then resolving the given network into appropriate A- and beta- circuits is the most crucial part of problem solving. The methodology of employing A-circuit and beta-circuit technique for solving problems involving feedback amplifiers is outlined in the following:

Identification of feedback network and type of feedback topology is the first step of solving any numerical problem on feedback amplifiers.

Feedback network It is easy to identify the feedback network in a given circuit simply by inspecting the circuit. The input port and the output port of the circuit are generally specified in the problem. Most of the time, input signal is fed between the input node and ground, while the output is taken between the output node and ground. Any circuit element or combination of elements that connect the output node directly or through any other part of the output loop to the input loop is the required feedback network. In most cases, feedback network is a combination of capacitor and resistances.

Feedback type Feedback can either be negative that reduces the input signal or positive that strengthens the input signal. The nature of feedback can be deciphered by finding the phase change of the input signal at each node in the given circuit while moving from input to the output node and back via the feedback network. If after this analysis it is found that the signal fed back to the input is 180^0 out of phase with respect to the input signal then the feedback is negative otherwise positive.

Feedback topology The way how the feedback network is connected with the output loop and to the input loop, decides the feedback topology. As a thumb rule, if the feedback network is directly connected at the output node, the sampling topology will be of **voltage sampling or Shunt sampling**. On the other hand, if the feedback network is joined at some other node (other than the output node) of the output loop, the sampling topology will be of **current sampling or Series sampling**. Similarly, if the feedback signal is directly connected to the input node, the mixing topology will be **Shunt mixing topology or Current mixing topology** otherwise it will be **Series mixing topology.** Let us consider the following circuits.

Figure 7.11a shows a typical feedback system in which the emitter of the npn transistor is the output node and the emitter resistance R_{Ei} serves as the feedback element. The voltage drop $V_f = (R_{Ei}.i_{Ei})$ constitutes the feedback voltage. Obviously, the feedback network samples the voltage V_f, and hence the **sampling topology is voltage sampling or shunt sampling**. The actual signal that is applied at the base of the transistor V_i is equal to the difference of source voltage V_s and the fed back voltage V_f, i.e. $V_i = (V_s - V_f)$. Since at the input the two voltages are added together, **the mixing topology is of series mixing**. Hence, the complete topology of feedback is **Voltage sampling–Series mixing or Series-shunt topology**.

In circuit of figure (b), collector which is the output node is directly connected to the base of the transistor. The feedback system samples voltage at the output, and therefore, the sampling is done using **voltage sampling (or shunt) topology**. However, the feedback network is connected at the base of the transistor. As a result, the source current Is splits up into two components one I_f and the other I_b such that $I_s = (I_b + I_f)$. Splitting of current or voltage in mixing is a clear sign of **Shunt mixing topology. The complete feedbacktopology may be defined as Voltage sampling** and **Shunt mixing topology.** It may also be called as **Shunt-Shunt topology.**

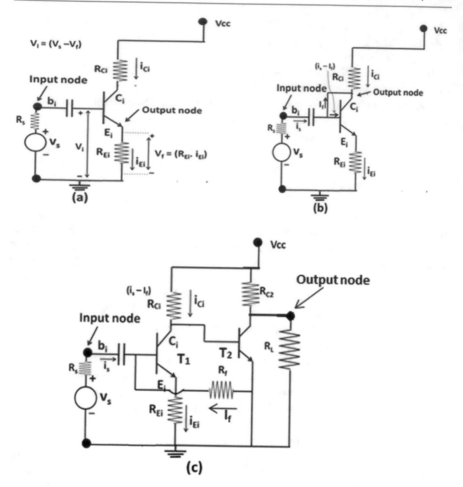

Fig. 7.11 Block diagram of a typical feedback system

Lastly, in circuit of Fig (c), feedback is achieved through resistance R_f which is connected from the emitter of the second transistor T_2 to the base of the first transistor T_1. At the output section, the feedbackelement is not directly connected at the output node. The output node is obviously the collector of transistor T_2, from where the load is connected. The sampling is of current; the sampling topology is **Current sampling or Series sampling**. Regarding the mixing of the feedback element, R_f is connected at the base or the input of transistor-one. This, therefore, is shunt mixing topology. The complete feedback topology is **Current sampling– Shunt mixing or Shunt-series topology**. Thus, using the above-mentioned thumb rules, it is possible to find the feedback topology of any given circuit.

Self-assessment question: Is it possible to achieve current mixing using series connection?

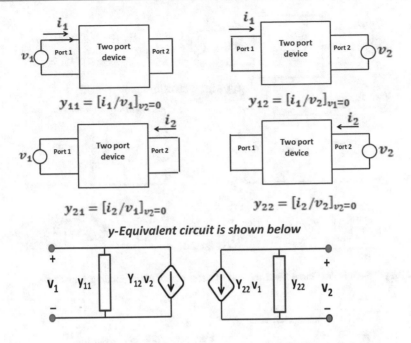

$$y_{11} = [i_1/v_1]_{v_2=0}$$

$$y_{12} = [i_1/v_2]_{v_1=0}$$

$$y_{21} = [i_2/v_1]_{v_2=0}$$

$$y_{22} = [i_2/v_2]_{v_2=0}$$

y-Equivalent circuit is shown below

Fig. 7.12 y-equivalent circuit and method of determining corresponding impedances

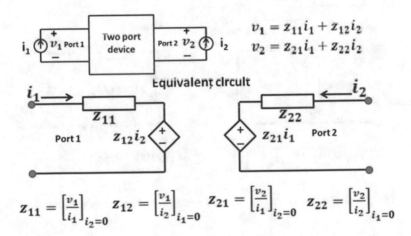

$$v_1 = z_{11}i_1 + z_{12}i_2$$
$$v_2 = z_{21}i_1 + z_{22}i_2$$

Equivalent circuit

$$z_{11} = \left[\frac{v_1}{i_1}\right]_{i_2=0} \quad z_{12} = \left[\frac{v_1}{i_2}\right]_{i_1=0} \quad z_{21} = \left[\frac{v_2}{i_1}\right]_{i_2=0} \quad z_{22} = \left[\frac{v_2}{i_2}\right]_{i_1=0}$$

Fig. 7.13 z-equivalent circuit and method of determining corresponding impedances.

Before developing the structures of the A-circuit and the beta-circuit for different topologies, let us summarize the features of different types of equivalent circuits which will be required for developing the A- and beta-circuits. The four types of equivalent circuits that will be required are: y-equivalent; z-equivalent; h-equivalent and g-equivalent which are summarized below.

Summary of small signal equivalents for two port networks.

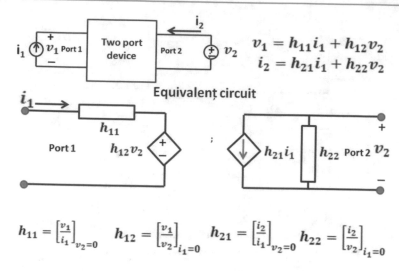

$$v_1 = h_{11}i_1 + h_{12}v_2$$
$$i_2 = h_{21}i_1 + h_{22}v_2$$

$$h_{11} = \left[\frac{v_1}{i_1}\right]_{v_2=0} \quad h_{12} = \left[\frac{v_1}{v_2}\right]_{i_1=0} \quad h_{21} = \left[\frac{i_2}{i_1}\right]_{v_2=0} \quad h_{22} = \left[\frac{i_2}{v_2}\right]_{i_1=0}$$

Fig. 7.14 h-equivalent circuit and method of determining corresponding impedances.

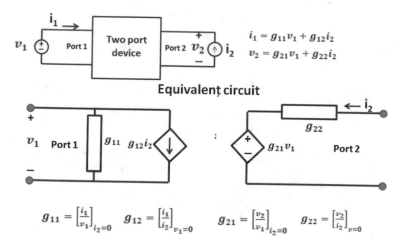

$$i_1 = g_{11}v_1 + g_{12}i_2$$
$$v_2 = g_{21}v_1 + g_{22}i_2$$

$$g_{11} = \left[\frac{i_1}{v_1}\right]_{i_2=0} \quad g_{12} = \left[\frac{i_1}{i_2}\right]_{v_1=0} \quad g_{21} = \left[\frac{v_2}{v_1}\right]_{i_2=0} \quad g_{22} = \left[\frac{v_2}{i_2}\right]_{v=0}$$

Fig. 7.15 g-equivalent circuit and method of determining corresponding impedances.

Amplifier is a two port device, the input port and the output port. To simplify problems based on two port networks it is required to draw their small signal equivalent circuits. We have already discussed some equivalent circuits in details in chapter-6. However, for the sake of ready reference a very brief description of the four types of equivalent circuits is included here.

A two port network is characterized by its port-1 and port-2 (not by input and output). In general there may be four types of two port networks: (i) that has voltage signals at both of its ports (ii) may have current signals at both ports (iii) has voltage signal at port-1 and current signal at port-2 and (iv) may have current signal at port-1 and voltage signal at port-2. There is another way of classifying two port networks, that is, on the basis of the type of connection with which port-1 and port-2 may be connected with some other network. For example, if it is required that port-1 is connected in series to some other network while port-2 in parallel to the other network, then h-equivalent is the correct choice as port-1 of h-equivalent gives a current source, while port-2 has a voltage source. In case of negative feedback amplifiers, as will be seen later, the feedback network will be replaced by an appropriate equivalent two port circuit. Therefore, the choice of the equivalent circuit will be decided by the nature of connections with which port-1 and port-2 be connected with the amplifier.

7.5.1 y-Parameter Equivalent

This equivalent corresponds to the case when the voltages v_1 and v_2, respectively, across port-1 and port-2 are known and currents i_1 and i_2 are unknown. Now a voltage source is normally connected in parallel or in shunt connection; therefore, this equivalent will be most suited to the case where both ports-1 and 2 are to be connected in parallel (shunt-shunt connection) like in the case of a CVT amplifier.

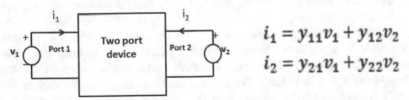

$$i_1 = y_{11}v_1 + y_{12}v_2$$
$$i_2 = y_{21}v_1 + y_{22}v_2$$

The currents i_1 and i_2 are related to voltages by the pair of equations given above. The components of y-parameters may be obtained as shown in Fig 7.12.

Y-parameter equivalent is also shown in Fig. 7.12. The same method may be used to specify the other three equivalents which are given below.

7.5.2 z-Parameters Equivalent

When the given network has current signals at both ports, it is convenient to use z-parameter equivalent. Details of z-equivalent and the way to determine the coefficients are explained in Fig. 7.13. Since current sources may be connected only in series, z-equivalent may be used to replace feedback network of series-series type topology (VCT).

7.5.3 h-Parameters Equivalent

If port-1 has a current signal and port-2 a voltage signal, then h-parameter is suitable for analysis. Hence, it may be used to replace the feedback network in Series-Shunt type (VVT) amplifiers. The h-equivalent circuit and method of determining corresponding impedances are shown in Fig. 7.14.

7.5.4 g-Parameter Equivalent

If port-1 has a voltage signal and port-2 a current signal, like a transconductance amplifier, it is convenient to use g-parameter equivalent to replace the feedback network. Details of the equivalent circuit and the method of determining corresponding impedances are shown in Fig. 7.15.

The specific structures of the A-circuit and of the beta-circuit for different feedback topologies will be discussed now. Before proceeding further, let us understand what are the A- circuit and the beta-circuit. The total network of a feedback amplifier may be divided into two parts; (i) the basic amplifier without feedback and (ii) the feedback network (FBN). Both these components of the circuit contain impedances, like the input and output impedances of the amplifier and source impedance, etc. As will be seen, the feedback network is finally replaced by its ac equivalent circuit which has some impedances and dependent current and/or voltage sources. After this replacement of the FBN by equivalent, the modified total circuit is manipulated such that **all impedances associated with the basic amplifier and also those associated with the FBN equivalent are included in the arms of the basic amplifier network and only dependent current and/or voltage sources are kept separately. The part of the manipulated network that contains the basic amplifier and all the impedances is called the A-circuit. The remaining component of the manipulated circuit that has dependent sources only is called the β-circuit.** In further analysis, the original FBN is used to find the values of impedances associated with the equivalent circuit and have been absorbed in the A-circuit. FBN is also used to determine the feedback factor β. The A-circuit is analyzed to obtain amplifier gain, input and output impedances with and without feedback and other required parameters.

7.5.5 To Resolve a Voltage Feedback Amplifier in A- and β-Circuits

A negative feedback voltage amplifier uses **Series-Shunt** (Voltage sampling-Series mixing) topology. The circuit diagram of an ideal voltage amplifier with feedback is given in Fig. 7.16. In an ideal case, it is assumed that the voltage source V_s does not

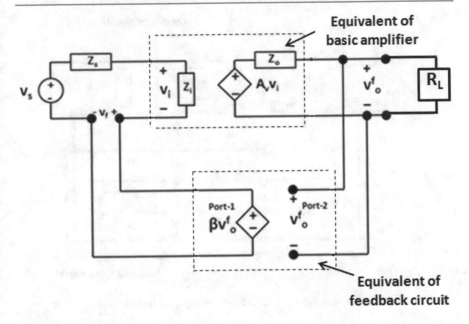

Fig. 7.16 Block diagram of a practical Series-Shunt feedback voltage amplifier with basic amplifier and feedback network replaced by their equivalent circuits

offer any impedance, however, in a practical situation, the voltage source does have an internal impedance say, Z_s, which must be included in a practical circuit. Figure 7.16 shows the block diagram of a practical Series-Shunt feedback amplifier where the basic amplifier and the feedback network are replaced by their equivalent circuits.

It may be noted that in Fig. 7.16, the source impedance, the input impedance, the output impedance and the load are all connected with the basic amplifier circuit either in series or in parallel connections. As the next step, the feedback equivalent is replaced by its h-parameter equivalent as shown in Fig. 7.16a. The reason why h-parameter equivalent has been selected for replacing the feedback network is that **only the h-equivalent circuit has a series connection at its port-1 (input port) and a shunt connection at its port-2, the output port. No other equivalent has such input and output connections**.

As may be seen in Fig. 7.16a, two new impedances h_{11} and h_{22} along with a voltage source ($h_{12} v_2$) on the input side and a current source ($h_{22} i_1$) on the output side have come into the picture. The h-equivalent terminology provides a prescription for calculating h_{11}, h_{22}, h_{12} and h_{21}. For example, h_{11} may be calculated from the feedback network (FBN) by shorting port-2 and looking into the FBN through port-1. Similarly, h_{22} may be obtained from the actual FBN by keeping port-1 open and looking into the FBN through port-2 as shown in Fig. 7.16b.

Amplifier circuit

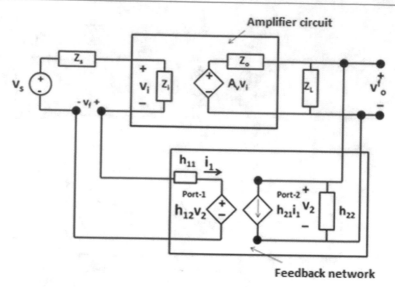

Fig. 7.16a Circuit after replacing the feedback network by h-equivalent

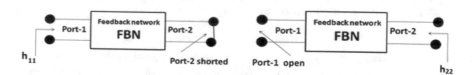

Fig. 7.16b Determination of impedances h_{11} and h_{22}.

Coming back to Fig. 7.16a, the term (h_{21} i_1) represents the forward transmission of the signal from port-1 to port-2. However, the basic assumption that has been made in this analysis is that both the amplifier and the FBN are unidirectional; amplifier transmits the signal only in the forward direction, while the FBN transmits only in the reverse direction, i.e. from port-2 to port-1. Under these assumptions, the forward transmission of FBN may be neglected so that (h_{21} i_1) is zero and port-2 becomes an open end. With these simplifications, it is now possible to absorb impedances h_{11} and h_{22} in the arms of the amplifier circuit. It is obvious that with this absorption of h_{11} and h_{22} in arms of the amplifier circuit, the strength of source v_s will change to a new value $v_s{}'$ and similarly the output voltage v_o^f will have a new value $v_o^f{}'$. The modified circuit is shown in Fig. 7.16c.

The A-circuit may be used to determine the gain, the input impedance, the output impedance of the network, while β-circuit to determine the value of the feedback factor β. Details as to how one may determine these parameters are shown in Fig. 7.16d.

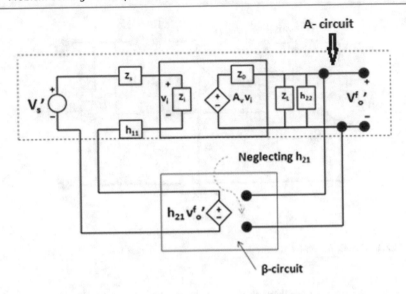

Fig. 7.16c Dissolution of feedback voltage amplifier into the A-circuit and the β-circuit

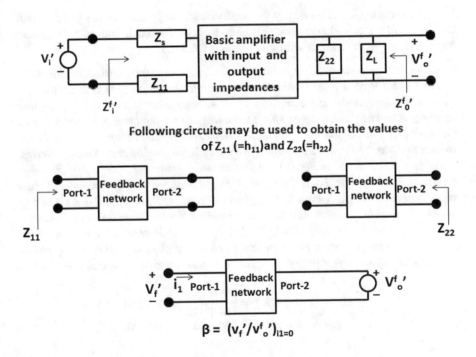

Fig. 7.16d A-circuit and method of determining some circuit parameters

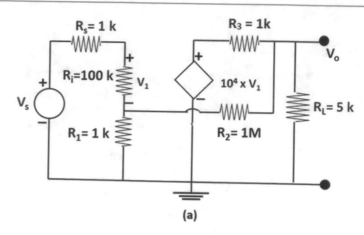

Fig. SE7.1 Circuit for problem (SE7.1)

The following example will illustrate how the A-circuit and the FBN can be used to solve the voltage amplifier problems (Figs. SE7.1, SE7.1.1 and SE7.1.2).

Solved example (SE7.1): Identify the feedback topology used in the given circuit and determine the feedback factor β, gain without feedback and with feedback; input and output resistances with and without feedback.

Solution: *It is simple to identify that the feedback network consists of the resistances R_1 and R_2. Since the feedback connection in the output side is taken directly from the output node, the sampling is of voltage and therefore the sampling topology is* **voltage sampling or shunt sampling.** *Since the feedback connection at the input is not directly at the input node, therefore, the* **mixing topology is series or voltage mixing.** *The complete feedback topology is* **Series-Shunt or Voltage-Series.**
As may be observed in Fig (b); while determining the impedance Z_{11}, the other end (port-2) of the FBN has to be short-circuited because port-2 is connected via a shunt connection to the A-circuit. This gives the value of $Z_{11} = (R_1||R_2)$. As a rule of thumb, the port in question must be short-circuited if it is connected to the A-circuit by a shunt connection and should be left open if it is connected with a series connection to the A-circuit. For example, in case of finding the value of Z_{22}, port-1 must remain open as this port is connected in series with the A-circuit.

$$Z_{11} = (1\,M \,||\, 1\,K) = \frac{1M \times 1K}{1M + 1K} \approx 1k \quad \text{and}$$
$$Z_{22} = (R_1 + R_2) = 1M + 1k = 1.001\,M$$

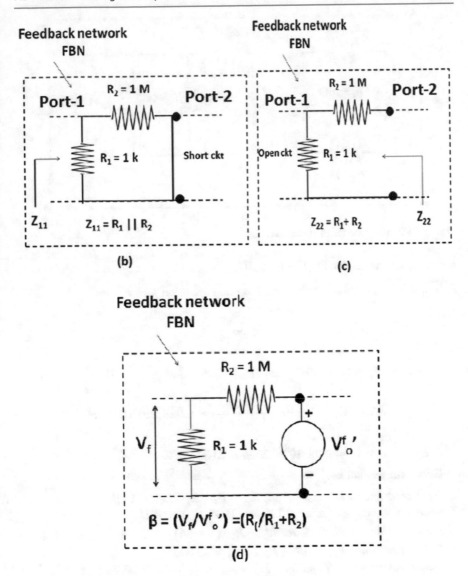

Fig. SE7.1.1 Circuits for finding the values of impedances Z_{11} and Z_{22}

To determine the value of the feedback factor β, a voltage source V^f_o' is connected across port-2 and the voltage that develops at port -1, V_f is measured. The ratio of $V_f / V^f o$' gives the value of β.

From Fig. (d) $\beta = \dfrac{V_f}{V^f_o} = \dfrac{R_1}{(R_1 + R_2)} = \dfrac{1\,k}{(1M + 1k)} \approx \dfrac{1k}{1M} \approx 10^{-3}$ V/V.

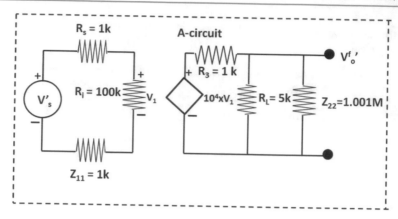

Fig. SE7.1.2   A-circuit for the given network

Drawing of the A-circuit (See Fig. SE7.1.2)

It is now easy to calculate the open-loop amplification factor A as

$$A = \frac{V_o^{f'}}{V_s'}$$

Now $V_1 = V_s'\left[\frac{100k}{102k}\right] = 0.98V_s'$ *and* $\left(R_L \parallel Z_{22} = \frac{1.001M \, x5k}{1.006 \, M} = 4.975 \, k\right)$,

$$Also \quad V_0^{f'} = V_1 x 10^4 \frac{4.975k}{4.975k + 1k} = 0.98V_s' x 10^4 x 0.8326$$

Therefore $A = \frac{V_o^{f'}}{V_s'} = 0.98 x 10^4 x 0.8326 = 8160 \ V/V$

Gain with feedback $A_v^f = \frac{A}{1 + \beta A} = \frac{8160}{1 + 10^{-3} x 8160} = \frac{8160}{9.160} = 890.83$

Input resistance without feedback $R_{in}' = R_s + R_i + Z_{11} = 102k$

Input resistance (excluding R_s*)* $R_{in} = (102 - R_s) = 101k$;

Input resistance with feedback $R_{in}^f = R_{in} x (1 + \beta A) = 101k x 9.16 = 925.16 k\Omega$

Output resistance $R_{out}' = [R_3 \parallel R_L \parallel Z_{22}] \approx 834\Omega$

Output resistance (without load R_L*)* $= R_{out} = \frac{1}{\frac{1}{R_{out}'} - \frac{1}{R_L}} = 1000.96\Omega$

Output resistance with feedback $= \frac{R_{out}}{1 + \beta A} = \frac{1000.96}{9.16} = 109.27\Omega$

7.5.6   To Resolve a Current Controlled Current Feedback Amplifier in A- and β-Circuits

A current feedback amplifier has a current source at its input and also delivers a current signal at its output. Since the feedback system samples current from the output, the sampling connection is series, similarly it mixes the fed back component of the current at its input, the mixing topology is a shunt. A current controlled current feedback amplifier, therefore, uses Shunt-Series (or Current sampling–

Current mixing, or Current-Shunt) topology. The block diagram of a practical current amplifier is shown in Fig. 7.17.

It is evident from Fig. 7.17 that the FBN-equivalent circuit needs a shunt connection on the input side or port-1 and a series connection on the output side (port-2). In case it is required to replace the FBN with an equivalent circuit, then only g-equivalent (which has a shunt connection on port-1 and a series connection on port-2) can be used. The circuit diagram of a current amplifier in which the feedback network is replaced by g-equivalent is shown in Fig. 7.17a.

The dependent voltage source $g_{22}V_1$ on the output side exhibits the forward transmission of the FBN. However, under the assumption that forward transmission of FBN is negligible, the g_{22} may be taken as zero and the dependent voltage source may be replaced by a short circuit. Finally, g_{11} and g_{22} (that are tranconductances and have units of Siemen or Ω^{-1}) may be absorbed, respectively, in the input and the output circuits of the amplifier. The A-circuit and the β-circuit for the current amplifier with negative feedback are shown in Fig. 7.17b.

Solved example SE7.2: In the following circuit, identify the type of feedback, feedback network and the kind of feedback topology used in the circuit. From the dc analysis obtain the operating points of transistors Q_1 and Q_2 and their hybrid-pi model parameters. Also, draw an ac equivalent circuit of the given network, and hence obtain the A- and the beta-circuits for feedback analysis. Assume that magnitudes of collector and emitter currents are equal for both transistors ($\alpha = 1$). β for both transistors is 100 and V_A is 75 V (Figs. SE7.2 and SE7.2.1).

Solution: *In the given circuit, feedback is provided from the emitter E_2 of the second transistor via resistances R_{E2} and R_f to the base of transistor Q_1. Therefore, FBN is the combination of resistances R_{E2} and R_f. To check the type of feedback, it may be observed that the base current i_{b1} of Q_1 will increase with the increase of the input signal, increase of i_{b1} will result in an increase of collector current I_{c1}, which in turn, will decrease the voltage V_{C1} at the collector of Q_1. But the collector*

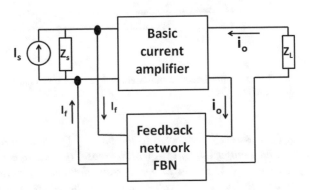

Fig. 7.17 Block diagram of a practical negative feedback current amplifier

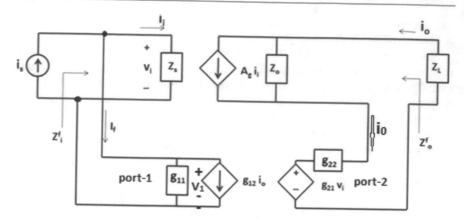

Fig. 7.17a Block diagram of a shunt-series amplifier with feedback network replaced by its g-equivalent

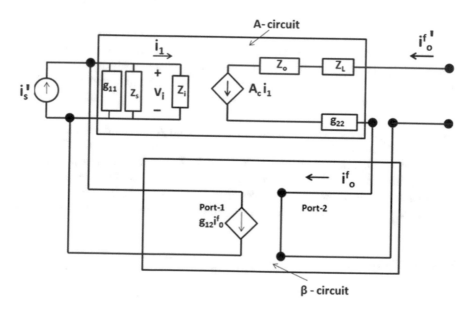

Fig. 7.17b Shunt-series amplifier equivalent after replacing the dependent voltage source by a short circuit

of Q_1 is directly connected to the base of Q_2, and therefore, Q_2 will conduct less, and the emitter current i_{E2} will decrease, this decrease in i_{E2} will increase emitter voltage V_{E2}, setting a larger current through R_f increasing larger voltage drop against R_f and a lower voltage signal at the base of Q_1. It may be observed that the increase of signal voltage V_s at base of Q_1 induces a signal of opposite phase at the

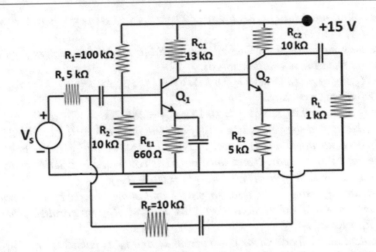

Fig. SE7.2 Circuit diagram for problem (SE7.2)

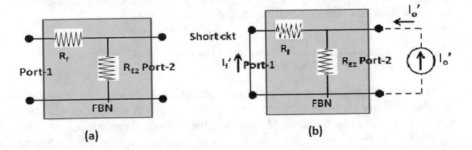

Fig. SE7.2.1 Feedback network (FBN)

base via the FBN, therefore, the feedback is negative. Since current is sampled and also current is mixed, therefore, the feedback topology is Current-Shunt or Shunt-Series.

The operating points of the two transistors may be found by doing dc analysis. Following calculations are self-explanatory

$$V_{B1} = 15x \frac{10\ k}{110\ k} = 1.36\ V; \quad V_{E1} = V_{B1} - 0.7 = 0.66\ V; \quad I_{E1} = \frac{0.66\ V}{660\Omega} = 1.0\ mA$$

Since it is given that $I_E = I_C$ for both transistors, $I_{C1} = 1.0\ mA$; $V_{C1} = 15$ $-R_{C1}I_{E1} = 15 - 13 = 2\ V$

Base voltage of Q_2 $V_{B2} = V_{C1}; = 2V$; hence $V_{E2} = V_{B2} - 0.7 = 2.0 - 0.7$ $= 1.3\ V$

And $I_{E2} = \frac{V_{E2}}{R_{E2}} = \frac{1.3\ V}{5k\Omega} = 0.26\ mA = I_{C2}$, $V_{C2} = 15 - R_{C2}I_{C2} = 15 - 10\ k\ x\ 0.26\ mA = 12.4\ V$

The operating point of Q_1 is ($V_{C1} = 2.0V$, $I_{C1} = 1.0$ mA); Operating point of Q_2 (12.4 V, 0.26 mA).

One may calculate the values of the hybrid-pi model parameters for the two transistors from the above dc analysis.

For Q_1: $g_{m1} = (I_c/25$ mV$) = 0.04$ A/V; $r_{\pi 1} = (\beta_{trd}/g_m) = 2.5$ kΩ; $r_{o1} = (V_A/I_c) = (75$ v/1 mA$) = 75$ kΩ.

For Q_2: $g_{m2} = 0.01$ A/V; $r_{\pi 2} = 10$ kΩ; $r_{o2} = 288$ kΩ.

Next, let us find the values of g_{11}, g_{22} and the feedback factor β. The FBN has already been identified and is shown in Fig. (7.2.1a). *To find g_{11}, one has to look into the FBN through port-1 and since port-2 is connected in a series connection, hence it is to be kept open. That gives $g_{11} = (R_f + R_{E2})$*

To obtain g_{22}, one looks through port-2 and short port-1 (because port-1 is connected to shunt connection) that puts R_f and R_{E2} in parallel. Therefore, $g_{22} = (R_f \| R_{E2})$

To calculate feedback factor β, a current source I_o' is connected at port-2 and port-1 is short-circuited (Why ?) as shown in Fig. (b). *The current I_f' through the short circuit lead is determined using the law of parallel connections, $I_f' = -(R_{E2}/R_f + R_{E2})I_o'$ The negative sign appears because of the fact that the actual current flows in a direction opposite to the assumed direction. Now*

$$\beta = \frac{I'f}{I'o} = -\frac{R_{E2}}{R_{E2} + R_f} = -(5/5 + 10) = -0.33(\Omega/\Omega)$$

In order to draw an ac equivalent circuit, the dc power source is grounded and all capacitors are short-circuited. As a result, R_1 and R_2 will come in parallel, R_{E1} will vanish as the capacitor across it will short it and R_{C2} will come in parallel with R_L. Further, in the problem, the signal source is given as a voltage source but the amplifier is a current amplifier, therefore, it will be appropriate to convert the given voltage source into a current source of strength $I_s = (V_s/R_s)$ in parallel to the resistance R_s. Also transistors Q_1 and Q_2 may be replaced by their r_π equivalents as shown in Fig. SE7.2.2. *Further, g_{11} (= $R_{E2} + R_f$) will appear in the input circuit in parallel connection and g_{22} (=$R_{E2}\|R_f$) in series to the ac equivalent circuit.*

The ac equivalent and A-circuit for the given current amplifier with negative feedback are given in Figs. SE7.2.3 and SE7.2.4.

7.5.7 To Resolve a Transconductance Feedback Amplifier in A- and β-circuits

A transconductance amplifier with negative feedback is characterized by series-series (current-series) feedback topology as the FBN samples current in the output and mixes voltage in the input. Since input has a voltage source, the source resistance is in series and similarly, the load resistance and the output resistance are in series. The FBN has series connections at both ends. If it is required to replace the FBN with an equivalent, then a small z-parameter equivalent that has series

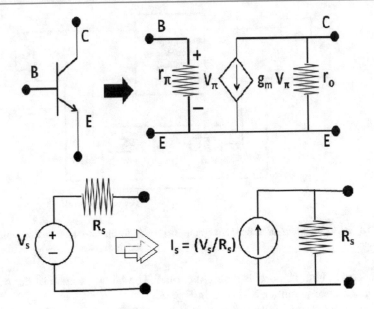

Fig. SE7.2.2 Equivalent circuits for the transistor and the voltage source

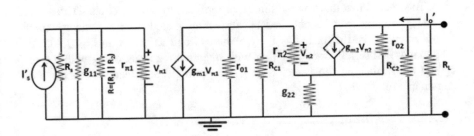

Fig. SE7.2.3 AC equivalent circuit

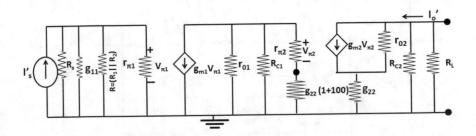

Fig. SE7.2.4 The A-circuit

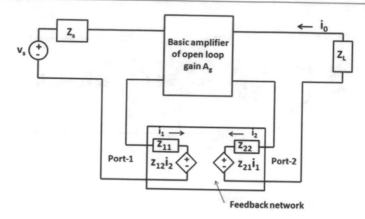

Fig. 7.18 Block diagram of series-series amplifier with feedback circuit being replaced by z-equivalent

connections at both its ports is the right choice. The block diagram of the practical transconductance amplifier is shown in Fig. 7.18.

The dependent voltage source $z_{21}i_1$ is due to the forward transmission of signal from port-1 to port-2 of the feedback network, which according to the assumption that FBN is unidirectional and does not transmit in the forward direction, must be zero. This means that this voltage source be replaced by a short circuit. The desired A- and β-circuits may be obtained by absorbing impedances z_{11} and z_{22} in the amplifier circuit and replacing $z_{21}i_1$ by a short circuit as shown in Fig. 7.18a.

Summary of series-series amplifier with negative feedback along with the procedure for evaluating the values of z_{11}, z_{22}, gain A_g and feedback factor β are shown in Fig. 7.18b.

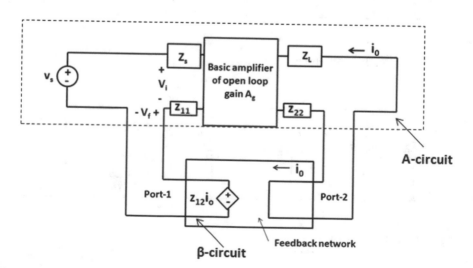

Fig. 7.18a A-circuit and the β-circuit for transconductance amplifier

Specific form of A-circuit for series – series feedback amplifier

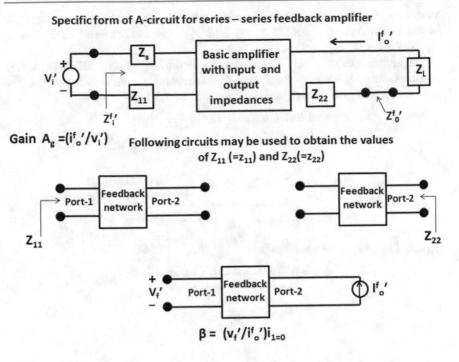

Gain $A_g = (i_o^f{}'/v_i')$ Following circuits may be used to obtain the values of Z_{11} $(=z_{11})$ and $Z_{22}(=z_{22})$

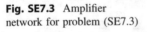

$$\beta = (v_f'/i_o^f{}')i_{1=0}$$

Fig. 7.18b Summary of Series-Series amplifier

Solved example (SE7.3) Draw A-circuit and beta-circuit for the current amplifier with negative feedback shown in the figure and calculate the feedback factor β and the z_{11} and z_{22} (Fig. SE7.3).

Fig. SE7.3 Amplifier network for problem (SE7.3)

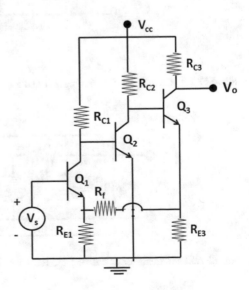

Solution: *The combination of three resistances R_{E1}, R_f and R_{E3} constitutes the feedback network. The values of z_{11} and z_{22} may be obtained as shown in Fig. SE7.3.1 a, b.*

The feedback factor β may be calculated with the help of Fig. SE7.3.1c , where a current source I_o is connected at port-2 that develops a voltage V_f at port −1. The ratio (v_f/I_o), will give the magnitude of β (Fig. SE7.3.1c).

$$\text{Now } V_f = R_{E1}.\text{current through } R_{E1} = R_{E1}\left[\frac{R_f + R_{E3}}{R_{E1} + R_{E3} + R_F}\right]I_o.$$

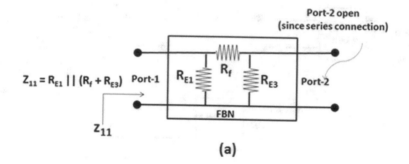

(a)

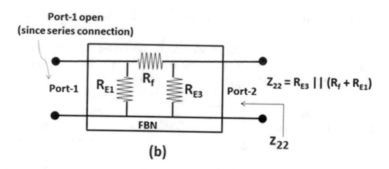

(b)

Fig. SE7.3.1 Circuits for determining Z_{11} and z_{22}.

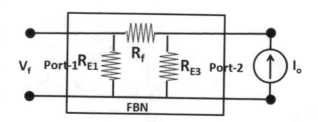

Fig. SE7.3.1c Circuit for calculating the feedback factor β

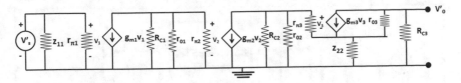

Fig. SE7.3.2 A-circuit for (SE7.3)

Therefore, $\beta = \dfrac{V_f}{I_o} = R_{E1}\left[\dfrac{R_{E3}}{R_{E1} + R_{E3} + R_F}\right]$ *(V/A).*

The A-circuit of the given network is shown in Fig. SE7.3.2.

7.5.8   To Resolve a Transresistance Feedback Amplifier in A- and β-Circuits

A transresistance amplifier amplifies an input current signal and delivers a voltage signal in the output. The feedback system of a transresistance amplifier samples voltage at the output using a shunt connection and delivers a current signal to be mixed at the input using a shunt connection. The feedback topology is, therefore, of Voltage-shunt or Shunt-Shunt. When it is required to replace the FBN with an equivalent circuit, only that equivalent circuit which has shunt connections at both ports may be used. Among equivalent circuits, only y-equivalent circuit is such that it has shunt connections at both ends, and therefore, may be used to replace the FBN in this case. The block diagram of a practical transresistance amplifier with negative feedback is shown in Fig. 7.19 after replacing the FBN with its y-equivalent.

As usual, the forward transmission component of FBN may be neglected and impedances y_{11} and y_{22} may be absorbed in the amplifier circuit constituting the A-circuit.

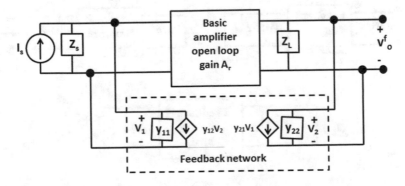

Fig. 7.19 Block diagram of a Shunt-Shunt feedback amplifier, FBN replaced by its y-equivalent circuit

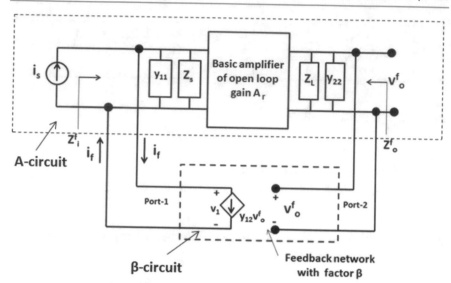

Fig. 7.19a A- and β- circuits for Shunt-Shunt amplifier

The A-circuit and the β-circuit corresponding to the Shunt-Shunt amplifier are shown in Fig. 7.19a.

Figure 7.19b shows how one can determine the values of y_{11}, y_{22} and feedback factor β from the FBN. Since both ports of the FBN are connected via shunt

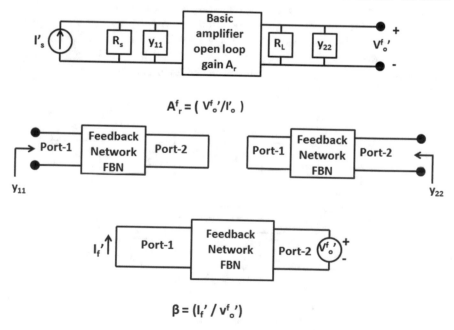

Fig. 7.19b Pictorial representation of obtaining y_{11}, y_{22} and β for a shunt-shunt amplifier with negative feedback

connections to the A-circuit, port-2 must be short-circuited while determining the magnitude of y_{11}, and similarly port-1 be short-circuited when finding y_{22}. The methodology of obtaining gain, feedback factor, y_{11} and y_{22} for shunt-shunt topology is shown in Fig. 7.19b.

Solved example (SE7.4) Identify the feedback topology and the type of feedback. With the help of FBN find the type of equivalent circuit, the corresponding impedances that need to be absorbed in the A-circuit. Calculate the feedback factor for the circuit.

Solution: *Circuit shown in the figure is a two stage amplifier, the first stage that uses transistor Q_1 is a CE amplifier, while the second stage is an emitter follower the output of which is connected to a load R_L. The feedback is achieved by the combination of resistances R_f and R_{E2}. Since feedback is taken directly from the output node, the sampling is voltage sampling (having shunt connection) and since the input feedback is connected directly to the base which is input node, it is shunt mixing. The feedback topology is, therefore, Shunt-Shunt.*

Any increase of the base voltage by the input signal will increase the base current of Q_1, which in turn, will increase the collector current of transistor Q_1. An increase of collector current of Q_1 will result in a decrease of potential at the collector C_1 of Q_1. Since collector C_1 is directly connected to the base of Q_2, decrease of base potential of Q_2 will reduce the emitter current of C_2. Reduction of emitter current of Q_2 will end up in increasing the potential at the upper end of R_{E2} from where the feedback is connected to the base of Q_1. As a result, larger current will flow through R_f and there will be more voltage drop across R_f, which will reduce the potential at the base of Q_1. The feedback network will thus feed a signal of opposite polarity to that of the source signal, hence the feedback is negative.

In shunt-shunt topology, both the ports of the FBN are to be connected with the amplifier circuit through shunt connections, the only equivalent network that has shunt connections at both ports is y-equivalent, hence y-equivalent circuit only may replace the FBN. The values of the two admittances y_{11} and y_{22} that ultimately have to be absorbed in the A-circuit may be calculated from the FBN.

As shown in Fig. SE7.4.1 , *the feedback factor is given by* $\beta = \dfrac{I'_f}{V_o}$

The   *current*

$I_o = \dfrac{V'_o}{(R_f \| R_{E2})}$ *and current I_f through the resistance R_f will be* $-\dfrac{R_{E2}}{(R_f + R_{E2})} I_o$.

The negative sign is because the assumed direction of the current is opposite to the actual direction of current flow (Fig. SE7.4.2).

Hence

$$I'_f = -\frac{R_{E2}}{(R_f + R_{E2})}\frac{V'_o}{(R_f \| R_{E2})} = -\frac{R_{E2}}{(R_f + R_{E2})}\frac{(R_f + R_{E2})}{R_f R_{E2}}V'_o$$

or

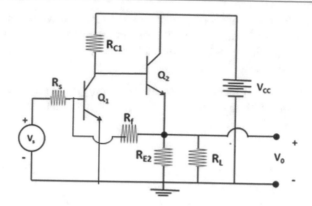

Fig. SE7.4.1 Network for solved example (SE7.4)

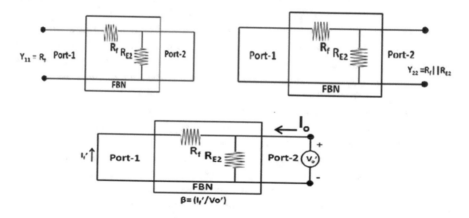

Fig. SE7.4.2 Calculation of y_{11} and y_{22}

$$\beta = \frac{I'_f}{V_o} = -\frac{1}{R_f}$$

Let us also draw the amplifier network without feedback, but retaining the loading of the amplifier by the feedback network. In the given network, a voltage source is provided at the input, but transresistance amplifier is driven by current, it is, therefore, required to change the voltage source into a current source. A current source of value (V_s/R_s) in parallel with a resistance R_s is included in the circuit at the input. The equivalent circuit of the given amplifier after removing feedback but retaining the loading of the original circuit by the feedback network is shown in Fig. SE7.4.3.

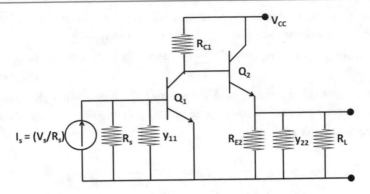

Fig. SE7.4.3 Equivalent circuit of the given amplifier network after removing feedback but retaining the loading of the network due to the feedback circuit

It may be seen in Fig. SE7.4.3 , the loading effect of the FBN may be included through the impedances y_{11} and y_{22}. The hybrid-pi equivalent circuit of the given network is shown in Fig. SE7.4.4 where transistors are replaced by their pi-equivalents in terms of resistances r_π's and dependent current sources $g_m V_\pi$'s.

In Fig. SE7.4.4 B_1, B_2, C_1, C_2, E_1 and E_2, respectively, stands for the bases, collectors and emitters of transistors Q_1 and Q_2. As has already been mentioned, the total given amplifier circuit may be divided into two parts, (i) a CE amplifier based on transistor Q_1, the output of which is given to (ii) the emitter follower based on transistor Q_2. Now the total gain of the circuit will be equal to the multiplication of the gain of the CE amplifier and the emitter follower. However, it is known that the gain of an emitter follower is ≈ 1, hence the total gain of the given network will be equal to the gain of the first stage of CE amplifier. Now it may be observed in the figure that the gain of the CE amplifier stage, A_r is given by

$$A_r = \frac{V_{\pi 2}}{I_b} \left(\frac{volt}{current}, i.e.ohm \right)$$

In order to calculate the gain of the amplifier let; $R_s \parallel y_{11} = R_1$ and $r_{01} \parallel R_{c1} = R_2$.

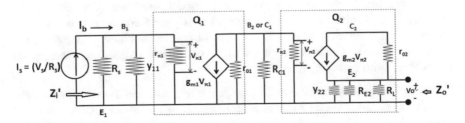

Fig. SE7.4.4 Hybrid-pi equivalent circuit of the Shunt-Shunt amplifier

Now, from the law of current division between two parallel resistances, the current I_1 through the resistance $r_{\pi 1}$ is given by; $I_1 = \frac{R_1}{(R_1 + r_{\pi 1})} I_b$.

Therefore $V_{\pi 1} = r_{\pi 1} I_1 = \frac{r_{\pi 1} R_1}{(R_1 + r_{\pi 1})} I_b$ *and   the   current* $I_{b2} = -g_{m1} V_{\pi 1} = -g_{m1} \frac{r_{\pi 1} R_1}{(R_1 + r_{\pi 1})} I_b$.

Here I_{b2} is the current through the base of Q_2. Again using the principle of current division, current through $r_{\pi 2}$ is given by $I_2 = \frac{R_2}{(R_2 + r_{\pi 2})} I_{b2}$ and the voltage drop against $r_{\pi 2}$ is given as,

$$V_{\pi 2} = r_{\pi 2} I_2 = r_{\pi 2} \frac{R_2}{(R_2 + r_{\pi 2})} I_{b2} = -r_{\pi 2} \frac{R_2}{(R_2 + r_{\pi 2})} g_{m1} \frac{r_{\pi 1} R_1}{(R_1 + r_{\pi 1})} I_b$$
$$= -g_{m1} \frac{R_1 R_2 r_{\pi 1} r_{\pi 2}}{(R_1 + r_{\pi 1})(R_2 + r_{\pi 2})} I_b$$

Hence gain $A_r = -g_{m1} \frac{R_1 R_2 r_{\pi 1} r_{\pi 2}}{(R_1 + r_{\pi 1})(R_2 + r_{\pi 2})}$

If following data is provided, $I_{c1} = 1$ mA, $I_{C2} = 0.5$ mA, $R_{C1} = 5$ kΩ; $R_{E2} = 5$ k Ω; $R_f = 10$ kΩ; $R_s = 2$ kΩ, $R_L = 5$ kΩ, $r_{01} = r_{02} = \infty$ and current gain of both transistors be ($\beta_{transistor} = 100$, then it is easy to show that the overall gain of the given amplifier is 101.2×10^3 (volt/Amp). Further, the feedback factor is given by $|\beta_{ffeedback}| = \frac{1}{R_f} = \frac{1}{10k\Omega} = 1x10^{-4}\Omega^{-1}$. The term $(1 + A\beta) = 11.12$.

The input impedance $Z_i' = R_1 \| r_{\pi 1} = 1.7k\Omega \| 2.5k\Omega = 1.01k\Omega$.
In Z is the actual input impedance looked from the source, then

$$\frac{1}{z_i'} = \frac{1}{Z} + \frac{1}{R_s} \quad or \quad \frac{1}{Z} = \frac{1}{z_i'} - \frac{1}{R_s} = \frac{1}{1.01} - \frac{1}{2} = 0.488; \quad Z = 2.05 \ k\Omega$$

The output impedance $Z_0 = y_{22} \| R_{E2} \| R_L = 6k \| 5k \| 5k = 1.76 \ k\Omega$
Input and output impedances will be further reduced by the factor $(1 + A\beta)$ due to negative feedback

$$Z_i^f = \frac{Z_i'}{1 + \beta A} = 0.184 \ k\Omega = 184 \ \Omega \ ; \ Z_o^f = \frac{1.76 \ k\Omega}{11.12} = 0.158 \ k\Omega = 158 \ \Omega$$

Solved Example (SE-7.5): Determine collector and base currents for the circuit given in Fig. SE7.5 and hence obtain hybrid–pi model parameters of the circuit. The current gain β_{tr} is given as 100 and $r_0 = \infty$. Identify the feedback network, type of feedback and feedback topology. Which type of equivalent circuit may replace the feedback network? Draw A-circuit for the network and obtain input and output impedances with and without feedback.

Solution: *The given network is a CE amplifier where feedback is provided from the collector to the base via a 47 kΩ resistance. Since one end of the FBN is connected at the output node, the sampling is of voltage with shunt connection. The FBN, i.e. the 47 kΩ resistance is also directly connected at the input node, the base and*

Fig. SE7.5 Circuit for solved
example (SE7.5)

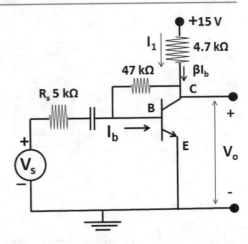

hence mixing is also of shunt connection. The feedback topology is, therefore, Shunt-Shunt or Voltage-Shunt type. Any increase in I_b due to the increase in the input signal will be accompanied by an increase of collector current I_c; increase in I_c will reduce potential V_c at the collector. This reduction in potential at collector will be communicated to the base via *the feedback resistance. Thus, any increase in the input signal will be opposed by the signal fed back to the input. Therefore, the feedback is of the negative type. The feedback network is the 47 kΩ resistance connected between base and collector in the circuit. Only y-parameter equivalent circuit that has shunt connections at both of its ports, may replace the FBN.*

   Figure SE7.5.1 shows how the admittances y_{11}, y_{22} and the feedback factor β for the circuit may be determined. Since the feedback element is only one resistance, $y_{11} = y_{22} = 47$ kΩ. The feedback factor.

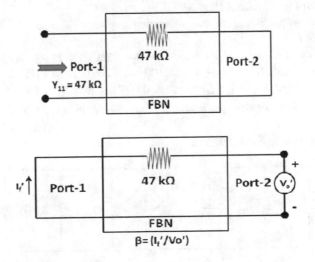

Fig. SE7.5.1 Circuits for determining y_{11}, y_{22} and feedback factor β

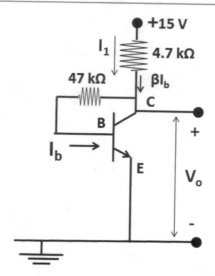

Fig. SE7.5.2 Circuit for dc analysis

$$\beta_{feed} = \frac{I'_f}{V'_0} = -\frac{(V'_o/47\ k\Omega)}{V'_o} = 0.021\ m\Omega^{-1}$$

To determine the base and collector currents, one has to carry-out dc analysis of the given network. While doing dc analysis, capacitors are taken as open circuits and ac voltage source is short-circuited. In the given network, there is one capacitor which if replaced by an open circuit, the modified network becomes the one shown in Fig. SE7.5.2 (Fig. SE7.5.3).

Since the transistor is in the operating region, $V_{BE} = 0.7\ V$. Also $V_C = 0.7V + 47k\Omega.I_b$ (a).

Further, current $I_1 = (\beta_{tr} + 1)I_b$ and $V_C = 15 - 4.7k\Omega x I_1 = 15 - 4.7k\Omega(\beta_{tr} + 1)I_b$ (b).

Equating (a) and (b); $15V - 4.7k\Omega(100 + 1)I_b = 0.7V + 47k\Omega.I_b$

Or $521.7x10^3 I_b = 14.3V$; Hence $I_b = 0.027\ mA$ and $I_C = 100x\ I_b = 2.7\ mA$.

The hybrid-pi model parameters: $g_m = 40I_C\ (in\ mA) = 40x2.7milli\Omega^{-1} = 108mSiemen$

And $r_\pi = \frac{\beta_{tr}}{g_m} = \frac{100}{108mS} = 0.926k\Omega$

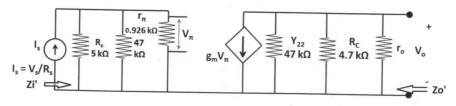

Fig. SE7.5.3 Hybrid-pi equivalent of the given network

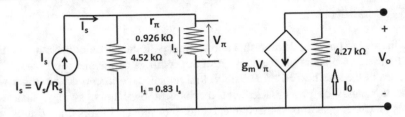

Fig. SE7.5.4 Reduced equivalent circuit

The reduced network of the equivalent circuit is shown in Fig. SE7.5.4. It is simple to calculate the current I_1 passing through r_π, which is given by $I_1 = \frac{4.52}{0.93 + 4.52} I_s = 0.83 \, I_s$.

The voltage drop against r_π $V_\pi = r_\pi I_1 = 0.93k \times 0.83I_s$

The current of the dependent current source $I_0 = -g_m V_\pi = -108x10^{-3} x(0.93x10^3 x0.83I_s)$.

Therefore, $I_o = -83.37I_s$; Hence $V_0 = 4.27x10^3 x[-83.37I_s]$

The gain $A_r = \frac{V_o}{I_s} = -356.0x10^3 \Omega$

The apparent input impedance *without feedback*

$$Z_i' = R_s \| y_{11} \| r_\pi = 5k \| 47k \| 0.93k = 0.77k\Omega$$

If actual input impedance looked by the voltage source V_s is Z_{in}, then

$$\frac{1}{Z_{in}} = \frac{1}{Z_i'} - \frac{1}{R_s} = \frac{1}{0.77k} - \frac{1}{5k} = 1.10 \, or \, Z_{in} = 1.1k\Omega$$

Output impedance $Z'_o = 4.27k\Omega$

Input impedance with feedback $Z_{in}^f = \frac{Z_{in}}{(1 + \beta_{fed}A_r)}$ (since there is shunt connection)

$$Z_{in}^f = \frac{1.1k\Omega}{(1 + 0.021x10^{-3}x356x10^3)} = \frac{1.1k\Omega}{8.48} = 0.13k\Omega = 130\Omega$$

$$Z_{Out}^f = \frac{4.27k\Omega}{(1 + 0.021x10^{-3}x356x10^3)} = 0.50k\Omega = 500\Omega$$

Amplifier gain with feedback $A_r^f = \frac{-356k\Omega}{8.48} = -42k\Omega$

Self-assessment question: what decides which equivalent circuit be used to replace the feedback network in a negative feedback amplifier when reducing it to an A- and beta-circuit?.

7.6 Oscillators

Oscillators are physical devices that show some periodic behaviour. Nature is filled with natural oscillators all the way from superstrings of the size of 1×10^{-33} to galaxies of the size $1 \times 10^{+33}$. Possibly, first human experiments on oscillator were done thousands of years back with mechanical oscillators, swings, then with acoustic (flute) and mixed-mechanical and acoustic oscillators like drums, etc. However, at present, we are interested in electronic devices that are capable of generating electric pulses (waveforms) of a desired shape periodically on a continuous basis. Such devices are called electronic oscillators. They may be divided into two broad categories; (i) the linear oscillators, which produce sinusoidal waves, and (ii) the relaxation or switching oscillators which produce waves of any other shape. Many topologies, like Colpitts, Hartley, phase shift, etc. have been proposed for sinusoidal or linear oscillators.

There are several ways of classifying oscillators, for example, on the basis of their frequency; there may be audio oscillators, radio oscillators, microwave oscillators, ultra high frequency oscillators, etc. on the basis of the physics behind their operation; positive feedback type, negative resistance type, crystal operated type, relaxation type, etc. on the basis of the shape of the output wave, like sinusoidal, sawtooth, square wave, oscillators, etc. However, only linear oscillators that generate sinusoidal waves of moderate frequency and use the principle of positive feedback will be discussed in this chapter.

7.6.1 Positive Feedback in Amplifiers

The essential components in a feedback amplifier are the basic amplifier, a positive feedback network, a frequency determining network and an amplitude limiting component. In most of the cases, the basic amplifier also performs the function of amplitude limiter and the frequency determining network, usually also serves as the feedback loop.

In literature on oscillators, one finds a mention of negative resistance oscillators, which employ an approach different from the positive feedback approach. Since the positive feedback oscillators also present impedance that assumes a negative resistance value at a particular time of the operation cycle, such oscillators may also be designed using the negative resistance formalism. However, in the present discussion, only the positive feedback approach will be used to describe the behaviour of the oscillator.

7.6.2 Transfer Function, Zeros and Poles

It may be useful to introduce at this stage the mathematical approach to the transfer function of a device. Any two port device has an input and an output. In the case of electronic devices, usually, both the input and the output signals are some sort of a pulse or a wave. Mathematically, any pulse or a waveform may be represented by a polynomial of the type $a_n s^n + a_{n-1} s^{n-1} + \ldots + a_1 s^1 + a_0$, or by a differential equation of the type $a_n \frac{d^n u}{dt^n} + a_{n-1} \frac{d^{n-1} u}{dt^{n-1}} + \ldots\ldots a_2 \frac{d^2 u}{dt^2} + a_1 \frac{du}{dt} + a_0$ where coefficients a_n etc. are real. Let the above polynomial represent the input to a device and another similar polynomial, say, $b_m s^m + b_{m-1} s^{m-1} + \ldots + b s^1 + b_0$ represents the output of the device. The transfer function of the device H(s), which is a rational function in the complex variable $s = \sigma + j\omega$, may be defined as

$$H(s) = \frac{b_m s^m + b_{m-1} s^{m-1} + \ldots + b s^1 + b_0}{a_n s^n + a_{n-1} s^{n-1} + \ldots + a_1 s^1 + a_0} \tag{7.48}$$

It is often convenient to factorize both the numerator N(s) and the denominator D(s) of (7.48) and put it in the form

$$H(s) = \frac{N(s)}{D(s)} = K \frac{(s - z_1)(s - z_2)(s - z_3)\ldots\ldots(s - z_{m-1})(s - z_m)}{(s - p_1)(s - p_2)(s - p_3)\ldots\ldots(s - p_{m-1})(s - p_m)} \tag{7.49}$$

The constant K in (7.49) is called the gain (or gain factor or gain constant) of the device and is given by $K = b_m/a_n$.

The transfer function of a system provides a basis for determining important system response characteristics without solving the complex differential equations. Much of this information may be obtained from the zeros ($s = z_1$, $s = z_2$,... etc.) and poles ($s = p_1$, $s = p_2$, ..., etc.) of the transfer function that are, respectively, the roots of equations;

$$N(s) = 0 \, and \, D(s) = 0 \tag{7.50}$$

It may be observed that when $s = z_i$, the transfer function vanishes (become zero), i.e.

$$\lim_{s \to z_i} H(s) = 0$$

Similarly, when $s = p_i$, the transfer function becomes unbound (approaches infinity), i.e.

$$\lim_{s \to p_i} H(s) = \infty$$

Since all the coefficients of the input and output polynomials are real, therefore, the poles and zeros must either be real or should appear in conjugate pairs. Which in the case of poles will mean that either, $p_i = \sigma_i \, or \, p_i = \sigma_i \pm j\omega_i$

The poles and zeros are the properties of the transfer function, and therefore, of the input, output differential equations describing the input- output dynamics. Together with the gain constant K, the poles and zeros completely characterizes the input–

output differential equations and provide the complete description of the total system. In order to appreciate the potential of poles and zeros of the transfer function, let us take an example. Suppose the transfer function of a system has a pair of conjugate poles. $p_1, p_2 = -1 \pm j2$, a single real zero $z_1 = -4$ and a gain $K = 3$ With this data, it is possible to find the differential equation of the system.

Now $H(s) = 3\dfrac{(s-(-4))}{(s-(-1+j2))(s-(-1-j2))} = 3\dfrac{(s+4)}{(s+1-j2)(s+1+j2)} = 3\dfrac{(s+4)}{(s^2+2s+5)}$

Therefore, the differential equation of the system is given by

$$\frac{d^2x}{dt^2} + 2\frac{dx}{dt} + 5 = 3\frac{dy}{dt} + 12$$

Thus, it may be observed that the knowledge of the zeros and poles of a system along with gain K provides complete information of the system dynamics and the differential equations representing input and output of the system.

Properties of the s-plane, zeros and poles

Transfer function can be mapped in a two-dimensional plane $s = (\sigma + j\omega)$ with σ-axis real and the ω-axis imaginary. The poles in this plane are generally represented by the symbol $\times$ and zeros by 0.

Figure 7.19c shows a typical pole-zero plot for the above example. Without going into the proof, the following properties may be assigned to the pole locations in the s-plane:

- If the complex conjugate poles of a system are located on the imaginary line, the system will oscillate with fixed frequency and amplitude under unforced or homogeneous response.
- If the poles lie on to the right of the imaginary line, the system will oscillate but go into an unstable condition of increasing amplitude with time (see inset (A) of Fig. 7.19d
- If poles lie on the left of the imaginary line, the system under homogeneous response will oscillate with decaying amplitude and move towards the stable condition of no oscillations. (Inset (B) of Fig. 7.19d.

Fig. 7.19c Pole and Zero plot for the above example in s-plane

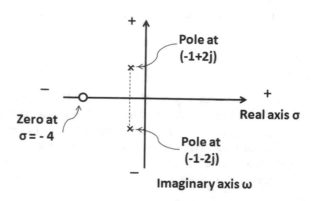

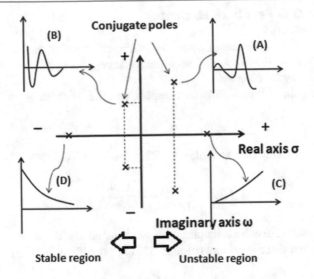

Fig. 7.19d System response for different locations of poles in s-plane

- If poles lie on the real line, there will be no oscillations instead (a) for positive real pole the system under unforced response will be unstable and will exponentially increase, (Inset (C)) (b) if pole is real but negative, the system under homogeneous response will exponentially decrease to attain the stable position. (see inset (D) in Fig. 7.19d.

It may, therefore, be said that the imaginary axis divides the s-plane into two distinct parts. The region on the right of the imaginary axis is the region of instability, the system whose conjugate poles lie to the right of the imaginary axis; under homogeneous response will be unstable and will move away from the imaginary line oscillating with increasing amplitude unless is forced to revert back. Similarly, a system that has a pair of conjugate poles on the left of the imaginary axis in s-plane, under unforced homogeneous response will move towards the imaginary axis, oscillating with diminishing amplitude to attain stability. The system the conjugate poles of which lie on imaginary axis will oscillate with fixed frequency and amplitude.

Self-assessment question: Which part of the s-plane is stable and why?

Self-assessment question: While designing an oscillator the poles *of the circuit should be in which part of the s-space?*

Self-assessment question An amplifier has its poles to the left of the imaginary axis in S-plane. Will it oscillate if positive feedback is provided to the amplifier?

7.6.3 Positive Feedback Oscillator

Now coming back to the case of a positive feedback amplifier, the block diagram of which is given in Fig. 7.19, the amplifier voltage gain is $A_v(jw)$, which in the general case may be complex.

Since the gain of the amplifier is complex and the feedback is positive.

$$(v_i + \beta(j\omega)v_0)A_v(j\omega) = v_0 \text{ or } v_iA_v(j\omega) = v_0(1 - \beta(j\omega))$$

Therefore, the amplifier gain after positive feedback is

$$A_v^f(j\omega) = \frac{v_0}{v_i} = \frac{A_v(j\omega)}{1 - \beta(j\omega)A_v(j\omega)} \tag{7.51}$$

For the positive feedback amplifier to work as an oscillator, an output signal must exist when there is no input signal. This is possible only when

$$1 - \beta(j\omega)A_v(j\omega) = 0$$

or

$$\beta(j\omega)A_v(j\omega) = 1 \tag{7.52}$$

Equation (7.52) expresses the fact that for oscillations to occur (without any input or self excited), the feedback loop gain must be equal to unity. This condition is called BARKHAUSEN criteria for self excited oscillations.

Suppose the open-loop gain $A_v(jw)$ is real and does not have any imaginary component, then one may represent the (real) open-loop gain by the symbol A_{v0}. Further, let us assume that the feedback factor $\beta(j\omega)$ is a complex of the form

$$\beta(j\omega) = \beta_r(\omega) + j\beta_i(\omega) \tag{7.53}$$

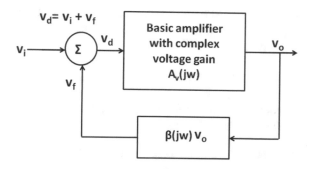

Fig. 7.20 Block diagram of an amplifier with positive feedback

Substituting the above value of $\beta(j\omega)$ in (7.52) and equating the real and imaginary parts one gets

$$\beta_r(\omega)A_{v0} = 1 \; or \; A_{v0} = \frac{1}{\beta_r(\omega)} \tag{7.54}$$

$$\beta_i(\omega) = 0 \tag{7.55}$$

Equations (7.54) and (7.55) jointly express Barkhausen criteria for the special case of open-loop gain being a real quantity. Further, (7.54) is called the gain condition and (7.55) as the frequency of oscillation condition. Equation (7.55) predicts the frequency at which the phase shift of the signal from the input of the amplifier to the output and back to the input via the feedback loop will either be 360^0 or integer multiple of it.

It is possible to write (7.52) in polar form (X $\angle\theta$) as

$$[|\beta(j\omega)A_v(j\omega)|]\angle[\beta(j\omega)A_v(j\omega)] = 1$$

This reduces to

$$[|\beta(j\omega)A_v(j\omega)|] = 1 \tag{7.56}$$

And

$$\angle[\beta(j\omega)A_v(j\omega)] = \pm n360^0 \tag{7.57}$$

So far we considered the case of a voltage amplifier, however, if there is a current amplifier with open-loop current gain $A_i(j\omega)$, and the feedback factor $\alpha(j\omega)$, the oscillation condition is given by

$$\alpha(j\omega)A_i(j\omega) = 1 \tag{7.58}$$

Another way of looking at the condition of oscillations is to treat A_v^f, the gain with feedback as a transfer function between the input signal v_i and the output signal v_0 of the system consisting of an amplifier plus a positive feedback network as shown in Fig. 7.21.

That is

$$A_v^f = H(s) = \frac{v_0}{v_i} = \frac{N(s)}{D(s)} = \frac{A_v(j\omega)}{1 - \beta(j\omega)A_v(j\omega)} \tag{7.59}$$

The condition for self excited oscillations is obtained by setting $D(s) = v_i = 1 - \beta(j\omega)A_v(j\omega) = 0$ that gives the poles of the transfer function $H(s)$.

As already mentioned in our discussion of poles, a stable oscillator has its conjugate poles on the imaginary axis. However, in electronic oscillators it is usually not possible to design an oscillator that has poles on the imaginary axis

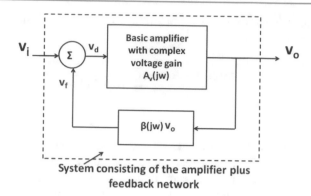

System consisting of the amplifier plus
feedback network

Fig. 7.21 Gain with feedback A^f_v may be treated as the transfer function for the system consisting of the amplifier and the feedback network

because of the non-linear nature of feedback loop. Broadly speaking, it is possible to design systems that have their conjugate poles either to the left or to the right of the imaginary axis. However, the design in which the conjugate poles lie to the left of the imaginary axis is not good as in this region the system will ultimately reach a stable state of no oscillation. The design in which poles lie to the right of the imaginary axis is good as long as the system remains in this region to the right, it will oscillate. The problem is of increasing amplitude of oscillations which may be tackled by incorporating some amplitude limiting mechanism. For example, if the system is initially only slightly away from the saturation region and moves to the saturation region with the increase of amplitude, the non-linear nature of gain in this region will reduce the gain, reduce the amplitude and will force the system to move towards the imaginary axis. Similar results may also be achieved by the transient signals that are generated when the power supply of the system is switched on and also by the noise signals generated by the heating of the device. Thus, an oscillator that has its poles towards the right side of the imaginary axis in s-plane attains a near constant amplitude and frequency, which are the average values of their variations due to non-linear effects that results in shifting of poles to left and the natural tendency of poles to shift to right, as shown in Fig. 7.22. It has been experimentally found that the Barkhausen condition of oscillation frequency given by (7.55), correctly predicts the frequency of oscillations.

(a) Hartley oscillator

Oscillatory processes are known for long, a swing is the simplest example of an oscillatory process. In case of the swing, the total energy of the swing is potential when it is at its two extreme positions and is totally kinetic when it is at the midpoint of the swing. The potential energy changes into kinetic energy while the swing moves from the position of maximum displacement towards the middle position where it becomes totally kinetic. As the swing moves further towards the other extreme, the kinetic energy gives way to potential, and finally at the extreme

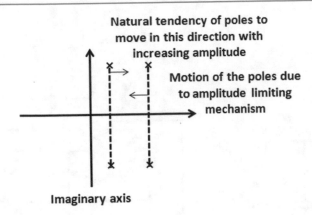

Fig. 7.22 Figure shows the effect of natural tendency of conjugate poles to move away from the imaginary axis with increasing amplitude and the effect of the amplitude limiting mechanism that makes the poles to move towards the imaginary axis with reduced amplitude. The oscillator stabilizes between the two extreme locations with average values of amplitude and frequency of oscillations

of motion, the energy of the swing again becomes totally potential. It may thus be observed that in an oscillatory motion, there are always two types of energies involved which keep transforming from one to the other type during the periodic motion. In case of electrical or electronic oscillatory motion also two types of energies take part. The simplest circuit that under suitable conditions may produce electric/electronic oscillations is a tank circuit that consists of an inductance and a capacitance in parallel connection as shown in Fig. 7.23a

To understand the operation of a tank circuit, let us start from the instant when switch K_1 is closed. A current flows through the capacitor C that charges it to the battery voltage with the upper plate of the capacitor having a positive polarity. During the charging of the capacitor, the battery supplies energy to the capacitor which gets stored in it in the form of electrostatic energy. Once the capacitor is fully charged, if switch K_2 is closed and K_1 made off, then the capacitor will discharge via the inductance L. During its discharge through inductance, the electrostatic energy stored in the capacitor converts into electromagnetic energy and gets stored around the inductor in the form of a magnetic field. Once the capacitor is fully discharged, total electrostatic energy gets converted into electromagnetic form and remain stored in the inductance. Since the magnitude of a capacitor discharging current keeps changing with time, it induces an emf of opposite polarity across the inductor, the magnitude of the induced back emf is maximum when discharging current becomes zero. This emf induced across the inductor has a polarity opposite to the potential difference across the capacitor that was causing the discharge current to flow. Now the induced emf sends current through the capacitor and charges it once again. However, the direction of flow of this charging current is opposite to the direction of discharge current, hence capacitor plates get charged to opposite polarities, upper place becoming negative and lower positive. Once again

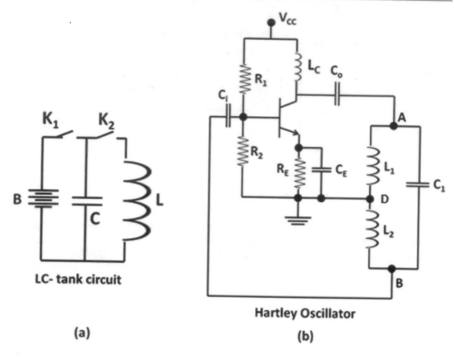

Fig. 7.23 a L-C tank circuit **b** Hartley oscillator circuit

electromagnetic energy contained in the inductor gets converted to electrostatic energy of the charged capacitor. This cycle of charging and discharging of capacitor through inductance will continue indefinitely if there is no loss of energy, that is both the capacitor and the inductors are ideal, do not dissipate energy. The charging and discharging sequences of capacitor will establish currents that will flow in one direction for half of the cycle and in the opposite direction in the next half cycle; thus generating sinusoidal current (or voltage) waves. These sinusoidal current or voltage waves are electrical/electronic oscillations. Actual or practical capacitors and inductors are both dissipative; particularly the inductance has a finite resistance and continuously converts energy into heat. As a result, the sinusoidal waves produced by the oscillations in the tank circuit have diminishing amplitude. The situation is comparable to the case of the swing, if left unattended, the amplitude of the swing goes decreasing because of the loss of energy against the friction. To overcome this loss of energy, in case of the swing a push at the right moment must be given. Similarly, in case of tank circuit, to compensate for the energy loss and to maintain the amplitude of oscillations at a constant value, the tank circuit must be supplied energy at regular intervals by closing switch K_1 and connecting it to the battery. Electronic devices, called oscillators often employ tank circuits to generate sinusoidal waves and to replenish the loss of energy by the tank circuit; they use the principle of positive feedback. A part of sinusoidal wave is taken from the output and is remixed with the input in such a way that it increases the magnitude of the input signal, i.e. the feedback part is in-phase with the input.

Figure 7.23b shows the circuit diagram of a Hartley oscillator, invented by Ralph Hartley and patented in 1915, while working at the research laboratory of the Western Electrical Company. As shown in this circuit, the BJT operates as an amplifier, the combination of resistances R_1, R_2, R_E and the bypass capacitor C_E provide stable dc bias. Capacitors C_i and C_o are used to isolate dc from the high-frequency ac generated by the tank circuit. An inductance L_C connects the collector of the transistor to the V_{CC} supply and works both as ac load and also isolates the high-frequency ac generated in the system from reaching the V_{cc} supply by offering high impedance. However, L_C behaves as a short circuit for dc supply. When the power is switched on to the amplifier circuit, current through the collector charges the capacitor C_1 developing a positive charge on the upper capacitor plate connected to terminal A. Once the capacitor is fully charged, it discharges through the series combination of inductances L_1 and L_2. Thus, a cycle of charging and discharging of capacitor C_1 through the series combination of L_1 and L_2 sets in across the tank circuit producing sinusoidal waves. In order to replenish the tank circuit of the energy dissipated by it, a part of the output sinusoidal signal is applied back at the base of the transistor amplifier through the lead connecting the terminal B of the tank circuit to the capacitor C_i, which blocks any dc at B but allow ac signals to reach the base. To make sure that the signal sent back to the base is in-phase with the signal at base, the central port of L_1 and L_2 is grounded. Grounding of the central point D of the junction of L_1 and L_2, ensures that the signal at port B is 180^0 out of phase with signal at port A. Suppose at a particular instant terminal A is positive, then terminal D will be at ground potential and polarity of the signal induced at B will be negative. It is known that in case of a BJT npn transistor amplifier in common emitter configuration, the signal at collector is 180^0 out of phase with signal at base. Therefore, the phase change of the signal from base to collector (terminal A) is 180^0 and from A to B another 180^0. As such the part of the signal feedback from the tank circuit to the base via lead BC_i, is $180^0 + 180^0 = 360^0$ out of phase (= in-phase) with the signal already existing at base. In this way, positive feedback is provided in this design of the oscillator.

The frequency of oscillations is approximately the resonant frequency of the tank circuit given by

$$f = \frac{1}{2\pi\sqrt{LC}} \tag{7.59}$$

Here, $L = L_1 + L_2$ and $C = $ the capacity of C_1.

Sometimes in the tank circuit, instead of using two different inductance coils, a single coil wound on a core is used and the central tapping of the winding is grounded. In such a situation, the two parts of the same coil have their individual inductances, L_1 and L_2 but have additional mutual inductance coming into the picture on account of the magnetic coupling of the two parts through core. In that case

$$L = L_1 + L_2 + k\sqrt{L_1 L_2} \tag{7.60}$$

As already mentioned, (7.59) gives only the approximate value of the frequency; the actual frequency will be lower than that because of the loading of the tank circuit by the transistor and also because of the parasitic capacitances present in the circuit.

Hartley oscillator is mostly used to produce sinusoidal waves of radio frequencies (in the range of kilo to mega H_z) used in signal transmission. It is not capable of producing low-frequency waves, primarily because of the requirement of inductances and capacitances of large values.

The main advantage of Hartley oscillator is that it requires very few components and that by using a variable capacitor it is possible to produce sinusoidal waves of variable frequencies.

The main disadvantage, apart from the limitation of low frequency, is that the sinusoidal waves produced by the circuit have a large component of higher harmonics unless some other frequency selecting circuit is coupled with the system.

Self-assessment question: As required, the poles of Hartley oscillator must be to the right of the imaginary axis in S-plane and will have the tendency to move further to the right with oscillations of increasing amplitude. What part of the Hartley circuit restricts this increase of amplitude? And how?

(b) Colpitts oscillator

Figure 7.24 shows the circuit diagram of a Colpitts oscillator. This oscillator is almost similar to Hartley oscillator and is, therefore, often referred to as the electrical dual of Hartley. The frequency producing element in the circuit is the tank network which has one inductance L_1 but two capacitances C_1 and C_2 joined in series. Often a variable dual gang condenser, in which the capacity of both condensers may be varied by a single rotating switch, is used in this circuit. The operation of the circuit is identical to that of the Hartley oscillator, the only difference being the feedback network which in this circuit is provided through the capacitor C_2. Since the midpoint of C_1 and C_2 is grounded, the phase of the signal at terminal B is opposite to that at terminal A. The total phase change from transistor base to terminal A, from A to terminal B and back to base is of 360^0. A phase difference of 360^0 ensures positive feedback.

The approximate value of the oscillator frequency is given by the frequency f of the tank circuit,

$$f = \frac{1}{2\pi\sqrt{LC}}$$

Here, $C = \frac{C_1 C_2}{C_1 + C_2}$ and $L = L_1$.

Self-assessment question: It may be observed that the expression that gives the frequency of oscillations provides only approximate value, why?

Fig. 7.24 Circuit diagram of Colpitts oscillator

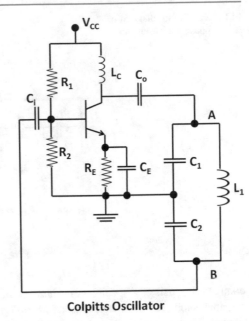

Colpitts Oscillator

(c) Phase shift oscillator

When any electronic circuit, and in particular, an amplifier is switched on, sinusoidal oscillations of many different frequencies get generated at its output. These oscillations are produced either because of the transients generated at the instant of switching-on or may be due to the noise generated as a result of current flow through circuit elements. In a normal situation, these sinusoidal oscillations die out with time. However, if there is a frequency selection network along with a feedback network that applies a part of these oscillations in the output back at amplifier input in-phase with the signal at the input,(i.e. positive feedback) sinusoidal waves of selected frequency may be obtained. A typical R–C network produces a phase shift of 60^0 and if three such combinations are connected in series a total phase shift of 180^0 may be obtained. Further, the same network has a resonance frequency given by $f = \frac{1}{2\pi\sqrt{6}RC}$ if all the three resistances are equal of value R and all the three capacitors are equal of value C. Circuit diagram of a R-C phase shift oscillator is shown in Fig. 7.25, where resistances R_a, R_b and R_c in combination with capacitors C_a, C_b and C_c provide both the positive feedback and frequency selection. In most of the practical circuits, the three resistances R_a, R_b and R_c are of the same value and so also the three capacitances. Apart from these feedback elements, the other components in the circuit constitute an npn BJT amplifier in CE configuration. Since there is an inherent phase difference of 180^0 between the base and collector signals in the CE amplifier, adding another 180^0 phase shift through the feedback network, makes a total phase change of 360^0, i.e. the feedback signal becomes in-phase with the signal at base. The resonance frequency of the oscillator may be

Fig. 7.25 Circuit diagram of a R-C phase shift oscillator

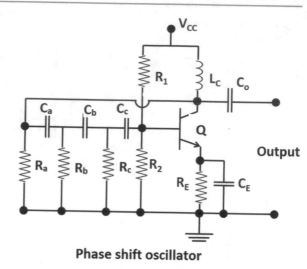

Phase shift oscillator

changed by using a special combination of three air core capacitors mounted together such that the magnitude of each capacitor may be changed by the same amount by the rotation of a single knob. Such capacitance combinations are called gang condensers.

The feedback network loads the transistor amplifier and reduces its gain. It can be shown that feedback network may reduce the amplifier gain by a factor of 29. Hence, for proper oscillatory action, the amplifier circuit must provide a gain of larger than 29, so that even after loading there is still amplification of the signal.

Special property of phase shift oscillator is its remarkable stability and produces well-shaped distortion free sinusoidal waves.

Self-assessment question: In phase shift oscillator apparently there is no tank or any other frequency generating network, only positive feedback is ensured. How the circuit then produces sinusoidal waves?

(d) Wien bridge oscillator

Max Wien in 1891, developed a bridge circuit for determining the magnitude of unknown impedance, later the same bridge circuit was used to make an oscillator. The bridge network has four arms, two of which have only resistances. In the remaining two arms, one has a series combination of resistance and capacitance while the fourth arm contains a parallel combination of resistance and capacitance. These series and parallel combinations of resistance and capacitance constitute a bandpass frequency filter which may be understood through the circuit given in Fig. 7.26a. Since the reactance X_c offered by a capacitance to a signal of frequency f is given by $X_c = \frac{1}{2\pi f C}$, therefore, for the series combination $C_1 R_1$, signals of all frequencies greater than a lower frequency, say, f_1 will find the reactance of the capacitance small and will pass through. The series combination, therefore, acts as a

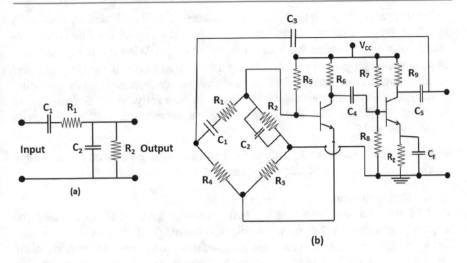

Fig. 7.26 **a** Band pass filter **b** Wien bridge oscillator

high pass filter. At this point, it may be mentioned that signals of different higher frequencies that pass through the high pass filter will have different amplitudes because the total impedance Z $(=X_c + R_1)$ faced by each high-frequency signal will have a different value. As such a combination of signals of higher frequencies $(f > f_1)$ having amplitude distribution will reach the low pass filter. The capacitor C_2 of the low pass filter will offer negligible reactance to all signals of frequencies larger than, say, f_2 and will work as a short circuit for these frequencies grounding them.

It may thus be observed that signals having frequencies between f_1 and f_2 will appear at the output. In this band of frequencies, there will be only one signal of frequency, say f, that will have the largest amplitude. The band pass filter, therefore, also works as a resonance frequency selector.

Figure 7.26b shows the circuit diagram of Wine bridge oscillator. As may be observed in this figure, it consists of a series combination of two CE amplifiers in cascade. Since each amplifier stage introduces a phase shift of 180°, the total phase shift from the base of the first amplifier to the out of the second is of 360°. The final output from the second amplifier is fed back to the input of the first amplifier through the bridge circuit. Since the total phase shift is 360°, the feedback is positive and the bridge circuit works as a frequency selection network. As it is clear from the circuit diagram of the oscillator, all other resistances, R_5 to R_8 and capacitor C_E bypassed resistance R_E provide load resistances to the two amplifier stages and establish stable transistor biasing. Capacitances C_3 to C_5 are coupling capacitors meant to isolate dc. It may be noted that the inclusion of two amplification stages not only ensures a positive feedback, but also takes care of output loading due to the bridge circuit.

In practical circuit designs, it is normal to keep $R_1 = R_2 = R$ and $C_1 = C_2 = C$, though in theory different values may be used. In the normal conditions, the frequency of the output sine wave is given by $f = \frac{1}{2\pi RC}$. Wien oscillators are very frequently used for generating sub audio and audio frequencies in the range of a few ten kilo hertz to few hundred kilo hertz. Wine oscillator is capable of providing stable and distortion free sinusoidal waves of relatively lower frequencies as compared to the Hartley and Colpitts oscillators that are suitable for radio frequencies.

Self-assessment question: Explain how does the band pass filter in the bridge circuit performs the task of frequency selection?

(e) Crystal oscillator

Most of the oscillators that employ tank circuit or phase shift mechanism for frequency generation suffer from two big shortcomings: (i) frequency instability; that is the frequency of the generated sinusoidal wave changes or drifts from the initial value with the heating of the system as time elapses. (ii) Frequency definition is not very precise, which means that the peak in the signal amplitude versus frequency graph is not very sharp. The broadness of the peak at its half maximum is a measure of the frequency selectivity of the oscillator and is related to Q, a parameter characterizing the frequency selectivity of the oscillator. A small Q-value means a good oscillator. A crystal oscillator is almost free from these two shortcomings.

A crystal oscillator uses a piezoelectric crystal for generating linear (sinusoidal) waves. Some crystalline materials have the special property that when pressed or distorted in a particular direction they develop an electric field (or a potential difference) along the direction of distortion. This property called the piezoelectric effect was discovered by Jacques and Pierre Curie in 1880. Initially, natural quartz crystals were found to exhibit this property. However, later synthetic quartz and some ceramic crystals have been developed that show piezoelectric effect. The reverse process called electrostriction occurs when a voltage difference is applied to a piezoelectric crystal in a specific direction, the crystal gets distorted. Thus, by applying an alternating potential difference across or on the two opposite faces of a piezoelectric crystal, it is possible to make the crystal vibrate with the frequency of the applied alternating electric field. This way of making a crystal to vibrate is called the method of forced vibrations. Now it is known that every material object including a piezoelectric crystal that possesses the property of elasticity has a natural frequency of vibrations that depends on several factors including the size and shape of the crystal. In case of forced vibration of a crystal, if the frequency of the applied alternating electric field matches the natural frequency of the crystal, the phenomenon of resonance occurs and the crystal vibrates with maximum amplitude. Since the natural frequency of each crystal is unique and precise, the resonance frequency is also unique and precise. Crystal oscillators use piezoelectric crystals like natural or synthetic quartz or ceramic crystals instead of the tank circuit or the phase shift mechanism for producing sinusoidal waves.

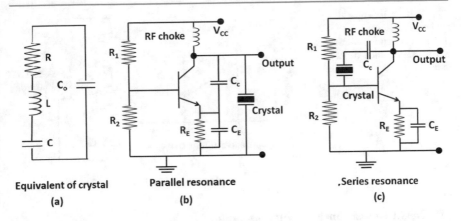

Fig. 7.27 a Electrical equivalent of the crystal b, c circuits, respectively, for parallel and series modes of operations

A quartz crystal can be modelled as an electrical network as shown in Fig. 7.27a. Such a crystal may have two resonance (natural) frequencies f_s (series) and f_p (parallel) given by

$$f_s = \frac{1}{2\pi}\frac{1}{\sqrt{LC}} \ and f_p = \frac{1}{2\pi}\sqrt{\frac{C+C_o}{LCC_o}}$$

The parallel resonance frequency of the crystal may be increased by adding an inductance across it, while it can be reduced by adding a capacitor across the crystal. In this way, it is possible to manage and regulate the parallel resonance frequency of the crystal and the oscillator. The crystal also exhibits high impedance if vibrating in parallel mode and low impedance in series mode. The circuit diagrams of crystal oscillators for parallel and series resonance mode are shown, respectively, in Fig. 7.27b, c. As may be observed in these figures, in a parallel mode circuit, the crystal resonator is connected at the output stage because of its high impedance, while in the series mode it is coupled between base and collector. Further, RF chokes are inductances that offer high impedance to the generated RF and do not allow them to pass to the V_{cc} power supply.

Crystal oscillators are used in digital electronic devices like computers, etc. for timing purposes. Digital watches also use crystal oscillators. Watches made using costly quartz crystal oscillators may have very small time errors of less than 1 s in 30 years or so. However, relatively cheap watches use ceramic crystals that are less precise but not very costly. Crystal oscillators are generally used to produce radio frequency sinusoidal waves.

Self-assessment question: Which circuit property restricts the amplitude of generated oscillations from increasing?

Fig. 7.28 a Example of
negative resistance
b Decaying current through a
series combination of R, L
and C

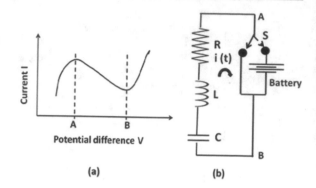

(a)   (b)

(f) Negative resistance or Gunn oscillator

Though negative resistance oscillators do not use positive feedback but brief discussion on it is included here for the sake of completeness. Let us start by introducing the property of negative resistance.

Resistance of a piece of a conductor is defined as the ratio of applied voltage V
to the current I that flows through the conductor, i.e. $R = \frac{V}{I}$. For most conductors I
increases on increasing V and such materials are called Ohmic materials. However,
some doped semiconductors behave differently. Figure 7.28a shows the current
variation with applied voltage in a typical piece of semiconductor GaAs or InP. As
may be observed in this figure, between points A and B current I decreases with the
increase of voltage. In such cases, it is better to define dynamic resistance r in
differential form as $r = \frac{dV}{dI}$, so that r assumes negative values between peak A and
the valley B. The peak at A is called the threshold potential at which the region of
negative dynamic resistance starts. Pieces of negative resistance semiconductors
having connecting terminals at the two ends are called Gunn diodes as it was John
Battiscombe Gunn, who in the year 1960, studied their properties in detail.

Figure 7.28b shows a typical series combination of an inductance L, resistance R
and a capacitance C connected through a switch to a battery. Initially, the switch
connects the series combination to the battery and allows the charging of the
capacitance. Later, when the switch is turned in the other direction, battery goes out
of the circuit and terminals A and B of the circuit are connected together. In this
condition when there is no battery in the circuit, some current, say, i(t) will flow
through the circuit at instant t, as capacitor C will discharge through R and L. Since
there is no external voltage source in the circuit, the sum of instantaneous voltage
drops across capacitor, inductor and resistance will be zero. Mathematically, it may
be expressed as

$$\int \frac{i(t)}{C} dt + L \frac{di(t)}{dt} + Ri(t) = 0 \qquad (7.61)$$

Equation (7.61) may be solved to obtain an expression for current i(t) which will be of the type

$$i(t) = e^{Kt} \qquad (7.62)$$

where

$$K = \frac{-R \pm \sqrt{R^2 - 4\frac{L}{C}}}{2L} \qquad (7.63)$$

The solution corresponding to the real values of K, when $(R^2 - 4\frac{L}{C}) > 0$ is well known, however, we shall explore the case when the under root term is negative, and therefore, K has two components; a real and an imaginary.

$$K = -\frac{R}{2L} \pm j\omega t \text{ where } \omega = \left| \frac{\sqrt{R^2 - 4\frac{L}{C}}}{2L} \right| \qquad (7.64)$$

Hence, current

$$i(t) = e^{-\frac{Rt}{2L}} e^{j\omega t} \qquad (7.64)$$

Equation (7.64) represents a sinusoidal current wave of frequency ω and amplitude $(e^{-\frac{Rt}{2L}})$. It may be observed that the amplitude of the wave depends on the value of factor $(-R/2L)$ and changes with time. In the case when R is positive, which is generally the case; the amplitude of the sinusoidal current waves will go on decreasing exponentially with time as shown in Fig. 7.29a. However, if R is negative, as in case of a Gunn diode; the amplitude of the sinusoidal current wave will increase with time as indicated in Fig. 7.29b. The reason for the exponential decrease of amplitude with time for R = positive is the negative sign of the exponent in $e^{-\frac{rt}{2L}}$. When R = -r, in case of a negative resistance device, the exponent becomes $e^{\frac{rt}{2L}}$ and the amplitude increases exponentially. If R = 0, there is no dissipation of energy by the resistance and the amplitude remains constant as shown in Fig. 7.29c.

Irrespective of the nature of resistance, ultimately the current in the circuit will die out since there is no source of energy (emf) in the circuit. To keep the current flowing, it is essential that energy is supplied to the circuit.

As shown in Fig. 7.30a, the circuit contains three usual elements R, L and C along with a Gunn diode. The system is energized by a dc potential source V_b which is kept isolated from the ac generated in the system by a high value inductance called isolation choke. The bias voltage V_b is adjusted in such a way that

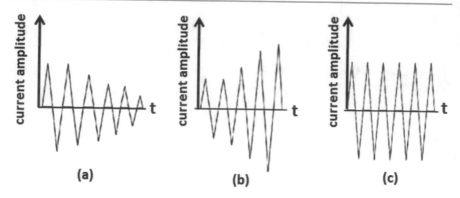

Fig. 7.29 Variation of the amplitude of sinusoidal current with time **a** R = positive (general case) **b** R = negative (as in Gunn diode **c** R = 0

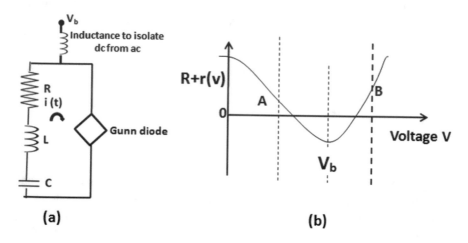

Fig. 7.30 a A combination of R, a Gunn diode, L and C being energized through an isolation choke by a dc bias voltage. **b** The amplitude of sinusoidal current varies between points A and B where power generation by the negative resistance is balanced by the power dissipation by the positive resistance

the total resistance of the combination of the Gunn diode and the resistance R is in the negative voltage region, as shown in Fig. 7.30b. On switching-on the bias supply, the system starts generating a sinusoidal current wave of exponential increasing amplitude which in the positive half cycle drags the operating point to A and to point B during the negative half cycle. At these points A and B, the energy generated by the negative resistance in the system is just balanced by the energy dissipated by the positive resistance, and hence there is no further increase in the amplitude of the generated ac signal. As a result, within a short time, sustained sinusoidal oscillations of (almost) constant amplitude are produced by the system. The cyclic frequency of these oscillations is given by (approx)

$$\omega = \sqrt{\frac{1}{LC}} \qquad (7.65)$$

Gunn oscillators are used to produce very high-frequency waves in the region of microwave to Terahertz. Actual Gunn oscillators use electromagnetic cavity and waveguides, etc. and are highly involved. The purpose of the present presentation was to illustrate the basic principle of operation of the device.

Self-assessment question: Is it true that current passing through a negative dynamic resistance generates energy?

Problems

P7.1 Identify the type of feedback, feedback network, and feedback topology for the circuit given in Fig. P7.1. Select the equivalent network that will be best suited to replace the feedback network and give reason for your selection. Calculate the feedback factor β and the values of **impedances that need to be absorbed in the A-circuit. Draw a hybrid-pi equivalent circuit for the network removing feedback but retaining the loading effect of feedback network.**

Answer: The feedback is negative, being provided by the combination of resistances R_3 and R_4 with Shunt-Shunt topology; y-parameter equivalent may replace the FBN as it has shunt connections at its both ends. $Y_{11} = 10$ kΩ; $y_{22} = (10 \text{ k} \parallel 5 \text{ k})$; $\beta = 0.1 mS$.

P7.2 If in Problem P7.1 each transistor has current gain β = 100, $r_\pi = 2.5$ kΩ, and $r_o = \infty$, calculate the gain of the amplifier and show that the gain with feedback is independent of the amplifier circuit and depends only on the feedback factor β.

P7.3 In the circuit given in Fig. P7.3 identify the type of feedback, its topology and the feedback network.

Answer: [Positive feedback using shunt-shunt topology, combination of $R_a C_a$, $R_b C_b$ and $R_c C_c$ constitutes the FBN].

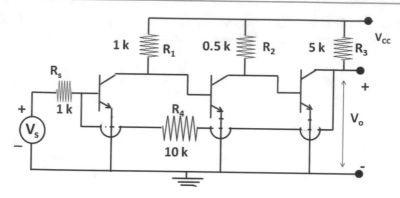

Fig. P7.1 Network for problem (P7.1)

Fig. P7.3 Network for problem (P7.3)

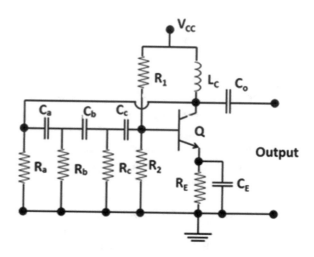

P7.4   A RC-phase shift oscillator uses three identical combinations of 100 Ω resistance and 0.01microfarad capacitance, calculate the approximate frequency of the oscillator and suggest the values of R and C to generate 6.5 kHz ac.

Answer: [65 kHz, 100 Ω, 0.1 µF].

P7.5   The resonant frequency of a Colpitts oscillator is 7.12 MHz. If two capacitors each of 1nF are used in the tank circuit, calculate the value of the inductance.

Answer: [1µH].

P7.6   An amplifier without any feedback has a gain of 10^6 with 10% stability. The gain stability of the amplifier got reduced to 1% using negative feedback. Calculate the magnitude of the feedback factor β and the gain with negative feedback

Answer: $[\beta = 9 \times 10^{-6}, A_f$ (with negative fee back) $= 10^5]$.

P7.7   Identify the nature of feedback and feedback topology for the circuit given in Fig. P7.7

Answer: [negative feedback; Shunt-Shunt topology].

P7.8   Figure P7.8 gives the ac equivalent circuit of a feedback amplifier when all dc sources are replaced by ground and all capacitors are shorted. Identify the feedback type, feedback topology and FBN. Also, calculate the feedback factor β.

Answer: [Negative   feedback,   Shunt-Series   topology, $\beta = (R_5/R_2 + R_5)A/A]$

P7.9   An amplifier has a gain of 40 dB and upper 3db frequency of 10^3 Hz. (i) At what frequency the gain of the amplifier will be 1? (ii) If a negative feedback circuit of $\beta = 0.01$ is included with the amplifier, calculate the new value of gain and the upper 3 dB frequency.

Answer: [(i) Unit gain frequency $= 1 \times 10^5$ Hz; (ii) Gain with feedback 50 Hz, upper 3 dB frequency $= 2 \times 10^3$ Hz].

Fig. P7.7 Circuit for
problem (P7.7)

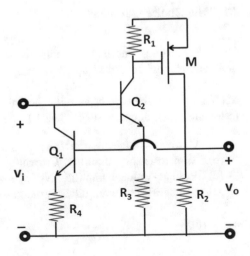

Fig. P7.8 Circuit for
problem (P7.8)

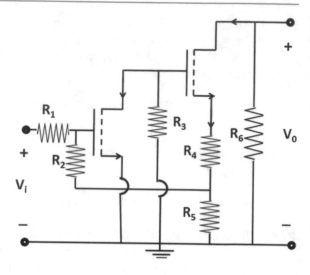

P7.10   A voltage controlled voltage amplifier has a gain of 10^3 V/V, input
impedance of 2.5 kΩ, output impedance of 5 kΩ and 10% distortion in the
output. It is required to reduce the distortion to 2% using Series-Shunt
negative feedback topology. Calculate the appropriate value of the
negative feedback factor β to achieve the desired reduction in distortion
and calculate the new values of the gain, input and output impedances.

Answer:

$$[\beta = 0.004; gain\ with\ feedback = 200\ \frac{V}{V},\ input\ impedance\ 12.5\ k\Omega; \\ output\ impedance = 1\ k\Omega].$$

Multiple Choice Questions

Note: Some questions may have more than one correct alternative; all correct
alternatives must be picked up for a complete answer.

MC7.1 Identify the type of amplifier and the associated feedback topology for the
following block diagram of Fig. MC7.1.

   (a) Current–Voltage amplifier; shunt-Shunt topology.
   (b) Transresistance amplifier; Current–Shunt topology.
   (c) Voltage-Voltage amplifier; Voltage-series topology.
   (d) Voltage-Voltage amplifier; Voltage-shunt topology.

ANS: [(c)].

Fig. MC7.1 Block diagram
for (MC7.1)

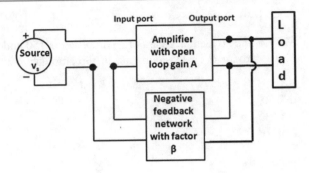

Fig. MC7.2 Block diagram
for (MC7.2)

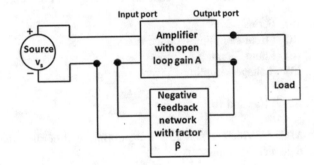

MC7.2 I dentify the type of amplifier and the associated feedback topology for the
following block diagram of Fig. MC7.2.

   (a) Voltage-Current amplifier; Series-shunt topology.
   (b) Transresistance amplifier; Current–Shunt topology.
   (c) Voltage-Voltage amplifier; Voltage-series topology.
   (d) Transconductance amplifier; Voltage-series topology.

ANS: [(b)].

MC7.3 Which of the following correctly lists the benefits of negative feedback in
amplifiers?

   (a) increased gain, better stability, and wider bandwidth
   (b) less noise, less distortion and higher gain
   (c) wider bandwidth, better stability and less distortion
   (d) reduced noise, increased gain and wider bandwidth.

ANS: [(c)].

MC7.4 The no feedback gain of an amplifier is 150 and the feedback factor is 0.05. The approximate closed-loop gain of the amplifier will be.

(a) 70
(b) 55
(c) 30
(d) 18

ANS [(d)].

MC7.5 It is required to increase both the input and output impedances of an amplifier, which of the feedback topology will do that?

(a) Series-Series
(b) urrent sampling–voltage mixing
(c) Shunt-Shunt
(d) Voltage sampling-Current mixing

ANS: [(a) and (b)].

MC7.6 Which of the following correctly describes the behaviour of a negative feedback amplifier?

(a) Upper corner frequency decreases, gain decreases and lower corner frequency increases
(b) Upper corner frequency decreases, gain increases and lower corner frequency increases
(c) Upper corner frequency increases, gain increases and lower corner frequency increases
(d) Upper corner frequency increases, gain decreases and lower corner frequency decreases.

ANS: [(d)].

MC7.7 Which of the following pole (s) is (are) invalid?

(a) $0 + 2J$
(b) $2 + j0$
(c) $2 \pm j5$
(d) $-5 - j1$

ANS: [(a) & (d)].

MC7.8 Tick the correct alternatives. The poles of an oscillator that generates linear waves may.

(a) lie on the left of the imaginary axis in s-plane of transfer function
(b) lie on the imaginary axis in s-plane of transfer function
(c) lie to the right of the imaginary axis in s-plane of transfer function
(d) lie only on the real axis in s-plane of transfer function

ANS : [(b), (c)].

MC7.9 The gain of an amplifier after 3 dB negative feedback is 97 dB. The open-loop gain of the amplifier is.

(a) 200 dB
(b) 150 dB
(c) 120 dB
(d) 100 dB

ANS: [(d)].

MC7.10 Negative feedback network of Shunt-Shunt topology may be replaced by an equivalent of.

(a) h-parameter
(b) g-parameter
(c) y-parameter
(d) z-parameter

ANS: [(c)].

MC7.11 Negative feedback network of Shunt-Series topology may be replaced by an equivalent of.

(a) h-parameter
(b) g-parameter
(c) y-parameter
(d) z-parameter

ANS: [(b)].

MC7.12 In a negative feedback amplifier, the feedback network may be replaced by h-parameter equivalent circuit. The feedback topology of the amplifier is.

(a) Series-Series
(b) urrent sampling–voltage mixing
(c) Shunt-Shunt

(d)   Voltage sampling-Current mixing

ANS: [(d)].

MC7.13 In Shunt-Shunt topology the unit of feedback fraction β is.

(a)   Siemen
(b)   Henry
(c)   Ohm
(d)   Volt per Volt.

ANS: [(a)].

MC7.14 Tick the correct statements about a Gunn diode.

(a)   May amplify an ac signal
(b)   May rectify ac into dc
(c)   Show the property of negative resistance
(d)   Used in making very high-frequency oscillators

ANS: [(c), (d)].

MC7.15 Which oscillator is best for producing highly stable and precise radiofrequency linear waves.

(a)   RC-phase shift oscillator
(b)   Wien Oscillator
(c)   Crystal Oscillator
(d)   Gunn oscillator

ANS [(c)].

Short Answer Questions

SA7.1   What is meant by feedback? Briefly describe negative and positive feedback and list the advantages of negative feedback.

SA7.2   Explain how positive feedback may produce sustained oscillations. What is a Barkhausen criterion for oscillations?

SA7.3   Explain how negative feedback improves the gain stability of an amplifier.

SA7.4   What is the main cause of non-linearity in amplifiers? How it can be improved by the use of negative feedback.

SA7.5 What is meant by the terms 'sampling' and 'mixing' with reference to feedback? Describe the four sampling-mixing topologies used in amplifiers to implement feedback.

SA7.6 Identify the amplifier in which h-equivalent circuit is used to replace *the feedback* network and discuss the method of determining parameters like h11, h22, etc.

SA7.7 What do you understand from the transfer function H(s) of an amplifier? Discuss zeros and poles *of* the transfer function and discuss their significance in the operation of an oscillator.

SA7.8 With the help of a block diagram, show how a negative feedback amplifier with Series–Shunt topology may be reduced to A-circuit and $\beta - circuit$.

SA7.9 Draw a neat circuit diagram of Hartley oscillator and explain how the total phase shift of the feedback component is 360^0.

SA7.10 With the help of the circuit diagram explain the role of the bridge circuit in producing oscillations in Wien bridge oscillator. Wien bridge oscillator is frequently used to produce oscillations of which frequencies?

SA7.11 Explain why the common connecting point of two capacitors is grounded in Colpitts oscillator? Also, elaborate on the role of the tank circuit in an oscillator

SA7.12 Write a brief note on crystal oscillators explaining the role of the piezoelectric crystal.

SA7.13 Outline the stepwise procedure you will adopt in analyzing a negative feedback amplifier to resolve it into A- and beta-circuits and to find the feedback factor β.

SA7.14 With the help of block diagrams and appropriate figures, summarize the method of resolving a negative feedback transconductance amplifier into its A- and $\beta-$ circuits.

SA7.15 Summarize the properties of the s-plane in *which transfer function* is mapped.

Sample Answer to SA7.5

SA7.5 What is meant by the terms 'sampling' and 'mixing' with reference to feedback? Describe the four sampling-mixing topologies used in amplifiers to implement feedback.

ANSWER *Feedback is the process by which a part of the output signal is applied back in the input of a system. Feedback network (FBN) in most of the cases is a passive network that is external to the main two port device like an amplifier. A typical block diagram of an amplifier and the FBN is shown in Fig. SA7.5*

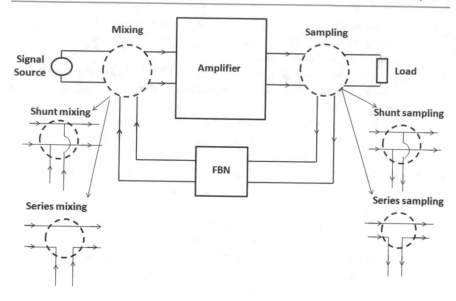

Fig. SA7.5 Block diagram of the amplifier feedback network

Amplifiers, according to the nature of the input and output quantities may be divided into four classes: (i) Voltage input and Voltage output, called Voltage-Voltage amplifier (VVT). (ii) Voltage input and Current output, called Voltage–current amplifier (VCT). ***(iii) Current input and Current*** *output, called Current-current amplifier (CCT) and (iv) Current input and voltage output or Current–voltage (CVT) amplifiers. The sampling process, that involves taking a part of the output signal, may have to sample either the voltage or the current. Sampling of voltage may be done only through shunt (or parallel) connections, while sampling of current may be done only using a series connection with the output. As such sampling topology may either be Shunt sampling topology or Series sampling topology. The connections of sampling for the two topologies are also shown in the figure. Similarly, for the mixing of the feedback signal with amplifier input, Shunt connection has to be used if the feedback signal is current and series connection if the feedback signal is voltage. Remember that in sampling, Shunt topology is used if the amplifier output is voltage and Series topology if the amplifier output is current. However, at the mixing end, Shunt mixing topology is used if the amplifier input is current and Series mixing topology if the amplifier input is voltage. As such there may be four different topologies that will be required for the implementation of feedback in amplifiers. They are:*

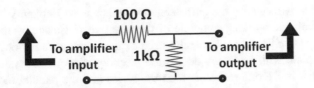

Fig. LA7.5 Feedback network for problem (LA7.5)

(a) *Shunt sampling–Series mixing topology that is used in VVT*
(b) *Shunt samplingShunt mixing topology that is used in CVT*
(c) *Series sampling–Shunt mixing topology that is used in CCT*
(d) *Series sampling–Series mixing topology that is used in VCT*

Long Answer Questions

LA7.1   Define positive feedback and explain how it may be used to generate sustained oscillations. Discuss Barkhausen criteria and its application in oscillators. With the help of a circuit diagram, explain the working of any one oscillator that may generate radio frequencies.

LA7.2   What are the essential components of an oscillator circuit? What role is played by the poles of the transfer function in explaining the production of sustained oscillations? Why is it essential that the poles must lie to the right of the imaginary axis? Explain the role of frequency selecting network in an oscillator.

LA7.3   Explain the role of negative feedback in amplifiers and list its advantages and disadvantages. Derive expressions for the gain stability with negative feedback in case of a voltage controlled voltage amplifier.

LA7.4   Draw the block diagram for a current controlled current amplifier with negative feedback and obtain expressions for the input and output resistances with feedback. Also, resolve the amplifier circuit into corresponding A and β circuits. How the values of the loading components of the feedback network and the feedback fraction β may be obtained?

LA7.5   What are the two important assumptions about the forward/backward transmission of signals through the amplifier and the feedback network? An amplifier uses Series-Series topology for negative feedback, identify the type of the amplifier and the type of the equivalent circuit that may replace the feedback network. Draw block diagrams for A- and the beta-circuits. Calculate the value of the feedback fraction β if the feedback network is as given in the Fig. LA7.5.

LA7.6 What is meant by the term 'negative resistance'? Explain with necessary details the working principle of the negative resistance oscillator. Why these oscillators are generally used in the microwave region?

LA7.7 Explain with necessary details the working of a quartz crystal oscillator. What is the difference between the series and the parallel modes of operation of the crystal? It is said that crystal oscillators are the most stable and precise, why?

Operational Amplifier

8

Abstract

The overall gain of a high-gain amplifier becomes independent of the amplifier circuit if large negative feedback is incorporated. The overall gain in such a case is decided by the feedback network. Amplifiers for which overall gain depends on the feedback network are called operational amplifiers. Operational amplifiers (Op amp) are extensively used in analog electronic circuits. In this chapter, working details of operational amplifiers and their important applications are discussed.

8.1 Introduction

The concept of operational amplifier (denoted in short as **op-amp** or **op amp**) was conceived by Harry Black in 1934 while working at Bell Labs in New Jersey, USA. He was confronted with the problem of designing an amplifier the gain of which may not change because of the changes in ambient temperature and the performance of active circuit elements. Such amplifiers were required for strengthening the audio signals at intermediate stations in long-distance telephone conversations. It is said that he got the idea of a very-high-gain negative feedback amplifier, the gain of which depends on the feedback network and is essentially free of changes in the performance of active circuit elements, while travelling by train /ferry from New York to New Jersey. Since the feedback network is generally made up of a combination of resistances which are much less sensitive to temperature changes, negative feedback amplifiers show extraordinary gain stability. Many other scientists including Harry Nyquist, who gave the generalized rule for avoiding instability in feedback amplifiers and Hendrick W. Bode who developed the systematic techniques of design whereby one could get the most of the specified situation and still satisfying Nyquist's criteria, played important role in developing an op-amp.

© The Author(s), under exclusive license to Springer Nature Switzerland AG 2021 677
R. Prasad, *Analog and Digital Electronic Circuits*, Undergraduate Lecture
Notes in Physics, https://doi.org/10.1007/978-3-030-65129-9_8

The first vacuum tube amplifier fitting the properties of an op-amp came about early in the 1940s wartime period and was used in Bell Labs designed the M9 gun director system.

In terms of op-amp details, Karl Swartzel Jr. of Bell Labs applied for a patent for his design of "Summing Amplifier" in May 1941, which remained pending during the war period, finally being issued in 1946. Further wartime op-amp development work was carried out in the laboratories of Columbia University of New York. Prof. John R. Ragazzini et al. were the first to coin the word "Operational amplifier" in their paper "Analysis of Problems in Dynamics by electronic circuits", Proceedings of the IRE, Vol. 35, May 1947, page 444–452.

Vacuum tube op-amps continued to flourish for some time into the 1950s and 1960s but their competition was eventually to arrive from solid-state developments. Three key inventions; invention of the transistor, invention of the integrated circuits and invention of the planer IC process opened way for the fabrication of solid-state modular and hybrid op-amp designs. The first generally recognized monolithic IC op-amp was from Fairchild Semiconductor Corporation (FSC), the μA 702 which was designed by the young Engineer Robert J. (Bob) Widlar. There were several improved versions of the device up to μA 748 etc. introduced in the market by 1969. Then came the precision AD 741 op-amp in the market.

As the name suggests, operational amplifier, often written as **Opamp, op-amp or Op-Amp** in short, is a very high-gain (open-loop gain tending to infinity) DC-coupled differential voltage amplifier with very large (tending to infinity) input impedance and extremely low output impedance. Since the operational amplifier under different configurations may perform various mathematical operations, like adding, multiplying, division, subtraction, integration, etc. of input analog signals, it is called an operational amplifier. Before the entry of digital computers, operational amplifier was the backbone of analog computers. Since the input signal may be directly coupled to the op-amp, it is possible to apply both DC and AC signals to the op-amp. An op-amp has two inputs; one non-inverting input is denoted by positive (+) sign and the other inverting denoted by (−) sign. In open-loop configuration, the op-amp circuit amplifies the difference in the instantaneous signal values at the non-inverting and inverting inputs by the open-loop gain; that is why it is called a differential amplifier. Since the input impedance of op-amp is very high, an ideal op-amp does not load the source of input signal. Similarly, on account of very low output impedance, it allows large currents to be drawn by the output load. In circuit diagrams, an op-amp is represented by a triangle with non-inverting (+) and inverting (−) inputs and a single output, Though it also has four other terminals; two where positive and negative power supplies (+V_{CC} and −V_{CC} or V+, & V−) are connected for proper biasing of various active inside components and two for null offset, but generally these terminals are not shown in circuit diagrams to avoid crowding of connections. Biasing power supply connections are shown by dotted lines in Fig. 8.1a. Arrays of Op-amps with several other circuits are generally fabricated on a single IC, however, the pin connections of an isolated 8-pin package of Op-amp AD 741 are shown in Fig. 8.1b. Null offset connections are used initially

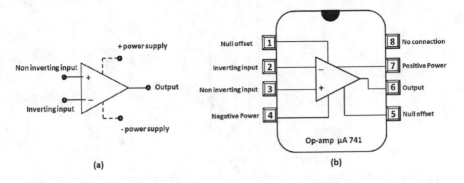

Fig. 8.1 **a** Symbolic representation of an Op-amp. **b** Pin connections of the Op-amp IC Op-amp AD 741

for zero setting of potentials at various points in the inner circuit for proper operation.

An operational amplifier is almost always used with negative feedback, which is provided by connecting the output terminal to the desired input terminal through a network of resistances (and maybe a few capacitances). With negative feedback of feedback fraction β the closed-loop gain of the amplifier (with feedback) becomes

$$A^f = G = \frac{A_{OL}}{1 + \beta A_{OL}}, \text{where } A_{OL} \text{ is the open loop gain}$$

Since A_{OL}, the open-loop gain of op-amp is very large $\approx 10^6$ or more, $\beta A_{OL} \gg 1$ and therefore, $G \approx 1/\beta$. The value of β depends on the resistance network external to the op-amp, hence **the close-loop gain of the amplifier may be adjusted to any desired value manipulating the resistance network, without interfering with the internal circuits of the Op- amp**. Moreover, because of the negative feedback, the overall operation of the op-amp gets stabilized. The distortion, etc. gets reduced and closed-loop gain becomes much more stable.

In principle, signals starting from DC to AC of any frequency may be applied at the input of the op-amp, such that the gain-bandwidth product remains constant.

The component-level circuit diagram of Op-amp 741 is shown in Fig. 8.2a, where it may be observed that there are around 20 BJT and about 12 passive circuit elements.

The block diagram of an op-amp is shown in Fig. 8.2b. As may be observed in this figure, the total op-amp may be divided into three main blocks; (i) the input stage of a differential amplifier, (ii) several stages of linear amplification and (iii) the output stage of a class B push-pull emitter follower.

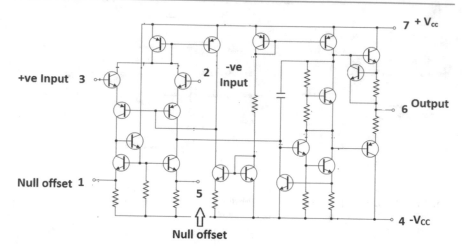

Fig. 8.2a Component-level circuit diagram of Op-amp 741

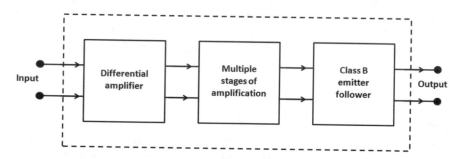

Fig. 8.2b Block diagram of an operational amplifier

8.1.1 Differential Amplifier

Basic circuit of a differential amplifier is shown in Fig. 8.2c, as may be observed, the circuit is symmetrical with respect to the two identical transistors T_1 and T_2. All elements of the two branches, like transistors T_1 & T_2 and resistances $R_{C1} = R_{C2}$, are exactly identical. It is easy to understand that in case two identical signals are applied at the two inputs, there will be no signal between the two outputs and in the case when two different signals are applied at the two inputs, an amplified signal proportional to the difference between the input signals will appear across the two outputs.

Fig. 8.2c Basic circuit of a differential amplifier

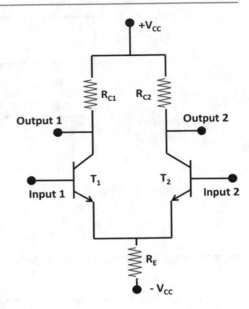

A differential amplifier may be configured in four different ways:

(i) **Single input unbalanced output** where the signal is applied at only one of the inputs, say on input 1, and the output is taken between output 1 and the ground. The other input is grounded.

(ii) **Single input balanced output** is the configuration where the input signal is applied at one input, say at input 1 but the output is taken between the two outputs. The other input (input 2 in this case) is grounded.

(iii) **Dual input unbalanced output** refers to the case when two input signals are applied one at each input but the output is taken between one of the two outputs and the ground.

(iv) **Dual input balanced output is** the configuration that is most frequently used. As obvious from the name, two signals are applied at the two inputs of the differential amplifier and the output is taken across the two outputs of the device.

It is customary to assign one of the two outputs of the differential amplifier as non-inverting output and the other as inverting output. Suppose output 1 is assigned as non-inverting output, then output 2 will be inverting output. Further, the non-inverting output is represented by the plus (+) sign and the inverting output by the minus (−) sign.

Since dual input balanced output configuration, shown in Fig. 8.2d, is the most used configuration we shall discuss it in some more details. The best way to understand the operation of the differential amplifier in dual input balanced output configuration is to carry out the AC analysis of the circuit. Let us assume that two sinusoidal signals v_1^i and v_2^i are applied, respectively, at input 1 and input 2 of the differential amplifier and we wish to determine the nature of the signal at output 1

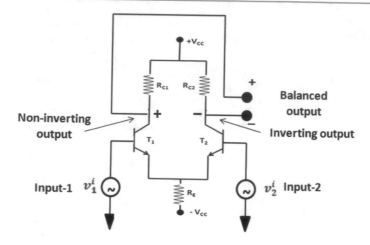

Fig. 8.2d Dual input balanced output differential amplifier

(non-inverting output or + output) with respect to output 2 (inverting output, or − output). The required AC analysis of the circuit may be carried out in two steps using the superposition theorem, according to which the desired output between + and − outputs may be obtained by algebraically adding the results obtained by taking source at one input at a time and grounding the other input.

Figure 8.2e shows a differential amplifier with a sinusoidal signal v_1^i at input-1 and the other input-2 grounded. It may be noted that signal v_1^i is with respect to ground. The transistor T1 with its collector resistance R_C and the common emitter resistance R_E constitutes a common emitter amplifier and, therefore, the output signal v_1^o at collector C_1 with respect to ground will be amplified and out of phase by 180° as compared to the input signal v_1^i, as shown in the figure.

One may, therefore, write

$$\left[v_1^o\right] \text{w.r.t ground} = -A_e\left(v_1^i\right)$$

In the above equation, w.r.t means that the signal v_1^o shown in the figure is with respect to ground and the negative sign on the right-hand side stands for the 180° phase change. Factor A_e is the gain of the common emitter amplifier.

It may be noted that the same signal v_1^i also appears at the common emitter (point A) of the two transistors. Now transistor T2 with grounded base and signal v_1^i at its emitter acts as a common base amplifier. The output signal v_2^o (with respect to ground) at collector C_2 will be A_b times the signal v_1^i, where A_b is the gain of the common base amplifier. Also v_2^o and v_1^i will be in phase as in a common base amplifier output signal is in phase with the input signal. Therefore,

$$\left[v_2^o\right] \text{w.r.t ground} = A_b\left(v_1^i\right)$$

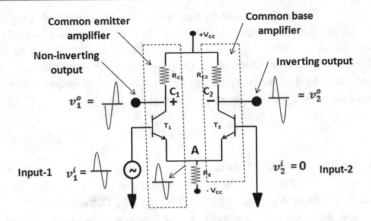

Fig. 8.2e Differential amplifier with signal at one input and other input grounded

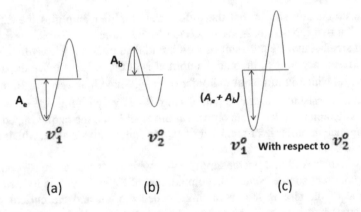

Fig. 8.2f **a** Signal v_1^o w.r.t. ground **b** signal v_2^o w.r.t ground **c** signal v_1^o w.r.t signal v_2^o

As has been mentioned, both signals v_1^o and v_2^o are with respect to ground with their amplitudes, respectively, A_e and A_b times the amplitude of the input signal v_1^i. However, a little thinking will tell that in case when signal v_1^o is measured with respect to signal v_2^o, the amplitude of the resultant signal (called the balanced signal) becomes the sum of the amplitudes of the two signals, which is shown in Fig. 8.2f. Summing up, it may be said that if a signal v_1^i w.r.t ground is applied at input 1, and input 2 is grounded, then a balanced output signal $v_{(bal-1)}^o$ appears between output 1 w.r.t output 2 which is given as

$$v_{(bal-1)}^o = -(A_e + A_b)v_1^i$$

In the second step, let us assume that input 1 is grounded and a signal v_2^i is applied at input 2 of the differential amplifier. It is easy to see that now an A_b time amplified signal (w.r.t ground) will appear at collector C_1, which will be in phase with v_2^i and an A_e time amplified signal (w.r.t. ground) will appear at C_2 which will be out of phase by 180° as compared to signal v_2^i. Balanced output signal $v_{(bal-2)}^o$, i.e. output signal at C_1 with respect to signal at C_2, is given by

$$v_{(bal-2)}^o = (A_e + A_b)v_2^i$$

Now, from superposition theorem, the balanced output when both signals v_1^i and v_2^i are present simultaneously, respectively, at input 1 and at input 2, may be obtained by algebraic addition of $v_{(bal-1)}^o$ and $v_{(bal-2)}^o$. Therefore,

The balanced output v_{bal}^o for simultaneous dual input signals is given by

$$v_{bal}^o = v_{(bal-1)}^o + v_{(bal-2)}^o = (A_e + A_b)\left[v_2^i - v_1^i\right]$$

The last equation shows that the differential amplifier amplifies the difference between the two input signals, as suggested by its name.

Using small-signal r-equivalent of a differential amplifier, it is easy to show that the input resistance of the differential amplifier is of the order of twice the dynamic resistance r_E, which in turn is of the order of $\approx \frac{V_T}{I_E}$. Since I_E is very small, the input impedance or resistance of a differential amplifier is very high in the order of few hundred kilo ohms or even mega ohms. On the other hand, the output impedance of the differential amplifier is of the order of R_c the collector resistance, which may be of few kilo ohms.

It may be observed that power supply $-V_{CC}$ with series resistance R_E constitutes a constant current source. Since this combination of R_E and $-V_{CC}$ is located almost at one end of the circuit, it is sometimes called the tail and the current passing through emitter resistance R_E as tail current denoted by I_T. Since each of the two identical transistors contributes a current I_E, tail current $I_T = 2\ I_E$.

8.2 Working of Operational Amplifier

The output of the differential amplifier is fed to several stages of linear amplifiers to obtain a very large value of overall gain. The output stage of the op-amp is usually a class B push–pull emitter follower. As is known, class B push-pull emitter follower has very small output impedance and may deliver considerable current to the output load. In the following sections, we shall discuss the operational details of operational amplifiers.

Self-assessment question: What property of an op-amp makes it so useful?

8.2.1 Feeding DC Power to the Op-Amp

As shown in Fig. 8.2a, an op-amp contains several active components (transistors, etc.), which need to be properly biased for the satisfactory working of the op-amp. Two terminals are provided in the op-amp for connecting DC power sources. These terminals are leveled as $V_{CC}+$ and $V_{CC}-$ (or V_S+, V_S-; $D_{SS}+$, $D_{SS}-$ etc.). The DC power to the op-amp may be supplied in two different ways. The first configuration called the **split power arrangement** is shown in Fig. 8.3, where two power supplies (or batteries) each giving the same potential (12.0 V in Fig. 8.3) are joined in series with the negative terminal of one with the positive terminal of the other. The common terminal is then grounded. As a result, one power supply gives $+V_{CC}$ (= +12.0 V) with respect to the common ground and the other $-V_{CC}$ (= −12.0 V) with respect to ground. The output of the combination is connected to the corresponding terminals of the op-amp.

Now suppose a source of voltage V_1 (with respect to ground) is connected at the non-inverting input and another source of potential V_2 (with respect to ground) is connected at the inverting input of the op-amp, as shown in the figure. The basic property of an op-amp is that it amplifies the **differential signal** $(V_1 - V_2)$ by the amount of its gain. When the op-amp is open loop, as in Fig. 8.3 the signal V_0 in the output is given by,

$$V_o = A_{OL}(V_1 - V_2) \tag{8.1}$$

Here A_{OL} is the open-loop gain of the op-amp which is generally very large in the order of 10^6. If it is assumed that $(V_1 - V_2) = 0.001 \ V$ and $A_{OL,} = 10^6$ then from (8.1) $V_o = 10^6 x (0.001) \ V = 1000 \ V$. The op-amp system cannot generate such a large voltage, therefore, the output voltage will get truncated or clipped at the $+V_{CC}$ (= +12 V), which is the maximum positive voltage of supply $+V_{cc}$. The maximum voltage that may be supplied by the DC power source is also called the line voltage or rail voltage. It may thus be realized that whenever $(V_1 - V_2) = + ve$; i.e $V_1 > V_2$ the output of the open-loop op-amp will be equal to the voltage of the positive power supply $V_{CC}+$. In this situation, the output of the op-amp will

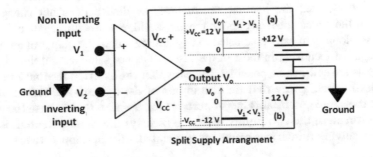

Split Supply Arrangment

Fig. 8.3 Op-amp as a comparator with split supply arrangement

be a positive step pulse with respect to the common ground terminal, as shown in inset (a) of the figure. Whenever the output of the op-amp is clipped at the V_{cc}, it is called the state of saturation.

In the other case when $(V_1 - V_2) = -ve$, i.e.; $V_1 < V_2$, output voltage V_0 will be negative and the negative step output pulse will get saturated at $-V_{CC}$. This is shown in inset (b) of Fig. 8.3. From this analysis, one may make two important conclusions: (i) **in case of split supply arrangement, all input and output signals are measured with respect to the common ground.** (ii) **A free-loop op-amp works like a voltage comparator; if the signal at the non-inverting input is larger in comparison to the signal at the inverting input, the output will be a positive saturated step pulse and in the reverse case a saturated negative step pulse. Thus a free-loop op-amp compares the magnitudes of the signals at its two inputs and delivers a positive step pulse at the output if the signal at the non-inverting input is larger than that at the inverting input and a negative step pulse if the signal at the inverting input is larger than that at the non-inverting input.**

Self-assessment question: What is the function of a comparator? Is it required to use a negative feedback network when using op-amp as a comparator?

Self-assessment question: Signals of -4 and -1 V are applied, respectively, at the non-inverting and inverting inputs of an open-loop op-amp. What will be the magnitude and nature of the output signal and how does the output will prove that the circuit is working as a comparator?

Let us calculate the maximum value of the differential signal (V_1-V_2) that will not saturate the output of the op-amp. Assuming that the open-loop gain to be 10^6, and $V_{CC} = 12$ V; differential signal (V_1-V_2) of slightly less than 12×10^{-6} V; i.e. $12 \ \mu V$ will give an output of less than 12 V and will not saturate the op-amp. It may be noted that $12 \ \mu V$ is negligibly small potential difference between the two inputs of the op-amp. One very important conclusion may be drawn from here and it is: **If the output of the op-amp is not saturated then the differential signal must be of a negligibly small magnitude which means that the two inputs of the op-amps will be almost at the same potential. This statement will remain true so long as the internal structure of the op-amps is not disturbed.**

When negative feedback is incorporated in an op-amp using an external network of resistances, etc., the internal structure of the device does not change and, therefore, the statement made above remains valid. Since with negative feedback the closed-loop gain can be adjusted to any desired value, the output of an op-amp with negative feedback, in most of the cases, does not saturate, and therefore, the two inputs are at the same potential. **It means that the external feedback network operates in such a way that the two inputs of the op-amp are forced to have almost the same potential. The property of the op-amp that in case the output is not saturated, the two inputs will have (nearly) the same potential, is often used for solving numerical problems associated with op-amp networks.**

Self-assessment question: What happens to the inherent (open-loop) gain of the amplifier when negative feedback is employed? Does it change?

In the other arrangements, power to the op-amp is provided by a single battery or a DC voltage source.

Let us assume that a positive output DC source of $+V_{CC}$ volt is used. In this case, the DC source is connected at $V_{CC}+$ pin of the op-amp and the inverting input is grounded. A series combination of two resistances R_1 and R_2 is also connected at the VCC+ pin as shown in Fig. 8.4a. The other end of the series combination of resistances is grounded and the junction point A of the resistance combination is connected to the non-inverting terminal of the op-amp. Since the input impedance of an op-amp is very high it is reasonable to assume that no current enters (or leaves) the non-inverting input from point A. The potential V_A of point A will depend on the ratio of R_1 and R_2 and may be adjusted to any desired value. If it is assumed that $R_1 = R_2$, then in that case $V_A = V_{cc}/2$. In this situation when $V_{CC}-$ is grounded **the potential V_A of point A becomes the new reference point for all signals applied to the op-amp or delivered at the output of the op-amp. The point A and the potential V_A behaves as if it is the new ground, and hence it is called the virtual ground**. It may be recalled that in the previous case of split power supply, when two DC sources one of $+V_{CC}$ volt and the other of $V-_{CC}$ volt were used the reference voltage with respect to which all input and output signals were measured was the ground, the potential of which is taken as zero. However, in the case of a single positive DC source operation the voltage at the non-inverting input works as the **virtual Ground**. In this example of positive DC source and equal values of the two resistances $V_A = V_{cc}/2$ and all input and output signals will now appear with reference to potential $V_{cc}/2$. **The main differences between a real ground and a virtual ground are: (i) the potential of the real ground is always Zero (ii) when any signal is applied at the real Ground, the signal dies out or vanishes. On the other hand, the potential of the virtual ground may not be Zero (ii) when some signal is applied at a virtual ground, the signal does not vanish, the signal level shifts to the potential at virtual Ground.**

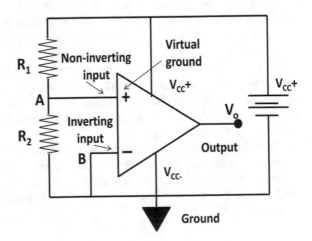

Fig. 8.4a DC power to Op-amp supplied through a single battery

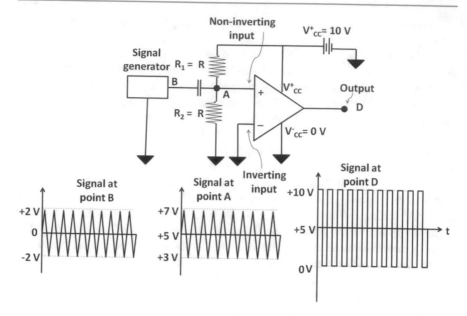

Fig. 8.4b Shift of reference level in case of single DC supply

Let a +10 V DC supply be connected at the $V_{CC}+$ pin of an op-amp and its inverting input be grounded, as shown in Fig. 8.4b. Further, the non-inverting input of the op-amp be connected to the middle point A and if $R_1 = R_2$; the potential at A will be $V_A = 5$ V. If now a signal, say sinusoidal pulses of 4 V peak-to-peak are applied at the non-inverting input of the op-amp then the profiles of the pulses at point A and at the output D (with respect to ground) will be as shown in the figure. At point A, the reference level of the pulse will shift from zero volt to +5 V and at point D, the op-amp output, there will be square pulses formed by the saturation of the signal due to very large gain, centred around the reference level of 5 V, extending up to 10 V on the positive side and up to 0 V on the negative side.

It is important to note that in the present case, there is no negative feedback, the gain of the op-amp is still very large; any positive signal with reference to the new reference level (+5 V) will derive the op-amp to saturation giving a step positive pulse while any negative pulse with respect to the new reference level will drive the op-amp to saturation in the negative direction delivering a negative saturated step pulse. Successive positive and negative step pulses will result in a series of square pulses as shown in the figure.

Since, in the present case, the op-amp remains in the state of saturation, the two inputs of the op-amp may be at two very different potentials. For example, the potential of the non-inverting input (point A) swings around the DC level +5 V; while the potential at the inverting input remains fixed at the ground potential of zero volts.

In the circuit of Fig. 8.4b, two external resistances R_1 and R_2 may be used to shift the level of the op- amp output to any desired value by choosing the values of the two resistances. However, **if no external resistances are used and two different power supplies one +E_1 volt and the other of −E_2 Volt are connected, respectively, to the power terminals $V_{CC}{}^+$ and $V_{CC}{}^-$ then the internal circuit of the op-amp will automatically set the output reference voltage to a value $V_{ref} = [\{+E_1 +(−E_2)\}/2]$ and this reference voltage will appear at the non-inverting input terminal**. For example, if $E_1 = 10$ V and $E_2 = −10$ V; the V_{ref} will be 0 V and the non-inverting input will be at zero volt. If $E_1 = 10$ V, $E_2 = 0$ V, then $V_{ref} = 5$ V and non-inverting input will be at this +5 V potential. The only condition for the proper operation of the op-amp is that power terminal $V_{CC}{}^+$ must always be at a higher potential than terminal $V_{CC}{}^-$.

Historically, the supply voltage for op-amp was typically ∓ 15 V and the dynamic range of the input and output was about ∓ 10 V. Lately, though there has been a trend towards lower supply voltage. This has happened for a couple of reasons, firstly high-speed devices work on lower voltages because of the slew rate limitation. Secondly, working with high voltages requires a larger heat dissipation capacity which becomes a major problem with miniature IC devices. It is for these reasons that modern-day fast op-amp require low-voltage power supplies of the order of $\mp 5V$ with dynamic rang of the order of ± 4.5 V or so.

Self-assessment question: An op-amp is biased using a +10 V and a +2 V power supplies, what will be the reference potential of the output signal?

8.2.2 Common-Mode and Differential-Mode Signals

An op-amp amplifies the difference between its two input signals. However, if one of the inputs of the op-amp is grounded, it will amplify the signal present at the other input. The input signal to an op-amp may be classified into two types: (i) common-mode signal and (ii) differential-mode signal.

(i) **Common-mode signals**: When the two input signals v_1 and v_2 to an op-amp are exactly equal in magnitude; $|v_1| = |v_2|$ and are also in the same phase, they are called common-mode signals. Since the difference $(v_1 − v_2)$ for common-mode signals is zero, no signal should appear at the output of the op-amp; **an ideal op-amp rejects common-mode signals**.

(ii) **Differential-mode signals**: When the two input signals to an op-amp are exactly equal in magnitude to each other; $|v_1| = |v_2|$ but they are opposite in phase; i.e. $v_1 = −v_2$, they are called differential-mode signals. An ideal op-amp amplifies the difference $(v_1 \sim v_2) = (2v_1$ or $2 v_2)$; corresponding to which a signal appears at the output of the op-amp.

It may thus be concluded that an ideal op-amp rejects common-mode signals and amplify the differential-mode signals.

In a practical op-amp, however, common-mode signals are not completely rejected, which means in case of common-mode signals at the inputs there is a small signal at the output. In other words, a practical op-amp has a finite but very small gain ($\approx 10^{-3}$) for the common-mode signals. The common-mode gain is often represented by A_{CM}. It is worth noting that the differential-mode open-loop gain (often denoted by A_{DM} or A_{OL}) of op-amps is very large $\approx 10^6$ or more.

Various kinds of noise signals are generated in all electronic devices including op-amps when they are in operation. In case of op-amps, however, nearly equal amounts of noise signals are picked up by the two inputs of the device. Moreover, the noise signals picked up by the two inputs are also in phase; therefore, noise signals constitute common-mode signals and are largely rejected; only a very small component of noise signals, much smaller than their strength at the inputs, reaches the output of the op-amp. As such the property of common-mode signal rejection helps in reducing the undesirable noise, hum and stray signals in the output of the op-amp.

8.2.3 Slew Rate

When a step signal is applied on any of the two inputs of an op-amp, the corresponding output is not a step pulse; rather the pulse rises (or falls) slowly with time. Slew rate is defined as the maximum rate of change of the output voltage of an op-amp and is given in unit of volts per microsecond.

Figure 8.5 shows the typical output pulse corresponding to a square pulse applied at the input of a practical op-amp with negative feedback (not saturated). As may be observed in this figure, the output pulse rises slowly and takes some time

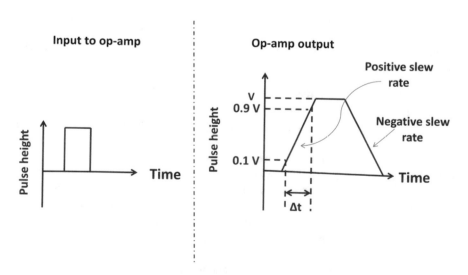

Fig. 8.5 Typical response of an operational amplifier to a square input pulse

($\approx$ few micro second (μs)) to attain the peak voltage V. If Δt is the time (in μs) that the output pulse takes in rising from 10% V to 90% V, then the slew rate is defined as;

$$\text{Slew rate} = \frac{\Delta V}{\Delta t} = \frac{(0.9 - 0.1)V}{\Delta t} \text{ volts per micro second} \tag{8.2}$$

An op-amp that has a large value of slew rate is considered to be a good op-amp that is why an ideal op-amp is supposed to have an infinite slew rate. Different practical op-amps available in the market have widely varying slew times, for example, op-amp OPA369 has a slew rate $\approx 0.005 \, V/\mu s$ while op-amps OPA 835 and OPA 847 respectively have slew times of 110 and 850 V per microsecond.

Sometimes, the positive slew rate (for the rising part of the output pulse) is different from the negative slew rate corresponding to the falling apart of the output pulse.

Since slew rate is a function of time, it may be used to determine the maximum operating frequency f_{max} of an op-amp as

$$f_m = \frac{\text{Slew rate}}{2\pi V_{(peak-to-peak)}}$$

The above expression tells that the maximum operating frequency of an op-amp of a given slew time depends inversely on the maximum amplitude (peak-to-peak value) of the applied input. Therefore, small-signal op-amps respond to higher frequencies of the input signal.

8.2.4 Common-Mode Rejection Ratio (CMRR)

Suppose that a differential input ($\Delta V = V_2 - V_1$) is applied at the input of an op-amp that delivers an output V_{out}. Now if the same voltage, say V (in the same phase) is added at both the inputs, then in principle the output of the op-amp should not change. However, in practice, it has been observed that the output changes on adding the same voltage at the two inputs. In other words, it may be said that the output of practical op-amp changes with the common-mode voltage. The common-mode rejection ratio (CMRR) of an op-amp is defined as the ratio of (A_{DM}/A_{CM}) the differential-mode gain to common-mode gain. For example, if a differential input change of Y volts (at the input) produces a change of 1 V at the output and a common-mode change of X volts (in the input) produces a change of 1 volt at the output, then CMRR is X/Y. It is obvious that X will be much larger than Y, since the common-mode gain is much smaller than the differential-mode gain. For a good op-amp, a very large change in common-mode voltage should change the output by 1 V, while a relatively much smaller differential voltage should change the output by 1 V. Therefore, a larger value of CMRR corresponds to a better op-amp quality. That is why the CMRR for ideal op-amp has infinite value.

It is possible to give CMRR either as a ratio or in dB. If represented in dB, CMRR is often called common-mode rejection (CMR); $CMR = 20\log_{10}(CMRR)$. In most cases, the CMR for practical op-amps varies between 70 and 120 dB, but it deteriorates with the frequency of the applied signal.

Self-assessment question: The common-mode gain of an op-amp is 10^{-2} and CMR is 80 dB; what is the differential-mode gain of the op-amp in dB?

8.2.5 Bandwidth and Gain-Bandwidth Product

Since an op-amp is basically an amplifier, its frequency response characteristic (or Bode plot) shows the typical flat gain response over a given frequency range and then falls off both at the upper and the lower corner frequencies (± 3 dB) at the rate of 20 dB for every decade of frequency. In case of op-amps manufacturer deliberately insert a dominant pole in the op-amp frequency response so that the output voltage versus frequency is predictable and the op-amp is stable. It is done because the op-amps are grown on the silicon die, have many active components each one with its own cut-off frequency. As a result, the frequency response of an op-amp is random, with poles and zeros of op-amps of the same family being different from one another, further for some values of poles the op-amp operation becomes unstable. To standardize the frequency response of a family of op-amps, manufacturers introduce a dominant pole at the same value in each op-amp of the family. Introduction of the dominant pole makes the operational amplifier behave like a single-pole system, which has a drop of 20 dB for every decade of frequency starting from the cut-off frequency.

When negative feedback is incorporated in any amplifier of high open-loop gain, it becomes possible to control the gain of the device at any desired value by adjusting the feedback network. In such a situation it is not the gain or the bandwidth but the gain-bandwidth product that becomes the important parameter. The gain-bandwidth product is always constant and fixed for a given device. For example, the gain-bandwidth product for the op-amp ADA4004 is 12 MHz. This means that at unit gain the bandwidth is 12 MHz. Since the open-loop voltage gain of this op-amp is $\approx 5 \times 10^5$, the open-loop band width of the op-amp is $\left(\frac{12 \times 10^6}{5 \times 10^5}\right) \cong 24$ Hz.

For a single-pole operational amplifier, the cut-off frequency f_c, the open-loop gain A_{OL}, and the gain A_ω at cyclic frequency ω are related by the expression

$$A_\omega = \frac{A_{OL}}{1 + j\frac{\omega}{2\pi f_c}} \qquad (8.3)$$

A typical frequency response graph for the op-amp of open-loop gain of $5 \times 10^5 = 114$ dB at 24 Hz, calculated using the above expression is shown in Fig. 8.6.

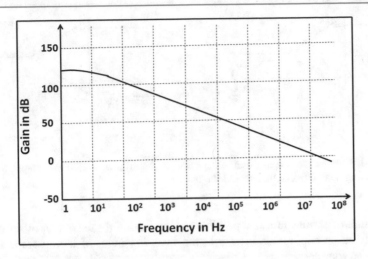

Fig. 8.6 Typical frequency response characteristic for an op-amp of open-loop gain of 500,000

As may be observed in the figure, at 24 Hz the gain in dB is 114 which falls down to 0 dB (gain = 1) at 12 MHz. For any intermediate value of gain, the gain-bandwidth product remains constant.

Self-assessment question: Use the graph shown in Fig. 8.6 to obtain the gain (in absolute unit, not in dB) of the amplifier at 5×10^4 Hz of frequency.

8.2.6 Output Offset Voltage

As already mentioned the backbone of an op-amp is the differential amplifier that is made up of two identical transistor branches as shown in Fig. 8.2c. It is only in an ideal situation that one may assume that the two-transistor branches are exactly identical; in practical case however, there is always some difference in the characteristics of two transistors of the same make and specifications. This miss-match in the characteristics of the two transistors, particularly the miss-match in the values of base-emitter voltages, V_{BE1} and V_{BE2} results in what is called the output offset voltage. As a result of the mismatch a voltage appears at the output of the op-amp even when both inputs are grounded. This voltage at the output is called the **output offset voltage**.

Figure 8.6a (i) shows how the mismatch of transistor characteristics in the differential amplifier circuit results in the output offset voltage $\pm V_{offset}$ to appear at the op-amp output. The problem of output offset voltage may be overcome in several different ways, the most common being the application of an appropriate voltage (V_{io}) at one of the inputs so that no voltage appears at the output of the op-amp, as shown in Fig. 8.6a (ii).

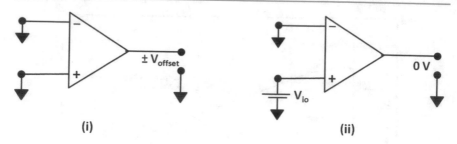

(i)   (ii)

Fig. 8.6a i Shows the output offset voltage and **ii** shows how to eliminate output offset by incorporating input offset in the input of the op-amp

Self-assessment question: Is it possible to compensate for the output offset voltage by applying an appropriate battery at the inverting input of the op-amp? Give a reason for your answer.

8.3 Ideal Op-Amp

An ideal operational amplifier has the following properties:

(i)   Infinite input impedance, zero input bias currents,
(ii) Zero output impedance (the output voltage is independent of the output current),
(iii) Infinite open-loop differential gain (independent of the input signal frequency),
(iv) Infinite slew rate,
(v)   Infinite common-mode rejection ratio (CMRR),
(vi) Infinite bandwidth.

However, the input impedance of a practical op-amp is usually > 10 MΩ; the output impedance is typically 50 to 75Ω. The open-loop differential voltage gain of a practical operational amplifier is of the order of 10^6.

8.4 Practical Op-Amp with Negative Feedback

Internal structure of a practical op-amp is quite complicated, has several active and passive elements as shown in Fig. 8.2. However, it is possible to analyse the operation of a practical op-amp without going into its structural details on the basis of two basic assumptions or rules which derive their justification from the properties of an ideal op-amp. These assumptions or rules are;

(i) **No currents can enter or leave the two inputs of the op-amp,**
(ii) **In normal case when op-amp output is not saturated, the two input terminals of the op-amp always remain at the same potential.**

The first rule is derived from the property of the ideal op-amp that its input impedance is infinite. Since input impedance is always in series, infinite input impedance means that no current may be drawn or delivered through the input terminals. In case of a practical operational amplifier, the input impedance though not infinite but is sufficiently large so that the input current, if at all, is only a few micro amp and may be treated as zero.

The second rule gets its justification from the fact that so long the output of the op-amp is not saturated; the magnitude of the differential signal between the two inputs is negligibly small. A practical op-amp is invariably used with negative feedback, which reduces the closed-loop gain of the amplifier such that the output signal (with negative feedback) never reaches the rail voltage. The negative feedback network is external to the structure of the op-amp and, therefore, in any operation of the op-amp with negative feedback, the potential difference between the two input ports (non-inverting and inverting) is negligibly small and may be taken as zero.

8.4.1 Negative Feedback Configurations

Negative feedback to an op-amp may be provided by connecting the output with a combination of resistances to the **inverting input** of the op-amp. There may be two possible configurations:

Self-assessment question: Why only passive circuit elements are used in the feedback network?

Self-assessment question: Is it possible to provide negative feedback to an op-amp by connecting the feedback network at the non-inverting input?

(a) **Non-inverting amplifier configuration**
When a positive or negative signal (with reference to the ground) is applied at the non-inverting (+) input of the op-amp and the output is connected through a combination of resistances to the inverting input, as shown in Fig. 8.7, the arrangement is called the **non-inverting amplifier configuration**.

Fig. 8.7 Non-inverting
amplifier configuration

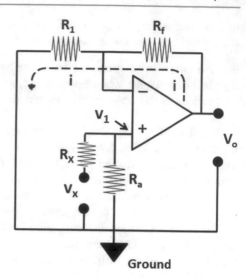

i. **Closed-loop gain**

As shown in Fig. 8.7, a signal of magnitude v_x (with respect to ground) is applied at
the non-inverting input of the op-amp using a combination of resistances R_X and
R_a. The actual signal reaching the non-inverting input of the amplifier is, however,
of the magnitude $v_1 = \frac{R_a}{(R_X + R_a)} v_x$. The op-amp output is connected through a series
combination of two resistances R_f and R_1 to the ground. The junction point of the
resistances is connected to the inverting input to establish the negative feedback.

Let V_0 be the voltage (with respect to ground) at the op-amp output (after
negative feedback). Since negative feedback is already in, V_0 must be less than the
rail voltage and the op-amp is not saturated.

From rule 1, no current may flow into the inverting input, therefore, the total
current 'i' from the output goes to the ground via the series combination of resistances R_f and R_1. Hence,

$$i = \frac{V_o}{\left(R_1 + R_f\right)} \tag{8.4}$$

Also the feedback voltage V_f applied at the inverting input = voltage drop across
the resistance R_1;

$$V_f = R_1.i = R_1.\frac{V_0}{\left(R_1 + R_f\right)} \tag{8.5}$$

Now according to op-amp rule 2, the potential at the two inputs of the op-amp
must be the same, therefore $V_f = v_1$ so equating (8.5) to V_1, one gets

$$v_1 = V_f = R_1 \cdot \frac{V_0}{(R_1 + R_f)}$$

Or

$$\frac{V_0}{v_1} = \frac{(R_1 + R_f)}{R_1} = 1 + \frac{R_f}{R_1} \tag{8.6}$$

But $\frac{V_0}{v_1}$ is the closed-loop gain G of the op-amp with feedback.

So, closed-loop gain G of the op-amp **in non-inverting amplifier configuration** is given by.

$$G = 1 + \frac{R_f}{R_1} \tag{8.7}$$

At this point, three important observations must be made: (i) The closed-loop gain G is the ratio of the output signal V_0 to the signal v_1 actually present at the non-inverting input; In practice, whenever a source of some signal is connected to the input of an op-amp it is invariably connected employing a network of resistances, and, therefore, the actual strength of the signal at the input may differ from the strength of the signal at the source. In the present case, for example, the strength of the actual signal at non-inverting input is v_1 which is different than the strength v_x of the source. (ii) It may be observed that the polarity or the phase of the amplified output signal in case of the non-inverting amplifier configuration does not change; that is if the input signal at non-inverting input is positive, the output amplified signal will also be positive and vice versa. It is essentially for this reason that this configuration is called the non-inverting configuration. (iii) In case R_f is zero or R_1 is infinite (open circuit), the closed-loop gain of non-inverting amplifier will become 1.

In the foregoing discussion, we considered only pure resistive elements in the circuit. However often capacitive and or inductive elements are also used for providing negative feedback, particularly when alternating signals are used. In such cases, the gain (8.7) of the non-inverting op- amp amplifier gets modified to,

$$G = 1 + \frac{Z_f}{Z_1}$$

Here, Z_f and Z_1 are, respectively, the impedances corresponding to the resistances R_f and R_1.

Self-assessment question: A closed-loop op-amp has a potential of 2.5 V at its non-inverting input, what will be the potential at its inverting input? Give a reason for your answer.

ii. **Finite open-loop gain**

In general open-loop gain of an op-amp is taken as infinite; however, it is possible to calculate the magnitude of the open-loop gain of a non-inverting amplifier if it has a finite value. For doing that one calculates the feedback factor β in the following way.

Let us calculate the feedback fraction β, which is the ratio of the feedback voltage V_f to the output voltage V_0. Hence, from (8.5);

$$\beta = \frac{V_f}{V_0} = \frac{R_1}{(R_1 + R_f)} \tag{8.8}$$

In case the open-loop voltage gain of the amplifier is **large but finite** and the finite value is denoted by A_{OL} then,

$$A_{OL} \cong \frac{1}{\beta} = \frac{(R_1 + R_f)}{R_1}$$

If A_{OL} is large, such that 1 may be neglected in comparison of $\beta.A_{OL}$, the closed-loop gain G is given by.

$G \approx \frac{1}{\beta} = \frac{(R_1 + R_f)}{R_1} = 1 + \frac{R_f}{R_1}$, the value obtained earlier (see (8.6)).

iii. **Input and output impedances**

The operational amplifier in the present case behaves as a voltage-controlled voltage amplifier for the differential signal, the equivalent circuit of the device is shown in Fig. 8.8.

The feedback topology employed in the present case is of voltage sampling and voltage mixing, i.e. Series-Shunt topology. As a result of series mixing the input impedance increases by the factor $(1 + \beta A_{OL})$ while the output impedance gets decreased by the same factor. If Z_{in}^{CL} and Z_{Out}^{CL}, respectively, denote the closed-loop input and output impedances then

$$Z_{in}^{CL} = Z_{in}^{OL}(1 + \beta A_{OL}) \, and \, Z_{Out}^{CL} = Z_{Out}^{OL}/(1 + \beta A_{OL}) \tag{8.9}$$

iv. **Effect of more than one signals at the input of non-inverting amplifier**

Figure 8.8a shows how three different signals, v_X, v_Y and v_Z, may be applied simultaneously at the input of a non-inverting amplifier. In order to find the magnitude of the output voltage, it is required to know the value of the signal that actually reaches the non-inverting input terminal of the op-amp. It is quite involved

Fig. 8.8 Equivalent circuit of the op-amp

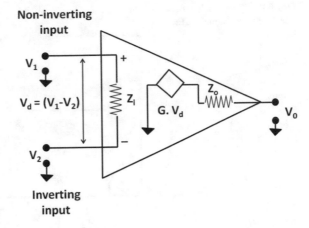

Fig. 8.8a Three signals simultaneously applied at the input of a non-inverting amplifier

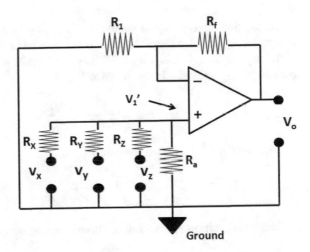

to determine the value of this actual signal at non-inverting input when all the inputs are present simultaneously. Therefore, usually what is done is that only one source at a time is considered, all other sources are replaced by short circuits and the potential at the non-inverting input is calculated separately for each source. Once the magnitudes of the potentials at non-inverting input for each source are known, corresponding output strength for each source may be obtained by multiplying the actual signal by the gain $G = (1 + R_f/R_1)$ which is the same in each case. One then uses the superposition theorem and takes the algebraic sum of output signal strengths obtained for each source individually. This sum gives the final value of the output when all the signals are present at the input simultaneously. This method of calculating the output is called the superposition method. Let us apply this method for the case of three inputs as shown in the figure.

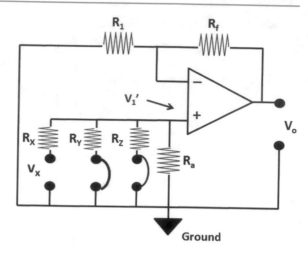

As the first step, let signal source v_X be retained and the other two sources v_Y and v_Z be replaced by short- circuits to the ground as shown in Fig. 8.8b.

It may be observed in Fig. 8.8b that three resistances R_Y, R_Z and R_a are in parallel giving an equivalent resistance R_{eq} as

$$R_{eq} = \frac{R_a R_Z R_Y}{R_a R_Z + R_a R_Y + R_Y R_Z} \qquad (8.9a)$$

And the current i_{eq} drawn from source as

$$i_{eq} = \frac{v_X}{R_{eq} + R_X} \qquad (8.9b)$$

The actual voltage v_1' reaching the non-inverting terminal due to source v_X alone is

$$v_1' = R_{eq}.i_{eq} \qquad (8.9c)$$

The amplifier output $V_o^{v_X}$ for the source v_X alone will, therefore, be

$$V_o^{v_X} = G v_1' \qquad (8.9d)$$

The same procedure may be adopted to obtain first the magnitudes of the actual signals v_Y' and v_Z' appearing at non-inverting input respectively when sources v_Y and v_Z alone are present in the circuit. And then corresponding outputs $V_o^{v_Y}$ and $V_0^{v_Z}$ maybe calculated. The final value of the output V_0 when all the sources are present in the input may then be obtained by adding all individual outputs as

$$V_0 = V_0^{v_X} + V_0^{v_Y} + V_0^{v_Z} \qquad (8.9e)$$

Following important observation may be made from the above derivation:

The magnitude of the actual signal reaching the non-inverting terminal not only depends on the magnitude of the source voltage and series resistance connected to the source but it also depends on the values of all other resistances present in the input network. Further, for the same source, the value of the actual signal will change if the number of sources in the input is changed. Often the term 'loading' is used to express such dependence of actual signal reaching the non-inverting terminal on the other sources present in the input.

In the case of three sources in the input, if.

$$R_X = R_Y = R_Z = R_a = R(say); \text{ (8.9a), (8.9b) and (8.9c) reduce to}$$

$$R_{eq} = \frac{R}{3}; i_{eq} = \frac{3v_X}{4R}; v'_X = \frac{v_X}{4}$$

Therefore, output $V_0^{v_X}$ due to a single source v_X is given as $V_0^{v_X} = G\frac{v_X}{4}$
And the output V_0 when all three sources are present is

$$V_0 = \frac{G}{4}(v_X + v_Y + v_Z) \tag{8.9f}$$

where gain $G = 1 + \frac{R_f}{R_1}$, the value of the gain may be set to any desired value by properly choosing the resistances R_f and R_1. If $\frac{R_f}{R_1} = 3$, Gain will become 4 and in that case.

$V_0 = (v_X + v_Y + v_Z)$, i.e. the op-amp non-inverting amplifier with three sources at its input works as a voltage adder, provided the gain is 4 and all resistances of the input network are of equal magnitude.

(b) Inverting amplifier configuration

The other negative feedback configuration refers to the case when the input signal (either negative or positive) is applied at the inverting input terminal. In this case, the non-inverting input is grounded and the negative feedback is provided by connecting the output terminal to the inverting input by a series combination of resistances R_f and R as shown in Fig. 8.9. **This configuration is called the inverter amplifier configuration**.

i. Closed-loop gain

Let a signal of magnitude V_2 with respect to the ground be applied at the inverting input through a resistance R and the inverting input be connected to the output of the op-amp via a resistance R_f to provide the negative feedback. Let us assume that a current i flows through resistance R, and since from rule 1 no current may enter or leave any input of the op-amp, the current i continues to flow unchanged through the other resistance R_f into the op-amp output.

Fig. 8.9 Negative feedback with input signal connected at inverting input

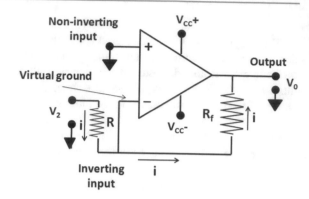

Since negative feedback is already included in the circuit, the op-amp output is not saturated and the magnitude of the output voltage V_0 is always less than the rail voltages. Invoking rule 2 leads to the fact that the potential at the inverting input must be equal to the potential of the non-inverting input, that is zero volts. As such the inverting input terminal behaves as a virtual ground. It means that current i passing through resistance R produces a voltage drop of magnitude V_2 so that the potential at the inverting terminal becomes zero. Hence,

$$R.i = V_2 \tag{8.10}$$

Therefore,

$$i = \frac{V_2}{R} \tag{8.11}$$

Further, applying KVL through the circuit path V_2–R–R_f–V_0 one gets

$$V_2 - R.i - R_f.i = V_0$$

or $V_2 - (R + R_F)i = V_0$ substituting the value of i from 8.16 one gets

$$-\frac{R_f}{R}V_2 = V_0$$

or

$$\frac{V_0}{V_2} = G = -\frac{R_f}{R} \tag{8.12}$$

It may thus be observed that in this configuration when the signal is applied at the inverting input, the closed-loop gain is given by $\left(-\frac{R_f}{R}\right)$, while in the other case when the signal is applied at the non-inverting input the closed-loop gain is given by $\left(1 + \frac{R_f}{R}\right)$.

For alternating signals and complex reactance's, the gain of an inverting op-amp amplifier is given by.

$G = -\frac{Z_f}{Z}$, where Z_f and Z are the impedances corresponding to R_f and R.

Self-assessment question: What is the feedback topology in case of the inverter amplifier configuration?

ii. Closed-loop gain calculations for inverting amplifier configuration when open-loop gain is finite

In the foregoing discussion, it was assumed that the open-loop gain of the operational amplifier was infinite. Now we consider the case when the open-loop gain is finite, say A. In this case, the potential.

V_x at point x will be (see Fig. 8.10).

$$V_A = -\frac{V_O}{A} \qquad (8.13)$$

And the current

$$i = \frac{[V_2 - (-V_0/A)]}{R} = \frac{(V_2 A + V_0)}{RA} \qquad (8.14)$$

But, $V_A - R_f i = V_0$ or $-\frac{V_O}{A} - R_f \frac{(V_2 A + V_0)}{RA} = V_o$

or $V_0 \left[1 + \frac{1}{A} + \frac{R_f}{RA}\right] = -\frac{R_f}{R} V_2$

or

$$\frac{V_0}{V_2} = G = -\frac{R_f/R}{\left[1 + (1 + R_f/R)/A\right]} \qquad (8.15)$$

Fig. 8.10 Inverting amplifier with finite open-loop gain A

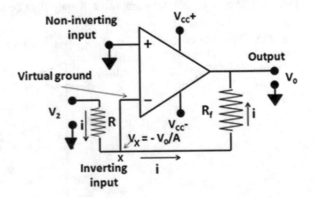

Here, G is the closed-loop gain for inverting op-amp amplifier in case the open-loop gain is not infinite. It may be observed that in case $A \to \infty$, from (8.15) $G \to -\frac{R_f}{R}$; the value obtained earlier (8.12).

iii. Input and output impedances

Controlled voltage source equivalent of an op-amp inverting amplifier is shown in Fig. 8.11. It may be observed in this figure that the inverting and the non-inverting inputs are joined by a dotted line shown as a virtual short circuit. It is because of the fact that though there is no connection between the two inputs since both of them are at the same potential of zero volt, it may be assumed that a virtual connection exists between the two. As such the input impedance, which is the effective impedance between the signal source and ground via the two inputs, is equal to the resistance R, through which signal V_2 is fed to the inverting input. Also, the output impedance in this configuration is zero.

An alternate method of calculating input impedance is to use the relation

$$Z_{in} = \frac{\text{Applied signal voltage at the input}}{\text{current drawn from signal source}} = \frac{V_2}{i} = \frac{V_2}{V_2/R} = R$$

That also gives the value of the input resistance as R.

The input impedance will depend on the frequency of the applied signal only in the case when input impedance Z_{in} contains reactive components. Also, the output impedance in this configuration is almost zero for signals of all frequencies.

iv. Effect of multiple sources at the input of an inverting amplifier

Figure 8.11a shows the circuit diagram of an op-amp based inverting amplifier that has three separate signal sources v_m, v_n and v_p connected at the input respectively with resistances R_m, R_n and R_p. It is important to note that in case of the inverting amplifier, the non-inverting input is always grounded. Since the two inputs of the op-amp must be at the same potential the inverting input terminal is also at ground potential in case of inverting amplifier. It may thus be observed that in case of the

Fig. 8.11 Controlled voltage source equivalent of op-amp inverting amplifier

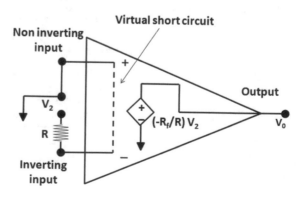

inverting amplifier both the inverting and the non-inverting input terminals are at ground potential. (This is not so in case of the non-inverting amplifier, the two input terminals of which are at the same potential but not necessarily at ground potential).

As shown in Fig. 8.11a, the currents i_m, i_n and i_p drawn, respectively, from signal sources v_m, v_n and v_p are given by

$$i_m = \frac{v_m}{R_m}, i_n = \frac{v_n}{R_n}, i_m = \frac{v_p}{R_p}, \tag{8.15a}$$

Now suppose source v_m is removed, i.e. $v_m = 0$. In this case, i_m becomes zero but there is no change in the values of i_n and i_p. This behaviour indicates that in case of inverting amplifier all input sources are totally independent of each other; there is no effect of absence or presence of any source on other sources. In technological terms, it is said that in case of inverting amplifier, the input is not loaded by signal sources. (This is a big difference in comparison with the non-inverting amplifier where input is loaded by sources and the presence or absence of each source affects the current through the remaining source).

Since no current is drawn by the input terminals of the op- amp, the sum of currents delivered by sources must be equal to the current $-i_0$ due to the output V_0.

Hence, $i_m + i_n + i_p = -i_0$ or $\frac{v_m}{R_m} + \frac{v_n}{R_n} + \frac{v_p}{R_p} = -\frac{V_0}{R_f}$

or

$$V_0 = -\left[\frac{R_f}{R_m}v_m + \frac{R_f}{R_n}v_n + \frac{R_f}{R_p}v_p\right] \tag{8.15b}$$

Fig. 8.11a Circuit diagram of an inverting amplifier with three signal sources at input

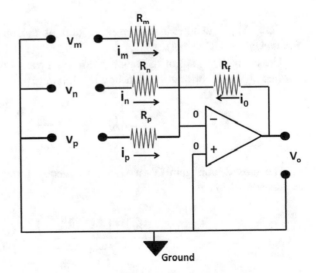

In the special case (i) $R_m = R_n = R_p = R$; (8.15b) reduces to

$$V_0 = -\frac{R_f}{R}\left[v_m + v_n + v_p\right]$$ (8.15c)

And for the case (ii) $R_m = R_n = R_p = R = R_f$;

$$V_0 = -\left[v_m + v_n + v_p\right]$$ (8.15d)

Equations (8.15c and 8.15d) show that an inverting amplifier may become a voltage adder that gives at the output the sum of the input signals with phase reversal.

As already discussed, a non-inverting amplifier under special conditions may also work as a voltage adder, but the output, in this case, is in phase with the input signal.

8.5 Frequency Dependence of the Gain for An Op-Amp

The gain, the input and the output impedances of an op-amp amplifier (both in non-inverting and inverting amplifier configurations) may depend on the frequency of the applied signal. The data sheet supplied by the manufacturer with the op-amp, generally does not specify the -3 dB frequency, called the cut-off or corner frequency (and denoted by f_c). Instead, datasheet provides the unit gain frequency, denoted by f_t which is related to f_c by the following relation:

$$f_t = A_{OL}^{(f=0)} f_c$$ (8.16)

Here, $A_{OL}^{(f=0)}$ denotes the open-loop gain of the op-amp for a signal of zero frequency (or DC signal).

The open-loop gain of the op-amp amplifier varies with the frequency f of the applied signal according to the following relation:

$$A_{OL}^f = \frac{A_{OL}^{(f-0)}}{\left(1 + j\frac{f}{f_c}\right)}$$ (8.17)

The closed-loop gain G^f at signal frequency f may be calculated using the expression (8.17) as

$$G^f = \frac{A_{OL}^{(f=0)} / \left(1 + \beta A_{OL}^{(f=0)}\right)}{1 + j\frac{f}{f_c(1 + \beta A_{OL}^{(f=0)})}} = \frac{G^{(f=0)}}{1 + j\frac{f}{f_c'}}$$ (8.18)

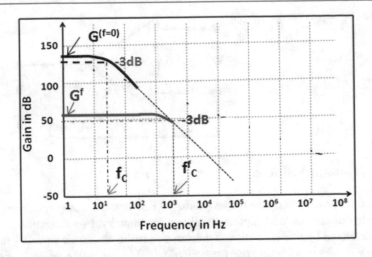

Fig. 8.12 Typical frequency response of an operational amplifier

Here, $G^{(f=0)} = A_{OL}^{(f=0)} / \left(1 + \beta A_{OL}^{(f=0)}\right)$ is the closed-loop gain for DC signal (f = 0). The new cut-off (or −3 dB frequency) f_c^f is given by

$$f_c^f = f_c \cdot \left(1 + \beta A_{OL}^{(f=0)}\right) \tag{8.19}$$

The new gain-bandwidth product is given by

$$G^{(f=0)} \cdot f_c^f = A_{OL}^{(f=0)} / \left(1 + \beta A_{OL}^{(f=0)}\right) \cdot \left[f_c \cdot \left(1 + \beta A_{OL}^{(f=0)}\right)\right] = A_{OL}^{(f-0)} \cdot (f_c) \tag{8.20}$$

Expression (8.20) tells that the gain-bandwidth product remains constant. Figure 8.12 shows the typical frequency response characteristics for an op-amp. The gain $G^{(f = 0)}$ for f = 0 and G^f at some frequency f along with corresponding corner frequencies f_C and f^f_C are shown in the figure.

8.6 Some Important Applications of Op-Amp

Because of their unique properties, considerably high input impedance, sufficiently low output impedance, adjustable closed-loop gain, etc. operational amplifiers are finding ever increasing applications in electronic circuits. Three of its basic applications, as voltage comparator, as non-inverting amplifier and as inverting amplifier, have been detailed in previous sections. Further, it has also been discussed how and under what conditions inverting and non-inverting amplifiers may behave as voltage adder. Some other important applications of op-amp will be discussed in the following.

Fig. 8.13 Op-amp as voltage
follower

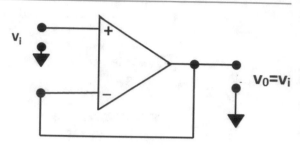

8.6.1   Voltage Follower

In electronics, very often one has to handle signals that are not only of very small amplitude but also have negligibly small current capacity. For example, electronic pulses delivered by nuclear radiation detectors or audio electronic signals generated when one speaks in a microphone are only of micro to millivolt amplitude and have pico to microamp current capacity. Such signals die out if they are directly fed to some low input impedance device like an amplifier or loudspeaker. In order to further process such feeble signals, they are first passed to a voltage follower, a device that has very high input impedance, almost unit gain, and very low output impedance. The output voltage of a voltage follower does not depend on the current drawn through the output. Circuit diagram of an op-amp-based voltage follower is shown in Fig. 8.13, where it is easy to realize that basically, it is a non-inverting operational amplifier (please refer to Fig. 8.7) with negative feedback resistance $R_f = 0$ and resistance R between the inverting input and ground being infinite $(R = \infty)$. The gain of the non-inverting amplifier (given as $G = 1 + R_f/R$, (8.6)) in case of voltage follower G_{VF} is given as

$$G_{VF} = 1 + \frac{(R_f = 0)}{(R = \infty)} = 1$$

Thus a voltage follower delivers at its output a signal identical to the input signal but of much higher current capacity.

Voltage follower is also called buffer circuit and is used to isolate one device connected at the input from the device connected at its output. A voltage follower matches the output impedance of the device connected at its input with the input impedance of the device connected at its output thus allowing the signal to pass from the input device to the output device without much attenuation.

Self-assessment question: Why do electronic signals get attenuated when they pass from one device to another the impedances of which do not match? How does voltage follower help in this regard?

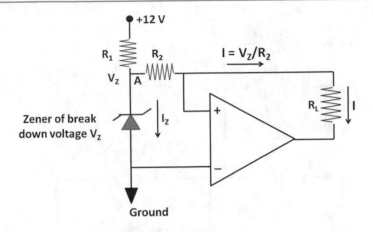

Fig. 8.14 Op-amp as a constant current source

8.6.2 Op-Amp as Constant Current Generator

Figure 8.14 shows the circuit arrangement of a constant current source. The 12 V power supply keeps the Zener diode in breakdown condition and, therefore, the voltage at point A in the circuit is always equal to V_Z, the Zener breakdown voltage. Since the inverting input of the op-amp is directly grounded, the non-inverting input is also at virtual ground. The constant current passing through the load is given by $I = V_Z/R_2$. In case the Zener used in the circuit is of 6 V breakdown voltage and resistance R_2 is of 200 Ω, then the constant current I of the source will be $I = 6/200 = 30$ mA.

Self-assessment question: What will happen to the current through resistance R_1 when load R_L is changed?

8.6.3 Voltage Adder

When a number of voltage sources with a series resistance each are connected in parallel to the inverting input of an op-amp the non-inverting input of which is grounded, the circuit behaves as a voltage scalar and adder. In Fig. 8.15 four voltage sources v_1, v_2, v_3 and v_4 each with a series resistance R_1, R_2, R_3 and R_4 respectively, are connected at the inverting input of the op-amp. Each voltage source contributes currents i_1 ($= v_1/R_1$), $i_2 (= v_2/R_2)$, $i_3 (= v_3/R_3)$, and $i_4 (= v_4/R_4)$ respectively to the network. Total current I through resistance R_0 is the sum of these individual currents, i.e.

$$I = i_1 + i_2 + i_3 + i_4 = \frac{v_1}{R_1} + \frac{v_2}{R_2} + \frac{v_3}{R_3} + \frac{v_4}{R_4}$$

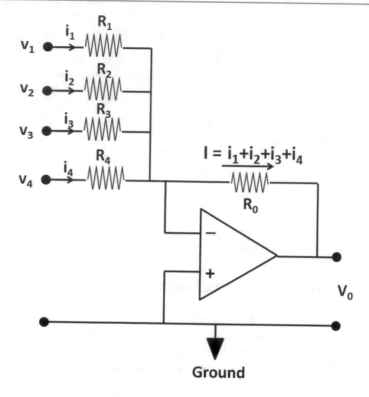

Fig. 8.15 Op-amp as a voltage adder

And the output voltage $V_0 = -R_0 I = \frac{R_0}{R_1}v_1 + \frac{R_0}{R_2}v_2 + \frac{R_0}{R_3}v_3 + \frac{R_0}{R_4}v_4$

For the case $R_0 = R_1 = R_2 = R_3 = R_4, V_0 = -(V_1 + V_2 + V_3 + V_4)$; the op-amp circuit behaves like a voltage adder.

For the case, $R = R_1 = R_2 = R_3 = R_4; V_0 = -\left(\frac{R_0}{R}\right)(V_1 + V_2 + V_3 + V_4)$; the op-amp circuit scales each input voltage by a scaling factor $\left(-\frac{R_0}{R}\right)$ and then add.

Self-assessment question: What is meant by virtual ground? Is there a virtual ground **in** Fig. 8.15?.

8.6.4 Voltage Adder and Subtractor

Let us consider the circuit shown in Fig. 8.16. Since the given circuit is quite complicated, therefore, we analyse it using the principle of superposition.

According to this principle, a complicated circuit may be broken down in to a combination of simpler circuits which may be analysed individually and the final result may be obtained by algebraic summing the results of the analysis of

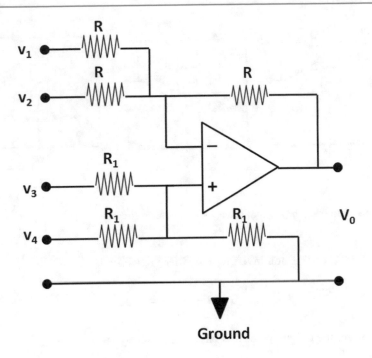

Fig. 8.16 Op-amp as adder and subtractor

individual circuits. The op-amp circuit shown in Fig. 8.16 may be broken down into a combination of three circuits as shown in Fig. 8.16a.

Circuit (i) corresponds to the situation when sources v_3 and v_4 are removed and replaced by ground. It is simple to analyse this circuit as an inverter adder. The non-inverting terminal is grounded and so the inverting input terminal is also at virtual ground. Since op-amp inputs do not draw any currents, the sum of currents delivered by the two sources v_1 and v_2 must be equal (but in opposite direction) to the current delivered by the output V_a. Hence,

$$\frac{v_1}{R} + \frac{v_2}{R} = -\frac{V_a}{R} \text{ or } V_a = -(v_1 + v_2) \tag{8.21}$$

Circuit given in figures (ii) corresponds to the cases when only source v_3 is retained and all other sources v_1, v_2 and v_4 are removed and replaced by ground. It is easy to identify this circuit as a non-inverting amplifier. However, the magnitude v_3' of the actual signal reaching the non-inverting input terminal (+) is given by

$$v_3' = v_3 \left(\frac{R_1/2}{R_1 + R_1/2} \right) = \frac{v_3}{3} \tag{8.22}$$

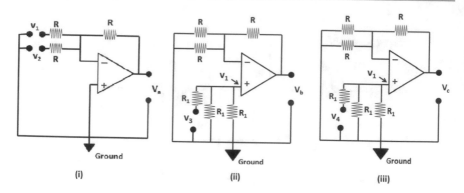

Fig. 8.16a Resolving the network of Fig. 8.16 into three circuits

The gain G of this non-inverting amplifier is given by

$$G = 1 + \frac{R_1}{R_1/2} = 3 \tag{8.23}$$

Therefore, the output V_b of this circuit is given by

$$V_b = G v_3' = v_3 \tag{8.24}$$

Circuit (iii) refers to the situation when all other sources are replaced by ground and only v_4 is retained. It may be analysed just like circuit (ii) and the result will be

$$V_c = G v_4' = v_4 \tag{8.25}$$

The algebraic sum of V_a, V_b and V_c gives the desired value of the output V_0 as

$$V_0 = -(v_1 + v_2) + v_3 + v_4 = (v_3 + v_4) - (v_1 + v_2)$$

The circuit may be called an adder and subtractor circuit.

8.6.5 Op-Amp as a Differentiator

The output signal of a differentiator circuit is proportional to the rate of change of the input signal with time. In order to make the output of an op-amp dependent on the time variation of the input signal, a reactive element like a capacitor is included in the feedback loop of the op-amp circuit, as shown in Fig. 8.17.

The non-inverting input B of the op-amp is directly grounded and is at 0 V potential, therefore, the inverting input A is also at virtual ground potential of 0 V.

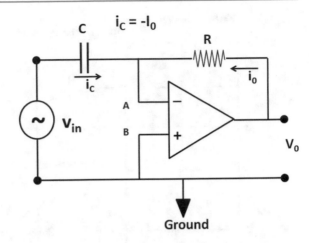

Fig. 8.17 Op-amp differentiator

If i_c and i_0 are, respectively, the values of the instant currents through the capacitor C and the feedback resistance R, then.

$i_C = -i_o$ as no current enters or leaves any input of the op-amp.

But $v_{in} = \frac{q}{C}$ and, therefore, $i_C = \frac{dq}{dt} = C\frac{dv_{in}}{dt}$

Also $i_o = \frac{V_0}{R}$ but $i_C = -i_o$

Hence.

$$\frac{V_0}{R} = -C\frac{dv_{in}}{dt} \text{ or } V_0 = -CR\frac{dv_{in}}{dt} \tag{8.26}$$

This shows that the instantaneous value of the output voltage is proportional to the time rate of the variation of the input voltage, $(-CR)$ being the constant of proportionality. It may be noted that there is a phase inversion in op-amp differentiation; a negative pulse appears in the output when the rate of change of the input pulse is positive. Figures 8.18a and 8.18b shows the typical differential outputs corresponding to a square input and for a triangular input, respectively. Negative spikes in case of square input wave appear in the op-amp differentiator output corresponding to the positive rise of the input wave, as expected. The output is zero whenever the input signal has a constant magnitude. The heights of the spicks are proportional to the numerical value of RC. The differential output of the triangular wave is easy to follow as the rate of change during the triangular rise and fall remains constant, corresponding to that square wave appears at the output.

8.6.6 Op-Amp as Integrator

As may be observed in Fig. 8.19, the feedback element in case of op-amp integrator is capacitance C. Since the non-inverting input is grounded, the inverting input is also at virtual ground potential, and since no current enters or leaves the inputs of the op-amp, the instantaneous currents i_R through resistance R and i_C through the

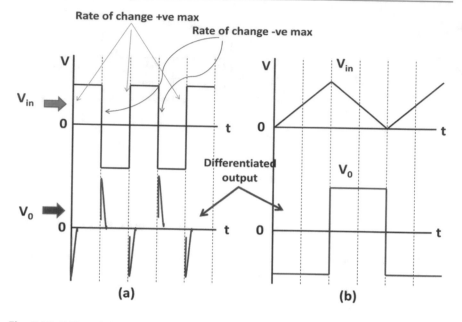

Fig. 8.18 Differential outputs of **a** a square wave; **b** a triangular wave

Fig. 8.19 Op-amp integrator

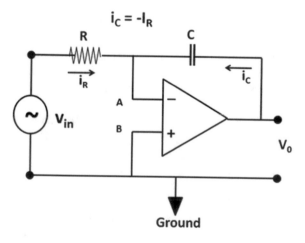

capacitance are equal in magnitude but are opposite in their direction of flow. Mathematically, one may write

$$i_R = -i_C \text{ or } \frac{v_{in}}{R} = -\frac{dq}{dt} = -C\frac{dV_0}{dt}$$

Therefore, $\frac{dV_0}{dt} = -\frac{1}{RC}v_{in}$

Hence,

$$V_0 = -\frac{1}{RC}\int v_{in}dt \tag{8.27}$$

The above expression shows that the output of an op-amp integrator is proportional to the time integral of the input signal. It may be noted that the output signal is 180° out of phase with the input and that the constant if proportionality is 1/RC.

8.6.7   Op-Amp Operated Precision Full-Wave Rectifier or Absolute Value Circuit

Operational amplifiers are often used to improve the performance of electronic circuits. For example, in rectifier circuits made by using ordinary semiconductor diodes, there are always voltage drops across diodes which make the circuit output non-ideal. Figure 8.20a shows the transfer characteristics of half and a full-wave rectifiers made using ordinary silicon diodes. As may be observed in these figures, the characteristic curves start after the 'on' voltage which is of the order of 0.7 V for silicon devices. Obviously, AC signals smaller than 0.7 V cannot be rectified using such ordinary rectifiers. Moreover, the voltage range of the output swings from ordinary rectifiers is less than that of the original input AC signal. All these problems make the rectified output of such circuit's non-ideal, particularly if signals

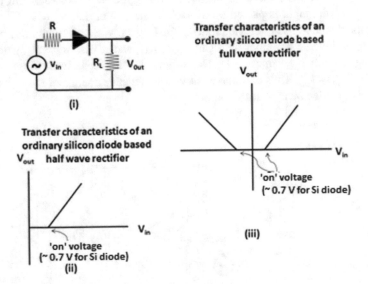

Fig. 8.20a Transfers characteristics of half and full-wave rectifier circuits based on ordinary diodes

are of low voltage. Problems associated with such voltage drops may be totally eliminated by employing op-amps in the circuit producing precision rectified output. Most op-amp-based rectifier circuits still use ordinary silicon diodes but they are used in a feedback loop so that the voltage drops across diodes get divided by the loop gain and are totally compensated.

Any full-wave rectifier circuit that changes the negative swings in the input signal into positive swings at its output may also be called an absolute value circuit as it produces the modulus or the absolute value of the input at its output.

An operational-amplifier-based precision full-wave rectifier is shown in Fig. 8.20b. The time-varying signal v_{in}, may be a sinusoidal wave, is applied at the non-inverting input of the first op-amp OP1.

As may be seen in the circuit diagram, resistances R_1, R_2 and R_3 constitute the feedback network while resistance R_4, generally of low value in the order of 50 Ω, provides the input impedance.

To understand the operation of the rectifier circuit let us denote the positive and negative swings of the input signal v_{in}, respectively, be v_{in}^+ and v_{in}^-. Further, consider the instant when the input v_{in} has positive swing. In this situation, the output of OP1 will be positive, diode D_2 will be forward biased (behaves like a short circuit) while diode D_1 will be reversed biased (behaves like an open circuit). The equivalent circuit for the positive input swing is shown in Fig. 8.20c.

As may be seen in Fig. 8.20c, for the positive input swing OP1 acts as a buffer or as a voltage follower and the positive swing of the input signal gets applied at the non-inverter input of OP2. Op-amp OP2 works as a non-inverting amplifier of gain $(1 + R_2/R_1)$. However, since D_1 is open, the equivalent value of R_1 becomes infinity and, therefore, the gain becomes $(1 + R_2/\infty) = 1$. As such the positive swing v_{in}^+ of the input is passed on as such at the output; i.e. $V_o = v_{in}^+$

For the negative swing of the input v_{in}^-, OP1 again works as the buffer, but OP2 works as an inverter amplifier with gain $-(R_2/R_1)$; and in case $R_1 = R_2$, gain of the inverting amplifier becomes -1. The output V_o for negative swing becomes $Vo = (-1)v_{in}^- = v_{in}^+$. Thus both the positive and negative swings of the input signal appear as positive swings at the output.

Fig. 8.20b Op-amp-based full-wave rectifier

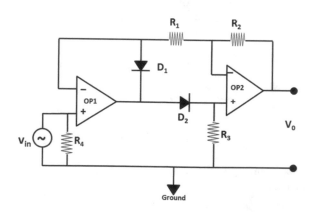

Fig. 8.20c Equivalent circuit for the positive swing of the input signal v_{in}

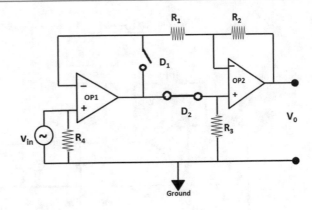

Fig. 8.20d Equivalent circuit for the negative swing of input signal v_{in}

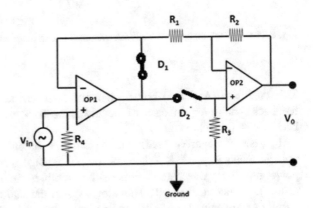

The precision full-wave rectifier can turn alternating current signals into single polarity signals. The two op-amps used in the circuit buffer the input signal and compensate for the voltage drops across D_1 and D_2 allowing the small-signal inputs. Further, the circuit may be used in applications that need to quantify the absolute value of the input signals which have both positive and negative polarities.

Self-assessment question: Draw transfer characteristics of a precision full-wave rectifier.

8.6.8 Op-Amp Operated RC-Phase Shift Oscillator

As shown in the block diagram of the feedback system (Fig. 8.21a (i)) oscillations are produced when transfer function $(F = \frac{V_{out}}{v_{in}})$ of the feedback system is in an unstable state; i.e.;

$F = \frac{V_{out}}{v_{in}} = \frac{A}{1 - \beta A}$ is unstable. This occurs when $(1 - \beta A) = 0$. Thus the key to designing an oscillator is that A $\beta = 1$, which is called the Barkhausen criterion.

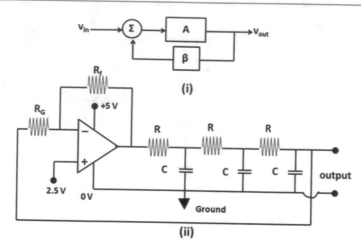

Fig. 8.21a i Block diagram of feedback system **ii** circuit diagram of an RC phase shift oscillator

This condition (A β = 1) holds when the feedback factor β is positive (positive feedback), however, in case of negative feedback the condition of oscillations becomes A β = −1.

In case of **negative feedback** system, the output voltage heads for infinite voltage when A β = −1. When the output voltage reaches the maximum of the power rail, the active components of the amplifier circuit change the gain A, as a result, the value A β ≠ −1. This slows down the rate of change of A β towards infinity and eventually it halts. At this stage one of the three things may happen:

(i) Non-linearity in the saturation or cut-off can cause the system to become stable and lockup.
(ii) The initial charge can cause the system to saturate/or cut-off, stay in that state for considerable period of time before it becomes linear and heads for the opposite rail voltage.
(iii) The system stays linear and heads for the opposite power rail.

In case (i) the system does not oscillate any more, in case (ii) system produces square waves, behaving like a relaxation oscillator. Option (iii) produces sinusoidal oscillations.

An RC phase shift oscillator may be built using a single op-amp and a single power supply, as shown in Fig. 8.21a (ii). It may be noted that the inverter amplifier configuration of an op-produces a phase shift of 180°, therefore, the additional feedback network must produce another 180° phase shift, so that the total phase shift becomes 360°. This means that each RC network should produce a phase shift of 60°. This occurs when $\omega = 2\pi f = \frac{1.732}{RC}$, as tan60 = 1.732.

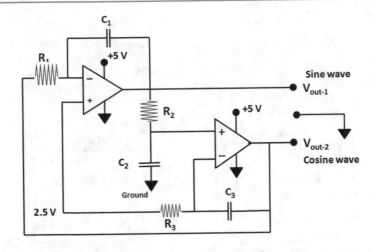

Fig. 8.21b Quadrature oscillator

It is worth noting that the virtual ground of the op-amp is at 2.5 V as a single power source of 5 V has been used.

Quadrature oscillator is another phase shift oscillator that has two outputs; one delivering sine wave and the other cosine wave. The circuit of Fig. 8.21b uses two low-voltage op-amps and three RC phase shift stages, each providing a phase shift of 90°. In the case $R_1C_1 = R_2C_2 = R_3C_3$ it may be shown that A $\beta = [1/(RC)^2]$ and oscillations occur at cyclic frequency $\omega = 2\pi f = 1/RC$.

Op-amp oscillators are often restricted to lower frequencies. It is because op-amps do not have the required bandwidth to achieve low phase shift at higher frequencies.

8.6.9 Op-Amp-Operated Active Filters

Filters are circuits that may allow or restrict the passage of signals of some desired frequencies through them. Filters may be classified into three categories; (i) low-pass filters that allow the passage of lower frequency signals up to a certain frequency f_c (called the cut-off frequency) and blocks all signals of frequencies larger than f_c. (ii) High-pass filters on the other hand allow the passage of all signals of frequencies larger than a certain frequency f_c (cut-off frequency) and block all signals with frequencies smaller than f_c. (iii) Bandpass filters allow a band of frequencies between f_{LC} and f_{HC} to pass through them and block signals of frequencies below f_{LC} (called lower cut off frequency) and signals with frequencies larger than f_{HC} (called the upper cut-off frequency). In fact, a combination of a low-pass filter followed by a high pass filter constitutes a bandpass filter.

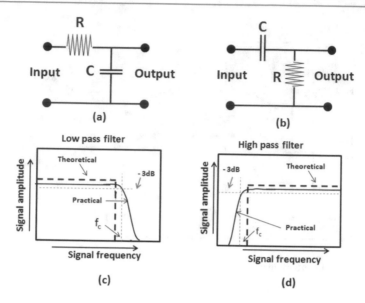

Fig. 8.22 Low-pass and High-pass passive filters and their frequency response characteristics

Filters made using passive circuit elements are called passive filters. For example, passive low-pass and high-pass filters made using resistance and capacitance are shown in Fig. 8.22a, b, respectively; their ideal and practical frequency response characteristics are shown respectively in Fig. 8.22c, d. The cut-off (or −3 dB) frequency, where the signal strength or gain falls to 0.707 times the maximum value, for both types of filters is given by.

$$f_C = \frac{1}{2\pi RC} \text{ or } \omega_c = \frac{1}{RC} \tag{8.28}$$

It may be observed in the frequency response characteristics of passive filters that the signal amplitude gets reduced on passing through filter. This attenuation occurs on account of resistive losses. Further, passive filters do not provide any mechanism for amplification of the signal and the output signal amplitude is also affected by the load in the circuit.

The transfers function; i.e. $\left(\left|\frac{v_{output}}{v_{input}}\right|\right)_{Low}$ for the low-pass filter may be given as

$$\left(\left|\frac{v_{output}}{v_{input}}\right|\right)_{Low} = \frac{|X_C|}{\sqrt{R^2 + X_C^2}} = \frac{\frac{1}{\omega C}}{\sqrt{R^2 + (1/\omega C)^2}} = \frac{1}{\sqrt{R^2\omega^2 C^2 + 1}} \tag{8.29}$$

Replacing $R^2 C^2 = \frac{1}{\omega_c^2}$ in (8.29), where $\omega_C = 2\pi f_C$ is the cyclic cut-off frequency (radians per unit time), one gets

$$(T_rF)_{Low} = \left(\left|\frac{v_{output}}{v_{input}}\right|\right)_{Low} = \frac{1}{\sqrt{1+\left(\frac{\omega}{\omega_C}\right)^2}} = \frac{1}{\sqrt{1+\left(\frac{f}{f_C}\right)^2}} \quad (8.30)$$

Similarly, the transfers function for the high-pass filter may be given as

$$(T_rF)_{High} = \left(\left|\frac{v_{output}}{v_{input}}\right|\right)_{High} = \frac{R}{\sqrt{R^2+X_C^2}} = \frac{\left(\frac{\omega}{\omega_C}\right)}{\sqrt{1+\left(\frac{\omega}{\omega_C}\right)^2}} = \frac{\frac{f}{f_c}}{\sqrt{1+\left(\frac{f}{f_C}\right)^2}} \quad (8.31)$$

Active filters made using active circuit elements like op-amp not only take care of the attenuation but also provides a mechanism to amplify the output signal. Three properties of an op-amp; high input impedance, low output impedance and the possibility of having signal amplification to the desired level, play important role in designing op-amp operated active filters of all types.

In principle, both non-inverting and inverting op-amp amplifier configurations may be used to amplify the output from low pass as well as from high pass passive filters resulting in corresponding active filters. Figure 8.23 shows three different arrangements of active low-pass filters that use non-inverting amplifier configuration. In Fig. 8.23 (i), op-amp is used as a voltage follower (voltage gain $A_v = 1$), in Figure (ii) and (iii) as an amplifier with voltage gain $A_v = (1 + R_f/R_1)$. Active filter output is obtained at the op-amp output and is given by

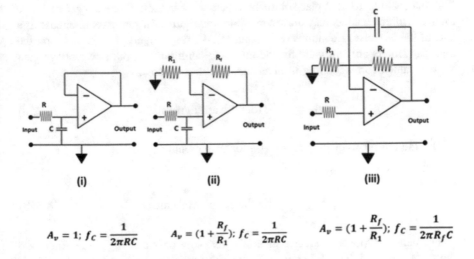

$$A_v = 1;\ f_C = \frac{1}{2\pi RC} \qquad A_v = (1+\frac{R_f}{R_1});\ f_C = \frac{1}{2\pi RC} \qquad A_v = (1+\frac{R_f}{R_1});\ f_C = \frac{1}{2\pi R_f C}$$

Fig. 8.23 Three methods of applying passive low-pass filter at input of an op-amp non-inverter amplifier

$$(V_{out})_{Low}^{Active} = A_v (T_r F)_{Low} (V_{in})_{Low}^{Passive} \tag{8.32}$$

And the cut-off frequency $f_c = \frac{1}{2\pi RC}$ for circuits showed in Fig. 8.23 i, ii, however, for the arrangement of Fig. 8.23 iii the cut off frequency is given as

$$f_c = \frac{1}{2\pi R_f C} \tag{8.33}$$

It is worth noting that while using formula given by (8.32) for active filters care should be taken for the type of arrangement as the value of voltage gain A_v and the cut-off frequency f_C have different values for the three different arrangements.

A high-pass passive filter may also be connected in the input of a non-inverting op-amp amplifier in the same three ways as a low-pass filter is shown in Fig. 8.23 i, ii, iii to get high-pass active filters. The circuit diagrams for high-pass filter are not given here and are left as an exercise for the reader. The active high pass filter output may also be given by an expression similar to (8.32) as follows:

$$(V_{out})_{High}^{Active} = A_v (T_r F)_{High} (V_{in})_{High}^{Passive} \tag{8.34}$$

Here also different arrangements have different values for A_v and f_C.

Inverting op-amp amplifier configurations may also be used for transforming passive filters into corresponding active filters. Two different arrangements where inverting op-amp amplifier configurations are used to convert high-pass passive filters into corresponding active filters are shown in Fig. 8.24. A passive low-pass filter may also be used in place of the high pass filter in these figures to get an active low-pass filter. Readers may draw corresponding figures for low-pass filter case as a part of the exercise. Equations (8.32) and (8.34), respectively, for the low and the high pass filters are still valid for the inverting configuration but now voltage gain A_v has values corresponding to inverting amplifier givens as

$$A_v = -\frac{R_f}{R_1} \text{ for both arrangments} \tag{8.35a}$$

But the cut-off frequency $f_C = \frac{1}{2\pi RC}$ for arrangment (a)
and

$$f_C = \frac{1}{2\pi R_f C} \text{ for the circuit of figure } (b) \tag{8.35b}$$

Frequency response characteristics of active low pass and high pass filters are shown in Fig. 8.25. The difference between the passive and active filters is evident. The cut-off frequencies, the frequency at which the signal strength falls to 0.707 times the maximum value or -3 dB value are indicated in these figures. In case of active filters, the gain or the signal strength falls off at the rate of 20 dB per decade of frequency, as shown in figure. In case of the active high pass filters, in principle

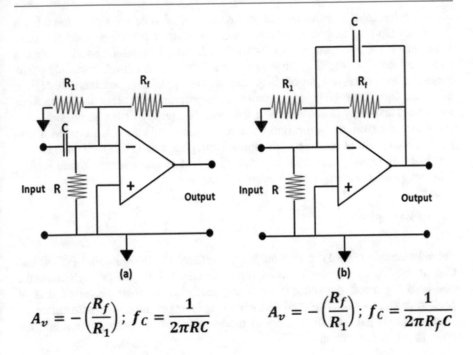

$$A_v = -\left(\frac{R_f}{R_1}\right); \; f_C = \frac{1}{2\pi RC}$$

$$A_v = -\left(\frac{R_f}{R_1}\right); \; f_C = \frac{1}{2\pi R_f C}$$

Fig. 8.24 Two different ways in which a high pass filter may be connected at the input of an inverting op-amp amplifier

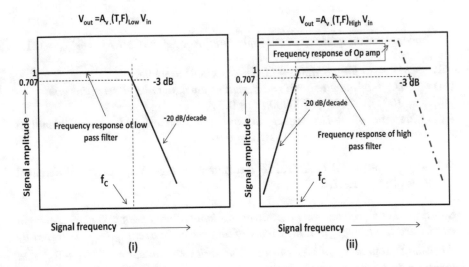

Fig. 8.25 Frequency response characteristics of **i** low pass **ii** high pass op-amp-based active filters

all higher frequencies above the cut-off frequency should be allowed to pass; but it is not so. Op-amp amplifier, whether non-inverting or inverting, used in active filters has a cut-off frequency of its own and, therefore, high-frequency signals passed by the high pass filter may still be restricted by the cut-off frequency of the amplifier. In general, the cut-off frequency of the amplifier is considerably higher, however, the active high pass filter will pass only those frequencies that lie between the cut-off frequency of the filter and the cut-off frequency of the amplifier.

In the last sections, some important applications of op-amps have been discussed, but the list of op- amp applications is very large and cannot be covered in a text like this. Op-amps are used in many other circuits like rate meter, rate limiter, time delay, selective amplifiers, current injector, etc.

Solved Examples

Solved example (SE8.1) An op-amp has a differential voltage gain of 5000 and CMRR of 50,000. A single-ended input signal of 100 μV is applied at the non-inverting input, the other input being grounded. A noise signal of 0. 1 V peak-to-peak, 50 cycles is picked up by both the inputs of the op-amp. Calculate the common-mode gain, CMR in dB and the strengths of the output signal and the noise signal at the output.

Solution:

(i) *Since CMRR = (Differential-mode gain/Common-mode gain),*

Hence, Common-mode gain = (Differential-mode gain/CMRR) = $\frac{5000}{50,000} = 0.1$.

(ii) *CMR = 20 log_{10}CMMR = 20log_{10}(50, 000) = 93.98.*
(iii) *Strength of output signal = differential input signal x differential gain*
       = 100 $\mu Vx5000 = 500000\mu V = 0.5V$.
(iv) *Strength of noise signal at output = common-mode noise signal x common-mode gain = 0.1Vx0.1 = 0.01 V.*

Solved example (SE8.2) Calculate the output voltage V_{out} for the differential amplifier of Fig. SE8.2.

Solution: *Let us consider the non-inverting input of the op-amp. Since no current can be drawn or given from the inputs of an op-amp; current $\frac{v_2}{(R_1 + R_o)}$ delivered by the voltage source v_2 passes through the series combination of $(R_1 + R_o)$ and therefore the voltage at non-inverting input is, say $= \frac{R_o v_2}{(R_1 + R_o)}$. Also, the two inputs of an op-amp are always at the same potential, the potential of the inverting input is also E. Further, no current goes into the inverting input, hence*

Fig. SE8.2 Circuit for
(SE8.2)

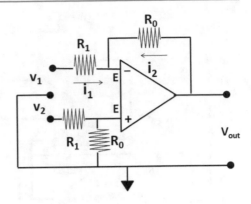

$$i_1 = -i_2 \text{ or } \frac{(v_1 - E)}{R_1} = -\frac{(V_{\text{out}} - E)}{R_o}$$

Substituting the value of $E = \frac{R_o v_2}{(R_1 + R_o)}$ *in the above equation one gets.*

$$V_{\text{out}} = \frac{R_0}{R_1}(v_2 - v_1).$$

Solved example (SE8.3) Analyse the following circuit to obtain output voltage in terms of the two input voltages.

Solution: *It may be observed in the given op-amp circuit the inputs of both the op-amps A and B are at ground potential. Circuit of op-amp B behaves like an inverting amplifier of gain -1, therefore, the voltage at the output of op-amp B is* $-v_2$. *A section of the circuit between points C, D and the output is shown in Fig. SE8.3b. Currents* i_1, i_2 *and* i_3 *are such that a zero potential is maintained at the tri-junction o. Further, these currents fulfil the following conditions:*

$$i_1 = \frac{v_1}{10k}; i_2 = \frac{v_2}{10k}, \text{ and } i_3 = \frac{V_{\text{out}}}{100k}; \text{Also}, i_2 - i_1 = i_3$$

$$or \frac{1}{10k}[v_2 - v_1] = \frac{1}{100k}V_{\text{out}} \text{ or } V_{\text{out}} = 10[v_2 - v_1].$$

Solved example (SE8.4) Obtain the value of output voltage for the circuit given in Fig. SE8.4.

Solution: *Let us calculate potential at the non-inverting input of the op-amp. Since no current is drawn by op-amp inputs, the current from 6 V battery* i_1 *is given by;*

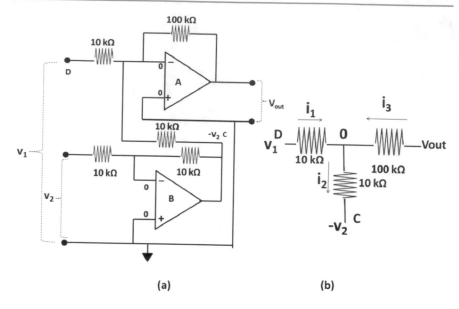

(a)   (b)

Fig. SE8.3 a given circuit **b** simplified part of the circuit

Fig. SE8.4 Circuit for
problem (SE8.4)

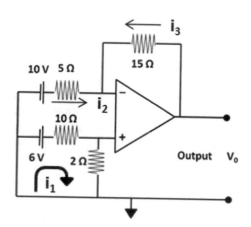

$$i_1 = \frac{6V}{(10+2)\Omega}$$
$$= 0.5A; \textit{Hence the potential at the non inverting}$$
$$\qquad - \textit{ potential at the inverting inputs}$$
$$= 2x0.5 = 1V$$

According to op-amp voltage rule, the potential at inverting input $= 1V$.
Currents i_2 *through 5 Ω resistance and* i_3 *through 15 Ω resistance are given as*

$$i_2 = \frac{(10-1)V}{5\Omega} = \frac{9}{5}A; i_3 = \frac{V_0}{15}A$$

However, $i_2 = -i_3$ or $\frac{9}{5} = -\frac{V_0}{15}$ and $V_0 = -\frac{15 \times 9}{5}V = -27V$

Output $V_0 = -27V$.

Solved example (SE8.5) Drive an expression for the output voltage V_0 in terms of the input voltages v_1 and v_2, for the op-amp circuit given in Fig. SE8.5.

Solution: *Though the given circuit appears to be very complex but it is easy to analyse it. The five op- amps in the circuits are designated as A, B, C, D and E and the signals at their outputs, respectively, v_a, v_b, v_c, v_D and V_0.*

It may be observed that the non-inverting inputs of both op-amps A and B are grounded and signals v_1 and v_2 are applied, respectively, at their inverting inputs via resistances of 2 k each. The two op-amps A and B operate as inverting amplifiers; A with gain $G_a = -8$ k/2 k = −4; and B with gain $G_b = -16$ k/2 k = −8. Therefore, Va = −4 v_1 and $V_b = -8$ v_2.

Output terminals of A and B are connected through 40 k resistances, respectively, to the non-inverting terminals of op-amps C and D. Since input terminals of op-amps do not draw any current, 40 k resistances are redundant, have no effect and signals V_a and V_b reach the non-inverting terminals of C and D. A careful look will tell that op-amps C and D are configured as voltage followers. Therefore, $v_c = v_a = -4$ v_1 and $v_D = v_b = -8$ v_2.

Signal v_D goes to the non-inverting terminal of op-amp E through a 0.5 k resistance, however, this resistance does not modify v_D as no current is drawn through the resistance.

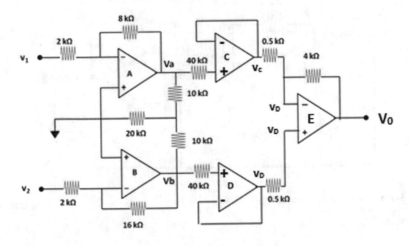

Fig. SE8.5 Circuit for problem (SE8.5)

Since signal v_D is present at the non-inverting input of E, the inverting input of E will also have v_D. The current i_1 through 0.5 k resistance, which is connected to the inverting input of op-amp E is given as

$$i_1 = \frac{v_c - v_D}{0.5k}$$

And the current i_0 through the 4 k resistance connected to Vo is given as.
$i_0 = \frac{V_0 - v_D}{4k}$, But $i_1 = -i_0$
Therefore, $\frac{v_c - v_D}{0.5k} = -\frac{V_0 - v_D}{4k}$ or $V_0 = -[8(v_c - v_D) - v_D] = -[8v_c - 9v_D]$
Substituting the values of $v_c = v_a = $ -4 v_1, and $v_D = v_b = -8$ v_2, one gets.
$V_0 = 32v_1 - 72v_2$.

Solved example (SE8.6) Find the amplitude and phase of the output signal with respect to the input signal (Fig. SE8.6).

Solution: *The circuit shown in the figure may be resolved into two components, the first to the left of the dotted line into a non-inverting amplifier and the other, towards the right of dotted line into an inverting amplifier. If A_1 is the voltage gain of the non-inverting amplifier and A_2 of the inverting amplifier, then the overall gain of the two cascaded amplifiers A will be given by $A = A_1$. A_2. Further, the input signal is given as $v_{in} = \sin(10^6 t)$, which means that the amplitude of the input signal is 1 and its cyclic frequency $\omega = 10^6$.*

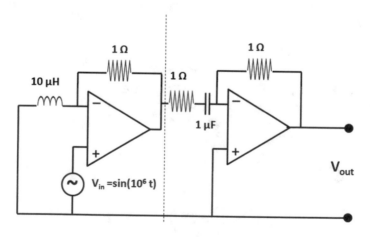

Fig. SE8.6 Circuit for solved example (SE8.6)

Now

$$A_1 = \left(1 + \frac{1\Omega}{X_L}\right) = \left(1 + \frac{1}{j\omega L}\right) = \left(1 + \frac{1}{j10x10^{-6}x10^6}\right) = (1 - 0.1j) \quad (8.6.8.1)$$

Also

$$A_2 = -\left(\frac{1}{1 - j\frac{1}{\omega C}}\right) = -\left(\frac{1}{1 - j\frac{1}{10^6 \cdot 10^6}}\right) = -\left(\frac{1}{1 - j}\right) \quad (8.6.8.2)$$

Therefore,

$$A = A_1 A_2 = (1 - 0.1j)\left[-\left(\frac{1}{1-j}\right)\right] = -(0.505 + 0.45j) = 0.46 \quad (8.6.8.3)$$

The amplitude of the output signal is 0.46 $\angle 41.7^0$. Further, the negative sign of A shows a phase change of $180°$ with respect to the input signal.

Solved example (SE8.7) Identify (i) the op-amp configuration (b) type of circuit and calculate the cutoff frequency of the circuit given in Fig. SE8.7.

Solution: *Since input signal is applied at the non-inverting input, the op-amp configuration is that of a non-inverting amplifier. Complete circuit represents an active high pass filter. The cut-off frequency is given by;* $f_C = \frac{1}{2\pi R_f C}$
$= \frac{1}{2\pi x10x10^3 x0.1x10^{-6}} = 159.16 Hz.$

Solved example (SE8.8) Identify the circuit shown in Fig. SE8.8. What do you expect to be the output of the circuit?

Fig. SE8.7 Circuit for problem (SE8.7)

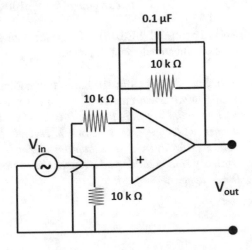

Fig. SE8.8 Circuit for
problem (SE8.8)

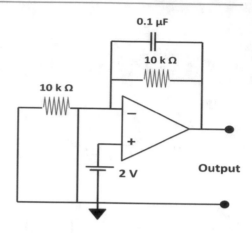

Solution: *As shown in the given circuit, the non-inverting input of the op-amp is at +2V and the inverting input is at ground potential, i.e. at 0 V. Since the two inputs are at different potentials, the op-amp is not unsaturated, it is saturated, and hence, it works as a comparator. Further, the potential at the non-inverting input is larger than the potential at the inverting input therefore, the output will be a step positive voltage of a value which is around 1 V less than the positive power supply $V_{CC}+$.*

Solved example (SE8.9) Unit gain-bandwidth of a non-inverting op-amp amplifier is 10 MHz. What will be the bandwidth when gain of the amplifier is 40 dB.

Solution: *A gain of 40 dB is equivalent to a linear gain of 100. The multiplication of bandwidth and gain at any instant must be equal to the unit gain bandwidth. Hence, bandwidth (Δw) at a gain of 100 is given as*

$$\Delta\omega = \frac{unit\ gain\ bandwidth}{linear\ gain} = \frac{10x10^6}{100}Hz = 100x10^3Hz = 100kHz.$$

Solved example (SE8.10) Discuss the nature of output voltage V_o with respect the applied sinusoidal input A sinωt.

Solution: *Figure (SE8.10) shows that a Zener diode of break down voltage $V_z = 6$ V and break down current $I_z = 1$ mA is connected between the inverting terminal and the output of the op-amp. It is important to remember that Zener break down occurs under reverse bias when reverse bias voltage exceeds the breakdown voltage. However, if a silicon-based Zener is forward biased, it behaves as an ordinary silicon diode and a current flows through it and a constant voltage drop of 0.7 V appears across it.*

Fig. SE8.10 Circuit for
problem (SE8.10)

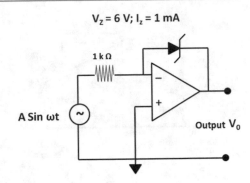

$V_z = 6\ V;\ I_z = 1\ mA$

A Sin ωt

Output V_0

Since the non-inverting input b of the op-amp is at ground potential, the inverting input 'a' will become virtual ground and its potential will also remain at 0 V throughout the operation of op-amp.

(i) Let us consider the case when the positive swing of the input sinusoidal signal is applied at the inverting input. As shown in Fig. SE8.10 (i), a current I_1 will flow along the direction of arrow that will force the Zener to operate as an ordinary diode with a forward current I_1 through it. A voltage drop of 0.7 V will appear against the conducting diode, with its 'a' terminal side as positive and output side negative. As such during the positive swing of the input–output will remain at a constant potential −0.7 V (Fig. SE8.10.1).

(ii) In the next half-cycle when a negative swing appears at the input, inverting input 'a' will be at zero volt and a current I_2 will flow in the direction of arrow shown in Fig. SE8.10 (ii). This current I_2 will force the Zener to undergo Zener breakdown and a constant voltage $V_z = 6.0$ V will appear against the Zener diode, with output side + ve and inverting input side 'a' at zero volt. Therefore, during the negative swing, a constant positive voltage of $V_0 = 6.0$ V will appear at the output.

The input sinusoidal and square output voltage signals are shown in Fig. SE8.10 (iii).

Self-assessment question: What are the sources for currents I_1 and I_2?

Solved example (SE8.11) Design a modified op-amp-based inverter integrator that has a time constant of 0.1 s and the cut-off frequency of 1 Hz. Calculate the circuit gain at a cyclic frequency of 10^4 rad/s.

Solution: *A normal or basic op-amp-based inverting integrator contains a resistance in the input and a capacitance in place of the feedback resistance R_f as shown in Fig. SA8.11 (a)). A modified inverting integrator Figure (SA8.11 (b)) has a parallel combination of R_f and C_f as feedback element.*

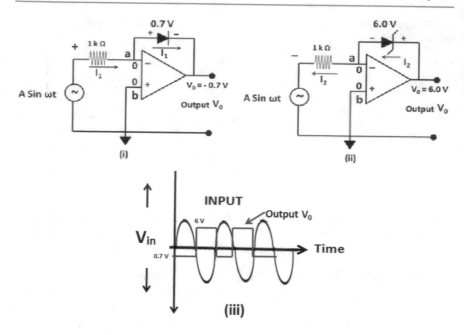

Fig. SE8.10.1 Current and output voltage during the negative and positive swings of the input signal

The time constant of the modified inverting integrator is given by $\tau = R_1 C_f$. It is given in the problem that the time constant has the value $\tau = 0.1s$. Suppose one chooses the capacitance C_f of the value $1\mu F$, and then the resistance R_1 should be of $0.1\ M\Omega$. The cut-off or pole frequency of the integrator will be given by

$$f_c = R_f C_f, \text{ which is required to be } = 1Hz, \text{ hence} f_c = R_f C_f = 1Hz.$$

But, for the required time constant $R_1 = 0.1M\Omega$.

Therefore, $R_f = \frac{1Hz}{2\pi C_f} = \frac{1Hz}{2\pi x 1.0\mu F} = 0.16M\Omega$.

It is now required to calculate the gain $(G)_{\omega=10^4}$ of the modified integrator at a cyclic frequency of 10^4 rad/s.

But $(G)_{\omega=10^4} = -\frac{Z_{eq}}{R_1}$, where Z_{eq} is the equivalent impedance of the parallel combination of R_f and C_f at $\omega = 10^4$ rad/s (Fig. SE8.11).

$$(Z_{eq})_{\omega=10^4} = \frac{Z_{R_f} Z_{C_f}}{Z_{Rf} + Z_{Cf}} = \frac{R_f \left[\frac{1}{j\omega C_f}\right]}{R_f + \left[\frac{1}{j\omega C_f}\right]} = \frac{(0.16x10^6)\left[\frac{1}{jx10^4 x1.0x10^{-6}}\right]}{(0.16x10^6) + \left[\frac{1}{jx10^4 x1.0x10^{-6}}\right]}$$

Therefore,

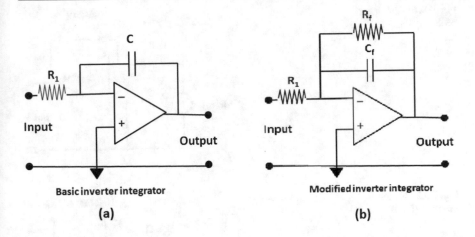

Fig. SE8.11   Inverter integrator

$$(G)_{\omega=10^4} = -\frac{Z_{eq}}{R_1} = -\left\{ \frac{(0.16x10^6)\left[\frac{1}{jx10^4x1.0x10^{-6}}\right]}{(0.16x10^6)+\left[\frac{1}{jx10^4x1.0x10^{-6}}\right]} \right\} \frac{1}{0.1x10^6}$$

$$= -\left\{ \frac{(1.6)\left[\frac{1}{jx10^4x1.0x10^{-6}}\right]}{(0.16x10^6)+\left[\frac{1}{jx10^4x1.0x10^{-6}}\right]} \right\} = -\{0 - j(0.001)\} = +j(0.001)$$

The gain at cyclic frequency 10^4 rad/s is 0.001. It may be observed that gain is less than 1 because the frequency is much more than the cut-off frequency (or pole frequency) and gain falls rapidly beyond it.

Problems

Problem P5.1 Discuss the nature of the output signal delivered by the circuit of Fig. P5.1. Two identical diodes used in the circuit have a breakdown voltage of 6.0 V.

Answer: [Phase-reversed square pulses of 6 V height of the same frequency as the applied signal].

Problem P5.2 Determine the input impedance, output impedance and gain of the inverting amplifier of Fig. P5.2 for a signal of 1 kHz.

Answer: [Input impedance: 1kΩ; Output impedance: infinity, Gain = −{10-j (1.592)}].

Fig. P5.1 Circuit for
problem (P5.1)

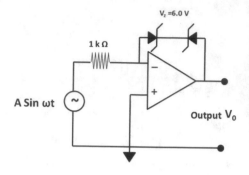

Fig. P5.2 Circuit for
problem (P5.2)

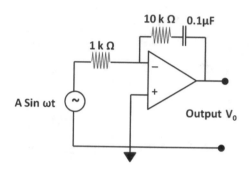

Problem P5.3 Obtain the value of output voltage V_0 (Fig. P5.3).

Answer: [−4.0 V].

Problem P5.4 What will be the output voltage for the circuit of Fig. P5.4? Give
reason(s) for your answer. Also, obtain the approximate value of the potential at
inverting input at steady state.

Answer: [Hint: Since the op-amp will get saturated the approximate value of
$V_0 \sim -10$ V; The circuit will behave as comparator, inverting terminal potential in
steady state will be +4.73 V].

Problem P5.5 As indicated in the circuit of Fig. P5.5 current through the dependent source is given by $I = A.V_x$, where V_x is the voltage drop against the 1 kΩ
resistor. Determine the value of the multiplicative factor A, if the output is 5.5 times
the input.

Answer: [$A = 1.1 \times 10^{-3}$].

Fig. P5.3 Circuit for
problem (P5.3)

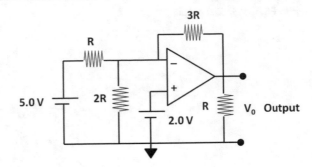

Fig. P5.4 Circuit for
problem (P5.4)

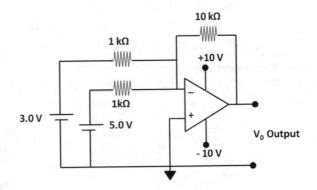

Fig. P5.5 Circuit for
problem (P5.5)

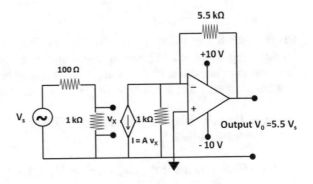

Problem P5.6 Design an op-amp based circuit that delivers an output $(V_1 + V_2 - 2 V_3)$ when voltages V_1, V_2 and V_3 are applied at its inputs.

Answer: Circuit given in Fig. P5.6 will perform the required operations.

Problem P5.7 What should be the ratio $R_1 : R_2$ if the gain of the circuit given in Fig. P5.7 is -12?

Answer: [1:16].

Fig. P5.6 Required circuit

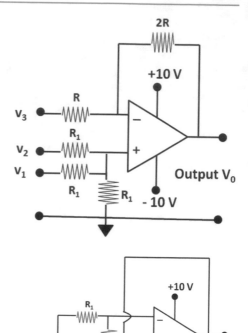

Fig. P5.7 Circuit for
problem (P5.7)

Problem P5.8 Design an op-amp based amplifier having following properties:
(i) input impedance = 2 kΩ; output impedance = ∞; gain = −12. Also, determine
the maximum amplitude of the input signal that may be amplified by the amplifier
without saturation if it is fed by a single 12 V power supply.

Answer: [Inverter amplifier with R_1 = 2 kΩ; and R_f = 24 kΩ. Maximum amplitude of input signal = 0.5 V].

Problem P5.9 Determine the magnitude of V_0 and the common-mode voltage gain
for the circuit of Fig. P5.9.

Answer: [20 ($V_1 - V_2$); 0].

Problem P5.10 Circuit given in Fig. P5.10 is called a differential output amplifier.
It has two outputs A and B, which give outputs V_A and V_B, respectively, corresponding to an input V_{in}. Find the voltage V_C at node C and the magnitudes of V_A
and V_B in terms of V_{in}, R_{f1}, R_{f2} and R_1. Obtain a relation between the resistances in
the circuit so that $V_B = -V_A$.

Answer: $\left[V_C = V_{in}; \ V_A = \left(1 + \frac{R_{f1}}{R_1}\right) V_{in}; \ V_B = -\frac{R_{f2}}{R_1} V_{in}; \ R_{f1} = R_{f2} - R_1 \right].$

Fig. P5.9 Circuit for problem (P5.9)

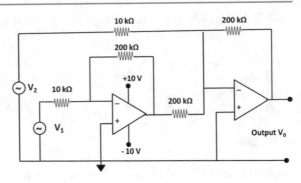

Fig. P5.10 Circuit for problem (P5.10)

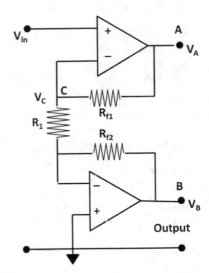

Multiple-Choice Questions

Note: More than one alternative may be correct in some of these questions. Selection of all correct alternatives will make a complete answer in such cases.

MC8.1 A differential amplifier ....

(a) Is a part of operational amplifier
(b) has two inputs
(c) has two outputs
(d) has two inputs and one output

ANS: [(a), (b), (d)].

MC8.2 If A_d and A_c represent, respectively, the differential and common-mode gains of an op-amp, then.

(a) $A_d < A_c$
(b) $A_d = A_c$
(c) $A_d > Ac$
(d) $A_d = 0$

ANS: [(c)].

MC8.3 Differential- and common-mode gains of an op-amp are, respectively, 3500 and 0.35. The CMMR and CMR of the op-amp are, respectively,

(a) 1000, 80 dB
(b) 80 dB, 1000
(c) 4, 10,000
(d) 10^4, 20 dB.

ANS: [(a)].

MC8.4 The most realistic value of the open-loop gain of an op-amp is.

(a) 10^5
(b) 10^2
(c) 10^1
(d) 10^0

ANS: [(a)].

MC8.5 Slew rate of an op-amp is 0.67 V/μs. The voltage at op-amp output will increase by 8 V in.

(a) 12 s
(b) 12 ms
(c) 12 μs
(d) 12 ps

ANS: [(c)]

MC8.6 Negative feedback is employed in op-amps to.

(a) Reduce the closed-loop gain
(b) make its linear operation possible
(c) to convert it into oscillator
(d) reduce noise

ANS: [(a), (b), (c), (d)].

MC8.7 A voltage follower

(a) has a voltage gain of 1
(b) does not change the phase of the input signal
(c) does not have a feedback resistance
(d) does invert the input signal

ANS: [(a), (b), (c)].

MC8.8 Value of $(V_A-V_B)/V_{in}$ for the circuit in Fig. MC8.8 is.

(a) 1
(b) 2
(c) 3
(d) 4

ANS: [(c)]

MMC8.9C8.9 If an op-amp powered by + 9 V and -9 V supplies delivers an output of 8.9 V when signals of V^+ and V^- volts are applied, respectively, at non-inverting and inverting inputs, then.

(a) op-amp is saturated
(b) $V^+ < V^-$
(c) $V^+ = V^-$
(d) $V^+ > V^-$

ANS: [(a), (d)].

Fig. MC8.8 Circuit for problem MC8.8

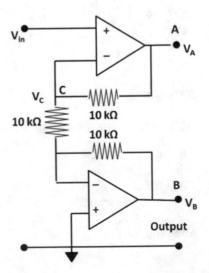

MC8.10 Two inputs of an op-amp are not at the same potential when.

(a) it is powered by a split supply arrangement
(b) non-inverting input is grounded
(c) inverting input is grounded
(d) it is saturated.

ANS: [(d)].

MC8.11 Feedback topology is generally used for negative feedback in op-amps:

(a) Series-series
(b) series-shunt
(c) shunt-series
(d) shunt-series

ANS: [(b)].

MC8.12 Typical characteristics of an ideal op-amp are.

(a) input impedance is zero
(b) output impedance is zero
(c) input impedance is infinite
(d) output impedance is infinite.

ANS: [(a), (d)].

MC8.13 An op-amp of voltage gain 3 has a signal of 2 V at one input while the output signal is of 12 V. The signal at the other input is.

(a) −4 V
(b) −2 V
(c) 0 V
(d) 2 V.

ANS: [(b)].

MC8.14 A voltage follower may also be called.

(a) difference amplifier
(b) integrator
(c) buffer
(d) unit gain amplifier.

ANS: [(c), (d)].

MC8.15 Which of the following circuit (s) are parts of an op-amp?

(a) Constant fraction pickup
(b) blocking oscillator
(c) differential amplifier
(d) Class B emitter follower.

ANS: [(c), (d)].

MC8.16 A triangular wave is applied at the input of an op-amp-based differentiator, the output wave will be.

(a) sine wave
(b) cose wave
(c) straight line
(d) a square wave.

Short Answer Questions

SA8.1 List the characteristic properties of an ideal op-amp.

SA8.2 Outline the method of determining the input and output impedances of a non-inverting op-amp amplifier.

SA8.3 What special arrangement changes the (almost) infinite open-loop gain of an operational amplifier to a manageable value independent of the op-amp circuit? Under what conditions the closed-loop gain becomes independent of the op-amp circuit.

SA8.4 Define slew rate and show that it is infinite for an ideal op-amp. How does the finite value of slew rate affect the performance of an op-amp?

SA8.5 What are CMMR and CMR? What is the advantage of having a large value for CMMR?

SA8.6 What is unit gain bandwidth and how it is related to the gain-bandwidth product?

SA8.7 Draw circuit diagrams for (i) non-inverting amplifier based low-pass filter (ii) inverting amplifier based integrator.

SA8.8 Draw the circuit diagram and the output waveforms for an op-amp based differentiator subjected to square waves.

SA8.9 Discuss how a voltage follower helps in cascading of circuits.

SA8.10 What is meant by the saturation of op-amp? Discuss the comparator operation of an op-amp.

SA8.11 What is a precision rectifier? Give a circuit diagram of a full-wave precision rectifier based on op-amp.

SA8.12   Draw circuit diagram for an inverting amplifier and drive expression for its gain.

SA8.13   NDraw a circuit to implement the operation $(V_1 - V_2)$.

SA8.14   Draw an op-amp-based circuit that may implement the operation $(V_1 . V_2)$.

Sample Answer to Short Answer Question 8.4

SA8.4   Define slew rate and show that it is infinite for an ideal op-amp. How does the finite value of slew rate affect the performance of an op-amp?

ANS: *When some time-varying voltage signal is applied at the input of an op-amp circuit, the output signal, in general, does not exactly follow the input signal. It is because the response of the op-amp to the voltage variations in the input signal is not fast enough. To understand this, let us refer to Fig. SA8.4a, where a square wave of duration 20 μs and amplitude 10 V is applied at the input of the op-amp based unit gain voltage follower. In the input square wave, the voltage rises instantaneously from 0 to 10 V. However, in the output of a practical op-amp the circuit takes some time for the voltage to rise from 0 to 10 V, say it takes 2 μs, as shown in the figure. This happens because of the slew rate of the device. Broadly speaking, the slew rate may be defined as the maximum rate at which the voltage may rise or fall in the output of the device. Strictly speaking, slew rate is the rate at which the voltage signal rises from 0.1% of its maximum value to 90% of its maximum value. The standard unit of slew rate is Volt/micro second = (V/μs). For the example*

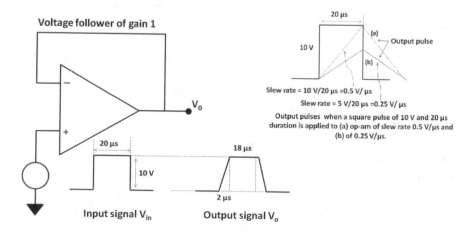

Fig. SA8.4a NExplaining the slew rate

shown in the figure, the slew rate $\approx \frac{10V}{2\mu s} = 5\ V/\mu s$.*It is obvious that the slew rate of a good device should be large so that it may respond to the input voltage changes. The slew rate of an ideal op-amp is, therefore, taken as infinite. In the practical case for proper and faithful reproduction of the signal at the output without distortions on account of slew rate, the slew rate of the device should be larger than the rate of change of voltage in the input signal.Distortion of the output pulse on account of slew time is apparent in the figure. If the slew rate of the op-amp is 0.5 V/μs, the input square pulse will appear as a triangular pulse of peak voltage 10 V in the output. Similarly, if the slew rate is still worst, say 0.25 V/μs, the output pulse will be a triangular pulse of peak voltage 5 V.In case of the sinusoidal signal at the input, the instantaneous AC voltage for which is given by.*$v(t) = A_m \sin \omega t$ *and the rate of change of the voltage as,*

$\frac{dv(t)}{dt} = A_m \omega \cos \omega t$.*The maximum value* $\left(\frac{dv(t)}{dt}\right)_{max}$ *of* $\frac{dv(t)}{dt}$ *occurs for* $\cos \omega t = 1$.

So $\left(\frac{dv(t)}{dt}\right)_{max} = A_m \omega = 2\pi f A_m$.*For distortion-free signal at the output, the slew rate of the op-amp should be greater than* $\left(\frac{dv(t)}{dt}\right)_{max}$,

i.e.Slewrate $\geq [\left(\frac{dv(t)}{dt}\right)_{max} = 2\pi f A_m]$.*Above expression may be used to obtain the maximum frequency* f_m *of the sinusoidal signal that may not be affected by the slew rate as*

$$f_m = \frac{Slewrate}{2\pi A_m}$$

It may also be mentioned that in some cases the slew rate of the same device has different values for the rising and the decreasing parts of the input signal.

Long Answer Questions

LA8.1 What is an operational amplifier and what are its basic properties? Discuss the technique of negative feedback and its implementation in an op-amp. With the help of a circuit diagram explain the working of a non-inverting amplifier and drive expression for its gain.

LA8.2 What is meant by the open-loop and closed-loop gains of an op-amp? Does incorporation of negative feedback ensure that the op-amp will not saturate? Write two important characteristics of a saturated op-amp and explain how it can be used as a comparator.

LA8.3 Discuss different techniques of providing power to an op-amp. An op-amp in inverting amplifier configuration of voltage gain 10 and input resistance 1kΩ is powered by a single supply of + 9 V; a sinusoidal signal of peak-to-peak value 0.1 V is applied at the input of the amplifier. (i) Draw a circuit diagram for the amplifier and (ii) draw a rough sketch of the input and the output signals showing their magnitudes, zero levels and time relation.

LA8.4 Draw circuit diagrams for op-amp-based integrator and differentiator circuits and explain their working. What will be the nature of the outputs if a square pulse is connected to the inputs of these circuits?

LA8.5 What are passive and active filters? With the help of circuit diagrams explain the working of active low pass and high pass filters, their time constants, cut-off (or pole) frequencies and transfer functions.

Part III
Digital Electronics

Electronic Signals and Logic Gates

9

Abstract

Digital data consists of sequences of 0 and 1, called binary code. Electronically, One (1) may be represented by the presence of a pulse of certain predefined height in a specified time interval, called the discrete time interval and Zero (0) by the absence of a pulse of that predefined height during the discrete time interval. Alphabets of any language may thus be coded into digital form assigning them different sequences of 0 and 1. Similarly, numbers may also be represented in digital form using different binary-based numeral codes. Analog data may be converted into digital form using analog-to-digital converter (ADC) while the digital data may be obtained back in analog form using digital-to-analog converter (DAC). Logical operations are digitally implemented using Logic Gates. Successive logic operations may be represented by Boolean expressions that follow laws of Boolean algebra. Details of discrete time digital signals, Numeral systems, principle of operation of ADC/ DAC and implementation of Logic gates, rules of Boolean algebra, methods of solving Boolean expressions, etc. are discussed in this chapter. A number of solved examples has been included to illustrate the implementation of Boolean algebra under different circumstances.

9.1 Electronic Signals

Electronic systems, in most cases, acquire information from some other source and convert it into 'electronic signals', processes these electronic signals, feed them to a computer and/or transmit them, if required. In general, information in any form is coded as a sequence of some symbols. For example, information in written form is coded as a sequence of alphabets and numbers. The alphabets and numbers are the symbols of written information. Similarly, basic sounds associated with different languages or more generally, consonants and vowels are the symbols for the

© The Author(s), under exclusive license to Springer Nature Switzerland AG 2021
R. Prasad, *Analog and Digital Electronic Circuits*, Undergraduate Lecture
Notes in Physics, https://doi.org/10.1007/978-3-030-65129-9_9

information that is presented as speech. Specific movements of fingers and other body parts are the symbols for the language designed for the deaf and dumb. As such, any piece of information may be viewed as a combination of signals and a sequence of signals may constitute the complete body of the information.

9.1.1 Discrete Time Electronic Signal

Signals in general may be looked as a sequence of symbols. However, an **'electronic signal'** may be viewed as a time sequence of **'electronic symbols'**. A voltage pulse (of specified voltage and duration) is very often used as an 'electronic symbol' to generate an electronic signal. In the case of **discrete time electronic signal**, the desired information is contained in the time sequences of these voltage pulses over a certain **discrete time duration**. Different time sequences of voltage pulses within the discrete time interval represent different pieces of information as shown in Fig. 9.1.

It is desirable to understand the concept of discrete time. Time is discretized by defining a time interval T which is repeated continuously. This time interval T is called the time period and reciprocal of T, i.e. 1/T is called the frequency, denoted by f. Electronic signal information may, therefore, be processed either in the time domain or in the frequency domain. At the first sight, it appears that the introduction of frequency domain is an unnecessary complication, but actually, some occurrences in nature are better represented in frequency domain than in the time domain; for example, one goes to school daily, eats twice a day, gets salary once a month, etc. are all those events that are better represented in frequency domain. Similarly, there are some electronic signal sequences that are easy to classify in frequency domain than in time domain. Obviously, those electronic signal sequences that occur several times in a time period T are represented easily in frequency domain, since the time duration of each sequence within the time period T is very small. However, it does not mean that there are two different classes of

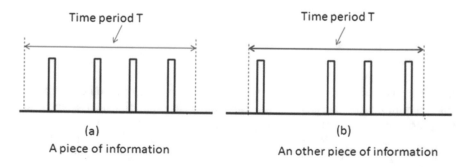

Fig. 9.1 Two different pieces of information **a** and **b** coded using different time sequences of electronic symbol square pulse in discrete time period T

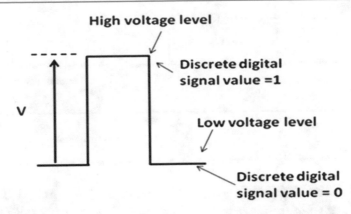

Defining the discrete digital signal values 1 and 0

Fig. 9.2 Defining the high and low levels of a pulse

events (or electronic signals): one that belongs to time domain and the other that belongs to frequency domain. This representation either in time domain or in frequency domain is just a matter of ease or convenience of representation.

Discrete time digital signals are used in information technology for transmitting information from one place to another. As already mentioned, the building block (or the electronic symbol) of discrete time electronic signals is a voltage pulse. A voltage pulse is characterized by two important levels: the high voltage level and the low voltage level, as shown in Fig. 9.2.

The highest (or simply high) and the lowest (or simply low) voltage levels of the voltage pulse are defined as '1' and '0', respectively. Thus, using discrete time, digital signals information acquired by an electronic system is coded in terms of 1 and 0 within a time period T.

The system of two digit representation where any piece of information may be coded in sequences of only two digits 0 and 1 is called the binary system. Each digit, 0 or 1, of the binary system is called a **bit**, the term obtained by joining the b of binary with it of digit. In the decimal system of numbers, there are ten different symbols: 0, 1, 2, 3, … 9, and any number can be represented as a sequence of these ten symbols. The decimal numbers may be added, multiplied, divide and subtracted following the rules of decimal number algebra. In a similar way, it is possible to generate binary numbers or 'binary word' using sequences of binary bits and all algebraic operations may be carried out on binary numbers using rules of a special class of algebra called Boolean algebra. It has been found that 8-bit binary numbers can represent most of the required information, in particular the alphanumeric text. An 8-bit or digit binary number is called a **'byte'**, while a 4-bit group of binary numbers is called a **'nibble'**.

Figure 9.3 shows the discrete time line at the top and below it three different pieces of information, represented with voltage pulses of different heights (and

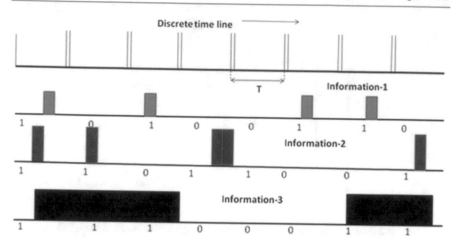

Fig. 9.3 Representation of three different pieces of information in 8-bit binary code

colours). If a voltage pulse falls within a time period T, it is represented by the binary digit 1, and if there no pulse in the time period T then, it is represented by 0. As such in 8-bit, binary code information-1 may be represented as: 10100110; information-2 as: 11011001 and information-3 as: 11100011. It is worth noting that the above-mentioned three different sets of information, each represented by a voltage pulse of a different height may be simultaneously coded using the same discrete time line without any interference from each other. It is because each piece of information may be distinguished from the other by selecting a different high voltage level for each piece of information. This method of simultaneously operating on several different forms of information is called **multiplexing** and makes discrete time electronic signal coding a very powerful tool for computer applications and communication.

Self-assessment question: Draw the discrete time line and show the voltage waveform for binary word 00110110

9.1.2   Signal Transmission

As has been said, discrete time information in binary code, consisting of 0 and 1, is used in operating computers and may be transmitted from one place to another. Although it is quite possible to transmit continuous time analog electronic signals from one place to another, as is done in radio transmission employing the amplitude or the frequency modulation techniques, such transmission suffers from two major defects: (i) signals cannot be transmitted to large distances without distortion and loss of power and (ii) signals pickup undesirable noise from the transmission medium. The content of noise depends on the environment and the distance of

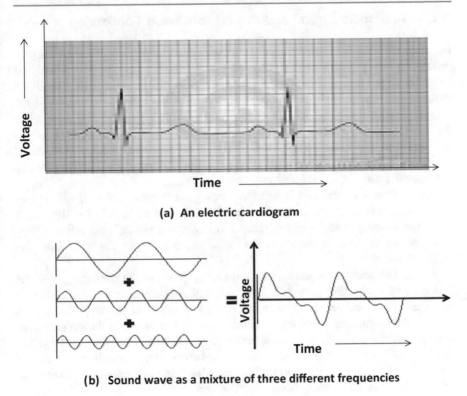

(a) An electric cardiogram

(b) Sound wave as a mixture of three different frequencies

Fig. 9.4 Figure shows two examples of continuous time analog information

travel. As a result, over a relatively large distance, the original signal becomes much too weak than the noise and the information transmitted through the continuous time analog signal is lost. Radio transmission becomes very noisy if the transmitting station is far off and if the weather is turbulent. On the other hand, when the signal is transmitted as discrete time digital signal, the transmission range increases considerably and there is almost no distortion in the signal due to noise. Moreover, noise, which consists of random signals, may be easily removed from the discrete time digital signals which are unidirectional and of fixed height. Therefore, for faithful and noise free transmission of information over large distances, the first requirement is that the information must be in discrete time digital form. Modern day computers also require that instructions and or information fed to them must be in discrete time digital binary code. However, most information's that we come across in our daily life are in continuous time analog form. Typical examples are electrocardiogram, electronic signals obtained from the heart using electrodes; sound signals of music (see Fig. 9.4), etc. In continuous time analog signals, the magnitude of the signal may assume any value at any time that changes with time continuously.

9.1.3 Analog to Digital and Digital to Analog Conversion

For long distance noise free transmission with minimum distortion, and also for processing of signals by the computer, continuous time analog signals are required to be converted into discrete time digital signals in binary code. This conversion requires two steps: (1) to make signals time discrete and (2) to digitalize the magnitude of the signal. This two-step process is performed by a system called Analog to Digital Converter or A to D (or A/D) converter. The block diagram of an A/D converter is shown in Fig. 9.5. The continuous time analog signal $X_a(t)$ is fed to the sampling system that consists of an oscillator of some known frequency f_{samp} that provides electronic pulses of duration $T = 1/f_{samp}$.

The continuous time axis is sampled through this oscillator to divide it into segments of time duration T as shown in Fig. 9.6a. It may be noted in this figure that a small time gap between two (successive) oscillator pulses is actually not there in the actual case but is shown in the figure to separate out the two successive oscillator pulses.

Figure 9.6b shows the function x[n], the magnitudes of the analog signal at the midpoint of each time period T. These magnitudes need to be digitalized or quantized. One method of digitalization of analog values of the magnitudes is to truncate the magnitude to the nearest whole number. For example, the magnitude of the signal at the midpoint of the first period is 13.21, it has been truncated to 13.0; similarly the value 18.14 is truncated to 18.0 and so on. It may thus be observed that truncation of the analog value to the nearest whole number amounts to digitalization. Now these digitalized values may be converted into corresponding binary numbers, for example, 13.0 corresponds to the binary number 1101 ($= 1 \times 2^3 + 1 \times 2^2 + 0 \times 2^1 + 1 \times 2^0$) and to the electronic code ▌▐(pulse, pulse, no pulse, pulse), that is shown in part (c) of the figure.

Truncation of analog data brings in approximations and introduces errors. However, these errors can be minimized by using high-frequency oscillator for time sampling, so that large number of data points becomes available. Hence, the use of high-frequency oscillators in ADC system is preferred.

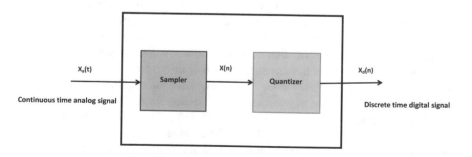

Fig. 9.5 Block diagram of Analog to Digital converter

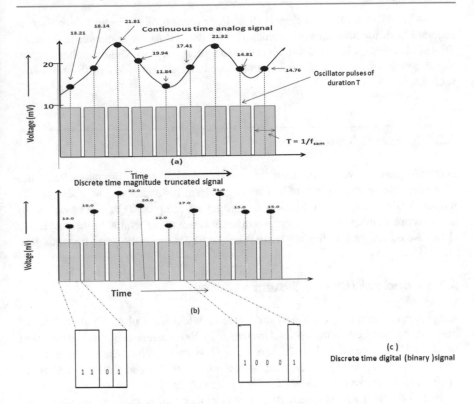

Fig. 9.6 a Shows how continuous time of an analog pulse may be digitalized with the help of oscillator pulses, **b** Shows the time discrete and magnitude truncated signal, **c** Shows the binary coding of discrete time digital signal

Reconstruction of the original analog information from the binary digital data is of prime importance for communication technology. In principle, it amounts to the reverse of the above-mentioned process of sampling and discretization. However, the reconstruction of the original analog signal, without ambiguity, from the corresponding digital signal puts some restriction on the frequency of the oscillator that provided the time sampling pulses at the earlier stage of sampling. It is known that analog signals are made up of signals of many different frequencies. For example, the sound signal shown in Fig. 9.4 is made of three different frequencies. If f_m represents the maximum frequency present in the original analog signal, then, according to Nyquist's theorem, it can be shown that the time sampling frequency of the oscillator f_{samp} should be at least two times the maximum frequency f_m. present in the analog signal, i.e. $f_{samp} \geq 2\, f_m$. In case the sampling frequency f_{samp} is much smaller than the maximum frequency f_m in the analog signal, then ambiguous signals called **'aliases'** are produced in the output. A detailed description of A/D and D/A conversion is provided in Chap. 11.

As already mentioned, analog to digital conversion of the input information is required both for information communication, as well as for computer operations. Digital information is coded into binary code which is a place value number system. Some details of binary and other number systems are provided in the following sections.

9.2   Numeral Systems

Numeral plays a very important role in measuring scientific parameters and recording data. Numeral system is a writing system for expressing numbers; that is a mathematical notation for numbers of a given set using digits or other symbols in a consistent manner. Many different numeral systems are available, however, we shall discuss only a few that are more relevant to data recording and transmission.

9.2.1   Decimal Number System

The first among them is the **decimal system**, which is a place value system of base-10. All numbers in the decimal system may be represented by the combination of 10 digits or symbols; 0, 1, 2, 3, 4, 5, 6, 7, 8 and 9. Numbers of the decimal system are specified by writing the subscript 10 at the end of the number, to distinguish them from numbers of other systems (Fig. 9.7).

In the decimal system, each digit has a cipher value and a place value, for example, let us consider the the decimal number 78901.89_{10}, which may be expanded as

Or $78901.89_{10} = 70000 + 8000 + 900 + 0 + 1 + 0.8 + 0.09$

It may thus be observed that in the decimal system each digit has a numerical value (or cipher value) that is multiplied by its place value. The place value of the first digit to the left of the decimal place is 10^0 of the second digit to the left from decimal is 10^1 and so on. On the right of the decimal, the first digit has a place value of 10^{-1}, the second digit has the place value of 10^{-2} and so on. Thus, the decimal number system has ten different digits to specify the cipher and the place values along with a base (or radix) of 10.

Fig. 9.7 Cipher and place values in a decimal number

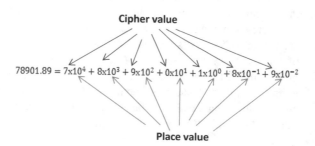

Cipher value

$$78901.89 = 7\times10^4 + 8\times10^3 + 9\times10^2 + 0\times10^1 + 1\times10^0 + 8\times10^{-1} + 9\times10^{-2}$$

Place value

9.2.2 Binary Number System

Like the decimal number system, binary number system is also a place value system, but has only two digits 0 and 1 to represent the cipher and place values. Further, the radix (or base) of the binary system is 2. In analogy to the decimal number system, in binary numbers, the place value of the first digit to the left of the decimal is 2^0, place value of the second digit to the left of decimal is 2^1, of the third digit 2^2 and so on, while on the right side after decimal the first digit has the place value 2^{-1}, the second digit has the place value 2^{-2}, the third digit has the place value 2^{-3} and so on. Binary numbers are specified by putting the suffix 2 at the end of the number.

Suppose a binary number 101011.01_2 is given, let us find the value of this binary number in the decimal system.

Now,

$$101011.01_2 = 1 \times 2^5 + 0 \times 2^4 + 1 \times 2^3 + 0 \times 2^2 + 1 \times 2^1 + 1 \times 2^0 + 0 \times 2^{-1} + 1 \times 2^{-2}$$
$$= 1 \times 32 + 0 \times 16 + 1 \times 8 + 0 \times 4 + 1 \times 2 + 1 \times 1 + 0.5 + 0.25$$
$$= 32 + 8 + 2 + 1 + 0.5 + 0.25 = 43.75_{10}$$

It may be observed that to express a rather small decimal number (43.75) of four digits, a binary number having 8 digit (or one byte) is required. This is because the base (or Radix) of binary system is only 2, while the base (or Radix) of decimal number is 10.

(a) Conversion of decimal number into a binary number

The above example shows that it is easy to convert a binary number into the corresponding decimal number. However, converting a decimal number into a corresponding binary number is somewhat involved. As an example, let us take the decimal number 4375.525_{10} and convert it into its corresponding binary number. Here we will have to convert the integer part (4375) and the fractional part (.525) of the given number separately to the corresponding binary components. We start with the integer part and convert it into the corresponding binary number, for that we make a three column table as shown below. In the first column of the table, the integer part of the given decimal number is written. The integer part in the first column is then divided by 2 and the integer value obtained by division is placed in column–II while the remainder is put in column-III. If the integer part of a decimal number is even, then the remainder will be 0 and if it is odd the remainder will be 1. Next, the quantity that is in column -II is put in column-I and divided by 2, the integer value obtained by division is put in column-II and the remainder in column-III. This procedure is continued till we obtain a zero in column-II. Arranging the remainders (from column-III) in reverse order, i.e. from bottom to top, gives the binary equivalent (Table 9.1).

Table 9.1 Method of converting integer part of a decimal number to corresponding binary number

Integer value	Integer value÷2	Remainder
4375 ÷ 2	2187	1
2187 ÷ 2	1093	1
1093 ÷ 2	546	1
546 ÷ 2	273	0
273 ÷ 2	136	1
136 ÷ 2	68	0
68 ÷ 2	34	0
34 ÷ 2	17	0
17 ÷ 2	8	1
8 ÷ 2	4	0
4 ÷ 2	2	0
2 ÷ 2	1	0
1 ÷ 2	0 Stop	1

Thus, the integer part of decimal number $4375_{10} = 1000100010111_2$ binary number

Next, we find the binary equivalent of the fractional part ($.525_{10}$). To do that, we **multiply** the fractional part by 2 and retain the integer part of the resulting value and repeat this procedure till only an integer value is reached. The following worked out example will make it clear.

$$.525 \times 2 = \boxed{1.500}$$
$$.500 \times 2 = \boxed{1.000}$$

Stop here as fractional value is zero

Hence, the decimal number fraction $0.525_{10} = 0.11_2$ and the complete decimal number

$$4375.525_{10} = 1000100010111.11_2$$

Self-assessment question: Verify that 1000100010111.11_2 is equal to 4375.525_{10}

9.3 Octal and Hexadecimal Numbers

Since the base or radix of the binary system is small, even a small decimal number of few digits is represented by a binary number of many bits. To overcome this difficulty, two other number systems called Octal and Hexadecimal systems are used.

9.3.1 Octal System

The Octal number system has a base (or radix) of 8 and may have eight different values, 0, 1, 2, 3, 4, 5, 6 and 7 for ciphers. The place value of the bit on the left hand side from the decimal increases from 8^0, to 8^1, 8^2, 8^3 and so on. On the right hand side from the decimal, the place value of bits reduces as 8^{-1}, to 8^{-2}, 8^{-3} and so on. For example, let us find the decimal equivalent for the following octal number

$$50234.12_8 = 5 \times 8^4 + 0 \times 8^3 + 2 \times 8^2 + 3 \times 8^1 + 4 \times 8^0 + 1 \times 8^{-1} + 2 \times 8^{-2}$$
$$= 5 \times 4096 + 2 \times 64 + 3 \times 8 + 4 + 1 \times 0.125 + 2 \times 0.015625$$
$$= 20480 + 128 + 24 + 4 + 0.125 + 0.3125 = 20636.4625_{10}$$

It may be observed that a decimal number of 9 digits is represented by an octal number of 7-bits.

(a) **To convert a decimal number into an octal number and vice versa**:
In order to convert a decimal number into an equivalent octal number, one follows exactly the same procedure as was done for the conversion of a decimal number into a binary number. The given decimal number is split into two parts: the integer part and the fractional part. The integer part is divided by 8 (in binary case it was divided by 2) and the integer part of the quotient is kept away in column-II for further division by 8, while the remainder is tabulated in the third column. The division of the integer part of quotient by 8 is continued till one gets a zero in the second column. The reverse sequence of remainders gives the octal equivalent.

To obtain the octal equivalent of the fractional part of decimal number, one multiplies the fractional part by 8, separate the integer (or the whole number) from the obtained value, again multiply the fraction by 8 and separate the integer. The process is continued till a whole number is obtained without any fraction. The forward sequence of integers or whole numbers gives the octal equivalent of the decimal fraction.

Table 9.2 Converting an integer decimal number into corresponding Octal number

Integer part	Integer part of quotient	Remainder
$824 \div 8$	103	0
$103 \div 8$	12	7
$12 \div 8$	1	4
$1 \div 8$	0	1

Stop as it has reached to zero

As an example, let us find the octal equivalent of decimal number 824.015625_{10}.

As the first step, the given number is broken into two parts; the integer part 824 and the fractional part 0.015625. To get the octal equivalent of an integer part, we make a table with three columns and divide the integer part of the decimal number and the whole number parts of subsequent quotients by 8, at each stage, keeping the remainder in column-III as shown in the table below (Table 9.2).

The integer part of the given decimal number $824_{10} = 1470_8$

Next, we find the octal equivalent for the fractional part $.015625_{10}$ of the given decimal number

To obtain that, we multiply the fractional part by 8 to get

$0.015625 \times 8 = 0.125$, the integer part of this is 0

Again, we multiply the fractional part by 8 to get $0.125 \times 8 = 1.00$

The integer part of 1.00 is 1 and the fractional part is 0, therefore, we stop here. The desired octal fraction equal to $0.015625_{10} = 0.01_8$

Therefore $824.015625_{10} = 1470.01_8$

9.3.2 Hexadecimal (or Hex) Number System

While working with computers, very often one comes across binary numbers that have large numbers of digits, may be 8, 16, 32 or even larger. That makes it difficult to read and write these numbers correctly, in particular, if working with many such binary numbers of 32 bits or so. To overcome this difficulty with the large-digit binary numbers, the digits are arranged in groups of 4-bits. These groups of 4-bits use another type of numbering system called **Hexadecimal numbers**.

Hexadecimal or simply hex, is a positional number system with base or radix of 16. The system uses 16 ciphers or digits that are: 0, 1, 2, 3, 4, 5, 6, 7, 8, 9, A, B, C, D, E, F. Symbols A, B, C, D, E and F (alternately, a, b, c, d, e and f) are used in place of 10, 11, 12, 13, 14 and 15 to avoid any confusion with corresponding binary bits.

Binary groups of four bits (or a nibble) produce a hexadecimal number. The smallest value of a 4-bit binary number 0000_2 is zero and the largest value of a 4-bit binary number 1111_2 is $(8 + 4 + 2 + 1 =)$ is 15; hence different combinations of binary digits in a 4-bit binary number can give 0, 1, 2, ... 15 different values, that is in all 16 different values including 0. Each of these sixteen values corresponds to a

Table 9.3 Equivalents of decimal numbers in binary and hexadecimal number systems

Decimal number	4-bit binary number	Hexadecimal number
0	0000	0
1	0001	1
2	0010	2
3	0011	3
4	0100	4
5	0101	5
6	0110	6
7	0111	7
8	1000	8
9	1001	9
10	1010	A
11	1011	B
12	1100	C
13	1101	D
14	1110	E
15	1111	F
16	0001 0000	10
17	0001 0001	11
18	0001 0010	12
19	0001 0101	13

Continue upwards in groups of four

hexadecimal number. Table 9.3 shows the decimal number, corresponding binary number and the corresponding hexadecimal number.

(a) Converting a many digit binary number into a hexadecimal number

Let us consider a binary number 1001001110001111.0010_2. The given binary number has in all 20 digits, 16 digits on the left of the decimal and four on the right. We group these digits in groups of four separately on each side of the decimal, as follows:

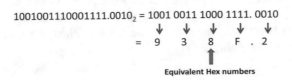

$$1001001110001111.0010_2 = 1001\ 0011\ 1000\ 1111.\ 0010$$
$$= 9 \quad 3 \quad 8 \quad F\ .\ 2$$

Equivalent Hex numbers

Therefore, $1001001110001111.0010_2 = \#938F.2_{16}$

Hexadecimal numbers are often written with # sign before the hexadecimal number followed by either h or subscript 16. Hexadecimal numbers may also be specified by writing 'Ox' before the number.

Let us now take another example of a binary number that has an odd number of digits both to the left and to the right of the decimal place like the number 101000100.011_2. In the given binary number, there are 9 digits to the left of the decimal place and three digits to the right. In order to convert it into hex, we must make groups of four digits each. This can be done by putting three extra zeros before the number and one additional zero at the end of the number, as shown below

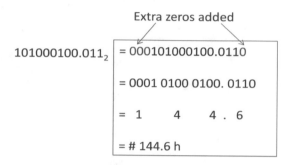

It may be observed that the additional zeros are put in such a way that the value of the given number does not change because of the zeros. The hexadecimal number # 144.6 h may also be represented as Ox 144.6_{16} or simply Ox 144.6.

(b) Converting a hex number into a binary and decimal number
What are corresponding decimal, binary and octal numbers for the hexadecimal number #D1CE h?

$$\#\text{D1CE h} = 13\text{x}16^3 + 1\text{x}16^2 + 12\text{x}16^1 + 14\text{x}16^0 = 53248 + 256 + 192 + 14 = 53710_{10}.$$
$$= 1101000111001110 = 1101000111001110_2$$

To convert the 16-bit binary number 1101000111001110_2 into an octal number, the first step is to make the total numbers of bits in the number a multiple of 3. The given binary number has 16-bits, the nearest number of bits that may be a multiple of 3 is 18. Therefore, the total numbers of bits in the given number are increased to 18 by including two zeros before the given number. Next, we group **three** digits each starting from the right, to get

Table 9.4 Converting a decimal number into an Octal number

Decimal number	Integer after division by 8	Remainder
$53710 \div 8$	6713	6
$6713 \div 8$	839	1
$839 \div 8$	104	7
$104 \div 8$	13	0
$13 \div 8$	1	5
$1 \div 8$	0	1

We stop here as zero is reached

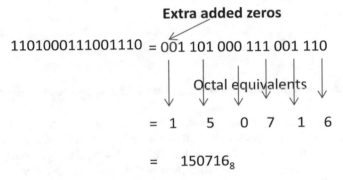

Extra added zeros

$1101000111001110 = 001\ 101\ 000\ 111\ 001\ 110$

Octal equivalents

$= 1 \quad 5 \quad 0 \quad 7 \quad 1 \quad 6$

$= \quad 150716_8$

The same result can also be obtained by dividing the decimal number 53710_{10}, by 8 as shown below in Table 9.4.

The Octal number is 150716_8. This is the same value as achieved by the grouping of three digits in a binary number.

Self-assessment question: Find the decimal, binary and octal equivalent of hex number $\#AB1_{16}$

9.3.3   Binary Coded Decimal Number (BCD)

Computer calculations generally give results in binary code. However, mostly it is required to present the result in decimal form. Conversion of binary numbers into decimal numbers is not only cumbersome but there is a possibility of errors if binary data contains large number of bits. It was, therefore, thought that binary to decimal conversion may become easy if a group of binary digits may represent decimal numbers from 0 to 9 and the next group 10 to 15. This has been made possible by the use of Binary coded decimal (BCD) number system.

In BCD, each of the ten decimal digits 0 to 9 is represented by a group of 4-binary bits. However, in BCD codes it is not necessary that binary equivalent of 10 decimal numbers are in consecutive order. Any group of 4-bit binary digits may represent any decimal number. Several BCD codes are, therefore, available, each of

Table 9.4a BCD_{8421} code for decimal numbers

Decimal number	BCD_{8421} MSD LSD			
	Place value 8	Place value 4	Place value 2	Place value 1
0	0	0	0	0
1	0	0	0	1
2	0	0	1	0
3	0	0	1	1
4	0	1	0	0
5	0	1	0	1
6	0	1	1	0
7	0	1	1	1
8	1	0	0	0
9	1	0	0	1

which uses a specific combination of 4-binary bits to represent a given decimal number.

A 4-bit binary number can represent 16 different characters, any ten of which may be chosen to represent 0 to 9 decimal numbers. Different BCD codes differ in this respect that they chose different combinations of 4-binary digits to represent the same decimal number. One code may represent decimal number 4 by 1110, other may be by 1010 and a third one by 1100. Out of theses several BCD codes, the most commonly used version is the one in which each decimal number from 0 to 9 is represented by a 4-bit binary word, that is the pure binary equivalent of the number and numbers are in consecutive order. This code is identified as BCD_{8421} (Table 9.4a).

It may be verified from the table that the BCD_{8421} codes for decimal numbers 8_{10}, 5_{10} and 3_{10} are, respectively, 1000; 0101 and 0011.

To represent two, three and higher digit decimal numbers another 4-bit binary number based on the same table is added for each digit, for example, $44_{10} = 01000100_{8421}$; $345_{10} = 001101000101_{8421}$.

Self-assessment question: What is the advantage of of BCD_{8421} code?

9.3.4   Alphanumeric Codes

Since it is convenient to convert binary numbers into digital form (since it has only two digits 0 and 1 which may be conveniently represented by the HIGH and the LOW value of a digitalized quantity like voltage), binary number system was readily accepted as a potential candidate for computer language. An added advantage of the binary system is that the digits of this system may also be used to answer logical questions; 1 or HIGH for yes and 0 or LOW for no. With the view to convert not only numbers but other necessary symbols, letters, alphabets, punctuation marks, etc. into binary and other digital number systems, some alphanumeric codes have been developed. These codes present data, including numbers, letters, punctuation marks, etc. in such a format that the computer may understand. With these codes, it is possible to interface input-output devices like printers, monitors, keyboards, etc.

The earliest alphanumeric code was used in telegraphy and was invented by Samuel F.B. Morse in 1837. The data in **Morse code** is made of a dot (.) and dash (–). For example, the letter A in this code is formed by a dot followed by a dash and number 3 by three dots in succession. In Morse code, Dot and Dash have different time durations.

Around 1860, another code called **Baudot code** was developed by E. Baudot, where all symbols were of the same length or duration but an alphabet in this code needed a minimum of 5 symbols.

Later in 1896, a new code called **Hollerith code**, developed by Herman Hollerith for tabulating machines of his company came into the market. This code was essentially used for punching cards that were used by the tabulating machines. It was this code that was extensively used for punching cards for mainframe computers in the early days. Since punch cards are no more used in computers, this code has become obsolete.

Present-day computers mostly use three codes: (i) American Standard-Code for Information Interchange (ASCII), pronounced as 'as-kee' (ii) Extended Binary Coded Decimal Interchange code (EBCDIC) pronounced as 'ebi-si disk' and (iii) Unicode

(a) American Standard-Code for Information Interchange (ASCII)

As-kee is a 7-bit code that has developed out of telegraphic code, and was first published in 1967. Since then, it has undergone several modifications and the latest version was updated in 1986. Being a 7-bit code, it may represent 128 characters. This list of 128 characters includes 26 upper case letters, 26 lower case letters (A to Z), 10 numbers (0 to 9), 33 special characters that include mathematical symbols, punctuation marks, spacing characters, etc. 7-bit characters of the code are made like $X_6\,X_5\,X_4\,X_3\,X_2\,X_1\,X_0$ where each X has the value either 0 or 1. A part of the ASCII table specifying the code is given in Table (9.5)

Most of the modern computers use an 8-bit (one byte) word, while the ASCII table has characters specified in 7-bit word. In order to increase the length of characters to 8-bit, a zero (0) is added before the Zone bit of the ASCII table. For example, the letter 'A' in this case will be specified as '0100 0001' and the '*' as '0010 1010'

(b) Extended Binary Coded Decimal Interchange Code (EBCDIC)

Ebi-si disk code is an 8-bit code, and therefore, can represent 256 different characters. The code is designed by computer manufacturing company IBM, therefore,

Table 9.5 A part of the ASCII code

Part of ASCII table of characters

Digits $X_3X_2X_1X_0$	$X_6X_5X_4$ (Zoned bits)					
	010	011	100	101	110	111
0000	SP	0	@	P	a	p
0001	!	1	A	Q	b	q
0010	"	2	B	R	c	r
0100	#	3	C	S	d	s
0101	$	4	D	T	e	t
0110	%	5	E	U	f	u
0111	&	6	F	V	g	v
1000	(	7	G	W	h	w
1001	)	8	H	X	I	x
1010	*	9	I	Y	J	y

It may be observed from the table that letter 'A' in ASCII code is specified as '100 0001', the sign of addition '*' as '010 1010' and so on

Table 9.5a Part of character table of EBCDIC code

Character	EBCDIC	HEX	Character	EBCDIC	HEX
A	1100 0001	C1	F	1100 0110	C6
B	1100 0010	C2	G	1100 0111	C7
C	1100 0011	C3	H	1100 1000	C8
D	1100 0100	C4	I	1100 1001	C9
E	1100 0101	C5	J	1101 0001	D1

it is used in all computers manufactured by the company. A table showing parts of EBCDIC code is given in (Table 9.5a).

(c) Unicode

Unicode is the latest alphanumeric code that has a unique presentation for each character. More and more users are getting to this code on account of its uniqueness. A small part of the Unicode character specification table is reproduced below (Table 9.5b).

Table 9.5b Part of Character table for Unicode

	0	1	2	3	4	5	6	7	8	9	0A	0B	0C	0D
0000	0													
0020		!	"	#	$	%	&	'	(	)	*	+	/	–
0040	@	A	B	C	D	E	F	G	H	I	J	K	L	M
0060	,	a	b	c	d	e	f	g	h	i	j	k	l	m

Self-assessment question: What is the purpose of having alphanumeric codes when there is already a code for representing binary numbers?

Self-assessment question: If a particular code uses 3-binary bits, how many different characters may be represented by the code?

Self-assessment question: What is the main difference between ASCII code and the ABCDIC code?

9.4 Logic Statement, Truth Table, Boolean Algebra and Logic Gates

Decimal numbers, text and mathematical expressions can be coded into digital code using ASCII tables. However, there are other kinds of statements that are often used in mathematical and other daily life situations. These are the **LOGIC** statements. By definition, a logic statement is a statement of some fact that may be true or false. For example, consider the following sentences:

(a) Delhi is the capital of India (a') Delhi is the capital of Iran
(b) 4.0 is less than (<) 5.0 (b') 4.0 is less than (<) 2.0
(c) Hindi is the best language
(d) Shut the door
(e) Is it raining?

Sentences (a) and (a') represent some facts about Delhi. The fact is that Delhi is the capital. The fact represented by sentence (a) is **true**; which in digital system may be represented by **1**. The fact represented by (a') is **false** and in digital language may be represented by **0**.

Next, consider sentences (b) and (b'). Again, both these sentences represent a fact that decimal number 4 is less or smaller than some other decimal number. However, the stated fact in sentence (b) is true (which in digital language may be represented by 1), while in sentence (b') it is false (which in digital terminology may be represented by 0).

It may thus be observed that (bivalent) logical statements are always associated with some value that may either be true or false, i.e. with 1 or 0.

Let us look to sentence (c), does it represents a statement of fact about the langue Hindi? No, it does not. A lover of Hindi may say that it is the best language but a lover of English will say that English and not Hindi is the best language. As such sentence (c) does not state a fact, it states an opinion. Opinions do not make a logical statement. Similarly, sentence (d) is a command or order that also is not a statement of fact, and hence not a logical statement. Sentence (e) is a question. Questions are also not logical statements. The answer to some questions may be logical statements, for example, answer to questions (e) may be: It is raining or It is not raining, both of which are statements of facts, and hence logical statements.

A large amount of research work was carried out in olden times by philosophers like Aristotle, etc. on bivalent logic that was associated with two possible values: true or false. In recent times, British Scientist George Boole laid the abstract mathematical foundation for (two valued) bivalent logic in 1847, in his book 'Mathematical analysis of logic' and later in 1854, wrote a more detailed book on the subject with the title 'An investigation of the laws of thought'. The branch of algebra based on two valued (or bivalent) logic is called Boolean algebra. In bivalent logic, there are only two possible values, true and false (or digital 1 & 0); however, another type of logic, called Fizzy Logic (or multi-valued logic), where a logical statement may have more than two possible values, each with a certain probability, is gaining applications in computer technology.

Boolean algebra provides the possible outcomes of those cases where more than one logical statement is operated either simultaneously or one after the other or in some other specified sequence. The possible results are generally recorded in a tabular form called the **truth table**.

Since every logic statement may have one of the two possible values, true or false; which may be represented by 1 or 0 (note that 1 & 0 used with reference to logic statements are neither decimal nor binary numbers but they are two digits to specify the true and false or two states; high voltage state and the low voltage state of an electronic circuit.), therefore, any algebraic operation on them will also give either 1 or 0. For example, particular bivalent logic operations may be written as: $1 + 1 = 1$ or $1.0 = 0$ and so on. It should not sound strange because in the vocabulary of binary numbers, the only possible symbols are 1 and 0. Nothing like '2' or '1/2', etc. exists in the digital world.

A similar situation also occurs in case when one determines the magnitudes of alternating current (ac) quantities. In the simple case of dc circuits, the addition of 3 V and 4 V gives 7 V. However, if the quantities refer to a.c, then 3 V(ac) + 4 V (ac) = 5 V(ac). It is because, in ac, the value of the voltage is determined by two factors: the magnitude and the phase. In other words, the dc quantities may be represented by real numbers, while ac quantities need complex numbers to represent them. Since the algebraic operations of real numbers and complex numbers are quite different from each other; the results of algebraic operations on real and complex numbers are also very different. In general, therefore, it may be said that the algebraic operations on a given set of numbers depend on how the numbers are defined.

Though Boolean algebra as a theoretical tool was available since 1854, however, its application to the analysis of electronic switching circuits was discussed for the first time by Claude Shannon, a Ph.D student at Massachusetts Institute of Technology (MIT), USA, in his thesis entitled 'A symbolic analysis of relay and switching circuits' in 1938. Since then Boolean algebra has acquired a central position in computer technology.

It is important to realize that symbols 1 and 0 used in Boolean algebra are **not** binary numbers. As a matter of fact, binary numbers are an alternative way or notation for representing decimal numbers. The 1 and 0 in Boolean algebra represent 'true' and 'false', that may also be associated with the **on** and the **off** states of a (electronic) switch. Further, in Boolean algebra 1 and 0 are only one bit numbers; the binary numbers may have many bits, depending on the magnitude of the decimal number and the system of representation (binary, octal or hexadecimal). Often in digital electronics, a voltage pulse of some defined height is associated with logic state 1 and the absence of any pulse that is a zero voltage is associated with logic state 0. These voltage pulses that define the logic states are called logic pulses.

9.5   Elements of Boolean Algebra and Logic Gates

Boolean algebra may be used to analyze the operation of (electronic) devices that have **only** two distinct stable states, one of which may be designated 1 and the other 0. A switch, generally shown by the symbol ⟿, is one such device. A switch can either be in the state 'on' (or closed) shown as ⟿ which may be designated as 1 or in the state 'off' (or open) shown as ⟿ that may be designated as 0. In Boolean algebra, devices that have only two distinct stable states are called variables and are generally represented by alphabets like A, B, ..., etc. As a rule, these variables can assume only two values; 0 or 1.

In ordinary mathematics, we have two basic operations: addition and subtraction; multiplication and division being the compound operations, respectively, of addition and subtraction. In Boolean algebra, there are three basic operations, called AND, represented by the symbol (.); OR represented by the symbol (+); Not or Inverse (or complement) represented by a bar over the variable. It is important to note that Boolean (.) and (+) symbols have entirely different meanings than the meaning these symbols have in normal arithmetic. To describe these Boolean operations, let us assume that there are two Boolean variables A and B and they are joined together through AND. Treating A and B as switches, the AND joining occurs when these switches are connected in series with a bulb that demonstrates the output of the system, as shown in Fig. 9.8. Figure 9.8a–d show that two switches may be joined in AND coupling in four different ways; in three cases when either one or both switches are off (in state 0) the bulb does not glow (state 0) but when both switches are in state 1 (on) the bulb goes to state 1(glows). These results may be tabulated as shown in Table 9.6a.

Table 9.6a is called the truth table of AND coupling. These results may also be expressed as a Boolean expression

$$A.B = X \quad \textbf{(Boolean expression for AND)} \tag{9.1}$$

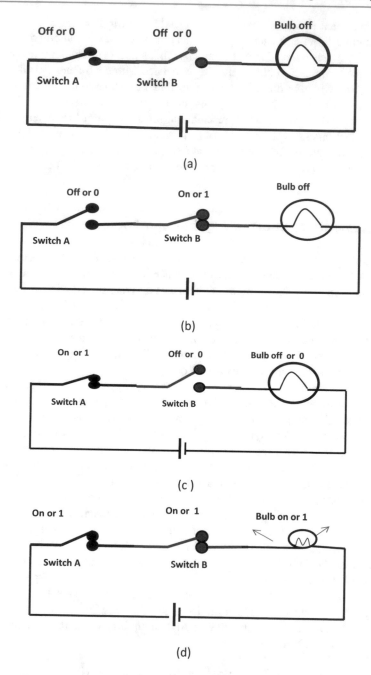

Fig. 9.8 Four different ways of coupling two switches in series

Table 9.6a Truth table for (logic gate) AND coupling of two variables A and B

State of variable A	State of variable B	State of the output X (bulb)
0	0	0
1	0	0
0	1	0
1	1	1

Variables A and B in (9.1) can have only two values, either 1 or 0. X, that represents the output, may also have two values; either 1 or 0. Table (9.6a) when translated in the form of (9.1), may be written as

$$0.0 = 0; \quad 1.0 = 0; \quad 0.1 = 0 \quad \text{and} \quad 1.1 = 1 \tag{9.2}$$

It is evident from (9.2), that AND of Boolean algebra which is represented by the symbol (.) behaves in the same way as the dot of normal algebra that stands for multiplication.

9.5.1 Logic Gate AND

An electronic device that may couple the two independent input digital signals such that the output is the same as given by Boolean AND coupling is called a logic AND gate. In electronic circuit diagrams, an AND gate is shown as in Fig. 9.9a. Table 9.6a is also called the truth table of logic gate AND.

As already mentioned, electronic pulse of some constant potential V corresponds to logic state-1, while the absence of any voltage (i.e. zero voltage) represents logic state 0 of Boolean algebra and both these pulses are called logic pulses. A logic gate receives logic pulse(s) at its input(s) and delivers a logic pulse at the output that bears some relation with the input pulse(s). Logic gates do not amplify or reduce the amplitude of input logic pulse(s), neither they change other parameters of the input pulse(s), they either allow the logic pulse to pass to the output or stops it or inverts it, hence they are called logic gates.

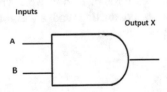

Fig. 9.9a Pictorial representation of AND coupling (Logic gate AND)

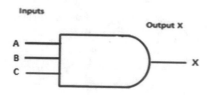

Input A	Input B	Input C	Output X X=A.B.C
0	0	0	0
1	0	0	0
0	1	0	0
0	0	1	0
1	1	0	0
0	1	1	0
1	0	1	0
1	1	1	1

Fig. 9.9b A three input Logic AND gate and its truth table

An AND logic gate may have two, three, four or any number (more than one) of inputs but only one output. A typical three input AND gate and its truth table is shown in Fig. 9.9b.

(a) Realization circuit for logic gate AND

A three input AND gate for actual use may be made using three NPN transistors connecting them in series as shown in Fig. 9.9c. The high voltage $+V_{cc}$ is supplied to the series combination of transistors Q_1, Q_2 and Q_3 through the load resistance R_L. The output voltage V_0 of the system is taken across the emitter resistance R_E. Also, $V_0 = X$, the logic gate output. The three inputs A, B and C are provided through the three base resistances R_{B1}, R_{B2} and R_{B3}, respectively, as shown in the figure. Each of the input terminals A, B or C may be connected either to the positive voltage V or to the ground potential. The resistances in the circuit are so adjusted that when the base terminal of any transistor is at the ground potential that transistor is cutoff and no current passes through it, however, when the base is at the +V potential, the transistor starts conducting and an emitter current is established through the transistor.

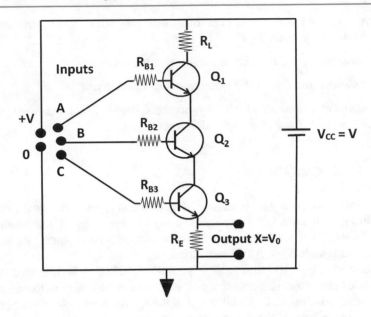

Fig. 9.9c A three input Logic AND Gate made using three NPN transistors

(b) **Working of the AND gate**

Assuming that potential +V corresponds to logic state 1 and ground potential as logic state 0; When inputs A, B and C are all connected to ground, i.e. A = 0, B = 0 and C = 0; all the three transistors are cutoff and no current passes through the emitter resistance R_E, the output voltage $V_0 = 0 = X$. Hence, the logic output is 0.

When the base of either one or any two transistors is connected to the ground and the base of the remaining one transistor is connected to potential + V; which means that if in the input, one has either 0,0,1 or 0,1,0 or 1,0,0; in any of these cases two of the three transistors do not conduct and are like open switches. Therefore, no current or very small current passes through the emitter resistance and the voltage drop across R_E is either zero or much smaller than V. The state of the output is equivalent to logic state 0.

In the case when any two transistors are conducting but one is cutoff; which corresponds to input conditions: 1,1,0; 1,0,1 or 0,1,1; no current or negligible current passes through emitter resistance and the potential drop across R_E is either zero or very small (<V) which corresponds to logic state 0.

In the only situation when all the tree inputs are connected to potential +V which corresponds to the input conditions A = 1, B = 1 and C = 1; all the three transistors conduct and a current I_E passes through the emitter resistance producing a potential drop $V_{out} = R_E \times I_E$. This value of V_{out} corresponds to the logic state 1. It may thus be observed that output X of the circuit has the value 1 only for the condition when inputs A = 1, B = 1 and C = 1. In all other cases the output X = 0. This confirms the truth table for the AND gate.

Self-assessment question: What is meant by a logic pulse?

Self-assessment question: Why logic gates are called gates?

Self-assessment question: Do binary numbers 0 and 1 are same as Boolean numbers 0 and 1? If not what is the difference?

9.5.2   Logic Gate OR

Another important logic gate is called OR gate and is represented by the drawing shown in Fig. 9.10a. An OR gate may have any number of inputs, minimum being two, and a **single output**. The mechanical equivalent of a three input OR gate using switches is shown in Figs. 9.10b and 9.10c.

As shown in Fig. 9.10c, the negative terminal of the battery is connected directly to the bulb, but its positive terminal is connected to a combination of three switches that are arranged in parallel. The other end of the parallel combination of switches is connected to the bulb. When some switch, for example, A is 'on' (as shown in the figure), its state is denoted by Boolean 1. It may be observed that when all the switches are 'off'; i.e. A = 0, B = 0 and C = 0, the bulb does not glow and the output of the gate is 0 (X = 0). Except for this case, in all other situations, when either one or two or all the three switches are 'on' the bulb gets connected to the battery and glows, which means X = 1. In the language of Boolean algebra, it may, therefore, be written as

A = 0, B = 0, C = 0; X = 0, A = 1, B = 0, C = 0; X = 1, A = 0, B = 1, C = 0; X = 1, A = 0, B = 0, C = 1; X = 1 and A = 1, B = 1, C = 1; X = 1 . These results are tabulated in the **truth table for OR gate** in Fig. 9.10b. The general Boolean expression for the OR operation is

Pictorial representation of an OR gate

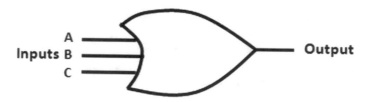

Fig. 9.10a Pictorial representation of a three input OR gate

Truth table for a three input Or gate

Variable A	Variable B	Variable C	Output X=A+B+C
0	0	0	0
1	0	0	1
0	1	0	1
0	0	1	1
1	1	0	1
0	1	1	1
1	0	1	1
1	1	1	1

Fig. 9.10b Truth table for a three input OR gate

Mechanical Equivalent of a three input logic gate OR

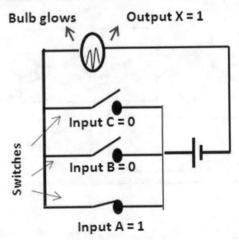

Fig. 9.10c Mechanical equivalent of a three input OR gate using switches

$$A + B + C \ldots\ldots = X \quad \textbf{(Boolean expression for OR)} \tag{9.3}$$

In (9.3), it may be noted that the plus sign (+) in the expression denotes coupling of variables A, B, etc. using an OR device. It is interesting to note that the Boolean (+) sign behaves like the plus sign of normal mathematics, with the difference that in this case, the **output cannot exceed 1**.

$$1 + 1 + 1 = 1, \; 1 + 0 + 0 = 1, \; 0 + 1 + 0 = 1; \; 0 + 0 + 1 = 1 \tag{9.4}$$

(a) Realization circuit for logic gate OR

A three input OR gate may be realized using three NPN transistors, the circuit diagram for which is shown in Fig. 9.10d.

As shown in the figure, the emitters of the three transistors are connected together and through resistance R_E to the ground. Similarly, the collectors of the three transistors Q_1, Q_2 and Q_3 are joined together and through the load resistance R_L is connected to the positive terminal of the power supply V_{CC}. The base connections of the three transistors are connected to a resistance each, R_{B1}, R_{B2} and R_{B3}, respectively. The open ends of these base resistances form the three inputs A, B and C. The resistances in the circuit are chosen in such a way that if the open base terminal A or B or C is grounded (logic state 0), the corresponding transistor goes to cutoff, it does not conduct, and hence no current passes through the transistor. However, if a positive potential +V is given (logic state 1) to the base of the transistor via the corresponding base resistance, the transistor goes into saturation

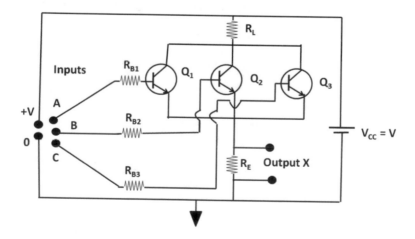

Fig. 9.10d Circuit diagram for a three input logic gate OR

and current I_E flows through the transistor. The output of the gates X is the potential drop across the emitter resistance R_E. If the potential drop across R_E is either equal to $R_E I_E$ or more than that, the logic state of the output X is taken as 1. When the potential drop across R_E is zero, it corresponds to logic state 0 for the output X.

(b) Working of the OR gate

Let us assume that all the three inputs A, B and C are grounded, which corresponds to the condition (A = 0, B = 0, C = 0). In this situation, all the three transistors will be cutoff and no current will flow through the emitter resistance R_E, therefore, the output X will be zero. On the other hand, when anyone or any two or all the three inputs A, B and C are connected to the positive logic potential +V, either one or two or all the three transistors, respectively, will go to saturation and conduct. Therefore, current either of magnitude I_E or 2 I_E or 3 I_E will flow through the emitter resistance, depending on how many transistors are conducting. The flow of emitter current will produce a potential drop of either $R_E I_E$ or $2R_E I_E$ or $3R_E I_E$ across the output terminals. In each of these cases, the output X will be in logic (or Boolean) state 1. These observations are in confirmation of the truth table of OR gate given in Fig. 9.10b.

9.5.3 Logic Gate NOT

Not gate or inverter gate is the only logic gate that has **one input and one output**. The essential property of the gate is that when a pulse corresponding to logic state 1 is applied in the input of the NOT gate, it produces a logic state 0 in its output; and when the logic state 0 is fed at the input of the NOT gate, it delivers the logic state 1 in the output. Since logic states 1 and 0 are complimentary to each other, NOT gate is also called the complement gate. In Boolean algebra, logic state 0 is the complement of logic state 1 and the vice versa, and this fact in algebraic form is expressed as $\bar{1} = 0$ and $\bar{0} = 1$. If a Boolean variable is represented by A, then its complement is indicated by $\bar{A}$.

The Boolean expression for NOT gate for the input A is

$$X = \bar{A} \quad \textbf{(Boolean expression for NOT gate)} \tag{9.5}$$

The symbol of the logic gate NOT is shown in Fig. 9.11a and the truth table for this gate in Table 9.7a. The annular ring or a 'bubble' is the characteristic sign of the NOT gate, often in circuit diagrams, the NOT gate is simply represented by putting a bubble (instead of the complete triangle and bubble) at the end of any other gate to produce the complement of the operation.

The switch equivalent of a NOT gate is shown in Fig. 9.11b, where an electromagnetic switch S_2 is connected to a battery via a simple switch S_1. In the condition when S_1 is off (Input state 0), the switch S_2 is on, the bulb glows, and the output X = 1. However, when switch S_1 is made on (corresponding to input

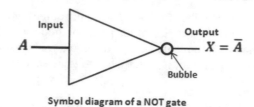

Symbol diagram of a NOT gate

Fig. 9.11a Symbol of NOT gate used in circuit diagrams

Table 9.7a Truth table for a NOT gate

Variable A	Output X
1	0
0	1

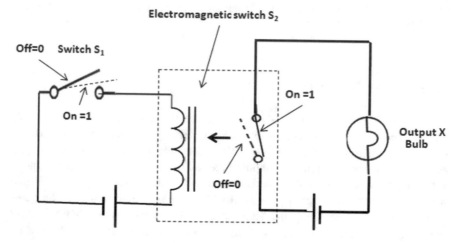

Fig. 9.11b Mechanical equivalent of NOT gate

state = 1), the electromagnetic switch S_2 gets off, breaking the bulb circuit, and making X = 0. Therefore, the output is always in the complement state of the input.

(a) Realization circuit of a NOT gate

The realization circuit for a NOT gate using NPN transistor is shown in Fig. 9.11c. As shown in this diagram, the collector C of the transistor is connected to the power supply V_{CC} through the load resistance R_L and the emitter E is directly grounded. The base is connected to a base resistance R_B, the other end of which serves as the

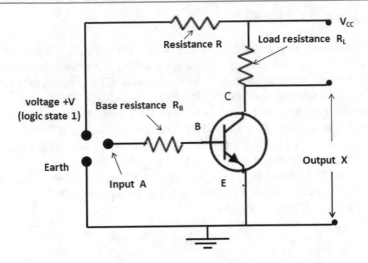

Fig. 9.11c Circuit diagram of the logic gate NOT using NPN transistor

input A. The output X of the system is taken between the collector C and the grounded Emitter E. The load resistance R_L and the base resistance R_B are so adjusted that the transistor remains in the cutoff state when input terminal A is connected to the ground, however, the transistor goes to saturation if a positive voltage +V is connected at terminal A. The potential +V (which corresponds to logic state 1) may be obtained from the power supply V_{CC} using a network of resistances being represented by R.

(b) Working of the circuit

When input terminal A is connected to ground (that corresponds to logic state 0 at the input), the transistor goes to cutoff and does not conduct. Therefore, almost negligible current passes through the load resistance R_L and the potential at the cathode terminal C with respect to the ground (= output voltage X) is almost equal to the V_{CC} which corresponds to logic state 1 in the output. However, in the case when the input terminal A is connected to the potential +V (that corresponds to state 1 at the input) the transistor goes to saturation and a high current passes through the load resistance R_L. As a result, a large potential drops across the load resistance and the potential difference between the cathode terminal C and the ground becomes negligible. A negligible small potential difference between cathode terminal C and ground corresponds to logic state 0 at the output, i.e. X = 0. In short, when input A is in logic state 0, the output X is in logic state 1 and when the input A in logic state1, the output X is in logic state 0. This verifies the truth table for the NOT gate.

9.5.4   Logic Gate NAND

The three basic logic gates AND, OR and NOT are the building blocks of digital electronics. However, some other useful logic gates may be constructed by coupling three basic gates. When the output of the logic gate AND is coupled to the input of a NOT gate, the combination is called the logic gate NAND. The NAND gate may be schematically represented in two ways, Fig. 9.12 shows the actual coupling of a AND gate and a NOT gate, while Fig. 9.12 shows a compact notation for the NAND gate. As may be observed in this figure, the NOT gate is often represented in circuit diagrams by the characteristic bubble alone.

It is known that the output X of a three input AND gate is given as: X = A.B.C, however, if this output of the AND gate is fed to the input of a NOT gate, the final output of the NAND gate becomes

$$X = \overline{A.B.C}. \tag{9.6}$$

The resulting truth table for a three input NAND gate is given in Table 9.7b.

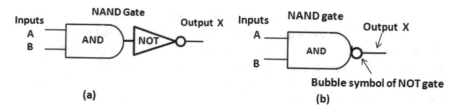

(a)   **(b)**

Fig. 9.12   a and **b** shows the two ways of drawing a NAND gate

Table 9.7b Truth table for a three input logic gate NAND

Input A	Input B	Input C	Output X $X = \overline{A.B.C}$
0	0	0	1
1	0	0	1
0	1	0	1
0	0	1	1
1	1	0	1
0	1	1	1
1	0	1	1
0	1	1	1
1	1	1	0

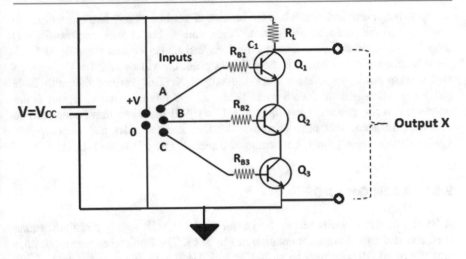

Fig. 9.12c Realization circuit for a three input NAND gate using NPN transistors

(a) Realization circuit for a three input NAND gate

The realization circuit for a three input NAND gate fabricated using three NPN transistors is shown in Fig. 9.12c. The three transistors are connected in series, and therefore, if any one of them is in cutoff, no current passes through the load resistance R_L which is connected to the power supply V_{CC}. The bases of the three transistors are connected, respectively, to three resistances R_{B1}, R_{B2} and R_{B3}. The open ends of these base resistances work as three inputs A, B and C. The positive potential +V that define the logic state 1 is obtained from the power supply V_{CC} and the ground potential defines the logic state 0. Each of the three inputs A, B and C may be connected either to the potential +V or to the ground, and thus they may be set in logic state 1 or 0, independently of each other. As shown in the figure, the output of the logic gate is taken between C_1 and ground, where C_1 represents the collector of transistor Q_1. Resistances R_L, R_{B1}, R_{B2}, R_{B3} and R are so designed that an individual transistor goes to cutoff state if its base input is connected to the ground and goes to saturation when the base input is connected to +V potential.

(b) Working of the circuit

Let us consider the condition when the three base inputs A, B and C of the three transistors are connected to logic potential +V and are, therefore, in logic state 1; i.e. A = 1, B = 1 & C = 1. In this situation, all the three transistors are in saturation condition and are conducting which sends a current, say, I_C through the load resistance R_L. Further, in saturation state the resistance of each transistor is almost zero, hence, in effect, the lower end C_1 of resistance R_L is at ground potential. It means that almost whole of the potential V_{CC} drops across the load resistance R_L and $R_L I_C \approx V_{CC}$. The potential between C_1 and ground is, therefore, zero and hence the output X of the NAND gate circuit is in the logic state 0. This verifies the last row of the NAND gate truth table.

However, in the situation when the base of either one or any two transistors are connected to the potential +V (goes to logic state 1) but at least one transistor is cutoff, no current will pass through the load resistance R_L and the potential difference between the ground and the collector point C_1 will be high ($\approx V_{CC}$) which will correspond to logic state 1 for the output X. This verifies the truth table statements contained in rows 3 to 7. When all the three inputs A, B and C are connected to the ground, (A = 0, B = 0, C = 0) all the three transistors become cutoff; again no current passes through the load resistance R_L and the output X again assumes logic state 1, confirming the second row of the truth table.

9.5.5 Logic Gate NOR

A NOR gate with three inputs is shown in Fig. 9.13. NOR gate in circuit diagrams is represented by the compact notation of Fig. 9.13. The Boolean expression for the output x of an OR gate may be written as $x = A + B + C$. When this output of OR gate is fed to the input of a NOT gate, the final output for the NOR gate becomes,

$$X = \bar{x} = \overline{A + B + C}. \tag{9.7}$$

The truth table for a NOR gate may be easily obtained from the truth table of the OR gate by replacing output logic states with their compliments.

(a) Realization circuit for a three input NOR gate

Circuit diagram for a three input NOR gate using three NPN transistors is shown in Fig. 9.13c. It may be observed in this figure that the three NPN transistors are arranged in parallel with their collectors connected together and through a load resistance R_L connected to the power supply V_{CC}. The emitters of the transistors are connected to earth or ground. Output X is taken between the common collector lead and the ground. Inputs to the three transistors are provided through their bases that are individually connected each to a base resistance R_{B1}, R_{B2} and R_{B3}, respectively.

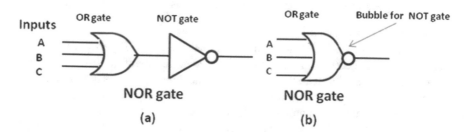

Fig. 9.13 **a** shows NOR gate as the combination of OR and NOT gates and **b** shows the compact notation for the NOR gate

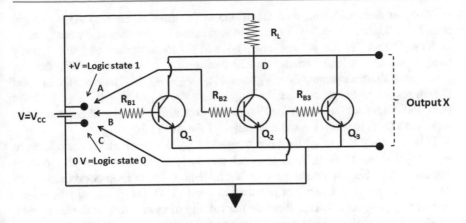

Fig. 9.13c Realization circuit for a three input NOR gate using NPN transistors

The open ends of base resistances A, B and C are the three inputs of the NOR gate. These inputs may be given +V volt potential (logic state 1) or the ground potential (logic state 0). The circuit elements are so arranged that each transistor goes to cutoff when its base is connected to the ground (logic state 0) via the base resistance and the transistor goes to saturation if the base is connected to potential +V corresponding to logic state 1.

(b) Working of the circuit

Let us consider the case when all the three inputs A, B and C are grounded (condition A = 0, B = 0, C = 0). In this case, all the three transistors will be in cutoff state, no current will pass through the load resistance R_L and the potential at the common collector point D in Fig. 9.13c will be high (as there will be no potential drop across resistance R_L and potential at D $\approx V_{CC} = +V$ corresponding to the logic state 1 (or True). Therefore, the output X = 1. This verifies row 1 of the truth table (Table 9.7c).

Table 9.7c Truth table for a three input NOR gate	Input A	Input B	Input C	Output X $X = \overline{A + B + C}$
	0	0	0	1
	1	0	0	0
	0	1	0	0
	0	0	1	0
	1	1	0	0
	0	1	1	0
	1	0	1	0
	1	1	1	0

Next, let us consider the case when one of the three inputs (say A) is in logic state 1 and the other two (say B and C) in logic state 0. The transistor whose base is in logic state 1 (in the present case Q_2) will go to saturation and will heavily conduct sending a load current (say) I_L through the resistance R_L. The flow of current will setup a potential drop V_L across R_L, $V_L = R_L \cdot I_L$. The potential of point D with respect to the ground, i.e. the output $X = V_{CC}\text{-}V_L = V_{CC}\text{-}R_L \cdot I_L$ ($< + v$). As such, the output will be much lower than the potential $+V$ and will correspond to logic state 0. This verifies rows 2, 3 and 4 of the truth table.

Finally, if any two or all the three transistors go to saturation, corresponding to input logic states (1, 1, 0; 1, 0, 1; 0, 1,1 and 1, 1, 1), either two or all the three transistors will heavily conduct producing a large potential drop across resistance R_L, that will result in a very low potential at point D. The output X will, therefore, assume the logic state 0. This confirms the validity of rows 5 to 8 of the truth table.

The NOR and the NAND gates are called universal gates as all other logic gates may be constructed using these universal gates.

9.6 Laws of Boolean Algebra

The rules that govern Boolean algebra are summarized in Table 9.8. In all, there are 16 rules that may be verified by substituting the values 0 and/or 1 for the variables A, B, C, etc. Applying these laws a complicated Boolean expression may be simplified.

NOTE: Rules R-15 and R-16 are known as DeMorgan theorem

As an example, let us verify the law

$$A + (A.B) = A.$$

We chose $A = 1$ and $B = 1$, then

Table 9.8 Laws of Boolean algebra

Boolean Identity or rule	Identity/Rule no.	Boolean Identity or rule	Identity/Rule no.
$A + A = A$	(R-1)	$A.(B + C) = (A.B) + (A.C)$	(R-10)
$A.A = A$	(R-2)	$A + (B.C) = A + C$	(R-11)
$A + \bar{A} = 1$	(R-3)	$\bar{A} + (B.C) = \bar{A} + C$	(R-12)
$A.\bar{A} = 0$	(R-4)	$A + (A.B) = A$	(R-13)
$\bar{\bar{A}} = A$	(R-5)	$A.(A + B) = A$	(R-14)
$A.B = B.A$	(R-6)	$\overline{A.B} = \bar{A} + \bar{B}$	(R-15)
$A + B = B + A$	(R-7)	$\overline{(A + B)} = \bar{A}.\bar{B}$	(R-16)
$A.(B.C) = (A.B).C$	(R-8)		
$A + (B + C) = (A + B) + C$	(R-9)		

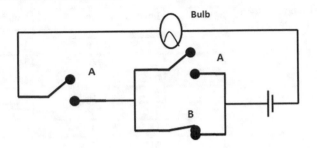

Fig. 9.13d Switch arrangement corresponding to Boolean expression A + (A.B)

A + (A.B) = 1 + (1.1) = 1 + 1=1, that is the value we chose for A.

Next let us take A = 0 and B = 1, in this case the expression A + (A.B) = 0 + (1.0) = 0+0 = 0 (which is the value we chose for A). As a matter of fact for any valid values of A and B, the Boolean product (A.B) can have the value either 1 or 0 and in both cases A + (A.B) = A + (1 or 0). But A + 1 = A and A + 0 = A. Hence for all possible values of A and B the rule may be verified.

If we treat A and B as switches then expression A + (A.B) corresponds to the arrangement of the switches as shown in Fig. 9.13d.

In Fig. 9.13d, it may be observed that the output of the circuit will be the same as the value of variable A. If A = 1, i.e. the switch A is 'on' (logic state 1) the bulb will glow or the output will be in the logic state 1, irrespective of the state of switch B. When A is in state 0 (Off), the output is also in state 0 (off) even when switch B is 'on' or 'off'. Therefore, the output is decided by the state of A.

All the above rules or laws of Boolean algebra may be verified using similar equivalent circuits.

Self-assessment question: Draw a logic diagram for the operation A + (B.C) using switches and show that A + (B.C) = A + C

9.7 Logic Gate Exclusive OR (XOR)

The exclusive OR gate (also written as XOR gate or EXOR gate) is a digital logic gate that performs exclusive disjunction. XOR gate may have two or more than two inputs and only one output. The output of the XOR gate will be in logic state 1 (high or true) only when an odd number of inputs are in logic state 1. If there is an XOR gate with only two inputs then the output will be in logic state 1 only for the inputs 0, 1 and 1, 0, for all other possible combinations 1, 1 and 0, 0 of the input states the output of the XOR gate will be in logic state 0 (Fig. 9.14a, Table 9.7d).

It is known that OR function is equivalent to Boolean addition, AND function is equivalent to Boolean multiplication and NOT function is equivalent to Boolean complementation (inversion), however, there is no such direct correlation between

Table 9.7d Truth table for a three input logic gate XOR

Input A	Input B	Input C	Output X $X = A \oplus B \oplus C$
1	0	0	1
0	1	0	1
0	0	1	1
1	1	0	0
1	0	1	0
0	1	1	0
1	1	1	1
0	0	0	0

XOR (Exclusive OR) operation and a corresponding Boolean operation. It is because of this reason that some time XOR gate is called a hybrid gate. The Boolean notation for XOR gate is the symbol $\oplus$ and the output X of a three input XOR gate may be written in expanded form as

$$X = A \oplus B \oplus C = A\overline{BC} + B\overline{AC} + C\overline{AB} \qquad (9.8)$$

For the case when A = 1, B = 0 and C = 0, the output X, using (9.8) becomes

$$X = 1\left(\overline{0.0}\right) + 0\left(\overline{1.0}\right) + 0\left(\overline{1.0}\right) = 1\left(\overline{0}\right) + 0\left(\overline{0}\right) + 0\left(\overline{0}\right) = 1.1 + 0.1 + 0.1$$
$$= 1 + 0 + 0 = 1$$

This confirms the statements given in the first three rows of the truth table. All other statements of the truth table may also be verified in the same way.

A two input XOR gate is shown in Fig. 9.14b along with its truth table and Boolean expression. A two input XOR gate operates like 'either/or', that is the output is 1 if inputs are different and output is 0 if inputs are identical. A three input XOR gate may be made from the combination of two XOR gates each with two inputs as shown in Fig. 9.14c.

A two input XOR gate may be constructed by combining different basic gates, some examples of which are given below.

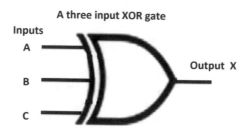

A three input XOR gate

Inputs

A

B

C

Output X

Fig. 9.14a Symbol for a three input XOR gate for use in circuit diagrams

A two input XOR gate

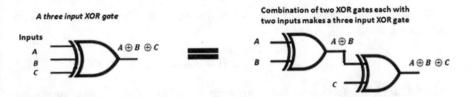

A $\quad$ $A \oplus B$

B

Input A	Input B	Output X X= $A \oplus B$
1	1	0
1	0	1
0	1	1
0	0	0

Fig. 9.14b A two input XOR gate and its truth table

A three input XOR gate

Inputs
A
B
C

$A \oplus B \oplus C$

Combination of two XOR gates each with two inputs makes a three input XOR gate

A $\quad$ $A \oplus B$

B

$A \oplus B \oplus C$

C

Fig. 9.14c A combination of two XOR gates each with two inputs is equivalent to a three input XOR gate

(a) A two input XOR gate from the combination of NOR gates

Figure 9.14d shows how a two input XOR gate may be made using five NOR gates.

$$X = \overline{(\bar{A}+\bar{B})} + \overline{(A+B)}$$

As indicated in Fig. 9.14d, the output X of the above combination of five NOR gates is given by

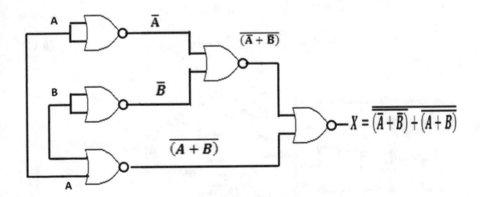

$$X = \overline{(\bar{A}+\bar{B}) + \overline{(A+B)}}$$

Fig. 9.14d Making a two input XOR gate using five NOR gates

$$X = \overline{(\bar{A}+\bar{B}) + (A+B)}\qquad(9.9)$$

Now we put $(\bar{A}+\bar{B}) = P\,and\,(A+B) = Q$. With this substitution (9.9) becomes $X = \overline{\bar{P}+\bar{Q}}$; We use Boolean law given by Boolean rule (R-15) and replace $\bar{P}+\bar{Q} = \overline{PQ}$ to get

$$X = \overline{\overline{P.Q}} = P.Q = (\bar{A}+\bar{B})(A+B) = A\bar{A} + A\bar{B} + B\bar{A} + B\bar{B} = A\bar{B} + B\bar{A}$$

We have used the Boolean identity $\bar{A}.A = 0 = \bar{B}B$ in the above expression. With these substitutions (9.9) reduces to

$$X = \bar{B}.A + \bar{A}.B = A \oplus B\qquad(9.10)$$

Equation (9.10) tells that the combination of five NOR gates shown in Fig. 9.14d is equivalent to a two input XNOR gate with Boolean variables A and B as the inputs.

(b) **A two input XOR gate from the combination of NAND gates**
Figure 9.14e shows how four NAND gates may be coupled to make a two input XOR gate. The output X of the above combination is shown to be equal to

$$X = \left[A.\overline{(A.B)}\right]\left[B.\overline{(A.B)}\right] = [A.(\bar{A}+\bar{B})][B.(\bar{A}+\bar{B})]$$

Or

$$X = [A.\bar{A} + A\bar{B}][B.\bar{A} + B\bar{B}] = [A\bar{B}][B\bar{A}] = \overline{(\overline{A}B)}\left(\overline{\overline{B}A}\right)$$

Or

$$X = \overline{(\overline{A}B)} + \left(\overline{\overline{B}A}\right) = A\bar{B} + B\bar{A}\qquad(9.11)$$

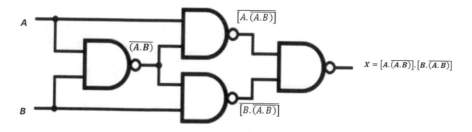

Fig. 9.14e Making a two input XOR gate using four NAND gates

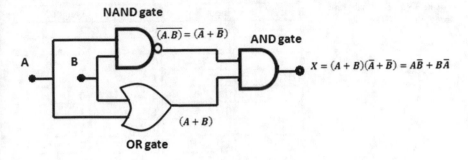

Fig. 9.14f A two input XOR gate using NAND, AND and OR gates

In deriving (7.11), we have used the laws of Boolean algebra given by (R-15) and (R-14), along with the identity $\bar{A}A = B\bar{B} = 0$.

It may be observed from (9.11) that the output X of the circuit shown in Fig. 9.14e is the same as that of a XOR gate

(c) **A two input XOR gate from the combination of a NAND, an OR and an AND gates**

Figure 9.14f shows how a NAND an OR and an AND gates can be coupled together to obtain an XOR gate. The value of the variable at each output is also indicated in the figure.

9.8 Logic Exclusive NOR or NXOR Gate

When the output of an exclusive OR gate is fed to the input of a NOT gate, the combination behaves as an Exclusive NOR or NXOR gate. The symbol of an exclusive NOR gate for use in circuit diagrams is shown in Fig. 9.14g, while Table 9.7e) shows the truth table for a three input XNOR gate.

Fig. 9.14g Symbol for an Exclusive NOR gate

Table 9.7e Truth table for a three input XNOR gate

Input A	Input B	Input C	Output X
0	0	0	1
0	0	1	0
0	1	0	0
0	1	1	1
1	0	0	0
1	0	1	1
1	1	0	1
1	1	1	0

As may be observed in the truth table, the output X of the XNOR gate is high (logic state 1) when either none or an even number of inputs are in logic state high, otherwise the output is in logic state 0. Symbolically the operation XNOR is indicated by $\odot$ and the output X of an XNOR gate with three inputs is given by the Boolean expression;

$$X = A \odot B \odot C = \overline{A \oplus B \oplus C} = \overline{ABC} + AB\bar{C} + A\bar{B}C + \bar{A}BC \qquad (9.12)$$

Any XNOR gate with more than two inputs may have the logic state 1 in the output in two situations; when all the inputs are in logic state 0 or when even number of inputs are in state 1. Therefore, an XNOR gate with more than two inputs may be used to detect even functions at the input and is called a MODULO-2 SUM (Mod-2-SUM).

Figure 9.14h shows the symbol for a two input XNOR gate along with its truth table and the Boolean expression for its output as

$$X = A \odot B = \overline{A \oplus B} = AB + \bar{A}\bar{B} \qquad (9.13)$$

It may be noted that a two input XNOR gate gives logic state 1 in its output only when the logic levels at its two inputs are identical, either 0, 0; or 1, 1. Hence, this gate works as a comparator of two input logic states; if they are identical then only high logic state 1 will appear at the output of the gate, otherwise the output will show a low logic state 0. Hence, a two input XNOR gate is often called a COMPARATOR or EQUIVALANCE gate.

Input A	Input B	Output X
0	0	1
1	0	0
0	1	0
1	1	1

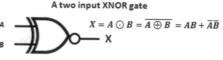

A two input XNOR gate

$$X = A \odot B = \overline{A \oplus B} = AB + \bar{A}\bar{B}$$

Fig. 9.14h A two input XNOR gate with its Boolean expression for the output X

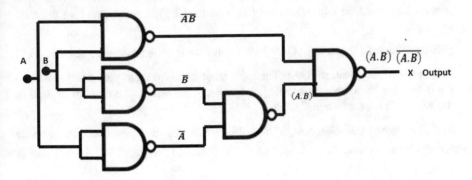

Fig. 9.14i Equivalent circuit for a two input XNOR gate using five NAND gates

A circuit that works like a two input XNOR gate can be assembled using five NAND gates

As shown in Fig. 9.14i, the Boolean function at the output of each NAND gate is also indicated in the figure. At the final output, the function X is given as

$X = (A.B)\overline{(A.B)} = \left\{\overline{\overline{AB}}.\right\}.\overline{(A.B)}$, here we have put $(A.B) = \left\{\overline{\overline{AB}}.\right\}$

But from Boolean algebra rule expressed by Eq. (7 g)

$$X = \{\overline{A}\overline{B}\} + (A.B) \qquad (9.14)$$

Thus, it is observed that the output of the circuit shown in Fig. 9.14i is equivalent to a two input XNOR gate.

Another combination of two NOT, two AND and one OR gates coupled as shown in Fig. 9.14j also results in a two input XNOR gate. The Boolean expression at each output stage of the circuit is shown in the figure. It may be said that in

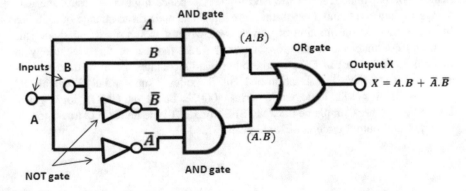

Fig. 9.14j Circuit diagram for a two input XNOR gate using two NOT, two AND and one OR gates

general any Boolean expression may be obtained by properly coupling different basic gates.

Self-assessment question SAQ: Which logic gate is called a comparator and why?

Self-assessment question SAQ: The truth table of a two input gate has 4 rows and the truth table for a three input gate has 8 rows. What will be the number of rows in the truth table for a N-input gate?

Self assessment question SAQ: Draw the circuit of logic gates that may generate the Boolean function $X = \overline{(A.B)} + \overline{(A+B)}$, starting from input variables A and B

9.9   Classification of Logic Technology

On the basis of technology and basic elements used to fabricate logic circuits in the manufacturing of integrated circuits (IC) and 'chips' on commercial scales, the logic circuits may be classified into five basic classes: (i) **D**iode–**T**ransistor **L**ogic (DTL) (ii) **R**esistance-**T**ransistor **L**ogic (RTL) (iii) **E**mitter **C**oupled **L**ogic (ECL) (iv) **T**ransistor-**T**ransistor **L**ogic (TTL) and (v) **C**omplementary **M**etal **O**xide **S**ilicon (CMOS) logic.

DTL and RTL technologies, respectively, used diodes and transistors and resistance-transistor networks to construct basic logic gates. These were the earliest technologies used in the initial stages when solid-state devices replaced tube circuits. Earlier, logic gates were fabricated using vacuum tubes. Being slow, requiring more power and cumbersome, these technologies for assembling logic gates were soon replaced by the faster circuits using ECL, TTL and CMOS technologies. ECL is not very different from TTL, bipolar junction transistors (BJT) either NPN or PNP type are used in both technologies, however, in ECL, the transistors never go into saturation regions and there is a small swing ($\approx 0.8V$) in the input/output voltages that may change the conducting state of the transistor very fast, the output impedance is low and input impedance high, as a result, the gate delays are low and fanning capabilities are high. The main disadvantage of ECL lies in its continuous consumption of power, even when it is in a quiescent state. This high and continuous consumption of power limits the use of this technology in modern IC and chips. In TTL, on the other hand switching action of transistors is exploited using them in saturation and cutoff modes. Commercially available IC series 7400 uses TTL logic while series 4000 is based on CMOS logic gates. CMOS uses either Complementary MOSFET or JEET type field effect transistors in the input and output circuits of logic gates.

The complexity of an integrated circuit (IC) or chip is decided by the number of transistors/logic gates are synthesized in a single package. The Small Scale Integrated Circuits (SSI) have up to 10 transistors or a few simple gates like AND, OR, NOT, etc. in a single packing. The Medium Scale Integrated (MSI) circuits accommodate from 10 up to 100 transistors or up to ten logic gate on a single silicon chip. Large Scale Integrated circuits may have 100 to 1000 transistors or up to 100 gates in a single packing. These (LSI) IC's are used as adders, decoder, counter, flip flop, etc. Very Large Scale Integrated (VLSI) circuits may contain from 1000 to 10, 000 transistors or from 100 to up to 1000 logic gates on a single package and are used as large memory arrays, programmable logic devices, etc. The Super Large Scale Integrated (SLSI) devices may have between 10,000 and 100,000 transistors or a few thousand to hundred thousand gates in a single package. These chips may be used as a microprocessor, calculator, microcontroller, etc. The Ultra Large Scale Integrated (ULSI) devices may contain more than one million transistors in a single package and are used in computer CPU, video processor, mobile phones, etc.

The ultimate level of integration is achieved in SYSTEM-ON-CHIP (SOC) devices where individual components like Input/ Output logic, memory, microprocessor, etc. are all fabricated in a single silicon chip. These chips may contain up to hundred million individual CMOS transistor gates on a single silicon chip and are used in digital cameras, higher level mobiles and in robotics.

9.10 Voltage Levels for the Two Logic States

The True/High/1 and False/Low/0 logic states in digital systems are specified by some prefixed voltage ranges. Most of the digital systems use positive voltages for defining the two logic states. The characteristic voltages depend on the type of logic; for example, in TTL IC's of 7400 series, the logic state 1(or High/True) is defined by a voltage of 2.0 V to 5.0 V and the logic state 0(or Low/False) by the voltage below 0.8 V to 0 V. It may also be mentioned that in this TTL system the supply voltage V_{CC} at which bipolar transistors work is 5.0 V. On the other hand, in CMOS 4000 series chips logic state 1 (High/True) is characterized by the voltage of 3.0 V to 18.0 V and the logic state 0 (Low/False) by the voltage 1.5 V to 0.0 V. The field effect transistors in CMOS logic works at the supply voltage V_{CC} of 18.0 V.

The defining voltages for logic states 1 and 0 in TTL and CMOS systems are shown in Fig. 9.15. The region of voltage between the lower limit of logic state 1 and the upper limit of logic state 0 is the voltage region where the logic is not defined. If the output or input of the gate shows a voltage in this undefined region, the response of the gate will not be defined and the behaviour of the system will be unreliable. Random noise pulse may derive the gate to the undefined region resulting in unreliable behaviour of the gate.

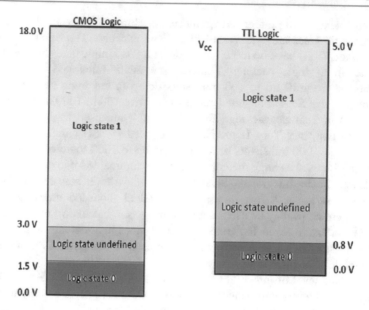

Fig. 9.15 Voltage levels for logic state 1 and 0 for TTL and CMOS devices

Self-assessment question: What is the advantage of ECL over TTL?

Self-assessment question: From the point of noise which of the logics TTL or CMOS is better and why?

9.11 Solving Problems Based on Logic Gates

Having studied the properties of all logic gates and the method of representing their performance using Boolean expressions, it is time to consider the method of solving problems based on the combination of several logic gates; this is called the combinational analysis. A special character of logic combinational circuit is that it has no memory, i.e. the output of the combination depends on the instantaneous values of the input states. In principle, combinational analysis of logic circuit may be done using two different techniques: (i) Using the laws/rules or theorems of Boolean algebra (ii) using Karnaugh mapping technique. Presently, we shall confine ourselves to the former; the applications of the laws/theorems of Boolean algebra tabulated in the Table (9.8). It is, therefore, important to remember the rules given in Table (9.8) by heart.

The first step in solving any combinational logic problem is to simplify the problem. Just by using the Boolean expression for each component logic gate in the combination, it is possible to write the Boolean expression for the final output. The

Boolean expression for final output or in general any Boolean expression may be written in two different forms **(a) as sum-of-products indicated by SOP in short or (b) product-of-sums indicated by POS.** For example

$$AB + (\bar{A}\bar{B}C) + \ldots + \ldots \qquad (9.15)$$

$$(AB + \bar{A}\bar{B}C)\left(\bar{A} + A + \overline{B}\overline{C}\right) \qquad (9.16)$$

Equation (9.15) represents the SOP form of a Boolean expression, while (9.16) gives the POS form. From the point of simplification, it is always convenient to convert the given Boolean expression in SOP form. It is important to remember that each multiplication in Boolean algebra represents an AND operation and each addition an OR operation. As such a Sum of Product Boolean expression consists of two or more AND operations that are ORed together. Further, each AND term consists of one or more variables individually appearing in either uncomplemented or complemented (inverted) form. It may be observed that in SOP form one inversion sign cannot cover more than one variable in the term. That is to say that in SOP for one cannot have terms like $\overline{XYZ}$ or $\overline{XY}Z$, etc. If such terms appear in the course of simplification, these terms may be reduced to simpler components using DeMorgan theorems given in Table (9.8)

Simplification of a given Boolean expression essentially means reducing the number of terms and/or the number of variables in different terms of the given expression without changing the outcome. Similarly, simplification of the given combinational logic gate network means reducing the number of logic gates used in the given network without changing the outcome of the network.

9.11.1 Simplifying Boolean Expression or Algebraic Simplification

Complicated and long Boolean expressions may be simplified using the rules of Boolean algebra as given in Table (9.8). The following examples may illustrate the procedure for simplifying both the Boolean expressions and complicated logic gate networks. Method of implementation of Boolean expressions is also demonstrated in these examples. It is important to remember that most of the Boolean expressions, particularly having a number of terms, may be implemented through different logic networks that give the same output.

Solved example (SE9.1) Simplify the following Boolean expression, and hence draw the logic circuit that may implement the given expression.

$$X = A\bar{B}\bar{C} + A\bar{B}C + ABC$$

Solution: *From the first two terms of the expression we take $A\bar{B}$ common to get*

$$X = A\bar{B}(\bar{C}+C)+ABC$$

However, from rule R-2 of Table (9.8), $(\bar{C}+C) = 1$, *therefore, the above expression becomes*

$X = A\bar{B}+ABC$, *A may be taken as common from this to give*
$X = A(\bar{B}+BC)$, *but* $(\bar{B}+BC) = (\bar{B}+C)$ *from R-11' of* Table (9.8)
Hence,

$$X = A(\bar{B}+C) \tag{SE9.1.1}$$

Equation (SE9.1.1) is the simplest form of the given expression. This expression may be implemented by the network of logic gates shown in (Fig. SE9.1.2).

Solved example (SE9.2) Design a combination of logic gates the output of which may be expressed by the Boolean expression $(\bar{A}+B)(\bar{B}+A)$. Test if the given circuit may be further simplified and if yes, draw the simplified logic gate network.

Solution: *To draw the logic gate combination that may deliver the desired output, it is advisable to start from the given expression. In the present case, the given Boolean expression is a multiplication of two brackets. Since multiplication in Boolean algebra is performed by the AND gate, therefore, the final gate in the combination will be an AND gate, the two inputs of which will be the brackets $(\bar{A}+B)$ and $(\bar{B}+A)$. It may be observed that each of the two brackets has the addition of two terms. Addition in Boolean algebra is performed by OR gate, therefore, there should be two or gates to perform additions, as shown in* Fig. SE9.2.1. *In order to perform inversion, two NOT gates need to be included in the network. The complete network of logic gates for the purpose is shown in this figure.*

For further simplification of the network, we once again consider the given expression

$X = (\bar{A}+B)(\bar{B}+A)$. *And open the brackets by term vise multiplication, to get*
$X - \bar{A}\bar{B}+B\bar{B} +\bar{A}A+B.A;$ *Now* $B\bar{B} = \bar{A}A = 0,$ *from Rule R-4.*

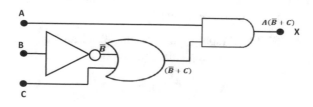

Fig. SE9.1.2 Combination of logic gates that may implement the Boolean expression (SE9.1.1)

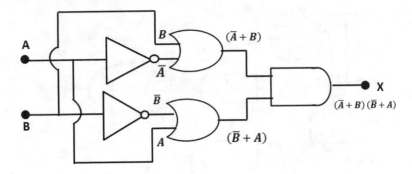

Fig. SE9.2.1 Logic gate combination that may deliver the desired Boolean expression

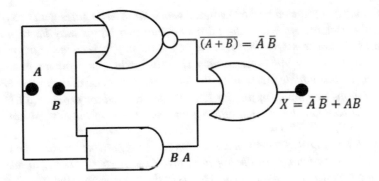

Fig. SE9.2.2 Logic network to realize the simplified Boolean expression

Hence $X = \bar{A}\bar{B} + B.A$, we see that the given expression may be simplified to this form and may be realized from the network given in Fig. SE9.2.2.

It may be observed from Figs. SE9.2.1 *and* SE9.2.2 *that only three gates are required to implement the simplified expression in comparison to the five logic gates used to implement the original expression.*

Solved example (SE9.3) Draw the logic gate combination that may implement the Boolean expression $X = \bar{A}BD + \bar{B}AC + \bar{C}\bar{D}$.

Solution: *Since the given expression for x contains three additives terms and in Boolean algebra additive terms may be generated through an OR gate, therefore, the last logic gate may be a three inputs OR gate as shown in* Fig. SE9.3.1. *Each input to the OR gate should produce one of the three terms, two of which have combinations of three components multiplied together and the one has a combination of two components multiplied together. Now multiplicative terms in Boolean*

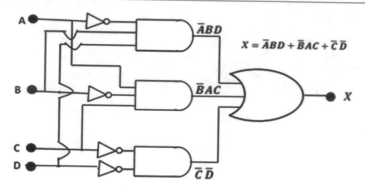

Fig. SE9.3.1 Logic circuit for implementing X

algebra may be produced by AND gates. As such three AND gates with two of them having three inputs each and the third one with two inputs may be required to generate the three terms, as shown in the figure. As may be seen in the figure, in the uppermost AND gate an inverter gate is included in one input to produce Ā; similarly, in the middle AND gate the inverter gate generates B̄. Further two inverters at the two inputs of the third AND gate are used for producing C̄D̄ at the output. The complete logic circuit for the implementation of the desired expression is shown in Fig. SE9.3.1.

It is relevant to ask if the given expression for X may be further simplified or not. A Boolean expression can be simplified in two ways or steps. First is to take some common components from the various terms of the given expression and then apply Rules of Boolean algebra. In the given expression for X, there are no common components in the given three terms and also no Rule of Boolean algebra can be applied. Hence, this expression cannot be reduced any further.

Solved example (SE9.4) Reduce the following Boolean expression for Y and draw the logic network to implement the reduced expression. What is the form of expression?

$$Y = (A\bar{B} + B\bar{C} \mid D\bar{A})(C\bar{D} + D\bar{B} + \bar{A}B)$$

Solution: *The given expression is in POS (product of sum) form. For simplification, it is required to be converted into SOP form which may be done by opening the brackets as follows:*

$$Y = A\bar{B}C\bar{D} + B\bar{C}C\bar{D} + D\bar{A}\,C\bar{D} + A\bar{B}D\bar{B} + B\bar{C}D\bar{B} + D\bar{A}D\bar{B}$$
$$+ A\bar{B}\bar{A}\,B + B\bar{C}\bar{A}\,B + D\bar{A}\bar{A}\,B \qquad\qquad \text{(SE9.4.1)}$$

The expanded Y contains 9 terms. Now all those terms that have factors like $A\bar{A}$; $B\bar{B}$; $C\bar{C}$ *and* $D\bar{D}$ *becomes zero as* $A\bar{A} = B\bar{B} = C\bar{C} = D\bar{D} = 0$, *from Rule R-4. As such terms No; 2, 3, 5 and 7 becomes zero. In the remaining 5 terms factors like A.A may be replaced by* $= A$ *an so on to get*

$$Y = ACD\bar{B} + AD\bar{B} + D\bar{A}\bar{B} + B\bar{C}\bar{A} + D\bar{A}B$$

In the above expression, terms III and V may be combined to get

$$Y = ACD\bar{B} + AD\bar{B} + D\bar{A}(\bar{B}+B) + B\bar{C}\bar{A}; \quad but \quad (\bar{B}+B) = 1\,from\,rule\,R-3$$

Finally, the expression becomes

$$Y = ACD\bar{B} + AD\bar{B} + D\bar{A} + B\bar{C}\bar{A} \qquad\qquad \text{(SE9.4.2)}$$

The simplified expression for Y may be implemented in many different ways. As a matter of fact, there is no unique method of implementing a Boolean expression that contains several terms. One possible way of implementing the expression given by (SE9.4.2) is shown in Fig. SE9.4. *One may use two AND gates with two inputs each in place of the four input AND gate used in the figure with the addition of another OR gate. Thus, there may be several different combinations of logic gates that may deliver the same output.*

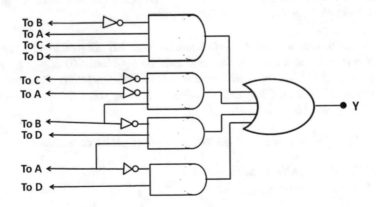

Fig. SE9.4 Logic network to implement Boolean expression Y

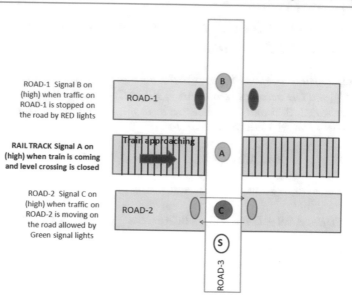

Fig. SE9.5 Figure for SE9.5

Solved example (SE9.5) Figure SE9.5 shows the layout of three roads and a railway track. Signals B and C on the two roads acquire high or 'on' state under the conditions given in the figure. Signal A of the railway track is in state high when the train is approaching and the level crossing is closed. A pedestrian on road-3 wants to go ahead. Discuss the status of signal S for the pedestrian and draw a logic gate network to implement the status of S.

Solution: *Let us assume that the pedestrian on road-3 is allowed to move on when signal S is in state high (or on). Now S should be high when (i) Signal C is LOW (off) (ii) Signal A is LOW (off) (iii) Signal B is HIGH (on). All these three conditions must meet simultaneously. It means that this is the case when the three conditions are logically joined by AND gate; i.e.*

$$S(HIGH) = (condition(i)AND(condition(ii)AND(condition(iii))$$

Or $S(HIGH) = \bar{C}AND\bar{A}ANDB = \bar{C}.\bar{A}.B$
The Boolean expression for S is;

$$S = \bar{C}.\bar{A}.B \qquad\qquad (SE9.5.1)$$

The truth table for S is

A	B	C	$S = \bar{C}.\bar{A}.B$
0	0	0	0
0	0	1	0
0	1	0	1
0	1	1	0
1	0	0	0
1	1	0	0
1	1	1	0
1	0	1	0

It may be observed that S has a high value only when A = 0, B = 1 and C = 0. The implementing logic for S is shown in Fig. SE9.5.1.

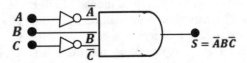

Fig. SE9.5.1 Logic gate for implementation of expression S

9.11.2 Karnaugh Map Technique

Karnaugh map or in short K-map technique is a powerful visual or graphical technique used to simplify logic expressions. The technique was introduced by Maurice Karnaugh in 1953. K-map is essentially a 2-D presentation of the truth table of the given logic expression. Boolean theorems/rules get automatically applied when the K-map technique is used to simplify logic expressions. Hence, there is no need to apply them separately.

Since the detailed theory of K-map technique is beyond the scope of the present discussion, the technique of applying K-map for reducing and simplifying logic expressions/truth tables will be discussed here, almost in a mechanical way, without going into its theoretical background

(a) Drawing of K-map
Drawing of K-map requires putting the values of the output variable in cells of a rectangular or square grid according to a defined pattern. The number of cells in the grid is determined by the number of variables in the given logic expression. If the number of variables is n, then the number of cells in the grid is equal to 2^n; the grid or K-map of an expression with two variables will have 4; with 3 variables 8 and with 5 variables 32 cells.

Each cell in K-map has a **place value or Code** which is represented by a number in binary code. The place values of cells are determined by using GRAY CODE. The special feature of this Gray code is that the place values of **two adjacent cells can not differ by more than 1 in binary notation**. It means that if a particular cell has the place value 01, then the two adjacent cells may have place values either 00 or 11 in any order but it cannot have the value 10. Similarly, a cell with code or place value 11 will have cells with code 01 and 10 on its two sides. Gray code values are assigned to cells both in the X-direction (or horizontally) and in Y-direction (or vertically). The final or ultimate place value or **address** of a cell is the binary word made by putting together its binary codes in y-direction and in X-direction. Suppose the binary code of a particular cell in Y-direction is y and in X-direction x. Then the address of the cell will be 'yx' if, y is the most significant digit. However, the address of the same cell will be 'xy' if x is the most significant digit. Thus, care must be taken while assigning addresses to cells. The binary address of each cell may be converted into the corresponding decimal number, which may be called the 'cell number'. Since binary numbering of cells in two directions is done using Gray code, the corresponding decimal numbering may not be in serial order. Thus each cell of the K-map has a binary address and a corresponding decimal number. Cells in the same row in K-map have the same value of binary code in y-direction and those in the same column, the same value for binary code in x-direction.

K-map is another way of representing the truth table of a logic (Boolean) expression. Let us start by taking a simple example of a two variable logic expression, where the function F is given by the Boolean expression

$$F = A\bar{B} + AB + \bar{A}B \tag{9.17}$$

Since function F has two variables, the K-map grid will have 2^2 = 4-cells. The four cell grid for the two variables case is shown in Fig. 9.16. It may be observed in this figure that variable B varies (takes different values) along the X-axis and the variable A along the Y or vertical axis. It is not necessary to take A along X-axis; it may be taken along the Y-axis, however, care is to be taken regarding the most significant (MSD) and the least significant digits (LSD). In the present case, the MSD is 'A' and LSD is 'B'. It can be shown that the final result is not affected by the choice of axes for the variables provided due care of MSD and LSD is taken. Now B can have only two values 0 and 1 that are shown at the top of the grid. Since these values are plotted along the X-axis, they may be called X-values as indicated in the figure. Similarly, A may have the two values 0 and 1 that are displayed in the vertical direction and represented as y. As shown in the figure, if A is taken as 1, then 0 represents $\bar{A}$. The same is true for B. The first cell of the grid, the top left cell is the one where both A and B have values 0, 0, therefore, the place value or the address for this cell is 00 (yx, in binary notation), which in the decimal system has the value 0. The decimal number for this cell is, therefore, 0 which is written in the lower right corner of the cell. The second cell, the top right cell is the one where A = 0 but B = 1, the code for this cell is 01, which corresponds to decimal number 1. The third cell, bottom left, has A = 1, B = 0, the code for this is 10, that corresponds to decimal number 2, written in the

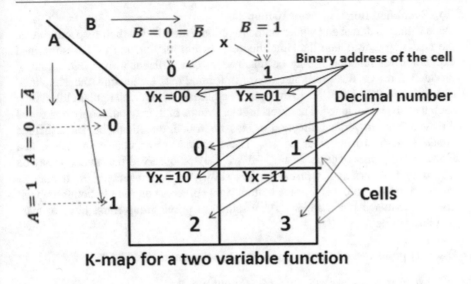

K-map for a two variable function

Fig. 9.16 K-map grid for a two variable logic function

right corner of the cell. The last cell, bottom left has the code 11 with the corresponding decimal number 3. In this case, since there are only two variables, Gray coding has no significant effect on the serial order of cells. However, for functions of more than two variables, as will be seen, Gray code plays a very significant role.

The truth table for function F is given in Fig. 9.7. It is now required to fill the values of function F from the truth table into the cells of the K-map. The first cell, code 00, corresponds to the case when both A and B are 0, in this case, F is also 0, hence 0 of the first row of truth table will go to the cell marked 00 with decimal number 0. When A = 0 and B = 1; that corresponds to cell code 01, F has value 1 (row-2 of truth table). This value of F will go to a cell designated 01. Similarly, F = 1 from row-3 of the truth table will go to cell designated 10 and the 4th value of F = 1 from row-4 will fill the cell coded 11. This is shown in Fig. 9.17.

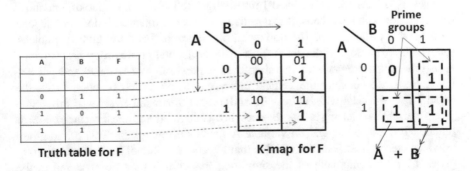

Truth table for F   K-map for F

Fig. 9.17 Truth table for function F and corresponding values in K-map cells

(b) **Simplified function from K-map**

Once K-map is drawn and values of the function F are filled in their respective cells, the next step is to obtain the simplified expression for function F. The simplified expression for function F may be obtained in two different forms: SOP (sum of product) form or POS (product of sum) form. First, let us find the simplified expression in SOP form. To do that, put all adjacent 1's in a rectangular closed loop such that no 0 is included. These rectangular groups of 1, may be in only horizontal direction, only in vertical direction or may be in both directions; but must include either 2, 4, 8, 16…number of 1's. These closed rectangular loops are called **Essential** or **prime groups.** In Fig. 9.17, it may be observed that two 1's may be ringed by two rectangular prime groups, shown by dotted rectangles. It may be observed in the figure that 1 in the bottom right corner is common to both groups. It does not matter if it so happens. An essential or prime group **must have at least one independent 1**.

Essential properties of a prim group may be summarized as

(i)   *It should enclose adjacent 1's without any 0.*
(ii)   *Total number of 1's in a prime group must either be 2, or 4, or 8, or 16 and so on. No prime group may enclose 1, 3, 5, 6, 7, 9, 10… numbers of 1's.*
(iii)   *Few 1's may be common to two or more prime groups but each prime group must have at least one independent 1 that is not common to other prime groups.*
(iv)   *Each prime group contributes an additive term to the simplified expression of the function.*
(v)   *Additive terms obtained from a prime group consist of the multiplication of those variables the value of which does not change in going from one cell to the next cell of the essential group.*
(vi)   *Sum all terms provided by prime groups plus a minimal set of other groups to get the final simplified Boolean expression in SOP form.*

Coming back to Fig. 9.17, the horizontal prime group encloses those 1's for which the value of variable $A = 1$ for both cells, but the value of variable B changes: B is 0 for the left cell and 1 for the right cell. Since the value of function F remains 1 for both these cells, it means that the value of function F does not depend on B for this prime group. The horizontal prime group, therefore, gives variable A (on which the value of F depends). Therefore, horizontal prime group has provided A as one of the elements of the simplified expression. We write A and put a + mark. Next, we consider the vertical prime group, for which function F has the value 1 for two cells with address 01 and 11. So, for member cells of this prime group variable, A changes but B does not change. It means that the value of function F does not depend on variable A for this prime group. This prime group provides the variable B as the variable that affects the value of F. This prime group gives B as the component of the simplified function. It may be observed in the K-map that all 1's in the map have been enclosed by the two essential groups and now no 1 is left without being included in any prime group, hence the terms B and A provided by two prime groups will constitute the simplified expression for F. The simplified expression for the function F is F = A + B. It is easy to verify that the given truth table is satisfied by this simplified expression. The original expression for F given by (9.17) also satisfies the truth table, but it contains three terms and each term has two variables, while the simplified expression obtained from the K-map has only two single variable terms. It is simple to visualize that a single OR gate with inputs A and B will implement the function F

i. Function with three variables

Next, we consider the case of a complex logic expression that has three variables A, B and C. It may be assumed that A represents the MSD and C the LSD. Since K-map is only in 2-dimensions, two variables out of the given three variables (A, B, C) must be put together on one axis in K-map. The MSD variable A is shown in the vertical direction and the product of variables BC in the X-direction in the K-map of Fig. 9.18a. The number of cells in 3-variable case will be $2^3 = 8$. The K-map will have 4-coulms and two rows as shown in this figure. In the horizontal direction, the pair of variables B C may be represented by two digits, the first representing B and the second C. As such, the horizontal binary code for successive cells (as per the guidelines of Gray code) will be **00** (when B = 0, C = 0), **01** (B = 0, C = 1), **11** (B = 1, C = 1), **10** (B = 1, C = 0). In the Y-direction that indicates the variation of variable A, the values that A can have are **0** (when A = 0) and **1** (when A = 1). These cell identifying codes are shown in Fig. 9.18a. The final identification address of each cell is constructed by prefixing the Y-direction code with the horizontal code; as such the final codes of cells are as follows:

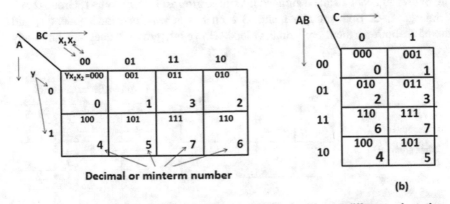

K-map for a logic function of three variables A,B,C in two different orientations

Fig. 9.18 K-map of a logic function of three variables in two different orientations

Row-1: 000, 001, 011, 010,; Row-2: 100, 101, 111, 110

Self-assessment question (SAQ): Why code of y is prefixed to get the final code of each cell?

These codes are written within each cell and the corresponding decimal numbers are also written at the bottom right of each cell. Decimal numbers are written to indicate the fact that ordering of cells does not follow the decimal numbering of cells because of the Gray code.

Figure 9.18b shows the grid structure for the same three variable logic function, but the variables AB are taken along the y-direction or vertical direction and the variable C is taken in horizontal or X-direction. The codes or addresses of cells and corresponding decimal numbers, in this case, are also written within each cell in Fig. 9.18b.

Let us assume that a logic function Z is given in the form $Z = \sum_{A,B,C}(1,3,6,7)$. This is a short form of representing a function in SOP. The subscript ABC to the summation sign indicates that the function has three variables A, B and C, with A being the MSD and C the LSD. The numbers given within bracket are the decimal or **minterm** numbers for which function Z has value 1; i.e. for decimal (or minterm) number 1 which corresponds to binary code 001, decimal number (or minterm) 3 (binary address = 011), minterm 6 (binary address = 110), and decimal number (minterm) 7 (binary address = 111) function Z has value 1 and for the rest other cells it is 0. With the help of this information, it is possible to develop the truth table of the function and also to fill the given function values in different cells of the K-map. The completed K-map and corresponding truth table for function Z is shown in Fig. 9.19.

The next step is to identify prime or essential groups in the K-map. As shown in the figure, adjacent 1 can be grouped into three groups: Prime group-1, Group-2 and Prime group-3. Both groups- 1 and -3 each has at least one independent 1, and therefore both of them are prime groups. However, group-2 does not have any

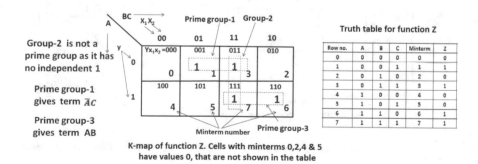

K-map of function Z. Cells with minterms 0,2,4 & 5 have values 0, that are not shown in the table

Fig. 9.19 K-map with function values filled in cells of minterms 1, 3, 6 and 7. The other cells contain 0 that are not shown to avoid confusion. Truth table of the function is also shown in the figure

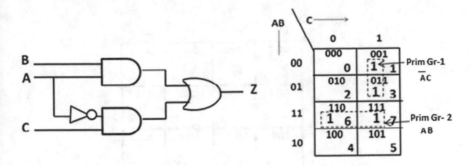

Fig. 9.20 Circuit to implement function Z

independent 1, hence it is not a prime group and need not to be considered any further.

For prime group-1, the variables on which the value of the function depends are: A = 0 and C = 1, that are common to both cells of the group. A = 0 means $\bar{A}$, hence $\bar{A}C$ is the term that prime group-1 provides to the SOP. Further, the value of variable B changes in going from cell 001 to cell 011, therefore, function Z does not depend on B and it will not contribute to SOP.

Similarly, prime group-3 provides term AB, which does not change in going from cell 111 to cell 110. Variable C does not contribute to SOP as the value of function Z for prime group-3 does not depend on it. Thus, from the analysis of K-map, the simplified expression for function Z is given as

$$Z = \bar{A}C + AB$$

The logic circuit that may implement this expression is shown in Fig. 9.20. K-map for the same function in other orientation is also shown in this figure with prime or essential groups and the terms they contribute to the SOP

ii. Function with four variables

A function F (A, B, C, D) with four variables ABCD, A being MSD and D being LSD will have a K-map with 16 cells as shown in Fig. 9.21. Digital address and decimal or minterm number of each cell is also shown in the map.

Figure 9.22 shows the truth table of a 4-variable function F and the corresponding K-map with function values put in respective cells. As may be observed in the K-map, there is only one essential or prime group in the map, digital addresses of the two cells of this group are 1000 & 1001. Digital values of these addresses actually refer to the variables A, B, C and D, respectively. The MSD in both these addresses is 1, which is common to both addresses and refers to variable A. The second and the third digits in the two addresses are 0 and common to both. These two digits refer to B and C. The last digit which is LSD has a different value for the two addresses; 0 &1. It means that the value of the function F does not depend on

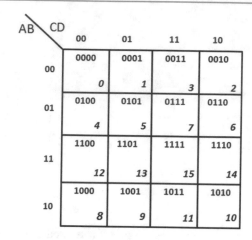

Fig. 9.21 16 cell K-map for a logic function of 4-variables

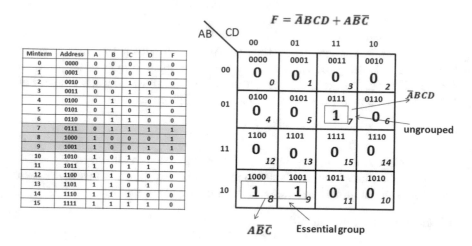

Fig. 9.22 Truth table and corresponding K-map

the value of the last digit which corresponds to variable D. Therefore, this essential group contributes the digital term 100 to the SOP. Now 1 means variable A, next zero (0) means $\bar{B}$ and the third 0 refers to the variable $\bar{C}$. The total term becomes $A\bar{B}\bar{C}$.

Having found the term contributed to the SOP by the essential group, we look to any other 1 left un-grouped in K-map. There is one such ungrouped 1. As a rule, all variables corresponding to the digital address of the cell that contains this ungrouped 1 contribute a multiplicative term to the SOP. The digital address of this cell is: 0111 that corresponds to $\bar{A}BCD$. Hence, the complete simplified function F in SOP form may be written as

$$F = A\bar{B}\bar{C} + \bar{A}BCD$$

iii. K-map with 'don't care'

Some truth tables contain entries denoted by X that means **'don't care'**. The phrase 'don't care' amounts to the fact that the value of the function does not depend on the particular set of variables specified in that row of the truth table. In other words, it may be said that it does not matter if X in truth table is replaced either by 1 or by 0. In such cases, X may be treated as 1 (or 0 as the case may be), while drawing K-map and may be included in making essential groups. This is indicated in Fig. 9.23, where the truth table of a function of four variables having 'don't care' X is shown.

As shown in the figure, now two essential (or prime) groups may be made by including X, which is treated as 1. Addresses of the two cells included in essential group-1 are 0111 and 1111. The last three digits are the same in the two addresses, and therefore, this group gives the term BCD to the SOP. Essential group-2 have 8-cells, it may be verified that except the first digit that is 1 (MSD), all other digits are different. Therefore, essential group-2 gives A as a term to SOP. The total SOP becomes

$$F = A + BCD$$

Minterm	Address	A	B	C	D	F
0	0000	0	0	0	0	0
1	0001	0	0	0	1	0
2	0010	0	0	1	0	0
3	0011	0	0	1	1	0
4	0100	0	1	0	0	0
5	0101	0	1	0	1	0
6	0110	0	1	1	0	0
7	0111	0	1	1	1	1
8	1000	1	0	0	0	1
9	1001	1	0	0	1	1
10	1010	1	0	1	0	X
11	1011	1	0	1	1	X
12	1100	1	1	0	0	X
13	1101	1	1	0	1	X
14	1110	1	1	1	0	X
15	1111	1	1	1	1	X

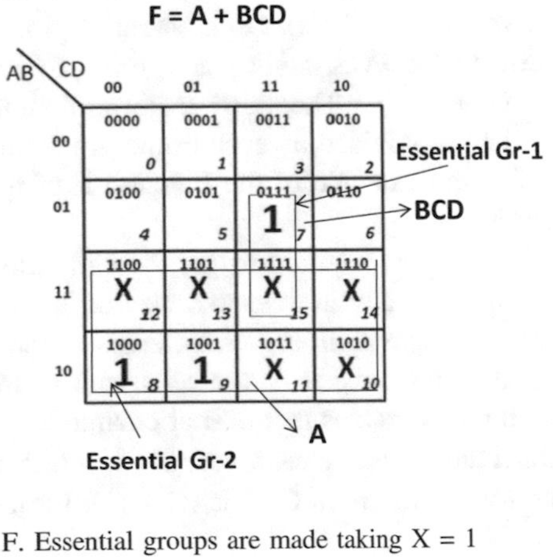

Fig. 9.23 Truth table and K-map of function F. Essential groups are made taking X = 1

Fig. 9.24 1's at the first and
the last cell of a row

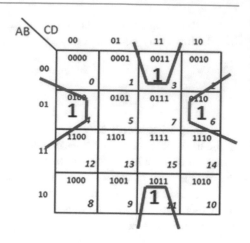

iv. 1's at the start and the end of a row or a column in K-map

Figure 9.24 shows a situation when 1's are in the first and the last cell of a row, as well as of a column in the K-map. These 1's are treated to have formed an essential group, assuming that if the K-map is folded-into form a cylinder in the X-direction and then in the Y-direction, the two 1's will become adjacent to each other and will form essential groups.

v. Output function in POS form using K-map
(a)   Given function in SOP form

Let us first consider the case when the given function is in SOP form. The K-map of the function is drawn in the same way as was done in the previous case of SOP. However, three changes in K-map are done to get simplified function in POS form, (i) the variables on the two sides of the K-map are written in the additive form (not in the multiplicative form as was done for SOP). (ii) Essential groups are made of zeros (not of 1's). Each essential group, in this case, provides a multiplicative term to the POS, while each term provided by the essential group has sums of variables). (iii) The third major difference is that compliments of non-variant variables provided by each prime group are taken. This will become clear from the following example of the function F of four variables that have been worked out earlier.

Since it is required to get the simplified value of the function in POS, 0's are grouped exactly in the same way as 1's were grouped to obtain the result in SOP form. As shown in Fig. 9.25, zeros in the K-map may be grouped into 4-groups.

Group-1 consists of the cells with addresses; 0010, 0110, 1110 and 1010. The common variables that do not change in moving across the cells of the group are: third digit 1 that means variable C and fourth digit 0 that represents $\overline{D}$. Common non-variant term in Gr-1 is $(C + \overline{D})$. In the next step, one **takes the complements**

Minterm	Address	A	B	C	D	F
0	0000	0	0	0	0	0
1	0001	0	0	0	1	0
2	0010	0	0	1	0	0
3	0011	0	0	1	1	0
4	0100	0	1	0	0	0
5	0101	0	1	0	1	0
6	0110	0	1	1	0	0
7	0111	0	1	1	1	1
8	1000	1	0	0	0	1
9	1001	1	0	0	1	1
10	1010	1	0	1	0	0
11	1011	1	0	1	1	0
12	1100	1	1	0	0	0
13	1101	1	1	0	1	0
14	1110	1	1	1	0	0
15	1111	1	1	1	1	0

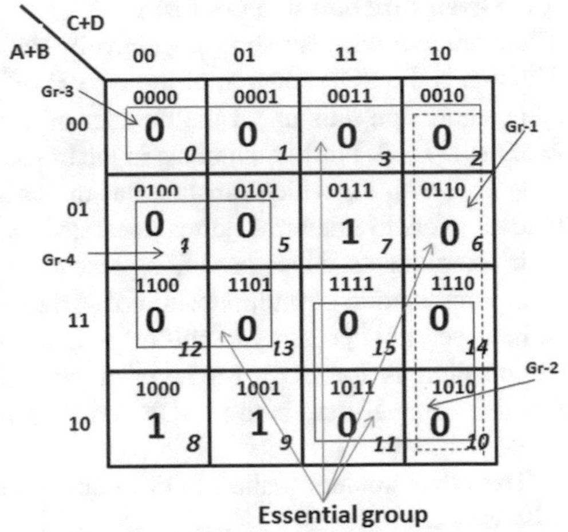

Fig. 9.25 Truth table and K-map of the function F; grouping of zeros to get the result in POS form

of the variables present in the common term to get the term which Gr-1 will contribute to POS. So, in this case, Gr-1 contributes the term $(\bar{C}+D)$

Group-2 also contains 4-cells of addresses, 1111, 1110, 1011 and 1010. The common variables that do not change in this group are; MSD 1 and third digit 1, which, respectively, correspond to A and C. The common term in this group is $(A + C)$ and the term Gr-2 will contribute will be $(\bar{A}+\bar{C})'$.

The next horizontal group-3 has common variables that do not change as 00, corresponding to $\bar{A}, \bar{B}$, hence this group will contribute the term $(A+B)$ to SOP.

The last group-4, in a similar way, contributes the term $(\bar{B}+C)$

Combining these terms, function F in POS form may be written as

$$F = F = (\bar{C}+D)(\bar{A}+\bar{C})(A+B)\,(\bar{B}+C)$$

Note: It is important to remember that in K-map simplification, if the required form is POS, then the following two steps are taken:

1. *Essential groups of 0's are made 2. Common variables that do not change in going from one cell to the other of the group are found and the complements of these common variables are added to get the multiplicative term that the group will give to the POS.*

(b) Given function in POS form

There are two different short notations by which logic functions may be specified. The first is the SOP notation of the type $F = \sum_{ABCD} m(1,5,7)$, where subscript to summation sign tells about the number of variables in the function and that A is MSD & D LSD. Further, numbers in the bracket refer to the decimal numbers of the cells in K-map for which function has the **value 1**. The small letter 'm' before the bracket refers to minterms, terms generated by the multiplication of all variables or their compliments. Since in SOP representation each cell of K-map corresponds to some combination of multiplication of variables, these terms are called minterms. It is, however, not necessary to put 'm' as the sign of summation is enough to tell that the notation is for SOP. Methods of obtaining a simplified expression for the function using K-map both in SOP and POS formats have already been discussed.

The other notation called POS notation may be used to define a function as follows:

$$F = \prod_{ABCD} M(1,4,9,12,14,15) \text{ or simply } F = \prod_{ABCD} (1,4,9,12,14,15)$$

In this notation also, subscript ABCD, etc. tells about the number of variables and that A is MSD while D is LSD. Capital 'M' in this case refers to Maxterm; that are additive combinations of all variables or their compliments. It is not always necessary to write 'M' as the sign of multiplication ($\prod$) immediately tells that the function is given in POS form. The most important thing is that in this case the numbers within the bracket refer to decimal numbers of cells in K-map where the **function has value 0**.

It is interesting to note that

$$F = \prod_{ABCD} M(1,4,9,12,14,15) = \sum_{ABCD} m(0,2,3,5,6,7,8,10,11,13)$$

Fig. 9.26 K-map of function F

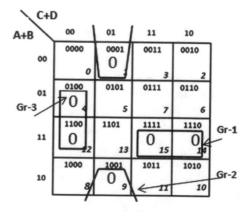

K-map of the given function is shown in Fig. 9.26. As may be seen in the figure, there are three essential groups. It is left as an exercise to verify that Gr-1 gives the POS term $(\bar{A}+\bar{B}+\bar{C})$, Group-2 term $(B+C+\bar{D})$ and Gr-3 $(\bar{B}+C+D)$. The POS form of the function is

$$F = (\bar{A}+\bar{B}+\bar{C})(B+C+\bar{D})(\bar{B}+C+D)$$

Following solved examples will further help in understanding the technique of K-map for simplifying logic functions.

Solved example (SE9.6) Obtain the simplified expression both in POS and SOP forms for the function F given as;

$$F = \prod_{ABC} M(1,3,5,7)$$

Solution: *Since the function is given in POS form, numbers in bracket refer to the decimal number of cells in K-map which contain zeros. K-map for the function is shown in Fig. SE9.6i. As may be observed in this figure, all four 0's may be enclosed in essential group-1. The binary terms for the four cells of the group are 001,011, 111 and 101. The common variable in all these cells is C. This group will contribute the term $\bar{C}$ to the POS. This there is only one term and the function in simplified form may be given as*

$$F = \bar{C}$$

Next, to simplify the given function in SOP form, the K-map is shown in Fig. SE9.6ii. Here all 1's can be grouped into an essential group as shown in the figure. The member cells contained in the group are: 000, 010, 110 and 100. The only common variable that does not change is the third digit 0, that corresponds to $\bar{C}$. Again from the SOP analysis also, the function may be given as

Fig. SE9.6 (i) K-map for POS form (ii) K-map for SOP form

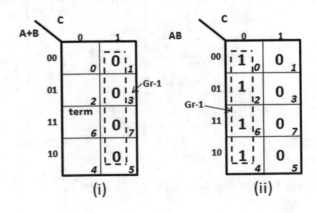

(i)   (ii)

$$F = \bar{C}$$

As expected the results obtained by the two methods of analysis are identical.

Solved example (SE9.7) Obtain simplified expressions both in POS and SOP forms for the function Z;

$$Z = \prod_{ABCD} M(0,1,2,4,5,7,10,15)$$

Solution: *The K-map corresponding to the given function is shown in Fig. SE9.7a. There are in all 8 zeros in the map that may be enclosed in three essential groups. Group-1 has cells of maxterms: 0000' 0001, 0100 and 0101. Common variable that does not change for this group are $\bar{A} + \bar{C}$. The sum of the compliments of these variables $(A + C)$ will make the multiplicative term for POS of the function.*

Group-2 consists of two cells of addresses 0111 and 1111. The common variables that do not vary are $(B + C+D)$; therefore, this group will provide the term $(\bar{B} + \bar{C} + \bar{D})$.

Similarly, Gr-3 has common variables $(\bar{B} + C + \bar{D})$ and the term it will contribute will be $(B + \bar{C} + D)$

The simplified POS form of the function is

$$Z = (A + C)(\bar{B} + \bar{C} + \bar{D})(B + \bar{C} + D)$$

To get the simplified expression in SOP, all 1's in the K-map are grouped into three essential groups as shown in Fig. SE9.7b.

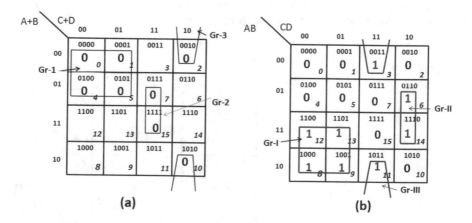

(a)   **(b)**

Fig. SE9.7 K-map for **a** POS form **b** for SOP form

Gr-I in figure (b) has common unchanged variables AC, Gr-II unchanged variables $BC\bar{D}$ and Gr-III unchanged variables $\bar{B}CD$. Simplified function in SOP form is

$$Z = AC + BC\bar{D} + \bar{B}CD$$

Problems

P9.1   Prove that $\bar{A} + (B.C) = \bar{A} + C$

Answer: [may be proved by taking different values for A, B and C]

P9.2   Write DeMorgan theorems and prove them

Answer: $[\overline{A.B} = \bar{A} + \bar{B}; \overline{(A+B)} = \bar{A}.\bar{B}$; may be proved by taking values 1,1; 1,0; 0,1 and 0, 0 for A and B, respectively]

P9.3   Identify SOP statements from the following

(i) $AB\bar{C} + B\bar{A}C + \bar{B}AC$;

(ii) $\bar{\bar{A}}B\bar{C} + BC + \bar{B}AC$;

(iii) $(AB\bar{C} + B\bar{A}C)(\bar{B})(A + B)$

Answer: [(i)]

P9.4   Identify the type of the Boolean expression X, simplify it if possible and draw a logic circuit to implement the original or the simplified statement (Fig. P9.4)

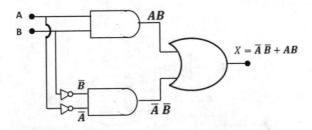

Fig. P9.4 Logic to implement X

$$X = (A + \bar{B})(B + \bar{A})$$

Answer: [Type POS, simplified $X = AB + \bar{A}.\bar{B}$

P9.5   Test the equivalence of logics (a) and (b) in Fig. (P9.5)

Answer: [Equivalence may be proved using DeMorgan theorem]

P9.6   Write Boolean expression for X and draw another logic circuit that may
implement the same logic as given in Fig. (P9.6)

Answer: [$X = \bar{C} + AB$, Equivalent logic may be implemented by a combination of an NOT, one AND and one OR gates as shown in Fig. (P9.7)]

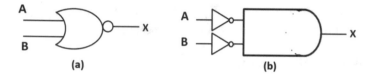

(a)   (b)

Fig. P9.5   Logics gates for Problem (P9.5)

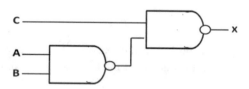

Fig. P9.6   Logic circuit for Problem P9.6

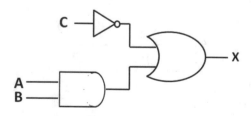

Fig. P9.7   Logic circuit for Problem P9.7

P9.7 Write Boolean expression for the output of the logic circuit given in Fig. (P9.7)

Answer: $[X = \bar{C} + AB]$

P9.8 Digitalized data on the height of wooden poles in the form of four bit binary number PQRS, where P is the most significant digit, is available. Heights vary from 5 ft to 9 ft in steps of 1 ft. It is required to generate a logic circuit that may select poles taller than 5 ft (equivalent to binary number 0101_2) and less than 9 ft for some purpose. Develop the truth table for the process and device the logic that may implement the truth table.

Answer: Hint: Let us assume that logic statement X represents the case when the height of the pole is equal or greater than 5ft (0101_2). The value of X remains LOW (or zero) for all heights up to 5 ft and X becomes High (or 1) when the height is equal or more than 5 ft. but less than 9 ft. The value of X again becomes LOW (zero) for heights equal or larger than 9 ft. The truth table of the selection process may look like the following:

P	Q	R	S	Height in ft	X	Remark
0	0	0	0	0	0	$X = 0$ $X = PQRS$
0	0	0	1	1 ft	0	$X = 0$ $X = PQRS$
0	0	1	0	2 ft	0	$X = 0$ $X = PQRS$
0	0	1	1	3 ft	0	$X = 0$ $X = PQRS$
0	1	0	0	4 ft	0	$X = 0$ $X = PQRS$
0	1	0	1	5 ft	0	$X = High$ $X = P\bar{Q}R\bar{S}$
0	1	1	0	6 ft	0	$X = High$ $X = \bar{P}QR\bar{S}$
0	1	1	1	7 ft	1	$X = High;$ $X = \bar{P}QRS$
1	0	0	0	8 ft	1	$X = High$ $X = P\bar{Q}\bar{R}\bar{S}$
1	0	0	1	9 ft	1	$X = High$ $X = P\bar{Q}\bar{R}S$
1	0	1	0	10	0	$X = 0$ $X = PQRS$

It may be inferred from the truth table that X should be the logical addition of all those terms that gives X = 1, i.e.

$$X = \bar{P}Q\bar{R}S + \bar{P}QR\bar{S} + \bar{P}QRS + P\bar{Q}\bar{R}S + P\bar{Q}\bar{R}S$$

It is now easy to devise a logic circuit for the implementation of the above expression for X.]

P9.9 Convert binary number 101010101010.0010_2 into Hexadecimal number

Answer: [#AAA.2_{16}]

P9.10 Convert binary number 10010100_2 into a decimal number

Answer: [148_{10}]

P9.11 Express hexadecimal number # $3CBA_{16}$ into decimal number and binary number

Answer: [15546_{10}; 1111101001010_2]

P9.12 Convert hexadecimal no. # $1A9_{16}$ into a binary number and decimal number

Answer: [110101001_2; 425_{10}]

P9.13 The memory location or address of a variable is at binary number 0001111101111100. What is the memory location address in hexadecimal notation?

Answer: [$1F7C_{16}$]

P9.14 If memory addresses are 32-bit values in hexadecimal, (i) Number of bytes memory addresses is? (ii) Number of hexadecimal digits in memory addresses is? (iii) Number of words in memory addresses is?

Answer: [(i) 4 (ii) 8 (iii) 1]

P9.15 Draw K-map for the function $F = \prod_{ABCD} M(1, 3, 4, 6, 9, 11, 14)$ and simplify it both in POS and SOP forms.

Answer: [$(B + \bar{D})(\bar{B} + D)$; $(BD + \bar{B}\bar{C}\bar{D} + \bar{B}C\bar{D}) = (BD + \bar{B}\bar{D})$]

P9.16 Express function $F = \sum_{ABC} m(1, 2, 4, 7)$ in SOP and POS forms

Answer: [$(\bar{A}\bar{B}C + \bar{A}B\bar{C} + ABC + A\bar{B}\bar{C})$; $(A + \bar{B} + \bar{C})(\bar{A} + B + C)$]

Multiple Choice Questions

Note: *More than one alternative may be correct in some of the questions. A complete answer in such cases requires picking of all correct options.*

MC9.1 What may be stored in a binary digit?

   (a) Two single digit values
   (b) one of the two values 0 or 1
   (c) one of the values from 0 to 9
   (d) a series of two 0/1 values as [1, 0]

ANS: [(b)]

MC9.2 Binary equivalent of 3F is;

   (a) 11001111
   (b) 1000 0000
   (c) 00111111
   (d) 01100111

ANS: [(c)]

MC9.3 Decimal number 15_{10} corresponds to;

   (a) 17_8
   (b) 1111_2
   (c) Ox F
   (d) # 0011 h

ANS: [(a), (b), (c)]

MC9.4 Alphabets A and Z in ASCII code are respectively,

   (a) 0111 101`0, 0100 0001
   (b) 0100 0001; 0111 1010
   (c) 1111 0000; 0000 1111
   (d) 1010 1010; 0101 0101

ANS: [(b)]

MC9.5 Bit, byte and nibble are respectively,

   (a) Digits of binary system, 4-bit binary number, 8-bit binary number
   (b) Digits of binary system, 8-bit binary number, 4-bit binary number

(c) 4-bit binary number, 8-bit binary number, digits of binary number

(d) 8-bit binary number, 4-bit binary number, digits of binary number

ANS: [(b)]

MC9.6 Boolean expression X = A + (A.B) is equivalent to

(a) A

(b) B

(c) A.A

(d) B $\bar{A}$

ANS: [(a); (c)]

MC9.7 If A and B are Boolean variables then $\overline{A.B}$ i.e. same as

(a) $\overline{\overline{A.B}}$

(b) A.B

(c) $\bar{A}.\bar{B}$

(d) $\bar{A}+\bar{B}$

ANS: [(d)]

MC9.8 For Boolean expression $X = A\bar{A} + A + A + \bar{A} + \bar{A}$, X is given by

(a) 0

(b) 1

(c) 2

(d) 3

ANS: [(b)]

MC9.9 Boolean variable A is given as X = 1+A; the value of X is

(a) 0

(b) 1

(c) 2

(d) A

ANS: [(b)]

MC9.10 Output X of the logic gate given in Fig. MC9.10 is

A ———⟮‾‾‾‾‾⟯o— X
B ———

Fig. MC9.10 Logic gate for MC9.10

(a) $\bar{A}+\bar{B}$

(b) $\bar{A}.\bar{B}$

(c) $\overline{A+B}$

(d) $\overline{A.B}$

ANS: [(d)]

MC9.11 Output X of the logic gate given in Fig. MC9.11 is

(a) $\bar{A}+\bar{B}$

(b) $\bar{A}.\bar{B}$

(c) $\overline{A+B}$

(d) $\overline{A.B}$

ANS: [(c)]

MC9.12 For the logic circuits of Fig. MC9.12i, ii

(a) X = Y
(b) Y = A + (A.B)
(c) X = A.(A + B)
(d) X = Y= $\bar{A}+(A.B)$

ANS: [(a), (b), (c)]

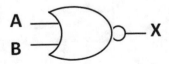

Fig. MC9.11 Logic gate for MC9.11

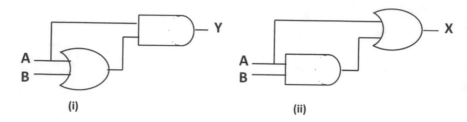

Fig. MC9.12 Logic circuits for MC9.12

MC9.13 X and Y in Fig. MC9.14i, ii will be simultaneously zero if;

    (a) A = 0
    (b) B = 0
    (c) A = B=0
    (d) for any all values of A and B

ANS: [(a), (c)]

MC9.14 Missing terms a and b in the truth table of the logic circuit of Fig. MC9.14 are

    (a) a = 0, b = 0
    (b) a = 1, b = 0
    (c) a = 0, b = 1
    (d) a = 1, b = 1

ANS: [(d)]

MC9.15 Output of a logic gate is HIGH when all its inputs are at logic LOW. The gate is

    (a) NAND or an EX OR gate
    (b) NOR or an EX-NOR gate

Truth table

A	B	Y
0	0	0
1	0	a
1	1	b
0	1	1

Fig. MC9.14 Logic circuit and truth table for MC9.14

    (c) an AND or an EX-OR gate

    (e) an OR an EX NOR gate

ANS: [(b)]

MC9.16 Logic function $Z = \prod_{ABC} M(0,3,7)$ is equivalent to

    (a) $\prod_{BCA} M(0,3,7)$

    (b) $\prod_{CAB} M(7,3,0)$

    (c) $\sum_{ABC} m(0,3,7)$

    (d) $\sum_{ABC} m(1,2,4,5,6)$

ANS: [(d)]

Short Answer Questions

SA9.1	What are hexadecimal numbers and what is their advantage over binary numbers?
SA9.2	What are discrete digital signals? Outline the process of A to D conversion.
SA9.3	Specify logic statements. What is Fizzy logic?
SA9.4	Draw notations for NAND gate and give its truth table
SA9.5	Which logic gates are called universal gates and why?
SA9.6	Discuss the EXOR gate and give its truth table
SA9.7	Explain how a binary number can be converted into a hexadecimal number.
SA9.8	Why the frequency of the time sampling oscillator of A to D converter must be at least twice the maximum frequency in the analog signal?
SA9.9	In what respect digital communication is better than analog communication and why?
SA9.10	Give the truth table and the realization circuit for a three input NOR gate using npn transistors
SA9.11	State and proof DeMorgan theorems of Boolean algebra
SA9.12	What minimum number of two input logic gates may implement the Boolean expression X = AB + CD + B? Draw the implementation logic.
SA9.13	Obtain the Boolean expression for X for logic circuit of Fig. SA9.13
SA9.14	Synthesize a EXNOR gate using AND, NOT and OR gates
SA9.15	List the commonly used alphanumeric codes. How these codes differ from each other?

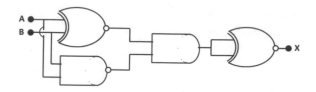

Fig. SA9.13 Logic circuit for Problem SA9.13

SA9.16   How decimal numbers 11_{10}, 111_{10} and 1111_{10} will be written in BCD$_{8421}$ code?

SA9.17   Give steps to draw K-map for a 3-variable logic function

SA9.18   What are the major differences in the K-maps of a function in SOP and POS formats?

Sample Answer to Short Answer Question SA9.2

SA9.2 What are discrete digital signals? Outline the process of A to D conversion.

Discrete time digital signal

Answer: *The term signal is generally applied to something that conveys information. Though signals may be represented in many different ways, in all cases, the information is contained in some pattern of variations of signals. Mathematically, signals are represented as functions of one or more independent variables. In most cases one of the independent variables of signals is time. The variation in the magnitude of the signal may be mapped continuously with time, the information generated in this way is called continuous time analog information and the signals as continuous time analog signals. However, it is also possible to divide the time axis into some convenient periods each of time T and map the signal over these successive discrete periods of time. The patterns of signal variations generated over successive discrete periods of time, each of duration T constitutes the discrete time information based on discrete time signals. It is easy to perform time discretization using an oscillator of frequency f_{osc} and time period $T = \frac{1}{f_{osc}}$.*

Digitalization of signal

The time axis of the signal can be digitalized using an oscillator as mentioned above. The signal magnitude may also be digitalized in the following way: In general, the magnitude of the signal will vary during the time period T, however, the magnitude of the signal may be taken as constant over a given digitalized time period T which is equal to the value of the signal at the midpoint of the period. Further, the midpoint magnitude of the signal is truncated at the nearest integer value. Obviously, these simplifications; replacing the signal variations over the

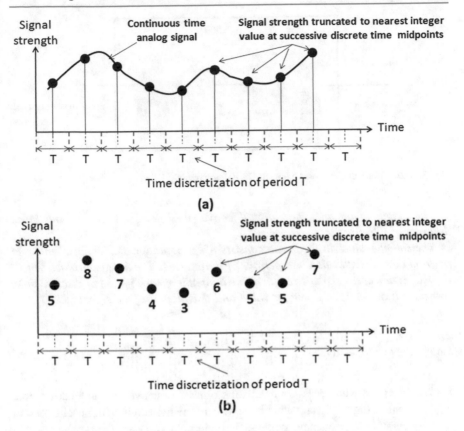

Fig. SA9.2 a,b Converting a continuous time analog signal into discrete time digital signal

discrete time T by a constant integer value brings in approximations which are likely to introduce errors, but these errors will be negligibly small if the time period T (for time digitalization) is small (or the oscillator frequency is large). Thus, discrete time digital signals may be produced by digitalizing time into small periods of time of period T and replacing the variations in the signal magnitude by an integer value corresponding to the magnitude of the signal at the midpoint of each discrete time period T.

Figure SA9.2a shows a continuous time analog signal. Discretization of time axis into successive time segments of period T is also shown in this figure. In order to digitalize the signal strength, the magnitude of signal strength at the midpoint of each period T is determined and is truncated at the nearest digital value, as shown in Fig. SA9.2b. Since signal strength may be measured in some arbitrary units, the midpoint values of signal strengths are nothing but some numbers. As such, time discrete digital signal is a sequence of numbers that remain fixed for each time period T. These numbers may be coded in any desired numeric code like binary, hexadecimal or any other code.

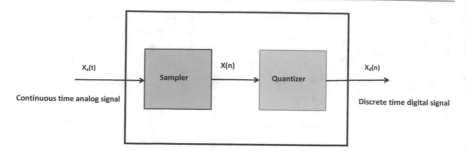

Fig. SA9.2c Block diagram of Analog to Digital converter

The layout of an analog to digital converter system is shown in the block diagram of Fig. SA9.2c.

The sampler block in the figure consists of an oscillator that samples time and perform time discretization into periods of durations T. The quantizer block senses the magnitude of the signal strength at the midpoint of each period of discrete time, truncates it at the nearest integer value and thus digitalize the signal.

Long Answer Question

LA9.1 Explain with necessary details the process of converting an analog signal into a digital signal. What is the role of the time digitalizing oscillator and why high-frequency oscillator is preferred? What are the typical characters of logic pulses used for defining binary states?

LA9.2 Describe AND and OR logic gates, give truth tables for three input OR and AND gates. With details of working draw realization circuit for one of these gates.

LA9.3 Describe the properties of an exclusive NOR gate with three inputs and give its truth table. Draw the logic circuit for an XNOR gate of two inputs using only NAND gates. How a NAND gate may be realized in practice?

LA9.4 A bank locker is shared by two business partners. The locker may be opened only when a bank representative and one representative of each partner simultaneously insert keys. Each business partner has two representatives any one of which may represent that partner. Develop a truth table for the operation of the locker and draw the corresponding logic circuit.

LA9.5 Give a detailed account of alphanumeric codes

LA9.6   List commonly used number systems and write the equivalent term for 21_{10} in all the listed systems explaining the method of writing.

LA9.7   Describe in detail how K-map can be developed for a given function and how the map can be used to simplify the function, both in POS and SOP forms

Some Applications of Logic Gates

10

Abstract

Logic gates are frequently used in digital electronics, particularly in implementing mathematical operations. Though in principle there are two basic mathematical operations; addition and subtraction, however, it may be only addition if numbers are defined as positive and negative. Addition of a negative number with a positive number will then represent subtraction. Addition of numbers is digitally implemented using half and full adder circuits that are combinations of logic gates. Logic circuits may be classified into: Combinatorial Logic circuit, where the output of the circuit depends only on the instantaneous value of the input, and Sequential logic circuits, in which the output not only depends on the instantaneous input value, but also on the previous history of the input. Thus, sequential circuits develop the property of retaining in digital form information of the past. These circuits work as elements of digital memory. Details of defining negative numbers, implementation of addition using adders, sequential circuits and their working along with their application in counters, multiplexers, parity coding and decoding, etc. are discussed in the following.

10.1 Introduction

Logic gates are essential components of digital electronic networks. They are frequently used in automation circuits. For example, a secured electronic lock may be controlled using the circuit given in Fig. 10.1.

As shown in the figure, the electronic lock can be opened if the output of AND gate-1 is high, that occurs when both switches S_1 and S_2 are in 'on' position. The two inputs to AND gate-1 goes to state-1 when S_1 and S_2 are set to 'on' and the output of this gate assumes state-1, which in turn opens the lock. The security of the lock is determined by the sequence in which the switches S_1 and S_2 are put to 'on' position. If switch S_1 is set 'on' first and then switch s_2 is put to 'on' there will be

© The Author(s), under exclusive license to Springer Nature Switzerland AG 2021
R. Prasad, *Analog and Digital Electronic Circuits*, Undergraduate Lecture Notes in Physics, https://doi.org/10.1007/978-3-030-65129-9_10

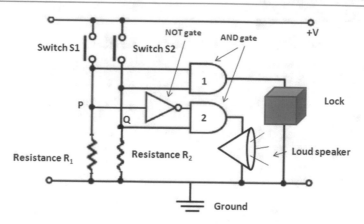

Fig. 10.1   Circuit diagram for a secured electronic lock

no problem and the lock will open without any alarm. However, in the case when switch S_2 is set to 'on' before switch S_1, an alarm sound will start coming from the loudspeaker. When S_2 is set to on before S_1, the potential at point P will be in logic state-0 (Switch S_1 is off), the inverter gate will feed a logic state-1 to one input of the AND gate-2, and since S_2 is closed the other input to AND gate-2 will also be in state-1. Therefore, both inputs of AND gate-2 will get logic state-1 signals that will result in a logic state-1 signal at the output of AND gate-2. A logic state-1 signal at the output of AND gate-2 will set the alarm on. Thus, putting the switch sequence wrong will result in the blowing of the alarm. For simplicity, only two switches have been shown in Fig. 10.1, but in actual practice, more than two switches may be used to make the chance switching of the correct sequence becoming more complicated. Similarly, automation of street lighting, watering of plants and similar other operations may be done using logic gates.

Logic gates are also the building blocks of digital computers. Every computer has a Central Processing Unit (CPU) where all mathematical and logic operations are performed. CPU is essentially made up of different combinations of logic gates. In mathematics, addition is the only operation on which other mathematical operations like subtraction (may be considered as addition of a negative number), multiplication (repeated addition), and division (repeated subtraction) are based. Combinations of logic gates may carry out addition of two binary numbers. Simplest binary number has one digit, like 0 or 1. The logic gate circuit that can add single digit binary numbers is called a HALF ADDER, while a combination of logic gate that may add more than one digit binary numbers is called a FULL ADDER.

10.2   Half Adder

Adders are digital electronic circuits that carry out addition of numbers. There may be different types of adders that may add numbers of many different representations like excess-3 or binary coded decimal. Adders are also used to calculate addresses,

Fig. 10.2 Block diagram of a half adder

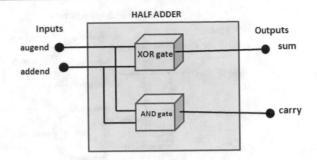

table indices in Arithmetic Logic Units (ALU), increment and decrement operators and similar other operations in different parts of processors.

A half adder circuit has two inputs and two outputs. It may add two input binary digits, called as **augend** and **addend**; and in the output gives two signals called **sum** and **carry**. Half adder consists of one XOR and one AND logic gates. The augend and addend inputs are fed in parallel to the XOR and the AND gates, the **sum** (in short denoted by S) output is provided by the XOR gate while the **carry** (in short denoted by C) appears at the output of the AND gate. The block diagram of a half adder is shown in Fig. 10.2.

The single digit binary numbers are 0 and 1. [It may, however, be remembered that 0_{10} (zero of decimal system) is equal to 0_2, the zero of binary system and similarly 1_{10} is equal to 1_2].These binary numbers may add in four different ways; (i) augend 0_{10}, addend 0_{10} (both of decimal system) which when added gives $0_{10} + 0_{10} = 0_{10} = 0_2$ (b) augend 1_{10} and addend 0_{10} on adding gives $1_{10} + 0_{10} = 1_{10} = 1_2$ (iii) augend 0_{10}, addend 1_{10} on adding gives $0_{10} + 1_{10} = 1_{10} = 1_2$ and (iv) augend 1_{10}, addend 1_{10} on addition gives $1_{10} + 1_{10} = 2_{10} = 10_2$. It may be noted that the output of adding two single digit binary numbers in first three cases is a single digit number either 0 or 1; however, in case-(iv) the output of addition gives a two digit binary number 10_2. Thus, it may be observed that for a complete representation of the sum of two one bit binary numbers, it is required to have two bits. In two bit representation, the above-mentioned four additions will be written as

Carry   SUM

(i) $0 + 0 = 00$

(ii) $1 + 0 = 01$

(iii) $0 + 1 = 01$

(iv) $1 + 1 = 10$

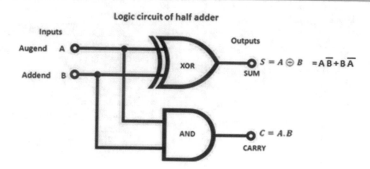

Fig. 10.3 Logic circuit of half adder

Table 10.1 Truth table of a half adder

Input A Augend	Input B Addend	Output Carry (C)	Output Sum (S)
0	0	0	0
1	0	0	1
0	1	0	1
1	1	1	0

In the final result of addition, the digit with **least place** value is called **sum** (**S**) and the digit with next higher place value is called **carry** (**C**). In the above example for the first three additions (i) to (iii), Sum has the value 0, 1, 1, respectively, and carry is 0, while in the (iv) addition sum is 0 and carry is 1. The truth table for a half adder is shown in Table 10.1

The logic circuit of a half adder is shown in Fig. 10.3. It is easy to verify that truth Table 10.1 represents the response of the circuit given in Fig. 10.3.

10.3   Full Adder

A half adder can add two binary numbers each of only one bit. To add bigger binary numbers with two, three and higher bits, full adders are used. A full adder has three inputs, A, B (usual Augend and Addend) and the third input is the carry output of the last half adder stage which is called the carry-in (denoted by C_{IN}). The full adder has two outputs, namely the SUM output and the carry-out output denoted by C_{out}. The block diagram of a full adder is shown in Fig. 10.4.

A single bit full adder may be assembled by cascading two half adders, as shown in Fig. 10.4. The binary numbers A and B, to be added, are fed to the two inputs of the first half adder which provides the sum S_1 and carry C_1 at its output. The sum output S_1 and the carry C_1 from the first half adder are then applied at the inputs of

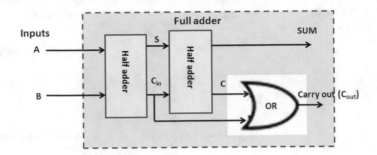

Fig. 10.4 Block diagram of a full adder

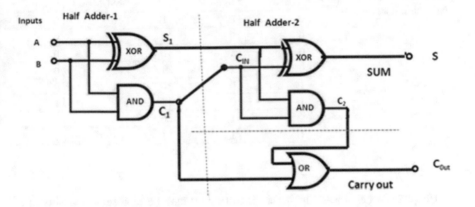

Fig. 10.4 Circuit diagram of a two bit full adder

the second half adder, carry C_1 becomes C_{IN} for the second half adder. The second half-adder delivers the final sum S and carry C_2. Carry C_1 and C_2 are then fed to the input of an OR gate that at its output gives the final carry out (C_{Out}). The truth table for the full adder shown in Fig. 10.4 is given in Table 10.1a.

Table 10.1a Truth table for the full adder of Fig. 10.4

A	B	S_1 $(A \oplus B)$ $(A\bar{B} + B\bar{A})$	$C_1 = C_{IN}$ $(A.B)$	S $(S_1 \oplus C_1)$ $(S_1\bar{C_1} + C_1\bar{S_1})$	C_2 $(S_1.C_1)$	C_{OUT} $(C_1 + C_2)$
0	0	0	0	0	0	0
0	1	1	0	1	0	0
1	0	1	0	1	0	0
1	1	0	1	1	0	1

Fig. 10.5 Schematic
representation of a single bit
full adder

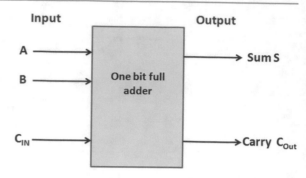

Table 10.2 Truth table for a
single bit full adder

Input A	Input B	C_{IN}	SUM	C_{OUT}
0	0	0	0	0
0	0	1	1	0
0	1	0	1	0
0	1	1	1	0
1	0	0	1	0
1	0	1	0	1
1	1	0	0	1
1	1	1	1	1

However, schematically a single bit full adder may be represented as shown in
Fig. 10.5, treating $C\text{-}_{IN}$ as an independent binary variable. In that case, the truth for
a single bit full adder may be written as shown in Table 10.2.

In computers for a multi bit addition, each bit is represented by a full adder and
is added simultaneously. In order to add two 8-bit binary numbers in a computer,
one needs eight full adders.

The Boolean expressions for the SUM and C_{out} of a full adder may be written as

$$\text{SUM } (S) = A \oplus B \oplus C_{IN} \quad \text{and} \quad C_{out} = (A.B) + [C_{IN}.(A \oplus B)] \qquad (10.1)$$

The layout of an 8-bit full adder is shown in Fig. 10.6, where inputs A and B are
8-bits wide and A (0) and B (0) are the least significant bits.

It is clear that an adder, which may add N-bit binary number must have
N-numbers of full adders cascaded in such a way that the C_{out} of the previous stage
becomes the C_{IN} for the next stage. This type of adders are called **ripple carry
adder**, because each carry bit ripples the next adder. The ripple carry adder is simple
in design but is relatively slow. It is because each adder has to wait for the previous
adder to calculate $C_{IN.}$. In a ripple carry adder that may add two N-bit binary
numbers, the C_{IN} must travel through N-XOR gates in the N-full adders and
N-carry generation blocks before it could affect the carryout (C_{out}).

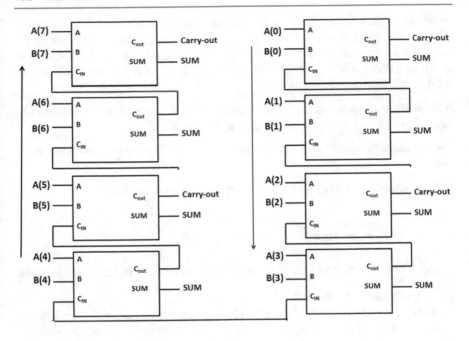

Fig. 10.6 Layout of an 8-bit full adder

In order to reduce computation time, faster adders have been designed. One class of such fast adders is called **carry lookahead** adder, in which two signals P and Q are generated for each bit position, based on whether a carry is propagated through from the less significant bit position (at least one of the inputs is in state-1), generated in that bit position (both inputs in state-1) or killed in that bit position (both inputs 0, 0). In most of the cases, P is the sum output of a half adder and Q is the carry output of the same adder. Once P and Q are generated, carries for each bit position are created.

In a different architecture of fast multi bit adders, the complete adder assembly is divided into blocks to optimize the delay time. In such block-based adders, signals P and Q are generated for each block rather than for each bit, and therefore, such adders are called 'carry skip/carry bypass' adders. In **carry select** adders, another variety of block adders, the sum and carry values are pre generated for the two possible carry inputs, 0 and 1 and the appropriate value is selected using a multiplexing network when the carry bit is known.

In digital electronics, it is often required to compress the data to save the storage space. Data compression mechanism may be of two types: lossy compression and lossless compression. In lossy compression, some part of the data may be lost when the data is uncompressed, while in lossless compression no part of the data is lost on uncompressing of the data. Adders may also be looked as lossy data compression systems. A full adder may be considered to be a **3:2 lossy compressor** since it sums three one bit inputs and returns a two bit number. For example, a binary input of 101 to a full adder results in an output of $1 + 0 + 1 = 10$; the sum representing 0 and carry representing 1. Thus, from three input bits, one gets in the output only two bits; a compression of the data in the ratio 3:2. Similarly, a half adder may be used as a 2 lossy compressor.

Self-assessment question SAQ: How many units of single bit full adders will be needed to add two binary numbers of one byte each?

Self-assessment question SAQ: What is the main disadvantage of ripple carry adder?

Self-assessment question SAQ: Differentiate between a bit and a byte.

10.3.1 Negative Numbers

Negative numbers in binary code are represented essentially in two different ways:

(a) **Sign-Magnitude method**
Add one more bit to the left of the given binary number and use this extra leftmost digit of the number as a special value to represent the sign of the number; 0 = positive, 1 = negative. For example, $+12_{10}$ may be written as $\mathbf{0}1100_2$. The leftmost zero which is written in bold, indicates that the number is positive. On the other hand, if the same number is written as: $\mathbf{1}1100$, then the leftmost 1(in bold) indicates that the number is negative and is equal to -12_{10}. An obvious problem with such representation is that one has to decide beforehand as the limit of the maximum value of the numbers that will be used. For example, if the fifth bit is assigned to the sign, then numbers that can use must not be greater than 4-bit length; the largest number will be $1111_2 = 15_{10}$.

(b) **Showing negative number by 'two's complement'**
In this scheme also an additional bit or digit is added to the left of the MSD which is assigned a value 0 if the number is positive and 1 if the number is negative, but in addition to this, the additional bit in case it is 1(means that it is negative number) is also assigned the place value of $((-2^{(x-1)}))$, where x is the number of the extra digit from the LSD. To understand this method, let us take the example of 4-bit binary number 0011_2

The LSD of this number is the rightmost 1 and MSD is the leftmost 0. Suppose that an extra bit is added to the left most after the MSD. If this extra bit has the value 0, the modified number becomes

$$00011_2 = +0011_2 = \left(0 \times 2^4 + 0 \times 2^3 + 0 \times 2^2 + 1 \times 2^1 + 1 \times 2^0\right)$$
$$= 0 + 0 + 0 + 2 + 1 = +3$$

However, if the extra bit is 1, the number comes

$$100011_2 = \left[-1\left(2^{(5-1)}\right) + 0 \times 2^3 + 0 \times 2^2 + 1 \times 2^1 + 1 \times 2^0 = -16_{10} + 0 + 2_{10} + 1_{10}\right]$$
$$= -13_{10}$$

Similarly,

$$111111_2 = [-1(2^{(5-1)}) + 1 \times 2^3 + 1 \times 2^2 + 1 \times 2^1 + 1 \times 2^0$$
$$= -16_{10} + 8_{10} + 4_{10} + 2_{10} + 1_{10} = -16_{10} + 15_{10} = 1_{10}$$

10.3.2   One's Complement of a Number

All binary numbers are made up of 0 and 1. The binary number obtained by replacing 0 by 1 and 1 by 0 of the original number, is called the 'one's complement' of that number. For example, consider the number 011001_2 the 'one's complement' of this number is: 100110. Similarly, one's complement of 1001100 is 0110011.

10.3.3   Two's Complement of a Number

If 1 is added to the LSD of the 'one's complement' of a number, the new number is called the two's complement;

Two'scomplement = one'scomplement + 1. One's and two's complements of some binary numbers are listed below (Table 10.3).

Table 10.3 One's and two's compliments of binary numbers

Binary number	One's complement	Two's complement
01010	10101	10101 + 1 = 10110
11001101	00110010	00110010 + 1 = 00110011
001	110	110 + 1 = 111
110	001	001 + 1 = 010
00010	11101	11101 + 1 = 11110

10.3.4   Subtraction of Binary Number Using 'Two's Complement'

The operation of subtraction of binary numbers may be converted into addition using the two's complement of the subtrahend. The operation may be carried out in the following steps:

(i)   As the first step, find two's complement of the subtrahend
(ii)   Then add the two's complement of the subtrahend to the minuend
(iii) If the final carryover of the sum is 1, it is dropped and the result is positive
(iv)   If there is no carryover, take two's complement of the sum which will be the result, and the sign of the result will be negative.

Let us explain it by taking two examples

(a) **Find the value of 110110 – 10110 in the decimal system**
(i)   It may be observed that subtrahend (10110) has 5-bits while minuend (110110) has 6 bits. The number of bits of the subtrahend is increased by adding one zero before the MSD so that subtrahend becomes: 010110. Now two's complement of subtrahend $010110 = 101001 + 1 = 101010$
(ii)   Adding two's complement of subtrahend to the minuend

$$110110$$
$$101010$$
$$\overline{1(\text{carry})\ 100000}$$

(iii) Dropping the carry over, the result comes out $= 100000_2 = 2^5 = 32_{10}$
Let us verify if the answer is correct: $\mathbf{110110 = 2^5 + 2^4 + 2^2 + 2^1 = 54_{10}}$

And $\mathbf{10110 = 2^4 + 2^2 + 2^1 = 22_{10}}$

$$110110 - 10110 = 54_{10} - 22_{10} = 32_{10}$$

(b) **Find the value of 10110 – 110110**
(i)   Again it is found that the number of bits in subtrahend and minuend are different, therefore, the number of bits in the two are made equal by increasing the zero on the left. The two numbers after adding a zero in minuend become,

$$010110 - 110110$$

(ii) The two's complement of subtrahend is $001001 + 1 = 001010$

(iii) Adding two's complement of subtrahend to the minuend

$$\begin{array}{r} 010110 \\ 001010 \\ \hline \textbf{(No carry)} \quad 100000 \end{array}$$

It may be noted that there is no carry in the adding, therefore, the addend is a negative quantity. To find the final value, the two's complement of 100000 is to be found. Which comes out to be $= 011111 + 1 = 100000$. The final result is

$$-100000_2 = -2^5_{10} = -32_{10}$$

It is important to realize that using the method of two's complement, the operation of subtraction may be converted into addition. The full and half adders may then be used to subtract binary numbers.

10.4 Sequential Logic Circuits: Latches and Flip-Flops

Logic circuits may be divided into two categories, namely, **Combinational logic circuits** and **Sequential logic** circuits. The main point of difference between the two types of circuits is that the outputs of combinational logic circuits at a given instant depends on the instantaneous value of the inputs; it does not depend on the past history of the output or the input of the circuit. On the other hand, the logic states of the outputs of sequential circuits may depend on three factors: (i) the logic states of instant inputs, (ii) the logic states of the inputs in the last state and/or (iii) the logic states of the outputs in the preceding or last state. In other words, it may be said that sequential logic circuits carry some memory of their past state. Therefore, they are used for storing digital data and are called memory elements. Memory circuits retain the memory of their preceding sequence, hence the name sequential circuits. A typical example of a sequential logic circuit is the 'channel-up/down' switch on the remote control of a television set. At any given instant the 'channel-up/down' switch knows the channel number to which the TV set is presently tuned, the number of the channel one unit less than the present channel and the number of the channel one unit above the present channel. Similarly, the switch designated by 'Back' on the remote controller of the TV keeps the memory of the channel that the TV set was tuned just before switching to the current channel.

This property of the circuit to retain the memory of the last state is achieved by employing in them feedback; that is by applying a part of the output back into the input. It is well known that feedback enhances the stability of the circuit, in turn, making it capable of retaining the memory of the preceding sequence.

Most of the sequential logic circuits are **bistable**, which means that the output or outputs of the circuit may remain stable in two different logic states. The circuit once set in a given stable state will remain **LATCHED** to that state till the time it is shifted to the other stable state by some external excitation/pulse. It is for this reason that the simplest memory devices that may store digital data of one bit size are either called a latch or a fillip-flop. There is, however, a little difference between a latch and a fillip-flop. A latch in general has two inputs and two outputs, one of the inputs is termed as SET input (in short represented by S) and the other input is called RESET (represented by R). The two outputs are termed as Q and $\overline{Q}$. As expected, $\overline{Q}$ is the complement of Q. A latch is set into a particular 'latched state' through the input S when the outputs Q and $\overline{Q}$ obtain particular logic state (either $Q = 1$ and $\overline{Q} = 0$ or $Q = 0\, \&\overline{Q} = 1$) and remain in the latched state for an indefinite time till a change in the logic state of input R resets it to the other stable state. The switching action of the latch from one stable state to the other is followed by the change in the logic state of R. On the other hand, a flip-flop, in general, has three inputs, the SET, the RESET and the CLOCK inputs, and two outputs Q and $\overline{Q}$.The circuit is put in the latched state by a signal at the SET input and remains there till a clock signal shifts it to the other stable state. The Clock is a device that generates rectangular pulses of adjustable frequency, duty cycle and amplitude. In case of the flip-flop, the output changes in accordance with the clock pulse, either at the rising edge or at the falling edge of the clock pulse. This makes the output of a flip-flop synchronous with the clock signal; hence, the flip-flop is called a synchronous system and the latch an asynchronous system. In general, those sequential logic circuits, the states of which get changed by a clock pulse, are termed as synchronous sequential circuits and the others where the clock pulse is not required to change the states are termed as asynchronous.

Latches and flip-flops can be made from cross coupled inverting elements like vacuum tubes, bipolar junction transistors, field effect transistors, inverters or logic gates. In digital circuits used in computers and other digital devices, latches and flip-flops are generally implemented using logic gates.

10.4.1 S-R Latch

A latch may be assembled using either two NOR or two NAND gates.

First, we discuss the operation of a letch implemented by two NOR gates as shown in Fig. 10.7a. As shown in this figure, one output of each NOR is connected back to one of the inputs of the other NOR. The cross connections of the outputs to the inputs of the other NOR provides the feedback to the system. The SET and RESET inputs of the latch are shown by S and R, respectively, while the outputs are

Fig. 10.7a SR latch using NOR gates

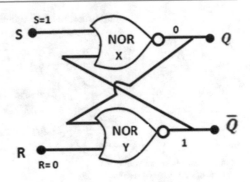

represented by Q and $\overline{Q}$ in this figure. One condition for the proper operation of the latch is that at **a given instant only one input of the latch can be in the logic state-1 (high)**. If both inputs are in logic state-1 (or High), the system goes to an undefined state. However, it is possible that both the inputs are simultaneously in the logic state-0 or Low.

In order to understand the operation of the NOR gate based letch, let us recall the properties of the NOR gate. It is known that the Boolean expression for the output Z of a NOR gate is given by $Z = \overline{(A+B)}$, where 'A' and 'B' are inputs to the gate. It is, therefore, clear that in case **one of the input to the NOR gate is in logic state-1, then the output of the gate will be in logic state-0, no matter what is the logic state of the other input. Also, if both inputs are in logic state-0, then the output of the NOR gate will be in logic state-1**. Now, in Fig. 10.7a, input S to the NOR gate X is in state-1, and therefore, irrespective of the logic state of the other input R, the output Q of this gate will be in logic state-0. The output of NOR gate X is fed to one of the input of the second NOR gate Y whose other input R is in logic state-0. Thus, both the inputs to NOR gate Y are in logic state-0, and therefore, the output $\overline{Q}$ of the NOR gate Y will be in logic state-1, as shown in the figure.

In the next step, let the logic state of the input S be changed from 0 to 1, without changing the logic state of the other input R. Let us now determine the logic states of the outputs Q and $\overline{Q}$. It may be noted that just before changing the state of S, the NOR gate Y, through its output, was feeding state-1 to the other input of NOR gate X, and when one input of a NOR gate has the logic 1, the output of the gate is 0, no matter what is the state of the other input. As such the output of NOR gate X remains in logic state-0 even after the logic state of input S is changed from 0 to 1. Obviously, the logic state of the other output $\overline{Q}$ also remains unaltered to the value 1 (High). This is shown in the upper half of Fig. 10.7b when the S-R latch moves from state marked by (I) to state marked by (II).

From the foregoing description, it may be observed that if the input S of the S-R latch which is in logic state-1 (High) is switched to the other logic state without changing the logic state of the input R which is in the state-0 (Low); the logic states of the two outputs do not change. This essentially means that in such switching of the S-R latch; the output of the latch remembers the logic

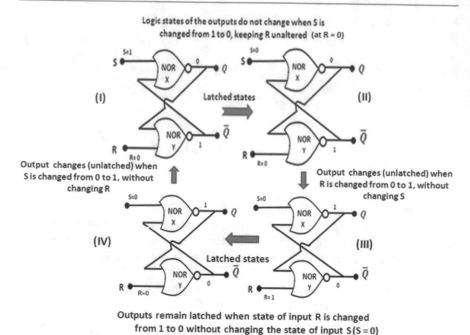

Fig. 10.7b Figure shows the two stable states of the S-R latch and how it retains the memory of its last state

states of the immediately preceding state or the output states before. The outputs of the circuit remain latched to the outputs of the preceding state. This is the reason why the circuit is called a latch.

So long as the S-R latch input that was in logic state-0 during switching (input R) is not disturbed, the logic states of its output Q and $\overline{Q}$ will remain latched to the values that they had in the immediately preceding state. In order to break this latching, it is required to change the logic state of the Reset input R (that remained in state 0 during the previous switching) without changing the logic state of the other input S. If the reset input R is switched from logic state 0 to logic state-1, the output $\overline{Q}$ of the NOR gate Y changes from 1 to 0, which in turn, changes the output Q from 0 to 1. It is to be noted that a change in the state of the reset input R unlatches the outputs. This unlatching is shown in Fig. 10.7b as the system moves from stage (II) to stage (III). In stage (III) the logic states of the S-R latch are: $S = 0, R = 1, Q = 1$ and $\overline{Q} = 0$. The circuit output will again get latched if the logic state of the input that has the value 1 (Reset input R) is switched to state 0 without changing the logic state of input S (which has logic state 0). It may be verified that when inputs S = R=0; the Q and $\overline{Q}$ outputs retain their previous values 1 and 0, respectively, as indicated by switching from stage (III) to stage (IV) in Fig. 10.7b.

The latching of output states may be removed by creating a new state of the S-R latch by switching the logic state of S from 0 to 1, without changing the logic state of R = 0. This change is shown in Fig. 10.7b as the system moves from stage (IV) to stage (I).

It may be observed that the outputs Q and $\overline{Q}$ may stay stable or steady (unaltered) in two different sets of logic states: (0, 1) or (1, 0). Therefore, the circuit is said to be bistable. The state of the latch when S = 1, R = 0, Q = 1 and $\overline{Q} = 0$ is called the **SET** state and the other stable state with S = 0, R = 1, Q = 0 and $\overline{Q} = 1$ is called the **RESET** state.

It has already been mentioned that if both the inputs of the latch are in logic state-1 then the latch does not work properly. It is because the logic states of the Q and $\overline{Q}$ outputs do not remain well defined. A slight difference between the characteristics of the two NOR gates may change the value of the output logic. It is important to note that switching time or the time that a gate takes in switching from one logic state to another logic state, plays an important role in the operation of latches. Since a latch contains two gates, both of which undergo switching from one to another state whenever the letch shifts from one steady state to another, therefore, switching times of both gates get involved in the operation. Although one may use identical gates, there is always some difference in the switching times of two identical gates. As a result, the change at the output of one gate takes place earlier than the other. This gives rise to what is called the 'racing' between the outputs of the two gates. Racing makes the logic of output states unsteady. It is for this reason that logic state-1 (high) simultaneously for both inputs is avoided. Sometimes the term 'metastable state' is also used to identify the unsteady state when S = R=1.

The truth table of an S-R latch implemented using NOR gates is given below in Table 10.4a

The operation of an S-R latch implemented by two NOR gates may be summarized using the Performance Table 10.4b and the Excitation Table10.4c given below. The sign X in these tables means that the value of the particular element does not matter, that is to say that the logic state of that particular element does not affect the result. Often, X is referred to as **'Don't care' condition**.

Performance of any logic circuit or gate may be summarized in tabular form in three different ways: (i) truth table (b) excitation table and (iii) performance table.

A truth table lists all possible but different combinations of circuit input variables and the corresponding outputs of the circuit. Listing of combinations of variables does not have any relation with the sequence of their occurrence in the actual operation of the circuit.

Table 10.4a Truth table of S-R latch

Set input S	Reset input R	Output Q	Output $\overline{Q}$	Comment
1	0	0	1	SET state
0	0	0	1	Latched
0	1	1	0	RESET state
0	0	1	0	Latched
1	1	Undefined	Undefined	Invalid state

Table 10.4b Performance table of an SR latch implemented using NOR gates

R	S	Q_{next}	Action
0	0	Q	Hold or Latch state
0	1	0	SET
1	0	1	RESET
1	1	X	Metastable

Table 10.4c Excitation table of an SR latch implemented using NOR gates

Q	Q_{NEXT}	R	S
0	0	0	x
0	1	1	0
1	0	0	1
1	1	X	0

The excitation table shows the minimum inputs that are necessary to generate a particular next state, in other words, to excite it to the next state, when the current state is known. In an excitation table, the current state and the next state are next to each other on the left hand side of the table and the inputs needed to make the state change happen are shown on the right side of the table.

The performance table tells about the states of the input variables for which the logic circuit attains stable states or go into an undefined state. Obviously, excitation and performance tables are more relevant for sequential logic circuits where the circuit may have stable states and there is a definite sequence of operation.

The first row of the performance table for S-R latch says that when both inputs S and R of the latch are in logic state-0, then the logic state of output Q is held to the present value if the logic state of S or R is changed, one at a time, keeping the other unchanged or constant. The above statement may be verified by referring to Fig. 10.7b, wherein the upper half of the figure, the logic state 0 of the output Q, is kept on hold when S goes from 0 to 1 and in the lower half the logic state of Q = 1 is kept at hold when R goes from 0 to 1. In the last row of this table, X in column-3 means that in the case S = R=1, the latch goes to the metastable state and the output of the latch has no definite relation with the logic of the last stage.

The first row of the excitation table states that if the output Q is in logic state-0 and the logic state of R is also 0 (Low), then the logic state of the output Q will not depend on the value (0 or 1) of the logic state of the input S. Thus, the corresponding rows of the performance and excitation tables states the same fact but in different ways.

The time sequence of S, R, Q and $\overline{Q}$ logic states are shown in Fig. 10.7c for the three valid states (S = 1, R = 0, Q = 0, $\overline{Q}$ = 1); (S = 1, R = 0, Q = 0, $\overline{Q}$ = 1); (S = 0, R = 0, Q = 0, $\overline{Q}$ = 1) and (S = 0, R = 1, Q = 1, $\overline{Q}$ = 0). The shaded part of the figure shows that the output of the latch remains unchanged when the state with S = 0, R = 0 is changed to S = 1, R = 0; the memory property of the latch.

The logic symbol of S-R latch synthesized using NOR-gate is shown in Fig. 10.7d(I).

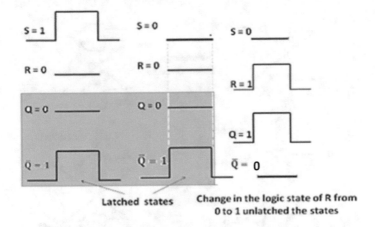

Latched states   Change in the logic state of R from
                      0 to 1 unlatched the states

Fig. 10.7c Time sequence of logic states of inputs S and R with outputs Q and $\overline{Q}$

Fig. 10.7d Logic symbol for
(I) NOR based S-R Latch,
(II) NAND based S-R Latch

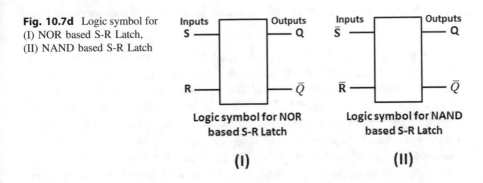

    Logic symbol for NOR   Logic symbol for NAND
      based S-R Latch   based S-R Latch

          (I)   **(II)**

Next, we consider the S-R Latch implemented using the universal NAND gates. The circuit connections are similar to the latch made by the two NOR gates, the output of each NAND connected to one of the inputs of the other, providing the feedback. The logic Symbol of a NAND gate latch is shown in Fig. 10.7d(ii). A visible point of difference between the NOR and NAND gate latches is that in the NAND gate latch the two inputs are $\overline{S}$ and $\overline{R}$ (inverse or complement) instead of S and R. This is because of the convention where it is assumed that the NAND gate devices operate with logic pulses of negative polarity, say, of -5 V, so that logic 0 corresponds to the logic state High (or True) and -5 V as logic state Low (or False).

In order to understand the operation of the S-R latch made using two NAND gates, it will be required to revise the properties of a NAND gate. The Boolean expression for the output X of a NAND gate is: $X = \overline{(A.B)}$, where A and B are the two inputs to the gate. The output will be 0 only when both inputs to the gate are in logic state-1, and will be 1 when either one or both inputs are 0. Figure 10.7e(i) shows an S-R latch made up of two NAND gates. The output of each NAND is coupled to one input of the other NAND to provide paths for feedback.

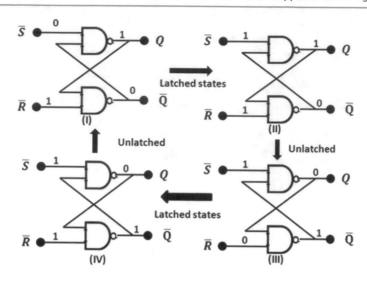

Fig. 10.7e Figure shows the sets of output logic states for different combinations of input logic states

As shown in Fig. 10.7e(i), when the inputs $\overline{S}$ and $\overline{R}$ are set, respectively, to logic state-0 and 1, the outputs Q and $\overline{Q}$ assume the logic states Q = 1 and $\overline{Q}$ = 0. These results can be easily arrived from the above-mentioned properties of NAND gates (if one input is 0, the output is 1 and if both inputs are 1, then output is 0). Figure 10.7e(II) shows that when the logic state of the input $\overline{S}$ is changed from 0 to 1 without changing the logic state of the reset input $\overline{R}$ the outputs Q and $\overline{Q}$ remains unaltered or latched to the previous settings. Thus, the latch outputs keep the memory of the immediately preceding state if the reset input $\overline{R}$ is not disturbed. However, on changing the logic state of the reset input R from 1 to 0 the logic states of the outputs also get changed (see Fig. 10.7e(III). It may be noted in Fig. 10.7e (iii) that input $\overline{S}$ is in logic state-1 and the input $\overline{R}$ is in logic state-0, now if the logic state of $\overline{R}$ is changed to 1 without disturbing the logic state of $\overline{S}$, (see Fig. 10.7e (IV)) the outputs Q and $\overline{Q}$ remain the same as they were in the immediately preceding state. Figure 10.7e essentially shows that outputs Q and $\overline{Q}$ retain the memory of their previous logic states when the S-R latch undergo transitions depicted in Fig. 10.7e from part (I) to (II) and part (III) to (IV).

Performance table for an $\overline{SR}$ latch is given below as Table 10.5.

Table 10.5 Performance table for an $\overline{SR}$ **latch**

Input $\overline{S}$	Input $\overline{R}$	Action
0	0	Metastable state
0	1	$Q = 1, \overline{Q} = 0$
1	0	$Q = 0, \overline{Q} = 1$
1	1	Hold Q

It may be noted in Table 10.5 that the undefined metastable state occurs for logic inputs 0, 0; that appears opposite to the condition for a normal S-R latch. However, in negative logic 0 stands for the state (High), and therefore, this is not different from the condition of metastable state in normal S-R latch. Time relationships between the logic states of inputs $\overline{S}, \overline{R}$, and outputs Q and $\overline{Q}$ are shown in Fig. 10.7f. As shown in this figure from time T_1 to T_2, the inputs $\overline{S}$ and $\overline{R}$ are, respectively, in logic state-0 and 1, while the output Q is in logic state-1 and $\overline{Q}$ in logic state 0. At time T_2, the input $\overline{S}$ is set to state-1 without changing the state of the other input $\overline{R}$. This change in the state of input $\overline{S}$ does not produce any change in the logic states of the two outputs as they are latched to the outputs of the immediately preceding configuration of the latch. At time T_3, the reset input $\overline{R}$ is set to logic state 0 keeping the logic state of the input $\overline{S}$ unchanged as 1. This changes the states of Q and $\overline{Q}$ from 1 to 0 and 0 to 1, respectively. Finally, if at time T_4 the state of input $\overline{R}$ is again switched to 1 (from 0), it may be observed that the logic states of the two outputs will not change, since they are latched to the outputs of the last configuration. The two stable states of the S-R latch are shown in the figure from T_1 to T_3 and T_3 to T_5. An important feature of Fig. 10.7f) is that at no time both the inputs are simultaneously in states 0, 0, though they are simultaneously in states 1, 1 during the time T_4 to T_5. This follows from the condition of the proper functioning of the latch. If the latch is operated with pulses of positive logic (as in the case of latch formed from two NOR gates), the two inputs can never be in state-1 simultaneously. However, if the letch operates with pulses of negative logic then in that case the two inputs cannot assume logic states 0, 0 simultaneously (in negative logic the True value is represented by 0 and the false value by −5 V). It is obvious that the NAND gate S-R latch may be run by positive logic if each input is fed through an inverter gate as shown in Fig. 10.7g).

Self-assessment question SAQ: Draw the *truth table for the NAND based* $\overline{SR}$ latch.

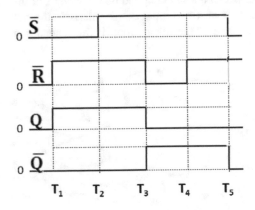

Fig. 10.7f Time relationship between the logic states of inputs $\overline{S}, \overline{R}$ and the outputs Q and $\overline{Q}$

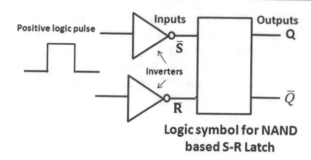

Fig. 10.7g NAND gate latch that may be operated with positive logic with inverters at each input

10.4.2   Gated Latch or Latch with Enable

The output of a latch depends both on its instant inputs, as well as the output of the previous stage. However, sometimes it is required that the latch is dictated when to latch the output and when not to be operative. This may be done by incorporating a pair of AND gates with a common Enable input as shown in Fig. 10.8a.

As shown in Fig. 10.8a, the S and R inputs to the NOR gate implemented latch are fed through AND gates X and Y. The other input of the AND gates are joined together to provide the enable input E. Input E may be kept at logic state 0 or 1 using the positive logic pulse.

Before discussing the operation of the circuit, let us briefly revisit the properties of an AND gate. If one input of a multi-inputs AND gate is in logic state 0 (low), the output of the gate is always 0, irrespective of the logic states of the other inputs. An AND gate with one input in state 0 is said to be **disabled** because the output of the gate does not depend on the logic states of its other inputs. If there is an AND gate with two inputs and one input is in logic state 0 and the other in logic state-1, the output will be 0, which does not depend on the value 1 of the other input. On the other hand, if one input of a two input AND gate is 1, the output of the gate will depend on the logic state of the other input, if the other input is also 1, then the

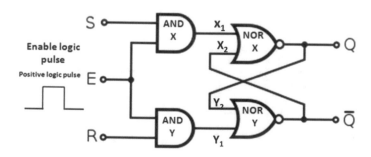

Fig. 10.8a Circuit diagram of a gated S-R latch implemented using NOR gates

Fig. 10.8b 'Disabled' and 'Enabled' states of an N-input AND gate

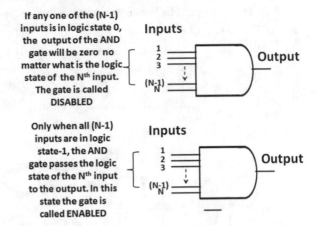

If any one of the (N-1) inputs is in logic state 0, the output of the AND gate will be zero no matter what is the logic state of the Nth input. The gate is called **DISABLED**

Only when all (N-1) inputs are in logic state-1, the AND gate passes the logic state of the Nth input to the output. In this state the gate is called **ENABLED**

output will be 1 and if the other input is 0, the output will be 0. In such a situation when the output of the gate shows the value given to the other input, the gate is said to be **enabled**. An enabled AND gate passes to its output the logic state of the other input, i.e. the gate becomes transparent to the other input. In short, a disabled gate stops the logic signal at its other input to pass to the output, while an enabled gate passes the signal at its other input to the output. It is simple to understand that an AND gate with multiple inputs may be disabled by setting only one of its input in state 0, while an AND gate if N-inputs will need that $(N-1)$ inputs be set at logic 1 so that it becomes enabled for the signal at its remaining input. The disabled and enabled states of an AND gate are shown in Fig. 10.8b.

Now we discuss the operation of the S-R latch circuit shown in Fig. 10.8a. Let us start by assuming that there is no pulse at the input E, which means that E = 0. Logic 0 at E disables both the AND gates X and Y, as a result, the signals at the other inputs of AND gates are blocked and cannot pass to the gate outputs. In this situation (when E = 0), the outputs Q and $\overline{Q}$ of the latch will be in some defined logic states, which may be (i) Q = 0, $\overline{Q}$ = 1 or (ii) Q = 1, $\overline{Q}$ = 0. When E = 0, no matter whatever be the logic states of S and R, the outputs Q and $\overline{Q}$ will remain latched to the state in which they are, i.e. either in state (i) or (ii). Remember that both Q and $\overline{Q}$ cannot be in the same logic state as they are complements of each other. Part of the truth table for a gated latch when Enable is in logic state-0 or Low is shown in Table 10.5a.

(I) **Next, let us consider the case, when E = 1, S = 1, R = 0 and (a) Q = 1,**
   $\overline{Q} = 0$

The logic states at different inputs and outputs of gates for this case are shown in Fig. 10.8c. Since the Enable input E is in logic state-1, the output X_1 of AND gate X assumes logic state-1. As such the two inputs of NOR gate X have logic state-1 & 0, output Q = $\overline{1+0}$ = 0. Therefore, the output Q switches from state-1 to state-0.

Table 10.5a Part of gated latch truth table when Enable is in logic state 0 (Low)

E	S	R	Q	$\overline{Q}$
0	0	0	Latched to whatever state it is	Latched to whatever state it is
0	1	0	-do-	-do-
0	0	1	-do-	-do-
0	1	1	-do-	-do

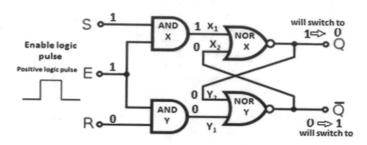

Fig. 10.8c Description of logic states of the inputs and outputs of case (I a) The case when E = 1, S = 1, R = 0 and (b) Q = 0, $\overline{Q}$ = 1 is shown in Fig. 10.8d

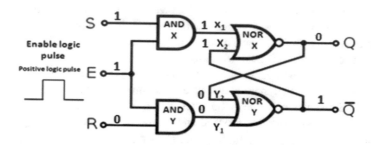

Fig. 10.8d Description of logic states of the inputs and outputs of case (I b)

With this switching, the two inputs y_1 and y_2 of NOR gate Y go to states 0 and 0. This switches the complementary output $\overline{Q}$ from state-0 to state-1.

As may be seen in this figure, X_1 = 1 & X_2 = 1 gives $Q = \overline{1+1}$ = 0. And since $\overline{Q}$ is a compliment of Q, it must be in logic state-1. It may be observed that if Q and $\overline{Q}$ are already in states 0 and 1, respectively, then they remain latched to these states on switching the Enable signal. In other words, one may say that [**Q= 0 and $\overline{Q}$= 1] are the steady output states when E = 1, S = 1 and R = 0.**

Logic states of X_1, X_2, Y_1 and Y_2 for the **case S = 0, R = 1, E = 1 and Q = 0,** $\overline{Q}$ = 1 are shown in Fig. 10.8e.

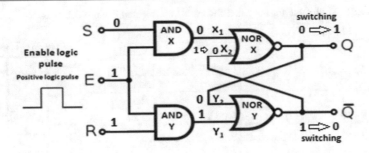

Fig. 10.8e Description of logic states for the case $S = 0$, $R = 1$, $E = 1$, $Q = 0$, $\overline{Q} = 1$

$E = 1$ enables both AND gates as a result Y_1 goes to state-1 while Y_2 is still in state-0; the output of NOR gate Y, therefore, becomes 0. It may be observed that logic states-1 for both E and R switches the logic state of $\overline{Q}$ from 0 to 1. This switching of the logic state of $\overline{Q}$, in turn, switches the logic state of X_2 from 1 to 0. Consequently, Q switches to state-1 from state-0. Obviously, the steady state for this configuration (when $S = 0$, $R = 1$, $E = 1$) is reached when Q goes to state-1 and $\overline{Q}$ to state-0.

It is left to the reader to draw a circuit diagram showing logic states of different inputs and outputs for the cases (II a): $E = 1$, $S = 0$, $R = 1$ and (b) $Q = 1$, $\overline{Q} = 0$; It will be easy to conclude from the circuit diagram that for the case $E = 1$, $S = 0$, $R = 1$ the steady state of outputs is [$Q = 1$ and $\overline{Q} = 0$]. The part of the truth table for the gated latch when Enable is 'on' ($E = 1$), is given in Table 10.5b. Tables 10.5a and 10.5b put together give the complete truth table for gated or enabled latch which is shown in Table 10.6a. The symbol of the gated latch is also given along with the table.

10.4.3 D (Data)-Latch or Transparent Latch

A look to the last row of the truth Table 10.6a of the gated S-R latch tells that the latch remains operative even in the state when it is undefined or in the metastable

Table 10.5b Part of gated latch truth table when Enable is in logic state 1 (High)

E	S	R	Q	$\overline{Q}$
1	0	0	Latched to whatever state it is	Latched to whatever state it is
1	1	0	0	1
1	0	1	1	0
1	1	1	Metastable state	Metastable state

Table 10.6a Truth table for the gated SR latch implemented using NOR gates

Input E	Input S	Input R	Output Q	Output $\overline{Q}$
0	0	0	Latched	Latched
0	1	0	Latched	Latched
0	0	1	Latched	Latched
0	1	1	Latched	Latched
1	0	0	Latched	Latched
1	1	0	0	1
1	0	1	1	0
1	1	1	Undefined	Undefined

Symbol for a gated SR latch

Table 10.6b Truth table for a D-latch implemented using NOR gates

Input E	Input D	Output Q	Output $\overline{Q}$
0	0	Latched to preceding state	
0	1		
1	0	1	0
1	1	0	1

Fig. 10.9 Circuit- diagram of a D-latch implemented using NOR gates

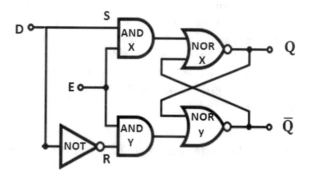

state with both S and R inputs in state-1. To avoid this unwanted condition in D-latch, two inputs are generated from a single input, one taking directly and the other by taking its inverted or complement signal. Figure 10.9 shows the circuit diagram of a D-latch or a transparent S-R latch that is implemented using NOR gates.

The truth table of a D-latch is given below in Table 10.6b. D-latches that are used in input/output of computers are available in the market as packages of many latches in a single IC, for example, IC 74HC75 contains four D-latches in a single packing.

It is interesting to study the time relation between the input D, Gate pulse E, outputs Q and $\overline{Q}$. It may be mentioned that signals at inputs D, E and at outputs Q

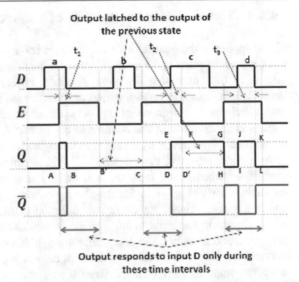

Time relation between the inputs D , E and the outputs Q and $\bar{Q}$

and $\bar{Q}$ are all digital signals of positive logic. As shown in Fig. 10.10 by a, b, c and d, the input signal at D stays in logic state-1 for different time durations and the enable pulses at input E have constant width and repetition rate. The circuit will respond to the input D-pulses only when E-input is in state-1, the corresponding time intervals are marked in the figure. During intervals of time when enabling input E is in logic state-0 (No pulse present at E), the pulses at the D-input will not affect the Q-output and the Q-output will remain latched to the value it had in the previous state when the enable pulse was present (E = 1) last time. Further, in case when input D is in logic state-1 and the E-input is also in state-1, then D = S = 1 and R = 0; and hence the latch output Q will be 1(High) and similarly, when E = 1, D = S = 0, the latch output Q will be in logic state-0. In other words, it may be said that the output Q follows the input D when Enable pulse is in state-1 (High).

Coming back to Fig. 10.10, it may be observed that during time t_1 both the D and the E pulses are simultaneously in state-1, and therefore, Q will follow D, i.e. during the time interval t_1, output Q will remain in state-1 (high), as shown in the figure (A to B). At the time indicated by point B, the D pulse attains logic state-0 and so also the output Q. During the time interval from B to B', E = 1 and D = 0. The output Q follows D and remains in state-0. The input E switches to 0 at time B' and the input D loses its control over Q. However, during the time interval from B' to C (when E = 0), output Q remains latched to its previous state-0. At C, the enable input E switches to 1 and Q starts responding to D till the point D' where E becomes 0. At D' (when E switched to 0), the Q was left in state-1 and it remains latched to this state up to G where E again switches to 1. From G up to K, the output Q follows D; becomes 0 between H and I and 1 between J and K. The other output $\bar{Q}$ is inverse or complement of Q as shown in the figure.

10.4.4   Signal Transmission Time of Logic Gate and Glitch

Signal applied at the input of a logic gate takes some time to transit through the gate and produce an effect at the output. Normally the transit time or the logic gate delay time is of the order of a few neno seconds, however, in case where the logic network contains many gates the total or overall transit time may become large and may affect the performance of the network adversely. Glitch produced in logic outputs is an example of the adverse effect of this delay time. To understand the phenomena of glitch, let us consider the simple circuit shown in Fig. 10.10a(i)

In a normal course, the output X of this circuit will have logic state-1, irrespective of whether the input A is 0 or 1. Let us assume that each logic gate responds to the change in its input value by a delay time of Δ. It may be observed that in the given circuit there are two different paths; Path-1 and Path-2 through which change in the input value may be communicated to the output. Path-2 has two gates, and therefore, the delay time through this path will be 2Δ, while Path-1 has only one gate and, therefore, a delay of only Δ. Now, let us consider the instant when the value of A is changed from 0 to 1; the output X will respond to this change after a delay time of Δ via Path-1 and the logic value 0 will appear at the output, after another time delay of Δ (total time delay $= 2\Delta$) the output will be activated via Path-2. Hence, the output X will first receive the signal from Path-1 and after a time gap of Δ, it will get information via Path-2; i.e. X = 0 + (time gap of Δ) 1. It may be observed that the output will hang with its value 0 for a time Δ, after which it will get information from Path-2 and will switch to 1. The time response of the output is shown in Fig. 10.10aii. After switching A from 0 to 1, the output X will switch from 1 to 0, retain a value 0 for a time Δ and will then switch to 1. This unwanted dip in the value of X for a time Δ is called glitch. It is needless to say that this glitch is the result of different transition times and would not be there if delay time Δ is zero. Glitches are unwanted signals and may adversely affect the operation of a logic circuit. In short, Glitches are the unwanted changes in output

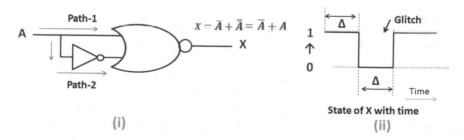

Fig. 10.10a **i** Circuit to show the effect of transition or delay time **ii** output x as a function of time

logic levels that occur for a short time due to different transmission times signals take to influence the output of a logic circuit.

10.5   Flip-Flops: The Edge Triggered Latch

Enabled D-latch, as shown in Fig. 10.10, responds to the D-input for the entire duration of time the E-pulse is in state-1. However, in many digital devices, it is required that the circuit should respond to the D-input for a much shorter time than the whole duration of the enable signal. Two types of circuits are available in this regard, one which allow the response of the D- input for a very short time when the enable signal goes from low level to high level and the other which allow the D-signal to affect the output for a very short time when the enable signal transit from high level to low level. The latch that employs the former technique is called 'leading edge triggered or positive edge triggered latch' and the other in which triggering pulse is generated when the enable pulse decreases from logic state 'high' to logic state 'low' is called the 'trailing edge or negative triggered latch'.

As has been mentioned, in edge triggered latches, a triggering pulse of very short duration is generated either when the enable pulse rises from zero to high level (leading edge trigger) or whenever it switches from high level to low level (trailing edge trigger). Figure 10.11 shows the leading edge and trailing edge triggering pulses. It is not very difficult to generate these triggering pulses using the delay time of logic gates. Every electronic device including logic gates has an inherent delay time τ, it is the time that a signal at the input of the device takes to reach the output terminal. Generally, these delay times are of the order of few neno seconds ($\approx 10^{-9}$). For generating positive trigger, the original E-input and the delayed inverted version of the signal are fed at the two inputs of an AND gate. The leading edge trigger pulses of time duration equal to the delay are produced at the output of the gate.

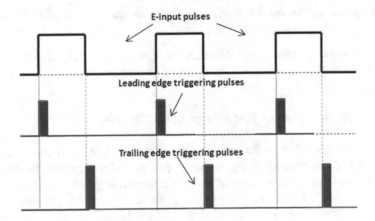

Fig. 10.11   Leading and trailing edge triggering pulses

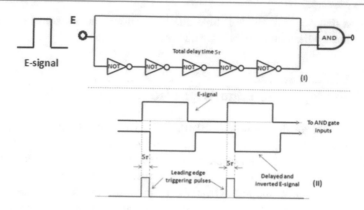

Fig. 10.12 **I** The circuit diagram of the network to produce leading edge triggers **II** Time relation between the two inputs to AND gate and the triggering pulses

Figure 10.12(i) shows the circuit diagram of the network that produces leading edge triggering pulses. A series combination of 5 (odd number of) inverter gates, each with delay time τ produces an inverted E-signal with a time delay of 5 τ with respect to the E-signal. One input of the AND gate is directly fed to the E-signal while at the other input inverted and time delayed E-signal is applied. The AND gate delivers pulses at its output when both inputs are in logic state-1.

As shown in part (II) of Fig. 10.12, both signals, E and time delayed inverted signal are in state-1 only during the 5 τ time just after the rising edge of signal E. Thus, the output of the AND gate delivers leading edge triggering pulses.

The circuit for producing trailing edge triggering pulses is shown in the upper part (I) of Fig. 10.13. In this case, a NOR gate is used instead of the AND gate of Fig. 10.12 (i). The NOR gate delivers the output only when both the inputs are simultaneously in logic state-0, which occurs at the trailing edge of the E-pulse, as shown in the lower part (II) of Fig. 10.13.

Self-assessment question SAQ: Why one uses an odd number of NOT gates in networks used to produce triggering pulses?

Self-assessment question SAQ: What are the logic states of S, R, Q and $\overline{Q}$ signals, respectively, in SET and RESET states of a S-R latch?

10.5.1   Working of an Edge Triggered Flip-Flop

The Q-output of a D-latch follows the input signal at D whenever there is a signal at E in state-1. Since edge triggering signals are of very short duration, the fillip-flop responds the D-signal at the instant there is a triggering pulse at E. As shown in Fig. 10.14, the first triggering pulse finds the D-signal in logic state-1 at A, therefore, output Q goes to state-1 at A and get latched to state-1 till another

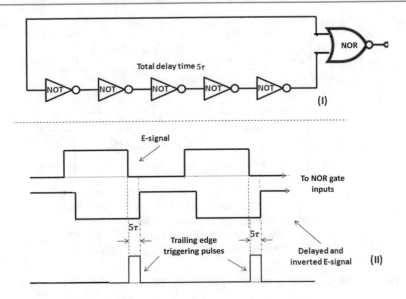

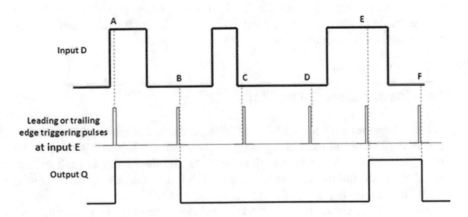

Fig. 10.13 I Circuit diagram of the network that produces trailing edge triggering pulses **II** Time relation between the pulses at NOR gate inputs and the trailing edge triggering pulses

Fig. 10.14 Signals at two inputs D and E in relation with the output Q of an edge triggered flip-flop

triggering pulse at B senses D in state-0. Output Q goes to state 0 at B and get latched to it (state-0) till the next triggering pulse which again senses D in state-0. The output Q remains latched to state-0 till the point E, where the triggering pulse senses the input signal D in state-1 and get latched to state-1 up to point F. At F, the triggering pulse finds the signal at D in state-0, and therefore, the output Q returns back to state-0. The signal at the other output $\overline{Q}$ is inverse or complement of Q.

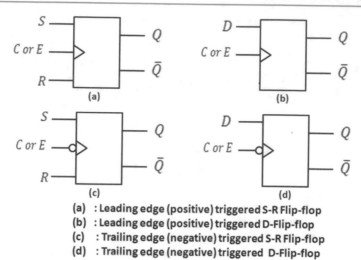

(a) : Leading edge (positive) triggered S-R Flip-flop
(b) : Leading edge (positive) triggered D-Flip-flop
(c) : Trailing edge (negative) triggered S-R Flip-flop
(d) : Trailing edge (negative) triggered D-Flip-flop

Fig. 10.15 Symbols of leading and trailing edge triggered S-R and D-flip-flops

The logic symbols for leading edge (positive) and trailing edge (negative) triggered S-R flip-flop and D-flip-flop are shown in Fig. 10.15. It may be noted in this figure that a triangle at C (or Clk clock) or at E (enable) port indicates that the circuit employs positive edge triggering, while a bubble and a triangle signifies the negative or trailing edge triggering.

10.6 Master-Slave D-Flip-Flop

It is a kind of series combination of two flip-flops, the first of which, called Master, captures the D-input at the leading edge of the clock (or E-input) pulse and the second, called Slave, picks up the signal captured by the Master at each trailing edge of the clock and passes it to the output. The block level circuit diagram of a Master-Slave D-flip-flop is shown in Fig. 10.16.

As shown in the circuit diagram, the D-signal is applied to the D- port of the first Flip-flop (FF1) and the Q_1 output of FF1 is fed to the D_2 input of FF2. The final output is delivered by the slave FF2 at its Q_2 output. The clock input is directly connected to the enable input E_1 of the master FF1. However, an inverted version of the signal at E_1 is fed to the enable input E_2 of FF2. The inversion of the clock signal is implemented using a NOT gate as shown in the figure.

A flip-flop is generally triggered by the triggering pulses produced using either the leading edge (positive) or trailing edge (negative) networks shown in Figs. 10.12 and 10.13.

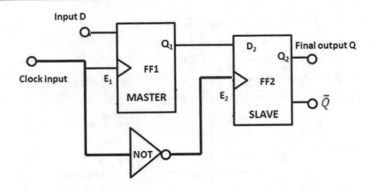

Fig. 10.16 Circuit diagram of a Master-Slave D Flip-flop

Before proceeding further, let us talk about the flip-flop triggering pulses. Flip-flops in general, if not specified otherwise, are triggered by the triggering pulse produced by the leading edge of the clock pulse. In the present case also, both the Master and the Slave flip- flops are triggered by the leading edge (the clock pulse edge where it raises from logic Low level to logic High level). The important point to remember is that the Master flip-flop in the present case is triggered by the **leading edge pulses of the clock directly connected at the E_1 input.** On the other hand, the Slave flip-flop is also triggered by leading edge pulses but of the **inverted clock signal applied at E_2.** In the inverted clock signal, the positive pulse of the original clock signal becomes negative and the negative pulse in the original clock signal becomes positive. The triggering pulse of the Slave flip-flop is derived from the negative pulse of the original clock signal which has become positive on inversion. This results in two facts: (i) both the Master and the Slave flip-flops are triggered by leading edge pulses. (ii) The triggering pulse of the Slave flip-flop occurs after a time gap of T/2 with respect to the triggering pulse of the Master flip-flop, where T is the time period of the clock pulse as shown in Fig. 10.17. As a result of this time delay between the enabling pulses of the two flip-flops, the two flip-flops never get enabled at the same time.

The Master is enabled first, say at instant 't' and passes the D-signal present at its input at instant 't' to output Q_1. At this instant, 't' Slave remains disabled. The state of the D-signal picked up by Master flip-flop remain hanging at Q_1 for a time duration T/2 from the instant 't' to instant (t + T/2). At instant (t + T/2), the Slave flip-flop gets an enabling pulse and passes the information hanging at D_2 (or Q_1) to the final output Q_2. At instant (t +T), the Master flip-flop again gets enabled, picks up the logic state of D-signal at instant (t + T), passes it to Q_1 where it remains hanging till Slave become enabled at the instant (t + 3T/2).

Clock signal, leading edge pulses that trigger Master flip-flop, pulses that trigger Slave flip-flop, signal at D-input, signals at Q_1 and at output Q_2 are shown in Fig. 10.18. As shown in this figure, the Master FF1 gets enabled by the triggering pulse marked 1 and picks the logic state Low or 0 of the D-signal at point 'a'. This value (0 or Low) of the D-signal gets latched at Q_1 till the next enabling pulse

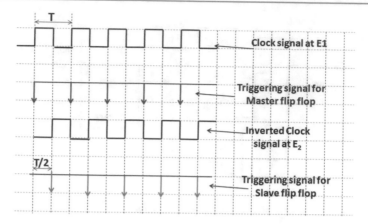

Fig. 10.17 Time relation between Master and Slave flip-flop triggering pulses

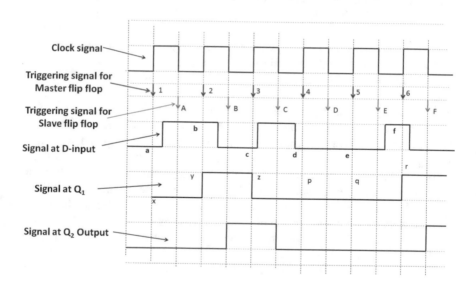

Fig. 10.18 Time relation between the triggering and the output signals of the Master and Slave Flip-flops with respect to D-input

marked 2 again enables the Master FF1. Now Master FF1 picks the logic state High (or 1) of the D-signal at the point marked b. As a result, the logic state of Q_1 which was Low (or 0) from x to y shifts to logic state High (or 1) at point y. Triggering pulse marked 3, which enables Master FF1, finds the logic state of D- signal Low (or 0) at point c. Consequently, the logic state of Q_1 which was latched to High at point b again shifts to Low at z.

Table 10.6c Truth table for Master-Slave D-flip-flop

Trigger	Input D	Output						Inference
		Present state		Intermediate		Next state		
CLK		Q	$\overline{Q}$	Q_1	$\overline{Q_1}$	Q	$\overline{Q}$	Set
↑	1	0	1	1	0	Latched		
↓		0	1	Latched		1	0	
↑		1	0	1	0	Latched		
↓		1	0	Latched		1	0	
↑	0	0	1	0	1	Latched		Reset
↓		0	1	Latched		0	1	
↑		1	0	0	1	Latched		
↓		1	0	Latched		0	1	

Triggering pulses marked 4 and 5 enable the Master FF1, but it senses logic state Low of the D-signal at points d and e. Therefore, the Q_1 gets latched to Logic level Low or 0 from points marked z to r. The 6th triggering pulse, however, finds the D-signal in logic High (1) at point f and the Q_1 which was latched to state 0 (or Low) shifts to High and remains latched to it till the next triggering pulse (not shown in figure).

Final output Q_2 follow or mimic the signal at Q_1 with a time delay of T/2. Therefore, the signal at final output Q_2 has the same shape as the signal at Q_1 but is shifted or delayed by a time T/2, as shown in the figure.

The truth table of a D-Master-Slave flip-flop is shown in Table 10.6c. Here it may be observed that the output of the Master flip-flop (data enclosed in shaded area) appear during the positive edge of the clock, indicated by an arrow pointing upwards. However, at this instant, the Slave outputs remain latched or unchanged. The same data is transferred to the final output pins of the Master-Slave system by the Slave during the Slave triggering pulse (shown by an arrow pointing down).

Self-assessment question SAQ: Fill the blanks in the following sentence."In a flip-flop, the D-input reaches the Q-output only when the enable clock signal is in logic state.......and when the clock signal is in logic state........ the output gets latched to ...."

10.7 Flip-Flop

The circuit diagram of a JK flip-flop is shown in Fig. 10.19. JK flip-flop is also called the universal flip-flop as it can mimic the performance of any other flip-flop. The alphabets J and K stands for **J**ack **K**ilby who invented this circuit. As shown in

Fig. 10.19 Circuit diagram
of a JK flip-flop

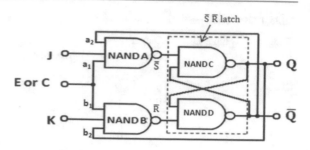

the figure, the JK flip-flop is basically made out of a $\overline{SR}$ latch by replacing its
$\overline{S}$ and $\overline{R}$ inputs by the outputs of two NAND gates A and B.

Another important change is that Q and $\overline{Q}$ outputs are fed back to one of the
three inputs, respectively, b_2 and a_2 of the two NAND gates A and B. One input (a_1
and b_1) each of NAND gates A and B is connected together to provide the enable
(or clock) input E or C. The remaining open input ports, named J and K serves as
the two inputs to the system.

Before proceeding into the analysis of the JK flip-flop one particular property,
the property of time delay produced by logic gates in receiving a signal at its input
and delivering the signal at its output must be understood. As a result of this time
delay, often it so happens that the signal fed back from some later stage of the
circuit keeps hanging at the input till the logic state of that stage gets effected by the
applied signal. This will become clear from Fig. 10.19a, where there are a number
of logic gates in the circuit and the output Q (= 1) of say, the N-th gate, is fed back
to the C-input of the initial OR gate. Suppose a signal S is applied at the B input of
the OR gate. If it is assumed that each of the gates takes a delay time of τ second to
deliver the signal at its output, the signal will take $((N - 1)\ \tau)$s to reach the input of
the N-th gate. During this period of $((N - 1)\ \tau)$ seconds, the C-input of the OR gate
will continue to have the logic state-1. With this background of time delay, we
move ahead to analyze the operation of the logic circuit.

To analyze the operation of the JK flip-flop circuit, we start by assuming that
$Q = 0$ **and $\overline{Q} = 1$**

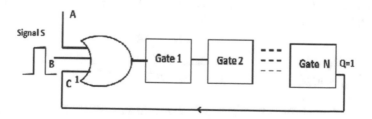

Fig. 10.19a Affect of time delay

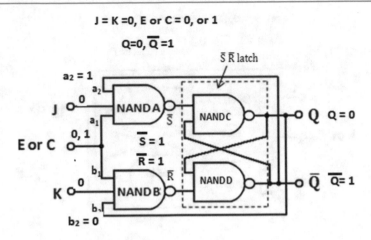

Fig. 10.19b Logic states for J = 0, K = 0 and E = 0

(i) Further, **let J = 0, K = 0 and E = 0**;
Since two inputs to NAND-A and B are in logic state-0 (low), therefore, both
$\overline{S}$ and $\overline{R}$ = 1. Thus, the two inputs of NAND D are at logic state-0 (feedback from
Q) and 1; as such the output $\overline{Q}$ is in logic state-1 (high) and since Q is complement
of $\overline{Q}$, $Q = 0$.

Hence, when there is no signal at inputs J and K and clock is in logic state-0 (no
pulse), the outputs Q and $\overline{Q}$ continue to be in the state they were, there is no change.
Also if the clock signal E is set to 1, (a positive pulse is applied at clock input E)
and inputs J and K are kept at level 0, then also there will be no change in the logic
states of Q and $\overline{Q}$, since in that case also both $\overline{S}$ and $\overline{R}$ = 1. The logic states at
different inputs and outputs are shown in Fig. 10.19b.

(ii) Next let **J = 1, K = 0 and E = 1**
Now the three inputs of NAND-A are in the states a_1 = 1; J = 1, a_2 = 1 (feedback
from the previous state of $\overline{Q}$ due to time delay), hence the output $\overline{S}$ goes to
state = 0. Now the two inputs of NAND-C have logic states 0 and 1 (feedback from
$\overline{Q}$). The output of NAND gate C, therefore, switches to 1 from its initial value 0.
Since the other output Q is complement of $\overline{Q}$, it switches to state 0. Thus, it may be
observed that when the enable clock signal is high (in state-1) and the input J is
high, while K is low, the Q output assumes the same logic state as that of the input
signal J, i.e. output Q follows input J as shown in Fig. 10.19c.

Similarly, it is simple to prove that if E = 1, J = 0 and K = 1; the output $\overline{Q}$
assumes the same state as that of input K. This is shown in Fig. 10.19d.

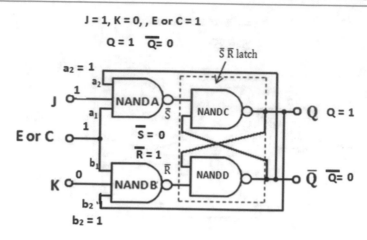

Fig. 10.19c For (J = 1, K = 0, E = 1) output Q follows J

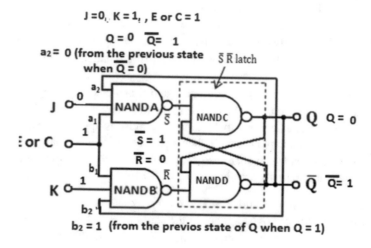

Fig. 10.19d For (J = 0, K = 1, E or C = 1) output $\overline{Q}$ follows K

(iii)   We now consider the case **when J = K = 1 and E = 1.**
In this case, two inputs of both NAND gates A and B have logic states 1, but the third input a_2 of gate-A is in state-1 (feedback from $\overline{Q}$) while input b_2 of gate B is in state = 0 (feedback from Q). Therefore, the outputs $\overline{S} = 0$ and $\overline{R} = 1$; Hence, the output Q switches from 0 to 1 and the output $\overline{Q}$ from 1 to 0. As has been mentioned earlier also, every electronic circuit has some delay time, and therefore, the switching of the logic states of outputs Q and $\overline{Q}$ is also not immediate, it takes some

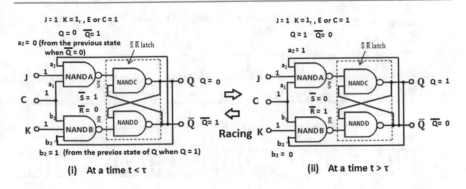

Fig. 10.19e Logic states of inputs and outputs at time t **i** t < τ and **ii** t > τ

time. Which means that as the clock pulse switches from low level to high (leading edge), the switching of the logic states of Q and $\overline{Q}$ occurs but with a time delay which is often represented by τ (Fig. 10.19e).

If the duration of the clock pulse is larger than the switching delay time τ, the output logic states again switch and this switching keeps going until the clock pulse switches from logic state −1 to 0. This repeated and automatic switching of the output levels during the time when the clock pulse is in state-1 is called RACING. It is important to distinguish between racing and toggling. **Racing is the switching of the output levels automatically, without any control when the clock pulse is in state-1.** On the other hand, **toggling is the controlled switching of the output levels; often in toggling the output levels undergo switching when some triggering pulse is applied.**

Figure 10.20 shows the time relation between the clock pulses, the logic states of inputs J and K and the logic states of outputs Q and $\overline{Q}$.

As is obvious, the racing occurs because of the feedback from the outputs Q and $\overline{Q}$, respectively, to the inputs b_2 and a_2. Racing which is often termed as racing hazard is undesirable and it is required to get rid of it.

In general, there are three methods of removing racing in the output. (i) To use clock pulses of short duration. If T is the time period of the clock pulse and τ the switching time of the circuit then for no racing to take place T/2 < τ. (ii) Another method is to use either the leading edge or the trailing edge pulses of short duration so that the E- pulse goes to state-0 before the racing occurs. (ii) Yet another method is to do away feedback from the outputs to the inputs of latch circuits. This may be achieved by using the Master-Slave flip-flop arrangement.

If the JK flip-flop is triggered by leading or trailing edge sharp pulses or by pulses of short duration, then each clock pulse switches the output levels and the flip-flop is said to be in the state of toggling. The truth table for JK flip-flop is given Table 10.7.

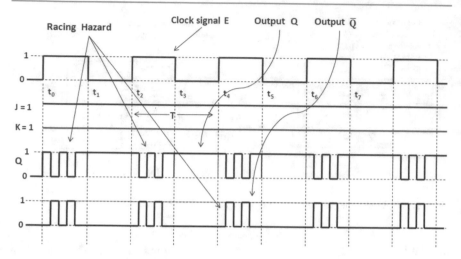

Fig. 10.20 Time relation between the clock pulses and the output signals

Table 10.7 Truth table of a JK flip-flop

Inputs		Output Q		Remarks
J	K	Before clock pulse	After clock pulse	
0	0	0	0	No change
0	0	1	1	
1	0	0	1	Q = J
1	0	1	0	
0	1	0	0	$\overline{Q} = K$
0	1	1	0	
1	1	0	1	Output toggle/race
1	1	1	0	

Self-assessment question SAQ: Differentiate between racing and toggling

Self-assessment question SAQ: What is the main reason for the racing hazard?

10.7.1 Master-Slave JK Flip-Flop

Figure 10.21 shows the circuit diagram of a master-slave JK flip-flop, where it may be seen that two flip-flops FF1 and FF2 are connected in series, with the two outputs Q_1 and $\overline{Q}_1$ of the master flip- flop (FF1) are connected to J_1 and K_1 inputs of the slave (FF2) flip-flop. Loops QA_2 and $\overline{Q}A_1$ provide the necessary feedback to the network. Positive logic clock pulses are fed to the master flip-flop at port E while

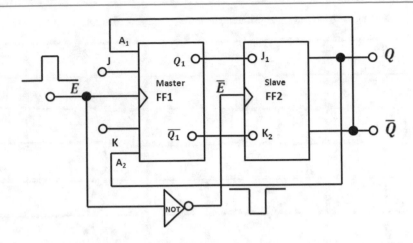

Fig. 10.21 Circuit diagram of Master-Slave JK flip-flop

inverted clock pulse, obtained by using the NOT gate, is applied to the clock input $\overline{E}$ (Or E_2) of slave flip-flop.

10.7.2 Working of the Master-Slave JK Flip-Flop

In this discussion of the working of a master-slave JK flip-flop, we assume that the signal at input K is a complement of the signal at input J. At the leading edge of the positive clock pulse, the master FF1 latches to the instantaneous logic state of input J, that is, the output Q_1 takes the value of the logic state of the signal at input J and the output $\overline{Q_1}$ switches to the logic state of the complement of J.

On assuming the logic state of J, the output Q_1 remains latched to this value till the full duration of the positive pulse (logic state-1). During the state-1 (clock pulse at level high) of the clock pulse, the Slave FF2 remains disabled since the input clock pulse $\overline{E}$ for slave FF2 is the complement of clock pulse E, and therefore, the state-1 of E is state-0 for $\overline{E}$. At the falling edge of the clock pulse E (which is the rising edge for $\overline{E}$) the slave FF2 gets enabled and pickups the value of the logic state of J that was waiting at input J_1 and passes it to the output Q. The state of Q remains latched to the logic state picked up by FF2 for the duration of the state-0 of the clock pulse (which is level 1 for $\overline{E}$) and also for the duration of the next level 1 of E when FF2 gets disabled. Thus, the logic state of J picked up by FF2 appears at Q after a time lag equal to the duration for which the clock pulse remained in state-1. This will become clear in Fig. 10.22, where the logic state-1 (high) is picked by FF1 at the rising edge of the clock signal at point 'a' and passes it to output Q at point 'd' after the time delay equal to the duration for which the clock signal stays in state-1. Q remains latched to this level (1) till point 'e', corresponding to the next rising edge of the clock pulse where the J-signal is again found to be in state-1

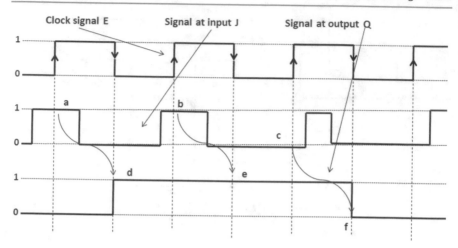

Fig. 10.22 Time relation between the clock signal, signal at input J and output signal Q

(high) at point 'b'. The output Q, therefore, again get latched to state-1 for the total duration corresponding to 'e to f' when the rising edge of the clock pulse finds the signal at input J in logic state-0 (or low) at point 'c'. The signal at Q, accordingly, switches to logic state-0 at f.

The master-slave JK flip-flop is a device which accepts the input at the rising edge of the clock pulse and delivers it at the output at the falling edge of the clock signal; as such it is a synchronous device.

In order to understand the toggling action of a master-slave flip-flop, we refer to its circuit diagram given in Fig. 10.23. We start with the initial conditions: $J = 1$, $K = 1$, $Q = 0$, $\overline{Q} = 1$, $E = 1$, $\overline{E} = 0$, when the clock pulse has just switched from state 0 to logic state-1.

At the rising edge of the clock pulse the three inputs of NAND gate A_1 will be in logic states; $a_1 = 1$, $J = 1$, $E = 1$, therefore, the output $q_1 = 0$, $q_3 = 1$. But since

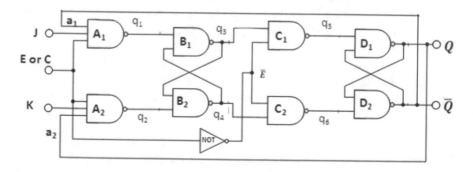

Fig. 10.23 Circuit diagram of a master-slave JK flip-flop

$\overline{E} = 0$, the NAND gate C_1 is disabled with the input q_3 at logic 1. Similarly, it is easy to show that NAND gate C_2 is also disabled with input q_4 at logic state-0.

The outputs continue to remain, respectively, in states $Q = 0$ and $\overline{Q} = 1$ till the falling edge of the clock pulse E, when $\overline{E}$ switches to 1 and enables the NAND gates C_1 and C_2 which pass the input data at q_3 and q_4 to the outputs Q and $\overline{Q}$. Thus, at the falling edge of the clock pulse, Q switches from 0 to 1 and $\overline{Q}$ from 1 to 0. The condition $Q = 1$, $\overline{Q} = 0$ continues for the duration of the logic state-0 of the clock pulse and also for the duration of the next logic state-1 of the clock pulse. Though the next leading edge of the clock pulse changes the logic states of inputs q_3 and q_4, these values are not passed through gates C_1 and C_2 which remain disabled for the duration of the clock pulse when it is in state-1. In nutshell, the new values for the logic states at q_1 and q_2 are generated at each leading edge of the clock pulse but they are passed on to the output at the trailing edge. As shown in Fig. 10.24, the logic states of outputs Q and $\overline{Q}$ alternate between 1 and 0 with a frequency half of clock the pulse.

It is interesting to note that for a master-slave JK flip-flop in toggling configuration, the two input signals at J and K remain in state-1, but the toggling is produced by the feedback signals at input a_1 and a_2 that switch their levels after each leading edge. Further, the uncontrolled racing is avoided because the inverted clock signal disables the slave flip-flop for the duration of clock pulse when it is in state-0. It may also be observed in Fig. 10.24 that the signals at Q and $\overline{Q}$ are complement of each other and have the frequency that is half of the frequency of the clock pulse. This property is used in making digital counters.

A detailed truth table of a J K Master-Slave flip-flop is shown in Table 10.7a. The arrow signs in the column marked 'Triggering pulses' specifies the triggering pulse; if the pulse triggers the Master flip-flop then it is indicated by the arrow

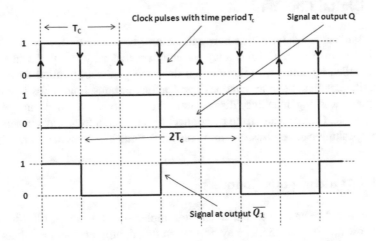

Fig. 10.24 Toggling of outputs of a master-servant JK flip-flop

Table 10.7a Truth table of a J K Master-Slave flip-flop

Trigger pulse	Input		Outputs						Remark
	J	K	Present state		Intermediate state		Next state		
			Q	$\overline{Q}$	Q_1	$\overline{Q_1}$	Q_2	$\overline{Q_2}$	
↑	0	0	0	1	0	1	Latched		No change
↓			0	1	Latched		0	1	
↑			1	0	1	0	Latched		
↓			1	0	Latched		1	0	
↑	0	1	0	1	0	1	Latched		Reset
↓			0	1	Latched		0	1	
↑			1	0	1	0	Latched		
↓			1	0	Latched		1	0	
↑	1	0	0	1	1	0	Latched		Set
↓			0	1	Latched		1	0	
↑			1	0	1	0	Latched		
↓			1	0	Latched		1	0	
↑	1	1	0	1	0	1	Latched		Toggle
↓			0	1	Latched		0	1	
↑			1	0	0	1	Latched		
↓			1	0	Latched		0	1	

Note:↑ Specify the pulse that triggers Master FF1 and ↓ specifies the pulse that triggers the Slave FF2

pointing upward and if it triggers the Slave flip-flop then it is represented with an arrow pointing downwards. Present state specifies the initial logic states of the outputs Q_2 and $\overline{Q_2}$. The intermediate state specify the value of the logic states at intermediate output Q_1 and $\overline{Q_1}$. Next state gives the logic levels of final outputs Q_2 and $\overline{Q_2}$. The table is quite general and can be used for any type of Master-Slave flip-flop with suitable modifications.

Self-assessment question SAQ: Why there is no problem of racing in JK flip-flop?

10.8   Digital Counters

Digital counters are devices that may calculate or record as to how many times a particular event occurred. There may be counters that may count the increasing number, called up counter, or those count decreasing numbers, called down counters. Digital counters have tremendous applications in household items like microwave oven, washing machine, dishwasher, digital clocks, laptops and personal computers, etc. Counters are made by connecting sequential logic units in series and broadly speaking can be classified into **Asynchronous** and **Synchronous**.

10.8.1   Asynchronous Counters

A typical layout of an asynchronous counter implemented using JK fillip-flops (JK FF) is shown in Fig. 10.25. As may be seen in this figure, the J and K inputs of all flip-flops are connected to logic state-1 (high) so that they operate in toggling mode.

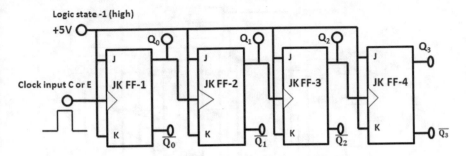

Fig. 10.25 Circuit diagram of an asynchronous counter

Further, the combination is essentially in series as the output Q_0 of the first flip-flop is fed to the clock input C or E of the second, and the output Q_1 of the second flip-flop to the clock input of the third flip-flop and so on. The event to be counted is fed in the form of clock signal at the input C or E of the first flip-flop. The Q_0, Q_1, ...Q_3 output signals and their relationships with clock signal is shown in Fig. 10.26.

The final output in an asynchronous counter is considerably delayed with respect to its input as it has to travel through all the flip-flops, each of which introduces some time delay. Hence, the name asynchronous counters.

10.8.2 Synchronous Counter

To overcome the problem of time delay in asynchronous counters, synchronous counters are implemented using JK flip-flops and triggering them with the same

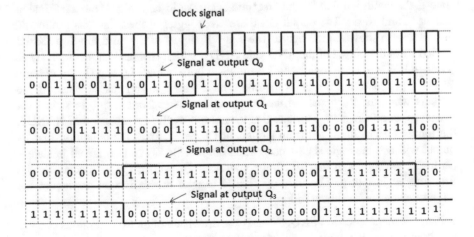

Fig. 10.26 Signals at different outputs of the asynchronous counter

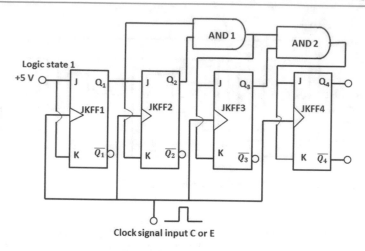

Fig. 10.27 Circuit diagram of a synchronous counter implemented using JK Flip-flops

clock signal, as shown in Fig. 10.27. The first flip-flop JKFF1 toggles at each clock pulse. The AND gates in the circuit are employed to ensure that JKFF2 toggles only when the signal at output Q_1 is in logic state-1, the JKFF3 toggles only when signals at Q_1, and at Q_2 are in logic state-1 and finally the JKFF4 toggles only when signals at Q_1, Q_2, and Q_3 are all in logic state-1.

Signals at different outputs along with the clock signal are shown in Fig. 10.28. Further, it may be kept in mind that Flip-flops are triggered by the leading edge of the clock signal.

As shown in the figure, the signal at the output Q_1 is the usual toggling waveform of a JK flip-flop with inputs J and K at logic state-1. The second JKFF2 toggles only when the clock pulse is rising from 0 to 1 and the output Q_1 is in logic state-1, the condition that for the first time occurs at instant 'a' and the next time at instant b, and so on. The output Q_2, therefore, toggles at these instants producing the waveform shown in the figure. The J and K inputs of the third flip-flop JKFF3 are fed from the output of the AND gate AND1, which ensures that the inputs (J & K) of JKFF3 will be in logic state-1 only when both outputs Q_1 and Q_2 are in logic state-1. This condition is met for the first time at instant m and next at instant n, the instants when the logic state of the output Q_3 switches from state-0 to state-1 (at instant m) and from state-1 to state-0 at instant n. Finally, the inclusion of NAND2, the inputs to which are fed from signals at Q_2 and Q_3 sets the condition that its output will assume logic state-1 only when outputs Q_1, Q_2 and Q_3 are all in state-1 (or high). As shown in Fig. 10.28, the condition $Q_1 = Q_2 = Q_3 = 1$ occurs for the first time at instant p. The flip-flop JKFF4 toggles at this instant producing the output waveform as shown in Fig. 10.28.

Self-assessment question SAQ: What is the drawback of an asynchronous counter and how is it overcome in a synchronous counter?

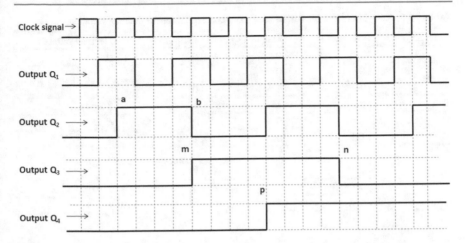

Fig. 10.28 Signals at different outputs of a synchronous counter

10.9   Four Bit Decade Counter

Since a single JK flip-flop can register a single bit of information, the total number of bits that a counter can register is decided by the number of flip-flops used in it. A four bit counter needs four JK flip-flops. Further, we have already seen that series combination of JK flip-flops connected in specific ways with other logic gates may produce at different outputs of the flip-flops signal patterns of different types. A decade counter that may count a sequence of ten numbers starting from 0, 1, 2, ...9 will be operational if waveforms shown in Fig. 10.29 are produced at the four successive outputs Q_1, Q_2, Q_3, and Q_4 of the combination of four JK flip-flops.

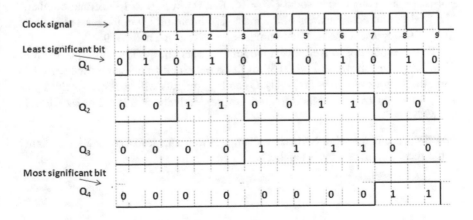

Fig. 10.29 Signal waveforms at four outputs of a 4-bit decade counter

Table 10.8 Ten different 4-bit output sequences corresponding to decimal numbers

Output Q_4	Output Q_3	Output Q_2	Output Q_1	Decimal number
0	0	0	0	0
0	0	0	1	1
0	0	1	0	2
0	0	1	1	3
0	1	0	0	4
0	1	0	1	5
0	1	1	0	6
0	1	1	1	7
1	0	0	0	8
1	0	0	1	9

The output at Q_4, the last (fourth in this case) flip-flop gives the value of the most significant bit while the output at the first flip-flop, represented by Q_1, provides the least significant digit. The ten different sequences of four bit each, obtained by collecting logic states of outputs in the order Q_4, Q_3, Q_2, Q_1, at a given instant within one clock time period, are given in Table 10.8.

Self-assessment question SAQ: On what factor(s) the number of bits in a counter depends?

Self-assessment question SAQ: If the time period of the clock signal is 2 ms, how much time it will take to generate the complete sequence of numbers?

10.10   4-Bit Binary Counter

A 4-bit binary counter can count ($2^4 = 16$; that is) from 0 to 15 decimal numbers. The 16 different combinations of the most significant bit Q_4 to the least significant bit Q_1 that represent the decimal numbers are given in Table 10.9.

The corresponding waveforms at the four outputs Q_1, Q_2, Q_3, and Q_4 of the four JK flip-flops are shown in Fig. 10.30. If the input clock signal has a time period of 1 ms, the periods of waves at Q_1, Q_2, Q_3 and Q_4 outputs will be respectively, 2 ms, 4 ms, 8 ms and 16 ms and their frequencies, respectively, 500 Hz, 250 Hz, 125 Hz and 62.5 Hz.

Self-assessment question SAQ: In a counter, the frequency of the least significant or the most significant bit is smallest?

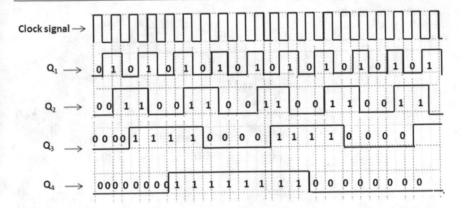

Fig. 10.30 Wave patterns at the four outputs of a 4-bit binary counter

Table 10.9 Sixteen sequences of 4-bits each representing decimal numbers

Q_4	Q_3	Q_2	Q_1	Decimal number	Q_4	Q_3	Q_2	Q_1	Decimal number
0	0	0	0	0	1	0	0	0	8
0	0	0	1	1	1	0	0	1	9
0	0	1	0	2	1	0	1	0	10
0	0	1	1	3	1	0	1	1	11
0	1	0	0	4	1	1	0	0	12
0	1	0	1	5	1	1	0	1	13
0	1	1	0	6	1	1	1	0	14
0	1	1	1	7	1	1	1	1	15

Self-assessment question SAQ: A binary counter has N-bits, how many count values in sequence it will have?

10.11   Characteristics of a Counter

(i)   **The counter modulus:** It is the total number of states or values generated by a counter as it progresses through its specific sequences. A decade counter has a modulus (or Mod) of 10, while a binary counter of N-bits has a mod of 2^N.

(ii) **The counter bits and stages:** The number of bits of a counter is equal to the number of flip-flops in the counter, each flip-flop storing one bit of information for the duration of clock pulse. An N-bit counter of modulus M generates M stages or specific sequences each of N-bit length. The output of the first flip-flop Q_1 provides the least significant bit (LSB) and the output Q_N of the last flip-flop the most significant bit (MSB).

Fig. 10.31 State transition
diagram of 4-bit decade
counter

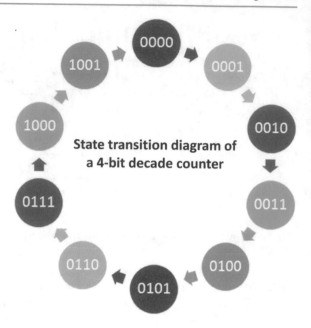

(iii)   **The state transition diagram:** states of a counter are the output numbers
generated by a counter at each clock pulse. The state transition diagram
shows the progression of counter states with successive clock pulses. The
counter modulus gives the total number of states of the counter. For example,
the transition diagram of a 4-bit decade counter may be represented by the
state transition diagram shown in Fig. 10.31.

(iv)   **Counter as a frequency divider:** Counters also work as frequency dividers
as they divide the clock signal frequency f_C by the modulus M of the counter
at each output. For example, in case of a binary counter, the frequency of the
signal at each successive output of the counter becomes half of the frequency
of the signal at the previous output; i.e. frequency of the signal at first output
$f_{Q1} = 1/2$ clock signal frequency f_C; $f_{Q2} = \frac{1}{2} f_{Q1} = 1/4$ fc, and $f_{QN} = 1/2$
$f_{Q(N-1)}$.

If $f_{(MSD)}$ denotes the frequency of the last output of a counter (that provides
the MSD), M the MODULUS or MOD of the counter and f_c the frequency of
the counter clock, then $f_{(MSD)} = f_c/M$. Counters are, therefore, often used to
produce signals that are synchronous with the clock signal but have lower
frequencies that are integer multiples of clock frequency.

Self-assessment question SAQ: The clock frequency of a 4-bit binary counter is
112 kHz, what will be the frequency of the signal at the last output of the counter?

Self-assessment question SAQ: What could be the maximum modulus of a 3-bit
counter?

Self-assessment question SAQ: what is the difference between a synchronous and an asynchronous counter?

Self-assessment question SAQ: What is meant by the positive edge triggering of a counter?

10.12 To Decode the Given State of a Counter

It is often required to identify a particular state of a counter. For example, in domestic microwave ovens, one may set a time for which the object placed in the oven is to be heated. The microwave is provided with a clock and a digital counter. When the oven is put 'on' the counter starts and produces states one after the other. If the counter is a 4-bit binary counter, it will generate states like 0000, 0001, 0010, ...and so on. Suppose it is required to heat for 6 s that corresponds to the counter state 0110. So to switch off the oven at the instant when the counter has generated the state 0110, it is required to identify or decode this state. This may be easily achieved by using a 4-input AND gate. The four inputs of the AND gate may be connected to the four outputs of the counter as shown in Fig. 10.32).

In state 0110, $Q_4 = 0$, hence $\overline{Q_4} = 1, Q_3 = 1,$ $Q_2 = 1$ but, $Q_1 = 0$ and $\overline{Q_1} = 1$. When inputs of the AND gates are connected as shown, all inputs go to state-1 (high) when the state 0110 is generated by the counter. For no other state of the counter, except the state 0110, the output of the AND gate will go to logic state-1. Thus whenever the output of AND gate goes to state-1, it will mean that state 0110 has reached. The output sequence $\overline{Q_4}, Q_3, Q_2, \overline{Q_1}$, that is required to decode a count state is called the **mini-term.** The output of the AND gate may be coupled to a relay which may switch off the microwave when the output attains the logic state-1. Count state decoding means that a signal may be generated for a particular count state. The type of decoding implemented using AND gate, the output of which goes to state-1 when the desired state is reached, is called the ACTIVE HIGH decoding.

It is easy to realize that identification of a counter state may also be done using a NAND gate instead of an AND gate. For example, to identify the counter state 0110 the four outputs of the counter, respectively, $\overline{Q_4}, Q_3, Q_2$ and $\overline{Q_1}$ be connected to the four inputs of the NAND gate, the output of the NAND gate in this case will be 0 (low). If the relay that switches off the microwave operates with logic state-0, then

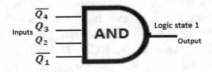

Fig. 10.32 Input connections for an AND gate to identify the given state of the counter

one may use the NAND gate for decoding. This type of decoding where the output of the decoding element (NAND gate) attains the logic state-0 on identifying the desired counter state is called ACTIVE LOW decoding.

When it is required to decode only a few count states, it is easy to use AND gates, but if many count states are to be decoded then it is better to use a digital IC decoder like 74LS138 or 74LS154. If the requirement is to decode a large number of count states and also to display them, then one may use a display decoder/driver like 74LS47 which has LED displays incorporated in the IC.

A decoder in nutshell has one input the state of which changes with time and it delivers an output only when a preset state of the input is reached.

Self-assessment question SAQ: An 8-bit binary counter is used in a washing machine and it is required to identify the counter state that corresponds to decimal number 105 using an AND gate. Indicate which outputs of the counter should be connected to the inputs of the AND gate

10.13   Multiplexer

In digital electronics Multiplexer, also called MUX or MPX in short, is a digital combinational circuit which is used to transfer a piece of data (in binary form) from a specified input line where there are several input lines, to a single output line. The input line that should be connected to the output is dictated by the controlling line or lines.

In data transmission and processing, it often happens that multiple data is provided by many different channels and it is required that data from a specified input line be displayed on a single screen or connected to a single device. A convenient example is the case of data provided by a metrological satellite; it sends to the ground controller, in digital form, several atmospheric parameters like Temperature, Wind direction, Wind velocity, Atmospheric density, Humidity, etc. simultaneously on different digital channels. These parameters received through different digital channels at the control lab are often displayed on a single screen in the ground control laboratory, one after the other in a particular sequence. The multiplexer circuit at the control lab connects one input channel, as specified by the controller say, the channel that gives Temperature, to the screen. In this way, a MUX circuit is a combinational digital logic circuit that connects a particular input data line (out of the several input data lines) to the (only one) output line.

The operation of a multiplexer can be best understood by the example of a mechanical rotary switch which may connect **any one** of the many input lines to the **single** output line. Which of the several input lines is to be connected with the output line is decided by the position of the shaft of the rotator switch; hence it may be called the control S, as shown in Fig. 10.33i. Depending on the number of input lines, a MUX is called a 2:1, 4:1 8:1, etc. multiplexer. Accordingly, Fig. 10.33i

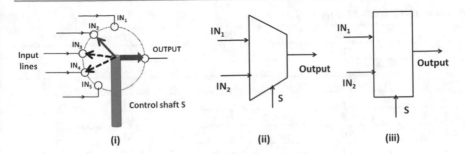

Fig. 10.33 **i** Mechanical equivalent of a MUX **ii** and **iii** alternate symbols of a digital 2:1 MUX

represents a 5:1 (mechanical) MUX. Figures (ii) and (iii) show the two alternate symbolic representations of a 2:1 digital MUX.

A digital MUX deals with digital data and, therefore, input lines carry data in the form of 1 (High) and 0 (Low). The control lines (also called source lines), that tells which of the input line is to be connected with the output, also instruct in digital format. Suppose there are only two input lines, then only one control line S will be sufficient to specify the input lines; if S = 0 (Low) it may mean that input line-1 may be connected to the output and if S = 1 (High) it may mean input line-2 may be connected to the output (or vice versa). However, if there are four input lines then a single control or source line will not be able to identify all the four input lines. The number 'N' of required control (or source) lines is related to the number 'n' of input lines by the equation,

$$n = 2^N;$$

where n = number of input lines and N is the required number of source or control lines.

As such a 4:1 MUX will need a minimum of N = 2 source lines to identify the 4-input lines as: 00, 01, 10 and 11.

The truth table of a 2:1 MUX that has two input lines IN_1 and IN_2 and only one source line S_1 is given in Table 10.10.

As may be observed in this table, when $S_1 = 0$ (control selects input channel IN_1) channel IN_1 is connected to the output and data of channel IN_1 is transferred to output channel. This is shown by the shaded part of the table. For $S_1 = 1$, input channel IN_2 is selected and is connected to the output. The Boolean expression for the output may be written as

$$Output = \overline{S_1}.IN_1 + S_1.IN_2$$

The above Boolean expression may be implemented using many different combinations of logic gates; two of which are given in Fig. 10.34.

Table 10.10 Truth table for a 2:1 MUX

S_1	IN_1	IN_2	Output
0	0	0	0
0	0	1	0
0	1	0	1
0	1	1	1
1	0	0	0
1	0	1	1
1	1	0	0
1	1	1	1

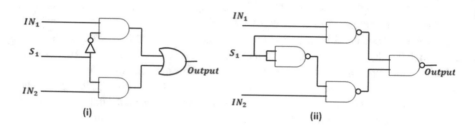

Fig. 10.34 Two alternate circuits that may implement 2:1 MUX

A 4-channel (4:1) MUX will require a minimum of two control lines and can be built using three (and not two) 2-channel MPX as shown in Fig. 10.35. Note that out of three control lines available with three 2-channel MUX, two (S_{M1} and S_{M2}) are connected together.

Built in 4-channel MUX are available in many commercially available IC's. For example, IC14052B has two 4-channel MUX in it. Pins marked X_0, X_1, X_2, and X_3 are the four input channels and pin marked X the corresponding output pin for each

Fig. 10.35 Block diagram of a 4-channel MUX constructed using three 2-channel MUX

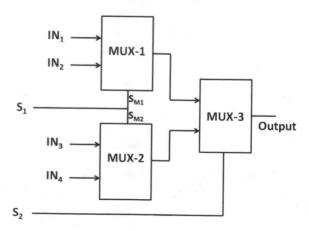

MUX. There are other pins which are meant for supplying V_{CC} etc. A special character of these MUX's is that input to the input pin may be an analog quantity like voltage, which is shown in the output. That essentially means that the analog signals are first digitalized using A/D converter before being applied to the MUX inputs.

Self-assessment question SAQ: What is the difference between a decoder and a MUX?

10.14 Parity of Binary Word and Its Computation

10.14.1 Parity of a Binary Word

Binary words are made of bits; each bit may have the value 0 or 1. The parity of a binary word is defined in terms of the number of one's in it. **If a binary word contains an even number of 1- bits, its parity is taken as even and represented by 0. If the given word contains an odd number of 1-bits, its parity is said to be odd and represented by 1.** For example, 1100110, an 8-bit binary word has 4-bits of value 1, hence its parity is even and is given as 0. Similarly, 10000101, another 8-bit word contains three 1-bits, hence has odd parity represented by 1.

10.14.2 Application of Parity

In the serial transmission of digital data, the data is transmitted and received bit by bit. Though digital transmission to a large extent is independent of environmental and ambient conditions, still there is a possibility that data may get corrupted during transmission and a bit which was 1 in the original word becomes 0 or vice versa. To detect such transmission errors, each word of the data is transmitted with an extra bit specifying its parity. For example, an 8-bit word in original data is transmitted as a 9-bit word, the first bit of which gives its parity. Generalizing, it may be said that an n-bit word is transmitted as an (n + 1) bits word and the first bit of the transmitted word contains the parity of the word. At the receiving end, the parity of the last n-bits of the received (n + 1) bit word is determined and compared with the value contained in the first bit. If the two values of parity are the same, there is no error in transmission and if they differ, the transmission is faulty.

10.14.3 Parity Generation and Checking

To generate parity bit for a word, it is required to break the word into segments of 4-bits each and check the parity of each 4-bit string. Suppose there is an 8-bit word like $X_0X_1X_2X_3X_4X_5X_6X_7$ which has parity S. It may be segmented into two strings

Y_1 and Y_2, of four bits each. If S_1 and S_2 are the parities, respectively, of Y_1 and Y_2, then the parity S of the 8-bit word is given by $S = S_1.S_2$.

The parity bit for a 4-bit word, say Y_1, may be generated by the logic circuit given in Fig. 10.36.

In Fig. 10.36, $Z_1 = 0$ when $X_o = X_1 = 0$, otherwise if $X_o \neq X_1$, $Z_1 = 1$. It means that Z_1 gives the parity of the 2-bit word X_0X_1. Similarly, Z_2 gives the parity of the sequence X_2X_3. If Z_1 and Z_2 are both 1, $Z = 0$, and if one of Z_1 and Z_2 is 0 and the other 1, then Z will be 1. In this way, Z gives the correct parity of the 4-bit word $X_0X_1X_2X_3$. The last NOR gate compares the parity Z of the 4-bit word with either 1 or 0 as set by the controller. If the comparing bit is 1, the parity bit generated at the final output is called a negative parity bit and if it is compared with 0, the positive parity bit. In this figure, the method of generating negative parity or positive parity bit for a 4-bit word is given which may be further extended to include words of larger bits.

Parity bit information with the type of parity bit, whether negative or positive is communicated to the receiver of the data, who can check the received word with the given information using the circuit given in Fig. 10.37. If the received 4-bit data corresponding to a four bit data $X_0X_1X_2X_3$ is $X'_0X'_1X'_2X'_3$ it may be fed to the circuit of Fig. 10.37 along with the parity bit sent by the transmitting centre. In case the parity data bit is generated with negative parity, then the output of the circuit in Fig. 10.37 will give 1, if data is faithfully transmitted, i.e. $X_0 = X'_0$; $X_1 = X'_1$ and so on. In case the output is 0, it will indicate that data has got corrupted in transmission.

Bit errors in digital data transfer may be of several types but the frequent ones are (i) Single bit error, where only one bit of the transmitted data got corrupted, (ii) Multiple bit error, where two or more bits are corrupted and (iii) Burst error, when a set of data bits are corrupted. In digital data transfer it is essential that error, if any, is detected at the first stage and corrected otherwise there is high probability that the data is lost. Several error detection codes, are, therefore, employed at each stage of transmission. Parity bit check is one of these error detection codes. **Hamming code** developed by R.W. Hamming at Bell Lab in 1950, is a linear code

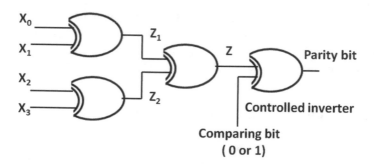

Fig. 10.36 Parity bit generating circuit

Fig. 10.37 Parity checking
circuit

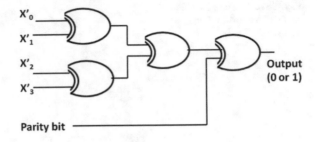

that is useful for detecting errors up to two immediate bits. Like parity bit, an extra
bit is attached with the data at a specified location that carries information required
to run Hamming code.

Solved Examples

SE 10.1 What is meant by count decoding? Determine the MOD of an **8-bit** binary
counter that is controlled by a 256 kHz clock. Also, determine the frequency of the
MSB signal. What will be the 'mini-terms' for the count states that correspond to
decimal numbers 19, 120 and 255?

Solution: *Count decoding means producing a signal when a preset count state is
reached.*

The MOD of an 8-bit binary counter = $2^8 = 256$;

The frequency of the MSB signal $= \dfrac{\text{Clock frequency}}{\text{MOD}} = \dfrac{256\,\text{kHz}}{256} = 1$ kHz.

Table SE10.1, *given below provides the mini-terms corresponding to different
decimal numbers.*

SE 10.2 Show the state transition diagram for a 4-bit binary counter.

Solution: *The MOD of a 4-bit binary counter is 16 which means that the counter
will generate 16 count states. The state transition diagram is shown in* Fig. SE10.2.

SE 10.3 A frequency divider binary network uses 10 JK flip-flops. Determine the
frequency of the signal at the output of the last flip-flop given that the clock
frequency is 1024 MHz.

Table SE10.1 Decimal numbers, corresponding binary numbers and mini-terms

Decimal number	Corresponding 8-bit binary number	Mini-term
19	00010011	$\overline{Q_8}\,Q_7\,\overline{Q_6}\,\overline{Q_5}\,\overline{Q_4}\,Q_3\,\overline{Q_2}\,Q_1$
120	01111000	$\overline{Q_8}\,Q_7\,Q_6\,Q_5\,Q_4\,\overline{Q_3}\,\overline{Q_2}\,\overline{Q_1}$
255	11111111	$Q_8\,Q_7\,Q_6\,Q_5\,Q_4\,Q_3\,Q_2\,Q_1$

Fig. SE10.2 State transition
diagram for a 4-bit binary
counter

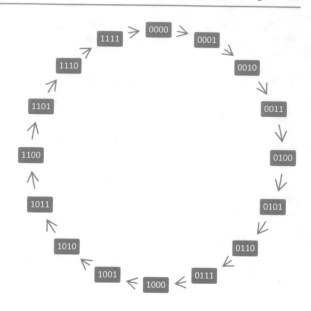

Solution: *Since the network uses 10 flip-flops and each flip-flop stores 1-bit of information, therefore, the total number of bits of the network is 10. Further, the network being binary; the MOD of the network* $\mathrm{MOD} = 2^{10} = 1024$. *The frequency of the signal at the output of the last flip-flop* $= \dfrac{\text{Clock frequency}}{\mathrm{MOD}} = \dfrac{1024\,\mathrm{MHz}}{1024} = 1\,\mathrm{MHz}$.

SE 10.4 Two 1's with a carry-in of 1 are added using a full adder. What are the outputs?

Solution: *Here the result is* $S = 1$ *and* $C_{out}= 1$; *(see* Table 10.1a, *row 4th). In decimal system* $1_{10}+ 1_{10}+ 1_{10}= 3_{10}$ *In binary system* $01_2+ 01_2+ 1(carry\text{-}in) = 11_2$.

SE 10.5 Design a circuit that counts the number of 1's present in three inputs X, Y and Z in two bit number AB, representing that count in binary. Assume active High logic. Also, give the truth table of the designed circuit.

Solution: *It will be easier to design the circuit if the truth table of the circuit is known. It is required to count the number of 1's in three inputs X, Y, Z each of which can assume a value of either 0 or 1. Therefore, the truth table must have a number of rows equal to the possible number of different combinations of X, Y and Z values. Total number of such combinations will be 8, as given in the truth table. Further, the number of 1's in a particular combination of X, Y and Z is to be represented by a two bit binary number AB. It means that AB should be 00 if there is no 1, be 01 if there is one, 10 if there are two 1's and 11 if there are three 1's. The truth table of the process will, therefore, look like as follows:*

In order to generate binary digit 'A', the desired values as represented in column under A in Table SE10.5 *may be produced by logic expression,*

Table SE10.5 Truth table for example (SE 10.5)

X	Y	Z	A	B
0	0	0	0	0
0	0	1	0	1
0	1	0	0	1
0	1	1	1	0
1	0	0	0	1
1	0	1	1	0
1	1	0	1	0
1	1	1	1	1

The last column of the truth table (i.e. B) can be easily identified as the output of the combination of two logic- EXNOR gates as are drawn below, and the truth table for $B = X \oplus Y \oplus Z$ is shown in Fig. SE10.5.

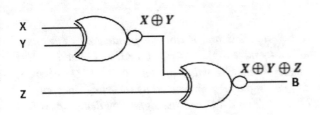

X	Y	Z	B
0	0	0	0
0	0	1	1
0	1	0	1
0	1	1	0
1	0	0	1
1	0	1	0
1	1	0	0
1	1	1	1

Fig. SE10.5 Circuit for example (SE 10.5)

$A = XY + XZ + YZ$ *and such logic may be generated by a combination of three AND and two OR gates as shown in* Fig. SE10.5.1, *where the truth table of A is also given.*

SE 10.6 What is the difference between a S-R latch and a D-latch? Write the truth table and the logic symbol for a D-latch. The clock pulses C (or E) and D-pulses applied to a D-latch are shown in the graph given below. Draw the signal at output Q on the same grid (graph).

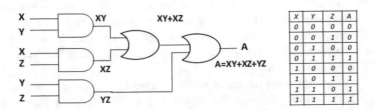

X	Y	Z	A
0	0	0	0
0	0	1	0
0	1	0	0
0	1	1	1
1	0	0	0
1	0	1	1
1	1	0	1
1	1	1	1

Fig. SE10.5.1 The desired circuit

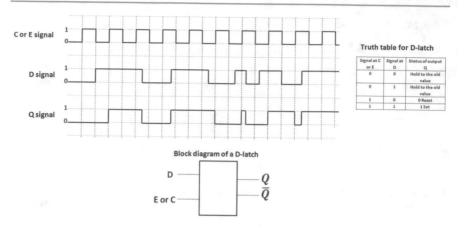

Truth table for D-latch

Signal at C or E	Signal at D	Status of output Q
0	0	Hold to the old value
0	1	Hold to the old value
1	0	0 Reset
1	1	1 Set

Block diagram of a D-latch

Fig. SE10.6   Circuit for example (SE 10.6)

Solution: *In a S R latch the two inputs S and R are independent of each other, which means that both S and R may have logic values 0 0 or 0 1, 1 0 or 1 1. This may produce a metastable state where the value of output Q (or $\overline{Q}$) becomes undefined. To avoid this in D-latch signal R is produced by inverting the signal S. The signal that is applied at input S is called D and the signal applied at input R is inverse of D, i.e. $\overline{D}$. With this there is no possibility that the latch may go to an undefined or metastable state. Block diagram, truth table and the time relation of signal Q are shown in* Fig. SE10.6.

SE 10.7 What is the main difference between a latch and a flip-flop? Draw the block level, as well as logic gate level circuit diagrams of a JK- Master-Slave flip-flop and give its truth table, as well the time relationship between the clock signal, the given J-signal and the Q-output.

Solution: *The main difference between a latch and a flip-flop is that a latch may change its state by sensing the level of the input which means that latches are level triggered. On the other hand, flip-flops change their state when the control signal either goes from low level to high level or from high level to low level. In other words, flip-flops are edge (either leading or the trailing) triggered. They change their state only either at the leading or the trailing edge of the control pulse. In general, a flip-flop contains two latches together; therefore, they are slow and also consume more power than a latch. The desired block level and the logic gate level circuit diagrams are given respectively in* Figs. 10.21 and 10.23. *The truth table for a J K Master-Slave is given in* Table 10.7a.

Time relationships between clock pulses, triggering signals and output signals for a JK Master-Slave flip-flop are shown in Fig. SE10.7. *As indicated in this figure, the Master FF1 is triggered by sharp pulses derived from the leading edges of clock pulses, while the Slave FF2 is enabled by the sharp pulses obtained from the rising edge of the inverted clock pulses. Master triggering signals marked as 1,*

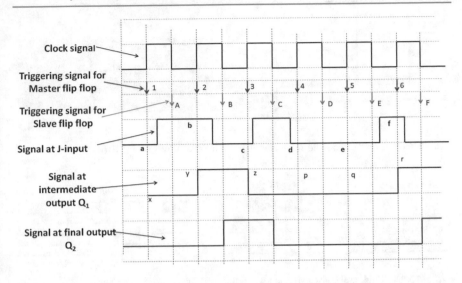

Fig. SE10.7 Time relationships of triggering and output signals for a JK Master-Slave flip-flop

2, 3.....and the Slave triggering pulses as A, B, C..... The Slave enabling pulses occur at a time lag of T/2 with respect to the corresponding Master triggering signal.

At each triggering pulse, Master FF1 senses the current logic state of J-signal and pass it to the intermediate output Q_1 where it remains latched (for a period of T/2 where T is the clock time period) till the Slave FF2 is enabled by its triggering pulse and passes it to the final output Q_2. The logic state recorded by Q_2 remains latched at Q_2 till the new value is passed by the Slave FF2. Thus, whatever logic state of J-signal is captured by the Master FF1 reaches the final output after a time delay of T/2 and remains latched Q_2 for a period of T/2 when a new logic state captured by the Master is passed to it. As such the Q_2 output mimics the intermediate output Q_1 with a time delay of T/2.

SE 10.8 Two half adders are connected as shown. Fill the blank columns of the accompanying table (Fig. SE10.8a).

Solution: *Truth tables for Half Adder-1 and Half Adder-2 are given, respectively, in Tables 1 and 2 of Fig. SE10.8b. It is state forward to write the values of SUM_1 from Table 1 for the desired combinations of X and Y. One then pick the value of C_1 from Table 1 for the chosen combination of X and Y and then look to Table 2 and select the corresponding value of SUM_2 from Table 2 to complete the desired table.*

For example, when X = 0 and Y = 1, then from row-2 of Table 1 SUM_1= 1. Also, for this combination of X and Y, C_1 from Table 1 has the value C_1= 0. Now use Table 2 to get the value of SUM_2 for Z = 1 & C_1= 0, which from the third column of Table 2 comes out to be SUM2 = 1. Thus, the first row of Table 3 may be

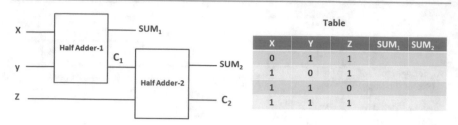

Fig. SE10.8a Logic circuit and the incomplete truth table of problem (SE 10.8)

Table				
X	Y	Z	SUM₁	SUM₂
0	1	1		
1	0	1		
1	1	0		
1	1	1		

Table-1

X	Y	SUM₁	C₁
0	0	0	0
0	1	1	0
1	0	1	0
1	1	0	1

Table-2

Z	C₁	SUM₂	C₂
0	0	0	0
0	1	1	0
1	0	1	0
1	1	0	1

Table-3

X	Y	Z	SUM₁	SUM₂
0	1	1	1	1
1	0	1	1	1
1	1	0	0	1
1	1	1	0	0

Fig. SE10.8b Truth tables for half adders and the final table

completed. The same method may be used for the completion of the complete table. Final complete table is given as fable 3 in the Fig. SE10.8b.

SE 10.9 It is required to synthesize a 4-bit adder using three full adders and one half adder. Draw a block level (not gate level) diagram of the **4-bit** adder, level all inputs and outputs and identify the MSB and LSB in the diagram.

Solution: *Block level circuit diagram of the desired 4-bit adder made using one half adder and three half adders is shown in Fig. SE10.9. The positions of the least significant and most significant digits have also been marked in this figure. The four bit binary number obtained by the addition of four two bit binary numbers* $X_1Y_1 + X_2Y_2 + X_3Y_3 + X_4Y_4$ *will be* $SUM_4SUM_3SUM_2SUM_1$

SE 10.10 Draw the time profile of signal X corresponding to the time profiles of the clock signal and signals at ports A and B. Assume that the initial logic state of Q is Low (or 0) (Fig. SE10.10).

Solution: *Essentially the logic circuit shows a D-flip-flop which is activated by the leading edge triggering pulses derived from the clock signal. Triggering pulses are marked as 1,2,3...5 in the figure.*

A typical property of a D-flip-flop is that whatever signal is present at the D-input at the instant when triggering pulse enables it, the same state of the signal is passed over to the Q-output and remains latched at Q-output till the next triggering pulse again enable the flip-flop. Therefore, during the time interval of two

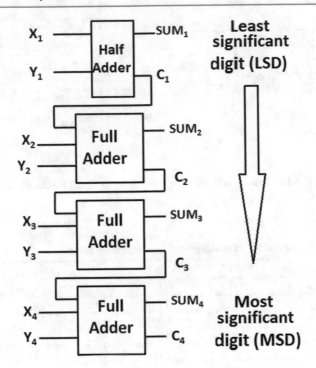

Fig. SE10.9 Four bit adder made using a half adder and three full adders

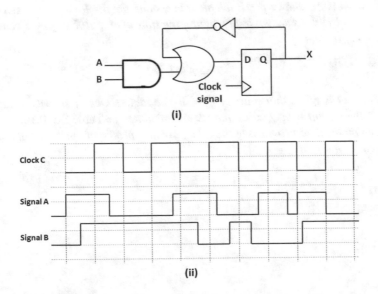

Fig. SE10.10 i Logic circuit **ii** Time profile of Clock pulse, A and B signals

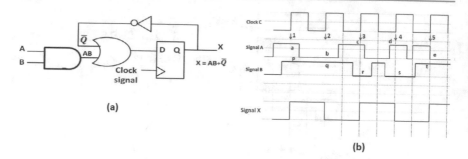

Fig. SE10.10.1 a Logic circuit with Boolean expressions for signal at each port **b** time profile of the logic signal at X

Table SE10.10 Logic states of different signals at triggering instants

Triggering pulse No.	A	B	$\overline{Q}$	$X = AB + \overline{Q}$
1	1	1	1	1
2	0	1	0	0
3	1	0	1	1
4	1	0	0	0
5	0	1	1	1

successive triggering pulses the Q-output remains latched to the previous value. Figure SE10.10.1a shows the Boolean expressions for the logic states at various ports of the circuit. As shown in this figure, the state of output X may be given by the Boolean expression

$$X = AB + \overline{Q}$$

Further, it is given that initially Q = 0. The logic states of signals at different ports at the instant of triggering pulses are tabulated in Table SE10.10.

Based on the data provided in the table, the time profile of the signal at output X is shown on the grid.

Problems

P10.1   Design a circuit that counts the number of zeros (0's) present in three
             inputs A, B and C in two bit number PQ, representing that count in
             binary. Give the truth table of the designed circuit.

Answer: [In Solved example SE 10.5 it was required to count the number of 1's,
while in the present problem it is required to count the number of zeros. If one
makes the truth table for the process, it will be found that $P = \bar{A}$ and $Q = \bar{B}$.
Therefore, the desired outputs P and Q may be obtained jus by including a NOT
gate in the outputs of the circuits given for A and B in the solved example]

P10.2   Time profile of the $\bar{Q}$ signal obtained from a SR latch made using two
             NOR gates is shown in Fig. P10.2). What will be the logic states of
             signals at S and R ports at instant t_1 and t_2?

Answer: [At t_1, S = 1, R = 0; at t_2 S = 0, R = 1]

P10.3   Figure P10.3 shows a D-flip-flop made of two D-latches, assuming that
             latches are enabled by leading edge of the clock pulse, discuss the
             working of the circuit if (i) Latch-1 is replaced by an AND gate
             (ii) Latch-2 is replaced by an AND gate. What will the value of Q in both
             cases?

Answer: [(i) if Latch-1 is replaced by AND: the And gate will pick up the
instantaneous value of D-signal when the clock is in state-1 but will register a value
0 when the clock is in state 0. Because of the inversion of the clock signal when
applied to latch-2 AND gate will pass on logic state 0 whenever Latch -2 is enabled
by the triggering pulse. As a result, the output Q will always be in state 0. Same is
true for (ii)]

P10.4   Clock signal and three times inverted clock signal are applied to an AND
             gate as shown in Fig. P10.4. Draw the time profile of the signal at B and

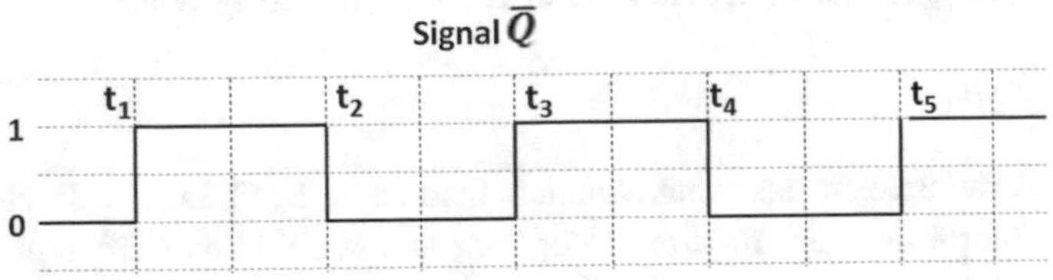

Fig. P10.2 Time profile of problem P10.2

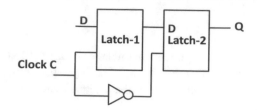

Fig. P10.3 Block diagram of circuit of problem (P10.3)

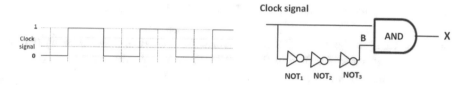

Fig. P10.4 Clock signal and logic circuit

the output X of the AND gate assuming that each NOT gate produces a time delay of Δ seconds.

Answer: [Profiles are given in Fig. P10.4.1. This is the method of producing leading edge trigger pulses of duration 3Δ)

P10.5 State of the two D-flip-flop circuit shown in Fig. P10.5) is represented by the two bit binary number $(Q_1 Q_2)$. If the initial state of the circuit is (0 0), determine the transition sequence (i.e. how values of $Q_1 Q_2$ change with each transition) of the circuit.

Answer: [The sequence is: 0 0; 1 1; 0 1; 1 0; 0 0....]

P10.6 Three D-flip-flops in the circuit of Fig. P10.6 are driven by leading edge signal of the clock. Before switching on the clock, the logic states of Q_1, Q_2 and Q_3 were, respectively, 0, 1 and 0. What will be the logic states of these outputs immediately after the first triggering pulse?

Answer: [0,1,1]

P10.7 A multiplexer has input channels marked A, B, C, D........P. How many control lines are required to operate this MUX? Give different one bit sequences that may be used to identify different input channels.

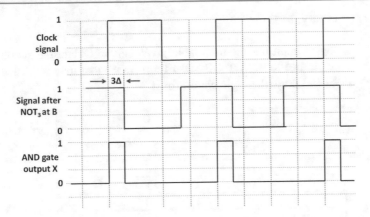

Fig. P10.4.1 Time profiles of logic pulses

Fig. P10.5 Circuit for problem (P10.5)

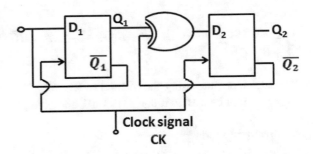

Fig. P10.6 Circuit diagram for problem (P10.6)

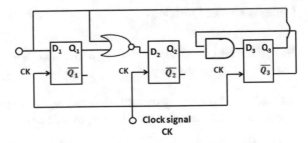

Answer: [The MUX has 16 input channels from A to P, each channel may be assigned a 4-bit digital number, like A = 0000, B = 0001, C = 0010 and so on; therefore, a minimum of 4 control lines are required]

P10.8 P10.8 Draw the logical circuit to implement $X = P\overline{Q} + PQ$. Next, simplify the Boolean expression and draw the corresponding circuit.

Answer: [It is easy, try]

P10.9   A logic circuit has the truth table given below. Obtain the simplified function in PRODUCT OF SUM form (POS) that may implement the table and draw the logic circuit for it.

A	B	C	X
0	0	0	0
0	0	1	1
0	1	0	1
0	1	1	1
1	0	0	1
1	0	1	1
1	1	0	1
1	1	1	0

Answer: [SOP form is: $(\overline{B}+B\overline{A}+A\overline{C}+C\overline{A}+B\overline{C}+C\overline{B})$, POS: $(A+B+C)$ $(\overline{A}+\overline{B}+\overline{C})$]

P10.10   Compute using the method of 2's complement the difference of binary numbers **1010100 − 11011001** and give you answer in decimal number.

Answer: [-133_{10}]

Multiple Choice Questions

Note: Tick all correct alternatives to a question. Some questions may have more than one correct alternative

MC10.1 In a 5-bit binary counter, the ratio of the frequency of signals at the third output to the frequency at the fifth output will be

- (a)   4:1
- (b)   2:1
- (c)   5:3
- (d)   3:4

MC10.2 A JK flip-flop toggles when the logic states are

- (a)   J = 1, K = 0
- (b)   J = 0, K = 1
- (c)   J = 1, K = 1

(d)   J = 0, K = 0

MC10.3 Racing is

(a)   Controlled toggling of the output of a latch
(b)   automatic switching of the output of a latch from 0 to 1 and 1 to 0.
(c)   due to the feedback from the output to the input of a latch
(d)   due to the long duration of the clock signal

MC10.4 Racing does not occur in JK Master–slave flip-flop because

(a)   There is no feedback
(b)   the slave flip-flop is disabled when the master flip-flop is enabled
(c)   both the master and the slave circuits are disabled simultaneously
(d)   clock pulses are of short duration.

MC10.5 In a High active input S-R latch, the inputs S and R are, respectively, in logic states 1 and 0. The state latched is

(a) $Q = 0, \bar{Q} = 1$
(b) $Q = 1, \bar{Q} = 0$
(c) $Q = 0, \bar{Q} = 0$
(d) $Q = 1, \bar{Q} = 1$

MC10.6 Which number system has the base of 16

(a)   Octal
(b)   Decimal
(c)   Binary
(d)   hexadecimal

MC10.7 The 9-bit number 111011000 is equivalent to

(a)   #138 h
(b) 472_8
(c) 730_8
(d) 472_{10}

MC10.8 16-bit number 1100 000 1001 0010$_2$ is equivalent to hexadecimal number

(a)   B092
(b)   A902
(c)   D092
(d)   C092

MC10.9 Which of the following pairs of logic gates give a logic state-1 in its output when all its inputs are set to logic state 0

(a) NOR and XNOR
(b) NAND and XOR
(c) OR and XNOR
(d) AND and XOR

MC10.10 Two switches are put at the two ends of a corridor both switches independently control the lighting of the bulb at the middle of the corridor. Which logic gate may represent the logic of the two switches

(a) NAND
(b) AND
(c) XOR
(d) OR

MC10.11 Minimum number of NAND gates required to implement the Boolean-function $Y = X + X\overline{A} + XB\overline{A}$ is

(a) 0
(b) 3
(c) 4
(d) > 5

MC10.12 Which of the following sentences is/are logic statement(s)

(a) Is it raining?
(b) It is raining
(c) Aligarh Muslim University is in Pakistan
(d) Go back!

MC10.13 Which set of logic states for A, B and C, respectively, will ensure that the output Y of the following circuit is in logic state-1 (MC10.13.1).

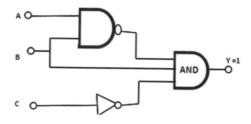

Fig. MC10.13.1   circuit diagram for problem MC-7.13

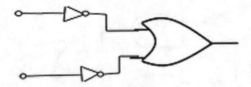

Fig. MC10.14.1 Circuit diagram for problem MC7.14

(a) 1, 1, 1
(b) 1, 1, 0
(c) 1, 0, 1
(d) 0, 1, 0

MC10.14 Circuit shown in Fig. MC10.14.1 is equivalent to

(a) NAND gate
(b) AND gate
(c) NOR gate
(d) XNOR gate

MC10.15 Number of valid entries in the truth table of a SR flip-flop are

(a) 2
(b) 3
(c) 4
(d) 5

MC10.16 The output of a circuit depends only on the input to the circuit, the circuit is

(a) Combinational circuit
(b) Sequential circuit
(c) Latch
(d) counter

MC10.17 A 4-bit adder will require a minimum of

(a) One half adder and three full adders
(b) Four full adders
(c) Four half adders
(d) two full adders

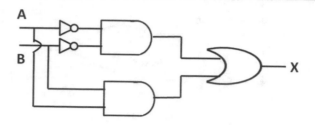

Fig. MC10.20 Circuit for problem (MC10. 20)

MC10.18 In Master-Slave D-flip-flop, Master and Slave flip-flops are activated,-respectively, by

    (a) Leading edge of clock pulse, trailing edge of clock pulse
    (b) Trailing edge of clock pulse, leading edge of clock pulse
    (c) Leading edge of clock pulse, leading edge of clock pulse
    (d) Trailing edge of clock pulse, trailing edge of clock pulse

MC10.19 Flip-flops and latches are triggered, respectively, by

    (a) Clock pulse level, Clock pulse level
    (b) Clock pulse edge, Clock pulse edge
    (c) Clock pulse level, Clock pulse edge
    (d) Clock pulse edge, Clock pulse level

MC10.20 Logic circuit of Fig. MC10.20 represents the logic equation

    (a) $X = \overline{A}B + AB$
    (b) $X = \overline{A}\overline{B} + AB$
    (c) $A \oplus B$
    (d) $X = \overline{A}B + AB$

MC10.21 Maximum counting speed of an 8-bit binary counter made up of flip-flops of transition delay of 25 ns is

    (a) 100 MHz
    (b) 50 MHz
    (c) 10 MHz
    (d) 5 MHz

MC10.22 Two's complement of binary number 1111 is

    (a) 1000
    (b) 0011

   (c) 0001
   (d) 1100

MC10.23 Parity of binary number 1100100_2 is

   (a) Even
   (b) Odd
   (c) 1
   (d) 0

Answers to Multiple Choice Questions

MC10.1: (a)
MC10.2: (c)
MC10.3: (b), (c), (d)
MC10.4: (b)
MC10.5: (b)
MC10.6: (d)
MC10.7: (a), (c), (d)
MC10.8: (d)
MC10.9: (a)
MC10.10: (c)
MC10.11: (a)
MC10.12: (b), (c)
MC10.13: (d)
MC10.14: (b)
MC10.15: (b)
MC10.16: (a)
MC10.17: (a), (b)
MC10.18: (c)
MC10.19: (d)
MC10.20: (b), (c)
MC10.21: (d)
MC10.22: (c)
MC10.23: [(b), (c)]

Short Answer Question

SA10.1   What is the difference in the operation of a half adder and a full adder?
          Draw a circuit diagram of a full adder.
SA10.2   Explain the terms (i) ripple adder (ii) carry lookahead adder (iii) carry
          skip and (iv) carry select adder.

SA10.3   Draw a block diagram (not gate level) of a four bit full adder with identification of least and most significant digits of the output.

SA10.4   Draw the logic circuit for an S-R latch and give its (i) truth table and (ii) performance table.

SA10.5   What is the difference between an S-R and a $\overline{SR}$ latch? Draw the logic gate level circuit diagram for a $\overline{SR}$ latch.

SA10.6   Show the time relationship between the inputs and the outputs of a $\overline{SR}$ latch.

SA10.7   What is the difference between a gated latch and a flip-flop? Explain your answer with the help of gate level circuit diagrams.

SA10.8   How many binary bits are necessarily required to represent 748 different numbers? Explain your answer.

SA10.9   Prove the following Boolean equation:

$$A + A.\overline{B} + A.\overline{B}.\overline{C} + A.\overline{B}.C + \overline{C}B.A = A$$

SA10.10   What is meant by the parity of a digital data? What is its significance?

SA10.11   Represent the decimal number (-10_{10}) as a negative binary number

SA10.12   Parity bit of a given data is 'even parity- 1'. What do you understand by this value of parity bit data?

SA10.13   Write the steps required to write the value of a given negative decimal number in binary form

SA10.14   Enumerate the steps for finding the difference of two binary numbers using the method of 2's complement.

SA10.15   Draw the logic circuit to generate the negative parity bit for a 4-bit digital word.

SA10.16   What is parity bit? why is it required? Differentiate between the negative parity bit and positive parity bit.

SA10.17   Draw the logic circuit to check the parity of a 4-bit digital word. What inference you will draw If the parity bit provided with the data is in the format of positive parity and the output of the checking circuit is 1.

SA10.18   What is the 2's complement of a binary number and what is its significance?

SA10.19   What is the function of a multiplexer? How many control lines a 4:1 MUX will need?

SA10.20   Give the mini-term for the decoder that wants to detect the tenth pulse of the clock counter.

SA10.21   What is the essential difference between a truth table and an excitation table?

Sample Answer to Short Answer Question SA10.10

What is meant by the parity of a digital data? What is its significance?

Answer: *In the serial transmission of digital data, data is transmitted bit by bit. Though digital transmission is almost free of noise and other environmental effects, it is still possible that the data gets corrupted during transmission; it may happen that a bit which was 0 in original data may get corrupted to 1 or vice versa. To detect if such error has taken place in transmission, each digital word of n-bits is transmitted with one extra bit, i.e. it is transmitted as a word of (n + 1) bits. The first bit (which is the extra bit) contains information regarding the parity of the word. The number of bits in the word that have value 1 decides the parity of the word. If the word contains an even number of bits with value 1, the parity of the word is called even and is denoted by 0, and if the number of bits with value 1 in the word are odd, the parity of the word is called negative and is denoted by 1. The first bit of the transmitted (n + 1) bit word contains the parity information. For example, an 8-bit word 11001001 that has 4-bits with value 1 has even parity which will be denoted by 0. While transmitting this word, it will go as a 9-bit word 011001001 with the leading zero indicating its parity. At the receiving end, the parity of the received word (i.e. of the last eight bits) is checked and if it is the same as it is given by the leading parity bit, then it is confirmed that the transmission is faithful. If the actual parity of the received word does not agree to the parity information provided by the leading bit then it means that the data has been corrupted in transmission. Thus, parity information plays a very significant role in detecting data transmission errors.*

Long Answer Questions

LA10.1   Discuss with necessary block and circuit diagrams the working of a half adder.

LA10.2   With the help of block level diagram, explain how 4-bit numbers may be added. Also, draw the logic circuit for one of the blocks used in the adder.

LA10.3   Explain the significance of the parity bit in linear data transfer and with necessary logic circuit discuss the process of generation and checking of parity bit.

LA10.4   What method is used to represent a negative number in binary code? How this representation is helpful in implementing subtraction using conventional adders?

LA10.5   With the help of a suitable example, explain the purpose of a multiplexer and with necessary details explain the working of a 2:1 MUX.

LA10.6   Explain the difference between a combinational and a sequential logic circuit. How does a sequential circuit acquire the property of retaining memory? Give logic circuit of an S-R latch and explain its working.

LA10.7   In what respect an enabled latch differ from an enabled flip-flop? Discuss the operation of a D-flip-flop and give its truth table.

LA10.8   Which flip-flop is called a universal flip-flop and why? Describe the operations of a JK Master-Slave flip-flop. What is the advantage of using a Master-Slave arrangement?

LA10.9   Distinguish between the phenomena of toggling and racing. What is the cause of racing in sequential circuits and how can it be avoided? Explain by taking a suitable example.

LA10.10   What is meant by the transition or delay time of a logic unit? Explain how glitch es may be generated because of the delay time. Explain with necessary logic circuits how the delay time may be used to generate leading edge and trailing edge triggering pulses from the clock signal.

LA10.11   Discuss the operation of a 4-bit decade counter and describe its characteristics.

LA10.12   Differentiate between an asynchronous and a synchronous counter. Give block level diagram of a 4-bit binary counter and its truth table.

Special Circuits and Devices

11

Abstract

Architectural layout and details of operation of some digital circuits/devices like semiconductor memories, analog-to-digital/digital-to-analog converters and arithmetic logic unit, are included in this chapter. These circuits/devices are essential components of computers and some of these are also used in digital cameras, watches, etc.

11.1 Semiconductor Memories

11.1.1 Introduction

Recording of information has played a crucial role in the development of human civilization. Different modes of recording information; carving on stones, writing on barks, on metallic plates, etc. have been used in past. With the invention of paper, pen, ink, printing press, etc. modern-day civilization achieved a landmark in recording of information or in generating data. With computers entering the lives of civilized World, the need for accumulating and storing a large amount of information or data increased by many folds. To keep pace with the advancement in computer sciences, new, fast and efficient means of data storage have been invented, namely, DRAM which may be looked as the counter part of pen, flash memory; a substitute for paper, disc as a conventional book, and cloud data centre as a library.

Since analog data requires unmanageable large memories, semiconductor memories are associated with digital data that may be managed with limited memory space. Memory is a semiconductor device capable of holding digital data or information in digital form. It is one of the most essential components of any computer, smartphones, digital cameras, etc. With ever growing demand for larger memory devices, new techniques are being developed to supplement the already existing technologies of ROM, RAM, EEPROM, flash memory, DRAM, SRAM, SDRAM, FRAM, etc.

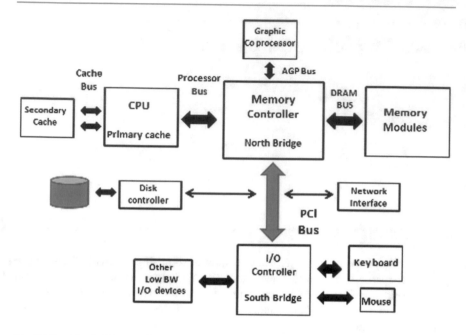

Fig. 11.1a Block diagram of a computer

Figure 11.1a shows a block diagram of a modern computer. CPU (central processing unit) is the heart of a computer where all operations are performed and controlled. Memory is an essential part of any computer. But computers use a hierarchy of memories, fasted within the CPU called cache L1, secondary cache next to CPU in communication with it via cache bus. Larger memory modules, basically DRAM modules are housed a little away from the CPU under control of memory controller block. All other systems, except the cache memory, communicate with the CPU via two connection bridges: the North Bridge and the South Bridge. Generally, all fast or high BW systems are connected through the North Bridge while Low BW systems through the South Bridge. Large memory modules are connected via North Bridge and Memory controller to the CPU. Some details of memory Banks kept in large memory modules of computers will be discussed here.

11.1.2   Memory Types

Broadly speaking, semiconductor memories may be divided into two classes: (i) program memory and (ii) data memory. Program memory stores instructions (in digital form) for running a device like microprocessor or a laptop, etc. Data memory stores the data used or generated by the device.

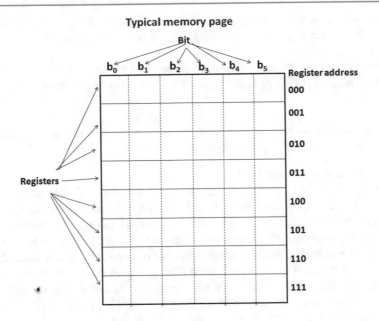

Fig. 11.1b Layout of a typical memory page

Before proceeding further, let us understand the elementary structure of a typical memory. In some ways, a digital memory may be compared to a notebook. Like a notebook, a memory may have pages. The structure of a memory page is like the page of a notebook, it is divided into rows, each row being called a register. Each register may contain several bits. For example, Figs. 11.1a, 11.1b shows a typical memory page with 8 registers, each register having, say, 6 bits. The sequence of bits in a given register is termed as bit-word.

Location of a bit in a memory stack is defined through the bit address. A bit address consists of three parts: (i) page address+(ii) register address+(iii) bit number. Suppose a memory stack has ten pages, each of these pages may be numbered from 0 to 9 using 4-bit address; 0000, 0001, 0010, 0011, 0100, 0101, 0110, 0111, 1000, 1001. The register address will depend on how many registers are accommodated on each page, if the number of registers per page is 8, then each registered on a page may be identified by a three-bit word, 000, 001, 010, 011, 100, 101, 110, and 111. Location of a bit in a register may be identified by the bit number, which in case of 6 bits per register may have values from 000, 001, 010, 011, 100, 101 and 110.

Each memory page may be looked as a combination of horizontal and vertical grids, the location of a bit on this grid may also be addressed by identifying the X (column number) and y (row number) coordinates. In most memories, a particular bit is identified by its X and Y addresses (coordinates) that are separated by a limiter. This way of addressing may not require memory page number if rows are numbered continuously across all pages.

Self-assessment question SAQ: What will be the address of 5th bit in third register of page 5 of a ten-page memory stack having 8 registers per page?

Self-assessment question SAQ: What will be the total number of bits required to address a bit in a two-page memory stack with 5 registers per page, each register having 4 bits?

Depending on the way of its operation semiconductor memories may be categorized in two broad classes: (a) ROM and (b) RWM.

(I) ROM: Read-only memory

These are memories in which data can be stored only once, though the same data may be retrieved a number of times, new data cannot be written after erasing the old data. Such memories are used to hold data (or instructions) to run a microprocessor or to boot the computer, etc. These read-only memories are very important from the point of getting corrupted. Since data is recorded only once in such devices, there is no chance of data getting changed/altered during operations. Another advantage of ROM is that they hold the data even when power is removed. It is for this reason that basic input–output system (BIOS) of a computer is kept in a ROM. ROM may also be used as a data-table and as a function generator.

Depending on the technology used, writing of data on ROM requires special hardware. In some ROMs, it is possible to erase the existing data, but that too requires special hardware.

'Mask-programmed' or classic ROM chips contain integrated circuits. A ROM chip sends current through a predetermined pathway and returns back via the output pathway determined by the manufacturer. Rewriting is functionally impossible and therefore data once written cannot be modified. It is difficult to produce the original template of a chip but once developed, it is very cheap to go for mass production.

Programmable ROM or **PROM** is essentially a blank version of ROM which can be programmed using a special programmer. Erasable Programmable ROM or **EPROM** chip allows one to write and rewrite on the chip. These chips use a specialised quartz window through which a specialized EPROM programmer emits a specific frequency of ultraviolet light that burns out tiny charges in the chip to reopen its circuit. This process essentially changes the chip into a blank ROM. **EEPROM,** Electrically Erasable Programmable Read-Only Memory, is like EPROM with the difference that in this the chip may be brought back to the original blank condition by the application of an electric field, instead of ultraviolet light. EEPROM is used in storing current time and date in different instruments. **Flash ROM** is like EEPROM with the difference that in EEPROM only one byte of data can be deleted or written at a particular time. However, in Flash ROM blocks of data, usually 512 bytes, can be deleted or written at a time; they are therefore much faster. Flash ROMs are used to store messages in mobile phones and photographs in digital cameras.

In general a ROM may have k-input (addresses) lines and n output (data) lines. The address lines specify a memory location. Such a ROM is called a $2^k \times n$ROM, the block diagram of which is shown in Fig. 11.2.

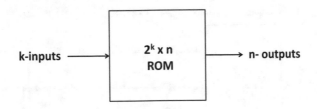

Fig. 11.2 Block representation of a $2^k \times n$ ROM

A $2^k \times n$ ROMXE"ReadOnlyMemory(ROM)" uses an address decoder such that the k address lines selects one **word** of the 2^k words of data stored in the ROM. Each of the $2^k \in n$bits inside of the ROM are programmable using opening/or closing switches. A ROM can implement multiple –input/multiple output logic functions inside it. Address lines are the logic function inputs and data outputs become logic functions. When an input or address is presented, the data stored in the specified memory location appears at the output.

As an example, let us implement the three input logic functions $f_0 = \sum(0, 1, 2, 6)$, $f_1 = \sum(2, 3, 4)$, $f_2 = \sum(0, 1, 5, 7)$, using ROM. First we make a table of inputs and corresponding outputs. Since there are three inputs $k = 3$, and three outputs, we require a 3 to $2^k = 8$ decoder. Figure 11.3 shows the implementation of the three functions using ROM.

In principle, a ROM is a combination of a decoder and OR gates. Fig 11.4 shows how two inputs and four functions may be implemented using a ROM. Since there are only two inputs, $k = 2$, therefore there will be 4 output leads (00, 01, 10 and 11). Further, each OR gate representing the output will have four inputs. Since $f_4 = \Sigma(1, 3)$, 1 and 3 inputs of OR gate will be connected to the corresponding horizontal outputs. In this way, all four functions may be implemented.

a	b	c	f_0	f_1	f_2
0	0	0	1	0	1
0	0	1	1	0	1
0	1	0	1	1	0
0	1	1	0	1	0
1	0	0	0	1	0
1	0	1	0	0	1
1	1	0	1	0	0
1	1	1	0	0	1

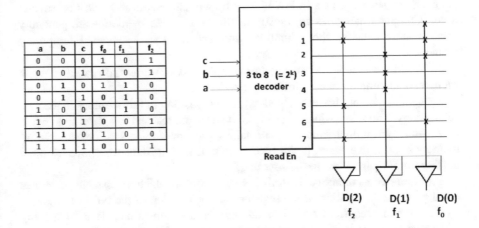

Fig. 11.3 Implementing functions using ROM

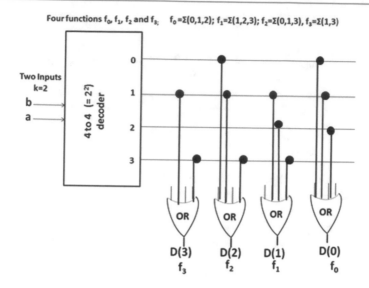

Four functions f_0, f_1, f_2 and f_3; $f_0=\Sigma(0,1,2)$; $f_1=\Sigma(1,2,3)$; $f_2=\Sigma(0,1,3)$, $f_3=\Sigma(1,3)$

Fig. 11.4 Block diagram of implementation

Advantages of ROM

(i)   Non-volatile,
(ii) Cannot be changed accidentally,
(iii) Easy to test,
(iv) More reliable than RAM,
(v)   Do not require refreshing,
(vi) Considerably lower coast.

(II) **RWM: Read and write memory**

As the name suggests, data in RWM may be written, erased and rewritten several times. Computer memories are mostly of this type. These memories are generally used to store data for short durations and are therefore also called temporary memories.

Read and write memories may be further specified as **sequential read and write memory** and **random-access memory (RAM).**

Sequential memory is relatively slow; the bit position in this memory is reached sequentially. Those bits which are near to the head are reached in a short time while those which are further from the head take more time to access. For example, in memory stack of five pages, it will take more time to write data or retrieve it from a register on 5th page than on the first page.

Random-access memory or RAM, is a memory in which assessment time does not depend on the location of the register and the location of the bit. Data may be written or retrieved from any data address in the same time. That is the big advantage of this type of read and write memories and they are fast also.

RAM may further be classified as **SRAM** and **DRAM,** standing, respectively, for static random-access memory and dynamic random-access memory.

SRAM uses bi-stable latches to store data and, therefore, data remains stored till the power is switched off. Memory of SRAM is stable, so long as power is there. On the other hand, DRAM is volatile; data stored in the memory slowly disappear unless it is rewritten in the memory at regular time intervals.

(i) SRAM Technology

Static random-access memory (SRAM) is meant to fulfil the two important requirements of computers. Firstly, it may provide a direct interface with the CPU at speeds that cannot be attained with DRAM. Secondly, SRAMs consume very little power as compared to DRAM, and, therefore, they replace DRAMs where ever possible. In its role as an interface, SRAM serves as an external cache, as shown in Fig. 11.5 that shows the memory configuration of a microprocessor.

On account of low-power consumption as compared to DRAM, which needs higher refreshing currents, SRAMs are used in most of portable devices.

(a) **SRAM cell** Three types of SRAM cells are in use.

(i) **4T or four-transistor cell**, this is the most common type of cell, very small in size and quite reliable. 4T cell uses two NMOS transistors as pass transistors. Gates of these transistors are connected to the Word line, thus connecting the cell to the columns. The other two transistors synthesize flip-flop inverters. Very high polysilicon resistors serve as inverter loads. Figure 11.6 shows the layout of a 4T cell. Despite the great advantage in size, a 4T cell has three main disadvantages;

Firstly, the cell has a standby current flowing through one transistor all the time, secondly because of very high polysilicon resistances, the cell is quite sensitive to noise and soft errors and thirdly, the cell is not as fast as the 6T cell

(ii) **6T-six-transistor SRAM cell** Six-transistor SRAM cell was made to overcome some of the problems associated with 4T cell, particularly, to reduce sensitivity to noise and soft errors. In T6 cell, two polysilicon resistances used in T4 cell that were the main cause of noise sensitivity,

Fig. 11.5 Memory configuration of a microprocessor

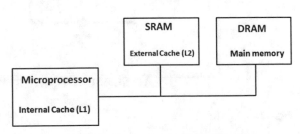

Typical PC Microprocessor memory configuration

have been replaced by two PMOS transistors. Thus there are in all six transistors, 4 NMOS and two PMOS. The structure of a 6T cell is shown in Fig. 11.7. Replacement of polysilicon resistive load by PMOS transistors reduces the standby current and improves noise performance, making cell faster and electrically superior to 4T cell. However, a considerable increase in size is a big disadvantage of this cell.

(iii)   **Thin-film transistor cell (TFT)** Standby current in an SRAM cell is essentially decided by the load attached to the flip-flop inverter circuit. With a view to reduce standby current, polysilicon resistances of 4T design were replaced by PMOS transistors. In TFT, the resistance of these PMOS transistors is further controlled by controlling the channel width of PMOS during operation. This is achieved by depositing several layers of polysilicon over the silicon surface of the PMOS. Thus a source/channel-drain is formed in the polysilicon layer. The structure a TFT SRAM cell is shown in Fig. 11.8. The overall performance of TFT is not very satisfactory.

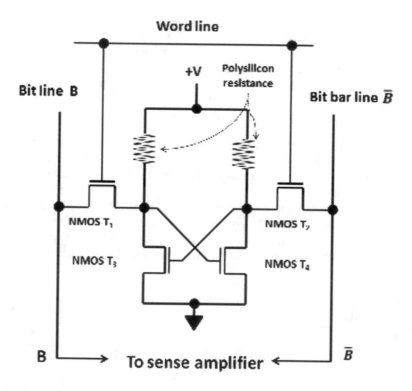

Fig. 11.6 4T SRAM Cell

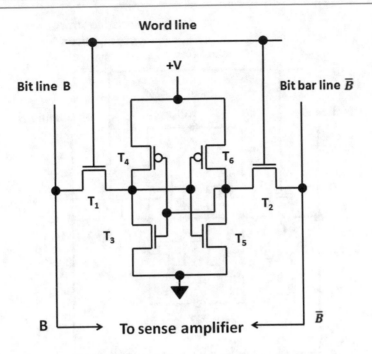

Fig. 11.7 Six-transistor SRAM cell

Working: A SRAM cell consists of a bi-stable flip-flop (Latch) that is connected to the internal circuitry through two transistors. When the SRAM cell is not addressed the two transistors remain closed and the data already latched to the bi-stable flip-flop remains stable. Since Flip-flop is a volatile device, it requires constant power to retain the data. But the data does not leak away if power is on, this is in contrast to the DRAM where data leaks even when the DRAM cell is powered.

Read operation

Let us first examine the read operation, for that let us refer to Fig. 11.9a, where two auxiliary capacitors C_1 and C_2 are connected at nodes A and B. Let us further assume that these capacitors are charged such that the voltage across them is say $V_{DD}/2$ V (in most cases $V_{DD}/2$ V defines logic state-1). Also, let us assume that the T_1 side of flip-flop is in logic state-1 and T_2 side in logic state-0. Now if the word line is set to logic state-1, i.e. the two transistors T_1 and T_2 are switched on, the two ends of the flip-flop will come in direct contact with nodes A and B. Node A will sense state-1 of the latch and since voltage at node A due to capacitor C_1 is same as the voltage on Latch 1, so no current will flow through the bit line. On the other hand, Node B will have higher voltage from capacitor C_2 but latch side will have 0 volt, therefore capacitor C_2 will discharge and a current will flow through the bit bar line. Now both bit and bit-bar lines will send signals to the sense amplifier. If the sense amplifier

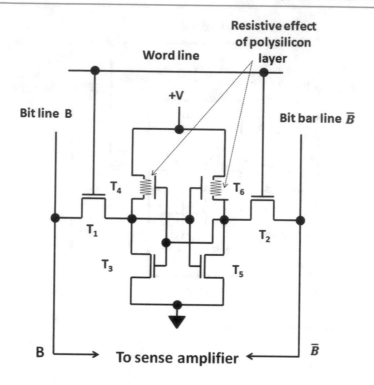

Fig. 11.8 Thin-film transistor (FT) SRAM cell

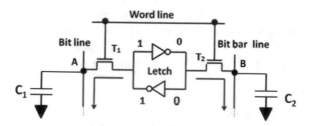

Fig. 11.9a Read operation of SRAM cell

gives out logic state-1, when no current comes from bit line and some current flows through the bit-bar line, then the output of the read operation will be 1.

In the other case if T_1 side of the flip-flop is in logic state-0 and the T_2 side in logic state-1 then on putting word line to logic state-1, current fill flow through node A, i.e. through the bit line and no current will flow through the bit-bar line. Signals from both the bit and the bit-bar lines will be compared by the sense amplifier, which in this case (current through B line and no current through $\overline{B}$ line) will be read as logic state-0.

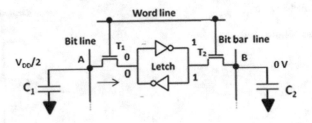

Fig. 11.9b Write operation on SRAM

Write operation

Write operation on SRAM chip is shown in Fig. 11.9b. Let us assume that at some instant T_1 side of the flip flop is in logic state-0 (zero volt) and T_2 side in logic state-1 ($V_{DD}/2$ V) .

If word line is in state-1, it means both transistors T_1 and T_2 are on, directly connecting the two sides of the flip-flop to notes A and B, respectively, then T_1 side of flip flop which was at zero volt will acquire voltage VDD/2 and will switch from state-0 to state-1. At the same time, T_2 side of flip flop which was in state-1 will switch to state-0 by discharging through node B.

Complete read and write operations are shown in Fig. 11.9c.

One of the main considerations in using the SRAM cell is its size. The size, in general, varies from manufacturer to manufacturer but it also depends on the memory size, for example, Samsung TFT of 4 Mbit has a size of 11.7 μm^2, while Sony 1Mbit TFT is of 20 μm^2. Widely used Intel Pentium 6T SRAM cell is of 33 μm^2 size.

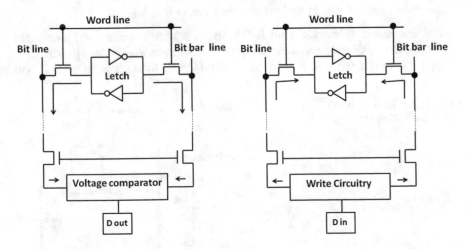

Fig. 11.9c Read and write operations of SRAM cell

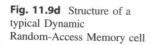

Fig. 11.9d Structure of a
typical Dynamic
Random-Access Memory cell

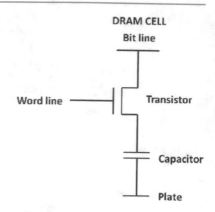

With the advancement of technology, some more SRAM cells with special applications have been developed, they include Synchronous, Asynchronous SRAMs, Burst mode SRAM and non-volatile SRAMs.

(ii) **DRAM Cell** Elementary structure of a Dynamic Random-Access Memory cell is shown in Fig. 11.9d. The cell contains a MOS transistor and a capacitor. Information in digital form (0, or 1) is stored on the capacitor. When the charge on capacitor is more than a preset value, it represents logic state-1; similarly, a charge less than a preset value corresponds to logic state-0.

Read and write operation

Refer to Fig. 11.9e to understand the read and write operations of a DRAM cell. As shown in the figure, capacitor C is the memory element while an auxiliary capacitor C_1 is connected at node A of the bit or data line. Capacitor C_1 is initially charged to some value, generally such that the potential across it is $V_{dd}/2$.

Read operation: When the word line is in logic state-1, transistor T is on and capacitor C gets directly connected to node A. Now there may be two possibilities, suppose C is charged (i.e. it is in logic state-1), then the voltage at node A will

Fig. 11.9e Read and write
operation for DRAM

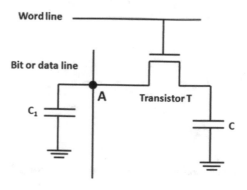

increase beyond $V_{dd}/2$. This information will be carried to the sense amplifier which will designate the status of C as 1 and this data will be transferred to the output buffer and from there to the output. On the other hand, if the potential across C is less than $V_{dd}/2$, the net potential at node A will decrease informing the sense amplifier that the logic state of C is 0 (zero). The bit line will transfer this information or data through the decoder to the output buffer, which in turn will pass on the data to the output. It may be observed that during read operation bit line performs data transferring to output. Another important aspect of read operation is that if C is in state-1, it gets discharged during read operation and its state changes to 0. However, the purpose of read operation is to know the present logic state of C and not to change it. Therefore, after every read operation, capacitor C is brought back to the logic state it has before read operation. This is called refreshing. Refreshing is automatically done after each read operation.

Write operation: During write operation, data input passes the data required to be loaded on capacitor C in terms of voltage. If the input voltage is more than $V_{dd}/2$, voltage of node A becomes more than $V_{dd}/2$. At this instant if transistor T is on, i.e. word line is in state-1, node A gets directly connected to C, and capacitor C gets charged to a higher value assuming logic state-1. On the other hand, if C was already in logic state-1, the voltage at node A does not affect the status of C.

The biggest disadvantage of the DRAM cell is the fact that the capacitor charge decreases with time even when the cell is properly powered. This happens because of the natural exponential discharge of the capacitance through lumped stray resistances. To overcome this shortcoming, the capacitor must be refreshed or rewritten with the same charge value that existed on it after each reading or at a regular interval. Rewriting existing data at regular intervals is called refreshing.

Figure 11.9f shows the simplified block diagram of DRAM. Block marked A in the figure stands for the sense amplifier. $\overline{RAS}$ and $\overline{CAS}$, respectively, represents Row Address Access Clock and Column Address Access Clock, bar over them indicating that both of them go active at a low level. As may be seen in the figure, gates of memory cells are connected to the rows. X and Y addresses are both presented at the same pad and multiplexed. Total operation cycle of DRAM may be divided into three segments; first segment or step consists of validating the row addresses internally by the $\overline{RAS}$ clock. The X addresses select one row through the row decoder, while the other rows not selected remain at 0 V. As may be seen in the block diagram, each cell of a row is connected to a sense amplifier, which senses if a charge has been loaded to the capacitor or not and translates it into digital code 1 or 0, as the case may be. This step is long as all the cells of a row are to be read by their respective sense amplifiers and their status (0 or 1) to be recorded. This step is also critical as capacitor charging time constant is large (resistances of all cells add up on account of their being in series) and maximum capacitance of cell capacitors is typically of the order of 25–30×10^{-15} F. Since charge accumulated on a capacitor is proportional to the capacitance of the capacitor, very weak charge gets stored on the capacitor which the sense amplifier has to read.

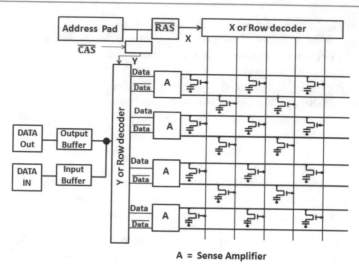

Fig. 11.9f Block diagram of DRAM

Second segment of operation is fast, column addresses are validated in this step internally through the $\overline{CAS}$ clock. Data present at each sense amplifier is then transferred through the column decoder to output buffer and eventually to the data output. The time taken in sequence one is of the order of 60 ns, while it takes only about 15 ns for second sequence.

The third step of sequences is called refreshing. As has been mentioned, the charge deposited on a capacitor of the cell leaks via the discharge of the capacitor. Further, the charge on a capacitor with state-1 is lost during Read operation. If left unattended the logic state-1 of the capacitor may become state-0, because of this charge leakage. In order to maintain the integrity of the capacitor logic state, it is necessary to refresh the memory of each cell. Refreshing or rewriting of the data existing on each cell memory (before any operation) is done in each operation cycle. This rewriting of data is done by the sense amplifiers that have the data, at the end of each cycle. When a particular row is selected all the cells of that row are read by the sense amplifiers and are refreshed, one at a time. Further, after a given time interval, of the order of 8–10 ms, burst refreshing is done, wherein all other commands are stopped and all rows, one after the other, are refreshed.

There are some parameters associated with the refreshing cycle. First is the refreshing time of each individual DRAM cell. It may be given either directly or in terms of clock cycles. For example, in case of a 133 MHz CPU, if 4 clock cycles is the refreshing time of each DRAM cell then in seconds this may be calculated as

$$1 \; clock \; cycle = \frac{1}{133 \times 10^6} s = 7.52 \times 10^{-9} \; s$$

$$\textit{Therefore refreshing time for 1DRAM cell is} = 4 \times 7.52 \times 10^{-9} \text{ s}$$
$$= 30.0 \times 10^{-9} \text{ s}$$

Suppose a DRAM contains 2^{13} rows then the time taken to complete one refreshing cycle $= 2^{13} \times 30.0 \times 10^{-9}\text{s} = 245.76 \times 10^{-6}\text{s}$

If 64 ms is the time after which burst refreshing is done, then refreshing over head is

$$= \frac{\textit{brust refreshing interval}}{\textit{time taken in complete one cycle of refreshing}} = \frac{64 \times 10^{-3}\text{s}}{245.76 \times 10^{-6} \text{ s}} = 260$$

In order to improve the performance of a DRAM cell sustained and continuous efforts are made to increase the capacitance of the cell capacitor. This has been further necessitated by the reduction of V_{CC} the maximum voltage applied to the device from 15 V to 5 V and now to 3.5 V. Since $Q = CV$, reduction of applied voltage reduces the charge accumulated at the capacitor unless the capacitance is increased. Two methods are used to increase the capacitance: change the shape of the capacitor and use dielectric between capacitor plates. Both these methods are used to further improve the cell performance.

Another concern associated with DRAM is the speed. Since the circuitry used for reading data from cells is inherently slow, DRAM could not match the speed of CPUs. It may be noted that speed of microprocessor systems has gone from 1MHz to 200 MHz or more, while the fastest DRAM units have reached the speed of 50 ns. To face this speed discrepancy problem, many different types of DRAM cells have been devised, mainly incorporating different interface circuitry. Each variation answers a particular situation and application. Without going into details names of some of the improved technologies are given here;

FPMRAM (fast page mode); EDORAM (extended data out), BEDORAM (burst EDO), EDRAM (enhanced DRAM), SDRAM (synchronous DRAM), Video DRAM, PSDRAM (pseudo-static RAM), RDRAM (Rambus), nDRAM (next generation), FRAM (ferroelectric RAM).

FRAM, introduced by Ramtron Corporation, uses a ferroelectric material, a ceramic film of lead-zirconium titanate (PZT) to provide non-volatile data storage. Since the memory is non-volatile, there is no need of refreshing at any stage. Also, the dielectric constant of the ceramic film is considerably large therefore the total charge accumulation is greatly enhanced.

11.2 Architecture of Analog-To-Digital and Digital-To-Analog Converter

The topic of analog-to-digital and digital-to-analog conversion has been touched in Chap. 9 while discussing discrete-time digital signals. We shall study it in some more detail in the following.

Most of the real-time signals are analog signals, be it the variation of temperature at a particular place, or the variation in wind velocity, the variation in the intensity of sound signals and so on. These analog signals in general cannot be sensed and processed by digital devices like computers, microprocessors, etc. It is, therefore, necessary to first convert analog information in digital form, process the digitalized information by digital devices, transport the digitalized information from one place to another using digital communication technology, store it, if required and again convert the digitalized final output into analog form for visual or any other form of display.

Block diagram of a typical system that uses Analog-to-Digital and Digital-to-Analog conversion is shown in Fig. 11.10a. An analog signal may be generated by any source; it may be a source of sound, varying temperature of some component in some industry and so on. The first step is to convert the analog signal into a corresponding (analog) voltage (or current, but mostly voltage) signal using an appropriate transducer. The voltage analog signal is then fed to pre-digitalization signal processing unit. Very often these units are frequency filters and Anti-aliasing circuits. The processed output from these units, which is still an analog voltage signal, goes to a Sample and Hold (S & H) unit. Sample and hold circuitry takes a sample or reads the voltage V_1 of the input analog signal at some instant of time t_1 and holds this value (V_1) for a time T_h at its output. During the time T_h, signal at the input of (S & H) circuit keeps changing but its output remains latched to the value V_1. The output of Sample and Hold circuit feeds the input of ADC. ADC processes the signal of constant amplitude V_1 and generates a digital code corresponding to the voltage. After the hold time T_h, the (S & H) unit again samples (or reads) the input analog voltage V_2 at time t_2 and holds it at its output for time T_h during which ADC generates the corresponding digital code. In this way (S & H) unit takes a sample of the analog voltage signal at regular intervals and holds the sample at its output for a fixed time during which ADC generates corresponding digital code.

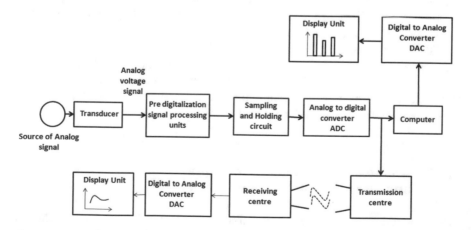

Fig. 11.10a Block diagram of a typical system using analog-to-digital and digital to-analog conversion

The digital output from the ADC may be fed to a computer that may further process the input digital signal and may record it in its memory or may pass it to a DAC which converts digital data into analog form for display.

The ADC output may also be passed on to a transmission centre, which may strengthen the signal, modulate the carrier with the signal and transmit. Since transmission of digital data is almost free from noise and environment disturbances, it may be received at a faraway receiving centre in good condition without much noise. At the receiving centre, the signal is retrieved using demodulation and, if required, may be fed to a DAC to convert into analog form for display.

11.2.1   Sampling and Hold Unit

Block diagram of Sampling and Hold unit (S & H) is shown in Fig. 11.10b. There are four basic components of the unit. First is an NMOSFET transistor Switch.

The operating condition of the transistor is such that when a positive voltage is applied to the Gate, the transistor behaves as a short circuit (switch on) and whatever signal is present at the Drain is passed to the Source. In case the voltage at Gate is zero, the transistor behaves as an open circuit and the contact between Drain and Source is broken. Signal V_{in} that is to be digitalized, is connected at the input of transistor switch. The second component is the control oscillator that generates square pulses of small duration and of adjustable frequency. That means the duration T_h, the time between two successive pulses can be set to the desired value. Square pulses from the control oscillator are applied at the Gate of transistor switch. The third component is a capacitor C, one end of which is connected to both, the switch and to the non-inverting terminal of the op-amp-based voltage follower. The other end of capacitor C is grounded. The last and the final component is an op-amp-based non-inverting voltage follower. Voltage follower circuit passes the signal present at its non-inverting terminal to its output. The output of voltage follower is connected to the ADC.

Self-assessment question: What is the purpose of an op-amp-based voltage follower at the output of (S & H) unit?

To understand the operation of the (S & H) unit, let us refer to Fig. 11.10c. The input signal V_{in} is shown in the upper part of the figure while pulses from the control oscillator are shown at the bottom of the figure. Suppose a positive square wave from the control oscillator turns the Gate potential positive turning the transistor switch on at time t_1. Once the switch is closed, signal V_{in} present at transistor input will pass on to capacitor C and will charge the capacitor. If V_1 is the voltage of V_{in} at instant t_1, the capacitor will charge to voltage V_1. Since Capacitor C is also connected to the input of the voltage follower, voltage V_1 will reach the input and then pass on to the output of the voltage follower.

During period T_h, when the voltage at the Gate is zero, the switch turns off, breaking the connection between the input signal V_{in} and the capacitor. However, during this period of time (T_h) voltage across the capacitor remains fixed to value

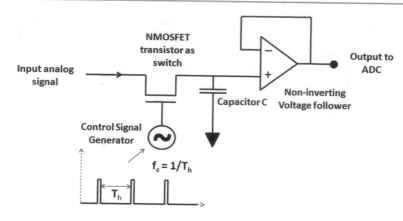

Fig. 11.10b Block diagram of Sampling and Hold unit

V_1. It is because the capacitor is cut off from the switch side so it can neither charge nor discharge through the open (off) switch. On the voltage follower side also, capacitor cannot discharge because of the infinitely high input impedance of op-amp. Thus voltage V_1 across the capacitor and at input and output of voltage follower remains at hold for the time duration T_h when control oscillator pulse has zero voltage. Control oscillator pulse becomes positive again at instant t_2, once again making the transistor switch on, connecting the capacitor to the input signal V_{in}. The capacitor will now sense the new instant value V_2 of the input signal and will get charged to potential V_2. This voltage V_2 will now hold across the capacitor, across voltage follower input and its output for a duration T_h when control oscillator voltage goes to zero. The signal sent by (S & H) unit to ADC is not the actual V_{in} but a histogram like signal shown shaded in Fig. 11.10c.

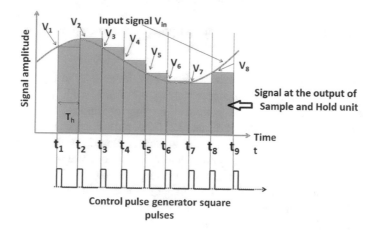

Fig. 11.10c Input and output of Sampling and Holding unit

In this way, the (S & H) circuit samples the input signal V_{in} at regular intervals when there is a positive pulse at transistor Gate and holds the sensed value for the time (T_h) when Gate voltage is zero. The (S & H) unit, sends a new value of the voltage to ADC after each time interval of duration T_h. The ADC processes this information in time T_h and generates a digital code corresponding to the voltage it has received from (S & H) unit.

It may be observed that (S & H) unit does not allow the original input signal V_{in} to reach the input of ADC. What actually reaches the ADC input is a collection of voltage samples (V_1, V_2, V_3, etc.) taking from the input signal V_{in} at different times. The question that often arises is how many samples are sufficient to convey the correct wave shape of the input signal. The number of samples essentially depends on the frequency f_C of the oscillator that generates square pulses. Actually, $f_C = 1/T_h$, where T_h is the time period of the square wave. After detailed analysis, Harry Nyquist concluded that the sampling frequency (the frequency f_c of the square wave oscillator) must at least be twice of the input signal V_{in}. If the sampling frequency is less than two times the frequency of input signal then falls signals called Aliases may be generated. The larger the frequency of the control oscillator as compared to the frequency of the input signal, more samples per cycle of the input signal will be taken by the (S & H) unit and more reliable reproduction of original signal will become possible.

The Nyquist Criteria

A continuous analog signal is sampled at discrete time intervals, $t_s = 1/f_s$. The discrete time interval must be carefully chosen so that the digitalized signals carry an accurate representation of original analog signal. It is clear that the more samples are taken (faster sampling rate), the more accurate will be the representation. If fewer samples are taken a point is reached where critical information about the original analog signal is lost. Harry Nyquist developed the mathematical basis for sampling and laid down the following rules for sampling, here f_s stands for the sampling frequency and f_a for the frequency of analog signal.

- A signal with a maximum frequency f_a must be sampled at a rate $f_s > 2 f_a$ or information about the signal will be lost because of aliasing.
- Aliasing occurs whenever $f_s < 2f_a$.
- A signal that has frequency components between f_a and f_b must be sampled at a rate $f_s > 2(f_b - f_a)$ in order to prevent alias components from overlapping the signal frequencies.

To demonstrate how aliasing or ghost signals, instead of the original signal, may get generated if the sampling frequency is less than two times the signal frequency, let us consider an analog sinusoidal signal of frequency f_a, shown in Fig. 11.10d. This analog signal is sampled with frequency f_s, which is only slightly larger than f_a. As shown in the figure, if one tries to reproduce the original signal by joining the sampled values one gets a sinusoidal signal of frequency $(f_s - f_a)$ not of frequency f_a. Signal of frequency $(f_s - f_a)$ is called the aliased signal or ghost signal that has developed because of the non-fulfilment of Nyquist criteria.

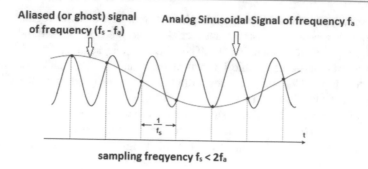

Fig. 11.10d   Generation of aliasing signal

11.2.2   Analog-To-Digital Conversion

The first step before undertaking A to D conversion is to convert the given analog information into corresponding voltage or current variation. This is required because most of the A to D converters, called ADC, accept analog signal at its input either in voltage or current form. Any analog signal may be converted into a corresponding voltage analog signal using an appropriate transducer. For example, an analog temperature variation may be converted into an analog voltage (or current) signal. The input of an ADC is analog signal and output is a digital signal, therefore, it is called a mixed-signal device. Since most ADC circuits accept analog signal in voltage form, we shall mostly refer only to voltage-activated ADCs.

An ADC, apart from having an input and an output port also has a reference voltage V_{Ref}. The input analog voltage is compared with this V_{Ref}. The output of the ADC provides information about the analog input (present at a given instant) in the form of a digital word. A word in digital language means a combination of bits, each bit is written with either 0 or 1. The word output from an ADC is also called the output code. The ADC compares the instant value of analog input with the reference voltage V_{Ref} and the output word tells (in digital language) what fraction of the reference voltage is the input voltage. Thus, an ADC may also be looked as a divider. Mathematically, the output from the ADC may be written as

$$ADC\ output(in\ the\ form\ of\ digital\ word\ or\ code) = \frac{G \times V_{in} \times 2^n}{V_{Ref}}$$

Here, G is the gain of the ADC, which in most cases is 1, and V_{in} is the input voltage. The power n appearing in the above equation is the bit size of the output code (or ward) of the ADC.

An ADC is characterized by the bit size 'n' of its output code. If suppose $n = 1$, then the output code will have only one bit and only two states 0 or 1 may be represented in the output. If the bit size is 2, $2^2 = 4$ different states, 00, 01, 10 and 11 may be represented by the output code. Similarly, for $n = 3$, eight different states, 000, 001, 010, 011, 100, 101, 110 and 111 may be represented in the output.

With the increase in the bit size of output code more and more different states may be represented by the output code.

The ADC divides the reference voltage V_{Ref} in to as many equal parts as is the possible number of states that may be represented by the output code. For example, a 2-bit ADC will divide V_{Ref} into 4 equal parts, a 3-bit into 8 equal parts, a 4-bit ADC into 16 equal parts and so on. Further, each least significant bit (LSB) in the output will represent a voltage change in the input equivalent to one small part obtained by the division of V_{Ref} by the number of possible output states. For example, for a 2-bit ADC, 1 LSB in output will represent a change of $V_{Ref}/4$ V, for a 3-bit ADC 1 LSB in output will represent an input voltage change of $V_{ref}/8$ V and so on. It is obvious from here that the larger the bit size of ADC output code, the change by one unit in the value of LSB in the output will correspond to a smaller fraction of voltage at the input. The bit size, therefore, decides the possible resolution that may be obtained from a given ADC, and the resolution of an ADC is given in terms of its bit size. An ADC of code size 3 bit is said to have a resolution of three bits, one with code size of 5 bit has a resolution of 5 bits and so on.

Self-assessment question: What will be the value (in volts) of 1 LSB of output for *an 8-bit* ADC of reference voltage 16 V?

Let us discuss the operation of a 3-bit ADC with reference voltage V_{Ref} of 8 V. Why this example? It is because it is manageable and one can easily draw graphs etc. for this case. A three-bit ADC can have 8 possible output states and each LSD in the output will correspond to $V_{Ref}/8 = 8/8 = 1$ V. Choosing this example makes life simple. In this example, whenever analog voltage at the input will change by 1 V, the LSD in the output code will be increased by 1. Suppose at any instant the input analog signal has a voltage of 3 V, the three-bit output code will display 011, now when input voltage will become 4 V, the output code will change to 100 (=011 + 001 binary addition).

It is clear from the above discussion that the overall resolution of an ADC depends both on the reference voltage and the bit length of the output code or word.

When one looks more carefully at this process of conversion of analog signal at input into output digital code, it becomes clear that the ADC is making approximations. Consider the situation when the input signal changes by an amount smaller than the resolution of the ADC. In the last example, where the resolution was 1 V if the input signal changes by say 0.4 V, the output code will remain unchanged. Thus any input signal change less than the resolution is not reflected in the output. For example of a 3-bit ADC and 8 V reference voltage, if an input signal of 4 V is coded as 100, then all input signals in the range 4.00000…1 V to 4.9999999… V will still be coded as 100; only when the input signal becomes 5.0 V, the output code will change to 101. It may be observed that the ADC has approximated all input signals larger than 4 V and smaller than 5 V to code 100.

Let us see the effect of reference voltage on the resolution, suppose V_{Ref} is reduced to 0.8 V (from 8 V). Obviously now the LSD of output code will correspond to a change of 0.1 V. One may infer from here that the resolution of the ADC system may become better by reducing the reference voltage, however, it is not completely correct. When resolution becomes too low, the electronic noise present in the system starts becoming more effective in interfering with the signal output.

Another way of improving (i.e. reducing) the overall resolution is to increase the bit length of the ADC output code. The resolution per LSD of a 4-bit ADC with a reference voltage of 8 V becomes 0.5 V.

Both methods; reducing the reference voltage and increasing the bit length of output code are used to obtain a better overall resolution, but increasing of bit length is restricted by the cost of the ADC while reduction of reference voltage beyond a limit is restricted by noise considerations.

(j) **Error of digitalization**

Figure 11.11 shows the graph of the analog input and digital output for a 3-bit, 8 V reference voltage ADC system. On the X- axis, $x = \frac{V_{Ref}}{Number\ of\ possible\ output\ states} = \frac{8V}{8} = 1$ V.

As may be seen in the figure, for the variation of input voltage from 0 V to any value less than 1 V, the output code remains 000; when input becomes 1 V, output

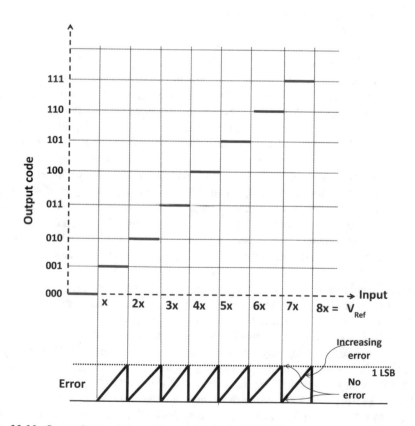

Fig. 11.11 Output–input relation and error distribution in ADC output

code changes to 001; again it remains fixed at 001 for the variation in input voltage from 1 V to any value less than 2 V, and it jumps to 010 when the input voltage reaches 2 V. Segments of dark lines in the graph show the input analog voltage intervals for which output code does not change. Vertically below each segment error associated with the segment is shown. It may be seen in the error graph, there is no error at the beginning and at the end of each segment, if input voltage is, say, 3 V, (point 3x on X-axis) the output code shows it accurately as 011 but for the interval from 3 V to just less than 4 V, the error (difference between the actual value and displayed value) increases linearly. However, when input reaches exactly 4 V, the output code also shows 100, which means the error now has become zero. In this way, error increases from zero at the beginning of each segment to a maximum of 1 LSB value at the end of the segment when it becomes zero again. The maximum error in the present case is of 1 LSB and it ranges from 0 to 1 LSB. This range of error is called **'quantization uncertainty'**. It is because in this range several different input voltages give the same code and one does not know the value of the actual input voltage. The **maximum value of quantization uncertainty is called the quantization error**. Quantization error is because of the finite resolution of the ADC, an n-bit ADC could resolve the input into 2^n discrete values. Each output code represents a range of input values, this range (for which output has the same code) is called a quanta and is denoted by **q**.

This error ranging from 0 to 1 LSB may be reduced to a value $+\frac{1}{2}$ LSB if an offset of $\frac{1}{2}$ LSB is introduced in the ADC.

Figure 11.12 shows how an initial offset of ½ LSB changes error from +1 LSB to ±½ LSB. With the introduction of the initial ½ LSB offset, for input voltage from 0 V to less than 0.5 V the output code will remain 000, it will become 001 when the input reaches 0.5 V. On further increase of the input voltage from 0.5 V to less than 1.5 V, output code will be 001, which at 1.5 V will become 010, and so on. With initial offset, the same error of 1 LSB has been divided into +1/2 LSB and −1/2 LSB.

In an ideal ADC, an input voltage of $q/2$ will just cause an output code to change from 000 to 001. Any deviation from this is called **zero scale error**.

(k) Differential and full-scale errors

An ADC is said to have completed a full cycle of operation when input voltage has changed by the amount of the reference voltage V_{Ref}. In one full cycle of operation, all possible output code values are displayed once. For the case of 3-bit ADC with a reference voltage of 8 V, maximum number of possible output codes is 8 and all these codes, starting from 000 to 111 will be displayed once each, when input voltage changes from 0 to 8 V. Thus in this particular example, one complete cycle may be divided into 8 steps, corresponding to unit change in LSB per step. When errors or any other parameter of the ADC system is computed, it may be evaluated over each step of operation (for the span when code changes from one value to the next) or it may be evaluated over one complete cycle of operation. Measurements

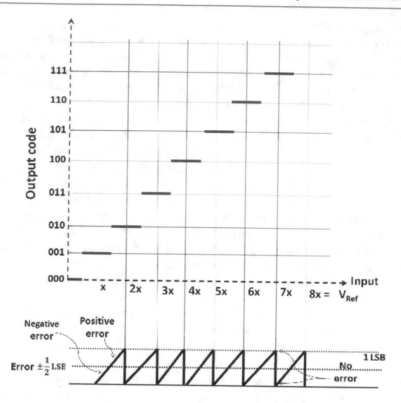

Fig. 11.12 Digitalization error reduces to $\pm\frac{1}{2}$LSB when an offset of $\frac{1}{2}$LSB is introduced initially

over one step of operation give the **distributed value** of the property while measurements over one full cycle give the **integral or full-scale value** of the property.

Self-assessment question: An ideal 5-bit ADC has a reference voltage of 3.2 V. (i) How many steps will make a full-scale transition, (ii) by what amount the input voltage will change in a full-scale transition and (iii) by what amount input voltage will change for each transition of output code?

(l) Gain error or Full-scale gain error
Theoretically speaking, the full-scale code transition, i.e. from the smallest value of the code to the highest value, should occur when input analog voltage changes by the amount of the reference voltage V_{Ref}. For the present example, code transition should occur when input voltage changes by 8 V. In practical case, however, full-scale code transition may happen before (or after) the input voltage changes by the amount of V_{Ref} (8 V in this case) or later. The **full-scale error is defined as the error in the actual full-scale output code transition point from the ideal value.** Full-scale error is made up of two components, one due to offset and the other due

to the error in the slope of the ADC. Full-scale error may be given as a percentage of the full-scale voltage (V_{Ref}) or in LSB.

Gain or Full-scale gain error is the deviation of the actual slope of the ADC from the ideal slope over one cycle. It may be recalled that the slope of an ideal ADC of gain 1, should be a straight line at 45° from the X- axis and passing through origin. In the case when gain is not 1, the slope line of an ideal ADC will still be a straight line inclined at some angle different from 45° but always passing through the origin. In actual ADCs it has been found that the actual slope line is neither parallel to the slope line of the corresponding ideal ADC nor it passes through origin even when compensated for the initial offset.

Gain error or full-scale gain error for a practical ADC with initial offset is shown in Fig. 11.13.

(m) Differential Non-Linearity

A staircase-like structure of output code versus input voltage is shown in Fig. 11.14. Horizontal steps in the graph refer to the input voltage spans over which output code does not change. In an ideal ADC, these horizontal steps remain constant, each of the same width equal to the overall resolution of the ADC system. In case of practical ADCs, it is found that actual step widths are not only different from the ideal width but also vary from one to the next. Difference in the step size of an ideal and an actual ADC is referred to as the **Differential Non-Linearity** written as DNL in short. It is worth noting that the magnitude of DNL for a given ADC is not constant, in general, it varies from one step to another step of the staircase.

Fig. 11.13 Illustration of gain error

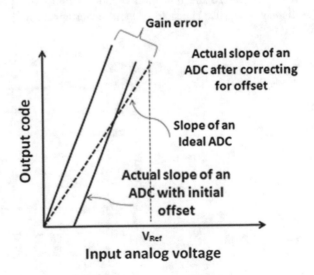

Gain error

Actual slope of an ADC after correcting for offset

Slope of an Ideal ADC

Actual slope of an ADC with initial offset

Output code

V_{Ref}

Input analog voltage

Fig. 11.14 Ladder like
structure of output code
versus input voltage graph

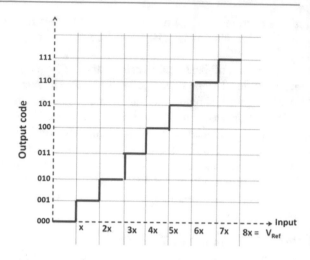

(n)   **Missing code**

Variations in DNL value from one step to the next may lead to the phenomenon of
missing code.

Figure 11.15 shows the transfer graph for an ADC that has DNL. As may be
seen in the figure, the width of the horizontal step ab is more than the ideal value
and at cd, it is still larger such that point d lies at a distance more than 0.5 times the
ideal step. Had point d be on the left of the midpoint of the horizontal step, the
transition of output code would have taken place in a normal way and output state
100 should have been displayed. However, in this case, point d lies to the right of
the midpoint so the transition of output code will jump the state 100 and will
display state 101. The state 100 will never appear in the output display in this case.

Fig. 11.15 DNL resulting in
missing code

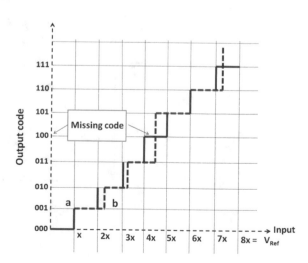

Fig. 11.16 Code jumping in
ADC with initial offset

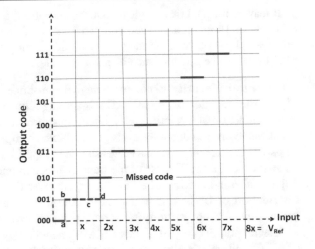

Figure 11.16 shows the transfer function for a 3-bit ADC that has 8 V reference voltage and in which an initial offset of +1/2 LSB has been introduced to reduce errors. In case the ADC is ideal, the output code transitions will take place at input voltages specified in Table 11.1. As usual the transfer function is made up of horizontal and vertical steps, the horizontal step representing the spread of input voltage for which code does not change. Now suppose the ADC is non-ideal and has DNL. The first code transition will take place at the right input voltage of 0.5 V which is the initial offset of ½ LSB. After first code transition, we assume that the width of the horizontal step gets increased by ½ LSB (i.e. by 0.5 V) because of the DNL error of ½ LSB. As a result of this increase the range of input voltage over which code transition will not take place has become 1.5 V (from original value of 1 V). As a result, code transition that was expected to take place at point c (input voltage 1.5 V) will now not happen instead next code transition will take place at input voltage of 2.5 V when code 011 will appear. Code 010 that was expected to appear when input voltage was 1.5 V will not appear ever in the output, because the range of input voltage from just above 0.5 V up to 2.0 V has now become the range where no code transition may take place.

Table. 11.1 Output code transition voltage

Number of code transition	Input voltage in units of x and in Volt
First code transition from 000 to 001	0.5 x = 0.5 V
Second code transition from 001 to 010	1.5 x = 1.5 V
Third code transition from 010 to 011	2.5 x = 2.5 V
Fourth code transition from 011 to 100	3.5 x = 3.5 V
Fifth code transition from 100 to 101	4.5 x = 4.5 V
Sixth code transition from 101 to 110	5.5 x = 5.5 V
Seventh code transition from 110 to 111	6.5 x = 6.5 V

It may be recalled that an ideal ADC has no offset, therefore, total DNL at point d in Fig. 11.16 including offset, compared to the ideal case is of 1 LSB (1/2 LSB offset + ½ LSB of DNL). It may be concluded from here that whenever DNL is 1 LSB or more, there is a chance of missing code.

(o) **Integral Linearity Error (or Integral non-linearity)**

The transfer function of an ideal ADC is a straight line passing through origin. The transfer function of a practical ADC after adjustment for offset and other errors, like quantization error and gain error, is generally not completely a straight line, it may be partly straight and partly curved as shown in Fig. 11.17. Integral linearity error is a measure of the straightness of the transfer function and it may be larger than the differential non-linearity. The amount of intergral linearity error depends on the **distribution of DNL**. It is important to note that integral linearity error does not include, quantization error, offset if any, and gain errors.

Integral non-linearity of a practical ADC may be specified in two different ways using: (i) end point method or (ii) best fit method.

In end point method, a straight line is drawn between the zero end point and the full-scale end point (See Fig. 11.18a) and maximum deviation from this line b and minimum deviation b' are given as a measure of full-scale non-linearity. In best fit method, a straight line best fitting the data is drawn, as shown in Fig. 11.18b and deviation 'a' from the best fit line is quoted as the full-scale or Integral non-linearity. The best fit line method gives the lower limit of INL while end point method gives both the maximum and minimum values for INL.

(p) **Total unadjusted error (TUE)**

Total unadjusted error (TUE) is a comprehensive specification that includes offset error, gain error and linearity error. It is a static parameter that has relevance when input signals change slowly. Because of different signs of errors, it is possible that one particular type of error may be larger than TUE.

Fig. 11.17 Transfer curves for an ideal and a practical ADC (adjusted for offset)

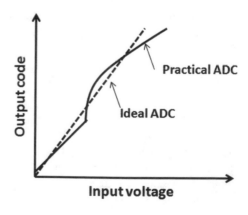

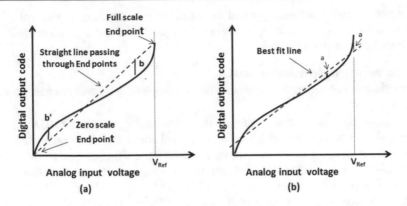

Fig. 11.18 a End point method. **b** Best fit method

(q) Resolution and Accuracy

Two terms, resolution and accuracy, are often interchangably used in ADC data-sheets. However, the two terms refer to two entirely different aspects. The overall resolution of an ADC is the smallest analog voltage that may increment the LSB of the ADC output by one. The overall resolution depends on both the resolution of the ADC and the reference voltage. Suppose an ADC has a resolution of 4 bits, i.e. the bit size of the output code is 4 bits. The total number of different output codes that may be generated by the ADC is $2^4 = 16$. That means that this ADC may divide the reference voltage V_{Ref} into 16 equal parts and each part will be represented by 1 LSB. Therefore, the overall resolution of the ADC $= \frac{Reference voltage V_{Ref}}{2^n}$, here n is the bit size of ADC output code which is also called the resolution of the ADC.

Accuracy of A/D converter determines how close the actual digital output is to the theoretically expected digital output for a given analog input. In other words, the accuracy of the converter determines how many bits in the digital output code represent useful information about the input analog signal. For example, in 16-bit output code of an A/D converter, it is possible that only 12 bits contain useful information and 4 LSB bits may represent random noise produced in ADC due to different error factors.

11.2.3 ADC Types

Analog-to-digital converters transform an analog voltage (or current) signal to a binary number and eventually to a digital number of base 10, for reading by a meter, or for display on a monitor. As already discussed, every ADC has a finite resolution decided by the bit size of the output code and the magnitude of the reference voltage. Different types of errors affect the output of a practical ADC. Practical ADCs are made using different techniques. Some of these techniques will be introduced in the following.

(a) Successive approximation ADCs

Block diagram of successive approximation analog-to-digital converter is shown in Fig. 11.19.

Surprisingly, it consists of a digital-to-analog converter (DAC), an op-amp-based comparator and a unit of control logic and few registers. Before going to the working of the system, let us understand the operation of the DAC.

A DAC converts input digital data into a corresponding analog voltage. Like reference voltage V_{Ref} in case of an ADC, a DAC has a voltage called full-scale voltage V_F. Setting a particular digital code at the input, the DAC delivers an analog output which is a fraction of V_F. Let us take the case of a DAC which has a 3-bit input code and the full-scale voltage V_F maybe, say, 10 V. The possible different codes that may be generated by a 3-bit code are $2^3 = 8$. The DAC will divide voltage V_F into as many equal parts as the number of codes generated by the bit size, in this case into 8 equal parts. As a result, when DAC input is 000, the analog output will also be zero; when DAC input is 001 the analog output will be $10/8 = 1.25$ V; Table 11.2 gives the value of the analog voltage at DAC output for each digital code at its input, for the example of 3-bit digital code and 10 V full-scale voltage.

It is important to note in this table that for input 100 the analog output of DAC has a voltage that is half of V_F. For input 010, the output voltage is ¼ of V_F and so on. The digital input of the DAC in our example has three bits, the leftmost bit is the most significant bit (MSB); therefore, code 100 means that the MSB is 1 and all other bits are 0. Similarly, 010 means that the next most significant bit (bit just before MSD) is 1 and all other bits are zero. These observations may now be generalized. If the bit size of digital input of DAC is different from 3 bit, maybe 5 bit say, and the full-scale voltage is V_F, then,

Fig. 11.19 Block diagram of successive approximation ADC

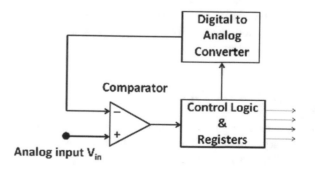

Table. 11.2 Analog voltage at DAC output for different input digital codes

DAC input code	Analog voltage at DAC output (V)	Output voltage as a fraction of V_F
000	0 0.00	0
001	1.25	1/8
010	2.50	2/8 $V_F/4$
011	3.75	3/8
100	5.00	4/8 $V_F/2$
101	6.25	5/8
110	7.50	6/8
111	8.75	7/8

Code 10000 at the input of DAC will deliver a voltage $V_F/2$ at the output;

Code 01000 at the input of DAC will deliver a voltage $V_F/4$ at the output;

Code 00100 at the input of DAC will deliver a voltage $V_F/8$ at the output;

And so on. With this understanding of DAC working, we proceed to discuss the operation of the successive approximation ADC block diagram of which is shown in Fig. 11.19.

As the first step, the control logic unit initially sets all input bits of DAC to zero. When an analog voltage signal is applied at one of the inputs of the comparator, the control logic unit forces the most significant bit (MSB) of DAC input to logic state-1. This forces the DAC output to ½ of the full-scale value. The analog voltage output from DAC is then compared by the comparator with the analog input signal V_{in}. Now there are two possibilities:

(i) analog input voltage $V_{in} > V_F/2$, the MSB of DAC input remains set at 1 and the next MSD is also set to 1; now MSD and next most significant digits are both 11 and all other bits are zero, at this setting the voltage delivered at the output of DAC will be $3V_F/4$, the input voltage V_{in} is now again compared with the current value $3V_F/4$. If V_{in} is still larger than the setting for MSD and the next to MSD bits is kept 11 and the next digit is set to 1, so that bit setting becomes 11100... Still higher voltage is delivered at its output by the DAC. This process of comparison continues till the LSD is compared.

(ii) In case $V_{in} < V_F/2$, then the control logic unit resets MSD to 0 and forces the next most siginficant digit to 1, the code will look like 01000... At this setting, the DAC will deliver ¼ V_F at its output which is again compared with V_{in}. If it is found that V_{in} is smaller than ¼ V_F the second most significant bit is reset to zero and the next MSD is forced to value 1, when code will look like 00100... and voltage $1/8V_F$ will appear at the DAC output. V_{in} will now be compared with $1/8V_F$. This process will go on till the Least Significant digit is compared.

At the end of the process, the output register will contain the digital code representing the analog input V_{in}.

Successive approximation ADCs are slow as the comparison process go serially and after each comparison the system has to pause for DAC to settle. Despite that conversion rates may easily reach over 1 MHz. Since 12- and 16-bit successive approximation ADCs are relatively less expensive they are often used in PC-based systems.

(b) Voltage to frequency ADCs

As shown in Fig. 11.20, voltage to frequency ADC employs a voltage to frequency converter that generates a train of digital pulses, the frequency of which is proportional to the amplitude of the analog signal. The output of the voltage to frequency converter contains a pulse train segment of which has pulses of the same size and shape but of different frequencies as shown in the figure. The frequency of pulses in a given segment is proportional to the amplitude of the input analog signal V_{in}. The pulse counter counts the pulses over a fixed period of time and thus determines the frequencies in different segments. Pulse counter output represents the digital voltage data.

Voltage to frequency ADCs are very good for use in noise environment. It is because the input pulse train with noise background is integrated into pulse counter to determine pulse frequency. Noise signals get cancelled during this integration process. It is because of this reason these ADCs are often used for remote sensing applications where uncontrolled background noise is present.

(c) Dual-slope Integrating ADCs

ADCs of this type use charging and discharging times of a capacitor for generating digital code for a given analog input voltage. The complete operation may be divided into two steps. In step-I, a capacitor is charged for a fixed period of time, by a current which is proportional to the amplitude of the input voltage. In step-II, the capacitor charged in step-I, is made to discharge at a constant current. Constant current is maintained using an auxiliary voltage source V_{ref}. The time required to

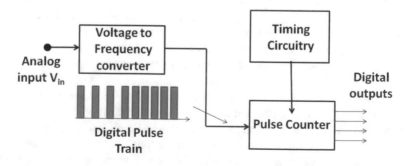

Fig. 11.20 Block diagram of a voltage to frequency ADC

Fig. 11.21 Integration and discharge times of the capacitor

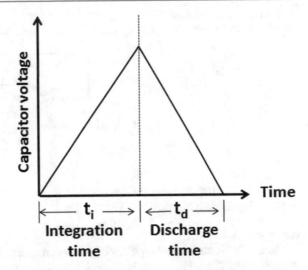

completely discharge the capacitor in step-II gives the value of the input voltage. Since slopes of charging and of discharging of the same capacitor are used to determine the input voltage, the method is called the dual-slope method. Also in the first step, the signal at the input of the ADC is integrated to obtain charging time, the ADC is, therefore, also called an integrating ADC. Integration and discharge times of a dual-slope ADC are shown in Fig. 11.21.

The magnitude of the input voltage V_{in} is given by

$$V_{in} = \frac{t_i}{t_d} V_{ref}$$

Dual-slope integration technique is quite accurate as it uses a ratio of two times and does not depend on the absolute value of the capacitor. Moreover, on account of integration in step-I, noise present at the ADC input is substantially reduced. The big disadvantage of the method is its speed that does not go up beyond 60 Hz, because several steps of integration and discharge are involved.

(d) Delta-Sigma ADC

It is one of the more advanced technologies employed in ADCs. In this type of ADC, the analog input voltage signal is fed to an integrator which produces a signal at its output the slope of which is proportional to the amplitude of the input signal V_{in}. This sampling voltage is then compared against ground potential (0 V) by an op-amp comparator. The comparator acts as a 1-bit ADC producing 1 bit of output (1, or 0) depending on whether the integrator output is positive or negative. The comparator's output is then fed back through a DAC (digital-to-analog converter) to the integrator. The purpose of mixing the comparator's output with integrator input

Fig. 11.22a Block diagram
of a Delta–Sigma ADC

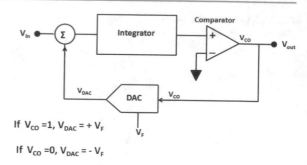

If $V_{CO} = 1$, $V_{DAC} = + V_F$

If $V_{CO} = 0$, $V_{DAC} = - V_F$

is to derive the integrator in the direction of zero volt output. The block diagram of
a Delta-Sigma ADC is given in Fig. 11.22a.

(e) Flash ADC

A flash ADC is also called a parallel A to D converter. It is built using a number of
comparators. The input signal is compared individually by each comparator with a
unique reference voltage. Outputs from comparators are connected to the inputs of a
priority encoder that in its output produces the digital code.

Since op-amp without feedback works as a comparator, they are often used for
the purpose of comparing the input signal with its unique reference voltage. Unique
reference voltage for each comparator is produced from a single precise and stable
voltage source (like an electronically regulated power supply) and a potential
divider arrangement. A combination of resistances in series constitutes the potential
divider. Voltages at different nodes of the potential divider circuit serve as a reference voltage for successive comparators as is shown in Fig. 11.22b.

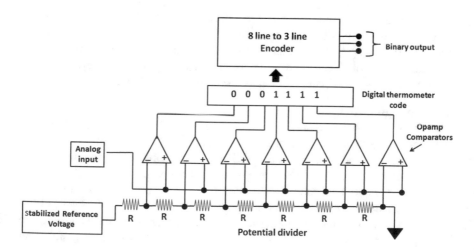

Fig. 11.22b A 3-bit Flash Analog-to-Digital converter

Fig. 11.22c Digital mercury code

Digital code shown in the figure is called Termometer code as High Logic states (1) fills the code like the murcury in a thermometer

If equal resistances are used in the potential divider chain, each comparator compares the input analog signal with a reference voltage that is higher than the next by an equal amount. This gives linearity to the system. One may use unequal resistances to make the output non-linear.

Outputs of those comparators for which analog signal equals or exceeds the reference voltage goes to logic state heigh (1) while those for which the input analog signal is lower than the reference voltage attains logic state low or 0. In this way, comparator output provides data in digital thermometer code. Thermometer code gets its name from the fact that high logic states in the code look as if they are like mercury in a thermomcter as shown in Fig. 11.22c. The digital thermometer code goes to the input of an 8–3 priority encoder. The priority encoder generates a digital code based on the highest order active input, ignoring all other active inputs.

In principle, flash ADC is the simplest and the fast ADC but it cannot be used for generating codes of large bit size. Even a 8- bit code will require $2^8-1 = 255$ comparators.

11.2.4 Digital-To-Analog Converter (DAC)

A digital-to-analog converter, often written as D/A, D2A, or D-to-A, is a device that converts a digital code into a corresponding analog signal. Figure 11.23a shows the symbol for a basic DAC.

A digital code specifies the binary weights or powers of 2, starting from the least significant bit with 2^0. A device that may generate voltages or currents of magnitudes proportional to the binary weights specified by successive input digital codes and adding them will produce the corresponding analog voltage or current signal in its output.

The principle of operation of a D/A converter may be understood with reference to Fig. 11.23b, where it is assumed that input to the system is fed by a 3-bit digital code that assumes different values at different instants. The digital input activates a 3-to-8 decoder that has 8 outputs corresponding to 8 possible states as shown in the figure. The reference voltage, say of 8 V is divided by a series chain of resistances

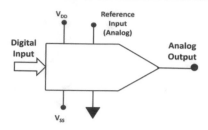

Fig. 11.23a Schematic symbol for a digital-to-analog converter (DAC)

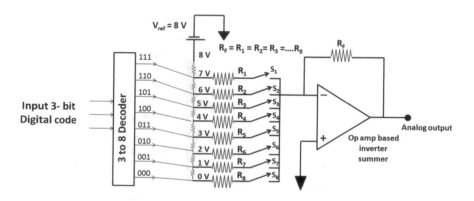

Fig. 11.23b Principal of operation of a digital-to-analog converter

into 8 equal parts of 1. 0 V each. The outputs of the decoder are connected to the corresponding nodes of the resistance chain or the potential divider. Potential divider nodes are connected to resistances R_1, R_2, R_3 ... R_8, which in turn are connected through respective switches S_1, S_2, S_3, ... S_8 to the input of an op-amp-based inverter summing circuit. When the value of resistances R_1, R_2, ... R_8 are all taken equal to the feedback resistance R_F the gain factor of inverter summer becomes 1 and different voltages connected to the inverter input add up without any scaling.

Let us assume that at instant t_1, the input digital code shows the value 101, the decoder will activate the corresponding output which will get connected to the potential divider node at 5 V. Let us assume that a control logic unit turns switch S_3 on so that 5 V signal is connected to the input of the inverter adder and output of adder shows a voltage $-[5V]$. Next at instant t_2, the 3-bit digital input changes to the value 111. The decoder will connect the corresponding output to the potentiometer node at 7 V. Control logic circuit this time will connect 7 V node to the input of inverter summer by putting switch S_1 in on position. The inverter summing output will now show $-[5V + 7V]$. In this way, voltage nodes corresponding to the values of successive digital input codes will get connected to the input of summing unit, one after the other, and the corresponding voltages will add up in the output. The inverter summing unit output will give the analog output. Though the analog ouput will have a negative sign it does not matter.

Above-mentioned principle of D/A conversion may be implemented in many different ways some of which are discussed here.

(i) Kelvin Divider or String DAC

Figure 11.23b without op-amp summing circuit, essentially shows the architecture of Kelvin Divider or String DAC. The n-bit version of this will have a resistance chain of 2^n resistors each of value R. It will also require 2^n switches, which most of the time are CMOS transistors operated in switch mode. This architecture is simple, has a voltage output and inherently monotonic, which means the output n cannot exceed output $n + 1$ even when a resistance gets short-circuited by any operational mistake. The system is also linear as it is made of resistances of equal value. The DAC analog output is in the form of voltage. The main drawback is its high impedance because of the large number of resistances. Use of an op-amp in the output usually overcomes this drawback. It may be noted that during code transition only two switches operate in the system, and the operation of a switch brings in disturbances called glitch, this architecture is a low glitch architecture. Before the invention of high-density IC technology, the use of Kelvin Divider was limited to low resolution (small n value) DACs. It was because even a 10-bit digital input required 1024 resistances and switches, which was difficult to fabricate on a single chip. However, now it is possible to use this architecture for high-resolution ADCs also.

(ii) Thermometer DAC

Figure 11.24 shows the block diagram for a 3-bit digital input current output DAC. As may be seen that, this DAC uses $2^n - 1$ switchable current sources connected to an output terminal. The analog current output may be converted into analog voltage output by the use of a resistance in output. All currents supplied by independent current sources are nominally equal. Input code 001 connects only one current source to the output, code 010 two current sources, code 011 three current sources and so on. This is achieved by adding some extra logic units to the decoder. Since currents are nominally equal from each source, the output is also linear. The inclusion of some complex logic in decoder may also correct the mismatch in currents from different sources to make the DAC fully linear.

(iii)   DAC with binary-weighted resistance

Circuit diagram of a 4-bit input, binary-weighted resistance DAC with an output coupled to an op-amp-based inverter summing amplifier is shown in Fig. 11.25. Terminals A, B, C and D correspond to the MSB down to LSB of digital input. As

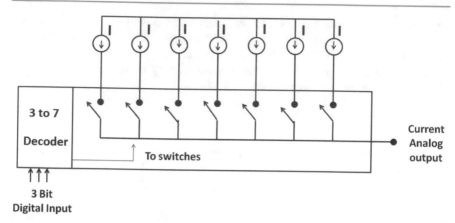

Fig. 11.24 Block diagram of Thermometer DAC

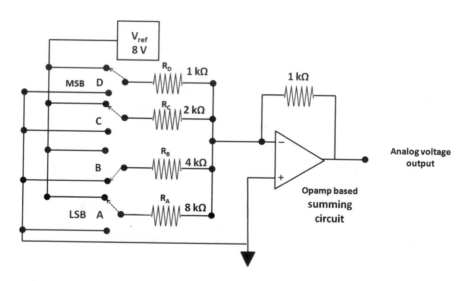

Fig. 11.25 Circuit diagram of a binary-weighted resistance DAC with op-amp inverter amplifier

may be observed in the figure, each of these terminals may be connected either to the reference voltage, say, of 8 V or to the ground. Terminal D is connected to the input of an op-amp-based inverter summing amplifier through a series resistance of 1 kΩ. Terminal C has a series resistance of 2 kΩ, terminal B of 4 kΩ and terminal A of 8 kΩ. These resistances are in the ratio of binary weights; i.e. R_D: R_C: R_B: R_A:: 1: 2: 4: 8.

Let us assume that at some instant t_1 the digital code communicated to DAC input is 1101. Then input signal will activate terminals of DCA and extra logic associated with the system (not shown in figure) will establish connections of terminals D, C and A with reference voltage (8 V) and of terminal B with ground. On account of different values of resistances, the gain of inverting amplifier for signals from different terminals will be different and the analog voltage at the output of op-amp will be given by

$$\textit{Analog voltage output} = -\left[\frac{R_f = 1k\Omega}{R_D = 1k\Omega} \times 8V + \frac{R_f = 1k\Omega}{R_C = 2k\Omega} \times 8V + \frac{R_f = 1k\Omega}{R_B = 4k\Omega} \times 0V + \frac{R_f = 1k\Omega}{R_A = 8k\Omega} \times 8V\right]$$

Or *Analog voltage output* $= -[8V + 4 + 0 + 1V] = -13V$ (zero appears as C is connected to ground).

It may be marked that if the output voltage is proportional to the decimal equivalent value of input binary code 1101 with a constant of proportinalty equal to -1.

It may be easily verified that for input code 0011, the output voltage will be $= -\left[0 + 0 + \frac{1}{4} \times 8 + \frac{1}{8} \times 8\right] = -3V$, which is the analog equivalent of 0011.

(iv) R-2R Ladder DAC

Figure 11.26 shows the circuit of a R-2R Ladder DAC (4-bit input) with its output connected to an op-amp inverter amplifier. The switching control logic connects bits with states-1 to the reference voltage and those with states-0 to the ground.

It is obvious that the input bit that is in the logic state-0 will not contribute to the output voltage. Only those bits that are in logic state-1 will contribute some specified fraction of reference voltage to the output. We calculate output voltage by keeping only one bit connected to the reference voltage and keeping all other bits

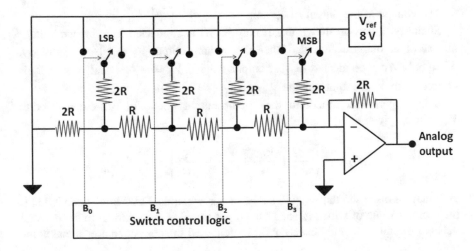

Fig. 11.26 Circuit of a 4-bit input R-2R Ladder DAC with op-amp inverter amplifier

Fig. 11.26a Equivalent
circuit when only MSB (B_3) is
connected to V_{ref} and all other
bits are grounded

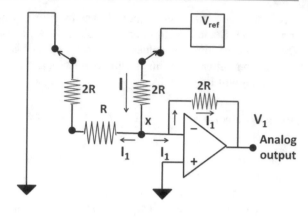

$$I = V_{ref}/2R \ , \ I_1 = I/2 = V_{ref}/4R$$

connected to the ground. One can then use the superposition theorem to get the
output voltage for any desired combination of input logic states by adding the
voltages of corresponding one-bit cases.

For example, when only B_3 (MSB) bit is connected to V_{ref}, and all other bits are
connected to ground, the equivalent circuit will look like Fig. 11.26a. Since node
X is at virtual ground, current I drawn from reference voltage V_{ref} is simply $I = V_{ref}/$
$2R$. Also at node X current I will be divided into two equal parts, $I_1 = I/2 = \frac{V_{ref}}{4R}$.
This current will pass through the feedback resistance of op-amp amplifier.

The voltage drop across feedback resistance due to current I_1 will be
$V_0 = 2RI_1 = \frac{V_{ref}}{2}$.

The voltage at the output of op-amp $= -V_1 = -\left[\frac{V_{ref}}{2}\right]$.

Similarly, it can be shown that if only bit B_2 is connected to reference voltage
and all other bits are grounded, then the op-amp output $V_2 = -\left[\frac{V_{ref}}{4}\right]$. Similarly,
when only B_1 is connected to V_{ref}, output voltage $V_3 = -\left[\frac{V_{ref}}{8}\right]$ and if only B_0 is
connected to V_{ref}, output voltage $V_4 = -\left[\frac{V_{ref}}{16}\right]$.

Using superposition theorem, the op-amp output when all bits are connected to
V_{ref} will be given as

$$V_0 = -\left[\frac{V_{ref}}{2} + \frac{V_{ref}}{4} + \frac{V_{ref}}{8} + \frac{V_{ref}}{16}\right] = \frac{15}{16}V_{ref}$$

It may be observed that V_0 will be the output when the DAC input code is 1111,
the decimal value of which is 15. Thus output signal is proportional to the decimal
value of digital code and hence DAC has delivered the correct analog signal at the
output.

Advantages of R-2R ladder DAC

(a) Easy to build as resistances of only two values are required and high accuracy may be achieved.
(b) Input digital code bit size may be easily increased by increasing the number of R-2R sections.
(c) On account of the small resistance spread of only 2, the R-2R ladder can be fabricated monolattical with high accuracy and stability

However, the real problem in R-2R ladders is to ensure that the temperature coefficient for all resistors is the same.

(v) DAC specifications

(a) **Resolution** The percentage resolution of a DCA depends on the number of bits its input can accept. A 16-bit DAC has better (finer or smaller) resolution than a 10-bit DAC. Resolution of a D/A converter is defined as the smallest change that can occur in the analog output as a result of a change in its digital input. Resolution is equal to the weight of the LSB which is also called the step size. A clock cycle or full-scale of operation consists of the time period in which all possible binary states of the system have been sampled. For example, in case of a 3-bit DAC, one cycle or full-scale operation means that all the possible 8 states from 000 to 111 have been sampled once. For a 4-bit DAC full scale will mean sampling of all possible 16 states and so on. Let us assume that a 4-bit DAC has a resolution of 0.1 V; when input digital code changes 1 LSB the output analog signal changes by 0.1 V. Now for a 4-bit DAC, the number of possible digital states is $2^4 = 16$ and in full scale all these states will be sampled once each, therefore, the total change of output analog voltage in full scale will mean a change of 16×0.1 V $= 1.6$ V. This is the full-scale output. A plot between digital code and corresponding analog voltage gives a staircase-like graph where voltage increases by a step equivalent to 1 LSB at each transition of logic code. In the example of 4-bit DAC and 0.1 V, resolution in one full-scale voltage will increase by 1.6 V as mentioned earlier. The step size in this case is 0.1 V.

(b) **Accuracy and error** Accuracy of any instrument means the closeness of the measured value with its theoretically expected value, often called ideal value. Since divergence between the ideal and actual value may arise due to several causes, or because of different types of errors, accuracy may also be specified with respect to different types of likely errors. In case of DAC (and also ADC), accuracy may be specified in terms of **full-scale errors and linearity errors**.

Full-scale error is the maximum deviation of DAC's output from its expected (ideal) value, expressessed as a percentage of full scale. Suppose a

DAC delivers a full-scale output of 9.834 V and it is specified that its full-scale error is 0.02%, then this percentage converts to,

$$error = \pm 9.834 \times 0.02\% = \pm 1.9668 mV$$

This means that the value of the output from this DAC may be off by as much as 1.9668 mV from its ideal value.

Linearity error is the maximum deviation of actual step size from the theoretical or ideal step size. If full-scale percentage error is specified then one can calculate linearity error from that.

Another type of error that is basically systematic error (not random error) is the offset error. Both an ideal DAC and an ideal ADC should display zero voltage when digital code has all zeros. However, in practical cases some small voltages either positive or negative are displayed for the all-zero code. This is called an off-set error and should be adjusted before using the device otherwise it will increase or reduce, depending on the sign of the offset error, the final result by the amount of offset error.

(c) **Settling time** Settling time of a DAC is the time in which one cycle of operation, i.e. sampling of all possible digital codes is completed. Smaller the settling time higher the operating speed of the DAC. Typical value of settling time is 1 0μs–50 ns.

(d) **Monotonicity** In a DCA, the output analog voltage increases with the increase in the value of digital code, this rise of output voltage with incremental digital code is termed as monotonicity. In terms of the input output graph, monotonicity means that the graph goes up at each step and does not fall down for any incremental successive input.

Solved example (SE11.1) A 10-bit DAC has a full-scale output of 0.1 V and a full-scale error of 0.05% of full scale. Calculate the step size and likely error in it. What will be the range of possible outputs for an input of 1000000000?

Solution:

$Stepsize = \frac{Full - scale\ output}{2^n - 1}$ where n is the number of bits in input code

Therefore, stepsize $= \frac{0.1V}{1024-1} = 9.775 \times 10^{-5}V = 97.75\mu V$

Full-scale error = 0.5 times full-scale value $= \frac{0.05 \times 0.1}{100}V = 0.00005V = 50\mu V$

Since code 1000000000 refers to ½ of maximum value, the value of output voltage for the given code = 0.1V/2 = 0.05 V. If errors are included the output value will be written as

$$0.05 \pm 00005V$$

that means that output voltage may have any value between 0.05005 V and 0.04995 V.

Error in step will also be of $\pm 50\mu V$.

Solved example (SE11.2) A 4-bit DAC uses binary-weighted resistances. The output of the DAC is fed to an op-amp-based inverting summing amplifier of feedback resistance 1 kΩ. If resistance connected in series with LSB node is 80 kΩ calculate the output analog voltage for input code of 1110 when the reference voltage is 1 V

Solution: *Since the series resistance at LSB node (R_4) is 80 k, resistances (R_3, R_2, R_1) connected to successively higher nodes will be 40 k, 20 k and 10 k. The resistance in series with MSB will be 10 k. The inverter summing amplifier will deliver an output of*

$$= -R_f\left[\frac{1}{R_1} \times V_{\text{ref}} + \frac{1}{R_2} \times V_{\text{ref}} + \frac{1}{R_3} \times V_{\text{ref}} + \frac{1}{R_4} \times V_{\text{ref}}\right]$$

Code 1110 means that R_3 is not connected to the reference voltage and therefore for the third term in the above expression V_{ref} will be zero.

The required output voltage is $= -1\text{ k}\left[\frac{1}{10\text{ k}} \times 1 + \frac{1}{20\text{ k}} \times 1 + \frac{1}{80\text{ k}} \times 1\right]$ V $= 0.1625$ V.

Solved example (SE11.3) A 8-bit DAC produces an output analog current of 10 mA for input code 11000000. What will be the output current for input code 00001111?

Solution: *For a DAC in general : Analog output = K × digital input*
Here K is constant of proporationality and is constant for a given ADC. In this example, it is given that output current is 10 mA for input code 11000000. This input code corresponds to decimal number ($1 \times 2^7 + 1 \times 2 + 0 + 0 + 0 + 0 + 0 + 0$) = 192
So 10 mA = K × 192; therefore K =(10/192) mA
Code 000011 = 03, hence the output current for code 000011 will be 3 × K = 3 × (10/192) mA
= 0.15625 mA.

Solved example (SE11.4) A 8-bit DAC has a step size of 10 mV. Determine the full-scale output and resolution of the device.

Solution: Total number of steps in a 8-bit DAC = $2^8 - 1 = 256 - 1 = 255$. Since step size is 10 mV, the total or full-scale output = step size × number of steps = 10 mV × 255 = 2.55 V.

$$Percent\ resolution = \frac{10\text{ mV}}{2.55\text{V}} \times 100 = 0.392\%.$$

11.3   Computer Organization and Arithematic Logic Unit (ALU)

Central processing unit (CPU) is like both, heart and brain, for a computer. All computer operations are performed here. It is like the human brain where all information and commands are received and different units of the CPU are instructed to carry out their required jobs. Airthematic Logic Units (ALUs) in a CPU carry out mathematical and logic operations as instructed by the CPU commands. A CPU may have more than one ALUs and they may be quite complex. Most of the operations of the CPU are performed by ALUs. To understand the operation of a CPU and of ALU, it is necessary to know about the basic components of a computer. A processor has.

- Clock module,
- Registers,
- Counter,
- RAM,
- Bus,
- Output,
- ALU.

Figure 11.27a shows some of the essential building blocks of a CPU.

Clock module: The function of a clock in a computer is to cycle forward the data or instructions. Clock module also determines the speed of operations of a computer, i.e. how fast the computer can execute instructions. The clock output is

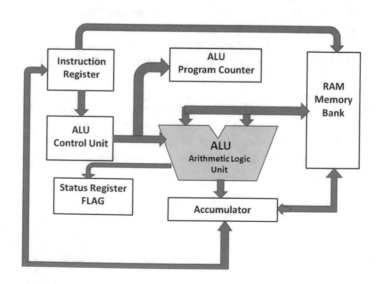

Fig. 11.27a Block diagram showing essential building blocks of CPU

voltage-driven outputted in pulses. Clock pulses enter all the units of CPU and make the unit work with after the clock outputs a pulse. Clock cycle is measured in terms of its frequency, the higher the frequency faster is the system.

Registers: These are the components where data or instructions are held in the form of voltages. A section of registers is used as fast memory, which holds values stored in them and which are parts of instructions. Digital data in the forms of bit words may also be stored in these registers and may be called when required.

Counter: Using a program counter is crucial as it often tells the computer with what to begin with. It is also used as an indicator of at what stage the computer is. It may also be used to feed in this information for longer program chains.

RAM: Random Access Memory is the main memory for the storage of programs. There are two types of memories, static and dynamic. RAM chips can do both Read and Write functions. In write mode, it will store data that is present at its I/O pins to the selected address line. In read mode, it will output the data at its I/O lines from the specified address line.

Bus: Bus, as the name suggests, is used for communication between and to different modules of the system. All computer modules can output some data to the bus and may also input data from the bus. In order to facilitate faster communication, the bus is generally located in the central part of the computer.

Output: Output of a computer is essentially a register that holds the results of the computer operation that are intended to be delivered. It may also contain some graphic information.

11.3.1 Airthematic and Logic Unit

It is the mathematical and logic processor of the CPU. The schematic symbol for an ALU is shown in Fig. 11.27b.

On getting command from the control unit, ALU may load two integer operands A and B from two specified registers. The ALU carries out operations as instructed

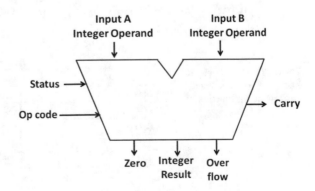

Fig. 11.27b Schematic symbol for ALU with inputs and outputs

by the control via Opcode and then delivers an integer output. In an ALU, there are other outputs that indicate the status of its operation and may give information regarding 'carryin', 'carryout' digits, overflow, division by zero, etc. during mathematical and logical operations being performed.

11.3.2 Design Architecture of ALU

While designing an ALU, the first consideration is of the 'word size' that ALU may handle. In case it is n-bit data, the ALU must have an output result of the same size. An ALU is characterized by the word size it can handle. An n-bit ALU can handle n-bit data at its two inputs A and B and will deliver a result in n-bit code. The **ALU opcode** tells it what operation it should carry out. For example, if ALU opcode size is 3-bit then it can give 2^3 different operational commands as given in the following table. Table. 11.3

It may be pointed out that in the present case only five operations have been defined and, therefore, only five out of possible 8 operation codes have been used. Above-mentioned facts are summarized in the following Fig. 11.27c.

A n-bit ALU may be built by coupling n-ALU's each of 1-bit. A 1-bit ALU is also called a slice. The following one-bit ALU may perform three logic functions AND, OR and ADD (Fig. 11.27d).

(i) **Implementing addition**: Addition in a 1-bit ALU is implemented using a full adder. A full adder is also called a (3, 2) adder as it has two inputs and three outputs. Table 11.4 specifies the inputs and outputs of the full adder in ith SLICE.

It may be shown that carry out $C_{(i+1)}$ is given by the logic equation

$$C_{(i+1)} = (b_i \cdot C_i) + (a_i \cdot C_i) + (a_i \cdot b_i) \tag{I}$$

And SUM S_i by logic equation.

$$\text{SUMS}_i = \left(a_i \cdot \bar{b}_i \cdot \bar{C}_i\right) + \left(\bar{a}_i \cdot b_i \cdot \bar{C}_i\right) + \left(\bar{a}_i \cdot \bar{b}_i \cdot C_i\right) + \left(a_i \cdot b_i \cdot C_i\right) \tag{II}$$

Table. 11.3 ALU control line (ALU op) code and corresponding function

ALU opcode	Function to perform
000	AND
001	OR
010	ADD
110	Subtract
111	Set-on-less-than

If operands A and B are each of n-bit wide, they must be connected through n-lines each. Arrow crossed by lines and n written near arrow indicates n-line connection. The result of operation carried out by ALU is also communicated via a n-bit line. An ALU with n-bit word is called a n-bit ALU

ALU control line (ALU op) provide command as to what operation the ALU should perform. In this case ALU op is a 3-bit word that may convey 8-different commands.

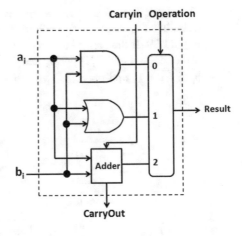

ALU op code	Function to perform
000	AND
001	OR
010	ADD
110	Subtract
111	Set-on-less-than

Fig. 11.27c ALU specification

A 1-bit ALU shown in the figure may perform two logic and one arithmetic operation. If the input data A (= $a_1, a_2,...a_n$) and B (=$b_1, b_2,....b_n$) is each of n-bits, then n-ALUs each of 1-bit may carry out these operations on different corresponding bits of A and B.

The 1-bit ALU will perform all the three operations and keep their outputs in its accumulator but will pass on to the result only that value that has been asked. If ALU is asked to add that means operation 2is to be preformed, then value of addition will be passed to Result

Fig. 11.27d A 1-bit ALU or SLICE

Table. 11.4 Input and output specifications for a full adder

Inputs			Outputs		Comments (Binary code)
a_i	b_i	Carry In C_i	Carry Out $C_{(i+1)}$	SUM S_i	
0	0	0	0	0	$0 + 0 + 0 = 00$
0	0	1	0	1	$0 + 0 + 1 = 01$
0	1	0	0	1	$0 + 1 + 0 = 01$
0	1	1	1	0	$0 + 1 + 1 = 10$
1	0	0	0	1	$1 + 0 + 0 = 01$
1	0	1	1	0	$1 + 0 + 1 = 10$
1	1	0	1	0	$1 + 1 + 0 = 10$
1	1	1	1	1	$1 + 1 = 1 = 11$

Figure 11.28 shows the hardware (logic circuits) to implement logic equations (I) and (II).

(ii) **Implementing subtraction**: Since $(A - B)$ is the same as $[A + (-B)]$, therefore subtraction can be implemented if B is converted into $(-B)$. To negate a two's complement binary integer, invert every bit in integer and add 1. Thus one can say

$$A - B = A + \overline{B} + 1$$

The above logical expression may be implemented in a 1-bit full adder by changing the carry in from 0 to 1 and using an XOR gate as shown in Fig. 11.29. If the control bit is 0, the XNOR gate allows b_i to reach the adder and operation $a_i + b_i$ is performed. When control bit is 1, XOR gate converts b_i to $(\overline{b}_i + 1)$ and operation $[a_i + \overline{b}_i + 1] = (a_i - b_i)$ is carried out.

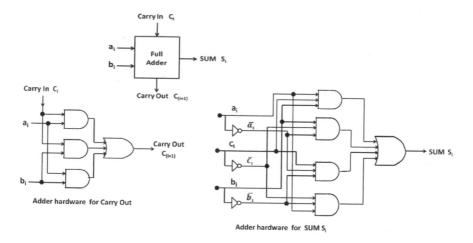

Fig. 11.28 (3, 2) Adder and logic circuits to implement carry out and sum

Fig. 11.29 Implementing
subtraction in a SLICE

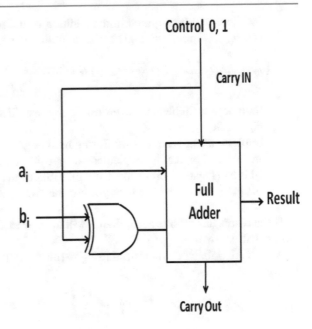

(iii) **Implementing Overflow:**

We have seen that a carry occurs when the result of addition or subtraction has number of bits more than the bit size of the ALU. For example, in case of a 4-bit ALU, if the number of bits in the result of addition or subtraction is more than 4 then a carry (or borrow) is generated. A related mathematical operation is **overflow, it tells that the sign of the result is different than the signs of two inputs.** For example, when 0111 $(7_2) + 0111$ $(7_2) = 1110$. Now this result (1110) in 4-bit two's complement means (-2_2). Here we see that the result of adding two positive integers has given a negative number. Overflow occurs when the bit size of the inputs is such that there is a carry that **changes the most significant sign bit. ALU will always output both carry and overflow, but both only make sense when the operation is addition or subtraction.** It is obvious that carry and overflow have no meaning in case of logic operations. Since overflow and carry both are the result of mathematical operations, overflow is considered in arithmetic operations only. In arithmetic operations also, if the sign of the two inputs is the same but the sign of the result is different then there is an overflow. Obviously overflow check is to be applied for the most significant bit, i.e. in case inputs A and B both are n-bit integers and are written as $A = [a_{(n-1)}$ $a_{(n-2)}$ $a_{(n-3)} \ldots$ $a_3 a_2 a_1 a_0]$; $B = [b_{(n-1)}$ $b_{(n-2)}$ $b - \ldots$ $b_3 b_2 b_1 b_0]$, then overflow will be required in the addition of the most significant bits $a_{(n-1)}$ and $b_{(n-1)}$ only. The logic for generating overflow will be

$$\left(a_{(n-1)} \; AND \; b_{(n-1)} \; AND \; \overline{Result_{(n-1)}}\right) \; OR \; \left(\overline{a_{(n-1)}} \; AND \; \overline{b_{(n-1)}} \; AND \; Result_{(n-1)}\right)$$

Hardwear to implement overflow in the last slice of the ALU is shown in Fig. 11.30.

(iv)   **Implementing zero output**: Zero output is given when all bits of all results are zero. This may be implemented by using an n-bit OR gate followed by 1-bit NOT gate. 1-bit results from all n-Slices may be fed to the input of the OR gate. Output of OR gate goes to the NOT gate as shown in Fig. 11.31.

If the result has all bits zero, then 1 will appear at the zero output, otherwise output will show a 0.

A n-bit ALU may be synthesized using n-1bit ALU units as shown in Fig. 11.32.

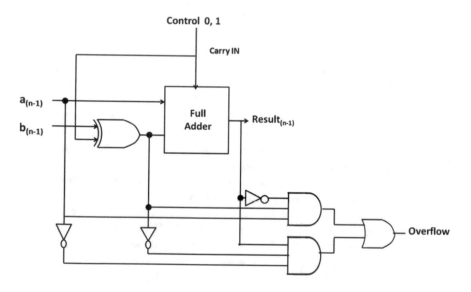

Fig. 11.30 Hardware to implement overflow in most significant bit slice

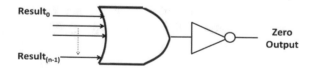

Fig. 11.31 Hardware for implementing zero output

Fig. 11.32 Layout of an n-bit ALU using n–1 bit slices

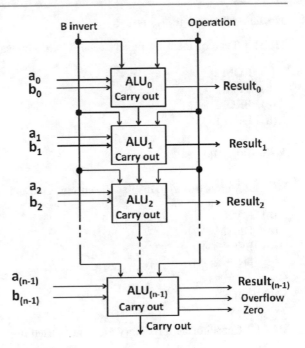

Problems

P11.1   Draw a block diagram for implementing of three functions $f_1 = \Sigma(0, 1, 3, 7)$; $f_2 = \Sigma(1, 2, 4, 7)$ and $f_3 = \Sigma(0, 1, 2, 5)$ with three inputs a, b and c using ROM.

P11.2   What will be the address of the 5th bit of the 5th register on a 10×5 memory page?

Answer: [0101 101].

P11.3   A ten-bit DAC has a step size of 10 mA. What are the full-scale output and percent resolution of the DAC?

Answer: [10.23 V; 0.0977%].

P11.4   A 10-bit DAC has a full-scale output of 10 mV. The device has a full-scale error of 0.5% of full-scale value. What will be the range of values for the output corresponding to the input code of 1000000000?

Answer: [5.0148 mV to 4.9852 mV].

Multiple-Choice Questions

MC11.1 Type of memory mostly used to store messages in mobile phones is

(a) ROM
(b) RAM
(c) EROM
(d) Flash ROM

ANS: [(a); (d)].

MC11.2 Memory used to display the current time and date in most digital devices is

(a) RAM
(b) DRAM
(c) EEPROM
(d) Flash ROM

ANS: [(c)].

MC11.3 Capacitor as a memory element is used in

(a) DRAM
(b) SRAM
(c) FRAM
(d) ROM

ANS: [(a), (c)].

MC11.4 Flip-flop with polysilicon resistance as the load is used as a memory element in

(a) DRAM Cell
(b) FDRAM Cell
(c) 4T SRAM Cell
(d) 6T SRAM Cell

ANS: [(c)].

MC11.5 Which of the following memory does not require refreshing?

(a) FRAM
(b) DRAM
(c) ROM
(d) synchronous DRAM

ANS: [(a), (c)].

MC11.5 Typical value of capacitance used in DRAM is of the order of

   (a) 30 F
   (b) 30 μF
   (c) 30 pF
   (d) 30 fF

ANS: [(d)].

MC11.6 Tick the properties of SRAM

   (a) Require less physical space
   (b) Faster access
   (c) hard to change data
   (d) High capacity

ANS: [(b), (c)].

MC11.7 Tick the properties of DRAM

   (a) Require less physical space
   (b) faster access
   (c) hard to change data
   (d) high capacity

ANS: [(a), (d)]

MC11.8 An op-amp driven voltage follower in the output stage of a sample and hold circuit is used to

   (a) Amplify the signal
   (b) attenuate the signal
   (c) avoid discharging of the capacitor through op-amp
   (d) to match the impedance of sample and hold circuit with the ADC

ANS: [(c), (d)].

MC11.9 An analog signal of frequency f_{in} is applied at the input of a sample and hold circuit with control frequency f_c. For satisfactory operation,

   (a) $f_C = f_{in}$
   (b) $f_c = 2 f_{in}$
   (c) $f_c < 2 f_{in}$
   (d) $f_c > 2 f_{in}$

ANS: [(d)].

MC11.10 An A/D converts

(a) voltage to current
(b) analog current to digital current
(c) digital voltage to digital current
(d) analog voltage to digital voltage

ANS: [(a), (d)].

MC11.11 Each digital input to a DAC is weighted according to their position in the

(a) decimal number
(b) binary number
(c) hexadecimal number
(d) octal number

ANS: [(b)].

MC11.12 In a 1-bit ALU, overflow occurs when

(a) A logic operation gives zero
(b) A logic operation gives infinity
(c) in adding operation sign of the result is different than the signs of two operands
(d) in adding operation signs of two operands are different

ANS: [(c)].

MC11.13 In a 1-bit slice operands 'a' and 'b' add to give result 'r'. The correct logic for the implementation of overflow is

(a) $(a \text{ AND } b \text{ AND } \bar{r}) \text{ OR } (\bar{a} \text{ AND } \bar{b} \text{ AND } r)$
(b) $(\bar{a} \text{ AND } b \text{ AND } \bar{r}) \text{ OR } (\bar{a} \text{ AND } \bar{b}, \text{ AND } r)$
(c) $(a \text{ AND } \bar{b} \text{ AND } \bar{r}) \text{ OR } (\bar{a} \text{ AND } \bar{b} \text{ AND } r)$
(d) $(\bar{a} \text{ AND } b \text{ AND } r) \text{ OR } (\bar{a} \text{ AND } \bar{b} \text{ AND } r)$

ANS: [(a)].

MC11.14 A 1-bit ALU adds integer 'a' and 'b'. If C_{in} is the value of carry in, the logic to generate carry out is

$$(\bar{a} \cdot C_{in}) + (b \cdot C_{in}) + (a \cdot b)(b)(a \cdot C_{in}) + (b \cdot C_{in}) + (a \cdot b)(a)(c)(\bar{a} \cdot + C_{in})$$
$$+ (b \cdot C_{in}) + (a \cdot b)(d)(\bar{a} \cdot C_{in}) + (b + C_{in}) + (a \cdot b)$$

ANS: [(b)].

Short Answer Questions

SA11.1 Explain how read and write operations are performed on an elementary DRAM cell.

SA11.2 With the help of a block diagram, give the outline of the three steps of operation of a DRAM assembly and explain the first step of operation.

SA11.3 Compare SRAM and DRASM. Why DRAM is used more often than SRAM?

SA11.4 Explain the advantages of a thin-film transistor SRAM.

SA11.5 What is FDRAM? Discuss its advantages over the other DRAMs.

SA11.6 What modifications were done to improve the noise and soft error response of SRAM?

SA11.7 Explain the read operation of a SRAM cell.

SA11.8 DRAM has high capacity, slow access and it is easy to change data in it. Explain each of these points.

SA11.9 Distinguish between resolution and accuracy of a DAC. What errors may cause a deviation of the actual output value from its ideal value?

SA11.10 What is the purpose of sample and hold circuit in A/D conversion?

SA11.11 Explain the terms: step size, percentage resolution, monotonicity of a DAC.

SA11.12 Most DACs have an oop-amp-based inverting amplifier summing circuit at its output. What is the purpose of this circuit?

SA11.13 A sample and hold circuit generally has an op-amp-based voltage follower as its output stage. What is the need for voltage follower when it does not provide any gain to the sampled signal?

SA11.14 What is the Nyquist criterion? Explain the production of aliased signals.

SA11.15 What is overflow? In which operation it is important?

Sample Answer of Short Answer Question SA11.15

What is overflow? In which operation it is important?

Answer: *In arithmetic operations carried out by the ALU of a computer, bits in a word are often treated as two's complement signed values. When 2 two's complement signed numbers are (digitally) added the final result may give a wrong value. For example, let us consider the addition of two 4-bit numbers (Two's complement notation) 1001 and 1010. In two's complement notation $1001 = -7$ and $1010 = -6$, when added these two negative numbers should give -13. However, their binary addition gives 3 as shown below*

$$
\begin{array}{r}
1001 \\
1010 \\
\hline
0011 \quad = 3
\end{array}
$$

Two important observations may be made here: (i) addition of two negative numbers has given a positive number (ii) since the word is of 4-bit, and the maximum negative number that can be represented by 4-bit word is (−8), while the correct addition gives (−13). Therefore, this error has occurred because of the limitation of word size.

Overflow occurs when the bit size of the inputs is such that there is a carry that **changes the most significant sign bit. ALU will always output both carry and overflow, but both only makes sense when the operation is addition or subtraction.** *It is obvious that carry and overflow have no meaning in case of logic operations. Since overflow and carry both are the result of mathematical operations, overflow is considered in arithmetic operations only. In arithmetic operations also, if the sign of the two inputs is the same but the sign of the result is different then there is an overflow. Obviously overflow check is to be applied for the most significant bit, i.e. in case inputs A and B both are n-bit integers and are written as* $A = [a_{(n-1)} a_{(n-2)} a_{(n-3)} \dots a_3 a_2 a_1 a_0]$; $B = [b_{(n-1)} b_{(n-2)} b - \dots b_3 b_2 b_1 b_0]$, *then overflow will be required in the addition of the most significant bits* $a_{(n-1)}$ *and* $b_{(n-1)}$ *only.*

Long Answer Questions

LA11.1   Discuss in detail DRAM technology and the improvements it has gone through.

LA11.2   What is the advantage/or disadvantage of DRAM over SRAM? Discuss with necessary details SRAM technology.

LA11.3   What is meant by ROM? What are important applications of this memory? Compare ROM with SRAM. With the help of a suitable example discuss the implementation of functions using ROM.

LA11.4   With the help of suitable diagrams explain the Read and Write operations on an SRAM cell. Is there any need of refreshing after every read operation in case of SRAM? If not, why?

LA11.5   With the help of a diagram explain the working of a sample and hold unit. Explain why the charge on the capacitor does decay during the hold period.

LA11.6   Write a detailed account of different types of errors that may affect the output of an ADC. Sometimes a ½ LSB offset is provided in ADCs. What is the purpose of this offset?

LA11.7   Discuss the working of an R-2R ladder DAC. What are the advantages of this type of ADC? List some of the applications of DACs.

LA11.8   Describe how a binary-weighted resistance DAC works. Define step size, settling time and monotonicity of a DAC.

LA11.9   Name different types of ADCs and discuss the operation of one of them in details.

LA11.10   Describe the working of a double slope ADC and explain why it is better than others.

LA11.11   With the help of a diagram discuss the working details of a successive approximation ADC. Why is it called a successive approximation ADC?

LA11.12   Design a slice that may add, subtract and carry out logic AND and OR operations. Discuss the implementation of subtraction using a full adder.

Index

© The Editor(s) (if applicable) and The Author(s), under exclusive license to Springer
Nature Switzerland AG 2021
R. Prasad, *Analog and Digital Electronic Circuits*, Undergraduate Lecture
Notes in Physics, https://doi.org/10.1007/978-3-030-65129-9

Printed in the United States
by Baker & Taylor Publisher Services